International Series in Operations Research & Management Science

Volume 307

More information about this series at http://www.springer.com/series/6161

Josef Kallrath

Business Optimization Using Mathematical Programming

An Introduction with Case Studies and Solutions in Various Algebraic Modeling Languages

Second Edition

Springer

Josef Kallrath
Department of Astronomy
University of Florida
Gainesville, FL, USA

ISSN 0884-8289 ISSN 2214-7934 (electronic)
International Series in Operations Research & Management Science
ISBN 978-3-030-73236-3 ISBN 978-3-030-73237-0 (eBook)
https://doi.org/10.1007/978-3-030-73237-0

Previously published under the imprint Palgrave

This Springer imprint is published by the registered company Springer Nature Switzerland AG.
The registered company address is: Gewerbestrasse 11, 6330 Cham, Switzerland

Dedication (2nd Edition)

To Albert & Diana

Dedication (1st Edition)

To those who increased my[1] pleasure in mathematics:

Wilhelm Braun (1970–1975)
Klaus Reusch and Wilhelm Gieselmann (1976–1979)
Gerard de Beuckelaer (1989–1992).

To[2] Helen, Alex, Tim, and Jack.

[1] Josef Kallrath.
[2] John M. Wilson.

Foreword

Mathematical optimization is an inherent paradigm of modern operational research (OR) for 80 years already. Although many subfields of OR got a kind of disciplinary independence over this period of time, "searching for the best" is still the main challenge for OR. Anyway, all sister disciplines of OR still benefit from more and more efficient methods of optimization developed within OR. Would machine learning have such a great trajectory without optimization? Would we hear about the success of deep learning if a proper optimization method would not be available? The present book reflects a continued interest in mathematical optimization. Its first edition written 20 years ago provided a faithful picture of optimization methods and their applications at the turn of the last century. An important progress in the methodology and applications of optimization, that we were able to observe over the last two decades, as well as a continuing interest in the first edition of the book, encouraged its first author to write the current update.

What one can learn from this book? A reader who remembers the first edition will quickly acknowledge that this edition goes far beyond the previous content. In particular, it includes new material on stochastic programming and nonlinear mixed integer optimization. Global optimization gets also a special place for its capacity of solving non-convex continuous or mixed integer nonlinear problems. The relation of global optimization to mixed-integer optimization, Lagrange relaxation, and polylithic modeling (sequence of models for one complex problem) is illustrated by interesting real-world examples from the academic publishing industry. Some topics, like phi-function techniques for cutting and packing, are not covered by other books on optimization known to me. A valuable feature of this book is also examples coded in popular algebraic modeling languages, and available online.

The book is written in full respect of the link existing between three necessary components of practical decision-making: "model-algorithm-software." In this way, the reader learns how real-world problems can be modeled in mathematical terms understood by optimization algorithms, and then solved with commercially available software. This holistic approach to solving complex problems with optimization methods makes this book self-contained. Moreover, because the content of theory, didactic examples, and real-world case studies are perfectly balanced in the book, the reader gets a convincing report on the practical relevance of contemporary mathematical optimization.

As a coordinating editor of the European Journal of Operational Research (EJOR) since 20 years, I can say that the character of this book is perfectly consistent with our editorial policy, in the sense that, similarly to EJOR, it underlines the link between the state-of-the-art methodology of optimization with practice of decision-making. Thus, I am pleased to congratulate Josef Kallrath on the excellent update of the first edition, and I am confidently recommending this book to students, teachers, researchers, and industrial practitioners who wish to learn how to effectively cope with complex decision problems using mathematical optimization.

Poznań, Poland
August 2020

Roman Słowiński

Preface to the 2nd Edition

With the continuing interest in my book, it is a pleasure, approximately 20 years after the completion of the first edition, to enhance it with some topics which, after two decades, have gained importance or led to a change in perspective.

This book, as the first edition, introduces business optimization using mathematical programming (optimization) not only from a practitioner's point of view but also from a researcher's perspective. It covers the entire process of solving a real-world decision problem by mathematical optimization: structuring and formulating the problem (free of mathematics) as well as collecting the input data, translating it into the mathematical language obtaining an optimization model, solving it, and validating the results. Linear programming case studies exemplify and showcase the learnings. Building on these concepts, the book examines mixed integer linear programming problems and presents problem formulations and case studies for these applications, and extends into the nonlinear optimization world, both continuous and discrete. Another series of case studies show optimization in practice employing integer programming, leading to larger practical examples, e.g., from production or supply chain planning, network planning, or cutting and packing. Hints are given on how users can control the optimization process and improve its efficiency. The book concludes with observations on the impact and implications of optimization in business. New to the second edition, many examples are coded in the algebraic modeling languages — `AMPL`, `GAMS`, FICO(R) Xpress `Mosel`, and `SAS/OR` — and are available *online*.

The second edition goes beyond linear optimization, i.e., beyond linear programming and mixed integer linear programming. Thus, the previous Chap. 11 has been renamed in *Beyond LP and MILP Problems*, and now includes fractional programming and its transformation into linear programming, successive linear programming as a special solution technique of nonlinear optimization, and optimization under uncertainty (especially stochastic programming). Also in Chap. 11, for quadratic programming, which is again a special case of nonlinear optimization, we provide an equivalent formulation based on special ordered sets. Chapter 12 provides an introduction into nonlinear continuous and mixed-integer optimization.

More attention is paid to the field of *Global Optimization*, as after 2002, strong commercial software using deterministic global optimization techniques became available for solving non-convex continuous or mixed integer nonlinear problems. The techniques used in this type of optimization are closely related to those used to solve mixed-integer optimization problems and are described in Chap. 12. They are illustrated with applications from paper and metals industry in Chap. 13. A further coupling of global optimization and mixed-integer optimization becomes apparent in the calculation of optimal breakpoint systems in connection with the modeling of nonlinear terms using SOS2 variables. In Chap. 14, *Polylithic Modeling and Solution Approaches*, we demonstrate how to use these SOS2 variables for obtaining good approximations for solving nonlinear problems. In the context of mixed-integer linear problems we also consider the topic of Lagrange relaxation for the improvement of lower bounds. Overall, as this second edition covers also the nonlinear optimization world, it is somewhat more demanding on the mathematical side. Therefore, it also contains Appendix C summarizing some of the fundamentals of linear algebra and calculus.

As cutting and packing is used in many examples throughout this book, and cutting and packing industry significantly contributes to the GDP of several countries, a full chapter on it has been added. Most of the *phi-function technique* material in Chap. 15 has been provided by Prof. Dr. Yuriy Stoyan and Prof. Dr. Tatiana Romanova.[3] This technique is a generic approach beyond and within mathematical programming, very suitable for solving large-scale real-world cutting and packing problems in 2D and 3D.

In *Tools around Optimization* in Chap. 17 we have added a section on algebraic modeling languages. Later in this chapter we address the importance of visualization of input data and output results.

Many examples in this book are coded in the algebraic modeling languages `mp-model`, `GAMS`, `Mosel`, and `SAS/OR` — online at www.springer.com — and referenced as MCOL (*M*odel *C*ollection *O*n*L*ine). This online access granted to readers of this book replaces the CD-ROM which came with the first edition. For compatibility with the first edition, we keep the `mp-model` files as well. In this second edition, we avoid using or explaining software specific syntax and keep implementation issues on a generic level.

[3]The National Academy of Sciences of Ukraine, Institute of Mechanical Engineering Problems, Department of Mathematical Modeling and Optimal Design, Kharkiv, Ukraine and Kharkiv National University of Radioelectronics, Department of Applied Mathematics.

Reading flow suggestions by target group and sequence of chapters:

flow	*target group*	*sequence*
F1	novice, beginners	1–17
F2	linear optimization	1–10
F3	nonlinear optimization	11–15
F4	practical optimization	1, 5, 10, 13–17
F5	advanced readers	10, 11, 13–15

In the first edition, a few larger case studies had been analyzed and solved with `XPRESS-MP`, the predecessor of what is now `FICO Xpress Mosel` and `FICO Xpress Optimization`, or `Xpress` in short. We have kept the text from the first edition in this second edition and continue referring to this modeling and solver `XPRESS-MP` from the 1990s in this context. We proceed similarly for currencies as well as the hardware or software equipment used. The real-world case studies from the 1990s remain unchanged. As `XPRESS-MP` and its modeling language `mp-model` use integer indices but not index sets, the first edition and its case studies are formulated using numerical indices. In this second edition, we leave the case studies untouched but use index sets in all material added. The formerly large MILP problems from 20 years ago have become toy examples for commercial solvers in 2020 — and the spreadsheet software `LOTUS-1-2-3` does not seem to be in use any longer. This should serve as a good warning and advise for the future. Only ASCII data has a reasonably long lifetime. Everything else strongly depends on market changes. The lifetime of software seems to be very limited — and even if the software still exists, backward compatibility is not guaranteed either. Fortunately, the first edition has been written in LaTeX, and is retained, largely, in the second edition.

Overall, we hope that the reader benefits from this second edition in various ways: providing a path to efficient modeling, and learning what is relevant nowadays, and which problems can be handled efficiently concerning mixed integer optimization and non-convex nonlinear optimization. If the reader has fun on the way reading this book, all the better.

Acknowledgment
It is a pleasure to again thank some friends and colleagues who have accompanied me for many years in my work or private environment and who in various ways have directly or indirectly contributed to the success of this book. This applies to everyone who has already been mentioned in the acknowledgment of the first edition. Between 1997 and 2020, I established close relationships or friendships with many new collaboration partners. These are, especially, Dr. Franz Nelißen and Dr. Michael Bussieck (GAMS Software GmbH, Frechen and Braunschweig, Germany), Prof. Dr. Christoudoulos A. Floudas (Princeton University, Princeton,

NJ, USA),[4] and Prof. Dr. Panos M. Pardalos (Center of Applied Optimization, University of Florida, Gainesville, FL). The contacts from my leadership of the GOR working group *Praxis der Mathematischen Optimierung* have also contributed indirectly to this expanded second edition.

A special word of thanks is directed to all the (new) project partners during the last 20 years: I have cherished many of them because of their enthusiasm and their in-depth knowledge of their application areas — and I have all of them in good memory. They all had in common that, for improving the benefits of their company, they wanted to enhance their methods in their applications areas by thorough mathematical modeling, which often led to mixed integer optimization. Interacting and communicating with them has been an important element during the projects and has some influence on this book. From the many German BASF project partners who have contributed indirectly to this book over several years, I would like to mention a few: Dr. Wolfram Schmidt and his team with Dr. Markus Klumpe and Bernd Heisel-Hoffmann, Norbert Vormbrock with a common history at Bonn University, and Dr. Gerd Fischer with his exciting rail car projects. It has been a pleasure to work with such people who feel the need to understand and solve a problem as deeply and as thoroughly as possible. High quality and sustainable solutions require deep *understanding*, dedication to detail, and the will to solve a problem on one's own initiative. I strongly hope that the project partners mentioned above will still have sufficient time left for focusing deeply on their projects and that they can enjoy their work.

For a thorough examination and proofreading of the manuscript as well as many constructive comments and suggestions that have improved the book, it is a pleasure to thank Dr. Jens Schulz and Dr. Susanne Heipcke (FICO, Berlin, Germany and Marseille, France); Jan-Erik Justkowiak (Siegen University, Siegen, Germany); Dr. Philipp M. Christophel (SAS Institute, Heidelberg, Germany); Dr. Johannes Schlöder (IWR, Heidelberg University, Heidelberg, Germany); Prof. Dr. Iiro Harjunkoski (Hitachi ABB Power Grids, Mannheim, Germany); Prof. Dr. Eugene F. Milone (University of Calgary, Calgary, Canada); Prof. Dr. Tapio Westerlund (Abo University, Finland); Prof. Dr. Ivo Nowak (Hochschule für angewandte Wissenschaften Hamburg, Hamburg, Germany); Prof. Dr. Alexandra Newman and Prof. Dr. Tulay Flamand and Phillip Bülow and Louis Kamga and Oluwaseun Ogunmodede (Colorado School of Mines, Golden, CO, USA); John Cox (US Air Force, Colorado School of Mines, Golden, CO, USA); Prof. Dr. David Morton (Northwestern University, Evanston, IL, USA); Dr. Joonghyun Ryu (Hanyang University, Seoul, Korea); Dominik Schweisgut who also wrote a first draft of Appendix C (Heidelberg University, Heidelberg, Germany); Prof. Eli V. Olinick (Southern Methodist University in Dallas, TX, USA); Prof. Dr. Ignacio E. Grossmann, Can Li, and Prof. Dr. Destenie Nock (Carnegie Mellon University, PA, USA);

[4]Unfortunately, for the whole community, Prof. Floudas passed away in August 2016. I lost a close collaborator and friend, we shared many common ideas and had joint activities since the early 1990s.

Dr. Michael Bussieck, Frederick Fiand, and Dr. Stefan Vigerske (GAMS Software GmbH, Braunschweig, Germany); Dr. Anna Schreieck (Neustadt a.d. Weinstraße, Germany); Prof. Dr. Stefan Helber (Leibniz University Hanover, Hanover, Germany); and Prof. Dr. Siegfried Jetzke (Ostfalia University of Applied Sciences, Salzgitter, Germany).

I thank Erwin Kalvelagen (www.amsterdamoptimization.com) for his kind permission to use his `GAMS` file *lagRel.gms* in `MCOL` and parts of his description on Lagrange relaxation in this book. From Fair Issac Corporation, I received the permission to use the FICO® Xpress Mosel modeling examples and FICO® Xpress Insight visualization examples. FICO is a trademark of Fair Isaac Corporation.

Finally, I thank my daughter Diana for producing the cartoon-like illustrations in this book and proofreading, and Christian Rauscher, the editor in charge at Springer (Heidelberg), with whom I had been working for many years — and who has promoted this second edition of the book.

Weisenheim am Berg, Germany Josef Kallrath
September 2020

Preface

This book arose from a realization that modeling using mathematical programming should be tightly linked with algorithms and their software implementation to solve optimization problems. Such linkage is necessary for a full appreciation of the methods used to model problems that will ensure they can be solved successfully. While there exist textbooks concentrating on the pure mathematics aspects of optimization, and others which just describe applications without providing sufficient technical background, we see our book as trying to provide a link between applications and the mathematics required to solve real-world problems. Few textbooks have integrated modeling with state-of-the-art commercially available software. Our book will also incorporate this missing link and will include the software to solve the models discussed.

Optimization using mathematical programming is an important subject area as it can determine the dramatic savings available to organizations that could not be achieved by other means. In this book, examples are cited where organizations are saving many millions of pounds (sterling) or dollars (US) by using optimization methods. Mathematical optimization models are part of tools that can help people in the process of making decisions concerning the use of resources and saving costs.

Mathematical programming also provides a way to solve problems that, because of their size or other features, would not otherwise be solvable by other methods. In major cities, e.g., London, mathematical programming models influence the control of the flow of domestic water through the city as the model is used to determine the most efficient strategy to move water from source to user as peaks and troughs in the usage pattern develop. Thus, the results from mathematical programming models are literally all around many of us.

The need for a source book of material on the subject was recognized while teaching at Heidelberg University and Loughborough University and while planning conference sessions on the practical relevance of mixed integer optimization.

Although there is an extensive literature on mathematical programming, the paucity of instructional materials in the area of efficient modeling and solving real-world problems is striking. The student, researcher, or industrial practitioner must read between the lines of material, usually only available in journal articles or similar, to glean the details of the modeling process and the "tricks of the trade". Yet the need is acute: as with many other areas of science, the computer revolution has given many modelers in industry as well as at universities the tools to attempt to solve realistic and complex models. In this work, we endeavor to provide a suitable background as an aid to the novice modeler, a useful reference book for the experienced modeler, and a springboard for the development of new modeling ideas. In particular, by tailoring this book around a commercially available software package we are able to illustrate some of the subtle details that determine the success or failure of the modeling efforts.

Readership

This book has been planned for use by more than one type of readership. Most of this book is designed to be used by readers who possess fairly elementary mathematical skills, i.e., the use of algebraic manipulation, and it is made clear which sections are not of this type. Further mathematical skills required are developed during the course of the book but the presentation should not prove too daunting. The material is suitable for use in courses in Business and Management Studies and operations research environments. Readers with stronger mathematical skills (e.g., linear and matrix algebra) and experienced practitioners in the field will still find much to interest them as the logic of modeling is developed. The book, therefore, will provide appropriate course material for lecture courses, short courses and self-teaching on the topics contained in it.

As some material is for the more advanced reader, or for the reader to use on a second pass through the book, certain sections in chapters have been marked as "advanced". These sections may be omitted on a first pass through the book. The more advanced parts of the book are written in such a way that it is sufficient if the reader is familiar with the basic concepts and techniques of linear algebra. A discussion of some foundations of optimization is provided at the end of some chapters, where it is helpful if the reader has familiarity with calculus techniques. It is also expected that the later advanced chapters will be read only once the reader has started to build models in earnest. A glossary at the end of the book will provide further help.

Scope

The focus of the book is primarily on models, model applications and individual case studies rather than algorithmic details. However, because the success of solution of complex problems requires efficient problem solving, it is important that models and algorithms are tightly connected. Therefore, we also concentrate on the mathematical formulation of models and the mathematical background of the algorithms. The understanding of the mathematics involved in a problem or model explains why certain model formulations work well while others do not. We have tried to present in this book a self-contained treatment of the subject where possible. The presentation of the material is not too far away from what real modeling in business looks like. Most of the case studies have a commercial or industrial background. For instance, some of the case studies in Chap. 10 stem from problems recently analyzed and solved in a mathematical consultancy group in the chemical industry.

Organization

Chapter 1 gives an introduction and overview of the field. Parts of this chapter, in particular the details on the software used in this book, can be skipped by the experienced practitioner. An overview on the history of optimization is presented in the appendix to Chap. 1. It is presented as an appendix because it requires some familiarity with the terminology of the subject. This chapter and parts of Chap. 2, illustrating how small linear and integer programming problems may be formulated, are kept on a very elementary level appropriate to the novice without a background in mathematics. We provide a systematic overview of mathematical solution techniques on both linear and mixed integer linear programming in Chap. 3, while Chap. 12 contains details on nonlinear optimization techniques. Exercises are included at the ends of chapters. These exercises can be tackled by hand or by using the software, where appropriate, included with the book.

Types of linear programming problems and their modeling are discussed in Chap. 4. Chapter 5 is a collection of case studies in the framework of linear programming. Chapters 6 and 7 cover foundations of integer programming while in Chap. 8 case studies are discussed. In Chap. 9 we consider how practitioners may best set up and solve their optimization problems and in Chap. 10 we consider examples of large cases. Then in Chap. 11 we consider other types of optimization, e.g., sequential linear, quadratic and mixed integer stochastic programming. New to the second edition are the chapters devoted to nonlinear optimization (Chap. 12), deterministic global optimization (Chap. 13), polylithic modeling and solution approaches (Chap. 14), cutting and packing (Chap. 15), impact of optimization and especially parallel optimization (Chap. 16). Finally, Chap. 17 reflects the author's view on mathematical optimization and modeling, how it is and should be used, and what is to be expected from it in the future.

Certain sections of chapters may be skipped by readers new to the area of optimization. They are marked by $\ominus$ in the section heading. These sections should be read through when required on a subsequent reading.

Instructors Manual

An Instructors Manual is available to *bona fide* lecturers. Please, contact the author Josef Kallrath.

Acknowledgments

We would to like to thank colleagues and mentors who have advised and/or inspired us over the years. These are too numerous to mention but we would like to single out Beate Brockmüller for providing material on the telecommunication network problem, Bob Daniel and Gunter Schnabel for the time they spent with us discussing the manuscript in great detail, Tom Horak, Gernot Sauerborn, Anna Schreieck, James Tebboth, Christian Timpe and Max Wagner for reading the manuscript (JK), and inspiration from the work of Peter Hammer, Ailsa Land, Gautam Mitra, Paul

Williams and the late Martin Beale (JMW). We also offer our thanks to Dash Associates for help, advice and cooperation over the inclusion of the XPRESS-MP software and related discussion material.[5] JK wants to express his special thanks to Marilyn Dalton for her kind hospitality during numerous visits to Blisworth House. Finally, JK wants to thank the clients involved in some of the real-world cases. The interaction and communication with the clients, most of whom were enthusiastic persons with deep knowledge of the business process they wanted to improve using mathematical optimization, was an important and irreplaceable resource which made the solution of challenging problems possible. Although, after all the years these people might have forgotten[6] the work and the exciting time we spent together and might not be aware how they indirectly contributed to this book I want to mention them: Peter Bassler, David DeSantis, Andy Hayter, Klaus Kindler, Jan Orband, Gunter Schnabel, Hubert Smuda and Eckhardt Schubert.

Ludwigshafen, Germany — Josef Kallrath
Loughborough, England — John M. Wilson
1997

[5]Dash holds a copyright on parts of the following Sects. 2.7, 10.2, 10.3 and 16.6.

[6]Of course, I hope they have not forgotten! All the special adventures involved in the time working on their problems could fill a book on its own.

Contents

About the Author

Prof. Dr. Josef Kallrath has studied mathematics, physics, and astronomy in Bonn, where he received a doctorate in 1989 for his dissertation in astrophysics on the dynamics of colliding double stellar winds. He has worked in companies (BASF SE, Ludwigshafen; 1989–2019), has been freelancing since 1998 as a scientific consultant and lecturer, and solving practical problems in industry with scientific computing and operations research techniques, most of it being modeling and solving mathematical optimization problems. His work focuses on mathematical optimization to support decisions in process, paper, metal, and energy industry, as well as transport infrastructure and the modeling of physical systems. He has taught at the University of Heidelberg (1991–2001) and at the University of Florida in Gainesville/USA (since 1997). Since 2002, Prof. Kallrath has been heading the Mathematical Optimization Practice Group of the German Operations Research society (GOR).

List of Figures

Chapter 1
Optimization: Using Models, Validating Models, Solutions, Answers

This chapter guides the reader into the field of mathematical optimization, distinguishes optimization from simulation, and introduces the key objects used in optimization models. It sketches how simple linear programming (LP) problems may be solved graphically. The chapter contains a survey of real-world problems and finally a review of the history of optimization. Those readers who want to use the optimization software attached to this book will find sufficient information in this chapter to do so. Readers less interested in implementation issues or already familiar with the use of commercial software may of course skip those parts.

1.1 Introduction: Some Words on Optimization

The title of this book contains the term *mathematical programming*. So let us first consider the question: "What is mathematical programming[1] or optimization and what are optimization problems?" Optimization problems arise in almost all branches of industry or society, e.g., in product and process design, allocation of scarce resources, production, logistics, scheduling, strategic planning, or traffic control. They can have different time scales: real-time (just now), operational (short-term), tactical (mid-term), and strategic (long-term) planning. Unfortunately, the word *optimization*, in nontechnical language, is often used in the sense of *improving* while the original meaning of the word is related to finding the *best*. In an optimization problem (OP), one tries to minimize or maximize an important

[1]The term *mathematical programming* has its historical roots in the first problems solved by *linear programming* in the Second World War. In these days, *programming* had more the meaning of *planning*. Nowadays, *mathematical optimization* seems more appropriate. In this book we will use the terms *mathematical programming* and *mathematical optimization* synonymously.

J. Kallrath, *Business Optimization Using Mathematical Programming*,
International Series in Operations Research & Management Science 307,
https://doi.org/10.1007/978-3-030-73237-0_1

characteristic of a process within the problem such as revenue, elapsed time or cost, by an appropriate choice of decisions which influence this process, e.g., budget allocations, manpower deployed, or quantities to be manufactured. Such decisions can be controlled, but are influenced and ultimately limited by a set of constraints, linked, for example, to physical limits, e.g., limits on the budget or manpower available. The results of the optimization process will suggest decisions that could be taken and, if appropriate, the levels of resources that should be utilized.

A traditional way to develop answers to decision problems is to propose a number of choices for the values of the decisions, using trial and error methods. The processes under investigation are then evaluated or simulated under these various choices, and the results are compared. This concept of *simulation* [cf. Pidd (1992,[439])] may involve varying certain input data or building scenarios from investigating the behavior of a model under various sets of conditions. Simulation can be carried out using sophisticated software systems, e.g., `SIMSCRIPT`[2], and is also prevalent as the "what if" facility provided by more general purpose software such as spreadsheets (e.g., `LOTUS`[3], `EXCEL`[4]). Analysts in charge of simulating these decisions have developed intuition and simple rules to select appropriate conditions, and simulation software exists to perform the evaluation of their performance. The "traditional" techniques may lead to useful and usable results, but there is no guarantee that the best solution (the optimal solution) or even a solution close to the optimal solution has been found. This is especially troublesome for large or complex problems, or those which require decisions with high financial impact. Below we will discuss the difference between simulation and mathematical optimization. This difference can be more easily discussed when we have answered the following question:

What do we need when we want to solve a real-world problem by mathematical optimization?

The first thing we need is to represent the real-world problem by a *mathematical model*. A mathematical model of a system is a set of mathematical relationships (e.g., equalities, inequalities, logical conditions) which represent an abstraction of the real-world problem under consideration [see Fig. 1.1]. Usually, a mathematical model in optimization consists of four key objects:

- data or parameters,[5]
- variables (continuous, semi-continuous, binary, integer),[6]
- constraints (equalities, inequalities),[7] and
- objective function.

[2] `SIMSCRIPT` is a trademark of CACI products.

[3] `LOTUS` is a registered trademark of Lotus Development Corp.

[4] `EXCEL` is a trademark of Microsoft Corp.

[5] Engineers often prefer the term *parameters* to refer to fixed data.

[6] Synonyms are *decision variables*, *unknowns*, or *columns*, depending on community.

[7] Sometimes also called *restrictions*.

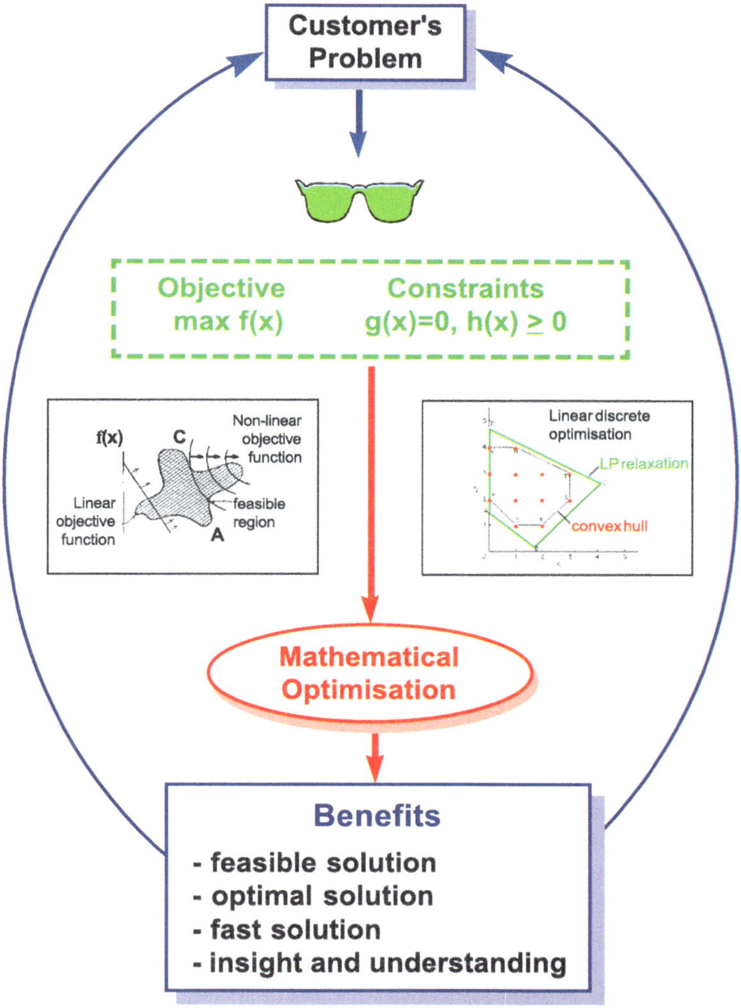

Fig. 1.1 Transforming a real-world decision problem to a mathematical optimization problem. The input data and parameters are hidden in the step from the customer's problem through the green glasses to the mathematical structure (objective function and constraints)

The data or parameters define an instance of a problem. They may represent costs or demands, fixed operating conditions of a reactor, capacities of plants, and so on. The variables represent the degrees of freedom, i.e., our decisions: how much of a certain product is to be produced, whether a depot is closed or not, or how much material should be stored in the inventory for later use. The constraints can be a wide range of mathematical relationships: algebraic, logic, differential, or integral; in this book, only algebraic and logic constraints are considered. They may represent mass balances, quality relations, capacity limits, and so on. The objective function, finally, expresses our goals or objectives in a mathematical form: minimize costs, maximize utilization rate, minimize waste, and so on. When building mathematical optimization models they usually lead to structured problems such as:

- linear programming (LP) problems,
- mixed integer linear programming (MILP) problems,
- nonlinear programming (NLP) problems, and
- mixed integer nonlinear programming (MINLP) problems.

Besides building a model and classifying the problem one needs a *solver*, i.e., a piece of software which has a set of algorithms capable of solving the problems listed above.

Above, the terms *integer* or *mixed integer* occurred. What is *mixed integer* or *discrete*[8]*optimization*? Classical optimization theory (calculus, variational calculus, optimal control) treats those cases in which the decision variables can be changed continuously, e.g., the temperature in a chemical reactor or the amount of a product to be produced. On the other hand, *mixed integer*, *combinatorial*, or *discrete optimization* addresses degrees of freedom which are limited to integer values, for example, counts (numbers of containers, ships), decisions (yes–no), or logical relations (if product A is produced, then product B also needs to be produced). This discipline, formerly only a marginal discipline within mathematical optimization, is becoming more and more important (Grötschel, 1992,[243]) as it extends the power of mathematical optimization to situations which are relevant to practical decision making where business success is at stake.

Above, simulation is also mentioned as a mean to improve a company's business. *What is the difference between simulation and mathematical optimization?* In contrast to simulation, optimization methods search directly for an optimal solution that fulfills all restrictions and relations which are relevant for the real-world problem, and prove that a certain solution is optimal (*proof of optimality*). In simulation, the effects on a model of selected solutions will be examined, but there will be no guarantee that any of the solutions under consideration is optimal. Besides optimality, notice the other very substantial difference between simulation and optimization: the existence of constraints. While in simulation somebody has to make sure that only those combinations of variables are evaluated which represent "appropriate conditions," in optimization models it has to be specified a priori what makes a feasible solution. Once the problem has been solved to optimality within the specified limitations or constraints, then it is known that no other set of decisions or values for quantities suggested by these decisions can provide any other solution which will give a "better" value to the characteristic that has been optimized. "Better" here will mean larger if the characteristic is to be maximized (e.g., discounted revenue being maximized) and smaller if the characteristic is to

[8]The terms *mixed integer optimization* and *discrete optimization* are used synonymously in this book. Sometimes, in the literature, the term *combinatorial optimization* is also used.

be minimized (e.g., total cost being minimized). The optimization process is a prescriptive one which tells its user what decisions should be taken and what levels of resources should be used. These levels might eventually not be accepted at face value in their entirety, for a variety of reasons, but they provide a yardstick against which all alternatively proposed decisions and levels of recommended resource can be compared. This is not to suggest that an organization will choose to implement a policy that is not the best for any frivolous reason, but rather it may choose to implement another policy which is very near to being optimal but which has certain other appealing features about it, e.g., it may only involve disturbing 10% of existing practice, whereas the optimal solution would disturb 90% of existing practice. Such features may be intangible and difficult to include directly in a representation of the problem. Based on the two fundamental differences (optimality proof, incorporation of constraints) between simulation and optimization we conclude: by using mathematical optimization it becomes possible to control and adjust complex systems even when they are difficult for a human being to grasp. Therefore, optimization techniques allow a better exploitation of the advantages inherent in complex systems.

What commercial potential is in mathematical optimization? To give some idea of the scope of mathematical programming in helping organizations we cite four recent examples of benefits achieved. First, at Delta Airlines it is reported in a paper by Subramanian et al. (1994,[536]) that use of an optimization model is expected to save the company (US)$300 million over a 3-year period. Similar models in use at other airlines around the world are likely to achieve comparable savings. Second, a comprehensive mathematical programming model used by Digital Equipment Corporation (DEC) is described by Arntzen et al. (1995,[33]). The model has helped the company to save (US)$100 million. Third, in the UK considerable savings in the electricity industry are attributable to the use of mathematical programming. Fourth, in some blending problems in the 1990s BASF-AG saved several hundred thousand DM/year. The potential saving in complex production networks in chemical industry is of the order of millions of DM. Pay-back time of projects is usually less than six months. Blackburn et al. (2014,[83]) have highlighted optimization-centered applications of Operations Research (OR) at BASF in the area of supply chain management (SCM) through the use of selected examples with annual cost savings ranging between several hundred thousand and several millions €. The role of mathematical optimization and its impact in industrial practice in chemical and petroleum industry is well described and reviewed by Grossmann & Harjunkoski (2019,[241]); they report about 30% energy savings in operations.

The reader can find more about the financial impact of optimization (and Operations Research in the broader picture) by searching for the Franz Edelman Award on www.informs.org, the webpage of the Institute for Operations Research and the Management Sciences (INFORMS). First awarded in 1972 as the INFORMS practice prize TIMS, in 1982 the prize has been renamed in *Franz Edelman Award* and is annually presented in honor of Franz Edelman, who founded the Operations Research division within Radio Corporation of America (RCA), one of the first corporations to embed operations research as a business imperative. Since its inception, nearly $250 billion in benefits have been tabulated among Franz Edelman Award finalist teams. Following the competition, INFORMS publishes papers by the Edelman finalists in the January issue of the INFORMS journal *Interfaces*.

1.2 The Scope of this Book

In the first edition of this book published in 1997, this section started with the following paragraph:

This book will consider the two types of mathematical programming problems introduced, namely the linear programming problem and its extension, the "mixed integer linear programming problem," and certain variations on these problems. We will consider how to set up these types of problem, how they are used, how they are solved, and how to use software to both model and solve practical situations that can be tackled using these two types of problems. Some other types of problems, principally those most closely related to these two, will be considered, but in very much less detail. That is not to say that problems which can be handled by techniques such as dynamic programming [cf. Bellman (1957,[59])] or nonlinear programming [cf. Chap. 12 or Bazaraa et al. (1993,[53])] are not important, but rather that most commercially available software has concentrated on the linear programming problem or the mixed integer linear programming problem because they are the most useful in applications and the most straightforward to solve for problems of large size. Thus we will tend to "follow the crowd," but this will have the advantage that we can benefit from the considerable effort that has gone into solving such problems over the last 30–40 years, and, in particular, from the developments in software to solve such problems.

The chapters of the book introduce the concepts necessary for an understanding of mathematical optimization problems. However, these topics are not introduced in a purely abstract form, but rather they will be related to using these techniques and approaches together with a software system, as would be done in practice.

In this second edition, the scope has been significantly extended. It covers nonlinear optimization as well as mixed integer nonlinear optimization. Polylithic modeling and solution approaches have been added and optimization under uncertainty has been treated in more detail. Readers interested in cutting and packing will find more specific material on this interesting topic with relations to computational geometry. Unless readers want to start with the nonlinear world right away, the easier path to follow is to start with linear programming, moving to mixed integer linear programming, and finally entering the world of nonlinear optimizations problems.

The software, i.e., the model files which come with this book, can be solved with the algebraic modeling systems (`AMSs`, hereafter) and the algebraic modeling languages (AMLs) contained therein: `GAMS`,[9] `FICO Xpress Mosel`, (short

[9] www.gams.com; see also Brooke et al. (1988,[100]) and Bussieck & Meeraus (2003,[106]).

form: Xpress Mosel or just Mosel),[10] or SAS/OR;[11] it should not be too difficult if the reader wants to use AMPL,[12] instead. Their modeling environment provides the user with an easy to use yet powerful language for implementing the problem. It also enables the user to gather the appropriate data from text files and a range of popular spreadsheet systems and database systems. The optimization software provides a solver that uses the problem description produced by the modeler and then solves it to provide the user with the optimal solution. AMLs and AMSs have been available and updated since the early 1980s and are innovative optimization systems widely used worldwide; for a timeline and a historical review see Kallrath (2004,[306]).

Coming with this book are coded models in GAMS, Fico Xpress Mosel, and SAS/OR hosted at www.springer.com. Some of these models are stand-alone models, while others will progressively introduce new features so that by the later stages of the book the reader should be able to formulate, model, and solve business problems of commercial size.

Apart from the classical AMLs we mention the recent Python based development Pyomo[13] as well as JuMP (a domain-specific modeling language for mathematical optimization based on Julia).[14] Matlab now also offers modeling based mathematical programming. Students might like to see optimization problems solved with software such as R or maybe Python as they are indeed very popular. We do not include R, Python, Pyomo, or JuMP code in this book, as we feel that especially the Python based approaches lack the stability required by industrial needs. That should, however, not discourage students from solving the exercises by using the tools.

Exercises will be presented at the end of most chapters. These will vary in difficulty and it is intended that these be worked through by the reader either by hand or by using an AMS — or by using R, Python, Pyomo, or JuMP if preferred. Complete solutions to exercises will be given with full instructions relating to the use of software on such exercises, where appropriate.

1.3 The Significance and Benefits of Models

The term *modeling* presupposes the term *model*, which first derives from the Latin word *modelus* (scale [diminutive of mode, measure] and then from the word *modelo*

[10] www.fico.com/xpress; see also Heipcke (2002,[259]), Guéret et al. (2002,[245]), Colombani & Heipcke (2002,[128]), Ciriani et al. (2003,[124]) and Colombani et al. (2004,[130]).

[11] http://go.documentation.sas.com; see also *SAS/OR User's Guide* (2018,[483]).

[12] www.ampl.com; see also Fourer et al. (1987,[198]) or Fourer et al. (1990,[199]) and Fourer et al. (2003,[200]).

[13] www.pyomo.org.

[14] www.juliaopt.org.

formed in the sixteenth century and which is used today in everyday life and in scientific language with various meanings, but mostly in the sense of a simplified, abstracted, or structured representation of an interesting part of reality. However, the idea and thus the terminology are much older. Even with Pythagoras around 600 BC, classical geometry consciously distinguished between wheel and circle; field and rectangle. Around 1100 A.D., a wooden model was made as an abstraction of the later Speyer Cathedral. Astrolabes and celestial spheres were used as manifested models of the motion sequences in the firmament to calculate the rising and setting of the sun, moon, and stars. Until the nineteenth century, mechanical models were understood as images of reality; they attributed all processes to the movement of the smallest particles, which followed the principles of classical mechanics. Further attempts were made to reduce all physical and other processes to the mechanistic model. Nowadays in physics and mathematical sciences one speaks of models, if

1. for reasons of simplification, the investigation is limited to certain phenomena which, in a given context, are regarded as important (*example*: in the movement of the planets, the spatial extent of these bodies is initially neglected),
2. for reasons of didactic illustration, one gives a picture based on classical ideas for phenomena which are not accessible in a descriptive way (*example*: the planetary model to illustrate the conditions in atomic nuclei), or
3. one studies situations in one range in analogy to known situations in another range (analogy models).

Let us generally define a model as an appropriate abstract representation of a real system [cf. Williams (1993,[580]) or Bossel (1994,[93])]. Naturally, a mathematical model of a process or a problem is formulated with the help of mathematical objects (variables, terms, relations). A (mathematical) model represents a real problem in the language of mathematics, i.e., using mathematical symbols, variables, equations, inequalities, and other relations.

A very important aspect, which is connected with modeling and precedes it, is the *model purpose*. It results directly in connection with the problem and very substantially influences the process of modeling. In science (e.g., Physics, Astronomy, Chemistry, and Biology) models are used to gain a deeper understanding of processes occurring in nature (an epistemological argument), e.g., to explain the movements of planets. The comparison of measurements and observations with the predictions of a model is used to determine the appropriateness and quality of the model. Sir Karl Popper (1980,[444]) uses the expressions *falsification* and *verification* in his famous book *Logic of Scientific Discovery* as tasks when accomplished deciding on whether a certain model is eliminated, slightly modified or accepted in improving the scientific process. Later, aspects and questions of accepting and improving global and fundamental models (e.g., general relativity or quantum physics) formed part of the discussion of the philosophy of science.

In the disciplines "*Scientific Computing*" and "Operations Research" models often have a rather local meaning; a special part of reality[15] is illustrated in great detail. Here, on the one hand, the motivation for model development is epistemological. On the other hand, we find military, pragmatic, and commercial aspects, from which operational instructions for actions can often be derived. What follows is a list of reasons and motives. The model maps most of the relevant features and neglects less important aspects to

- provide insight into the problem,
- permit experimentation but avoid expensive and/or dangerous experiments,
- avoid the production of unwanted side products,
- optimize some objective,
- propose careful use of resources.

Let us now have a closer look on how to develop a mathematical model for a real-world problem, or alternatively, how does one transform a given real-world problem into a mathematical model? The associated process of modeling is by no means easy nor unique; its difficulties somewhat resemble the solution of text problems, in school, cf. Polya (1979,[443]), but the advantage was that from the teaching content it was always known which mathematical technique was the most suitable one to solve the problem at hand, e.g., the law of Pythagoras, quadratic equations, linear systems with several variables, to name a few. In practice, however, one usually does not know which mathematical technique is needed. Therefore it is important to be open and creative in the first approaching attempts to understand and solve the problem. It is helpful to know as many problem solving strategies and heuristics [cf. Michalewicz & Fogel (2000,[397])] as possible and to be able to use them or develop new ones from them. The following points are useful to remember when starting to build a model:

1. There is no precise recipe telling the user how to build a model,
2. Experience and judgment are two important aspects of model building,
3. There is not necessarily a *correct* model,
4. There may not necessarily be a *unique* model, as different models focusing on different aspects may be appropriate. Co-existence of models can be useful to investigate different aspects (*example*: wave-particle dualism in physics). It should be noted that different models for the description of the same facts may require different data input based on very different databases.
5. Model building should follow the principle "as simple as possible, as complex as necessary" or in a free translation of the words of Antoine de Saint-Exupéry: "A

[15]Reality itself is already as compact as possible to be able to present itself lossless. Consequently each self-image on a small section (individual human being) is inevitably connected with a loss of information. So we shape ourselves and our view of the world only by a very small part of reality which is filtered by our sensory organs and learned behavior. Our terms and languages in general are products of limited perception — and therefore highly subjective. Modeling then occurs as reduction in already limited worldviews.

model is not ready when you can't add anything anymore, but when you can't leave anything out."

However, there are some aspects that should be considered when building a model:

1. *Definition of the problem and the purpose of the model*: A clearly formulated task should be used as a basis for defining the purpose of the model.
2. *Problem scope*: It is necessary to clearly define what belongs to the problem or model and what does not (*example*: in a production planning system for medium- and long-term planning we may neglect detailed planning aspects).
3. *Identification of important model objects* (*examples*: products, production sites, warehouses, transport links, time periods, etc.) and relations between these objects.
4. *Acceptance* and *users*: The model formulation should be adapted to the problem, the model purpose, and the potential user. For instance, the planner who needs to accept the model for his planning purposes. It helps if the modeler knows all relevant aspects — and the motivations of people involved or designated to use the model. Models for vehicle routing and their use can encounter acceptance problems among truck drivers, if the client, the routing planner, has named the minimization of travel costs as a target for the modeler, but each individual truck driver is paid on the basis of the delivery value of goods. So a driver would rather take a trip with one expensive piece of furniture to be delivered and mounted than a trip with many destinations, where only cheap and perhaps even time-consuming set-up work has to be done.

These points may already indicate that modeling is one of the most important and critical tasks in optimization. This might suggest the question: is it possible to learn good and efficient modeling? The answer is both "Yes" "No." For good model building, experience is essential; cf. Williams (1993,[580]) or Ashford and Daniel (1992,[37]). The choice of the right decision variables to include the model and a good model formulation correlate with one another. It is not necessarily the number of decisions or restrictions that reflects the complexity of a problem [concerning complexity, see remarks on page 555]. The criterion that separates a good formulation from a bad one has a mathematical foundation which is difficult for a modeler to establish, if at all possible. Nevertheless, rigid approaches exist which try to improve the model formulation by exploiting certain mathematical features [see Chaps. 8 and 9, and examples]. The effect is that the computing time needed to solve the problem is reduced significantly. For further discussion see Jeroslow & Lowe (1984,[288]) and Williams (1993,[580]). It is important that a model developer needs to be knowledgeable about both: The "model" (the right level of abstraction...) and the "mathematical formulation," which is also often referred to as the "mathematical model." We will come back to the importance of good modeling again and again in this book. At this stage, let us first focus on the question (Fig. 1.2):

Fig. 1.2 A powerful modeler. Sometimes, modeling done right seems like magic. Produced for this book by Diana Kallrath, Copyright ©2020

What is a model good for and how can we benefit from it? A well formulated and documented model is useful for deduction, i.e., it allows us to derive consequences and results. The results from the model have to be interpreted with respect to their consequences in reality. The model building process increases our understanding of a real-world problem. It documents structural features of that problem and thus provides the know-how that others can use later on. The evaluation of a model is coupled to the degree of reality contained in the model and the modeler's power. While within mathematics a strict mathematical *proof* is the ideal measure, and one speaks of natural scientific models being subjected to *falsification* or *verification*, in economic or business orientated models *validation* is what comes next to falsification and verification, whereby the relations

$$\boxed{\text{Validation} \leq \text{Falsification/Verification} \leq \text{Mathematical Proof}}$$

are fulfilled. However, one should always be aware that with a well-validated model, the model is only a partial aspect of reality, and hence the model is less than reality. However, it does not necessarily have to be a lot less. Electrodynamics, for example, reproduces macroscopic electromagnetic phenomena so accurately that it is possible to speak of a match between model and reality within the scope of measurement accuracy. Furthermore, for the real-world problems and models we are dealing with, we have to keep in mind that today's reality (and model) can be quite different from tomorrow's reality. Let us conclude this section with a statement found in Box & Draper (1987,[95]) "... all models are wrong, but some are useful."

1.4 Mathematical Optimization

Four fundamental problem types are considered in this book. The first is linear programming (LP), introduced now, the second is mixed integer linear programming (MILP), to be introduced in Chap. 2, the third and fourth are nonlinear (NLP) and mixed integer nonlinear programming (MINLP) to be treated in Chap. 3.

1.4.1 A Linear Optimization Example

We now turn to an example where we will identify the main features of this linear optimization problem. Sunshine Cruises is a national company specializing in renting out river-going boats to tourists. The company makes use of two types of boats — the Premier and the Standard. The Premier is the more expensive boat to hire and each Premier boat brings in an average net revenue of 800 per week, while for the Standard the corresponding figure is £600 per week. The company is about to scrap its older boats and lease new ones and has asked for some advice about decision making. Decisions to be taken will be based on a scenario of "a typical week."

The company has maximum berthing accommodation around the country for 350 boats in total, of which not more than 200 can be Premier boats. The company also wants to have no fewer Premier boats than Standard. Maintenance of boats is a further restriction. Each Premier craft requires 4 h maintenance per week while each Standard craft requires 3 h. The maximum number of maintenance hours available per week is 1400 (Fig. 1.3).

Fig. 1.3 Boat problem illustrated. The decision maker from Sunshine Cruises pondering the number of Premier boats and Standard boats. Produced for this book by Diana Kallrath, Copyright ©2020

The problem introduced so far is simple and structured. Many aspects of complexity[16] have been removed in advance and much information in the problem is in an easily digestible form.

What are the degrees of freedom and what are the decisions?

Berthing and maintenance are consequences of having boats and are not fundamental decisions. The fundamental decisions are:

How many Premier boats should be leased?

How many Standard boats should be leased?

Once we have the answers to these questions, most other aspects of the problem fall into place. We now endeavor to express the problem in terms of variables representing the decisions (decision variables) and subject these decisions to a series of conditions (constraints). We also associate a "driving force" with the problem termed the objective (objective function) and we aim to optimize this objective.

[16]The advanced reader or those who have had already a glance at Chap. 3 notice that this problem is technically an integer linear programming problem. But for the moment, we avoid the difference between Linear Programming involving only continuous variables and Integer Programming involving integer variables.

Let p be the desired number of Premier boats, and s the desired number of Standard boats. For practical sense it is clear that neither p nor s should ever take a negative value, but otherwise we can expect large or small values for p and s - at the moment we are uncertain. It might be more realistic if p and s could avoid taking values which include fractions. The fractional solution of 18.5 Premier boats might not be appropriate, even if it were acceptable mathematically. However, we will assume that for the moment such an answer will be regarded as appropriate, if it arises, as such an answer is considered admissible in a linear programming problem.

For the example introduced, the variables can now be linked together in expressions to model the relationships of the problem.

Firstly, there is a limit on accommodation for berthing. The total number of boats in our model is

$$p + s \tag{1.4.1}$$

and this total is limited by 350. We therefore introduce the inequality

$$p + s \leq 350 \tag{1.4.2}$$

to model this restriction. Notice that the restriction takes the form of an inequality rather than an equation as the total number of boats which is desired to be used may turn out to be less than 350.

The restriction that not more than 200 can be Premier boats is modeled by the inequality

$$p \leq 200. \tag{1.4.3}$$

An inequality as (1.4.3), or any constraint (equality or inequality) containing a right-hand side coefficient and only a single variable with coefficient 1 in front of it is called a *bound*. It turns out that mathematical programming software handles bounds particularly efficiently.

To model the maintenance restriction we first calculate the total number of maintenance hours required. This is obtained by taking the number, 4, of maintenance hours required by each Premier boat and multiplying it by the number, p, of Premier boats used giving $4p$ and adding to that maintenance time a similar expression for the Standard boats, $3s$, giving $4p + 3s$. We then construct the inequality

$$4p + 3s \leq 1400 \tag{1.4.4}$$

to show that the total number of maintenance hours used is limited by 1400.

To model the restriction that there must be no fewer Premier boats than Standard we introduce the constraint

$$p \geq s,$$

which may be rewritten as

$$p - s \geq 0. \tag{1.4.5}$$

Finally we have the non-negativity conditions mentioned at the start of the definition of the variables

$$p \geq 0 \quad , \quad s \geq 0. \tag{1.4.6}$$

The inequalities (and any equations, if they are required) which restrict the solution space are called *constraints*.

Notice that different units are involved in different constraints. The first two constraints are expressed in terms of numbers of boats, but the third is measured in terms of hours. It is very important and always required that there is consistency between all terms of a constraint, in particular between the left-hand side of a constraint and the right-hand side of a constraint — these two sides must be expressed in the same units. Clearly, quantities with dimensions *kg* and *cm* cannot be added, but care is also needed when quantities of similar type are added (e.g., cm and km, or liter and m^3) which need to be converted to the same units before addition takes place. In general a model contains several units in different constraints: e.g., energy, mass flow, number of personnel at work, money, and so on.

All of the restrictions on the problem have now been modeled. For the problem the obvious objective seems to be to maximize net weekly revenue. As each Premier boat gives us revenue of £800 per week, if we have p Premier boats, then the weekly revenue from these is $800p$. Similarly, the revenue from s Standard boats is $600s$, so the revenue function expressed in £ units is

$$800p + 600s. \tag{1.4.7}$$

This function is called the *objective function* (or objective) and is to be maximized. In mathematical programming problems an objective function has to be maximized or minimized (e.g., minimization may be appropriate if the objective function was a measure of cost). The optimal solution is characterized by the optimal objective function value and the solution values of the variables leading to that value. Let us summarize our problem as follows:

$$\max \quad 800p + 600s \tag{1.4.8}$$

subject to harbor capacity, limit on the number of Premier boats, maintenance capacity, more-premier-than-standard boats, and non-negativity:

$$\begin{array}{rcr} p + s & \leq & 350 \\ p \phantom{{}+3s} & \leq & 200 \\ 4p + 3s & \leq & 1400 \\ p - s & \geq & 0 \end{array} \quad , \quad p \geq 0 \quad , \quad s \geq 0. \tag{1.4.9}$$

The problem just introduced, along with other problems containing many more constraints and variables, will be solved using optimization software. However, problems containing only two variables may be represented graphically as follows, and this representation gives some "feel" as to the operation of the optimization process.

How might we solve such a problem? We might start with a trial and error approach. We could try values of 50, 100, 150, 200, 250, 300, 350 where appropriate for the variables and we might hit on $p = 150$ and $s = 150$ and the revenue = 210,000 or $p = 200$ and $s = 100$ and the revenue = 220,000.

This problem may be represented graphically as shown in Fig. 1.4. As s and p are non-negative, we need only concern ourselves with a graph of non-negative values. In the graph the boundary (excluding the axes) of the constraint

$$p + s \leq 350 \tag{1.4.10}$$

is represented by the line corresponding to

$$p + s = 350 \tag{1.4.11}$$

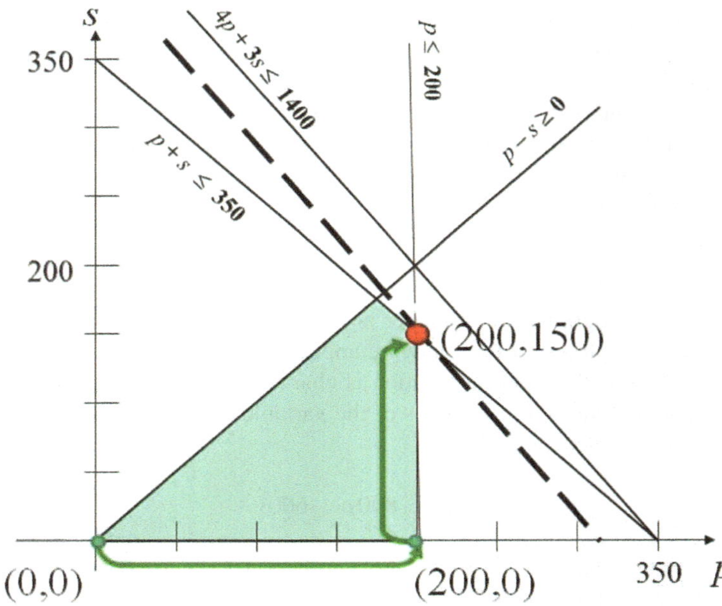

Fig. 1.4 Graphical solution of an LP problem in two variables, p and s. The colored area represents the feasible region established by harbor capacity $p + s \leq 350$, limit on the number of Premier boats $p \leq 200$, maintenance capacity $4p + 3s \leq 1400$, more-premier-than-standard boats $p \geq s$, and non-negativity $p \geq 0$ and $s \geq 0$. The dashed line represents the contour line of the objective function through the optimal point (in red) at (200, 150)

and as all values of p and s such that

$$p + s < 350 \tag{1.4.12}$$

are also valid, the region lying between the line and the axes is also valid. Other constraints are represented similarly. The shaded area denotes all the possible values of p and s satisfying the constraints. The objective function must lie parallel to any line of form

$$800p + 600s = k \tag{1.4.13}$$

where k is a constant. Such a contour line may be represented on the graph by choosing a typical value for k, say by taking values of 200 for s and p and then determining the slope of the line.

As we are maximizing, we wish to move the contour line such that the constant k increases, while still lying within the shaded area, or just touching it. This occurs at $p = 200$, $s = 150$.

In the small example the optimal solution turns out to be $p = 200$, $s = 150$ and objective function value = $250, 000. Thus the recommendation being made is to lease 200 Premier craft and 150 Standard craft, which can be accommodated, and will result in average net weekly revenue of £250,000. The solution is not entirely intuitive in the sense that the higher contribution is given by Premier craft, but leasing some Standard craft has also been recommended. Even on such a simple problem we see that mathematical programming appears to be able to provide us with information which we could not acquire in other ways. It should be noted that in this example it is fortunate that the decision variables took integer values in the solution, and so "made sense" as practical decisions. This will not happen in all such problems and we will see in Sect. 3.3 how to ensure that we do not end up with values such as "25.3 boats."

1.4.2 A Typical Linear Programming Problem

The problem of linear programming (LP) is to optimize (maximize or minimize) a linear function, see Eq. (1.4.14) below, of decision variables subject to a series of constraints, which are also linear, on these variables. A decision variable is denoted by a symbol, such as p, x, or y, used to represent a quantifiable decision, such as "how much" (continuous) or "how many" (integer valued) in a decision making problem. A new example is introduced here to illustrate some features absent from the example of the previous section. Considering a simple situation involving only the three decision variables x, y, and z, then a typical linear programming problem in these variables might be

$$\max \quad 2x + 4y + 3z \tag{1.4.14}$$

subject to

$$\begin{aligned} 3x + 2y + \quad\; z &\le 38.5 \\ x + \;\; y + 1.3z &= 10.0 \\ 4y - \quad\; z &\ge \;\; 6.2 \end{aligned} \tag{1.4.15}$$

with the additional requirement that all variables must take non-negative values (symbolically written as $x,\ y,\ z\ \geq\ 0$). It is very important to *specify the basic domain of variables*. In LP problems, the variables can usually take all real values, but often only non-negative values. For the boat problem in the previous section we should have explicitly required that the variables representing the "how many boats" decision can only take integer values.

A number of features of the LP problem described by (1.4.14) and (1.4.15) are typical of all linear programming (LP) problems.

1. There is an objective function (the first expression) which is to be maximized (in this case) or to be minimized. This function is linear in the (three) variables, i.e., the function contains only variables multiplied by a constant and each expression *Constant · variable* is separated from an adjacent pair by a + or − sign. There is no multiplication of variables such as $x \cdot y$ or introduction of mathematical functions such as power (e.g., x^3), exponential (e.g., e^x) or trigonometric (e.g., $\cos x$) functions of variables, which would make the problem nonlinear. Independent of the optimization sense, it is the objective function which drives the process of optimization.
2. The variables are constrained by a number of expressions introduced by the phrase "subject to" or the abbreviation *s.t.* for short. These are termed "constraints," and, like the objective function they are also linear expressions on the left-hand side. They are linked to a constant right-hand side by a relation which will be one of $\leq$, $\geq$, or $=$. No other relations are permitted.
3. There is also a requirement that the variables cannot take negative values (this requirement will be reconsidered subsequently).
4. In the linear expressions the constants are whole numbers and decimal fractions. GAMS allows symbolic expressions such as $3/7$, $\sqrt{2}$ or π, but approximates them to a given number of digits.
5. Inequalities $<$ or $>$ are not permitted[17] in the constraints, but this turns out not to be a drawback. Usually practical models will only require $\leq$ or $\geq$ constraints. However, if the condition

$$2x + 3.1y > 18.5 \tag{1.4.16}$$

[17]The reason is that continuous functions do not necessarily assume their optimum on non-compact sets. Using $\leq$ or $\geq$ ensures that the feasible set is compact, even a closed set in this case, which guarantees that the optimum is realized for some point contained in this set; see also Appendix C.7.

had to be included, then it could probably be replaced by a constraint such as

$$2x + 3.1y \geq 18.5 + \varepsilon, \tag{1.4.17}$$

where $\varepsilon = 0.000125$ is a sufficiently small quantity for the purpose of the application. This can present difficulties when integer values are required for variables and such a "sufficiently small quantity" is of equivalent magnitude to the degree of tolerance the software uses to determine whether a decimal number is actually an integer. Care should be taken in implementing $<$ or $>$ using this approach.

The frequently used convention that the right-hand side of a constraint has to be a constant, in order to represent a constraint in a standard form, leads to a constraint initially expressed in the form

$$x + 2y \leq 1 + z. \tag{1.4.18}$$

Simply collecting variable terms over to the left-hand side leads to the desired form

$$x + 2y - z \leq 1. \tag{1.4.19}$$

1.5 Using Modeling Systems and Software

The problem introduced in Sect. 1.4.1 was comparatively easy to handle. It had only two variables and four constraints, one of which was a simple bound. Realistic problems are much larger and may involve hundreds or thousands of variables and hundreds of constraints. Such problems must be handled systematically — in terms of modeling and solving the problem. To solve a larger problem some software will be required and this will take the form of a modeling system (sometimes called a Matrix Generator, or even a Matrix Generator) and an Optimization System which will optimize the model.

1.5.1 Modeling Systems

A *modeling system*[18] allows the user to specify variables, constraints, objective functions, and other details of the problem in a systematic way. The modeling system is then used to generate a mathematical programming matrix by systematically reorganizing this information into a standard form.

[18]The modeling system associated with the first edition of this book was `XPRESS-MP`, a typical representative of many commercially available systems in the late 1990s. `XPRESS-MP` contained the AML `mp-model` which appeared in 1983 on the market.

Modeling systems offer a number of advantages, such as the general aspects of software system discipline applied to modeling. Once a model is built using a Modeling System it is easy to modify the model to accommodate changes required by the client, particularly if it is necessary to change the scope or size of the model when moving from the broad-brush approach to the detailed approach. It is rare that a model is only solved once and its results then implemented as decisions. Clients will want to ask "what if" questions, data may undergo small changes and new facts will become available during the life of a project. Hence, the problem will be run and re-run. This is tedious to do if substantial retyping of a model is required to accommodate small changes. A Modeling System will facilitate the incorporation of small changes very easily, and large changes at modest effort, and allow re-running to take place. Once clients are "comfortable" with a model they will be able to handle much experimental and developmental work for themselves using the Modeling System [see Sect. 2.8].

Once a model has been formulated algebraically, in what Fourer (1983,[197]) called a modeler's form and the data obtained, it is then converted into a form suitable for a solver. The industry standard input form of LP and MILP problems is MPS (Mathematical Programming System), a file format originally developed by IBM and used by almost all systems. This format goes back to the early days of computers with its reliance on data being read in sequentially as a fixed format list. Although its origins are in Fortran programming it has remained the standard since the 1960s. The style is an ordered list of constraints (rows) followed by the matrix of non-zero coefficients: each is specified by an 8-character variable (column) name, a similar styled constraint name and the element value, e.g.,

```
xprod    fact1con    5.6
```

where `xprod` is a variable name and `fact1con` is one of the constraints in which it appears with a non-zero coefficient. At most two elements can be specified per line and the line has to conform to a strict layout. A model with 1000 constraints and 5000 variables might require an MPS file of 10,000 lines which is clearly impractical to prepare or maintain with a standard data preparation and editing facility. Because of the presence of sparsity, which will be discussed in Sect. 2.11.2, data preparation should also be performed in an efficient way to avoid duplication of effort by continually retyping some frequently occurring coefficient. What is required is a procedure to generate the coefficient matrix easily and efficiently. This can be achieved using matrix generation programs.

1.5.2 A Brief History of Modeling Systems

Here we present just a brief history of how matrix generation programs of the early days developed into algebraic modeling languages for optimization; more details and more recent developments are provided in Sect. 17.2.4. Initially matrix generators were purpose built programs designed by users for specific purposes written in high-level languages and were task specific. Programs were not user friendly, were hard to maintain, update, debug, and were rarely well documented.

An early advance was the introduction of the (special purpose) matrix generator generators OMNI/PDS MAGEN by Haverly Systems and MGG by Scicon, software written in a high-level language but offering flexibility across models and applications. The software allowed the user to provide a model and data in the modeler's form, albeit fairly structured and not in a particularly user friendly manner and generated MPS format matrices from the specification and data provided. Checking within the system was provided but this did not necessarily avoid the problem of creating totally unreasonable matrices. Such software encouraged self-documentation and was clearly a dramatic advance on user written programs. There were disadvantages in that the software could be slow to use and somewhat awkward.

The next development was the introduction of *general purpose* matrix generation programs that interpreted the modeler's form to produce the MPS matrix. Examples are mp-model within XPRESS-MP in 1982 [Ashford & Daniel (1987,[35])], MAGIC [Day (1982,[147]], GAMS [Meeraus (1983,[392])], LINGO, AMPL [Fourer et al. (1990,[199])], and MPL [Kristjansson (1992,[342])]. Systems such as OMNI, DATAFORM[19], and XPRESS-MP allowed the user in 1990s to read data from files, spreadsheets or databases. These latter developments have been particularly influenced by the development of small, inexpensive but increasingly powerful small computers, either PCs or workstations. The development of matrix generation systems has thus been influenced by the move away from large computers to smaller ones, which perhaps interface with other machines. Data is now more commonly picked up from existing sources already stored on the machine or downloaded onto it, so flexibility of format styles is important in matrix generation software. Modeling languages have been developed by several authors, including Anthonisse (1970,[26]), Bisschop & Meeraus (1982,[80]), Fourer (1983,[197]), Bisschop & Kuip (1993,[81]). Systems now offer more flexibility for use on smaller machines. Among such systems is also EASY MODELER [IBM (1993,[277])]. A review of systems available appears in Sharda (1993,[501]), a comparison of several systems appears in Greenberg & Murphy (1992,[238]), and Kallrath (2004, 2012;[306],[312]).

1.5.3 Modeling Specialists and Applications Experts

Once a model has been built by a *modeling specialist* it is often used by somebody — here called *application expert*[20] — possibly without any background in math-

[19] DATAFORM is a trademark of Ketron Inc.

[20] The term "application specialist" here refers to a person who does not necessarily have familiarity with management science, LP, or modeling but may be a specialist in another field, e.g., accounting or engineering, or may perform a purely clerical or administrative role.

ematics or modeling. The modeling specialist will be responsible for formulation, validation, and debugging of the model but then needs to present it in a sufficiently robust manner for the non-specialist to use. The model must therefore be usable by people other than the designer, and so must be easily accessible and user friendly. In particular, data input facilities need to be straightforward to initiate and modify. Further, the management of the model building and optimization process must be transparent and pictorial displays of model and solution must be meaningful to the environment in which they will be used. Graphical and Windows based interfacing will be used frequently. Many models will need to interface with commonly used spreadsheets such as `LOTUS` (in the 1980s and 1990s) or `EXCEL`. Data might also be recovered from SQL-queries to corporate or departmental databases. The state-of-the-art modeling software provides this functionality.

Thus the user of the model and its results will usually be insulated from mathematical programming and optimization and will be able to treat results as if coming from a "black box" system. This may sound oversimplified as it would be unwise for an end user to employ an optimization model indiscriminately, but a compromise between the two extremes of naïve user and expert user is necessary.

1.5.4 Implementing a Model

Let us illustrate how to formulate the problem of Sect. 1.4.1 using an `AMS`. As the problem is small a simplified approach to modeling can be taken.

After installing the `AMS`, and calling up the `AMS` icon in the file manager, we see a graphical modeling environment. This should allow us to choose the `File` option from the menu and click with the mouse on it. Then click on `New` and enter a Model Name, in this case `boats` in the box and click on `OK` (we can provide information in the other boxes by moving around with the mouse and typing if we wish, but it is not mandatory). The model has two essential parts `VARIABLES` and `CONSTRAINTS`. In the first part, the `VARIABLES` section, the decision variables are specified. In the `CONSTRAINTS` section, the constraints and objective function are specified. The first step is to identify the decision variables. The keyword `VARIABLES` is typed followed by the names of the decision variables. Any information on a new line following the `VARIABLES` keyword is treated as variable definition information until the next keyword is read. The main structure of the model, i.e., the constraints and objective function, is now described in the `CONSTRAINTS` section.

The `CONSTRAINTS` section requires some explanation. AMLs require that each constraint be given a name. For this problem the constraint names used are `berthing` for berthing accommodation available, `prem_lim` for the limit on Premier boats, `maintenance` for the maintenance hours restriction, and `mixture` for the "no fewer Premier boats" restriction. To enter these constraints, the keyword `CONSTRAINTS` is typed and each constraint specified. The constraint name is followed by a colon and then by the inequality. If we look carefully we see that a constraint is made up from:

a constraint name,	followed by
a separator [colon (:), colon (;)]	followed by
a linear expression,	followed by
a constraint direction,	usually followed by
a right-hand side value.	

A linear expression was alluded to in Sect. 1.4.2. Constraint directions are, for instance, in `GAMS` :

$$\begin{aligned}
&= \mathrm{L} = \text{ meaning a } \leq \text{ constraint,} \\
&= \mathrm{G} = \text{ meaning a } \geq \text{ constraint, and} \\
&= \mathrm{E} = \text{ meaning an } = \text{ constraint.}
\end{aligned}$$

As the objective function has a similar structure to a constraint, it is coded as a constraint. In keeping with constraints, the objective function is given a name. We use the name `revenue`.

By default, most AMLs assumes that all variables are constrained to be non-negative unless stated otherwise. Therefore, there is no need to specify non-negativity constraints on variables. Also by default, all variables not specified in particular constraints are assumed to have the coefficient zero. Thus, variables which do not occur in particular constraints or the objective function do not need to be specified with zero coefficients.

1.5.5 Obtaining a Solution

Now we wish to maximize this problem. So, either in the GUI or by some language commands, we tell the system the sense of the optimization problem: *max* or *min*. Finally, we call the optimizer or solver to take care of the problem and to solve it for us. The solver call may return with the indications on problems, or it may return the solution.

How to identify particular values or extract the solution values of variables, depends on the `AMS` at hand. For the number p of Premier boats we find $p = 200$. We can also display shadow prices/dual values (on the constraints) and reduced costs (on the variables) discussed in Sects. 3.4.4 and 3.4.3. To exit from each solution dialogue, click on the top left corner of its box and choose `Exit`. Save the model by clicking on `File` on the menu and then `Save`. It is advisable to save a model every time you make changes that you wish to keep. If we are finished with the `AMS` we could now choose `Exit`. If we wish to continue with the `AMS` we should `Close` the current `File`.

What happens when things go wrong in our model? Suppose we have typed in an erroneous problem which, when we attempted to compile and run it, resulted in two errors. Ideally, the indicated type of error and the line of the model on which it occurs helps us to identify the problem. Both these errors can be corrected by over-typing on the appropriate lines of the model and then the problem can be `Run` again. Note that correcting all obvious errors in the model and achieving a working model does not ensure that the model is "correct." There may be other errors of logic in the model that are harder to disentangle.

1.5.6 Interpreting the Output

The names of the decision variables and their optimal values have been obtained above. We saw that the optimal value of p is 200, i.e., 200 Premier boats should be leased in order to maximize profit. Similarly, the optimal value of variable s is 150, i.e., 150 Standard boats should be leased.

`We` also see the optimal revenue produced. The optimal revenue obtained is £250,000. In the present example it is fortunate that the numbers of boats required of each type have turned out as whole numbers. If a problem requires that integer numbers be produced in the optimal solution, then we normally require the techniques of (mixed) integer linear programming. Further details of this approach can be found in Chap. 3.

Although we have found an optimal solution to the problem given, this may not always be possible. If the constraints were insufficient to constrain the problem it may be possible to improve the objective function indefinitely. In this case the problem is said to be *unbounded*. This usually means that a relevant constraint has been omitted from the problem or that the data are in error rather than indicating an infinite source of profit. Conversely, if the constraints are too restrictive it may not be possible to find any values for the decision variables that satisfy all constraints. In this case, the problem is said to be *infeasible*.

We have now seen that AMSs may have two components, a modeling component and a solver. We have just built a simple model, and put it into the language format expected. We then solved the problem. Finally we displayed the optimal solution values. In the next section we shall see what our client has to say about the solution and extend the boat problem to incorporate more details.

1.6 Benefiting from and Extending the Simple Model

We are now ready to present our promising results to the owner of Sunshine Cruises. She is very impressed by the information she receives from the results of the linear programming calculations, but now something happens which is quite common when modeling real-world problems. She realizes that the problem has not been modeled in sufficient detail.

After discussion, we learn further that Sunshine Cruises has a seasonal business. Their year runs from the end of March to the end of October and is divided into three parts:

early season	high season	late season

Different numbers of boats will be required for each season and essential maintenance is performed in early or late season making some boats unavailable. As the type of customer varies in different seasons, the profile of the demand for the different types of boat alters. In high season most customers are families who require the larger Premier boat. Pricing policies are also different in the different seasons. Thus Sunshine Cruises would normally lease different numbers of boats at different seasons.

The simplified and aggregated version of the model was too simple and now needs to be disaggregated. We can do this by introducing three pairs of variables:

$$\begin{array}{ccc} p_e & p_h & p_l \\ s_e & s_h & s_l \end{array} \tag{1.6.1}$$

to denote the numbers of Premier and Standard boats required in each season.

Restrictions on requirements for boats are: in early and late seasons more Standard then Premier boats are required, while in high season more Premier than Standard boats are required. Thus we require

$$p_e \le s_e \quad , \quad p_h \ge s_h \quad , \quad p_l \le s_l \tag{1.6.2}$$

Giving the new revenue structure the objective function becomes

$$\max \quad 500p_e + 400s_e + 700p_h + 600s_h + 400p_l + 350s_l \tag{1.6.3}$$

subject to

$$\begin{array}{ll} p_e + s_e \le 300 & , \; p_e \le 150 \\ p_h + s_h \le 350 & , \; p_h \le 200 \\ p_l + s_l \le 300 & , \; p_l \le 150 \end{array} \tag{1.6.4}$$

the upper limits accounting for availabilities of boats. Other constraints for maintenance require

$$\begin{array}{l} 4p_e + 3s_e \le 1000 \\ 4p_h + 3s_h \le 1400 \\ 4p_l + 3s_l \le 1000 \end{array} \tag{1.6.5}$$

As before all variables are non-negative.

This problem, which you find under the name *boats2*, can now be solved using a MILP solver and values for the variables obtained. We find the new policy is

$$\begin{array}{ccccccc} p_e & s_e & p_h & s_h & p_l & s_l & objective \\ 100 & 200 & 200 & 150 & 100 & 200 & 470,000 \end{array} \tag{1.6.6}$$

Notice that revenue has risen, but revenue must be regarded as the total revenue for three weeks with a typical week from each season included.

This model is now more realistic than the first simple model and provides Sunshine Cruises with much more information. Further elaboration to the model can now be considered by bringing in constraints involving these factors:

- fuel is supplied by a company part-owned by Sunshine;
- breakdowns require a boat to replace the broken down boat, conduct repairs, compensate and perhaps swap customers on to a spare boat;
- other aspects of cost;

and so on.

Further major analyses are possible as the fuel company has its own mathematical programming model to determine what is best for its operations, of which arrangements to provide fuel to Sunshine Cruises is only one part. The two models can now be compared, or indeed amalgamated, to determine if jointly further gains to revenue are obtainable.

Thus the Sunshine Cruises model has illustrated the typical use of mathematical programming to aid a company in a series of stages:

1. a simple model is developed initially;
2. once the capabilities of the analysis are grasped, the model is made more realistic;
3. the model is expanded in scope;
4. further gains obtainable from modeling are appreciated and the analysis is further extended.

1.7 A Survey of Real-World Problems

The problem just considered was small but illustrated a typical situation in which mathematical programming might be used. In this section we describe the areas in which applications of mathematical programming are to be found.

The brief survey of real-world problems given in this section is typical for many businesses and industries but many other applications also occur in other areas.

- allocation problems (allocating people to tasks);
- production planning (production, logistics, marketing);
- scheduling problems (production of goods using processes such as machines);
- sequencing problems (putting production into order);
- blending problems (production and logistics);

- process design (process industry);
- engineering design (all areas of engineering);
- selection problems (strategic planning);
- network design (planning, strategic planning);
- refinery planning and scheduling (refineries, chemical process industry)
- timetabling problems (in schools, universities);
- timetabling or scheduling and routing for aircrafts and trains;
- distribution and logistics problems;
- purchasing;
- long term resources acquisition/disposal;
- economic problems;
- problems in the financial service industry.

Specific skills often mean that only certain people can perform specialized tasks. Thus optimization methods can solve problems of *allocating* personnel to tasks, see Chap. 4. Manpower modeling is discussed in Schindler & Semmel (1993,[489]).

Companies which are in a situation to utilize the advantages of a complex production network as in Kallrath (1995,[300]), often because they operate at several sites, may greatly benefit from *production planning* and *production scheduling*, see Chap. 10. Of course, scheduling problems occur also in other branches of industry. Production planning and scheduling require detailed answers to the questions: when is the production of a specific product on a specific machine to be started? What does the daily production sheet of a worker look like? Scheduling problems belong to a class of the most difficult problems in optimization. Typical special structures, which can be tackled by optimization, are minimal production rates, minimal utilization rates, minimal transport amounts. Items of production must also be *sequenced*.

Blending problems occur in a wide variety in the chemical process industry, but in modified form also in the mineral oil or food industry. In Kallrath (1995,[300]) a model is described for finding minimal cost blending which simultaneously includes container handling conditions and other logistic constraints, see also Chap. 10.

In planning it is often required to make a *selection* from a series of projects in an optimal manner. This is clearly another type of optimization problem.

Questions of how a telecommunication network should be structured and designed when the annual demand is known, or what the traffic infrastructure should look like for a given traffic demand, lead to *network design problems*.

People often need to be *timetabled*, i.e., allocated to sessions with conflicts being avoided. These applications lead to complex optimization problems. They can be classified as resource constrained scheduling problems with time windows. Such a case is discussed in Sect. 10.5.

Problems of *distribution and logistics* provide a fruitful outlet for optimization — choosing vehicles, routes, loads, and so on, see Chap. 7.

While the problems listed above can be solved with linear mixed integer methods, problems occurring in process industry very often lead to *nonlinear discrete problems*. Production planning or scheduling problems in the refinery or petrochemical industry include blending problems involving many *pooling* problems [see Sect. 11.2.2]. If, in addition, we have to consider that plants operate in discrete modes, or that connections between tanks and cracker or tanks and vacuum columns have to be chosen selectively, then mixed integer nonlinear optimization problems need to be solved. Process network flow or process synthesis problems usually fall into this category too. Examples are heat exchanger networks, distillation sequencing, or mass exchange networks.

Problems in the financial services industry are related to financial optimization [see Sect. 8.4] and are concerned with risk management and financial engineering and lead usually to nonlinear or even mixed integer nonlinear programming problems. Models in financial optimization are used, for example, by investment and brokerage houses.

It should not be inferred from this list of examples that optimization is confined to large sophisticated problems occurring in large organizations.[21] Small problems are equally important, but it may be that the role of techniques to solve problems is less important. Some small problems lack the structure required for the use of optimization techniques and for others the use of techniques would be seen as using a sledge hammer to crack a nut. A balance is required where gains to be obtained by finding optimal answers to a problem outweigh the effort required to obtain these answers.

1.8 Summary and Recommended Bibliography

After reading this chapter the modeler should be able to:

- recognize the features of mathematical programming and, in particular, linear programming (LP) problems;
- identify decision variables, constraints, and an objective function for small LP problems;
- draw a graph of a 2-variable LP problem;
- solve a small LP problem using an `AMS`;
- appreciate some of the extensions possible to small LP problems;
- be aware of some of the range of applications of mathematical programming.

The book *Introduction to Applied Optimization* by Diwekar (2008,[163]) may be a good starting point for the reader new to optimization. *50 Years of Integer Optimization* by Jünger et al. (2010,[295]) is a comprehensive (superb combination of theory, algorithms, and applications with both linear and nonlinear problems) collection of superb, high quality chapters by many distinguished scientist of optimization.

---

[21] The use of mathematical optimization in small or medium size business [in the German-speaking part of the world: kleine und mittelständige Unternehmen] is discussed in Sect. 16.5.

1.9 Appendix

1.9.1 Notation, Symbols, and Abbreviations

Below we list the most important symbols and abbreviations used in this book.

Symbol	Explanation	Example
$\{\}$	Set	$I = \{1, 2, 3, 4, 5, 6\}$
x_i	Subscripted variable	$x_1, x_2, x_3, x_4, x_5, x_6$
$\mathbf{x}^T$	Transposed vector, row vector	$\mathbf{x}^T = (x_1, x_2, x_3, x_4, x_5, x_6)$
A^T	Transposed matrix	
$\in$	Is an element of	$i \in I$
$\forall$	For all (universal quantifier)	$\forall i$
$\sum$	Summation	$\sum_{i=1}^4 x_i = x_1 + x_2 + x_3 + x_4$
Π	Product	$\Pi_{i=1}^4 x_i = x_1 \times x_2 \times x_3 \times x_4$
$\subset$	Is strictly contained in	$\{2, 3, 4\} \subset \{2, 3, 4, 5\}$
$\subseteq$	Is contained in or equal to	$\{2, 3, 4\} \subseteq \{2, 3, 4\}$
$\leq$	Less than or equal to	$x \leq y$
$\geq$	Greater than or equal to	$y \geq z$
$:=$	Defined equal to	$x := y^2$
$\|\mathcal{I}\|$	Number of elements in a set; card($\mathcal{I}$)	$\|\{1, 2, 3, 4, 5, 6\}\| = 6$
$\mathbb{N}$	The set of positive integers	$\mathbb{N} = \{1, 2, 3, 4, \ldots\}$
$\mathbb{N}_0$	The set of non-negative integers	$\mathbb{N}_0 = \{0, 1, 2, 3, 4, \ldots\}$
∞	Infinity	
exp	Exponential function	$\exp(x) = e^x, e = 2.7182\ldots$
ln	Natural logarithm function	$e^{\ln x} = x$
$\lfloor x \rfloor$	Down-rounding function	$\lfloor 4.3 \rfloor = 4$
$\lceil x \rceil$	Up-rounding function	$\lceil 4.3 \rceil = 5$
AML	Abbreviation: *algebraic modeling language*	
AMS	Abbreviation for *algebraic modeling system*	
MCOL	Abbreviation for *model collection online*	
e.g.	The Latin abbreviation *exempli gratia* meaning:	"For example"
i.e.	The Latin abbreviation *id est* meaning:	"That is to say"
s.t.	Abbreviation for *subject to*	
w.r.t.	Abbreviation for *with respect to*	

Furthermore, we keep the following convention throughout the book:

Object	Sets	Vectors	Matrices	Software products	Solver names	File names
Print font	Calligraphic	Bold face	Sans serif	Typewriter	Small caps	Italics
Example	$\mathcal{I}$	$\mathbf{x}$	A	`GAMS`	BARON	*optGrid.gms*

In mathematical model descriptions decision variables/unknowns are always denoted by small letters where Greek letters refer to integer, binary, semi-continuous or variables belonging to special ordered sets. Capital letters indicate known input data and parameters of the optimization. Equations are numbered and cross-referenced by (1.2.3) giving chapter, Section, and equation number in that section.

1.9.2 A Brief History of Optimization ⊖

The notion of *optimization* — used in a mathematical sense — means the determination of the maximum or minimum of a special function, which is defined through a (limited) region or series of states. Optimization has become more important, because of remarkable improvements in techniques and the dramatic increase of calculation (computer) power. However, the benefits of optimization have been around for longer as will now be shown.

The discipline *Operational Research (OR)* developed during the Second World War and in the immediate post-war period; cf. Nemhauser (1994,[420]) and Jünger et al. (2010,[295]). The problems to be solved in the period following the Second World War were closely linked to the military operations to be carried out. Teams were put together, combining interdisciplinary or multi-disciplinary professionals with others with a mathematical-natural science background, who then worked together on problems. Physicists were often in these groups. In the 1950s these approaches were more and more used for civilian problems and a variety of techniques, methods, and algorithms were developed. Developments were paralleled on both sides of the Atlantic. In the USA the term Operations Research was favored rather than Operational Research, but to a degree the concepts were similar and fortunately both terms have the same initials (OR) as an abbreviation. Operational Research professionals now form a sizeable group, with Professional Societies linked internationally by IFORS (the International Federation of Operational Research Societies). In Britain the Operational Research Society has approximately 3000 members and in the USA its counterpart, INFORMS, has over 10,000 members. These professionals are not licensed practitioners in the way that lawyers or dentists are required to be, so people other than professional society members may practice OR.

The idea of producing a best solution to a given problem has usually been central to OR. However, not until the 1960s, did a sub-discipline within OR emerge to put these procedures on a well-defined mathematical footing. The notion of *mathematical programming* developed as a generalization of the types of optimization being conducted in OR. (The word *programming* here means planning rather than computer programming, as problems were being considered before the advent of computers.) Nowadays *optimization* has returned as the more suitable term, which itself can be understood as a part of a practically orientated discipline of mathematics, namely *scientific computing*, which has developed since the 1980s. The catalyst in all this has been the development of cheaply available computer power to solve problems.

Both the disciplines of scientific computing or operational research try to solve practical problems of high complexity by mapping given *real-world problems* onto *mathematical models*. The methodology is to formulate the problem in the language of mathematics using variables, mathematical functions, and relations. Next techniques (*algorithms*) are developed for solving the mathematical problem. Eventually these algorithms are implemented on a machine (*computer*) which leads to a piece of *software*.

Already in the early nineteenth century certain optimization problems played a role and were solved with the help of a technique known as Lagrange multipliers. In this period the French mathematician Fourier investigated constrained optimization problems and proposed an outline of a technique to solve them but did not make the final breakthrough. People continued to be interested in optimization problems even though they did not know precisely how to solve them. From the field of economics there was interest in optimal diet problems [Stigler (1945,[519])] and other problems of resource allocation. Economics, with its interest in scarce resources, was a natural field to be interested in the optimization of the utilization of these resources. However, for the birth of the first solved optimization problems in OR, 1947 is the key, being the year in which Dantzig developed his Simplex algorithm for solving a known optimization problem from OR (LP) and used it in an air-force-planning model (Dantzig et al., 1954,[145]).

The Simplex algorithm of Dantzig for solving LP problem exploits the geometrical shape of the feasible region which in that case is a polyhedron. It searches from corner (vertex) to corner progressively. An optimal corner can be found and the optimal solution actually lies at this corner. As early as 1954 the first commercial code for solving LP problems was written by Orchard-Hays (1969,[426]).

In the 1950s the first ideas on an optimization problem known as *dynamic programming* were introduced by Bellman (1957,[59]). Ford & Fulkerson (1962,[195]) looked at and solved *network problems*. Kuhn & Tucker (1951,[346]) established the principles of solving *optimization problems* where the conditions and/or objective of the problem cannot be described by the type of rigid conditions enunciated by Dantzig et al. (1954,[145]). Gomory (1958,[230]) presented the first general algorithm for the solution of an extension of Dantzig's LP approach to what are termed *integer optimization problems*, and finally, Dantzig & Wolfe (1954,[143]) examined *decomposition procedures* and their application on large-scale problems which related the solution technique back to fundamental ideas on decentralization arising in economics. The development of solution techniques went hand-in-hand with the development of large mainframe computers to which industry, commerce, and government departments had access through new data-processing departments. Further information on the history of the development of optimization in OR is contained in Orden (1993,[428]) and Bodington & Baker (1990,[89]).

Nowadays OR optimization problems with several hundred thousand decision variables and conditions to be satisfied, for example, in the civil aviation area, are routinely solved. The best Simplex algorithms compete with the newer approaches termed interior-point methods, whose original development goes back to Dikin (1967,[161]). Initiated by the work of Karmakar (1984,[325]) interior-point methods have undergone further development involving the best researchers in operations research. Depending on the structure of the problem one or other procedure is more suitable. In some cases a mixture of several methods is suited for the solution of problems with several million decision variables.

Essential aspects which lead to an improvement of the algorithms are under continual investigation by researchers, but these components could only be improved significantly because in the meantime computer systems have improved (architecture, memory and software surroundings) and make fast testing easy and thereby sensible experimenting possible.

The improvement within LP has, on the one hand, consequences for the size and complexity of the problems which can be solved. On the other hand the technique of LP is often used within other optimization processes for more complex problems (e.g., in mixed integer programming, see Chap. 3).

The developments in optimization techniques facilitated the solution of problems in the areas of logistics, transport, production planning, finance, communication, and yield management; cf. Grötschel & Lovasz (1982,[244]). An important characteristic of such problems is the need to model discrete decisions and very complex structures in the optimization. In addition, due to more efficient algorithms, hardware and software, problems with several million decision variables can be solved in reasonable time on a simple PC laptop, removing the earlier need for access to large computers. Among the improvements in optimization techniques are:

- improved algorithms;
- good model formulation;
- preprocessing (investigating the problem intensively before the solution technique is applied, see Sect. 9.1);
- approximate methods (heuristics);
- efficient use of resources on a computer (e.g., storage, access and processing).

Algorithms are under continual review and development. The principal recent development has been of hybrid algorithms. These algorithms combine the strengths of several different techniques and either automatically or by user intervention permit to solve problems by switching from method to another method.

With the help of a facility in optimization software known as automatic preprocessing which examines the constraints and variables of a model, it is possible to eliminate redundancies in the model and to improve it logically. Difficult problems, which could hardly be solved previously, can now be solved and previously solvable problems can be solved more efficiently.

Heuristics provide ways of approximately solving problems. They may be used as a starting point by an existing technique or may be used to modify a solution provided by a technique when it is not possible to solve the problem easily.

Computer resources can now be dedicated to problem solving allowing modelers dedicated use of storage or machinery or the hard-wiring of aspects of the software into the machine. Turnkey systems which allow naive users to solve problems with simple menu prompting are now cheap with the advent of inexpensive small computers.

Chapter 2
From the Problem to its Mathematical Formulation

This chapter will cover the fundamentals of modeling problems such as the one introduced in the previous chapter but now with a view to modeling and solving much larger problems. By modeling we mean taking all the features of the original industrial or commercial problem and encapsulating the ideas in it as a mathematical programming problem. We aim for a one-to-one correspondence between the original problem and the model.

2.1 How to Model and Formulate a Problem

If we understand a *model* as an *abstraction of the real world*, it is clear that it cannot cover all details. The main reasons are:

1. Data availability (in most part of this book we develop and discuss deterministic models assuming that the data are deterministic, although, in practice, input data may be uncertain, a situation covered in Sect. 11.3).
2. Effort to collect and maintain additional data.
3. Complexity to understand the interactions.
4. Finally the results.

Although a one-to-one correspondence between the original problem and the model would be ideal, we will normally have to simplify aspects of the real problem in order to make the modeling manageable (Fig. 2.1). This is not intended as some

Supplementary Information The online version contains supplementary material available at https://doi.org/10.1007/978-3-030-73237-0_2.

J. Kallrath, *Business Optimization Using Mathematical Programming*,
International Series in Operations Research & Management Science 307,
https://doi.org/10.1007/978-3-030-73237-0_2

Fig. 2.1 A modeler thinking about a model of a real-world problem. Produced for this book by Diana Kallrath, Copyright ©2020

kind of opt out, but the simplifications may arise for good reasons as the following three possibilities (scale, visualization, simplicity) suggest:

1. *Scale*: When we start to model it may be convenient and easier to commence with a scaled down version of the real problem so that we can develop a "feel" for what is going on. This will be particularly useful as an aid for discussions between the modeler and the client for whom the model is being produced. For example, if there are 100 vehicles delivering from 12 depots, it may be convenient to temporarily simplify the problem to 10 vehicles and 2 depots. We start to examine what conditions are affecting these vehicles and depots. We then can discuss with the client whether we have taken the correct view of the problem in all its aspects. What we are trying to do here is to marry together "client's world" and "modeler's world." The model now becomes a focus for debate between modeler and client.
2. *Visualization*: In our own modeling it may be beneficial to visualize what is going on by trying to model parts of the problem first. Once we have confidence that we understand the problem, we can attempt the full model.
3. *Simplicity*: Parts of a client's problem may be genuinely difficult to model, e.g., some complex conditions within an industrial process. Or the client cannot provide data on that detailed granular level required. Thus, it may be best to simplify or even ignore these difficulties and see what progress can be made with a model that is not complete, but copes with the main thrust of the client's requirements. Once the client sees the results of some solved problems, it may be that the information is sufficient to make decisions, e.g., if aspects of a problem are treated in an overgenerous way, but the results suggest no gain can be made, then the client may be satisfied that the proposed ideas are not worth further pursuing. Alternatively, the provisional results may be sufficiently tempting to suggest gains are possible, and now the client requires a much more precise and complete model to be built. The moral here is that it may not be worthwhile

aiming for a high degree of precision on the first attempt at building a model as a broad-brush approach may provide considerable insight.

In the three suggestions made it is important that full communication is retained between modeler and client so that the client is aware of simplifications being proposed and limitations imposed. These can then be agreed as a suitable starting point.

To model a problem as a mathematical programming problem three key parts of the problem must be identified:

- Variables
- Constraints
- Objectives

The identification of each of these three will form the next three sections of this chapter. In addition to these three parts we have:

- Data structures (clarify what are basic objects and what are attributes to objects).
- Indices (they represent the basic objects)

In more sophisticated modeling these would be the first two parts to be identified, but in simple examples with few variables and constraints they can be ignored. This will be done in this chapter, but the importance of data structures and indices will be stressed as part of the good practice of modeling.

2.2 Variables, Indices, Sets, and Domains

Mathematical problems revolve around variables. Variables are used to model the decisions "how many," "what kind," "where to," "where from," "how much," and so on. Variables are often thought of in algebraic terms as representing "unknown quantities." Symbols such as x, y, and z may be used to denote variables. These are commonly used in textbooks on algebra for "unknowns." However, in our work we want to make a distinction between what is *unknown* and what is the *direct result* of a decision. Therefore, the variables or unknowns are often thought of as "decision quantities" and are also called *decision variables*.[1]

For example, if a manager of a factory has to make a decision as to how many trucks to dispatch at fixed 5 min intervals, starting with the earliest at 08.30, then initially the planned departure time of the last truck is unknown. However, the final departure time is not a direct decision, but the total number of trucks to dispatch is. Once we know that 20 trucks are to be dispatched, the planned departure time of the last truck is known. Thus the key to identifying variables will be to identify decisions.

Sometimes it is difficult to identify exactly which decisions exist in a problem situation. It may often be useful to pose questions such as "what does the manager

[1] In the LP community variables are also called *activities* or just *columns*.

need to know?" in order to identify what are the necessary decisions and hence the variables of the model. In this part of the process we are re-emphasizing the earlier distinction made between the "client's world" and the "modeler's world." The client wishes to know what decisions to take but may wish to remain indifferent about which variables are in the model. The variables are part of the modeler's world.

Hence we start to recognize that variables can be classified according to type. In mathematical programming the usual types of variables are:

- *Continuous* variables, $x \geq 0$

These are non-negative variables that can take any value between zero and some specified upper limit X, $0 \leq x \leq X$, and X may take the value infinity. Note that we use lower-case characters for continuous variables.

A continuous variable might for example be used to model the amount of oil measured in tons flowing through a pipeline.

- *Free* variables, $-\infty \leq x \leq \infty$

These are continuous variables unrestricted in sign. They can take any positive, zero, or negative value. They are also called *unconstrained* variables or *unrestricted* variables.

In contrast to the earlier remark about variables usually being non-negative, free variables exist to model varying quantities such as "growth," "change," or "net profit" where the desired value of a variable could be positive or negative, or indeed zero. A free variable can be thought of as or represented by the difference between two non-negative variables, e.g., $x^+ - x^-$, and it is possible to avoid the use of free variables in models, but this will be unnecessary within most modeling languages. However, they are often useful to express directly a quantity where conventional negative values are not unusual.

So we keep in mind, free variables can describe growth, change, or account balance. On page 400, for example, they are also called unconstrained variables to solve special nonlinear optimization problems by sequential linear programming.

- *Integer* variables $\{0, 1, 2, 3, 4, \ldots\}$

These are non-negative variables that can take any integer (whole number) value between zero and some specified finite upper limit. Throughout the book we use lower-case Greek characters, e.g., $\alpha, \beta, \gamma, \ldots$, for all integer variables or other non-continuous variables, e.g., the binary variables introduced below.

The number of Premier boats, for example, could be represented by an integer variable.

- *Binary* or $\{0,1\}$ variables.

These are a special form of integer variables that can take only the values 0 or 1. Thus a binary variable has an upper limit 1. Note that a binary variable, typically named δ, always obeys the weaker restriction

$$0 \leq \delta \leq 1. \tag{2.2.1}$$

Such a restriction is called *relaxation*, or to be more precise, a *domain relaxation*, and is a very useful concept to treat optimization problems containing integer or binary variables.

Binary variables are used to model decisions that have exactly two outcomes, e.g., do commence a project or do not commence it. In this instance we may choose to associate the value "1" with the commencement of the project and the value "0" with non-commencement. These associations are arbitrary and are decided on by the modeler. Notice that a binary variable models the logic of a decision, not a value that occurs in the client's world. In many models continuous variables occur alongside binary variables, the binary variables being used to model major dichotomies in the decision making, and the continuous variables modeling the consequences of these decisions.

- *Discrete* variables

can only take certain isolated or discrete values, i.e., $\sigma \in \{W_1, W_2, \ldots, W_n\}$. The values $W_1, W_2, \ldots, W_n$ do not necessarily have to be integer or binary, e.g., $\sigma \in \{-3, 0, +0.25, +1, +17.1\}$ is also possible. Discrete variables contain integer variables as special cases and can be described with the help of binary variables δ_i, the equation $\sigma := \sum_{i=1}^{n} W_i \delta_i$, and the constraint $\sum_{i=1}^{n} \delta_i = 1$ or the special ordered sets of type 1 described in Sect. 6.8,, respectively.

- *Semi-continuous* variables (and also *semi-continuous integers*), σ

These are variables that can take the value zero or any continuous value equal to or greater than a constant. In many software packages the constant is predefined as 1. Thus the variables need to be scaled accordingly, e.g., if a variable x is to be zero or any continuous value greater than $X^- = 3$, then $\sigma = x/3$ is a semi-continuous variable taking the value zero or any continuous value greater than 1. Formally, this variable fulfills the relation

$$x = 0 \quad \text{or} \quad X^- \leq x \leq X^+, \tag{2.2.2}$$

where the lower limit X^- is any positive number, $X^- > 0$, and the upper limit X^+ may take the value infinity. If scaling to $\sigma^- = 1$ is necessary, then a semi-continuous variable

$$\sigma = \frac{x}{X^-} \quad \text{which is equivalent to} \quad x = X^- \sigma \tag{2.2.3}$$

is introduced. Note that σ is scaled to unity, i.e.,

$$\sigma = 0 \quad \text{or} \quad 1 \leq \sigma \leq X^+/X^-. \tag{2.2.4}$$

These variables are used to model aspects of industrial processes where either nothing is produced or if anything is produced, it has to be above a minimal level. Semi-continuous variables are also used in transport problems. Imagine a company with production sites on different continents. They can ship goods from one site to another. But if goods are shipped, then the requirement is that at least, say, 20 tons are shipped. Otherwise no transport is possible. Such a situation occurs in the case study in Sect. 10.4 where semi-continuous variables are also used to include certain plant utilization rates.

- *Partial integer* variables

Partial integer variables, also known as semi-integer variables, are variables that must be integral if they lie below some stated value; otherwise they are considered to be continuous variables. They are a generalization of semi-continuous variables. The reason for having such variables is that once a variable is larger than some stated value any fractional part associated with it will be sufficiently small compared to the integer part of its value that it may be regarded as relatively insignificant. Thus the variable can be regarded as continuous once it is relatively large. As continuous variables are easier to deal with when solving mathematical programming problems, one should always check whether it is possible to replace integer variables by partial integer variables. Conversely, such a variable taking a low value would mean that if the value were not integral then the fractional part formed a significant part of the overall value and could not be neglected. There will be no standard breakpoint that can be used to form the "stated value" — it will depend on the individual decision modeled by the variable; see also the example on page 307.

Note that apart from what we have learned above about the various types of variables, binary, integer, semi-continuous, and partial integer variables are often referred to as *discrete variables*.

All work in linear programming (LP), integer linear programming (ILP), or Mixed Integer Linear Programming (MILP) variables are assumed to be non-negative unless otherwise stated. Sometimes, however, it is necessary to have variables unrestricted in sign.
It is not obligatory to use single letters to denote variables. In fact it may be useful to use mnemonic names to denote them, e.g., $xcash$, $prod1$, $prod2$, as this can help to document the model (but one should bear in mind that it makes mathematical relations more difficult to read). However, there is a trade-off between mathematical elegance and the use of mnemonic names that might improve the understanding of a model, particularly to those people not involved in building it in the first place.

So far in this section we have thought of variables as individual items whose individual values may range widely. In more realistic modeling it will be useful to think of variables collectively in sets and to define their range.

2.2.1 Indices, Sets, and Domains

If we wish to model quantities of a product produced at 12 different time intervals and 5 different locations, it would be quite cumbersome or impractical, to introduce 60 individual variables with names $x_{01}, \ldots, x_{60}$ and to remember which index value corresponds to which combination. It is much easier to think of a generic variable, x, depending on the time slice t and location l. The variable x now becomes an array of variables. The indices are linked with x and are used to describe each member of a set of variables. Therefore, we introduce the index $t = 1, 2, \ldots, 12$ for time and the index $l = 1, 2, 3, 4, 5$ used for the locations. More precisely, t is used as an element (index) in the *index set* $\mathcal{T} = \{1, 2, \ldots, 12\}$, in mathematical writing $t \in \mathcal{T}$; and similarly for l and $l \in \mathcal{L} = \{1, 2, \ldots, 5\}$. Now, x_{tl} or $x_{t,l}$ represents production at time t at location l, e.g., $x_{4,5}$ for a quantity of the production quantity in time interval 4 at location 5; in general

$$x_{tl} \quad , \quad t \in \mathcal{T} \quad , \quad l \in \mathcal{L}.$$

This is an alternative, more general, and therefore preferred formulation to

$$x_{tl} \quad ; \quad t = 1, 2, \ldots, 12 \quad , \quad l = 1, 2, \ldots, 5.$$

If it is clear that all elements of an index set are meant, we write this with the universal quantifier $\forall$ (for all) and demand for example

$$x_{tl} \geq 10 \quad , \quad \forall t \in \mathcal{T} \quad , \quad \forall l \in \mathcal{L},$$

$$x_{tl} \geq 10 \quad , \quad \forall t \quad , \quad \forall l,$$

or, if no confusion is possible,

$$x_{tl} \geq 10 \quad , \quad \forall \{tl\}.$$

The symbol $\in$ is read as "is an element of," while the symbol $\forall$ means "for all." Thus $t \in \mathcal{T}$ is read as "t is an element of set $\mathcal{T}$," and $\forall l$ means "for all l." If we refer, for instance, to a previous time period, it is safer to separate the indices by comma, e.g., $x_{t-1,l}$.

The set of index values that a particular variable can take will be referred to as its *domain*, e.g., {2,3,4}. Integers used as subscripts are typical for time periods or hours during the day, e.g., in energy models. However, when using index sets it is not necessary to restrict the elements of the domain to integers, names of products, or cities are also valid. So, instead of numeric index sets like $\mathcal{L} = \{1, 2, \ldots, 5\}$, we could have an index set for locations that appears as $l \in$ {Athen, Rom, Paris, London, Bonn} with alphanumeric or string-valued index quantities. That results in much more flexibility and readability.

Index sets make it possible to easily create model instances of small size for test purposes, by only creating a part of the indices to be used, e.g., instead of $t = 1, 2, \ldots, 12$ only $t = 1, 2, 3$. Another advantage is the extension of index sets to subsets explained below. The use of discrete-time slices (hours, days, weeks, months, quarters, or years) leads to the so-called *discrete-time models*. In scheduling problems, especially, in the process industry, also *continuous-time model formulations* are promising alternatives; cf. the examples in [280], [279], [281], [252], [364], and [366] by Floudas and his collaborators. In these models, the start and end times can take any values on the continuous timeline.

Sometimes the convention $x(t, l)$ is used to avoid the need for subscripts because most computer languages do not use subscripts, but frequently the modeler uses the subscripted notation as a shorthand way of writing. Perhaps in the future two-dimensional formula input and output, e.g., x^2 or y_{ij}, will become the standard as is already the case in `Maple`[2] and `MathCAD`.[3]

In Sect. 2 the use of scaled down models was advocated. This can be handled easily when indices are used. The ranges over which the indices are allowed to vary are scaled down for the prototype model and then ranges can be extended later. More than two indices can be associated with a variable, but it is often found that notation becomes very cumbersome when more than four subscripts and associated indices are used, particularly by a non-expert modeler. As an upper limit on indices is approached it may be beneficial to rethink the nature of the variables. In any model the choice of which variables to use (not simply the choice of names or characters to use to denote them) will not be unique and may depend on personal style.

2.2.2 *Summation*

In Sect. 2.2.1 indices were introduced. Indices are used to reference a variable, e.g., x_{14}. Indices also allow sets and subsets of variables to be added together and their sum to be referenced. To do this the summation symbol $\sum$ is introduced, where $\sum$ means "sum of." For example, if we have defined the variables y_i $(i = 1, 2, \ldots, 6)$ or $i \in \mathcal{I} = \{1, 2, \ldots, 6\}$, then the expression

$$\sum_{i=1}^{6} y_i \text{ or } \sum_{i \in \mathcal{I}} y_i \text{ or } \sum_{i} y_i \qquad (2.2.5)$$

means the sum

$$y_1 + y_2 + y_3 + y_4 + y_5 + y_6 .$$

In the $\sum$ expression the part "$i = 1$" under the $\sum$ is the starting value of the index i to be used in the sum and the part "6" above the $\sum$ is the ending value of that index,

[2] `Maple` is a registered trademark of Waterloo Maple Inc.

[3] `MathCAD` is a registered trademark of MathSoft, Inc.

and

$$\sum_{i=1}^{6} \text{ or } \sum_{i \in \mathscr{I}} \text{ or } \sum_{i} \tag{2.2.6}$$

means "add all values which have subscripts between 1 and 6, inclusive" or "add all values which have subscripts contained in index set $\mathscr{I}$." Thus

$$\sum_{i=1}^{3} y_i \tag{2.2.7}$$

means

$$y_1 + y_2 + y_3$$

and

$$\sum_{i=4}^{6} y_i \tag{2.2.8}$$

means

$$y_4 + y_5 + y_6.$$

We may use further sophistication, e.g.,

$$\sum_{i=1 \wedge i \neq 3}^{6} y_i \tag{2.2.9}$$

means

$$y_1 + y_2 + y_4 + y_5 + y_6$$

and

$$\sum_{i=1}^{3} y_{2*i} \tag{2.2.10}$$

means

$$y_2 + y_4 + y_6.$$

In gAML the notation $\sum$ could be replaced by `SUM ( )` and the expressions (2.2.5), (2.2.7), (2.2.8), (2.2.9), and (2.2.10) are written, respectively, as

$$\texttt{SUM}\ (i = 1:6)\ y(i)\ \text{ or }\ \texttt{SUM}\ \{i \text{ in } \mathscr{I},\ y(i)\}$$

$$\texttt{SUM}\ (i = 1:3)\ y(i)$$

$$\texttt{SUM}(\ i = 4:6)\ y(i)$$

$$\texttt{SUM}\ (i = 1:6 \mid i.NE.3)\ y(i)$$

$$\texttt{SUM}\ (i = 1:3)\ y(2*i).$$

The summation notation provides a convenient shorthand for complex expressions and is particularly useful when we want to express something like

$$A_1y_1 + A_2y_2 + A_3y_3 + A_4y_4 + A_5y_5 + A_6y_6,$$

which can be written as

$$\sum_{i=1}^{6} A_i y_i \ \text{ or } \sum_{i \in \mathscr{I}} A_i y_i$$

and could be implemented in gAML as

$$\texttt{SUM}\ (i = 1:6)\ A(i) * y(i)\ \text{ or }\ \texttt{SUM}\ \{i \text{ in } \mathscr{I},\ A(i) * y(i)\}.$$

A more complex summation such as

$$\sum_{t=1}^{12} \sum_{l=1}^{5} A_{tl} x_{tl}$$

is handled as

$$\texttt{SUM}\ (t = 1:12, l = 1:5)\ B(t,l) * x(t,l),$$

or

$$\texttt{SUM}\ \{(t \text{ in } \mathscr{T}, l \text{ in } \mathscr{L}),\ B(t,l) * x(t,l)\}.$$

The benefit of using sets becomes even more obvious when the models contain variables depending on several indices. As a simple example, consider a variable s_{hc} representing the amounts of shipment from pairs of countries $\mathscr{C}$ ={Jamaica, Haiti, Guyana,Brazil,Germany} and harbors $\mathscr{H}$ ={Kingston, SanDomingo, Georgetown,

Belem, Bremen, Hamburg}. The complete tuple set $\mathcal{H} \times \mathcal{C}$ would contain 30 elements, but in reality, only the six combinations

$$\begin{aligned}\mathcal{HC}(h,c) = \{&(\text{Kingston, Jamaica}); (\text{SanDomingo, Haiti});\\ &(\text{Georgetown, Guyana}); (\text{Belem, Brazil});\\ &(\text{Bremen, Germany}); (\text{Hamburg, Germany})\}\end{aligned}$$

really exist in this harbor–country relationship; note that a country can have more than one harbor, but each harbor is uniquely assigned to a country. Most AMLs would allow to write a sum over such a tuple — also called subset — as

$$\texttt{SUM}\,\{\mathcal{HC}(h,c), s(h,c)\},$$

which means that the loop within the SUM runs only over six index combinations and that this term contains only six variables sent to the solver instead of 30 for the full combination set. For this small example, the difference in time for model generation is small. But if we talk about problems with millions of variables, it makes a great difference. This handling of indices, by the way, is also one of the strong arguments for AMLs as they do index and set handling very efficiently. Free approaches such as `Pyomo`, `JuMP`, etc. may have long running times when problems reach this size.

2.3 Constraints

The model developed in Sect. 1.4.1 illustrated some types of constraints, e.g., $4p + 3s \le 1400$. In general, constraints take the form

$$\text{(left-hand side) “relation” (right-hand side)}\,. \tag{2.3.1}$$

The left-hand side must be a linear expression in the variables of the model. By this we mean it must take the form

$$2x + 3.5y - 8.1w \ldots \tag{2.3.2}$$

with each variable being multiplied by a positive, negative, or zero constant (coefficient)[4] and the constant·variable items added together. Products or powers

[4]Conventionally variables with zero coefficients are omitted from the expression, so not all variables are necessarily present in the expression. Also a coefficient of 1.0 is conventionally not shown next to its associated variable, i.e., $1x$ is written as x.

(e.g., xy or x^4) of variables, or even functions of variables such as $\exp(x)$ are not permitted in linear models (but see Chap. 6).

The relation can be $\leq$, $=$ or $\geq$. The relations $<$, $>$ cannot be used because of mathematical restrictions arising from the solution approach to problems. Recall, as mentioned earlier, $<$ and $>$ are used in software coding of models, as $\leq$ and $\geq$ are not on keyboards.

Thus typical constraints include

$$x + y + w = 10, \tag{2.3.3}$$

$$2x - 3y + 4.2w \geq 5.6, \tag{2.3.4}$$

$$x + 2y - z \leq 1, \tag{2.3.5}$$

or also

$$x + 2y \leq 1 + z. \tag{2.3.6}$$

In the last constraint it is not necessary to collect together all the non-constant terms on the left-hand side of the constraint for input into the optimization software. In fact it is often convenient to use this style of constraint to clarify modeling, e.g., in the constraint

$$\text{new stock} = \text{old stock} + \text{purchases} - \text{sales} . \tag{2.3.7}$$

What about other constraints? Using binary variables it is possible to model some very sophisticated relations as linear constraints. This will be discussed in Sect. 6.5. In particular, it becomes possible to model the absolute value function $|x|$, i.e., to handle expressions of the form $|r_1 - r_2|$ where r_1 and r_2 may refer to output rates of plants in consecutive time intervals.

2.3.1 *Types of Constraints*

As in the small problem, constraints can be used to model a variety of sets of circumstances. Constraints will now be considered according to type and several different types introduced and discussed. Identification of the type of constraint will help the modeler to formulate the problem. We have tried to make the list as comprehensive as possible, but inevitably it is not possible to cover every single type of constraint. Once the reader has seen a variety of different types of conditions that can be formulated as constraints, then it becomes easier for them to identify different conditions that will be required in a model. Knowing how to formulate such conditions aids the modeling process and avoids mistakes.

One of the most frequently occurring sets of conditions gives rise to the type of constraints known as *availability constraints*. Such constraints may be used when modeling purchase of raw materials needed for production. Often raw materials are available only in limited amounts. To illustrate this type, let r_1 and r_2 represent two raw materials needed and let there be a total limit of 20 units over all periods. This is modeled as

$$r_1 + r_2 \le 20. \tag{2.3.8}$$

Quite similar to availability constraints are *capacity constraints*. They are more related to the capacity of machine for processing raw materials and producing products. Let x and y be quantities produced, and 3 and 8 their unit requirements (kg/kg) of raw materials. It is required to model the fact that 200 units is the maximum amount of raw material available. The constraint is

$$3x + 8y \le 200. \tag{2.3.9}$$

Although capacity and availability look mathematically alike, there is a difference regarding the business process. Capacity constraints are subject to changes or improvements within the company. If it turns out (by inspection of marginal effects) in a machining problem that a certain constraint has a strong effect on the solution, it might be possible to tune the machine differently and thus increase the capacity. Availability constraints related to the purchase of raw material are not that easy to change because often it is not possible to buy more materials on the market. Available work hours per week, people available for working during nights or weekends, and hiring personal are further examples of even harder business constraints.

Other typical restrictions are flow or mass *balance constraints*. They may appear in several disguises as we will see below. Let r_1, r_2, r_3 indicate flows into a process and y_1 and y_2 represent flows out. A constraint is required to model the fact that flow must be conserved in the sense that the total flow out cannot exceed the total flow in. The constraint is

$$r_1 + r_2 + r_3 \ge y_1 + y_2. \tag{2.3.10}$$

A tighter form of (2.3.10) and flow conservation is of course

$$r_1 + r_2 + r_3 = y_1 + y_2, \tag{2.3.11}$$

which ensures that flow out and flow in are exactly equal to each other. Even if (2.3.11) is the correct formulation of flow conservation there is an advantage of using (2.3.10) discussed in Chap. 9, on page 299.

Some models also include *capacity balances*. It may be required to generalize the flows of a balance constraint into other expressions that balance. This may give rise to a constraint to be associated with constraint (2.3.10). Let 2, 4, 10, 3, and 8 be the unit costs for raw materials r_1, r_2, r_3, y_1, and y_2, respectively. Consider the requirement that the total cost of items flowing into the process must not exceed the total cost of items flowing out. The required constraint is

$$2r_1 + 4r_2 + 10r_3 \leq 3y_1 + 8y_2. \tag{2.3.12}$$

In many practical problems we need *quality constraints*.[5] Consider the following example. A process consists of mixing two ingredients, X and Y, without loss of volume. Let the quantities of the two ingredients be x and y. Let X and Y contain 5% and 4% of alcohol, respectively. Then $5x+4y$ is the total contribution to quality, i.e., alcohol in mass units, from mixing the ingredients. Let 4.8 be the minimum quality level (in percent) required in the mixture of the ingredient. The required constraint (the left-hand side measures alcohol concentration) on quality is

$$\frac{5x + 4y}{x + y} \geq 4.8, \tag{2.3.13}$$

which, assuming $x, y \geq 0$ and $x + y > 0$, may be rewritten as

$$5x + 4y \geq 4.8x + 4.8y \tag{2.3.14}$$

after cross-multiplication and may be further rearranged as

$$0.2x - 0.8y \geq 0. \tag{2.3.15}$$

In models where there are quality constraints there are often *recipe constraints*. Such constraints occur when there are rigid requirements in the composition of a product that is composed of raw materials blended or mixed together. Let us assume that in a chemical process two liquids, Xyrene and Zilcene, are blended together without loss of volume to form the liquid Ailene. Ailene must comprise exactly 60% Xyrene and 40% Zilcene. Let x and z represent the quantities used of Xyrene and Zilcene, respectively, and let a represent the quantity of Ailene produced. Then we require the recipe constraints

$$0.6a = x, \quad 0.4a = z. \tag{2.3.16}$$

Another group of constraints ensures *requirements satisfaction*. Let x, y, and w be quantities produced of three products; 85.3 a revenue target that must be met; and 2, 3.5, and 8.9 the unit contributions to the revenue. The revenue requirement is

[5] Modelers in the refinery industry call these constraints *quality constraints*. Quality constraints describe concentrations, sulfur content in streams and similar quantities. These constraints, which are always related to properties of streams or materials, might also be called *property constraints*.

modeled as

$$2x + 3.5y + 8.9w \geq 85.3. \tag{2.3.17}$$

In multi-period modeling (see Sect. 8.2) we need additional *multi-period flow constraints*. These constraints connect adjacent time intervals and ensure that stocks and production balance the way they should. We must ensure that in each period final stock comprises previous stock plus stock acquired less stock used. If the index t denotes times such a constraint typically appears as

$$s_t = s_{t-1} + b_t - u_t - w_t, \tag{2.3.18}$$

where s represents stock, b some purchase of raw material, u what we use in production, and w what we sell or ship. When formulating multi-period flow constraints attention has to be paid to the correct connection between initial and final stock.

Finally, we encounter a simple group of constraints called *bounds*. These constraints contain only one variable and a right-hand side. Their simple structure allows more efficient numerical treatment. For instance, we may have a direct limitation placed on variables such as stock level s, e.g., s must be at least 10 units (stock security level) and at most 20 units (stock capacity), i.e.,

$$s \geq 10 \tag{2.3.19}$$

and

$$s \leq 20. \tag{2.3.20}$$

Constraint (2.3.19) is called a *lower bound* and constraint (2.3.20) an *upper bound*.

A further type of constraint, a *logical constraint*, will be discussed in Chap. 6, together with the use of discrete variables for formulation purposes. Another type of constraints used for tightening the model formulation is *cuts*, sometimes also called *valid inequalities*; see Chap. 3 and examples in Chap. 10.

2.3.2 Example

A timber company has resources of two types of wood, pine and birch. These woods can be used to manufacture two products, ply or board. To produce 100 m^2 of ply requires 300 m of pine and 300 m of birch, while to produce 100 m^2 of board requires 100 m of pine and 500 m of birch. The company has access to 5000 m of pine and 6000 m of birch each working period. Costs and revenues can be estimated for the different raw materials and products and the company wishes to build an LP model to maximize net revenue per period. The company wants to produce at least $500\,\mathrm{m}^2$ of each product per period.

First we identify the decision variables p_y and p_d to represent the quantities (in m^2) of ply and board produced, respectively, and the variables $u_{py}, u_{pd}, u_{by}, u_{bd}$ (in units of m) to represent the quantities of pine and birch used, respectively. We need to "subdivide" the quantities of pine and birch used for each type of product in order to organize the recipes correctly.

Now we identify constraints. There are two simple bounds given by the requirements that are modeled as

$$p_y \geq 500 \quad , \quad p_d \geq 500. \tag{2.3.21}$$

There are then two availabilities of raw materials modeled as

$$\begin{aligned} u_{py} + u_{pd} &\leq 5000 \\ u_{by} + u_{bd} &\leq 6000. \end{aligned} \tag{2.3.22}$$

There are also recipe constraints

$$\begin{aligned} 3p_y &= u_{py}, \quad & 3p_y &= u_{by} \\ p_d &= u_{pd}, \quad & 5p_d &= u_{bd}. \end{aligned} \tag{2.3.23}$$

The model would now be completed by an objective function.

In this example we have seen how we can identify structure in a problem and turn conditions into known types of constraints. Initially it would have been tempting to have single variables to represent pine and birch, but as soon as we try to model the recipe aspects we notice that it will be impossible to ensure correct proportions of raw materials required by the products unless we associate separate raw material variables with each product.

2.4 Objectives

In practical problems it is not always clear how to choose the objective function associated with a model. Profit maximization or cost minimization may be appropriate but may be over-simplistic or too difficult to measure. More sophisticated accountancy measures such as "return on capital employed" may have to be used as objective functions. A client may in fact have a number of objectives in mind, each of which could lead to the formulation of an objective function. Some of these may involve maximization and some minimization and are likely to conflict, e.g., sales maximization and stock minimization may be desirable in a retailing environment but are in direct conflict. It will be best to elicit from a client some principal objectives and test how the solution to a model impinges on other objectives.

Attempts at combining objectives and handling multi-objective problems have been made and will be discussed in Chap. 4. However, in LP or MILP models, ultimately an objective function must be a linear expression in variables and must be maximized or minimized.

Two particular types of objective functions deserve special consideration: *Minimax* and *Maximin*. To construct a typical Minimax expression, let

$$t_i \quad , \quad i \in I \tag{2.4.1}$$

represent times by which events numbered $i \in I$ must be completed. Each t_i may have been defined in a linear expression. If it is required to minimize the time t by which all events $i \in I$ will have been completed, i.e.,

$$\min \quad \left\{ \max_i \{t_i\} \right\} \tag{2.4.2}$$

this may be modeled using the constraints

$$t_i \le t \quad , \quad i \in I \tag{2.4.3}$$

and minimizing the objective function

$$\min \quad t. \tag{2.4.4}$$

The objective function and the constraints will have the effect of ensuring that t is larger than the completion time of all events but is the minimal time value that satisfies that requirement.

In another problem we may wish to have an objective function that is, loosely speaking, the opposite of the Minimax objective just introduced. Assume that we want to maximize the time available before any of the events are completed (Maximin). This may be modeled, in the notation just introduced, by the constraints

$$t \le t_i \quad , \quad i \in I \tag{2.4.5}$$

and the objective function

$$\max \quad t, \tag{2.4.6}$$

which is equivalent to

$$\max \quad \left\{ \min_i \{t_i\} \right\}. \tag{2.4.7}$$

We now consider in detail how to model variables, constraints, and objectives.

2.5 Building More Sophisticated Models

We shall consider two further models, one in this section and one in Sect. 2.7. They are a little more complex than the ones we have already introduced but will illustrate some of the more powerful features of algebraic modeling languages. It should be noted that when a model is entered into an AMS and then saved, it can be modified later. This saves having to re-type a model when it is changed and is fundamental to any model management scheme.

In this simple production planning example we build up a model piece by piece in preparation for saving it to a model file (a *.mod* file in a generic AML). We also illustrate the use of data tables and how to enter data directly into these tables. We then generate the problem matrix, input it to the solver, and solve the problem.

2.5.1 *A Simple Production Planning Problem: The Background*

A firm is processing three products A, B, and C on two machines M_1 and M_2. Each day there are 8 h available on M_1, and 10 on M_2. The machine hours required per 100 units of the products are given in the following table:

	A	B	C
M_1	2.2	1.8	1.9
M_2	2.4	2.0	2.1

Currently the profit contributions per 100 units of the products are:

A	B	C
24.7	22.4	19.7

Letting a represent the number (in hundreds) of units of product A to be made, and b and c similarly, our LP maximizing profit is

$$\max \quad 24.7a + 22.4b + 19.7c \tag{2.5.1}$$

subject to

$$\begin{aligned} 2.2a + 1.8b + 1.9c &\le 8 \\ 2.4a + 2.0b + 2.1c &\le 10 \end{aligned} \tag{2.5.2}$$

and a, b, c are non-negative.

2.5.2 Developing the Model

We require three decision variables a, b, and c to represent the quantity of the products A, B, and C, respectively, to produce. The following input to gAML will define the variables required.

```
VARIABLES
   a
   b
   c
```

Now we define data tables REQ(2,3) to store the machine time requirements for each product–machine combination and PC(3) to store the profit contributions from each product. These data tables are just like matrices or arrays in mathematics. When we define REQ(2,3) in a TABLES section we create a 2-dimensional array with 2 rows and 3 columns. Defining PC(3) we create a 1-dimensional array with 3 entries. If we wish to refer to the entry in the second row and first column of REQ(2,3), then we specify REQ(2,1).

In gAML we create tables, which initially have all zero entries, by the lines:

```
TABLES
    REQ(2,3)
    PC(3)
```

We can put some data into these tables using the simplest of the methods available. More flexible and powerful techniques will be considered later. First we handle the machine requirements. We can enter these row-by-row as

```
DATA
    REQ(1,1)  = 2.2, 1.8, 1.9
    REQ(2,1)  = 2.4, 2.0, 2.1
```

The DATA section lets you specify a starting position in a table and then a stream of numbers that will be put into the table starting at the specified position and carrying on into cells to the right. The lines starting with REQ(1,1)=. . . and REQ(2,1)=. . . have the following effect on the elements of the array REQ:

REQ(1,1) gets the value 2.2
REQ(1,2) gets the value 1.8
REQ(1,3) gets the value 1.9
REQ(2,1) gets the value 2.4
REQ(2,2) gets the value 2.0
REQ(2,3) finally gets the value 2.1

Similarly for the profit contributions table. To initialize this we have

```
    PC = 24.7, 22.4, 19.7
```

If the coefficient of the starting position in a table is not specified it is assumed to be the first row and column in the table.

We have now defined the required decision variables whose values are to be decided by the optimization process and seen the commands to enter the required data into tables. The constraints can now be defined as follows:

```
CONSTRAINTS
    PROFIT: PC (1)    *a + PC (2)    *b + PC (3)    *c
    M1    : REQ(1,1)*a + REQ(1,2)*b + REQ(1,3)*c <  8
    M2    : REQ(2,1)*a + REQ(2,2)*b + REQ(2,3)*c < 10
```

Models are usually assembled in files with an extension of *.mod*, where the first part of the file name can be selected by the user to reflect the problem being modeled. You should now assemble all the input items together, in the order that we have developed them, and save the model in a file *prdx.mod*. Here is the text:

```
VARIABLES
  a
  b
  c

TABLES
  REQ(2,3)
  PC(3)

DATA
  REQ(1,1) =  2.2,  1.8,  1.9
  REQ(2,1) =  2.4,  2.0,  2.1
  PC       = 24.7, 22.4, 19.7

CONSTRAINTS
  PROFIT: PC (1)    *a + PC (2)    *b + PC (3)    *c
  M1    : REQ(1,1)*a + REQ(1,2)*b + REQ(1,3)*c <  8
  M2    : REQ(2,1)*a + REQ(2,2)*b + REQ(2,3)*c < 10
```

Now we run this model. If you made any errors typing in the model, gAML will return an error message, correct any errors, and resubmit the corrected problem. You should be able to obtain the optimal objective value 99.5556.

2.6 Mixed Integer Programming

Let us now introduce a new feature that has very significant consequences: variables that are restricted to discrete values. When an optimization problem contains integer variables, or certain other types of variables, it becomes a *mixed integer programming* (MIP) problem, usually containing both integer and continuous variables. The special case in which only integer variables occur is IP. In Sect. 12.6 we refer briefly to Mixed Integer Nonlinear Programming (MINLP) but in most parts of the book we just concentrate on linear problems.

When an LP contains integer variables, this leads us to a Mixed Integer Linear Programming problem (abbreviated as an MILP problem). Of course, the structure of MILP problems retains the linear objective and constraints but adds an integrality attribute to the non-negativity requirement on some, or all, of the variables. The term integer linear programming problem (ILP) is normally used to denote an LP problem where all variables are required to take integer values, sometimes called a pure integer linear programming problem, but the term ILP is sometimes also used as an all-embracing term to describe both pure integer linear programming problems and MILPs. Thus an ILP may look like the problems introduced in the previous sections, with an additional requirement on the nature of the variables. While the differences between an LP problem and an MILP problem may look to be minor, they turn out to be major in terms of the difficulty of solving, with MILP being substantially more difficult.

One might try to solve an MILP problem by solving the LP problem neglecting the integrality on the variables and then fixing those variables to nearby integer variables. Sometimes this technique may work but the following example shows why rounding does not really help. Consider

$$\max \quad x_1 + x_2 \tag{2.6.1}$$

subject to

$$-x_1 + x_2 \geq 0.5 \quad , \quad -0.8x_1 + x_2 \leq 1.3 \quad , \quad x_1, x_2 \in \mathbb{N}_0. \tag{2.6.2}$$

In Sect. 1.4.1 we saw how we can solve this problem graphically. The straight lines

$$S_1 : x_2 = x_1 + 0.5 \quad , \quad S_2 : x_2 = 0.8x_1 + 1.3 \tag{2.6.3}$$

and the x_1- and x_2-axes establish the feasible region (draw this as an exercise). If we neglect the integrality conditions the solution is

$$x_1 = 4 \quad , \quad x_2 = 4.5 \quad , \quad Z^{LP} = 8.5. \tag{2.6.4}$$

However, the best integer feasible solution is

$$x_1 = 1 \quad , \quad x_2 = 2 \quad , \quad Z^{IP} = 3. \tag{2.6.5}$$

This example shows that the best integer solution may be far away from the solution in which the integrality condition is not forced. Thus simple rounding techniques are not sufficient.

Because ILP and MILP introduce the idea of a variable that takes a discrete set of values, they also allow for the possibility of using the values of variables to denote decisions that are not strictly quantitative. Problems using this idea are frequently termed discrete optimization or discrete programming problems and will be discussed later in the book (see Chap. 7 for a systematic overview on types of such problems). Here we just briefly list some features and properties in models that can be formulated using MILP:

- Counting (e.g., Sects. 2.6.1 and 7.1.2)
- Representation of states and yes–no decisions (Chaps. 7 and 10)
- Logical implications (Sects. 6.1 and 6.2)
- Simple nonlinear functions (Sects. 6.5, 6.7, and 6.8)

The next section provides an MILP example that will later in Sect. 3.3.1 also serve to illustrate a solution method for MILP problems.

2.6.1 Example: A Farmer Buying Calves and Pigs

Tom, a British farmer keeping calves (baby cows) and pigs, wants to increase his animal stock. Tom has the chance to buy calves at the price of 1000 £ per animal and young pigs at the same price. Unfortunately, at the moment, Tom can only invest 3500 £. Keeping calves and pigs for 2 years, the anticipated net profit is 3000 £ per cow and 2000 £ per pig. As Tom is already a fairly successful farmer, his farm can only take another two calves and two pigs. It is not possible to enlarge the animal housing. Of course, the farmer wants to increase his total profit. The question is how many calves and pigs Tom should buy.

Obviously, the best solution is to buy two calves and one pig resulting in a total profit of 8000 £. In this simple case it was easy to guess the solution. But imagine if the farmer could also have chickens and horses, more money to invest and so on. In that case mathematical optimization could help. Let us try to translate the problem into mathematics and to solve it as a mixed integer programming problem.

The task of the model builder is to transform the real-world problem into a mathematically suitable form. As we learned in Sect. 2.2, the modeler first identifies variables that allow him to formulate constraints and the objective function. The variables are usually chosen in a natural way such that the basic questions are answered. In the present case it seems appropriate to introduce variables c and p to represent the number of calves and pigs, respectively, to be purchased. Our choice of the set of valid solutions must reflect the fact that it is not possible to buy living animals in fractions. Our model builder, experienced in farm optimization, chooses 1,000 £ (k£) as the basic unit for money in this model.

The advantage of doing so is related to scaling: to have a model that behaves numerically well, all quantities, data and variables, should ideally be within a range of 0.1 and 10. This is not always possible but it is desirable, and therefore one should pay some attention to this issue. Scaling is discussed in Sect. 9.2.2.

Using this unit the objective function takes the form

$$\max \quad Z = 3c + 2p. \tag{2.6.6}$$

The constraints are

$$0 \leq c \leq 2 \quad , \quad c \quad \text{integer} \tag{2.6.7}$$

$$0 \leq p \leq 2 \quad , \quad p \quad \text{integer} \tag{2.6.8}$$

$$c + p \leq 3.5, \tag{2.6.9}$$

where we will refer to the constraints (2.6.7) and (2.6.8) as *integrality conditions*.

In Sect. 3.3.1 we show how this problem can be solved. At this stage, we can already use an AMS, e.g., AMPL, GAMS, Mosel, or SAS/OR to answer our farmer's questions. We input the model as shown in Fig. 2.2. A model implementation of this problem is contained as *calves* in MCOL coming with this book.

Now we Run and solve the model, remembering to maximize. In Sect. 3.3.1 we learn how to use a Branch-and-Bound algorithm to solve this problem exploiting *nodes* of a *tree*. In Fig. 3.3 we see that an integer solution was obtained at node 3 and a total of 5 nodes were explored; in Sect. 3.3.1 it will become clear what *nodes* are. The objective function value obtained at node 3 was 8.0 and this was never bettered subsequently. So 8 is the best objective function value we can get, and the decision variables take the values $c = 2$ and $p = 1$ telling our farmer to buy two calves and one pig with an expected profit of 8000 £.

```
* define the budget
  Scalar BUDGET budget /3.5 / ;

  Integer variable x1 the number of cows
                   x2 the number of pigs ;

  Variable profit ;

  Equations
       Objective          "maximize profit"
       Budget_restr       "budget constraint"
  ;

  Objective..          profit =e= 3*x1 + 2*x2 ;

  Budget_restr..       x1 + x2 =l= BUDGET ;

  model cows_pigs /all/ ;

* upper bounds on the number of animals
  x1.up = 2 ;
  x2.up = 2 ;
```

Fig. 2.2 Integer model coded in GAMS. This screenshot is just to get the flavor of how a model could look like in an AML. No need to understand the syntax in detail

2.6.2 A Formal Definition of Mixed Integer Optimization

With the introduced terms we are now able to give a formal definition of an integer or mixed integer optimization problem. The optimization problem class we have in mind is mixed integer programming (MIP) problems specified by the augmented vector $\mathbf{x}_{\oplus}^{\mathrm{T}} = \mathbf{x}^{\mathrm{T}} \oplus \mathbf{y}^{\mathrm{T}}$ established by vectors $\mathbf{x}^{\mathrm{T}} = (x_1, \ldots, x_{n_c}) \in X \subseteq \mathbb{R}^{n_c}$ and $\mathbf{y}^{\mathrm{T}} = (y_1, \ldots, y_{n_d}) \in V \subseteq \mathbb{Z}^{n_d}$ of n_c continuous and n_d discrete variables, a contiguous subset $X \subseteq \mathbb{R}^{n_c}$, and a discrete and restricted subset $V \subseteq \mathbb{Z}^{n_d}$, an objective function $f(\mathbf{x}, \mathbf{y})$, n_e equality constraints $\mathbf{h}(\mathbf{x}, \mathbf{y})$, and n_i inequality constraints $\mathbf{g}(\mathbf{x}, \mathbf{y})$. The problem[6]

$$\min \left\{ f(\mathbf{x}, \mathbf{y}) \left| \begin{array}{l} \mathbf{h}(\mathbf{x}, \mathbf{y}) = 0, \ \mathbf{h} : X \times U \to \mathbb{R}^{n_e}, \ \mathbf{x} \in X \subseteq \mathbb{R}^{n_c} \\ \mathbf{g}(\mathbf{x}, \mathbf{y}) \geq 0, \ \mathbf{g} : X \times U \to \mathbb{R}^{n_i}, \ \mathbf{y} \in U \subseteq \mathbb{Z}^{n_d} \end{array} \right. \right\} \tag{2.6.10}$$

is called *Mixed Integer Nonlinear Programming* (MINLP) problem, if at least one of the functions f, $\mathbf{g}$ or $\mathbf{h}$ is nonlinear. The vector inequality, $\mathbf{g}(\mathbf{x}, \mathbf{y}) \geq 0$, is to be read component-wise. If instead of $\mathbf{y} \in V \subseteq \mathbb{Z}^{n_d}$ the basic area $\mathbf{y} \in Y$ is any set with discrete elements, e.g., $Y = \{-3, 0, +0.25, +1, +17.1\}$, then (2.6.10) is a discrete optimization problem, which can be formulated as an integer optimization problem with the help of binary variables like on page 37. Any vector $\mathbf{x}_{\oplus}^{\mathrm{T}}$ satisfying the constraints of (2.6.10) is called a *feasible point* of (2.6.10).[7] The set of all feasible points forms the feasible set S. Any feasible point whose objective function value is less or equal than that of all other feasible points is called an *optimal solution*, or more formally: The vector pair $(\mathbf{x}, \mathbf{y})$ is called *optimal solution*, if it is allowed and applies to all allowed points $(\mathbf{x}', \mathbf{y}') \in S$: $f(\mathbf{x}, \mathbf{y}) \leq f(\mathbf{x}', \mathbf{y}')$. It follows from this definition that a problem may have several, not uniquely defined optimal solutions. In addition, a typical difficulty in the nonlinear non-convex optimization is that it is usually only possible to determine local optima, but it is much more difficult to prove that a global optimum has been determined. In simple terms, a global optimum is the point whose objective function value is no worse than any other valid point, while a local optimum $(\mathbf{x}_*, \mathbf{y}_*)$ has this property only in a neighborhood of that point $(\mathbf{x}_*, \mathbf{y}_*)$. The formal definition for this is: In a given minimization problem, a point $(\mathbf{x}_*, \mathbf{y}_*) \in S$ is a local minimum with respect to the environment U of $(\mathbf{x}_*, \mathbf{y}_*)$ (or simply local minimum) if

$$f(\mathbf{x}_*, \mathbf{y}_*) \leq f(\mathbf{x}, \mathbf{y}), \quad \forall\, (\mathbf{x}, \mathbf{y}) \in U(\mathbf{x}_*, \mathbf{y}_*)$$

applies. If $U(\mathbf{x}_*, \mathbf{y}_*) = S$, then $(\mathbf{x}_*, \mathbf{y}_*)$ is called *global minimum*.

[6] A minimization problem has been chosen as the standard form; since any maximization problem can be formulated as a minimization problem, this is not a limitation.

[7] Often a *feasible point* is also called *feasible solution*. We do not want to do this in this book because the aim of optimization is to determine the optimal point, or in case of ambiguity also optimal points, and therefore only those can be regarded as a solution.

Now an example: The integer optimization problem in the variables y_1 and y_2

$$\min_{y_1, y_2} \left\{ 3y_1 + 2y_2^2 \,\middle|\, \begin{matrix} y_1^4 - y_2 - 15 = 0 \\ y_1 + y_2 - 3 \geq 0 \end{matrix} \quad , \quad y_1, y_2 \in V = \mathbb{N}_0 = \{0, 1, 2, 3, \ldots\} \right\}$$

has, for example, the feasible point $(y_1, y_2) = (3, 66)$ and the unique optimal solution $\mathbf{y}^* = (y_1, y_2)^* = (2, 1)$ mit $f(\mathbf{y}^*) = 8$.

Depending on the functions f, $\mathbf{g}$, and $\mathbf{h}$ in (2.6.10) we get the problem types

Acronym	*Type of optimization*	$f(\mathbf{x}, \mathbf{y})$	$\mathbf{h}(\mathbf{x}, \mathbf{y})$	$\mathbf{g}(\mathbf{x}, \mathbf{y})$	n_d
LP	Linear programming	$\mathbf{c}^T\mathbf{x}$	$\mathsf{A}\mathbf{x} - \mathbf{b}$	$\mathbf{x}$	0
QP	Quadratic Programming	$\mathbf{x}^T\mathsf{Q}\mathbf{x} + \mathbf{c}^T\mathbf{x}$	$\mathsf{A}\mathbf{x} - \mathbf{b}$	$\mathbf{x}$	0
QCQP	Quadratically Constrained QP	$\mathbf{x}^T\mathsf{Q}\mathbf{x} + \mathbf{c}^T\mathbf{x}$	$\mathsf{G}\mathbf{x} - \mathbf{r}$	$\mathbf{x}^T\mathsf{P}\mathbf{x} + \mathbf{d}^T\mathbf{x} + q$	0
SOCP	Second Order Cone Program	$\mathbf{c}^T\mathbf{x}$	$\Vert\mathsf{A}_i\mathbf{x} + \mathbf{b}_i\Vert_2$	$\leq (\mathbf{q}_i^T\mathbf{x} + d_i)$	
NLP	Nonlinear Programming				0
MILP	Mixed Integer LP	$\mathbf{c}^T\mathbf{x}_\oplus$	$\mathsf{A}\mathbf{x}_\oplus - \mathbf{b}$	$\mathbf{x}_\oplus$	≥ 1
MIQP	Mixed Integer QP	$\mathbf{x}_\oplus^T\mathsf{Q}\mathbf{x}_\oplus + \mathbf{c}^T\mathbf{x}_\oplus$	$\mathsf{A}\mathbf{x}_\oplus - \mathbf{b}$	$\mathbf{x}_\oplus$	≥ 1
MINLP	Mixed Integer NLP				≥ 1

with a matrix $\mathsf{A} \in \mathcal{M}(m \times n, \mathbb{R})$ of m rows and n columns, a matrix $\mathsf{Q} \in \mathcal{M}(n \times n, \mathbb{R})$ of n rows and n columns, $\mathbf{b} \in \mathbb{R}^m$, $\mathbf{c} \in \mathbb{R}^n$, and $n = n_c + n_d$. MIP problems without continuous variables are integer programs (IPs). Sometimes, the problems can be equivalently formulated in several classes. A binary variable δ in an MILP problem can also be used as a continuous variable in the equality constraint

$$\delta\,(1 - \delta) = 0$$

in an NLP problem; unfortunately, this is numerically not efficient. Another example for several classes is found, e.g., Rebennack et al. (2009,[457]) who derive an exact algorithm from a continuous, bilinear formulation of the fixed charge network flow problem. The authors reformulate this classical MILP problem with a continuous QP programming formulation. In LP and MILP one usually writes the constants on the right side, i.e., instead of $\mathsf{A}\mathbf{x} - \mathbf{b} = 0$, one writes $\mathsf{A}\mathbf{x} = \mathbf{b}$. Since some of these optimization problems listed above occur as subproblems when solving others, it is very important to understand the solution algorithms well and to implement them as efficiently as possible.

2.6.3 *Difficult Optimization Problems*

While LP problems can be solved relatively easily (the number of iterations, and thus the effort to solve LP problems with m constraints grows approximately linearly in m), the computational complexity of MILP and MINLP grows exponentially with n_d but depends strongly on the structure of the problem. Numerical methods to solve NLP problems work iteratively and the computational problems are related to questions of convergence, multiple local optima, and availability of good initial solutions. Global optimization techniques, cf. Horst & Pardalos (1995,[271]), Floudas (2000,[190]), or Floudas & Gounaris (2009,[191]), can be applied to both NLP and MINLP problems and their complexity increases exponentially in the number of all variables entering nonlinearly into the model.

From a practical point of view, we call optimization problems difficult when they cannot be solved to optimality, or within a reasonable integrality gap or any guaranteed bound by any standard solver within a reasonable time limit. Problem structure, size, or both can produce difficult problems. From a theoretical point of view it is relevant to observe that solving MILP problems and the problem of finding appropriate bounds is often $\mathcal{NP}$-hard, which makes these problems hard to solve. A consequence of this structural property is that these problems scale badly; we must fear the worst. Even if we can still solve a certain problem instance well, slightly changed input data that only slightly increases the problem can lead to the fact that we can no longer solve the problem at all in a reasonable time. But these classifications are based on worst case behavior: For instance, knapsack problems are (weakly) $\mathcal{NP}$-hard but not really hard to solve for 100 items. *Dynamic programming* is suitable for such problems. Even $\mathcal{NP}$-hard problems can be computationally tractable. A good example is the cutting stock problem in Sect. 14.2.1 when using *column generation* following the ideas from Gilmore & Gomory (1961,[220]). They exploit the problem-specific structure of the cutting stock problem leading to a decomposition into a column-generating model (also called pricing model), a master model generating the input data for the continuous pricing model, and a partitioning model solving the original MILP problem with all columns dynamically generated. A recommended overview on column generation is by Lübbecke & Desrosiers (2005,[376]). The meaning of what are columns varies from problem to problem:

1. In the cutting stock problem the columns are patterns generated by solving knapsack problems, i.e., in this case, the pricing model is a knapsack problem.
2. In vehicle routing problems, cf. [376], the columns are tours in a network generated by solving shortest path problems.
3. In a multiple container loading problem, Eley (2003,[172]) provides a set partitioning approach to minimize the number of containers used to host a given set of rectangular items (boxes) by using box packing to individual containers — generated by a heuristic — as columns; as the pricing problem is solved by a heuristic, no optimality is guaranteed though.

Fig. 2.3 Simple spreadsheet

	A	B	C	D
1				
2				
3				
4				

2.7 Interfaces: Spreadsheets and Databases

The use of spreadsheets has become a very popular way for managers to model problems, store data, model variables, and investigate scenarios. In many ways spreadsheets such as `LOTUS 1-2-3`[8] in the older days and `EXCEL`[9] nowadays have replaced the use of certain programming languages for the more casual user. Familiarity with spreadsheets has led many modelers to expect to use the row and column matrix layout shown in Fig. 2.3 as a standard for data input.

An existing modeling application may exist as model and data in a spreadsheet. The modeler may now intend to move on to optimization of the model that is contained in the spreadsheet, or parts of that model. Hence, the modeler is moving from "what if" modeling, for which the spreadsheet is very useful, to "what is ideal, and what are the means to achieve it." Many spreadsheet systems offer limited optimization facilities, but this may not be suitable for realistic models. The system `What's Best?`[10] was able to amalgamate spreadsheet style modeling with optimization capability, making use of the established `LINDO`[11] optimization system. `EXCEL`[12] even includes nonlinear programming and Mixed Integer Nonlinear Programming features for small problems. These systems use very simple algorithms to solve the problems and they are thus restricted by memory and computing time, and sometimes they lack stability and robustness. The UT Austin "Paul Jensen Excel Add-ins" (found at https://pypi.org/project/ormm/) can be a good source for "robustifying" Excel.

The more useful form of incorporating spreadsheets into optimization processes is to use the spreadsheet to store data and then to draw on this data for a model

[8]`LOTUS 1-2-3` is a registered trademark of Lotus Development Corp.

[9]`EXCEL` is a registered trademark of Microsoft Corp.

[10]`What's Best?` is a registered trademark of Lindo Systems Inc.

[11]`LINDO` is a registered trademark of Lindo Systems Inc.

[12]`MS-EXCEL` is a registered trademark of Microsoft Corp.

built by the modeling facility of optimization software. Thus the data is held in the spreadsheet, but the logic model is held by the optimization software. The spreadsheet provides a simple to use interface, particularly for the less experienced user, and helps the user to organize the data or to produce plots.

A model will now be developed, which makes use of spreadsheet interfacing. A feature of this example is the use of data tables. These contain the data to be used with a model. Data for these tables can be entered from a variety of sources such as text files and many spreadsheets and databases. In the following example a more advanced model is described, which uses data input from these sources. It assumes that you are familiar with the use of `LOTUS 1-2-3` to the extent that you can create and save a new spreadsheet and name ranges within it. Readers who are less familiar with these areas may wish to postpone the rest of this section until later.

2.7.1 Example: A Blending Problem

This example illustrates how data may be read into tables from a text file and a `LOTUS 1-2-3` spreadsheet. It also shows how we can create arrays of variables, i.e., subscripted variables. It assumes that you are able to set up named ranges and edit tables within `LOTUS 1-2-3`. Let us first set up the model background.

A mining company has two types of ore available: Ore 1 and Ore 2. Denote the amount of these ores to be used by o_1 and o_2, and introduce an index set $\mathcal{O}(2)$ to indicate that there are two different types of ore: the set itself reads $\mathcal{O} = \{Ore1, Ore2\}$. These ores can be mixed in varying proportions to produce a final product of varying quality. For the product we are interested in, the "grade" (a measure of quality) of the final product must be between the laid-down limits of $L_1 = 4$ and $L_2 = 5$. The grades of the ores are given by G_j, $j \in \mathcal{O}$. The final product sells for $S = \$125$ per ton. The costs C_j of the two ores vary, as do their availabilities A_j, $j \in \mathcal{O}$. Maximizing net profit z (i.e., sales revenue less cost of raw material) gives us the objective function:

$$z = \sum_{j\in\mathcal{O}} (S - C_j) o_j . \tag{2.7.1}$$

We then have to ensure that the grade of the final ore is within certain limits. The type of constraint required was shown in (2.3.13) to (2.3.15). Assuming that the grades of the ores combine linearly, the grade of the final product is

$$\sum_{j\in\mathcal{O}} G_j o_j \Big/ \sum_{j\in\mathcal{O}} o_j . \tag{2.7.2}$$

This must be greater or equal to L_1; so cross-multiplying and collecting terms, we obtain the constraint:

$$\sum_{j \in \mathcal{O}} (G_j - L_1) o_j \geq 0. \tag{2.7.3}$$

Similarly the grade must not exceed L_2, so we have the further constraint:

$$\sum_{j \in \mathcal{O}} (L_2 - G_j) o_j \geq 0. \tag{2.7.4}$$

Finally there is a limit to the availability of each of the ores. We model this with the constraints:

$$o_j \leq A_j \quad , \quad \forall j \in \mathcal{O}. \tag{2.7.5}$$

As we have seen before, these constraints explicitly bound exactly one variable and are called bounds.

2.7.2 Developing the Model

The problem description above sets out the relationships that exist between variables but contains few explicit numbers. Focusing on relationships rather than figures makes the model much more flexible. In this example only the selling price and the upper/lower limits on the grade of the final product are fixed.

In developing the model we assume that grades of the ores are available in a data file called *quality.dat*. Using a text editor, create a data file called *quality.dat* and insert a single line representing the grades of ores 1 and 2. Assuming that these values are 2.1 and 6.3 we just type:

```
2.1,6.3
```

We will also assume that the varying costs and availabilities are in a `LOTUS` spreadsheet, the costs being held in a named range called `COSTS`. We shall access the availabilities directly — they will be stored in cells A7 .. B7. Now set up a `LOTUS` 1-2-3 version 3 spreadsheet and call it: *ore_data.wk3*.

Enter the costs 85.00 and 93.00 in adjacent cells of the spreadsheet and give a name `COSTS` to the range containing these two cells, using/`Range Name Create`. In cells A7 and B7 enter the ore availabilities (in tons) for the coming week. We have specified 60 tons of ore 1 and 45 tons of ore 2, but you may wish to experiment with different values. Save the data and exit `LOTUS`.

The next step is to rewrite the algebraic model we began with a format that `gAML` can interpret. Note that comments can be included if preceded by an exclamation mark "!"

Enter the following model script[13] into a model file (*blendx.mod* contains a version in MCOL):

```
INDICES
  oretype(2)

VARIABLES
  o(oretype)              ! Quantity of each ore to purchase

TABLES
  COST (oretype)          ! Cost per tonne of ores
  AVAIL(oretype)          ! Availability of ores
  GRADE(oretype)          ! Quality of ores

DATA
 oretype = "t1", "t2"

DISKDATA
  GRADE = quality.dat                 ! Input grades from file

DISKDATA -l                           ! Input from Lotus 1-2-3
  COST  = ore_data.wk3 (costs)        ! costs from a named range
  AVAIL = ore_data.wk3 (A7 .. B7)     ! availability from range
                                      ! description

 CONSTRAINTS

! Maximize (Revenue - Costs)
    Max: SUM(j = oretype) ( 125.0 - COST(j) ) * o(j) $

! Grade < upper limit
    QualMax: SUM(j = oretype) ( 5.0 - GRADE(j) ) * o(j) > 0

! Grade > lower limit
    QualMin: SUM(j = oretype) (GRADE(j) - 4.0) * o(j) > 0

! limit on oresl < avail
    Limit(j = oretype): o(j) < AVAIL(j)
```

In this model file, the two ore variables have not been identified uniquely by name. Instead we have defined a table, or array, where the variables can be found. This makes the model more general and means that it can be applied to problems involving many types of ore if required. The array is written as

```
VARIABLES
  o(2)
```

[13]The DISKDATA command for reading text data is supported by mp-model's successor FICO Xpress Mosel by FICO. Just LOTUS is probably not much used these days.

This creates a one-dimensional array of variables, o_1 and o_2. An array written as

```
VARIABLES
  y(5,8)
```

would create a five row by eight column array of y variables. The variable in the second row, third column would be referred to as y_{23}.

When data are input from a spreadsheet file, then gAML must be told the filename using the appropriate syntax. We shall call the model *blendx.mod*. Having set up the data files as required we can run, and the ore costs, availabilities, and qualities will be extracted from the various input files as necessary by gAML. The model can now be maximized and the optimal objective value 3618.947 is obtained.

2.7.3 Re-running the Model with New Data

Now suppose you want to run the model again with different costs and availabilities associated with the two ores. Simply call up the spreadsheet software and make the required changes to the appropriate cells. Make sure you save the new spreadsheet and then re-start gAML.

gAML will extract the revised data from the spreadsheet. It will then generate a new matrix file and you can go on to optimize the new model, just as before.

By now you should be able to see the advantages of storing the model in a text file — there is no need to type in the model each time you wish to run it. As long as the model's structure does not change, the model can be used again and again to obtain a new optimal solution whenever the data producing a particular model instance change.

As well as using a spreadsheet to provide the data for an optimization model, we would expect results from runs of the model to be returned to the spreadsheet and stored. This can be accomplished and details are available from the manual for the AML at hand. Similarly, a proprietary database system may be used to provide the data for the model and results returned to be stored in the database. Most commercial mathematical programming systems offer the capability to interface with database systems.

2.8 Creating a Production System

In the previous section the idea of using links between mathematical programming modeling and spreadsheets and databases was introduced, but this only touched upon the one important part of delivering a friendly production system to the end user.

Once a large practical model has been developed and is presenting sensible and useful results, it will frequently be developed into a production system. It is quite likely that the user of this production system will be someone who is not versed in mathematical programming and will expect to see a system that uses terms and concepts that are comprehensible to him or her. The typical output of a modeling or optimization system will not be acceptable to this sort of end user. The detailed design of such a system is beyond the scope of this book but it is worthwhile pausing to see what facilities are provided by modern modelers and optimizers that will aid in the construction of friendly, easy to use systems.

It is an absolute prerequisite that the modeling system supports reading from and writing to spreadsheet and database systems. Such data repositories may reside not only on the end user's machine but may also be part of the corporate database, living on departmental servers or perhaps even on the company's mainframes. Usually, the modeling system must be able to gather data from several data sources, some local, some from within the user's department, and some perhaps from distant machines. In a planning model it is likely that data will be required from many different sources, including the Production, Purchasing, and Sales departments. It is imperative that the data are up-to-date and are reliable, or else the results will not be believed. Getting the data directly from their source is the only way to guarantee its freshness.

The requirement for outputting data, including all sorts of solution values and possibly numbers derived from them, is important as the report writing facilities in mathematical programming systems are very poor compared to those cheaply available in spreadsheets, databases, and commercial report generating packages. It is almost certain that the end user will either be to able write reports in the preferred package or will have access to someone with data processing skills to do this.

The important questions of data verification and validity are probably best answered in the data repository, rather than in the modeling system. They have richer facilities for inspecting data and it is likely that the end user will have some familiarity with using them anyway.

We shall now briefly consider various aspects of integrating modeling and optimization into a complete system. The easiest and most obvious one is to use the modeler and optimizer in a batch system (for instance, on a PC we might create a *.bat* file with a set of commands that include a call to a stand-alone modeler and a stand-alone optimizer). For various aesthetic reasons this is now considered to be inelegant as it does not give the end user the idea of an integrated system.

Modern optimization systems now also come in the form of a library of subroutines, which can be linked into the data capturing/result presentation software so that the modeling and optimization process is effectively hidden from the end user.

Under Windows subroutine libraries generally take the form of Dynamic Link Libraries (DLLs) that can be called from a variety of different languages. The system integrator can choose the language for developing the software that interacts with the end user. 1990s favorites included `Visual Basic` and `Delphi`, as it was very easy to put together a pleasing and familiar interface in a relatively short time. These programs then call the DLLs to perform modeling and optimization, check for correct data, and display results.

It is important to note that though the core of the application is the optimizer, the end user only sees the surrounding software and the acceptability of the system is very often dominated by his or her perspective of this. Thus it is a (perhaps unfortunate) fact that the optimization modeler must be aware of the rapid developments taking place in graphical user interfaces and whatever happens to be the current fashionable fad as a development system because the credibility of the model may lie not in the results themselves but in the way they are presented.

Various forms of presentation software may be employed, ranging from simple report writing programs through to animations and even videos of the results.

2.9 Collecting Data

This section contains some fundamental remarks. So far it has been assumed that data are readily available, accurate, and reliable, for the models to be built. If data are to be drawn from an existing database, then it may not be possible to scrutinize their origins; however, it is advisable not to take too much on trust. A number of elements of good practice exist, which are in fact part of good practice within operational research, of which mathematical programming may be considered a subset. Details of the practice of operational research may be found in Mitchell (1993,[408]). Some suggested procedures are the following:

1. The origins of data should be checked where possible and it should be established if data are appropriate.
2. The accuracy levels to which data have been measured should be established and it should be decided if these were appropriate for the model.
3. The limitations of data collected should be noted.
4. Assumptions made when the data were collected should be established and these should be checked against what is required for the model.
5. Limitations such as the timeliness of the data should be established and these should be checked against what is required for the model.
6. The extent to which estimation was used in data collection should be established.

In addition to the above checks and balances whose level needs to be decided upon, if data are to be collected specifically for a new model, then the following considerations should be made. Note that the points listed above and also below are both the modeler's and the client's responsibility:

1. It should be established with the client what definitions of terms such as "cost" and "revenue" are to be used and data should be collected in accordance with these definitions.
2. The level of accuracy required by the client should be agreed upon and data should be collected to support this level.
3. The client should agree upon assumptions and simplifications (e.g., aggregation of data) that may be used with data.

4. The delegation of responsibilities for data collection and subsequent supply of data when the model is used in the future should be established.
5. Procedures should be laid down as to how data should be collected in the future in accordance with practice developed for the development of the model.

While in this section we gave a recipe how to collect data, in Sect. 5.5.5 we discuss the problem of consistent and obtainable data when modeling real-world problems. In Sect. 14.1.2.3 we provide a technical approach for data consistency checks.

2.10 Modeling Logic

When a model of a problem is produced, it may "work" and produce answers, but that will not necessarily mean that the model is an accurate representation of the original problem. The ideas that the modeler has tried to convey may not be what was intended.

To illustrate this point, consider the following example: in a beer production planning model it may be intended to include the following relationship into a model: producing one pint of beer requires eight pints of water. It is easy to translate this relationship incorrectly within the model into the form that eight pints of beer require one pint of water in the production process. With this erroneous relationship present the model may have nothing apparently wrong with it, but the answers it produces may be very suspect.

Hence, two types of checks are necessary in modeling logic. The first is to ensure that the modeler has correctly appreciated the client's problem and is modeling the agreed problem. The second is to ensure that once the model has been produced, what it contains stands up to scrutiny as a valid representation of reality as it applies to the client. Thus the problem both implies the model and is implied by it.

It will often be difficult to capture the logic of a problem from discussions with a client. A structured approach similar to a flow chart in computer programming or the approach computer scientists take in producing pseudo-code will be useful. This allows the problem to be broken up into manageable portions, each connected logically in some way to the rest. Thus the model will be broken down into submodels and then each submodel may be more directly transferable into mathematical programming constraints and variables. Each submodel must stand up to scrutiny. Thus a carefully structured approach to modeling is to be recommended.

2.11 Practical Solution of LP Models

The smaller examples of LP do not illustrate many of the features of the problems that are solved in practice. Therefore, we summarize some fundamental facts that the modeler should have in mind when trying to solve real-world problems.

LP models are typically large, involving hundreds or thousands of constraints (rows) and thousands or tens of thousands of variables (columns). Such problems are solved by software in times that are often less than 1 min. It is usual that the problem has many more variables than constraints. It is quite common that there are ten times more variables than constraints. Some commercial problems involve 1.5 million variables. We now address the twin problems of: Why are LP problems so large? Why are LP problems comparatively easy to solve?

2.11.1 Problem Size

A typical model will incorporate a number of dimensions, e.g., a production problem may have products, factories, and time periods. Thus an organization may be interested in production planning over several products that it makes at several locations and production varies in each time period. If there are 50 products, 5 locations, and 10 periods and we require decision variables of the form: "how much of product A is to be produced at factory Z in period 4?"

In order to compute the number of variables we have to multiply the number of products times the number of factories times the number of time periods. Thus, we require $50 \cdot 5 \cdot 10 = 2500$ variables. We can see immediately why a problem will easily have many variables. A similar argument applies to the number of constraints.

2.11.2 Ease of Solution

Four particular reasons allow the Simplex algorithm to perform well on problems, even when the numbers of variables and constraints are substantial, and allow them to be solved in a routine and reliable way. The reasons concern the algebra of the problems as well as their structure.

1. As LP problems increase in size, the time taken to solve them will be expected to increase. There is a class of mathematical problems, known as $\mathcal{NP}$-hard problems (see the remarks on complexity theory on page 555) that are problems where the difficulty of solving rises exponentially (i.e., faster than any simple power such as squared, cubed, etc.) as problem size grows. It turns out that although some optimization problems similar to LP are in this category, LP is in an easier category. Instead of a precise definition of problem size, cf. Padberg (1996,[431]), we might take the sum of rows and columns, $n + m$, as a measure. LP is known to be solvable by a polynomial algorithm (Khachian (1979,[332]) and later Karmarkar (1984,[325])), i.e., one that requires computation time in proportion to a power of problem size. However, the Simplex algorithm, the commonly used algorithm for LP, in the worst case has behavior that is exponential in problem size; Klee & Minty (1972,[334]). Thus there would seem

to be some loose justification why LP problems are known to be solvable in polynomial time but when being solved by an algorithm that is not polynomial rarely encounter their worst case behavior. Thus it is not so surprising that the Simplex algorithm of Dantzig (1963,[142]) should have stood the test of time.

2. Most LP problems are *sparse* although that may be not so obvious from the examples and smaller case studies in the book. This means that if we draw a table of the constraints as rows and the variables as columns marking down the values of the coefficients of each variable, then this matrix has few non-zero coefficients. Typically in a large problem each variable only occurs in 5–10 constraints and each constraint only contains 5–10 variables. There will be common constraints in which many variables are present, e.g., financial or resource base, but most constraints will be localized. Hence, vast sections of the coefficient matrix will be empty, i.e., have zero coefficients. As far as computation is concerned, calculations that involve zero can be conducted easily, so the Simplex algorithm moves rapidly on a sparse problem. Can we understand these features? Why should this be so? If we think back to the production problem considered earlier, we will find that many constraints will be special for each product, each factory, and each time period and will not interact as each block of constraints represents a submodel for local production of certain products in a certain period. So, for example, a variable that refers to London in June does not appear in a constraint that restricts Amsterdam in May.
3. Some LP problems are even *supersparse*. Not only are there few non-zeros in the matrix of LP coefficients but the non-zeros tend to come from a small set of distinct numbers, e.g., 1,2,2.5,8 may appear in many places and few other numbers. This is called supersparsity. Because of supersparsity [297] it is easier to pack the problem into the main memory of the computer and avoid slow disk transfer operations during the running of the software. The set of distinct non-zeros is called the element pool. Computer operations on coefficients are performed by using pointers that point to an element in the pool, rather than actual coefficients. Efficiency arises when many pointers point to the same pool element. Best examples of exploiting supersparsity are the huge set covering problems with 10 million variables occurring when modeling airline problems. Every entry in the matrix is 1 so there is just one distinct element (with value 1.0). Since we know a non-zero is 1, we do not even have to store the value or pointers to this value, truly a supersparse representation.
(With computer memory becoming cheaper all the time, supersparsity is becoming less of an issue and most vendors have removed this feature from their software.)
4. A further reason that makes the Simplex algorithm perform well when successive problems are being solved is that it can make advantageous use of the previous optimal solution when starting to solve the next problem.

2.12 Summary and Recommended Bibliography

In this chapter we started with a very simple problem and showed how to model it by identifying variables, constraints, and an objective function. We investigated how to solve such a model using software and then we moved on to consider more complicated models. We saw how models might require variables that were not continuous. Later in this chapter we investigated the care required when we model, formulate, collect data, and present results. Thus by the end of this chapter the reader should be able to:

- Model more complex LP problems by combining together logic and data
- Be able to identify different types of variables and to make use of sets of variables
- Be able to identify different types of constraints and objectives
- Be able to formulate and solve more complex LP problems involving indices and files of data
- Be aware of the need for Mixed Integer Linear Programming (MILP)
- Be aware of issues affecting the size of practical LP problems and the ability to solve them

For modeling problems in Operations Management we strongly recommend the humorous book by Helber (2014,[261]). The book by Suhl & Melluoili (2007,[537]) is for readers interested in optimization systems, models, methods, software, and applications.

2.13 Exercises

1. Consider the example given in Sect. 2.3.2.

   If it is also required that at least twice as much ply as board is required, show how this can be modeled as a constraint.

   In addition, revenue is 10 units per m^2 of ply produced and 8 units per m^2 of board produced. The cost per m of pine is 4 and for birch is 10. If it is required that the ratio between revenue and cost must be at least 1.5, show how this may be modeled as a constraint.

2. A company is able to manufacture 4 products. The profit margin on product X is GBP 3, on product Y is GBP 2, on product Z is GBP 5, and on product W is GBP 4. Let w, x, y,and z be the quantities manufactured of the respective products. The company sets up the problem of manufacture under limited resources and establishes the following three constraints:

$$
\begin{aligned}
2w + x + 2y + z &\le 500 \\
2w + 3x + \quad + 2z &\le 460 \\
x + 4y \quad &\le 420.
\end{aligned}
\tag{2.13.1}
$$

(a) Use an AMS to solve the problem of finding the maximum profit that can be made by manufacturing quantities of the products.
(b) By comparing z with w in the constraints establish why the values in your solution occurred.
(c) An additional constraint is added to the problem

$$w + 2x + y + z \leq 340.$$

Re-run AMS to see what effect this will have on the problem.
(d) Investigations on the model reveal that two coefficients were inaccurate. These are: the coefficient in the second constraint for x should be 2 rather than 3, and the coefficient for w in the second constraint should be 1.5 rather than 2. Re-run AMS to see what effect these changes will have on the optimal solution.

3. Using AMS solve the following linear programming problem, where all variables are non-negative.

$$\max \quad 15x_1 + 6x_2 + 9x_3 + 2x_4$$

subject to

$$\begin{aligned} 10x_1 + 5x_2 + 25x_3 + 3x_4 &\leq 100 \\ 12x_1 + 4x_2 + 12x_3 + x_4 &\leq 96 \\ 7x_1 + x_4 &\leq 70. \end{aligned} \tag{2.13.2}$$

(a) In formulating the problem your advisers were uncertain about the coefficient of 25 attached to x_3. What analysis can you perform on this from your answer?
(b) A new variable x_5 is to be considered. This will have coefficients of 10 in the objective function and 4 and 1 in the first and second constraints, respectively. Using AMS, establish what will be the maximum coefficient the variable could have in the third constraint and still result in a non-zero value for the variable.

Chapter 3
Mathematical Solution Techniques

This chapter provides some of the mathematical and algorithmic backgrounds to solving LP and MILP problems.

3.1 Introduction

In Sect. 1.4.1, the simple "Boat Renting" problem gave rise to the model

$$\max \quad 800p + 600s \tag{3.1.1}$$

subject to

$$\begin{aligned} p + s &\le 350 \\ p &\le 200 \\ p - s &\ge 0 \\ 4p + 3s &\le 1400 \end{aligned} \tag{3.1.2}$$

with the additional implicit condition that both p and s are non-negative variables.

Supplementary Information The online version contains supplementary material available at https://doi.org/10.1007/978-3-030-73237-0_3.

J. Kallrath, *Business Optimization Using Mathematical Programming*, International Series in Operations Research & Management Science 307, https://doi.org/10.1007/978-3-030-73237-0_3

3.1.1 Standard Formulation of Linear Programming Problems

How would such a problem be solved by the software? General mathematical solution methods exist for such LP problems and for the extension to the case of MILP problems. Since algorithms and software usually require some standard notation, we will also formulate LP problems using the standard notation[1]

$$\max \quad c_1x_1 + c_2x_2 + \ldots + c_nx_n \tag{3.1.3}$$

subject to

$$a_{11}x_1 + a_{12}x_2 + \ldots + a_{1n}x_n = b_1 \tag{3.1.4}$$

$$a_{21}x_1 + a_{22}x_2 + \ldots + a_{2n}x_n = b_2 \tag{3.1.5}$$

$$\vdots \tag{3.1.6}$$

$$a_{m1}x_1 + a_{m2}x_2 + \ldots + a_{mn}x_n = b_m \tag{3.1.7}$$

$$x_1, x_2, \ldots, x_n \geq 0 \tag{3.1.8}$$

or in matrix–vector notation (see Appendix C for precise definitions of vectors and matrices)

$$\begin{array}{ll} \max & \mathbf{c}^{\mathrm{T}}\mathbf{x} \\ s.t. & \mathsf{A}\mathbf{x} = \mathbf{b} \quad , \quad \mathbf{x} \geq 0. \end{array} \tag{3.1.9}$$

Note that $\mathbf{x}$ and $\mathbf{b}$ are column vectors, while $\mathbf{c}^{\mathrm{T}}$ is a row vector, the product of a row and column vector is the scalar product

$$\mathbf{c}^{\mathrm{T}}\mathbf{x} := \sum_{j=1}^{n} c_j x_j, \tag{3.1.10}$$

and the matrix–vector product $\mathsf{A}\mathbf{x}$ follows the usual rules known in linear algebra; see also Appendix C.3.

Let us illustrate the matrix notation by inspecting the "Boat Renting" problem [Eqs. (3.1.1) and (3.1.2)] and identifying

$$\mathbf{x} = \begin{pmatrix} p \\ s \end{pmatrix}. \tag{3.1.11}$$

---

[1] Other standard notations are possible. One that is also used very often is $\max \mathbf{c}^{\mathrm{T}}\mathbf{x}$ subject to $\mathsf{A}\mathbf{x} = \mathbf{b}, \mathbf{l} \leq \mathbf{x} \leq \mathbf{u}$. Note that instead of the non-negativity constraints, we have lower and upper bounds on the variables. See Sect. 3.8.3 for how this case is treated.

The objective function then appears as

$$\max \quad (800, 600) \begin{pmatrix} p \\ s \end{pmatrix}. \tag{3.1.12}$$

The left-hand side of (3.1.2) is

$$\begin{pmatrix} 1 & 1 \\ 1 & 0 \\ 1 & -1 \\ 4 & 3 \end{pmatrix} \begin{pmatrix} p \\ s \end{pmatrix}, \tag{3.1.13}$$

and the right-hand side is just the vector

$$\begin{pmatrix} 350 \\ 200 \\ 0 \\ 1400 \end{pmatrix}. \tag{3.1.14}$$

Note that we have expressed the left and right sides of the constraints in vector notation, but we have not yet said anything about the relation between both sides. The original problem is not yet in the standard form as it still has less-than-or-equal and greater-than-or-equal constraints being present.

Although in Chap. 2 different types of constraints and objectives were introduced, we will see that every LP problem can be written in the standard form.

An objective function that is "minimize" can be converted to the one that is "maximize" because

$$\min f(x) = -\max(-f(x)). \tag{3.1.15}$$

What can we do about inequalities? That question is answered in the next section.

3.1.2 Slack and Surplus Variables

Inequality constraints can all be converted into equality constraints by introducing additional variables. For instance, the berthing inequality (1.4.2)

$$p + s \leq 350 \tag{3.1.16}$$

will be converted into an equation

$$p + s + t = 350. \tag{3.1.17}$$

In the equation, the variable t is called a *slack variable*. Note that from (3.1.16), we have

$$t = 350 - p - s \geq 0, \tag{3.1.18}$$

i.e., the slack variable t is chosen such that it is a non-negative variable. Sometimes the slack variable has no physical meaning, and it merely accounts for any difference between the left and right-hand sides of the inequality. But often, as in our case, it accounts for unused resources, unused capacities, unsatisfied demands, and so on. In our case, t models the amount of spare (or slack) berthing capacity available. Slack variables are always introduced in such a way that they become non-negative variables. Thus, t conforms to the same rule for variables as p and s.

In an inequality that is of the "$\geq$" type, a *surplus variable*, rather than a slack variable, may be inserted to lead to a new non-negative variable, e.g.,

$$3x + 2y \geq 500 \tag{3.1.19}$$

becomes

$$3x + 2y - u = 500, \tag{3.1.20}$$

where u is a non-negative surplus variable. Again, the new variable indicates the difference, if any, between the left-hand side and the right-hand side of an inequality. Note that a different slack or surplus variable would be required for each inequality. If slack and surplus variables are inserted into each inequality, then all constraints of the problem take the form of equations as in the LP standard form (3.1.9).

3.1.3 Underdetermined Linear Equations and Optimization

The linear equation part of LP problems is usually *underdetermined*, i.e., has more variables (degrees of freedom) than equations or has redundant[2] equations (see example below); this enables optimization. Equation (3.2.2), which describes the boat problem in the standard notation and already makes use of slack and surplus variables, is an example with more variables than equations. The linear system

$$\begin{aligned} 2x + 3y &= 1 \\ 4x + 6y &= 2 \\ 6x + 9y &= 3 \end{aligned} \tag{3.1.21}$$

[2]We call a certain equation $f_i(\mathbf{x}) = 0$ of a system $\mathbf{f}(\mathbf{x}) = \mathbf{0}$ of equations *redundant* if it does not provide any further information with respect to the solution of that system. All information carried by $f_i(\mathbf{x}) = 0$ is already contained in the other equations.

is an example with 2 variables and 3 equations, but in which two equations are redundant. Although the equations look different, they just carry the same information:

$$y = \frac{1}{3} - \frac{2}{3}x. \tag{3.1.22}$$

The reason for the redundancy in this case is that the equations depend linearly on each other. If we multiply the first equation by 2, we get the second one, and if we multiply the first equation by 3, we get the third one.

Let us first consider a simple system that is not underdetermined:

$$\begin{aligned} x + y &= 4 \\ 8x - y &= 5 \end{aligned} \tag{3.1.23}$$

and that involves only two equations and also only two variables. Adding the equations gives

$$9x = 9, \tag{3.1.24}$$

which shows uniquely $x = 1$ and consequently $y = 3$. Thus, here two equations allow us to determine two variables uniquely.

In underdetermined systems, the given equations do not allow us to determine all variables uniquely. That gives us the freedom to choose a solution that maximizes the objective function, rather than just find the unique solution that satisfies the constraints.

We may thus think of methods that solve LP problems as procedures that solve equations with the goal of optimizing an objective function. Such methods involve a sequence of steps that get repeated and are termed *algorithms*. It should be noted that the algorithm of solving LP problems will be different from the more familiar process of solving a set of simultaneous equations. However, the algorithm is related to the concepts of solving systems of linear equations. The principal difference between solving LP problems and solving sets of equations comes from the fact that LP problems have in most cases more variables than constraints. Let n and m denote the number of variables and constraints. Since there are too many variables, $n - m$ of them can be treated as *independent* (we can fix these variables to zero, or in the case of general bounds (see Sect. 3.8.3) on the variables, to one of their bounds), while the other m variables are *dependent* variables derived from a linear system with m constraints.

3.2 Linear Programming

In this section, we will consider two algorithms for solving LP problems: the Simplex algorithm and interior-point methods.

3.2.1 *Simplex Algorithm — A Brief Overview*

One of the best known algorithms for solving LP problems is the *Simplex algorithm* developed by George B. Dantzig in 1947 and described in Dantzig (1963,[142]) or, e.g., Padberg (1996,[431]). We illustrate the key ideas of this algorithm at an elementary level in Sect. 3.2.2. The first step is to compute an initial feasible solution (see Sect. 3.8.2) as a starting point, possibly by using another LP model that is a variant of the original model but allows us easily to determine an initial feasible solution. The Simplex or the revised Simplex (a more practical and efficient form for computer implementation) algorithm finds an optimal solution of an LP problem after a finite number of iterations, but in the worst case, the running time may grow exponentially, i.e., for large problems, we should be prepared that running time is an exponential function of the number of variables and constraints. Nevertheless on many real-world problems, it performs better than the so-called polynomial time algorithms developed in the 1980s, e.g., by Karmarkar (1984).

In most commercially available software systems, the Simplex algorithm provides the foundation of a method that will comfortably produce rapid solutions to problems involving ten thousands of variables and thousands of constraints. When a problem is formulated as an LP, the formulation will not be unique, e.g., some modelers may prefer to introduce certain variables to represent intermediate stages in operations, while the others will avoid these concepts. However, provided that the models are valid representations of the problem, the resulting LP problems will all be essentially equally easy to solve and will provide equivalent solutions.

3.2.2 *Solving the Boat Problem with the Simplex Algorithm*

In order to demonstrate how the Simplex algorithm works, let us express the boat problem introduced in previous chapters in the standard formulation. Since we want to reserve the symbol s for slack variables, we rename the original variables, p and s, to x_1 and x_2. Then, the problem reads as follows:

$$\max \quad 800x_1 + 600x_2 \tag{3.2.1}$$

subject to

$$\begin{array}{llllll} x_1 + & x_2 + s_1 & & & & = 350 \\ x_1 & & + s_2 & & & = 200 \\ -x_1 + & x_2 & & + s_3 & & = 0 \\ 4x_1 + & 3x_2 & & & + s_4 & = 1400. \end{array} \tag{3.2.2}$$

Note that in addition to the non-negative variables x_1 and x_2 denoting the number of Premier and Standard boats, we have introduced the non-negative slack or surplus variables $s_1, \ldots s_4$. The third equation is derived from $x_1 - x_2 \geq 0$ leading to

$x_1 - x_2 - s_3 = 0$. In order to have the auxiliary variables (slack or surplus variables) appearing with coefficient +1 (we see below why that is an advantage), we multiplied the equation by −1.

Let us for a moment neglect the objective function. Then, we are facing the problem of finding a solution of a system of linear equations. How can we solve the (underdetermined) system of equations (3.2.2)? You can easily check that

$$x_1 = x_2 = 0, \quad s_1 = 350, \quad s_2 = 200, \quad s_3 = 0, \quad s_4 = 1400 \tag{3.2.3}$$

is a solution. How could we obtain this solution? In this case, it is easy as each slack or surplus variable appears in exactly one equation with coefficient +1, and each row contains exactly one slack or surplus variable. We will call variables with this property (appearing in only one equation, having coefficient +1, each equation has only one of them) *canonical variables*. Canonical variables are a special case of *basic variables*[3] and allow us to apply the following heuristic: assign the value 0 to all other variables; these are the so-called *non-basic* variables. Then, the right-hand side of the equations gives the values of the basic variables. The solution is called a *basic solution*, and a *feasible basic solution* if all basic variables have non-negative values. However, the value of the objective function associated with this solution (which we refer to as $\mathbf{x}^0 = (0, 0, 350, 200, 0, 1400; 0)$) is 0, which will not keep our boat owner too happy.

Our solution $\mathbf{x}^0$ suggests that we should not lease any boats. How about slightly increasing the non-basic variables from 0 to 1? We do not want to lose feasibility. Let us further introduce a quantity, d_j,

$$d_j := Z^{new} - Z^{old}, \tag{3.2.4}$$

which measures, for each non-basic variable, the effect on the objective function when that variable is increased from 0 to 1 while considering the system of linear equations. Note that Z^{new} and Z^{old} are the values of the objective function after and before the increase of a non-basic variable. To give an example, we consider d_1 associated with x_1 :

$$Z^{new} = 800+0\cdot(350-1)+0\cdot(200-1)+0\cdot(0+1)+0\cdot(1400-4) = 800 \tag{3.2.5}$$

and $Z^{old} = 0$ and so $d_1 = 800$.

---

[3]To readers with a background in linear algebra, the concept of basic variables stems from linear algebra. Consider the constraint matrix A and a collection of m columns, $\mathbf{A}_j$. If these columns are linearly independent, then the corresponding variables x_j are called *basic variables*. Since there may exist different sets of m linearly independent column vectors, we also have different sets of basic variables. Thus, to say a variable is a basic variable makes sense only if this variable is seen as a member of an ensemble of variables. The situation with canonical variables is different. It can easily be seen whether a variable is a canonical one or not. If we inspect the columns of A associated with a set of m canonical variables, we see that, after ordering the columns, we get the identity matrix, which is a very special example of a set of linearly independent columns.

Let us derive a general formula from this computation: the first term is obviously the objective function coefficient, c_j, of the non-basic variable with index j that we are interested in. The coefficients in front of the brackets are the objective function coefficients of the current basic variables. The terms in the brackets are the values of the current basic variables minus the coefficient of the non-basic variable in the corresponding row. Using the standard form (3.1.9) with matrix A and vectors $\mathbf{b}$ and $\mathbf{c}$, as well as a little knowledge of linear algebra, we can simplify the expression by introducing the scalar products and get a simple formula that helps us to compute the quantities d_j in general:

$$d_j := \left[c_j + \mathbf{c}_B^{\mathrm{T}} \left(\mathbf{x}_B - \mathbf{A}_j\right)\right] - \mathbf{c}_B^{\mathrm{T}} \mathbf{x}_B, \tag{3.2.6}$$

where $\mathbf{c}_B^{\mathrm{T}}$ is a row vector containing those coefficients in the objective function associated with the basic variables, $\mathbf{x}_B$ is a column vector containing the values of all basic variables, $\mathbf{A}_j$ is a column vector established by the jth column of the current[4] matrix A. Simplifying (3.2.6), we get the expression for the *relative profit* d_j

$$d_j = c_j - \mathbf{c}_B^{\mathrm{T}} \mathbf{A}_j. \tag{3.2.7}$$

In the example above, we have

$$\mathbf{c}_B^{\mathrm{T}} = (0, 0, 0, 0) \quad , \quad c_1 = 800 \quad , \quad \mathbf{A}_1 = \begin{pmatrix} 1 \\ 1 \\ -1 \\ 4 \end{pmatrix}, \tag{3.2.8}$$

and thus

$$d_1 = 800 - (0, 0, 0, 0) \begin{pmatrix} 1 \\ 1 \\ -1 \\ 4 \end{pmatrix} = 800 \tag{3.2.9}$$

again.

The term relative profit reflects the fact that it describes a profit per unit increase of a non-basic variable and mutual decrease of a basic variable (the term $\mathbf{x}_B - \mathbf{A}_j$ accounts for that in (3.2.6)). If we were solving a minimization problem, we would use the term *relative cost* or *reduced cost*. For an economic interpretation of the reduced cost, see also Sect. 3.4.4.

Applying (3.2.7) to the non-basic variable x_2, we find $d_2 = 600$.

[4]In the original Simplex algorithm, the matrix A is updated in every iteration.

What about the value of d_j if x_j is a basic variable? In that case, we have $d_j = 0$ as the column $\mathbf{A}_j$ corresponding to x_j is, by definition of a basic variable, a column vector containing exactly one entry equal to unity, while all other entries are zero. Thus, the scalar product $\mathbf{c}_B^{\mathrm{T}}\mathbf{A}_j$ gives just the coefficient of the objective function corresponding to x_j and that is of course c_j, i.e., $\mathbf{c}_B^{\mathrm{T}}\mathbf{A}_j = c_j$ and thus $d_j = 0$.

What are the d_j useful for? We can use them to aid selection of which variable to put into the basis, i.e., increase from zero to a positive value. Because $d_1 > d_2$, we consider x_1 as the candidate for becoming a basic variable. The procedure leading to this decision is called *pricing* or *pricing-out* the variables, where the d_j are the prices. Choosing the non-basic variable with biggest d_j is a heuristic procedure, and the other procedures exist. Any non-basic variable with $d_j > 0$ would suffice. Concerning numerical performance, we would of course pick the one guaranteeing the fastest solution. Unfortunately, there is no rule for that.

From the definition of basic variables, or especially canonical variables, we conclude that we can have at most m basic variables, where m denotes the number of constraints, i.e., linear equations. Thus, if we select a new basic variable from the set of non-basic variables, we need to eliminate one of the existing basic variables. How can we determine which one to eliminate? In order to take this decision, let us inspect Eq. (3.2.2) once more. We consider only the basic variables $s_1, \ldots, s_4$ and the new basic variable x_1, i.e., x_2 is zero throughout. Then, (3.2.2) reduces to

$$\begin{aligned} x_1 + s_1 \qquad\qquad\qquad &= 350 \\ x_1 \qquad + s_2 \qquad\qquad &= 200 \\ -x_1 \qquad\qquad + s_3 \qquad &= 0 \\ 4x_1 \qquad\qquad\qquad + s_4 &= 1400. \end{aligned} \tag{3.2.10}$$

The question of which basic variables we are going to eliminate is answered by determining by how much we can increase x_1. The requirements $s_1, \ldots, s_4 \geq 0$ lead to the inequalities

$$\begin{array}{rcl} 350 - \ \ x_1 \geq 0 & & x_1 \leq 350 \\ 200 - \ \ x_1 \geq 0 & \Leftrightarrow & x_1 \leq 200 \\ 0 + \ \ x_1 \geq 0 & & x_1 \geq \ \ 0 \\ 1400 - 4x_1 \geq 0 & & x_1 \leq 325. \end{array} \tag{3.2.11}$$

The second inequality becomes active[5] first, namely when the value of x_1 increases to 200. The third inequality never becomes active and is always fulfilled; it is an empty condition. We are now able to select the variable to be eliminated from the basis. $x_1 = 200$ implies $s_2 = 0$, which means that we want to eliminate s_2 from the basis. The so-called *minimum ratio rule* following from our example reads as

[5]An inequality is said to be *active* for some values of the variables if the left- and right-hand sides are equal, i.e., for these values, the inequality is fulfilled as an equality. Example: $x + y \leq 5$ becomes active for $x = 3$ and $y = 2$.

follows: eliminate that basic variable x_e for which the minimum value

$$\frac{b_e}{A_{ej}} = \min_i \left\{ \frac{b_i}{A_{ij}} \middle| A_{ij} > 0 \right\} \tag{3.2.12}$$

is taken (the corresponding constraint e for which the minimum ratio is realized is the first one that becomes active and would become infeasible if the new basic variables is further increased); note that j is the entering variable index. The symbol | starts defining a logical condition, i.e., it can be read "for which." If all column entries, A_{ij}, corresponding to the new basic variable x_j are non-positive, the LP problem is unbounded.

While we already know the value of the new basic variable x_1, we do not yet know the modified values of the other ones. In addition, we also want to see the fact that x_1 is a basic variable (in our case, even a canonical variable) in the system of linear equations, i.e., x_1 appears in only one equation and with unit coefficient.

In order to achieve this goal, we borrow some knowledge from linear algebra. Two systems, $\mathsf{A}_1\mathbf{x} = \mathbf{b}_1$ and $\mathsf{A}_2\mathbf{x} = \mathbf{b}_2$, of linear equations are *equivalent* if they have the same set of solutions. To illustrate this property, consider the systems

$$\begin{aligned} x_1 - 2x_2 + x_3 - 4x_4 + 2x_5 &= 2 \\ x_1 - x_2 + x_3 - 3x_4 - x_5 &= 4 \end{aligned} \tag{3.2.13}$$

and

$$\begin{aligned} x_1 \quad + x_3 - 2x_4 - 4x_5 &= 6 \\ x_2 \quad + x_4 - 3x_5 &= 2. \end{aligned} \tag{3.2.14}$$

Instead of checking whether both systems have the same solutions, we can show that they are connected by the so-called *elementary row operations*. These operations include multiplication of rows by a non-zero factor and addition and subtraction of rows; they do not change the set of solutions. In the example above, the second system is derived from the first one, by applying the following elementary row operations to (3.2.13) subtract the first row from the second one and then add twice the second row to the first one.

Let us now apply this technique to (3.2.2) and remember that x_1 replaces s_2 as a basic variable. This decision was based on the minimum ratio rule that considered the rows of the system. Thus, the index e is used to refer to the row in which an existing basic variable is to be eliminated. This row is transformed according to the following formula:

$$A_{ec} \rightarrow A'_{ec} = \frac{A_{ec}}{A_{ej}} \quad ; \quad c = 1, \ldots, n+1, \tag{3.2.15}$$

where A_{ej} is the coefficient of the new basic variable in that row, and column $n+1$ is the right-hand side, i.e.,

$$A_{r,n+1} = b \quad , \quad A'_{r,n+1} = b'. \tag{3.2.16}$$

The other rows, $r \neq e$, are transformed according to the slightly more difficult rule

$$A_{rc} \rightarrow A'_{rc} = A_{rc} - \frac{A_{rj}}{A_{ej}} \cdot A_{ec} \quad ; \quad \begin{matrix} r = 1, \ldots, m \\ r \neq e \end{matrix} \quad ; \quad c = 1, \ldots, n+1. \tag{3.2.17}$$

If we apply the formulas (3.2.15) and (3.2.17) to (3.2.2), we get the system of equations again in the canonical form, i.e., each equation contains exactly one canonical variable:

$$\begin{aligned} x_2 + s_1 - s_2 &= 150 \\ x_1 \qquad + s_2 &= 200 \\ x_2 \quad + s_2 + s_3 &= 200 \\ 3x_2 \quad - 4s_2 \quad + s_4 &= 600 \end{aligned} \tag{3.2.18}$$

and

$$\mathbf{x}_B = \begin{pmatrix} s_1 \\ x_1 \\ s_3 \\ s_4 \end{pmatrix} = \begin{pmatrix} 150 \\ 200 \\ 200 \\ 600 \end{pmatrix} \quad , \quad \mathbf{c}_B^{\mathrm{T}} = (0, 800, 0, 0). \tag{3.2.19}$$

Now our boat owner is much happier because the objective function, i.e., profit, has increased to £160,000. The questions remain: Is that all we can get? Is this the best solution? How can we know answers to the questions posed here? To answer this question, let us do the pricing again, yielding

$$\mathbf{d} = (0, 600, 0, 0, 0, 0). \tag{3.2.20}$$

There is a candidate to improve the solution, namely x_2. According to the minimum ratio rule (3.2.12), we need to eliminate the variable corresponding to the first equation in (3.2.18), i.e., s_1. Transforming the linear equations again yields

$$\begin{aligned} x_2 + s_1 - s_2 &= 150 \\ x_1 \qquad + s_2 &= 200 \\ - s_1 + 2s_2 + s_3 &= 50 \\ - 3s_1 - s_2 \quad + s_4 &= 150 \end{aligned} \tag{3.2.21}$$

and

$$\mathbf{x}_B = \begin{pmatrix} x_2 \\ x_1 \\ s_3 \\ s_4 \end{pmatrix} = \begin{pmatrix} 150 \\ 200 \\ 50 \\ 250 \end{pmatrix} \quad , \quad \mathbf{c}_B^{\mathrm{T}} = (600, 800, 0, 0) \tag{3.2.22}$$

and a profit of £250,000. The boat owner gets excited about optimization. But are we finished now? Let us do pricing again. We get

$$\mathbf{d} = (0, 0, -600, -800, 0, 0). \tag{3.2.23}$$

This result implies that we cannot improve the objective function any more. This is quite fundamental, and therefore let us be clear about how we could derive this conclusion from **d**. The vector **d** is actually referring to changing the non-basic variables. The basic variables cannot be altered without altering the non-basic variables. By non-negativity, the non-basic variables are only allowed to increase. The reduced prices **d** listed above show that this can only reduce the optimum value. Thus, we have proved that $x_1 = 200$ and $x_2 = 150$ is the optimal solution, and we learn that pricing is good for two things: *proof of optimality* and *selecting a new basic variable*.

Let us summarize what characterizes the complete output of the Simplex algorithm: the values of the variables, evaluation of the constraints, their slacks if present, the value of the objective function, the reduced costs, and a proof of optimality. There is one further piece of relevant information: the shadow prices or dual variables discussed in Sects. 3.4.3 and 3.5 and in Appendix 3.8.1.

Further geometrical and algebraic properties of the Simplex and revised Simplex algorithm, as well as questions related to degeneracy, are reviewed in Appendix 3.8.1 in this chapter.

3.2.3 Interior-Point Methods — A Brief Overview

The region defined by linear constraints gives rise to a feasible region that is a polyhedron. While the feasible region can be easily visualized if there are only two variables present, and even possibly in the case of three variables, it is not possible in a problem involving hundreds of variables. The Simplex algorithm, in the non-degenerate case, i.e., no optimal solutions on a line connecting two vertices, can be thought of as an algorithm that moves from one corner (vertex) of the polyhedron to a "new" corner by advancing along one edge at a time (see Fig. 3.1a).

In contrast to the idea of a vertex-following method to solve an LP problem, more recently developed methods have concentrated on moving through the interior of the polyhedron (see Fig. 3.1b). Such methods are called *interior-point methods* and first received widespread attention after work by Karmarkar (1984,[325]). Since then,

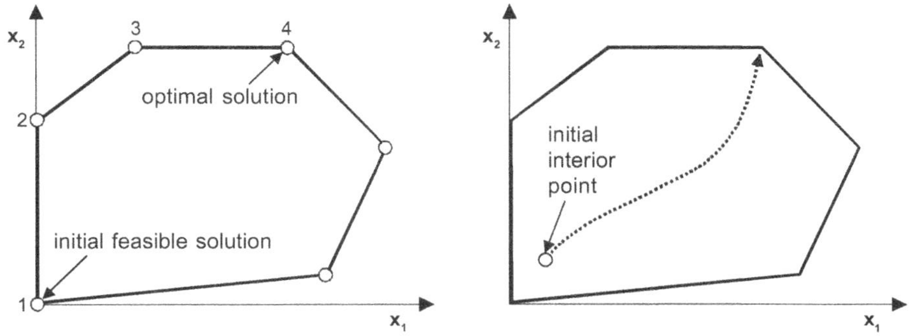

Fig. 3.1 Illustration of the (**a**) Simplex algorithm compared to (**b**) interior-point methods

about 2,000 papers have been written on the subject and research in optimization experienced the largest boom since the development of the Simplex algorithm; cf. Freund & Mizuno (1996,[201]). The idea of interior-point methods is intuitively simple if we take a naive geometric view of the problem. However, first, it should be noted that in fact the optimal solution to an LP problem will always lie on a vertex, i.e., on an extreme point of the boundary of the feasible region. Second, the shape of the feasible region is not like, say, a multi-faceted precious stone stretched out equally into many dimensions but more likely to resemble a very thin pencil stretched out into many dimensions. Hence, an algorithm that moves through the interior of a region must pay attention to the fact that it does not leave the feasible region. Approaching the boundary of the feasible region is penalized. The penalty is dynamically decreased in order to find a solution on the boundary. Interior-point methods will in general return an approximately optimal solution that is strictly in the interior of the feasible region. Unlike the Simplex algorithm, no optimal basic solution is produced. Thus, "purification" pivoting procedures from an interior point to a vertex having an objective value no worse have been proposed and cross-over schemes to switch from interior-point algorithm to the Simplex method have been developed [21].

The commercially available solvers are the so-called *logarithmic barrier*-type solvers (see the Appendix in Chap. 3), but the trend is toward *homogeneous* and *self-dual solvers*.[6] The barrier method is usually attributed to Frisch (1955,[202]) and is formally studied in Fiacco & McCormick (1968,[181]) in the context of nonlinear optimization, i.e., well before Karmarkar. Gill et al. (1986,[218]) have shown that there exists a formal relationship between Karmarkar's new interior-point method and the classical logarithmic barrier method.

[6]Such solvers are discussed in the review article by Freund & Mizuno (1996,[201]). Further details are found in Andersen & Ye (1995,[22]) and Vanderbei (1996,[560],2014,[561]).

Since interior-point methods contain complicated mathematics and use advanced mathematical concepts, we postpone the description of these methods to Appendix 3.8.5.

3.2.4 LP as a Subroutine

The solution of LP problems is also relevant for solving other optimization problems. LP problems occur as subproblems when solving the optimization problems listed in the following table:

Abbrev.	Optimization problem type	Using ...
NLP	Nonlinear programming	sequential LP
ILP	(Pure) integer LP	LP relaxations (Sect. 3.3.1)
MILP	Mixed integer LP	LP relaxations (Sect. 3.3.1)
MINLP	Mixed integer NLP	MILP master problems (Sect. 12.6.3)

For all these problems, the LP algorithms are used as subroutines and called several thousands or millions of times. In this book, we will concentrate on MILP problems and their formulation, but whatever we do in order to solve MILP problems, we should have a clear concept in mind of what is needed to solve the subproblem, i.e., the LP problems. By now, we already have an idea of how the Simplex algorithm works and what are the main ideas:

- pricing-out,
- eliminating a basic variable, and
- linear algebra aspects (pivoting).

The model building process can influence these tasks enormously. Here, we will briefly mention a few key ideas one should have in mind:

- efficient pricing is supported by a strong driving force in the objective function,
- equations with many zeros in right-hand side entries can create difficulties when applying the minimum ratio rule, and
- sparsity and density of the model, to exploit sparsity aids the linear algebra.

Most commercial solvers also provide presolvers (i.e., systems for analyzing the model before attempting optimization). Such systems eliminate fixed variables, attempt to tighten bounds on variables, and try to perform other operations that enable the solvers to operate more efficiently. Presolving is considered in more detail in Sect. 9.1.1.

The advances in state-of-the-art hardware and software have enabled the inexpensive, efficient solution of many large-scale linear programs previously considered intractable. Still large LPs can require hours, or even days, of runtime and are not guaranteed to yield an optimal (or near-optimal) solution. Klotz & Newman (2013,[338]) present practical guidelines for solving difficult LPs and suggestions for diagnosing and removing performance problems in state-of-the-art linear programming solvers and guidelines for careful model formulation, both of which can vastly improve performance.

3.3 Mixed Integer Linear Programming

A *mixed integer linear programming* (MILP) problem is an LP problem, where some variables are enforced to be integers. In other words, MILP problems include both integer and continuous variables.

Neither ILP (an MILP problem without continuous variables) nor MILP problems can be solved by the same approach as LP problems (except in certain unusual circumstances), but the LP solution technique does form the foundation of most algorithms for solving the IP or MILP. An intuitive idea for solving MILP problems leads to *explicit enumeration* as demonstrated in Sect. 3.3.1; our farmer from Sect. 2.6.1 has used this approach. All combinations of binary or discrete variables are noted down. That yields LP problems in which the discrete variables have fixed values. If we solve all LP problems (some might be infeasible, of course), we will pick out the best solution, which is also the optimal solution to our original MILP problem. However, as the number of combinations grows exponentially,[7] this approach is not practical and an *implicit enumeration* method (where all combinations are considered without enumerating them one by one) is preferable. Surely, there are also methods that are based on neither explicit nor implicit enumeration.

3.3.1 Solving the Farmer's Problem Using Branch and Bound

Recall the farmer's problem, i.e., the relations (2.6.6) to (2.6.9) from Sect. 2.6.1:

$$\max \quad Z = 3c + 2p \tag{3.3.1}$$

subject to

$$c + p \leq 3.5 \tag{3.3.2}$$

$$0 \leq c \leq 2 \quad , \quad c \quad \text{integer} \tag{3.3.3}$$

$$0 \leq p \leq 2 \quad , \quad p \quad \text{integer.} \tag{3.3.4}$$

[7]Imagine the case that we have n binary variables, δ_i, we have 2^n possible combinations. While $2^3 = 8$ and $2^{10} = 1024$, 2^{50} is already a number that starts with leading digit 1 and then has 15 zeros. Note that we have approximately $2^n \approx 10^{0.3n}$ or to be precise $2^n \approx 10^{M \cdot n}$ with $M = \ln(2)/\ln(10) \approx 0.30103\ldots$.

Our farmer, ill-educated in optimization, tries to solve the problem by *explicit enumeration*, i.e., by identifying all the budget- and housing-compatible combinations of calves and pigs, computing the value of the objective function, Z, and finally, selecting the pair (c, p) with the highest value of Z. As $c \leq 2$ and $p \leq 2$, we have only 9 possible pairs in $\{0, 1, 2\}$, i.e., at most $3^2 = 9$ combinations. The values of Z are given in the following table:

p\c	0	1	2
0	0	3	6
1	2	5	8
2	4	7	–

The combination $(2, 2)$ is infeasible because it violates the constraint (3.3.2). As expected, the result is $(c = 2, p = 1)$ and $Z = 8$. For small-sized problems,[8] this enumeration method can be successful. If the farmer had enough investment money to buy 9 calves or pigs instead of only 2, then we would have $10^2 = 100$ pairs out of $\{0, 1, 2, 3, 4, 5, 6, 7, 8, 9\}$. A more complicated situation would arise if the farmer had additional kinds of animals on the farm (Exercise 3.7.8). For that reason, it would be desirable to have an optimization algorithm available, capable of determining the optimal solution without using complete enumeration.

Our general method, which is called *Branch and Bound* (B&B) with LP relaxation, is the most common method used for solving integer programming problems originated by Land & Doig (1960,[349]). The method proceeds as follows: we first solve the LP relaxation **LP-1** of the original problem given above by ignoring the integrality conditions. Obviously, **LP-1** is an LP problem as there are no integer variables. In Sects. 1.5 and 3.2.2, we have learned how to solve such problems. The result is

$$\mathbf{LP-1}: \qquad c = 2 \quad , \quad p = 1.5 \quad , \quad Z = Z^1 = 9. \tag{3.3.5}$$

Although this solution includes fractional values and does not provide an optimal solution to the original MILP problem, it will still provide some useful information for us. Since we relaxed the original problem and gave more freedom to it, the solution Z^1 puts an upper limit on what we can anticipate. We have formally proven that the profit cannot be larger than 9 kGBP. In the formal description of the B&B algorithm, we will keep track of the upper bound denoted by z^{LP}, initialized at $z^{LP} = Z^1 = 9$. We will also consider a lower bound z^{IP}, which is initialized at $z^{IP} = -\infty$; we will see below how this bound is improved. The bounds restrict the objective function value, z^*, of the optimal solution to the range

$$z^{LP} \geq z^* \geq z^{IP}. \tag{3.3.6}$$

[8]The *problem size* of a model may be expressed by the number of variables and constraints.

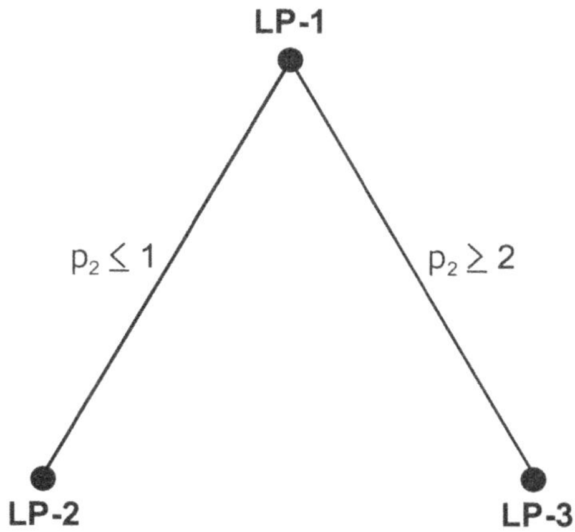

LP relaxation at node LP-1 and the first two subproblems of a B&B tree.

Fig. 3.2 LP relaxation and the first two subproblems of a B&B tree

What can we do now about the integrality condition? We produce two new problems (we will call them *subproblems*) as shown in Fig. 3.2

$$\begin{array}{llll} \mathbf{LP-2}: & LP-1 & \text{with} & p \leq 1 \\ \mathbf{LP-3}: & LP-1 & \text{with} & p \geq 2 \end{array} \tag{3.3.7}$$

by taking the original (relaxed) problem and adding some bounds to it. These are again linear programming problems, but using the additional constraints, we exclude the set of values between 1 and 2 for the variable p. Note that $p = 1$ and $p = 2$ are not excluded. Can you figure out graphically how these constraints change the feasible region? We leave that as an exercise! The subproblems are also called *branches*, and we have to decide on which problem to branch first. Let us solve **LP-2** first. The solution is

$$\mathbf{LP-2}: \qquad c = 2 \quad , \quad p = 1 \quad , \quad Z = Z^2 = 8. \tag{3.3.8}$$

This solution satisfies all of the constraints (3.3.3–3.3.2), and it gives the optimal answer that becomes obvious from our farmer's table.

But without the table, how can we be sure that this is the optimal answer? We have a solution with $Z = 8$, thus we update $z^{IP} = 8$, and we know that we cannot do better than $z^{LP} = 9$. In order to prove the optimality of $Z = 8$, we investigate the other subproblem (or node):

$$\mathbf{LP-3}: \qquad c = 1.5 \quad , \quad p = 2 \quad , \quad Z = Z^3 = 8.5. \tag{3.3.9}$$

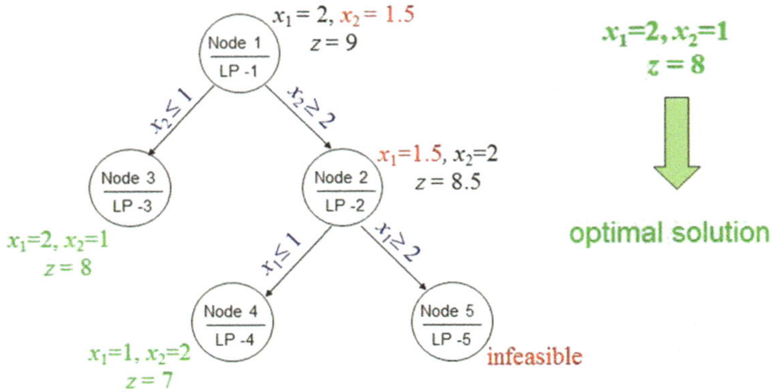

Fig. 3.3 B&B tree for the cows and pigs problem. Five nodes suffice to prove global optimality. Node 3 shows the maximal profit solution

If we are going to investigate this branch, we know that we cannot obtain a solution better than $Z = 8.5$. Proceeding as above, we generate two subproblems

$$\mathbf{LP-4}: \quad \mathbf{LP-3} \quad \text{with} \quad c \le 1 \tag{3.3.10}$$

and

$$\mathbf{LP-5}: \quad \mathbf{LP-3} \quad \text{with} \quad c \ge 2. \tag{3.3.11}$$

The whole B&B tree is depicted in Fig. 3.3. Now the solution of **LP-4** follows as

$$\mathbf{LP-4}: \qquad c = 1 \quad , \quad p = 2 \quad , \quad Z = Z^4 = 7 \tag{3.3.12}$$

is integral but smaller than our current best $Z = 8$. The other subproblem, **LP-5**, is obviously an infeasible problem, as it originates from **LP-3** and adds the new bound $c \ge 2$, which cannot be satisfied simultaneously with $p \ge 2$. So we are done. This example shares another nice property with other mixed integer problems. Sometimes it is possible to tighten the bounds[9] and produce a solution much faster. The key is the constraint (3.3.2), i.e., $c + p \le 3.5$. Due to the integrality of c and p, we can replace this constraint by

$$c + p \le 3. \tag{3.3.13}$$

As adding two integer values gives an integer and if an integer value has to be no more than 3.5, then it has to be no more than 3. If we then compute the LP relaxation,

[9]Tightening of bounds and other presolving operations are further described in Sect. 9.1.1.2.

we get at once the solution $c = 2$ and $p = 1$ and the optimal value $Z = 8$. This nice case in which the LP relaxation gives the optimal integral solution is said to have zero integrality gap (a concept to be discussed in Sect. 6.9).

Another remark is useful. After we have inspected the solution **LP-3**, we could have taken the objective function values of **LP-2** and **LP-3** concluding

$$z^{LP} = \max\{z(\mathbf{LP-2}), z(\mathbf{LP-3})\} = 8.5. \tag{3.3.14}$$

At this stage, we also had $z^{IP} = 8$ available and thus according to (3.3.6) the chain

$$8 \leq z^* \leq 8.5. \tag{3.3.15}$$

As in the original MILP problem, the variables c and p can take only integer values and the objective function has only integer coefficients, and the objective function value is always integral. Since there are no further integral values between 8 and 8.5, (3.3.15) proves already the optimality of the solution found at node **LP-2**. Thus, it is not necessary to evaluate the nodes **LP-3** and **LP-4**. So, we can see how bounds can be used to reduce the B&B tree.

3.3.2 Solving Mixed Integer Linear Programming Problems

A great variety of algorithms to solve mixed integer optimization problems has arisen during the last decades. The following methodologies are well-known exact algorithms for solving ILP and MILP problems:

- explicit enumerative algorithms,
- implicit enumerative algorithms,
- cutting-plane algorithms,
- Branch-and-Cut (B&C) algorithms, and
- dynamic programming.

Only small MILP problem can be solved by *explicit enumeration*: we make a list of all combinations of values the integer variables could have, solve the corresponding continuous LP problems, and select the best one with the best objective function. If the problem size growths, the prohibitively large number of combinations makes this approach impossible for solving most real-world problems.

The most common algorithm used by software for IP and MILP is the Branch-and-Bound (B&B) algorithm, originated by Land & Doig (1960,[349]) and described in its implemented form by Dakin (1965,[137]). It is the most important representation of *implicit enumerative algorithms*. Such algorithms include *pruning criteria* so that not all feasible solutions have to be tested for finding the optimal solution and for proving optimality.

In Sect. 3.3.1, we have already outlined the concept of a B&B algorithm based on LP relaxation. The key idea in B&B is that the relaxation leads to easier problems, which can be solved in a relatively short time and provide useful bounds for the original problem. These bounds reduce the size of the search tree significantly. In general, relaxation means we are weakening some restrictions of the problem. Assuming that our original problem is feasible, the relaxed problem always has a feasible set of solutions, S^R, of which the original feasible set, S, is a subset, i.e., $S \subseteq S^R$. The relaxation mostly used in MILP solvers is variable domain relaxation, i.e., binary variables $\delta \in \{0, 1\}$ are relaxed to $\delta \in [0, 1]$, and integer variables $\alpha \in \mathbb{N}$ are relaxed to $\alpha \in \mathbb{R}_0^+$. Other relaxations are such that some constraints are weakened, moved into the objective function for penalization (cf. *Lagrange Relaxation* in Sect. 14.1.3.3 or *Benders Decomposition* in Sect. 14.1.3.2), or removed completely. The B&B algorithm with LP relaxation is discussed in more detail in Appendix 3.8.6.

While the B&B algorithm works efficiently to solve integer programming models, it still requires solving many subproblems using certain branching and node selection schemes. Therefore, solving an ILP or MILP is still not as easy as solving an LP model because of subtleties within the mathematics of MILP models and solutions. Typically, two problems of identical size in terms of constraints and variables, where one is LP and the other is MILP, may differ by a factor of up to 2^n (n denotes the number of binary variables in the problem) in computer time needed to solve the MILP compared to computer time taken to solve the LP. Similarly, when data changes are made to models, but the numbers of constraints and variables remain the same, the time taken to solve an LP will remain fairly constant, but the MILP solution time may be subject to dramatic fluctuations. Fortunately, real-world problems are not always such worst-case scenarios that necessarily show this strong increase in computing time. Thus, the preceding does not imply that MILP models should never be built, but rather that caution should be exercised as regards their ease of solution.

What are the points a novice modeler should pay attention to? It is difficult to answer this question without some mathematical background. Technically speaking, the model formulation should be based on wisely chosen variables such that the LP relaxation should be as close as possible to the *convex hull*. At several places in the book, we refer back to the concept of the convex hull. The *convex hull* of an MILP is shown in Fig. 3.4. It is the polyhedron with the smallest volume which still contains all MILP feasible solutions. It can be mathematically shown that the convex hull can always be described by a set of linear constraints. Unfortunately, in most cases, it is not possible to write down the system of constraints explicitly. In cases in which it is possible, the LP relaxation of the original problem restricted to the convex hull as the feasible region gives the solution of the original MILP problem.

The remark made about equivalent LP models at the end of Sect. 3.2.1 is not appropriate for MILP models. Although equivalent MILP models will give rise to equivalent solutions, the ease of achieving these solutions may vary dramatically. Thus, modeling using MILP is very much an art and a number of "tricks of the trade" exist. These include the choice of which decisions to model as variables and certain additional constraints, which though apparently superfluous, may make a

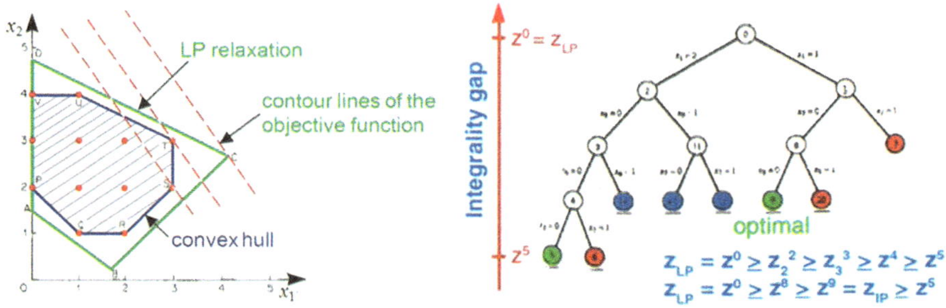

Fig. 3.4 LP relaxation (green: vertices A, B, C, and D) and convex hull (blue: vertices P, Q, R, S, T, U, and V) for a problem with two variables (left figure). A B&B tree for a maximization problem with nine binary variables x_1 to x_9 (right figure). The first integer feasible solution is found at node 5, and the optimal is node 9

substantial difference to the ease of solving the MILP. These and other related issues are discussed in Chap. 6.

The reader should be aware that presolving already mentioned in Sect. 3.2.4 becomes even more important in the context of MILP. In some circumstances it is possible that a simple analysis of the problem before using B&B, i.e., in the root node, may result in improvements that hold throughout the search. For example, using presolve immediately strengthened a constraint $1.62 \leq x \leq 3.73$ by converting it into $2 \leq x \leq 3$ for an integer x variable. Further details on MILP, presolving can be found in Sect. 9.1.1.

There are two further methods for solving MILP problems, which are also based on solving LP problems as a fundamental building block. The first is the *cutting-plane method*, and the second is the *Branch-and-Cut* (B&C) algorithm.

A number of other methods exist for solving ILP problems that contain no continuous variables, particularly problems with binary variables, which are not based on solving LP problems as an integral part. One of the oldest methods is the *additive algorithm* due to Balas (1965,[43]). Other methods are those of Granot & Hammer (1972,[233]) and Wolsey (1971,[586]). These methods rarely feature in commercial software systems, but ideas from these specialist algorithms are sometimes incorporated as subsidiary steps in methods such as B&B.

Dynamic programming as described in Nemhauser & Wolsey (1988,[421]) or Ravindran et al. (1987,[453]) is not a general purpose algorithm but originally was developed for the optimization of sequential decision processes. This technique for multi-stage problem solving may be applied to linear and nonlinear optimization problems which can be described as a nested family of subproblems. The original problem is solved recursively from the solutions of the subproblems. Examples can be found in Nemhauser & Wolsey (1988,[421]).

Finally, there exist *heuristic methods*, local and global search algorithms, e.g., *simulated annealing* and *tabu search*. However, these methods do not provide any proof of optimality or give an estimation of the quality of a feasible point found.

3.3.3 Cutting Planes and Branch and Cut (B&C)

Cutting-plane methods operate by starting in the same way as B&B but then moving toward a solution by restricting the feasible region defined by the constraints of the relaxed problem. The optimal solution is normally found before the end of the process which has to establish if any better solution exists. The cutting-plane method was introduced by Gomory (1958,[230]) and further refined by Glover (1968,[224]) and Young (1968,[592]). With the B&B algorithm, a number of unfruitful paths toward a solution are normally explored and abandoned subsequently. In contrast to B&B, the cutting-plane algorithm moves gradually toward the optimal solution by a route that ensures that at the last step in the process the optimal solution is found.

The *B&C algorithm* combines features from both the B&B algorithm and the cutting-plane approach. In each iteration, constraints (*cuts*) are added in order to exclude fractional solutions. B&C algorithm still branches on a variable (as in B&B) but also adds a set of cuts to tighten the bound. These cuts, also called *valid inequalities*, cut off parts of the LP feasible region without cutting off any valid integer solution (see Fig. 3.5). An early introduction to the subject appeared in Padberg & Rinaldi (1987,[432]). B&C may operate in two main ways:

(a) only cuts that are always valid for the problem may be generated;
(b) cuts that are only valid at particular nodes (because of branching already conducted) may be generated.

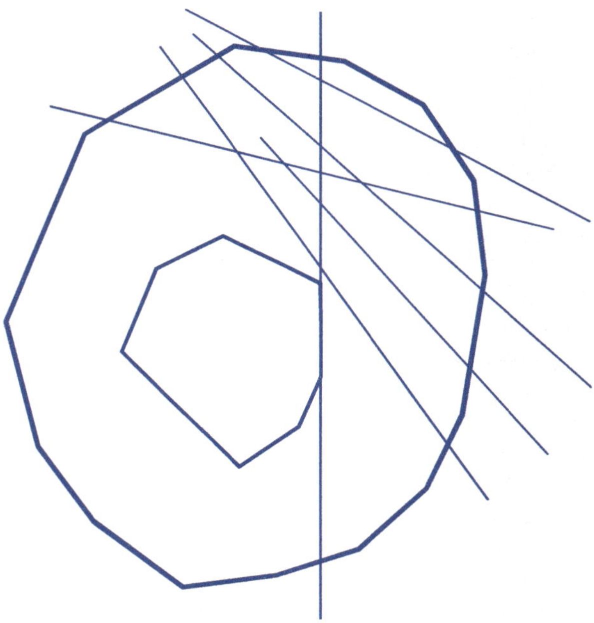

Fig. 3.5 Illustrating the idea of Branch and Cut

The reasons for the difference may not seem apparent at first. The difficulty is that in an IP problem many cuts will be available, but it will be tedious to try to generate them all, and if they were all generated, the resulting LP problem would be very large in terms of the number of constraints. Thus, although an integer solution would be guaranteed, it would be difficult to obtain it. Hence, the subtlety of B&C lies in generating violated cuts, i.e., ones that are violated by the current LP solution. Thus in (a), violated cuts can be added progressively, while in (b), only violated cuts valid at a particular node need to be added. It may also be possible to drop cuts progressively in either approach. Research is currently active in the field of B&C, but some early successes have been reported in Padberg & Rinaldi (1987,[432]) and Savelsbergh et al. (1994,[487]).

Although in 1997, B&C was promising and used in some commercial software, B&B remained as the most popular choice for commercial software systems in those days. While B&B required certainly a well-experienced modeler, B&C asked for even more mathematical insight because the construction of cuts (valid inequalities) was very difficult (see Sect. 10.6.3.2 for an example how valid inequalities are established).

Cutting-plane algorithms and B&C are standard nowadays. Commercial solvers such as CPLEX, GUROBI, or XPRESS-OPTIMIZER support B&C by checking automatically whether the model allows certain *standard cuts*. Compared to the 1990s, the current state-of-the-art optimizers outperform the user when it comes to tuning solver parameters. Nevertheless, it may happen that for a given problem, the lower or upper bound may not change at all or may not even find a feasible solution. For such cases, Klotz & Newman (2013,[339]) propose suggestions for appropriate use of solvers and guidelines for careful formulation, both of which can vastly improve performance. Overall, as stressed by Klotz (2014,[337]) , the numerics of solving MILP problems is by far not trivial and always challenging.

3.3.4 Branch and Price: Optimization with Column Generation

Branch-and-Price (B&P) methodology is a different form of a B&B algorithm, where at each node, the LP relaxation is solved by *column generation* as briefly introduced in Sect. 2.6.3. B&P is a useful extension of B&B for large-size MILP problems with many, i.e., several million variables as in Vanderbeck & Wolsey ([559]). Because of this large number of variables, most variables are not taken into account during the LP relaxation; this can be a good start heuristic, because most variables in large problems assume the value zero anyway. To prove the optimality, a pricing optimization problem is solved by identifying new variables to be included in the base and recalculating the LP relaxation if necessary. Branching is performed if the LP relaxation does not meet all integer conditions. Column generation as introduced in Chap. 2 finally takes place in every node of the B&B tree. Once you have a method available that can handle a large number of variables, this can be very useful, because the following is often observed in practical problems:

1. Compact formulations of MILP problems, such as in scheduling and in vehicle routing problems, often have weak LP relaxations. Often the efficiency of the model can be enhanced considerably by reformulating the model.
2. Compact formulations of MILP problems often contain symmetrical structures that lead to inefficient performance in standard B&B procedures; a much larger number of variables or reformulations in general can tackle this symmetry.
3. Column generation decomposes the original problem into a main and a subproblem, which in some cases leads to significant improvements in terms of solution quality and computing time. Column generation provides the optimal solution for the LP problems only, while it solves the LP relaxation for integer models. It is usually followed by solving the master problem as an MILP problem.

The basic idea of B&P as outlined, for instance, by Barnhart et al. (1998,[49]) — see also section "Column Generation and Branch-and-Price Methodology", seems simple, especially if the problem is classically decomposed according to Dantzig & Wolfe (1960,[143]). Technical difficulties, e.g., compatible branching rules, often only occur if the problem is not decomposed and the model formulation, e.g., set-partioning problems, has many integer variables. Desrosiers et al. (1995,[158]) explain this for route planning and scheduling problems, Savelsbergh (1997,[485]) applies B&P to the Generalized Assignment Problem, which we solve in section "Example: Generalized Assignment Problem" by means of Lagrange relaxation.

3.4 Interpreting the Results

In Chap. 2, any LP solver could have been used to solve a number of LP models. In this section we shall consider in further detail the kind of information with which a solved LP or MILP can provide us and how that information may be interpreted. It will be helpful if you refer back to Chap. 2 or examine a run of on LP solver on the boats problem of that chapter.

3.4.1 LP Solution

When an LP solver solves an LP problem, it often invokes the revised Simplex algorithm. Some summary information is produced, as the algorithm progresses, which can be examined in the *log* by invoking some kind of a `View Log` command. Each complete operation of the steps of the algorithm, which then repeats, is called an iteration ("`Its`" in the *log*), and after a group of iterations, some information on progress toward solution is given. Four columns will be particularly useful to observe.

`Its`	This tells us the total number of iterations performed so far
`Ninf`	The number of infeasibilities, this tells us how many of the constraints are not satisfied by the trial solution; `Ninf` needs to fall to zero before the optimal solution can even be reached; however, it does not decrease monotonically
`Obj Value`	This tells us the objective value of the current trial solution and we can observe this value growing as the iterations progress if we are maximizing the objective function (and decreases if the problem is minimize); if `Ninf`$\geq$ 0, it shows the value of the penalty objective function
`Time`	This tells us how much time has elapsed.

The section commences with some information about the *crash basis*. Crashing is a heuristic method which the software uses to try to make sensible mathematical "guesses" about which variables will be important in the optimal solution and is akin to the idea of a "hot start." Crashing is discussed by Maros & Mitra (1996,[386]) and in Chap. 9.

When we are experimenting with a new model, we may get one of the two troublesome occurrences that give rise to unexpected solutions and to messages in the *log*:

- an infeasible problem
  meaning that there is no solution possible to the problem. This may come as a surprise and may be remedied by checking the formulation for errors, e.g., $\leq$ and $\geq$ confused or + and − confused, or more fundamental difficulties;
- an unbounded problem
  meaning that the optimal solution to the problem will give an infinitely large (or negatively infinitely large) value to the objective function, i.e., in this case there is no optimal solution. Again this will probably not be what is desired and should be remedied in a similar way to "an infeasible problem." Usually, in this case one or more constraints have been omitted, e.g., a restriction on the machine capacity.

3.4.2 Outputting Results and Report Writing

All AMLs provide a function to retrieve the "Status" after a solve statement has been executed. When an acceptable solution has been reached (or even an infeasible problem, as we may get useful information about why it seems infeasible), we can display the details of the solution. After some initial statistics on the size of problem and the number of iterations required, the optimal objective function value will be given. Then follows the tables where variable and constraint information can be viewed. The constraint value tells us the value of the left-hand side of a constraint. Since the objective function is included under this section, the objective function "constraint" gives us the optimal value of the objective function. In the variables section, we find details of values of variables and their reduced costs. For constraints, we find their (left-hand side) values and shadow prices; reduced costs and shadow prices are discussed in the next two sections. Note that the concept of shadow prices and reduced costs strictly holds only for LP but not for MILP problems.

Although this paragraph might seem to be out of place in a chapter on solution techniques, it is worth making a few comments on report writing as it also helps to interpret the solution. Users of a model want to see the results from the model being solved expressed in the context of their world, with its associated jargon, styles and symbols. For them it will be far preferable to see "increase James's target to 1,000" as a solution value, rather than the bare `XJAMES(1)=1000`. Thus, it will be incumbent on modelers to make good use of report writing facilities.

At the same time, as one set of results from a model is being produced, the user may wish to vary the conditions under which the results were produced and explore the model further. This will also be performed under the umbrella of the report writing commands.

Beyond textual reports, nowadays in 2019, reporting tools have become such a standard that you should refer to them as well. Plain numbers do not tell much. Users want to get aggregated numbers and drill down into the specific details — this can be accomplished with nowadays Business Intelligence (BI) tools that apply automatic filtering upon selection and can be highly customized to the need of the application.

3.4.3 Dual Value (Shadow Price)

In a constraint, the optimal values of the variables will be such that the left- and right-hand sides of the constraint are either equal or unequal. For an equality constraint (or for the objective function), equality will be guaranteed. If for a $\leq$ constraint, the left-hand side of the constraint is strictly less than the right-hand side, then the resources to which the constraint refers are not restrictive, and so there is no inherent value in obtaining additional resources. The *dual value* of a constraint is the increase, for a maximization problem, obtainable by a one unit increase in the right-hand side of that constraint. Thus for the constraint "berthing" (1.4.2), a unit increase in berthing capacity will yield an increase of 600 units to the objective function. Note that the units of increase are unit of objective function over unit of the constraint. The dual value can also be interpreted as the unit decrease in the objective function caused by a unit decrease in the right-hand side of a constraint. As can be shown mathematically, the shadow price is the partial derivative of the objective function with respect to the right-hand side of a constraint evaluated at the solution.

The dual values can be computed with the help of formula (3.8.8), and for our "Boat Renting" problem, they turn out to be £600/boat, £200/boat, £0/boat, and £0/hour for the four constraints (this is left as Exercise 3.7 for the reader).

The dual value of the `maintenance` constraint is zero, indicating that the restriction is not "binding" or active, i.e., it is not restricting values of variables because there is some slack (100 units) available.

For a $\geq$ constraint, the dual value for a minimization problem is the unit increase in the objective function achieved by a unit increase (decrease) in the right-hand side value of the constraint. A positive dual value would mean that the objective would be increased, i.e., worsened, by that amount per unit increase of the right-hand side. For a maximization problem, the interpretation needs to be adjusted accordingly.

Note that gains indicated by the dual values (shadow prices) are optimistic; in most cases, they cannot fully be realized. As dual values are marginal values, their extent may be reduced by other factors in the problem, so 1 extra unit of .... may give 10 units more to the maximum output, but 2 extra units of .... may only give 10.5 units more to the maximum output. Thus the shadow price is only a local property.

Dual values can be interpreted as values of resources, and through LP a mathematical explanation of some aspects of the economics of scarcity can be achieved. This work is discussed further in Young & Baumol (1960,[593]).

3.4.4 Reduced Costs

If a variable takes a positive value in the optimal solution, it is apparent that it can make a useful contribution in a maximization problem. If a variable takes the value zero in the optimal solution, its contribution to the optimal solution could be regarded, in some sense, as unsatisfactory.

If a variable has solution value zero in a problem, its *reduced cost* introduced in Sect. 3.2.2 tells us by how much (additively) its objective function coefficient must be increased for it to achieve a non-zero value in the optimal solution. Alternatively, the reduced cost may be viewed as telling us by how much a variable is "underpriced" (maximization problem) or "overpriced" (minimization problem) compared to other variables. In a production problem, underpricing will mean that revenue is less than production cost, caused by the constraints of the problem.

Reduced cost is a marginal concept and the caveats regarding marginal values indicated in the previous section apply.

An interesting description of many aspects of investigating the solution to problems is contained in four articles by Greenberg (1993a[234], b[235], c[236]; 1994,[237]). The reader is referred to these useful articles for further details.

3.5 Duality ⊖

The existence of shadow prices and reduced costs suggests that when we solve an LP problem using the Simplex algorithm, we are not simply interested in the values of variables. The marginal values we investigated in the previous sections are evidently values of other varying quantities, in an algebraic sense. To find out their origins, we have to introduce a second LP problem (the *dual* LP problem), which is directly associated with any given LP problem. Knowledge of the behavior of this problem also aids our understanding of the concept of optimality of an LP problem and provides some reasons why an optimal solution can be found and identified.

The formal mathematical foundations addressing the topics above are related to the concept of *duality*. Duality is a fruitful concept for theoretical considerations, improving numerical properties of algorithms for nonlinear, linear, and also discrete optimization. A dual version of the Simplex algorithm is essential when using B&B techniques based on LP relaxations. Efficient interior-point algorithms (see Appendix 3.8.5) cannot be understood without the concept of duality.

Finally, duality allows us to perform *sensitivity analyses*, i.e., to investigate how small changes in our input data affect the solution. In particular, it becomes possible to consider how the constraints affect our solution in an optimization problem and which financial implications they cause.

3.5.1 Constructing the Dual Problem

If we just consider LP problems, it is very simple to get to the dual problem. Consider an LP problem (there is no need to do this in the standard form)

$$\mathbf{LP}: \quad \max \quad 2x_1 + 3x_2 + x_3 \tag{3.5.1}$$

subject to

$$\begin{array}{l} 2x_1 + x_2 + x_3 \leq 20 \\ x_1 + 2x_2 \leq 30 \end{array} , \quad x_1, x_2, x_3 \geq 0. \tag{3.5.2}$$

The problem, **DP**,

$$\mathbf{DP}: \quad \min \quad 20y_1 + 30y_2 \tag{3.5.3}$$

subject to

$$\begin{array}{l} 2y_1 + y_2 \geq 2 \\ y_1 + 2y_2 \geq 3 \\ y_1 \geq 1 \end{array} , \quad y_1 \geq 0 \quad , \quad y_2 \geq 0 \tag{3.5.4}$$

associated with the problem **LP** above is called the *dual* of **LP**, which is itself referred to as the *primal* problem. Both problems are related to each other according to a few properties that can easily be seen:

1. **LP** is maximize and all constraints are of the $\leq$ type. **DP** is minimize and all constraints are of the $\geq$ type.

2. The coefficients in the objective function of **LP** (in order) are the same as the right-hand sides of the constraints of **DP**.
3. The right-hand sides of the constraints of **LP** (in order) are the same as the coefficients in the objective function of **DP**.
4. The coefficients of the left-hand sides of the constraints of **LP**, read horizontally in order, are the same as the coefficients of the left-hand sides of **DP**, read vertically in order. Readers familiar with linear algebra will recognize this as the transposed matrix; otherwise, see Appendix C.
5. **LP** has 2 constraints and 3 variables; **DP** has 3 constraints and 2 variables. This follows of course from property 4 above. The variables of the dual problem, the dual variables, are the shadow prices we encountered already in Sect. 3.4.3.

It should be noted that the optimal values of x_1, x_2, and x_3 in an **LP** problem do not correspond to the optimal values of y_1 and y_2 in the **DP** problem. This is perhaps not surprising as an LP problem and its dual usually have very different numbers of variables. If the **LP** had m constraints and n variables, then the **DP** will have n constraints and m variables.

An **LP** problem and its dual problem, **DP**, are closely connected. A very important result (see Sect. 3.5.3) is that, if optimal solutions exist, then the optimal values of the objective function of both problems are the same.[10] This concept of relationship between the problems is known as *duality* and is expressed in the *strong duality theorem*.

It should be further noted that **DP** can be rewritten in any standard form of choice. If we then "take the dual of **DP**," we will obtain a problem that is equivalent to **LP**. Hence, **DP** is the dual of **LP** and **LP** is the dual of **DP**. Thus, we may consider the operations of "take the dual" as those required to move from **LP** to **DP** or those required to move from **DP** to **LP**.

Let us consider the following examples to get familiar with the construction of dual problems. The first example is

$$\begin{array}{ccccc} \text{primal problem} & & \text{dual problem 1} & & \text{dual problem 2} \\ \\ \begin{array}{rr} \max & \mathbf{c}^{\mathrm{T}}\mathbf{x} \\ s.t. & \mathsf{A}\mathbf{x} \le \mathbf{b} \\ & \mathbf{x} \ge 0 \end{array} & \longleftrightarrow & \begin{array}{rr} \min & \mathbf{b}^{\mathrm{T}}\mathbf{y} \\ s.t. & \mathsf{A}^{\mathrm{T}}\mathbf{y} \ge \mathbf{c} \\ & \mathbf{y} \ge 0 \end{array} & \Longleftrightarrow & \begin{array}{rr} \min & \mathbf{y}^{\mathrm{T}}\mathbf{b} \\ s.t. & \mathbf{y}^{\mathrm{T}}\mathsf{A} \ge \mathbf{c}^{\mathrm{T}} \\ & \mathbf{y}^{\mathrm{T}} \ge 0. \end{array} \end{array} \tag{3.5.5}$$

Some care is necessary with the multiplication of row and column vectors. Vectors like $\mathbf{c}^{\mathrm{T}}$ and $\mathbf{y}^{\mathrm{T}}$ are row vectors, while $\mathbf{c} = \left(\mathbf{c}^{\mathrm{T}}\right)^{\mathrm{T}}$ is a column vector. The dual problem is sometimes formulated in two different but equivalent ways.[11]

[10]Note that this property is not valid in MILP problems.

[11]Using the results $\mathbf{u}^{\mathrm{T}}\mathbf{v} = \left(\mathbf{u}^{\mathrm{T}}\mathbf{v}\right)^{\mathrm{T}} = \mathbf{v}^{\mathrm{T}}\mathbf{u}$ and $\left(\mathbf{y}^{\mathrm{T}}\mathsf{A}\right)^{\mathrm{T}} = \mathsf{A}^{\mathrm{T}}\mathbf{y}$ from linear algebra, it is easy to see that both formulations of the dual problem are equivalent.

In the first version of the dual problem, the vector of dual variables, $\mathbf{y}$, is treated as a column vector and some data of the primal problem are transposed to formulate the dual problem in the column vector space (primal vector space); see Appendix C.4 for a precise definition of a vector space. The second version keeps the vector of dual variables and all relations in the row vector space (dual vector space) and leaves the original data unchanged.

The next example is the primal–dual formulation for the standard form introduced in Sect. 3.1.1.

$$
\begin{array}{rlcrlcrl}
\multicolumn{2}{c}{\text{primal problem}} & & \multicolumn{2}{c}{\text{dual problem 1}} & & \multicolumn{2}{c}{\text{dual problem 2}} \\
\max & \mathbf{c}^{\mathrm{T}}\mathbf{x} & & \min & \mathbf{b}^{\mathrm{T}}\mathbf{y} & & \min & \mathbf{y}^{\mathrm{T}}\mathbf{b} \\
s.t. & \mathsf{A}\mathbf{x} = \mathbf{b} & \longleftrightarrow & s.t. & \mathsf{A}^{\mathrm{T}}\mathbf{y} + \mathbf{s} = \mathbf{c} & \Longleftrightarrow & s.t. & \mathbf{y}^{\mathrm{T}}\mathsf{A} + \mathbf{s}^{\mathrm{T}} = \mathbf{c}^{\mathrm{T}} \\
 & \mathbf{x} \geq 0 & & & \mathbf{s} \geq 0 & & & \mathbf{s}^{\mathrm{T}} \geq 0.
\end{array}
\qquad (3.5.6)
$$

Note that the dual variable $\mathbf{y}$ is a free variable, i.e., it is unrestricted in sign. The dual problem contains the surplus variable $\mathbf{s}$ so that the dual problem also appears in the standard form (only equality constraints).

After having seen some examples of primal–dual pairs, we can summarize the rules for constructing the dual from the primal problem:

	primal (maximize)	$\longleftrightarrow$		*dual* (minimize)
A	coefficient matrix	$\leftrightarrow$	A^{T}	transposed matrix
$\mathbf{b}$	right-hand side vector	$\leftrightarrow$	$\mathbf{c}$	price vector
$\mathbf{c}^{\mathrm{T}}$	price vector	$\leftrightarrow$	$\mathbf{b}^{\mathrm{T}}$	transposed rhs vector
ith	constraint is an equation	$\leftrightarrow$	y_i	y_i is a free variable
ith	constraint is an $\leq$ ineq.	$\leftrightarrow$	y_i	y_i is non-negative
ith	constraint is an $\geq$ ineq.	$\leftrightarrow$	y_i	is non-positive
x_j	is a free variable	$\leftrightarrow$	jth	dual constraint is an eq.
x_j	is a non-negative variable	$\leftrightarrow$	jth	dual constraint is a $\geq$ ineq.
x_j	is a non-positive variable	$\leftrightarrow$	jth	dual constraint is a $\leq$ ineq.

3.5.2 Interpreting the Dual Problem

To illustrate the idea of duality and to interpret its meaning, let us consider the dual of the "Boat Renting" problem [see (3.1.1) and (3.1.2)]:

$$\min \quad 350y_1 + 200y_2 + 1400y_4 \qquad (3.5.7)$$

subject to

$$\begin{array}{l} y_1 + y_2 - y_3 + 4y_4 \geq 800 \\ y_1 \qquad\;\;\; + y_3 + 3y_4 \geq 600 \end{array} \quad ; \quad y_1, y_2, y_3, y_4 \geq 0. \tag{3.5.8}$$

Note that we changed the third relation in (3.1.2) to $-p + s \leq 0$ to get all $\leq$ inequalities and to apply the rule expressed in (3.5.5). Further note that the third relation in (3.1.2) has a zero right-hand side, which explains why y_3 does not appear in the objective function. Let us now try to interpret the meaning of the dual problem. The inequalities (3.5.8) tell us that y_1, y_2, and y_3 have the unit *money/boat*. If we want to find out about y_4, we have to keep in mind that the coefficients 4 and 3 in the original (primal) problem had the unit hours/boat. Thus, the unit of y_4 is clearly *money/boat*. Let us also remember that the values 350, 200, 0, and 1400 had the units *boat*, *boat*, *boat*, and *hours*. The conclusion is that while the primal problem involved maximizing revenue, the dual is a problem of minimizing costs; the objective function of the dual problem has the unit *money*. What sort of costs we are minimizing in the dual problem? The cost terms in the dual objective function are given by the right-hand side of original constraint. The strong duality theorem (in an optimal solution, both primal and dual objective functions take the same value, see Sect. 3.5.3) tells us that if we relax our original constraints, say from 350 to 351, the costs in the dual problem increase by y_1, and so does the revenue in the primal problem. Thus, the dual variables y_i carry the information on what constraint i is going to cost us. They are the shadow prices introduced earlier. We can also learn that if a constraint in the primal problem is not active, then the associated dual variables are zero; otherwise, the dual value can be different from zero. This is known as the *complementary slackness condition* (see Sect. 3.5.3):

$$y_i \cdot s_i = 0, \tag{3.5.9}$$

where y_i and s_i are the dual value and slack variable of constraint i. The complementary slackness condition (see also Sect. 3.5.3) we encounter here in linear programming is a special case of nonlinear programming where a similar relation holds. In nonlinear programming, the dual values are usually called the *Lagrange multipliers*.

If we briefly refer to Appendix 3.8.1, we can also try to interpret the constraints in the dual problem above. The sufficient conditions (3.8.10) in a maximization problem in vector notation read

$$\boldsymbol{\pi}^{\mathrm{T}}\mathsf{A} \geq \mathbf{c}. \tag{3.5.10}$$

Thus, if we identify $\mathbf{y} = \boldsymbol{\pi}$, the constraints in the dual problem just represent the sufficient condition for the optimality of the primal solution.

Eventually, let us consider the solution of the dual problem: if we solve, we find out that y_1 = \$600/boat and y_2 = \$200/boat, y_3 = \$0/boat, and y_4 = \$0/hour, which leads to the objective function value \$250, 000, which is, and should be the same, as the objective function value of the primal problem.

3.5.3 Duality Gap and Complementarity

To consider the connection between the primal and the dual problem in more detail, let us consider the difference of their objective functions. We take the primal–dual pair constructed in example (3.5.5) to derive some important result from duality theory.

If $\mathbf{x}$ and $\mathbf{y}$ are feasible points of the primal and dual problems, the quantity

$$\Delta(\mathbf{x}, \mathbf{y}) = \mathbf{b}^T\mathbf{y} - \mathbf{c}^T\mathbf{x} = \mathbf{y}^T\mathbf{b} - \mathbf{c}^T\mathbf{x} \tag{3.5.11}$$

is called *duality gap*. If the primal problem is a maximization problem then we can state the *weak duality theorem* as

$$\Delta(\mathbf{x}, \mathbf{y}) \geq 0. \tag{3.5.12}$$

To see this, let us choose the second formulation of the dual problem in example 2, and let $\mathbf{x} \geq \mathbf{0}$ and $\mathbf{y}^T \geq \mathbf{0}$ denote feasible points of the primal and the dual problem. The chains of inequalities

$$\mathsf{A}\mathbf{x} \leq \mathbf{b} \quad \Longrightarrow \quad \mathbf{y}^T\mathsf{A}\mathbf{x} \leq \mathbf{y}^T\mathbf{b} \tag{3.5.13}$$

and

$$\mathbf{y}^T\mathsf{A} \geq \mathbf{c}^T \quad \Longrightarrow \quad \mathbf{y}^T\mathsf{A}\mathbf{x} \geq \mathbf{c}^T\mathbf{x} \tag{3.5.14}$$

lead directly to $\mathbf{c}^T\mathbf{x} \leq \mathbf{y}^T\mathbf{b}$, which proves $\Delta(\mathbf{x}, \mathbf{y}) \geq 0$.

We can exploit the weak duality theorem in the following sense. If the primal problem is a maximization problem, the dual problem provides an upper bound for the primal problem; vice versa, the primal problem provides a lower bound for the dual problem. This property again leads to the following conclusion: if the primal problem is feasible and unbounded, then the dual problem is infeasible; vice versa, if the dual problem is feasible and unbounded, then the primal is infeasible.

The *strong duality theorem* states that if the primal and the dual problem are both feasible, then optimal solutions $\mathbf{x}^*$ and $\mathbf{y}^*$ exist and the duality gap in the optimal solution vanishes, i.e.,

$$\Delta = \Delta(\mathbf{x}^*, \mathbf{y}^*) = 0. \tag{3.5.15}$$

The existence of the optimal solution follows from the compactness of the feasible region and that continuous functions take the optimal value on compact sets. To show that $\Delta(\mathbf{x}^*, \mathbf{y}^*) = 0$ is more difficult to prove; cf. Padberg (1996,[431], p.87).

Vice versa, if $\Delta(\mathbf{x}', \mathbf{y}') = 0$ for some feasible points $\mathbf{x}'$ and $\mathbf{y}'$, then these points are the optimal solution, i.e., $\mathbf{x}^* = \mathbf{x}'$ and $\mathbf{y}^* = \mathbf{y}'$. This direction of the strong duality theorem is more difficult to prove, but it has an important consequence for interior-point algorithms. Feasible points $\mathbf{x}'$ and $\mathbf{y}'$ with non-zero duality gap are interior points (points not on the boundary or at a vertex) and the value of $\Delta(\mathbf{x}', \mathbf{y}')$ provides a measure for how far we are away from the optimal solution. Thus $\Delta(\mathbf{x}', \mathbf{y}')$ or some related quantities can be used as a termination criterion in interior-point algorithms.

Another property, that of complementary slackness, follows easily from duality. If we define the primal and dual slack variables $\mathbf{s}$ and $\mathbf{w}$

$$\mathbf{s} = \mathbf{b} - \mathsf{A}\mathbf{x} \quad , \quad \mathbf{w}^{\mathrm{T}} = \mathbf{y}^{\mathrm{T}}\mathsf{A} - \mathbf{c}^{\mathrm{T}}, \tag{3.5.16}$$

then for feasible points $\mathbf{x}$ and $\mathbf{y}$, we have the following complementary slackness property:

$$\mathbf{w}^{\mathrm{T}}\mathbf{x} + \mathbf{s}^{\mathrm{T}}\mathbf{y} = \mathbf{0} \quad \Longleftrightarrow \quad \Delta(\mathbf{x}, \mathbf{y}) = 0, \tag{3.5.17}$$

which relates the complementary slackness condition

$$\mathbf{w}^{\mathrm{T}}\mathbf{x} + \mathbf{s}^{\mathrm{T}}\mathbf{y} = \mathbf{0} \tag{3.5.18}$$

and the zero duality gap at an optimal point with each other.

Let us interpret (3.5.17). If the complementary slackness condition (3.5.18) holds for a feasible primal–dual pair $\mathbf{x}$ and $\mathbf{y}$, then these points are the optimal solution because zero duality gap implies optimality according to the strong duality theorem; vice versa, zero duality gap implies that the complementary slackness condition holds.

Since all variables involved are non-negative, $\mathbf{w}^{\mathrm{T}}\mathbf{x} + \mathbf{s}^{\mathrm{T}}\mathbf{y} = \mathbf{0}$ further implies

$$\mathbf{w}^{\mathrm{T}}\mathbf{x} = \mathbf{0} \quad , \quad \mathbf{s}^{\mathrm{T}}\mathbf{y} = \mathbf{0}, \tag{3.5.19}$$

and this again implies the component-wise complementary slackness property

$$w_j x_j = 0 \quad , \quad \forall j \tag{3.5.20}$$

and

$$s_i y_i = 0 \quad , \quad \forall i, \tag{3.5.21}$$

which we encountered already in Sect. 3.5.2.

The proof of (3.5.17) is easy as by inserting all terms we see that

$$\mathbf{w}^{\mathrm{T}}\mathbf{x} + \mathbf{s}^{\mathrm{T}}\mathbf{y} = \left(\mathbf{y}^{\mathrm{T}}\mathsf{A} - \mathbf{c}^{\mathrm{T}}\right)\mathbf{x} + (\mathbf{b} - \mathsf{A}\mathbf{x})^{\mathrm{T}}\mathbf{y} = \mathbf{y}^{\mathrm{T}}\mathbf{b} - \mathbf{c}^{\mathrm{T}}\mathbf{x} = \Delta(\mathbf{x}, \mathbf{y}) \tag{3.5.22}$$

and thus the following useful property between the *complementarity gap*, $g(\mathbf{x}, \mathbf{y})$,

$$g(\mathbf{x}, \mathbf{y}) = \mathbf{w}^{\mathrm{T}}\mathbf{x} + \mathbf{s}^{\mathrm{T}}\mathbf{y} \tag{3.5.23}$$

and the duality gap $\Delta(\mathbf{x}, \mathbf{y})$ for any feasible points

$$g(\mathbf{x}, \mathbf{y}) = \Delta(\mathbf{x}, \mathbf{y}). \tag{3.5.24}$$

3.6 Summary and Recommended Bibliography

In this chapter we have considered the different ways of solving LP problems and have investigated certain aspects of software which will solve such problems. Moving on from LP problems, we considered MILP problems and how they could be difficult to solve. Techniques used in software systems for solving MILP problems were then considered. Thus the reader should now be familiar with:

- a standard form of an LP problem;
- the use of slack and surplus variables;
- how the Simplex algorithm works and some very general principles of interior-point methods;
- how B&B works for solving MILP problems and some very general principles of B&C;
- solution information from LP solvers and the basic concept of dual values and reduced costs; and
- the basics of duality theory.

For further reading on LP, MILP, and polyhedral theory, we refer the reader to Nemhauser & Wolsey (1988,[421]), Vanderbei (2014,[561]), and Lancia & Serafini (2018,[348]). Combinatorial optimization is well covered by Schrijver (2003,[493]) or Korte & Vygen (2018,[341]). *Algorithms for Convex Optimization* by Vishnoi (2021,[566]) is strongly recommended for convex optimization.

3.7 Exercises

1. Solve the problems **LP** and **DP** from Sect. 3.5, page 98, and show that they have the same optimal objective value.
2. Construct the dual of **DP** from Sect. 3.5, page 98, using the rules given in (3.5.5), and show that this dual problem is equivalent to **LP** after a simple rearrangement.
3. Solve the dual of the "Boat Renting" problem given on page 100, and compare its solution with the solution of the primal "Boat Renting" problem.
4. When a free variable x occurs in an LP model, it is replaced by $x^+ - x^-$ where $x^+, x^- \geq 0$. Inspection of the solution shows that at most one of x^+ and x^- is non-zero in the optimal solution to the problem. This result is expected. Can you explain this feature and prove it?

5. Consider the production planning problem in Sect. 2.5.1. How many products will there be at most in the optimal portfolio yielding maximum profit? Answer this question without solving the problem explicitly. Sections 3.2.2 and 3.8.1 might help you to do so.
6. How many basic solutions and basic feasible solutions can at most exist in the boats problem (1.4.8,1.4.9) described in Chap. 1? Derive all basic solutions, i.e., give the basic variables, their values, and the basic matrices. Hint: Fig. 1.4 and (3.2.2) might help. Note that you should have read Appendix 3.8.1 before you tackle this exercise.
7. Calculate the shadow prices involved in the boat problem. Appendix 3.8.1 might help you to do so.
8. Investigate the complexity of the "Calves and Pigs" problem in Sect. 3.3.1. Assume that the farmer has n different kinds of animals and enough investment money and space to purchase m_i animals of each kind.

   (a) How many combinations would the farmer have to check when using explicit enumeration to solve the problem?
   (b) What is the solution if there is no money constraint?

3.8 Appendix

3.8.1 *Linear Programming — A Detailed Description*

Consider the linear program in the standard form

$$\mathbf{LP}: \max \quad \left\{ \mathbf{c}^{\mathrm{T}}\mathbf{x} \mid \mathsf{A}\mathbf{x} = \mathbf{b}, \quad \mathbf{x} \geq 0, \quad \mathbf{x} \in \mathbb{R}^n, \quad \mathbf{b} \in \mathbb{R}^m \right\}, \tag{3.8.1}$$

where $\mathbb{R}^n$ denotes the vector space of real vectors with n components.

In Sect. 3.2.2, we introduced the concept of basic variables collected into the basic vector $\mathbf{x}_B$. The algebraic platform is the concept of the basis $\mathscr{B}$ of A, i.e., a linearly independent collection $\mathscr{B} = \{\mathbf{A}_{j_1}, \ldots, \mathbf{A}_{j_m}\}$ of columns of A. Sometimes, just the set of indices $\mathscr{J} = \{j_1, \ldots, j_m\}$ referring to the basic variables or linearly independent columns of A is referred to as the basis. The inverse $\mathscr{B}^{-1}$ gives a *basic solution* $\bar{\mathbf{x}} \in \mathbb{R}^n$, which is given by

$$\bar{\mathbf{x}}^{\mathrm{T}} = \left(\mathbf{x}_B^{\mathrm{T}}, \mathbf{x}_N^{\mathrm{T}}\right), \tag{3.8.2}$$

where $\mathbf{x}_B$ is a vector containing the basic variables computed according to

$$\mathbf{x}_B = \mathscr{B}^{-1}\mathbf{b}, \tag{3.8.3}$$

and $\mathbf{x}_N$ is an $(n-m)$-dimensional vector containing the non-basic variables:

$$\mathbf{x}_N = \mathbf{0} \quad , \quad \mathbf{x}_N \in \mathbb{R}^{n-m}. \tag{3.8.4}$$

If $\bar{\mathbf{x}}$ is in the set of feasible points $S = \{\mathbf{x} : \mathsf{A}\mathbf{x} = \mathbf{b},\ \mathbf{x} \geq 0\}$, then $\bar{\mathbf{x}}$ is called a *basic feasible solution* or a *basic feasible point*. If

1. the matrix A has m linearly independent columns $\mathbf{A}_j$,
2. the set S is not empty, and
3. the set $\{\mathbf{c}^{\mathrm{T}}\mathbf{x} : \mathbf{x} \in S\}$ is bounded from above,

then the set S defines a convex polyhedron P and each basic feasible solution corresponds to a vertex of P. Assumptions (2) and (3) ensure that the LP is neither infeasible nor unbounded, i.e., it has a finite optimum.

It can be shown that in order to find the optimal solution, it is sufficient to consider all basic solutions (sets of m linearly independent columns of A), check whether they are feasible, compute the associated objective function, and pick out the best one. In this sense finding an optimal solution for an LP is a combinatorial problem. In the worst case, there are

$$f_1(n,m) := \binom{n}{m} = \frac{n!}{m! \cdot (n-m)!} \tag{3.8.5}$$

basic solutions. If the problem is not degenerate (this term will be explained below), the optimal solution is among the finite set of basic feasible solutions. Therefore, an LP problem can have at most m positive variables in the solution. At least $n-m$ variables, these are the non-basic variables, must take the value zero. McMullen (1970,[391]) has shown that there can exist at most[12]

$$f_2(n,m) := \binom{n - \left\lfloor \frac{m+1}{2} \right\rfloor}{n-m} + \binom{n - \left\lfloor \frac{m+2}{2} \right\rfloor}{n-m} \tag{3.8.6}$$

basic feasible solutions. Let us consider an example with $n = 16$ and $m = 8$. In that case we have $f_1(16,8) = 12870$ and $f_2(16,8) = 660$. Although f_2 gives a much smaller number than f_1, we get a feeling that this purely combinatorial approach is not attractive in practice.

Geometrically, the (primal) Simplex algorithm can be understood as an *edge-following* algorithm that moves on the boundary of a polyhedron representing the feasible set, i.e., from vertex to vertex of the polyhedron (see Fig. 3.6). In each move corresponding to a linear algebra step (technically, a Pivot step), the objective function value is either improved or does not change. Klee & Minty (1972,[334]) provide a set of examples of LP problems with 2^n vertices which all

[12] This result is only true if no upper bounds (see Sect. 3.8.3) are present.

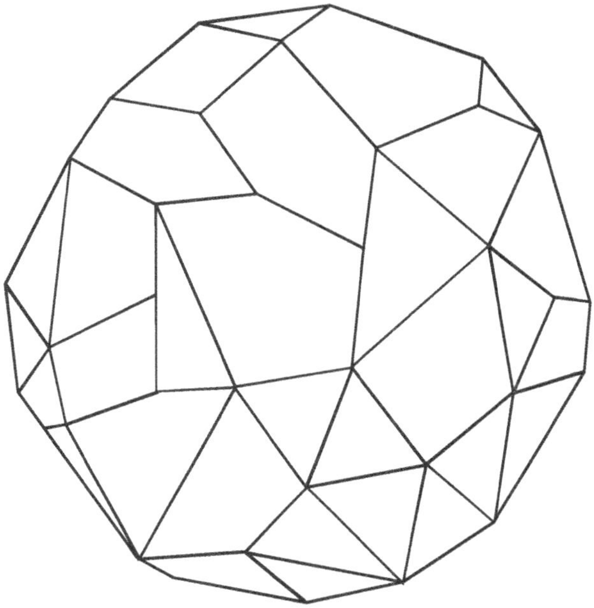

Fig. 3.6 Feasible region of an LP problem

need to be investigated, i.e., in which the algorithm requires exponentially many steps. Algebraically, in each iteration, one column of the current basis is modified; according to this exchange of basic variables, matrix A and vectors $\mathbf{b}$ and $\mathbf{c}$ are transformed to matrix A' and vectors $\mathbf{b}'$ and $\mathbf{c}'$. Technically, this procedure is called a *pivot* or *pivot operation* or *pivot step*.

Now we can also understand degenerate cases in LP. A purely algebraic concept is to call an LP problem degenerate if the optimal solution contains basic variables with value zero. If we combine the algebraic and geometric aspects, we can interpret a degenerate problem as one in which a certain vertex (usually we are considering the one leading to optimal objective function) has different algebraic representations, i.e., two vertices are co-incident with an edge of zero length between them.

Instead of keeping and computing the complete matrix A based on the previous iteration, the revised Simplex algorithm is based on the initial data A, $\mathbf{b}$, and $\mathbf{c}^{\mathrm{T}}$ and on the current basis inverse $\mathscr{B}^{-1}$. Let us now summarize the computational steps of the revised Simplex algorithm (see also Fig. 3.7).

The first step is to find a feasible basis $\mathscr{B}$ as described in Sect. 3.8.2. Note that this problem is, in theory, as difficult as solving the optimization problem itself.

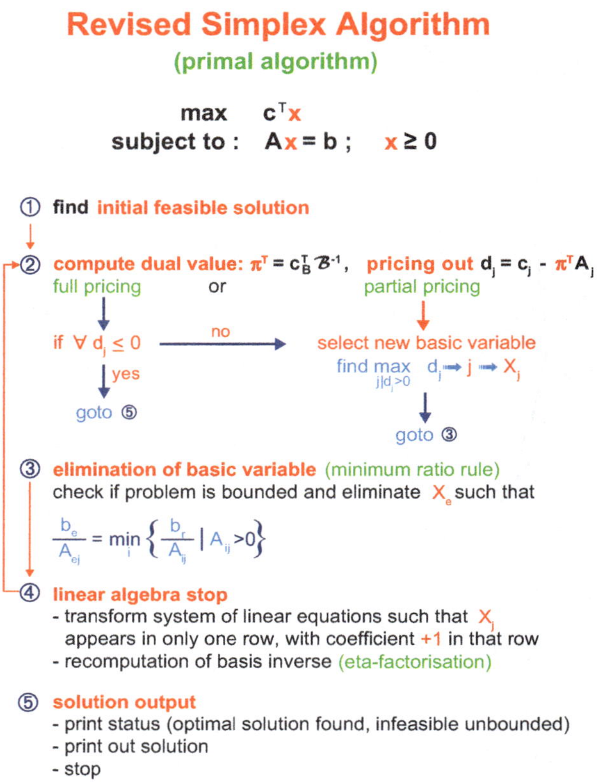

Fig. 3.7 The revised Simplex algorithm

Once the basis is known, we can compute[13] the inverse, $\mathscr{B}^{-1}$, of the basis, the values of the basic variables

$$\mathbf{x}_B = \mathscr{B}^{-1}\mathbf{b} \tag{3.8.7}$$

and the dual values, $\boldsymbol{\pi}^{\mathrm{T}}$,

$$\boldsymbol{\pi}^{\mathrm{T}} := \mathbf{c}_B^{\mathrm{T}}\mathscr{B}^{-1}. \tag{3.8.8}$$

Note that $\boldsymbol{\pi}^{\mathrm{T}}$ is a row vector. Now we are in the position to compute the reduced costs, d_j,

$$d_j = c_j - \boldsymbol{\pi}^{\mathrm{T}}\mathbf{A}_j \tag{3.8.9}$$

[13] As further pointed out on page 110, the basis inverse is only rarely inverted explicitly.

for the non-basic variables (the reduced costs of the basic variables are all equal to zero $\mathbf{d}(\mathbf{x}_B) = \mathbf{0}$). Note that formula (3.8.9) computes the reduced costs by using only the original[14] data c_j and $\mathbf{A}_j$ and the current basis $\mathscr{B}$. If any of the d_j are positive, we can improve the objective function by increasing the corresponding x_j. So, the problem is not optimal. The choice of which positive d_j to select is partly heuristic: conventionally, one chooses the largest d_j, but commercial solvers are different from textbook implementations. They use the so-called *partial pricing* and also *devex pricing* [255]. The term partial pricing indicates that the reduced costs are not computed for all non-basic variables. Sometimes, the first reduced cost with positive[15] sign gives the variable to become the new basic variable. Other heuristics choose one of the non-basic variables with positive sign randomly. And there must be a heuristic device that tells the algorithm when to switch from partial to full pricing. Only full pricing can do the optimality test. A sufficient optimality criterion for an optimal solution of a maximization problem is

$$d_j = c_j - \boldsymbol{\pi}^{\mathrm{T}}\mathbf{A}_j \leq 0 \quad , \quad \forall j. \tag{3.8.10}$$

In a minimization problem, the criterion is

$$d_j = c_j - \boldsymbol{\pi}^{\mathrm{T}}\mathbf{A}_j \geq 0 \quad , \quad \forall j.$$

Note that we said a sufficient but not necessary condition. The reason is that in the case of *degeneracy*, several bases define the same basic feasible solution and some can violate the criterion. If non-degenerate, alternative optimal solutions exist (this case is called *dual degeneracy*), then necessarily the reduced cost for some of the non-basic variables is equal to zero. If $d_j < 0$ for all non-basic variables in a maximization problem, then the optimal solution is unique.

If we have not yet reached optimality, we check whether the problem is unbounded. If the problem is bounded, we use the minimum ratio rule to eliminate a basic variable. Both steps are actually performed simultaneously: the minimum ratio rule fails precisely when the incoming vector gives infinite improvement. The data needed for applying the minimum ratio rule are also derived directly from $\mathscr{B}^{-1}$

$$\mathbf{A}'_j = \mathscr{B}^{-1}\mathbf{A}_j. \tag{3.8.11}$$

After the minimum ratio rule has been applied, we have the new basis, i.e., a set of indices or linearly independent columns.

What needs to be done is to get the current basis inverse $\mathscr{B}^{-1}$. There are several formulas to do this, but all of them are equivalent to computing the new basis inverse although the inverse is never computed explicitly. To be correct, the basis is only rarely inverted explicitly. Elementary row operations carry over the existing basis

---

[14]From now on $\mathbf{A}_j$ denotes the columns of the original matrix A corresponding to the variable x_j.

[15]In this case, we are solving a maximization problem.

inverse to the next iteration.[16] However, every, say 100 iterations, the basis inverse is refreshed by inverting the basis matrix taken from the original matrix A. Through this procedure, rounding errors do not accumulate. In addition, in most practical applications A is very sparse, whereas after several iterations the transformed matrix A′ becomes denser so that, especially for large problems, the revised Simplex algorithm usually needs far less operations.

The algorithm continues by computing the values of the basic variables, dual values, and so on until optimality is detected.

Let us come back to the basis inverse. Modern software implementations of the revised Simplex algorithms do not calculate $\mathscr{B}^{-1}$ explicitly. Instead commercial software uses the product form

$$\mathscr{B}_k = \mathscr{B}_0 \eta_1 \eta_2 \dots \eta_k \tag{3.8.12}$$

of the basis to express the basis after k iteration as a function of the initial basis $\mathscr{B}_0$ (usually a unit matrix) and the so-called rather the *eta-matrices* or *eta-factors* η_i. The η_i-matrices are $m \times m$ matrices

$$\eta = \mathbb{1} + \mathbf{u}\mathbf{v}^{\mathrm{T}} \tag{3.8.13}$$

derived from the dyadic product of two vectors $\mathbf{u}$ and $\mathbf{v}$ leading to a very simple structure (“1” on the diagonal, and just non-zeros in one column). To store the η-matrices, it is sufficient to store the η-vectors $\mathbf{u}$ and $\mathbf{v}$. Computing equations such as $\mathscr{B}\mathbf{x}_B = \mathbf{b}$ yielding $\mathbf{x}_B = \mathscr{B}^{-1}\mathbf{b}$ are then solved by

$$\mathbf{x}_B = \eta_k^{-1} \eta_{k-1}^{-1} \dots \eta_1^{-1} \mathbf{b}. \tag{3.8.14}$$

The inverse of the η-matrices (under appropriate assumptions) can be computed very easily according to the formula

$$\eta^{-1} = \mathbb{1} - \frac{1}{1 + \mathbf{v}^{\mathrm{T}}\mathbf{u}} \mathbf{u}\mathbf{v}^{\mathrm{T}}. \tag{3.8.15}$$

[16] The “Boat Renting” problem shows this property very well. Let us inspect the system of linear equations in each iteration. The columns associated with the variables $s_1, \dots, s_4$, the original basic variables, give the basis inverse associated with the current basis. The reason is that elementary row operations are equivalent to a multiplication of the matrix representing the equations by another matrix, say M. If we inspect the first iteration, we can understand how the method works. The initial matrix and, in particular, the columns corresponding to the new basic variables are multiplied by M and obviously give $\mathscr{B} \cdot \mathsf{M} = \mathbb{1}$, where $\mathbb{1}$ is the unit matrix. Thus we have $\mathsf{M} = \mathscr{B}^{-1}$. Since we have multiplied all columns of A by M, and in particular also the unit matrix associated with the original basic variables, these columns just give the columns of the basis inverse $\mathscr{B}^{-1}$. In each iteration k we multiply our original matrix by such a matrix M_k, so the original basic columns represent the product of all matrices M_k, which then is the basis inverse of the current matrix.

Note that we need to store all η_i-vectors. As the iterations proceed, the amount of storage for the factors increases. So a *re-inversion* of the basis occurs not only for reasons of numerical accuracy but also due to a "storage versus computation" trade-off. Readers more interested in the details of the linear algebra computations, LU factorizations, η-vectors, and conserving sparsity may benefit from reading Gill et al. (1981,[217], p.192), Padberg (1996,[431], Section 5.4), and Vanderbei (1996,[560] & 2014,[561]).

Let us illustrate some of the new things we learned in this section and apply them to the "Boat Renting" problem. Some care is needed with the correct use of the indices. The list of basic variables in the solution on page 82 is (x_2, x_1, s_3, s_4), and thus we have

$$
\mathbf{c}_B^{\mathrm{T}} = (600, 800, 0, 0) \quad , \quad \mathscr{B} = \begin{pmatrix} 1 & 1 & 0 & 0 \\ 0 & 1 & 0 & 0 \\ 1 & -1 & 1 & 0 \\ 3 & 4 & 0 & 1 \end{pmatrix} . \tag{3.8.16}
$$

Note that the columns of the basis matrix $\mathscr{B}$ need to coincidence with the sequence of basic variables and that it is taken from the original equation (3.2.2). Applying the formalism known in linear algebra, we can compute the basis inverse as

$$
\mathscr{B}^{-1} = \begin{pmatrix} 1 & -1 & 0 & 0 \\ 0 & 1 & 0 & 0 \\ -1 & 2 & 1 & 0 \\ -3 & -1 & 0 & 1 \end{pmatrix} , \tag{3.8.17}
$$

which gives us the shadow prices

$$
\pi^{\mathrm{T}} = (600, 200, 0, 0). \tag{3.8.18}
$$

If we have a look on page 82 again we can see that the basis inverse can be read off immediately from the final system of linear equations (the columns corresponding to the first basic feasible solution, i.e., the columns 3 to 6 associated with the slack variables s_1 to s_4 contain the current basis inverse).

Earlier in Sect. 3.5 we introduced the idea of the dual problem and its corresponding primal problem. When the dual problem is solved, the optimal values of its variables (and slacks) correspond to the values of the reduced costs and shadow prices of the primal problem. Thus the operation of the Simplex algorithm on the primal problem is governed by the updating of the solution values of the dual problem, which provides current values of the reduced costs on variables. Thus the Simplex algorithm is an algorithm that is implicitly moving between the primal and dual problems, updating solution and reduced cost values, respectively.

Understanding the concept of dual values and shadow prices, we can also give another interpretation of the reduced costs in terms of shadow prices. While the dual values, or Lagrange multipliers, give the cost for active constraints, the reduced cost of a non-basic variable is the shadow price for moving it away from zero, or, in the presence of bounds on the variable, to move the non-basic variable fixed to one of its bounds away from that bound. That also explains why basic variables have zero reduced costs: in non-degenerate cases, basic variables are not at their bounds.

3.8.2 Computing Initial Feasible LP Solutions

The Simplex algorithm explained so far always starts with an initial feasible basis and iterates it to optimality. We have not yet said how we could provide an initial solution. There are several methods, but the best known are *big-M methods* and *phase I and phase II* approaches. Less familiar are heuristic methods usually referred to as *crash methods*. To discuss the first two methods, consider the LP problem with n variables and m constraints in standard form (here it is advantageous to consider a minimization problem)

$$\min \quad \mathbf{c}^{\mathrm{T}}\mathbf{x} \tag{3.8.19}$$

subject to

$$\mathsf{A}\mathbf{x} = \mathbf{b} \quad , \quad \mathbf{x} \geq \mathbf{0}. \tag{3.8.20}$$

By multiplying the equations $\mathsf{A}\mathbf{x} = \mathbf{b}$ by -1 where necessary, we can assume that $\mathbf{b} \geq \mathbf{0}$. That enables us to introduce non-negative *violation variables* $\mathbf{v} = (v_1, \ldots, v_m)$ and to modify the original problem to

$$\min \quad \mathbf{c}^{\mathrm{T}}\mathbf{x} + M \cdot \sum_{j=1}^{m} v_j \tag{3.8.21}$$

subject to

$$\mathsf{A}\mathbf{x} + \mathbf{v} = \mathbf{b} \quad , \quad \mathbf{x} \geq \mathbf{0} \quad , \quad \mathbf{v} \geq \mathbf{0}. \tag{3.8.22}$$

M is a "big" number, say 10^5, but it is very problem dependent as to what big means. The idea of the big-M method is as follows. It is easy to find an initial feasible solution. Can you see this? Check that

$$\mathbf{v} = \mathbf{b} \quad , \quad \mathbf{x} = \mathbf{0} \tag{3.8.23}$$

is an initial feasible solution. Now we are able to start the Simplex algorithm. If we choose M to be sufficiently large, we hope that we get a solution in which none of our violation variables is basic, i.e., $\mathbf{v} = \mathbf{0}$. What if we find a solution with some positive variables v_j? In that case either M was too small, or our original problem is infeasible. How do we know the right size of M? One could start with small values, and check whether all violation variables are zero. If not, one increases M. Ultimately, M must become very large if the problem appears infeasible and one is essentially doing the two-phase method described in the next paragraph.

There is an alternative approach that does not depend on a scaling parameter such as M: the two-phase method. The idea is the same, but it uses a different objective function, namely

$$\min \quad \sum_{j=1}^{m} v_j, \tag{3.8.24}$$

i.e., just the sum of the violation variables. In this case we are certain that if the objective function has a value different from zero, our original problem is infeasible. Actions to be taken in that case are discussed in Sect. 8.5.1.

Why do we have two methods? Would not one be enough? If you inspect both methods carefully, you will notice that they have different advantages or disadvantages. If one takes the limit $M \to \infty$, the big-M methods become the two-phase method. Using the big-M method, the software designer has to ensure and to worry that M is big enough. Often M is adapted dynamically when trying to find an initial solution. It is just good enough to find a solution with $\mathbf{v} = \mathbf{0}$. Keeping the original variables in the objective function may provide an initial solution that is closer to the optimal solution. In practice the number of artificial variables is kept to a minimum. If a certain row already has a slack or surplus variable, there is no need to introduce an additional one. Mixtures of big-M and two-phase methods are also used.

In addition, commercial LP software employs the so-called crash methods (see Sect. 9.2). These are heuristic methods aiming at finding an extremely good initial solution very close to the optimal solution.

Although it has only marginally to do with initial feasible LP solutions it is worth mentioning, the reuse of a basis saved from a previous related run. This approach produces good initial feasible solutions quickly if the model data have only changed a bit or if only a few variables or constraints have been added.

New methods for computing initial solutions or good starting candidates are *hybrid methods*. Such methods, combining the Simplex algorithm and interior-point method, are described in Sect. 3.8.5.

3.8.3 LP Problems with Upper Bounds

So far we have considered the LP problem with n variables and m constraints in standard form (it is not really important whether we consider a minimization or a maximization problem)

$$\min \quad \mathbf{c}^{\mathrm{T}}\mathbf{x} \tag{3.8.25}$$

subject to

$$\mathsf{A}\mathbf{x} = \mathbf{b} \quad , \quad \mathbf{x} \geq \mathbf{0}. \tag{3.8.26}$$

In many large real-world problems, it is advantageous to exploit another structure that occurs frequently, upper and lower bounds on the variables. This is formulated as

$$\min \quad \mathbf{c}^{\mathrm{T}}\mathbf{x}' \tag{3.8.27}$$

subject to

$$\mathsf{A}\mathbf{x}' = \mathbf{b}' \quad , \quad \mathbf{l}' \leq \mathbf{x}' \leq \mathbf{u}'. \tag{3.8.28}$$

Since we can always perform a variable substitution $\mathbf{x} = \mathbf{x}' - \mathbf{l}'$ and observe that the new variables have the bounds $\mathbf{0} \leq \mathbf{x} \leq \mathbf{u}' - \mathbf{l}' = \mathbf{u}$, it is sufficient to consider the problem

$$\min \quad \mathbf{c}^{\mathrm{T}}\mathbf{x} \tag{3.8.29}$$

subject to

$$\mathsf{A}\mathbf{x} = \mathbf{b} \quad , \quad \mathbf{0} \leq \mathbf{x} \leq \mathbf{u}. \tag{3.8.30}$$

Of course we could reformulate this problem by introducing some slack variables $\mathbf{s} \geq \mathbf{0}$ in the standard way

$$\min \quad \mathbf{c}^{\mathrm{T}}\mathbf{x} \tag{3.8.31}$$

subject to

$$\begin{array}{r} \mathsf{A}\mathbf{x} = \mathbf{b} \\ \mathbf{x} + \mathbf{s} = \mathbf{u} \end{array} \quad , \quad \mathbf{x} \geq \mathbf{0} \quad , \quad \mathbf{s} \geq \mathbf{0}. \tag{3.8.32}$$

Since in many large real-world problems we have $n \gg m$ (there are often between three and ten times as many variables as rows), a straightforward application of the Simplex algorithm would lead to a very large basis of size $(m+n) \times (m+n)$. Exploiting the presence of the upper bounds will lead to a modified Simplex algorithm that still is based on a basis of size $m \times m$ only.

The idea is to distinguish between non-basic variables, x_j, $j \in \mathcal{J}_0$, that are at their lower bound of zero (the concept we are familiar with) and those x_j, $j \in \mathcal{J}_u$, that are at their upper bound (the new concept). With the new concept no explicit slack variables are necessary. Let us now try to see how the Simplex algorithm works when performing a basis exchange. Pricing now tells us that a current basis is not optimal if one of these two situations occurs:

$$\begin{aligned}&1)\text{there exist indices } j \in \mathcal{J}_0 \text{ with } d_j < 0\\ &2)\text{there exist indices } j \in \mathcal{J}_u \text{ with } d_j > 0.\end{aligned}$$

In case 1) we could increase a non-basic variable, in case 2) we could decrease it, and in both cases the effect would be a decreased value of the objective function. A new consideration, when increasing or decreasing a non-basic variable, is that we need to calculate whether it could reach its upper (respectively, lower) bound.

In Sect. 3.2.2 we learned that the minimum ratio rule controlled how we could change a non-basic variable. We had to stop increasing a non-basic variable when one of the basic variables became zero. The minimum ratio rule in the presence of upper bounds on variables gets a little bit more complicated because variables might hit their upper bounds.

Case 1) leads to two sub-cases:

1a) the non-basic variable x_j can be increased to its upper bound, while no basic variables reaches zero or its upper bound, or
1b) while increasing the non-basic variable x_j, a basic variable reaches zero, or a basic variable reaches its upper bound.

Case 1a), called a flip for obvious reasons, is easy to handle: one just moves the index j into the new set $\mathcal{J}_u$. Reaching zero in 1b) is handled as in the standard Simplex algorithm: variable x_j enters the basis and the index of the variable leaving the basis is added to the new set $\mathcal{J}_0$. When an upper bound is hit in 1b) the variable x_j enters the basis and the index of the variable leaving the basis is added to the new set $\mathcal{J}_u$.

Case 2) can be analyzed by considering the slack $s_j = u_j - x_j$. If the non-basic variable x_j is at its upper bound, then s_j is at its lower bound (zero). s_j plays the same role (note that its upper bound is u_j) as the non-basic variable x_j considered in 1a) and 1b), and thus the argument is the same.

The linear algebra involved in the iteration is similar to the standard Simplex, and essentially no extra computations are required. Readers more interested in the subject are referred to Padberg (1996,[431], pp.75–80).

We have now seen why exploiting bounds on variables explicitly leads to better, i.e., faster numerical, performance: there is a little more testing and logic required in the algorithm but no additional computations. This has significant consequences for the B&B algorithm that adds only new bounds to the existing problem.

Note that as the bounds are treated explicitly and not as constraints, no shadow prices are available on these "constraints." But the shadow prices can be derived from the reduced costs of the non-basic variables fixed at their upper bounds. The idea is as follows: the shadow prices are the change in objective function per unit change in the right-hand side. For a particular constraint $x + s = u$, changing the right-hand side is the same as changing the slack variable s (or, x) by the same amount. Thus, we have $d_j = \pi$. To see this in more detail, let us start with the example

$$\begin{array}{rlcrl} & \text{problem (constraint)} & & & \text{standard formulation} \\ \min & -x-y & & \min & -x-y \\ s.t. & 2x+y \le 3 & \Longleftrightarrow & s.t. & 2x+y+s_1=3 \\ & x+2y \le 3 & & & x+2y+s_2=3 \\ & x \le 0.5 & & & x+s_3=0.5 \\ x \ge 0 & , \quad y \ge 0 & & x \ge 0 & , \quad y \ge 0. \end{array} \tag{3.8.33}$$

This problem gives the solution

$$\begin{pmatrix} x \\ y \end{pmatrix} = \begin{pmatrix} 0.5 \\ 1.25 \end{pmatrix}, \quad \begin{pmatrix} s_1 \\ s_2 \\ s_3 \end{pmatrix} = \begin{pmatrix} 0.75 \\ 0 \\ 0 \end{pmatrix}, \tag{3.8.34}$$

and[17]

$$\begin{pmatrix} d_x \\ d_y \end{pmatrix} = \begin{pmatrix} 0 \\ 0 \end{pmatrix} \quad , \quad \begin{pmatrix} \pi_1 \\ \pi_2 \\ \pi_3 \end{pmatrix} = \begin{pmatrix} 0 \\ -0.5 \\ -0.5 \end{pmatrix}. \tag{3.8.35}$$

Note that both x and y, as well as s_1 and s_4 are basic variables. The second and third constraints are active. The problem (3.8.33) can be formulated with the third constraint as a bound leading to

$$\begin{array}{rlcrl} & \text{problem (bound)} & & & \text{standard formulation} \\ \min & -x-y & & \min & -x-y \\ s.t. & 2x+y \le 3 & \Longleftrightarrow & s.t. & 2x+y+s_1=3 \\ & x+2y \le 3 & & & x+2y+s_2=3 \\ 0 \le x \le 0.5 & , \quad y \ge 0 & & 0 \le x \le 0.5 & , \quad y \ge 0. \end{array} \tag{3.8.36}$$

Of course we get the same solution with respect to x and y but only y is a basic variable, x is a non-basic variable at its upper limit, and the reduced costs and shadow prices (formally, there are only two of them) are different (note, the we use the superscript B to refer to the formulation using bounds)

$$\begin{pmatrix} x \\ y \end{pmatrix} = \begin{pmatrix} 0.5 \\ 1.25 \end{pmatrix}, \quad \begin{pmatrix} d_x^B \\ d_y^B \end{pmatrix} = \begin{pmatrix} -0.5 \\ 0 \end{pmatrix}, \quad \begin{pmatrix} \pi_1^B \\ \pi_2^B \end{pmatrix} = \begin{pmatrix} 0 \\ -0.5 \end{pmatrix}. \tag{3.8.37}$$

---

[17]Note that LP solvers may have a different sign convention for the shadow prices and may print out $\pi_2 = \pi_3 = +0.5$.

However, note that in the example we have $\pi_3 = d_x^B$. If we could show that this relation holds in general, then we could solve LP problems exploiting explicit upper bounds and then derive the shadow prices associated with those bounds. Let us do a little algebra to show that such a relation indeed is valid. Below we formulate the same problem with m constraint (not counting the bounds) and n variables once with constraints considering the upper bound and once treating the bounds explicitly.

$$
\begin{array}{rrcrr}
 & \text{problem (constraint)} & & & \text{problem (bounds)} \\
 & \min \quad \mathbf{c}^{\mathrm{T}}\mathbf{x} & & & \min \quad \mathbf{c}^{\mathrm{T}}\mathbf{x} \\
s.t. & \mathsf{A}\mathbf{x} + \mathbf{s} = \mathbf{b} & \longleftrightarrow & s.t. & \mathsf{A}\mathbf{x} + \mathbf{s} = \mathbf{b} \\
 & \mathbf{x} + \mathbf{s} = \mathbf{u} & & & \\
 & \mathbf{x} \geq 0 & & & 0 \leq \mathbf{x} \leq \mathbf{u}
\end{array}
. \tag{3.8.38}
$$

Let us assume that the solution has the first n_u variables at their upper limits (by appropriate change of columns in matrices involved, this is always possible). In the formulation using the constraint, the basis can be written as the block matrix

$$
\mathsf{B} = \begin{pmatrix} \mathsf{U} & \mathscr{B} \\ \mathbb{1} & \mathsf{0} \end{pmatrix}, \tag{3.8.39}
$$

where is an $(n_u + m)(n_u + m)$ matrix, U is an $m \times n_u$ matrix, $\mathbb{1}$ is the $n_u \times n_u$ unit matrix, $\mathsf{0}$ is an $n_u \times m$ matrix of zeros, and $\mathscr{B}$ is (under appropriate assumptions) an $m \times m$ regular matrix, which at the same time is the basis of the solution of the problem in which bounds are treated explicitly. In our little example above we have $m = 2$ and $n_u = 1$ and

$$
\mathsf{B} = \begin{pmatrix} 2 & 1 & 1 \\ 1 & 2 & 0 \\ 1 & 0 & 0 \end{pmatrix} \quad , \quad \mathsf{U} = \begin{pmatrix} 2 \\ 1 \end{pmatrix} \quad , \quad \mathbb{1} = (1) \quad , \quad \mathscr{B} = \begin{pmatrix} 1 & 1 \\ 2 & 0 \end{pmatrix} .
$$

To show that $\boldsymbol{\pi}_U^{\mathrm{T}}$, the shadow prices corresponding the upper bounds formulated as constraints, equals the reduced costs, $\mathbf{d}_U^{\mathrm{T}}$, of the non-basic variables fixed at their upper limits, we invert the matrix B to derive $\boldsymbol{\pi}_U^{\mathrm{T}}$ and compute $\mathbf{d}_U^{\mathrm{T}}$ based on $\mathscr{B}$.

The inverse B^{-1} can be written down explicitly

$$
\mathsf{B}^{-1} = \begin{pmatrix} \mathsf{0} & \mathbb{1} \\ \mathscr{B}^{-1}\mathbb{1} & -\mathscr{B}^{-1}\mathsf{U} \end{pmatrix} . \tag{3.8.40}
$$

The shadow prices $\boldsymbol{\pi}_U^{\mathrm{T}}$ follow from evaluation of

$$
\boldsymbol{\pi}^{\mathrm{T}} = \left(\mathbf{c}_U^{\mathrm{T}}, \mathbf{c}_B^{\mathrm{T}}\right) \begin{pmatrix} \mathsf{0} & \mathbb{1} \\ \mathscr{B}^{-1}\mathbb{1} & -\mathscr{B}^{-1}\mathsf{U} \end{pmatrix} = \left(\mathbf{c}_B^{\mathrm{T}}\mathscr{B}^{-1}\mathbb{1}, \mathbf{c}_U^{\mathrm{T}} - \mathbf{c}_B^{\mathrm{T}}\mathscr{B}^{-1}\mathsf{U}\right)
$$

and give

$$\boldsymbol{\pi}_U^{\mathrm{T}} = \mathbf{c}_U^{\mathrm{T}} - \mathbf{c}_B^{\mathrm{T}} \mathscr{B}^{-1} \mathsf{U}. \tag{3.8.41}$$

The reduced costs of the non-basic variables $\mathbf{x}_U$, which are at their upper bounds (in the formulation exploiting explicit upper bounds), are exactly [cf. formula (3.8.9)]

$$\mathbf{d}_U^{\mathrm{T}} = \mathbf{c}_U^{\mathrm{T}} - \mathbf{c}_B^{\mathrm{T}} \mathscr{B}^{-1} \mathsf{U}. \tag{3.8.42}$$

So we have shown that $\boldsymbol{\pi}_U^{\mathrm{T}} = \mathbf{d}_U^{\mathrm{T}}$.

3.8.4 Dual Simplex Algorithm

The *(primal) Simplex algorithm* concentrates on improving the objective function value of an existing basic feasible solution. In contrast, there is also the *dual Simplex algorithm* that solves an LP problem by taking a dual optimal basic solution that is optimal in the sense that the reduced costs computed according to (3.8.9) have the correct sign but are not basic feasible. The dual Simplex algorithm tries to achieve feasibility of the solution while retaining its optimal properties. The two approaches can be seen as "dual" to each other in the sense introduced in Sect. 3.5, while the primal algorithm makes the choice of the new basic variable first and then decides on which existing basic variable should be eliminated, the dual algorithm eliminates first an existing basic variable and then selects a new basic variable.

The dual Simplex algorithm is often used within the B&B algorithm. When a branch is made, the subproblem just differs from the original problem in having a different bound on the branching variable; remember from Sect. 3.8.3 that bounds can be treated very efficiently in the Simplex algorithm. So the LP solution obtained at the parent node could now be considered optimal but not feasible. In most cases, the dual Simplex algorithm restores feasibility quickly to the solution, say, within a few iterations. By using the dual Simplex algorithm we are able to take advantage of the earlier work done by the Simplex algorithm and then move to a new optimal solution quickly. However, there is no guarantee that this will always happen. If the (primal) Simplex algorithm was used after each branch the problem would need to be solved from the start again and this would be burdensome.

3.8.5 Interior-Point Methods — A Detailed Description

As has already been said in Sect. 3.2.3, initiated by the work of Karmarkar (1984,[325]), a large variety of *interior-point methods* (IPMs) has been developed [cf. Gonzaga (1992,[231], Lustig et al. (1992,[379])], and Mehrotra's so-called *primal–dual second-order predictor–corrector methods*, Mehrotra (1992,[393]), have already been integrated into some LP solvers. In linear programming, IPMs are well suited especially for large, sparse problems or those, which are highly degenerate. Here considerable computing-time gains can be achieved. With respect to the solution strategy most of these algorithms can be classified as *affine scaling methods*,[18] *potential reduction methods*,[19] *central trajectory methods*;[20] cf. Freund & Mizuno (1996,[201]).

The idea of IPMs is to proceed from an initial interior point $\mathbf{x} \in S$ satisfying $\mathbf{x} > 0$, toward an optimal solution without touching the boundary of the feasible set S. The condition $\mathbf{x} > 0$ is (in the second and third methods) guaranteed by adding a penalty term to the objective function.

To explain the essential characteristics of central trajectory interior-point methods, let us consider the *logarithmic barrier method* in detail when applied to the primal–dual pair

$$\begin{array}{rrcrr} & \text{primal problem} & \longleftrightarrow & & \text{dual problem} \\ & & & & \\ & \min \quad \mathbf{c}^{\mathrm{T}}\mathbf{x} & & & \max \quad \mathbf{b}^{\mathrm{T}}\mathbf{y} \\ s.t. & \mathsf{A}\mathbf{x} = \mathbf{b} & & s.t. & \mathsf{A}^{\mathrm{T}}\mathbf{y} + \mathbf{w} = \mathbf{c} \\ & \mathbf{x} \geq 0 & & & \mathbf{w} \geq 0 \end{array} \tag{3.8.43}$$

with free, dual variable $\mathbf{y}$, the dual slack variable $\mathbf{w}$, and the solution vectors $\mathbf{x}^*$, $\mathbf{y}^*$ and $\mathbf{w}^*$. A feasible point $\mathbf{x}$ of the primal problem is called strictly feasible if $\mathbf{x} > 0$, and a feasible point $\mathbf{w}$ of the dual problem is called strictly feasible if $\mathbf{w} > 0$. The primal problem is mapped to a sequence of nonlinear programming problems

$$P^{(k)} : \min \left\{ \mathbf{c}^{\mathrm{T}}\mathbf{x} - \mu \sum_{j=1}^{n} \ln x_j \;\middle|\; \begin{array}{l} \mathsf{A}\mathbf{x} = \mathbf{b} \\ \mathbf{x} > 0 \end{array} \quad , \quad \mu = \mu^{(k)} \right\} \tag{3.8.44}$$

with *homotopy parameter* μ where we replaced the non-negativity constraint on the variables with the logarithmic penalty term; instead of using x_j in the penalty

---

[18]*Affine scaling methods* are based on a series of strictly feasible points (see definition below) of the primal problem. Around these points the feasible region of the primal problem is locally approximated by the so-called *Dikin ellipsoid*. This procedure leads to a convex optimization problem which is in most parts linear except for one convex quadratic constraint.

[19]Potential reduction methods are those that have been introduced in Karmarkar's famous 1984 paper. The objective function is the potential function $q \ln\left(\mathbf{c}^{\mathrm{T}}\mathbf{x} - \mathbf{b}^{\mathrm{T}}\mathbf{y}\right) - \sum_{j=1}^{n} \ln x_j$ built up by the logarithm of the duality gap and a logarithmic term designed to repel feasible points from the boundary of the feasible region. The constraints are the linear constraints for primal and dual feasibility.

[20]*Central trajectory methods* are primal–dual methods based on the concept of the *central path* or *central trajectory*. They are discussed in detail below. In their primal–dual predictor–corrector version, they are the most efficient ones, and according to Freund & Mizuno (1996), they have the most aesthetic qualities.

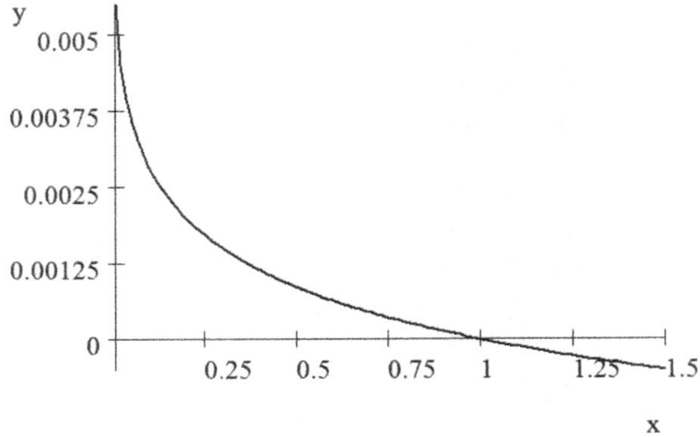

Fig. 3.8 Logarithmic penalty term for $y = -\mu \ln x$ with $\mu = 0.0125$. Note that for constant μ, $\lim_{x \to 0^-} y(x) = +\infty$

term, we could have considered the inequality $\mathbf{A}_i \mathbf{x} \leq b_i$ through terms of the form $\ln(b_i - \mathbf{A}_i \mathbf{x})$.

At every iteration step k, μ is newly chosen as described in Sect. 3.8.5.4. As shown in Fig. 3.8 the penalty term, and therefore the objective function, increases to infinity. By suitable reduction of the parameter $\mu > 0$, the weight of the penalty term, which gives the name logarithmic barrier problem to this methods, is successively reduced and the sequence of points obtained by solving the perturbed problems converges to the optimal solution of the original problem. So, through the choice of $\mu^{(k)}$, a sequence $P^{(k)}$ of minimization problems is constructed, where the relation

$$\lim_{k \to \infty} \left(\mu^{(k)} \sum_{j=1}^{n} \ln x_j \right) = 0 \tag{3.8.45}$$

has to be valid, viz.

$$\lim_{k \to \infty} argmin(P^{(k)}) = argmin(LP) = x^*, \tag{3.8.46}$$

where the function *argmin* returns an optimal solution vector of the problem.

We have replaced one optimization problem, namely (3.8.43), by several more complex NLP problems. So, it is not a surprise to learn that interior-point methods are special homotopy algorithms for the solution of general nonlinear constrained optimization problems. Applying the Karush–Kuhn–Tucker (KKT) conditions [Karush (1939,[329]) and Kuhn & Tucker (1951,[346])], these are the necessary or the sufficient conditions for the existence of local optima in NLP problems, we get a system of nonlinear equations that can be solved with the Newton–Raphson algorithm as shown below. The good news is that the problems $P^{(k)}$ or systems of nonlinear equations they produce need not to be solved exactly in practice, but one is satisfied with the solution achieved after one single iteration in the Newton–Raphson algorithm.

3.8.5.1 A Primal–Dual Interior-Point Method

So far we have considered the primal problem. To get to the dual and finally the primal–dual[21] version of interior-point solvers (motivation will be given below), the Karush–Kuhn–Tucker (KKT) conditions are derived from the Lagrangian function (a common concept in NLP)

$$L = L(\mathbf{x},\mathbf{y}) = \mathbf{c}^{\mathrm{T}}\mathbf{x} - \mu\sum_{j=1}^{n} \ln\, x_j - \mathbf{y}^{\mathrm{T}}(\mathsf{A}\mathbf{x} - \mathbf{b}). \tag{3.8.47}$$

Note that the constraints are multiplied by the dual value (Lagrange multipliers), $\mathbf{y}^{\mathrm{T}}$, and then are added to the original primal objective function. If we introduce the abbreviations[22]

$$\mathbf{e} = (1,\ldots,1)^{\mathrm{T}} \in \mathbb{R}^n\ ,\ \mathsf{X} = diag(x_1,\ldots,x_n)\ ,\ \mathsf{X}^{-1} = diag(x_1^{-1},\ldots,x_n^{-1}), \tag{3.8.48}$$

where $\mathbf{e}$ is the n-vector of all ones, and X is a diagonal matrix with the elements x_j, the KKT conditions read

$$\frac{\partial L}{\partial \mathbf{x}} = \mathbf{c} - \mu\mathsf{X}^{-1}\mathbf{e} - \mathsf{A}^{\mathrm{T}}\mathbf{y} = 0 \tag{3.8.49}$$

$$\frac{\partial L}{\partial \mathbf{y}} = \mathsf{A}\mathbf{x} - \mathbf{b} = 0 \tag{3.8.50}$$

$$\mathbf{x} \geq 0. \tag{3.8.51}$$

The system (3.8.49–3.8.50) can be rewritten as the primal feasibility constraint

$$\mathsf{A}\mathbf{x} = \mathbf{b}, \tag{3.8.52}$$

and in addition the pair of the dual feasibility constraint

$$\mathsf{A}^{\mathrm{T}}\mathbf{y} + \mathbf{w} = \mathbf{c}, \tag{3.8.53}$$

and a *perturbed* complementary slackness condition

$$\mathsf{XW}\mathbf{e} = \mu\mathbf{e}\quad ,\quad \mathsf{W} := diag(w_1,\ldots,w_n). \tag{3.8.54}$$

---

[21] The reason for using this expression is that the method will include both the primal and dual variables.

[22] $diag(x_1,\ldots,x_n)$ denotes a diagonal matrix with diagonal elements $x_1,\ldots,x_n$ and zeros at all other entries.

So, the KKT conditions derived from the Lagrangian function (3.8.47), which reproduces our original equations of the primal–dual pair (3.8.43) and gives us in addition the nonlinear equation (3.8.54) depending on the barrier parameter.

Let us try to interpret these equations: the system of equations (3.8.52)-(3.8.53) is identical to the KKT system for the original LP problem in which the complementary slackness conditions are perturbed by μ. A non-negative solution of (3.8.52)–(3.8.53) is called an *analytic center* and depends on the barrier parameter. The set of solutions $[\mathbf{x}(\mu), \mathbf{y}(\mu), \mathbf{w}(\mu)]$ defines a trajectory of centers for the primal and dual problems, respectively, and is called a *central path* or *central trajectory*.

For points on the central path and barrier parameter μ we note the property

$$\mathbf{w}^{\mathrm{T}}\mathbf{x} = g(\mu) = \Delta(\mathbf{x}, \mathbf{y}) = 2\mu\mathbf{e}^{\mathrm{T}}\mathbf{e} = 2n\mu. \tag{3.8.55}$$

Equation (3.8.55) shows that in the absence of the barrier problem, the duality gap vanishes. If μ approaches zero, the complementarity gap and duality gap also approach zero, which implies according to the strong duality theorem that we are approaching the optimal point of the LP problem (3.8.43).

The relation (3.8.55) enables us by inserting into (3.8.54) to derive a non-parameterized representation of the central path. The central path is the set of all points satisfying the system of equations

$$\mathsf{A}\mathbf{x} = \mathbf{b} \quad , \quad \mathsf{A}^{\mathrm{T}}\mathbf{y} + \mathbf{w} = \mathbf{c} \quad , \quad \mathsf{XWe} = \frac{\mathbf{w}^{\mathrm{T}}\mathbf{x}}{2n}\mathbf{e} \quad , \quad \begin{matrix} \mathbf{x} > \mathbf{0} \\ \mathbf{w} > \mathbf{0}. \end{matrix} \tag{3.8.56}$$

Let us now concentrate on how we can solve the system of equations (3.8.52)–(3.8.54). Such systems $\mathbf{f}(\mathbf{z}^*) = 0$ are solved by the Newton–Raphson algorithm based on an initial point $\mathbf{z}$, Jacobian matrix J, and the linearization

$$0 = \mathbf{f}(\mathbf{z}^*) = \mathbf{f}(\mathbf{z} + \Delta\mathbf{z}) = \mathbf{f}(\mathbf{z}) + \mathsf{J}(\mathbf{z})\Delta\mathbf{z}\ ; \ \mathsf{J}(\mathbf{z}) = \frac{\partial \mathbf{f}}{\partial \mathbf{z}}(\mathbf{z}) \tag{3.8.57}$$

eventually, yielding

$$\Delta\mathbf{z} = -\mathsf{J}^{-1}\mathbf{f}(\mathbf{z}) \tag{3.8.58}$$

and

$$\mathbf{z}^* = \mathbf{z} + \Delta\mathbf{z}. \tag{3.8.59}$$

In (3.8.59) the point $\mathbf{z}^*$ used in next iteration is computed from the current point $\mathbf{z}$ by adding the full step size $\Delta\mathbf{z}$. In practice, one often uses a factor, δ, $0 < \delta \leq 1$, which may cut down the step size in order to prevent the new point from leaving the feasible region and computes the new point according to

$$\mathbf{z}^* = \mathbf{z} + \delta\Delta\mathbf{z}. \tag{3.8.60}$$

Applying the Newton–Raphson method to (3.8.52)–(3.8.53) leads to the system of linear equations

$$\begin{pmatrix} 0 & \mathsf{A} & \mathbb{1} \\ \mathsf{A}^{\mathrm{T}} & 0 & 0 \\ \mathsf{W} & 0 & \mathsf{X} \end{pmatrix} \cdot \begin{pmatrix} \Delta\, \mathbf{x} \\ \Delta\, \mathbf{y} \\ \Delta\, \mathbf{w} \end{pmatrix} = \begin{pmatrix} \mathbf{c} - \mathsf{A}^{\mathrm{T}}\mathbf{y} - \mathbf{w} \\ \mathbf{b}-\mathsf{A}\mathbf{x} \\ \mu \cdot \mathbf{e}-\mathsf{X}\mathsf{W}\mathbf{e} \end{pmatrix}, \tag{3.8.61}$$

which is called Newton equations system; note that $\mathbb{1}$ denotes the unit matrix of appropriate dimension. The third equation can be eliminated leading to

$$\Delta\, \mathbf{w} = \mathsf{X}^{-1} \cdot (\mu \cdot \mathbf{e}-\mathsf{X}\mathsf{W}\mathbf{e}-\mathsf{W}\cdot\Delta\, \mathbf{x}) \tag{3.8.62}$$

and the reduced Newton equations system

$$\begin{pmatrix} -\mathsf{X}^{-1}\mathsf{W} & \mathsf{A} \\ \mathsf{A}^{\mathrm{T}} & 0 \end{pmatrix} \cdot \begin{pmatrix} \Delta\, \mathbf{x} \\ \Delta\, \mathbf{y} \end{pmatrix} = \begin{pmatrix} \mathbf{c} - \mathsf{A}^{\mathrm{T}}\mathbf{y} - \mu\mathsf{X}^{-1} \cdot (\mu\mathbf{e} - \mathsf{X}\mathsf{W}\mathbf{e}) \\ \mathbf{b}-\mathsf{A}\mathbf{x} \end{pmatrix}. \tag{3.8.63}$$

Now it is also possible to eliminate $\Delta\, \mathbf{x}$ yielding

$$\mathsf{A}^{\mathrm{T}}\mathsf{W}^{-1}\mathsf{X}\mathsf{A}\Delta\, \mathbf{y} = \mathbf{b}-\mathsf{A}\mathbf{x}+\mathsf{A}^{\mathrm{T}}\mathsf{W}^{-1}\mathsf{X}\cdot \left[\mathbf{c} - \mathsf{A}^{\mathrm{T}}\mathbf{y} - \mu\mathsf{X}^{-1} \cdot (\mu\mathbf{e} - \mathsf{X}\mathsf{W}\mathbf{e}) - \mathsf{A}\Delta\, \mathbf{y}\right] \tag{3.8.64}$$

and

$$\Delta\, \mathbf{x} = -\mathsf{W}^{-1}\mathsf{X}\cdot \left[\mathbf{c} - \mathsf{A}^{\mathrm{T}}\mathbf{y} - \mu\mathsf{X}^{-1} \cdot (\mu\mathbf{e} - \mathsf{X}\mathsf{W}\mathbf{e}) - \mathsf{A}\Delta\, \mathbf{y}\right]. \tag{3.8.65}$$

Solving (3.8.63) or (3.8.64) is the major computational effort within the iterations of interior-point methods. Note that the matrix $\mathsf{A}^{\mathrm{T}}\mathsf{W}^{-1}\mathsf{X}\mathsf{A}$ in (3.8.64) is positive definite, which would allow us to use a Cholesky solver. However, the disadvantage is that by doing so we would destroy sparsity to some extent. Solving (3.8.63), we would conserve sparsity, but the linear algebra is less efficient. Nevertheless, for most problems, this is the most efficient approach.

Once (3.8.64) has been solved, $\Delta\, \mathbf{x}$ is computed by (3.8.65), and finally, $\Delta\, \mathbf{x}$ is computed by (3.8.62). Then, the maximum step sizes are computed such that the non-negativity of the variables is conserved. In practice the primal and dual step sizes are computed separately:

$$\mathbf{x}^{(k+1)} = \mathbf{x}^{(k)} + \delta_P \cdot \Delta\, \mathbf{x} \quad , \quad \begin{pmatrix} \mathbf{y} \\ \mathbf{w} \end{pmatrix}^{(k+1)} = \begin{pmatrix} \mathbf{y} \\ \mathbf{w} \end{pmatrix}^{(k)} + \delta_D \cdot \begin{pmatrix} \Delta\, \mathbf{y} \\ \Delta\, \mathbf{w} \end{pmatrix}. \tag{3.8.66}$$

Here the damping factors δ_P and δ_D are computed according to the following heuristic:

$$\delta_P = \min\left\{1, (1-\varepsilon) \cdot \delta_P^{'}\right\} \quad , \quad \delta_P = \min\left\{1, (1-\varepsilon) \cdot \delta_D^{'}\right\} \tag{3.8.67}$$

with

$$\delta'_P := \min_i \left\{ \frac{x_i}{-\Delta x_i} \middle| \Delta x_i < 0 \right\} \quad , \quad \delta'_D := \min_i \left\{ \frac{s_i}{-\Delta w_i} \middle| \Delta w_i < 0 \right\} . \tag{3.8.68}$$

The step sizes are slightly reduced by the factor $1-\varepsilon$ to prevent hitting the boundary, with a small positive parameter, ε, say, $\varepsilon \approx 10^{-4}$.

3.8.5.2 Predictor–Corrector Step

What needs to be explained is the predictor–corrector part involved in most interior-point solvers. Predictor–corrector techniques belong to the class of higher-order methods used to solve nonlinear systems of equations. As we have already pointed out, the most expensive computational step is the solution of (3.8.63) or (3.8.64) requiring a matrix inversion or factorization of the matrix. If we factorize the matrix $\mathsf{A}^T\mathsf{W}^{-1}\mathsf{XA}$, we can solve for $\Delta\mathbf{y}$ and $\Delta\mathbf{x}$ if the right-hand side of the terms is known. The idea of predictor–corrector techniques is to reuse the factorization and to solve for different right-hand sides with the objective to find a better "search direction." Mehrotra's second-order predictor–corrector strategy [393] solves the Newton equations in the first step for $\mu = 0$ yielding the *affine scaling (predictor)* search direction Δ_α. If a step of size δ is taken in that direction, the complementary gap is used to estimate the new barrier parameter μ_e. Next, the higher-order component of the predictor–corrector direction is computed and based on the requirement that the next iteration is perfectly centered, i.e.,

$$(\mathsf{X} + \Delta\mathsf{X})(\mathbf{w} + \Delta\mathbf{w}) = \mu_e \mathbf{e} \tag{3.8.69}$$

or equivalent equation

$$\mathsf{W}\Delta\mathbf{x} + \mathsf{X}\Delta\mathbf{w} = -\mathsf{X}\mathbf{w} + \mu_e\mathbf{e} - \Delta\mathsf{X}\Delta\mathbf{w}. \tag{3.8.70}$$

Instead of setting the second-order term on the right-hand side equal to zero, Mehrotra proposes to estimate $\Delta\mathsf{X}\Delta\mathbf{w}$ using the affine scaling direction $\Delta\mathsf{X}_\alpha\Delta\mathbf{w}_\alpha$. It is the Eq. (3.8.70) with the estimated barrier parameter μ_e, which defines the predictor–corrector direction.

3.8.5.3 Computing Initial Points

By definition interior-point methods operate within the interior of the feasible region and thus need strictly positive initial guesses for the vectors $\mathbf{x}$ and $\mathbf{w}$. How can such feasible points be obtained? This is a very difficult task in IPM research. The method described in Sect. 3.8.2 does not help this time because we do not want to use the Simplex algorithm. Since interior-point methods are path-following methods, one would like to have an initial point as close as possible to the central path and to be as close to primal and dual feasibility as possible. How that goal is reached can be read in Andersen et al. (1996,[23]).

The reader interested in this topic might find it astonishing that the so-called primal–dual infeasible interior-point methods proved to be successful. These methods start with initial points $\mathbf{x}^{(0)}$ and $\mathbf{w}^{(0)}$ but do not require that primal and dual feasibility is satisfied. This is quite typical for nonlinear problems that use the Newton-type algorithms. Feasibility is attained during the process as optimality is approached.[23] Therefore, a primal and dual feasibility test is part of the termination criterion (see Sect. 3.8.5.5). It needs to be mentioned that the primal–dual infeasible methods have difficulties to detect possible primal or dual infeasibility of the LP problem. This problem also promoted the development of homogeneous self-dual methods.

3.8.5.4 Updating the Homotopy Parameter

The control of the homotopy parameter μ determines how efficient the interior-point method works. If one wants to stay close to the central path, then small changes of μ are recommended (*short step methods*). However, this might result in slow convergence. The opposite, *large step methods*, might leave the vicinity of the central part and provide numerical difficulties or require several Newton steps in each iteration.

The relation (3.8.55), which is valid for the central path, suggests that the barrier parameter should be connected to the complementarity gap or the duality gap. Note that because the method does not require primal and dual feasibility, the complementarity gap might be different from the duality gap.

A common choice of the $\mu^{(k)}$ is the heuristic given in Lustig et al. (1991,[378])

$$\mu = \frac{\mathbf{b}^{\mathrm{T}}\mathbf{y} - \mathbf{c}^{\mathrm{T}}\mathbf{x}}{\varphi(n)}, \tag{3.8.71}$$

where n represents the number of the variables and $\varphi(n)$ is an auxiliary function

$$\varphi(n) = \begin{cases} n & , n \le 5000 \\ n^3 & , n \ge 5000. \end{cases} \tag{3.8.72}$$

In Mehrotra's predictor–corrector method [393], the barrier parameter is chosen as

$$\mu = \left(\frac{g_a}{g}\right)^2 \frac{g_a}{n}, \tag{3.8.73}$$

[23]The screen output of an interior-point solver may thus contain the number of the current iteration, quantities measuring the violation of primal and dual feasibility, the values of the primal and the dual objective function (or the duality gap) and possibly the barrier parameter as valuable information on each iteration.

where g_a is the predicted complementarity gap used in that method. Another approach that is particularly useful in parallel IPMs determines $\mu^{(k)}$ by exploiting extrapolation techniques (Bock & Zillober, 1995,[88]) known from the theory of differential equations.

3.8.5.5 Termination Criterion

For the given μ viz. μ^k, the algorithm proceeds until relative primal feasibility

$$\Delta^P(\mathbf{x}) := \frac{\left\|\mathsf{A}\mathbf{x}^{(k)} - \mathbf{b}\right\|}{\max\left\{1, \left\|\mathsf{A}\mathbf{x}^{(0)} - \mathbf{b}\right\|\right\}} < \varepsilon^P \tag{3.8.74}$$

and relative dual feasibility

$$\Delta^D(\mathbf{y}) := \frac{\left\|\mathsf{A}^{\mathrm{T}}\mathbf{y}^{(k)} + \mathbf{w}^{(k)} - \mathbf{c}\right\|}{\max\left\{1, \left\|\mathsf{A}^{\mathrm{T}}\mathbf{y}^{(0)} + \mathbf{w}^{(0)} - \mathbf{c}\right\|\right\}} < \varepsilon^D \tag{3.8.75}$$

are achieved up to some specified accuracy and the relative *duality gap* $\Delta^R(\mathbf{x}, \mathbf{y})$ is such that

$$\Delta^R(\mathbf{x}, \mathbf{y}) := \frac{\mathbf{c}^{\mathrm{T}}\mathbf{x} - \mathbf{b}^{\mathrm{T}}\mathbf{y}}{1 + \mid \mathbf{b}^{\mathrm{T}}\mathbf{y} \mid} < \varepsilon^G. \tag{3.8.76}$$

Typical values for the accuracies are $\varepsilon^P = \varepsilon^D = 10^{-10}$ and $\varepsilon^G = 10^{-8}$. It was already mentioned in Sect. 3.5.3 that the size of the duality gap gives a measure of how far a feasible point is away from being an optimal point. Thus we are not surprised to see $\Delta^R(\mathbf{x}, \mathbf{y})$ as a part of a termination criterion.

3.8.5.6 Basis Identification and Cross-Over

Interior-point methods produce an approximation to the optimal solution of an LP but no optimal basic and non-basic partition of the variables. Since the optimal solution produced by the interior-point method is strictly in the interior of the feasible solution, there are many more variables not fixed at their bounds than we would expect in a Simplex solution. The concept of basic solutions is, however, very important for sensitivity analyses and the use of LP problems as subproblems in B&B algorithms. The availability of a basis facilitates warm-starts.

Therefore, in many cases, it is advantageous to be able to purify an optimal interior-point solution into a basic solution. Such procedures are called *basis identifications procedures*. Such procedures derive a basis from any feasible solution of an LP problem such that this basic solution gives the same objective function value as the feasible point it is derived from. Details on these methods are found in Andersen & Ye (1994,[21]) and Andersen (1996,[19]).

Commercial implementations of interior-point methods use *cross-over techniques*, i.e., at some time controlled by the termination criterion described in Sect. 3.8.5.5, the algorithm switches from the interior-point method to the Simplex algorithm. Crossing-over starts with the basis identification providing a good feasible initial guess for the Simplex algorithm to proceed. The Simplex algorithm improves this guess quickly and produces an optimal basic solution.

Let us conclude with the remark that cross-over is not a trivial task and the experience of software developers may greatly influence the efficiency.

3.8.5.7 Interior-Point Versus Simplex Methods

At the moment the best Simplex algorithms and the best interior-point methods are comparable. It depends on the problem which procedure is the more efficient. For some benchmarks and test runs both with interior-point methods and sophisticated Simplex algorithms, we refer the reader to Arbel (1994,[30]). Some general characteristics can be given, Nemhauser (1994,[420]) in the following points.

The Simplex algorithm needs many iterations, but these are very fast. The number of the iterations grows approximately linearly with the number of constraints and logarithmically in the number of variables.

While for the Simplex algorithm, there are a theory and a clear understanding of the complexity of the method, there is no clear concept on the worst-case complexity of interior-point methods. A theoretical bound for the number of iterations

$$\mathcal{O}\left(\sqrt{n}\ln\frac{1}{\varepsilon}\right) \tag{3.8.77}$$

is not in good agreement with the number

$$\mathcal{O}\left(\ln n\right) \quad \text{or} \quad \mathcal{O}\left(n^4\right) \tag{3.8.78}$$

observed in practice. Interior-point methods usually need about 20 to 50 iterations; this number grows weakly with the problem size as described by (3.8.78). Every iteration requires the solution of an $n \cdot n$ system of nonlinear Equations, which is quite costly. This system is linearized. Thus the central computing consumption for a problem with n variables is the solution of an $n \cdot n$ linear equation system. That is why it is essential for the success of the IPM and that *this* system matrix is sparse.

Although problem dependence plays an essential role for the valuation of the efficiency of Simplex algorithms and IPMs, the IPMs seem to have advantages for large, sparse problems.

Especially for big systems, hybrid algorithms seem to be very efficient. In the first phase, these determine a nearly optimal solution with the help of an IPM, viz. determine a solution near the polyhedra edge. In the second phase, "purification" pivoting procedures are used to create a basis. Finally, the Simplex algorithm uses this basis as an initial guess and finally iterates to the optimal basis.

As with the B&B method, IPMs still have a disadvantage, because they do not allow an efficient warm-start (see Sect. 9.2.1 for definition) iterating in a few steps from an initially infeasible point to an optimal feasible one. So for that reason, their use is also limited to hybrid techniques in which they are used to provide good initial points and then cross-over to the Simplex algorithm.

3.8.6 Branch and Bound with LP Relaxation

Although the idea of the Branch-and-Bound (B&B) algorithm first developed by Land & Doig (1960,[349]) has already been briefly sketched at several places in this book, in this appendix, we summarize the method and present a more technical approach explaining some relevant strategies and control parameters being used in B&B.

The *branch* in B&B hints at the partitioning process used to produce solutions or to prove the optimality of a solution. Lower and upper *bounds* are used during this process to avoid an exhaustive search in the solution space. The B&B idea or *implicit enumeration* characterizes a wide class of algorithms which can be applied to discrete optimization in general.

Figure 3.9 summarizes the computational steps of the B&B algorithm. After some initialization, the LP relaxation — that is that LP problem that results if we relax all integer variables to continuous ones — establishes the first node. The node selection is obvious in the first step (just take the LP relaxation), but later on it is based on some heuristics (see Sect. 9.4.2.2). A B&B algorithm of Dakin (1965,[137]) with LP relaxations uses three *pruning criteria*: infeasibility, optimality, and value dominance relation. In a maximization problem the integer solutions found lead to an increasing sequence of lower bounds, z^{IP}, while the LP problems in the tree decrease the upper bound, z^{LP}. Note that α denotes an *addcut*, which causes the algorithm to accept a new integer solution only if it is better by at least the value of α. If the pruning criteria fail branching starts, the branching in this algorithm is done by variable dichotomy: for a fractional y_j^*, two child nodes are created with the additional constraint $y_j \leq \lfloor y_j^* \rfloor$, respectively, $y_j \geq \lfloor y_j^* \rfloor + 1$. Other possibilities for dividing the search space are, for instance, generalized upper bound dichotomy or enumeration of all possible values, if the domain of a variable is finite [104, 421]. The advantage of variable dichotomy is that only simple lower and upper bounds are added to the problem. In Sect. 3.8.3 we have shown why bounds can be treated much easier than general constraints.

The *selection of nodes* plays an important role in implicit enumeration; widely used is the *depth-first* plus backtracking rule as presented above. If a node is not pruned, one of its two sons is considered. If a node is pruned, the algorithm goes back to the last node with a son that has not yet been considered (backtracking). In linear programming only lower and upper bounds are added, and in most cases the dual Simplex algorithm (see Sect. 3.8.4) can reoptimize the problem directly without data transfer or basis re-inversion [421]. Experience has shown [104] that it is more

Branch & Bound
(with LP relaxation)

(✱) max $c^T x + d^T y$; x continuous
s.t. $Ax + By = b$; $x, y \geq 0$; y integer

① **Initialization**
$k = m = 0$, $z_k^{IP} = -\infty$, $z_m^{LP} = +\infty$, node-list = {0}

② **Define LP relaxation** (call this node 0)
as (✱) but y is relaxed: y continuous

③ **Node selection**
if node-list = { } goto ⑥
if node-list ≠ { } apply heuristic to select node n

④ **Solve LP problem** associated with node n
↳ check status, z^n, x^n, y^n
if status = "infeasible" then remove node from node-list; goto ③
if status = "feasbile" then
if y^n integral then
if $z^n > z_k^{IP} + \alpha$ then $z_{k+1}^{IP} = z^n$, set k = K+1
remove node from node-list, goto ③
end if
if $z^n < z_k^{IP} + \alpha$ then remove node from node-list ; goto ③
update z_m^{LP}: if $m = 0$ then $z_1^{LP} = z^n$, $m = 1$
if $m > 0$ then $z_{m+1}^{LP} = \max z\,(LP \mid LP \in \text{node-list})$
set m = m+1
goto ⑤

⑤ **Variable selction**
- insect current LP solution and choose a fractional y_j
- generate two subproblems (nodes)
$LP_{current} \oplus y_j \leq \text{int}(y_j)$ and $LP_{current} \oplus y_j \geq \text{int}(y_j) + 1$
- goto ③

⑥ **Solution Output**
- no integer solution found
- or print-out the solution
- stop

Fig. 3.9 The Branch-and-Bound algorithm

likely that feasible solutions are found deep in the tree. Nevertheless, in some cases the use of the opposite strategy, breadth-first search, may be advantageous.

Another important point is the *selection of the branching variable*. A common way of choosing a branching variable is by user-specified priorities, because no robust general strategy is known. Degradations or penalties may also be used to choose the branching variables, both methods estimate or calculate the increase of the objective function value if a variable is required to be integral, especially penalties are costly to compute in relation to the gained information so that they are used quite rarely [421].

The B&B algorithm terminates after a finite number of steps. Termination occurs when the node list becomes empty. In that case the result is either the optimal integer feasible solution or the message that the problem does not have any integer feasible solution. In practice, it happens very often that the user does not want to wait until

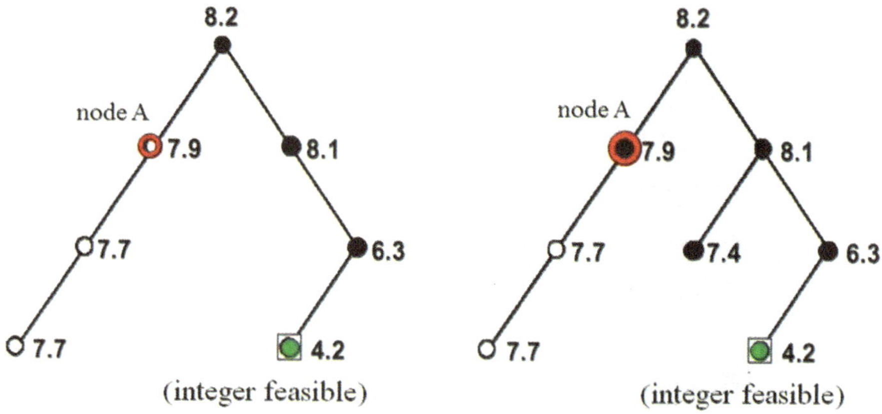

Fig. 3.10 Branch-and-Bound tree: Updating the value of the LP relaxation. The number beside each node denotes the value of the LP relaxation; open circles represent active nodes. The nodes of the tree are evaluated in the sequence 8.2-8.1-7.4-6.3-4.2-7.9 (node A). The absolute integrality gap δ before evaluating node A is $\Delta = 8.2 - 4.2 = 4$ (left part of the figure). Once node A has been evaluated (right part), we get $\Delta = 7.9 - 4.2 = 3.7$

the node list becomes empty but wants to stop after one, or several integer solutions have been found. If an integer feasible solution has been found, the upper and lower bounds mentioned above may be used to estimate the quality of the solution. Let us see how that works: during the B&B for a maximization problem, we know that

$$z^{LP} \geq z^* \geq z^{IP}, \tag{3.8.79}$$

where z^* is the (unknown) value of the best integer solution, z^{IP} (possibly $-\infty$) is the value of the best integer solution found so far in the search and $z^{LP} = \max_i\{z_i^{LP}\}$, where z_i^{LP} is the optimal value of the LP relaxation at *active* node i (nodes that have been fathomed are not considered). For example, in Fig. 3.10 (left part) when an integer feasible solution has been found, $z^{IP} = 4.2$, and $z^{LP} = 8.2$, and thus $\Delta = 8.2 - 4.2$. Once the value of the LP relaxation of node A is known (right part of Fig. 3.10), the maximum of all LP relaxations at active nodes is 7.9, yielding a reduced gap of $\Delta = 3.7$.

We have found the optimal solution to our MILP problem if $\Delta = z^{LP} - z^{IP} \leq \Delta^{\mathrm{a}}$, with some tolerance Δ^{a} on the absolute gap. So it can be seen that a criterion for node selection in B&B is to reduce the integrality gap Δ. One way to do this is to select a node which has a good chance of yielding a new integer feasible solution better than the current z^{IP}. Another way is to branch on the node having the largest z_i^{LP} on the grounds that a descendant of this node will certainly have no higher a value of z_i^{LP} and probably will have a lower value, in which case z^{LP} will be smaller.

Chapter 4
Problems Solvable Using Linear Programming

In this chapter a number of standard LP problems will be formulated. These are problems that occur frequently and may provide ways of formulating parts of the users' problems, to which other conditions, e.g., integrality conditions or nonlinear terms, will be added, and secondly they will give an indication as to the ease or difficulty of solving a particular problem when it becomes apparent that it is similar to some standard problem. For some of the problem types a case study is also provided and a model relating to the case is supplied with the software included with the book.

4.1 Cutting Stock: Trimloss Problems

Cutting stock or trimloss problems in various industries consist of cutting a fixed number of patterns defined by the length and width of certain materials from continuous, finitely long rolls or finitely long pallets of paper, textile, glass, wood, or metal of known width at right angles, thereby minimizing the unusable and superfluous material. Models for solving trimloss problems were developed by Gilmore & Gomory (1961,[220]) and subsequently by numerous authors, especially Dyson & Gregory (1974,[167]). Applications in the paper industry are, e.g., described by Vajda (1961,[556]) or Kallrath et al. (2014,[319]), and applications in the glass industry by Dyson & Gregory (1974,[167]). In Farley (1991,[176]) a model for cutting photo paper is described; finally in Sect. 13.2 an example can be

Supplementary Information The online version contains supplementary material available at https://doi.org/10.1007/978-3-030-73237-0_4.

J. Kallrath, *Business Optimization Using Mathematical Programming*, International Series in Operations Research & Management Science 307, https://doi.org/10.1007/978-3-030-73237-0_4

found in which not only the number of patterns but also the design of the patterns is determined by the optimization model.

4.1.1 Example: A Trimloss Problem in the Paper Industry

A paper roll of infinite length and a width of 20 meters is given as well as customer demand of 30×5, 30×7 and 20×9 m as defined by $L \times B$. With regard to length, no interruptions are possible, with regard to width, very well (Fig. 4.1).

To model this problem, it must first be decided which combinations can be cut from the roll. For example, it would be possible to cut two layers of width 7m and one layer of width 5m at the same time; the rest of width 1m cannot be used any further. The determination of the set of all combinations is a problem in itself and should not be described here; all permissible combinations c have in common that they contain a waste W_p that cannot be used further. In the present case the six patterns result to be read column-wise (permutations of these patterns — e.g., 7,5,7 — are not listed):

p	1	2	3	4	5	6
	5	5	5	5	7	9
	5	5	5	7	9	9
	5	7	9	7		
	5					
W_p	0	3	1	1	4	2

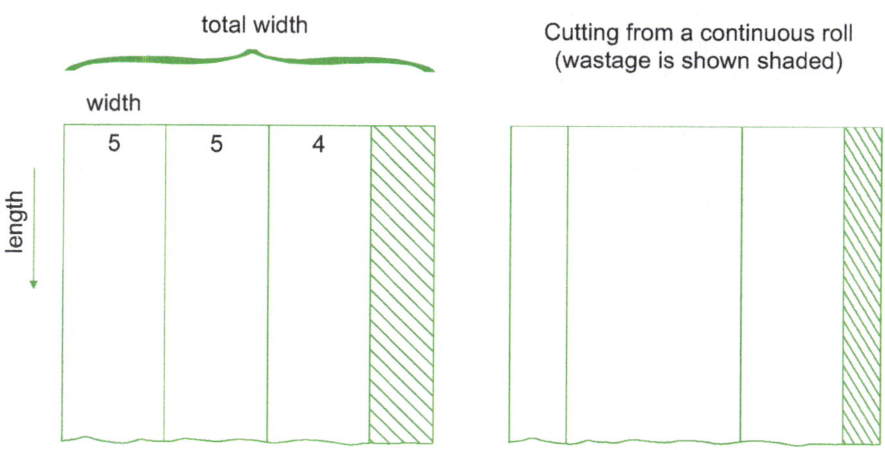

Fig. 4.1 Geometry of a trimloss problem. The length of the rolls is assumed to be infinite. The figure shows two patterns cut from the master roll of width 17 dm

The occurring widths 5, 7, and 9 are indicated below with w ($w = 1, 2, 3$), the patterns with p ($p = 1, 2, \ldots, 6$). First of all, we introduce the non-negative, continuous variables

$$x_p \geq 0 : \text{Length of roll to be cut when using pattern } p$$
$$s_w \geq 0 : \text{Excess length of cut when using width } w.$$

The objective function is the minimization of the total waste

$$\min \quad (3x_2 + x_3 + x_4 + 4x_5 + 2x_6) + (5s_1 + 7s_2 + 9s_3). \tag{4.1.1}$$

The terms in the first parenthesis describe the direct cut given by the combination, and the terms in the second parenthesis the cut resulting from the excess length; the first two combinations must be cut with a length of 30,000 m, combination $c = 3$ with a length of 20,000 m.

As the first condition we note: The total length of the 5 m wide cuts (these are only possible in the first four combinations) must be at least 30,000 m. Since some combinations result in several pieces of a certain length and surplus lengths may occur, this is expressed by the equality

$$4x_1 + 2x_2 + 2x_3 + x_4 = 30000 + s_1.$$

Analogously, the conditions regarding the total length of the 7 m widths (here also 30,000 m are required) and the 9 m widths (here 20,000 m are required) are as follows:

$$x_2 + 2x_4 + x_5 = 30000 + s_2$$

and

$$x_3 + x_5 + 2x_6 = 20000 + s_3.$$

This model formulation is suitable for the current, relatively small problem with solution $z = 35000$, $x_1 = 3750$, $x_4 = 15000$, and $x_6 = 10000$ or also $x_3 = 7500$, $x_4 = 15000$, and $x_6 = 6250$; however, it leads to difficulties in real industrial applications if the number of combinations and thus the number of variables become very large.

Instead of (4.1.1), we could also minimize the total cut length

$$x_1 + x_2 + x_3 + x_4 + x_5 + x_6 \quad ; \tag{4.1.2}$$

this leads to the same solution. The reason for this is the observation that the total cut length must cover the demand plus the waste. Since the required quantity is fixed, minimizing the total cut length of all patterns is equivalent to minimizing the waste; in the next section we show this algebraically for a similar situation.

4.1.2 Example: An Integer Trimloss Problem

This chapter is dedicated to issues that can be addressed within the framework of the LP. However, as the integer problem at hand is very similar to the one in Sect. 4.1.1, we allow an exception and treat it here; it also shows how quickly an LP problem can become an MILP problem. In the present case, 70 cm wide paper rolls in three widths of 12, 20, and 22 cm are to be cut; permitted patterns are known again. The following weekly demand for rolls is to be satisfied:

w		1	2	3
B_w	width [cm]	12	20	22
D_w	demand	10	13	8

How should the master roll be cut if one wants to minimize waste? In the model, the indices

$$w \in \mathcal{W}\ ,\ \mathcal{W} = \{1(12\text{ cm}), 2(20\text{ cm}), 3(22\text{ cm})\} : \text{set of widths}$$
$$p \in \mathcal{P}\ ,\ \mathcal{P} = \{1, 2, \ldots, P\} \qquad : \text{set of patterns}$$

are used. The model requires the following input data

D_w : demand for rolls of width w
W_p : trimloss (in cm) of pattern p
N_{wp} : the number of rolls of width w, we obtain from a single 70 cm roll. master roll when cut to pattern p
B_w : width (in cm) of roll w

How should the master roll be cut if you want to minimize waste? In the model, the indices are

	p	1	2	3	4	5	6	7	8	9	10	D_w
B_w	w											
12	1	5	4	4	2	2	2	0	0	0	0	10
20	2	0	1	0	2	1	0	3	2	1	0	13
22	3	0	0	1	0	1	2	0	1	2	3	8
W_p		10	2	0	6	4	2	10	8	6	4	

The patterns — they are constructed that way — fulfill the condition

$$\sum_{w \in \mathcal{W}} N_{wp} B_w \leq B_{\text{MR}} \quad , \quad \forall p,$$

where B_{MR} denotes the width of the stock rolls, in this case 70 cm.

Now two formulations seem reasonable: (A) If unused widths can be returned to the warehouse (this is usually the case for standard sizes that are frequently used), the waste to be minimized is $\sum_{p\in\mathscr{P}} W_p\alpha_p$, and the requests are to be fulfilled

$$\sum_{p\in\mathscr{P}} N_{wp}\alpha_p \geq D_w \quad , \quad \forall w$$

with the integer variables $\alpha_p \in \{0, 1, 2, \ldots\}$ that describe how many rolls are cut according to pattern p.

(B) If unused widths cannot be reused, they shall also be considered as waste and the objective function to be minimized may be reduced simply to the number

$$z = \sum_{p\in\mathscr{P}} \alpha_p .$$

The master rolls to be used follow as

$$\begin{aligned}\sum_{p\in\mathscr{P}} W_p\alpha_p + \sum_{w\in\mathscr{W}} B_b\left[\sum_{p\in\mathscr{P}} N_{wp}\alpha_p - D_b\right] &= \sum_{p\in\mathscr{P}}\left(W_p + \sum_{w\in\mathscr{W}} B_w N_{wp}\right)\alpha_p - C \\ &= \sum_{p\in\mathscr{P}} B_{\mathrm{MR}}\alpha_p - C \\ &= B_{\mathrm{MR}}\sum_{p\in\mathscr{P}} \alpha_p - C .\end{aligned}$$

The constants — B_{MR}, the width of the master roll, and $C := \sum_{w\in\mathscr{W}} B_w D_w$ — can be neglected as they lead to the same optimal solution. This has the advantage that the objective function value is integer-valued, which allows us to select a minimal improvement of $\alpha = 1$ as described in Sect. 10.2.2, which accelerates the B&B algorithm significantly. For this problem and the data specified above, we obtain the objective function value $z = 9$ and, for instance, the pattern multiplicities $(\alpha_1, \alpha_7, \alpha_9) = (2, 3, 4)$, $(\alpha_1, \alpha_7, \alpha_9, \alpha_{10}) = (2, 4, 1, 2)$, $(\alpha_1, \alpha_7, \alpha_8, \alpha_{10}) = (2, 1, 5, 1)$, or $(\alpha_3, \alpha_4, \alpha_7, \alpha_{10}) = (2, 1, 4, 2)$. The solution is not unique. From a practical point of view, the first solution with only three patterns is preferable because it minimizes waste *and* the number of pattern changes (resulting in less work in changing the knife positions).

4.2 The Food Mix Problem

The food mix problem concerns the blending together of quantities of ingredients to make a final blend. Restrictions are placed on the relative composition of the ingredients in the final product. Let us illustrate this by the *Manufacturing Foods* model based on Williams & Redwood (1974,[582]) who describe an LP model used to optimize the production of foods by blending oils. Each food type is required to be no more than a certain proportion of the total of all foods produced. The following formulation of the constraints was used with indices

$$
\begin{aligned}
i &= 1, 2, \ldots, n \ : \text{the set of ingredients} \\
f &= 1, 2, \ldots, m \ : \text{the set of food constituents}
\end{aligned}
$$

data

$$
\begin{aligned}
A_{fi} &: \text{proportion of ingredient } i \text{ that provides food constituent } f \\
B_f &: \text{maximum proportion of the total food that can} \\
&\quad \text{be food constituent } f
\end{aligned}
$$

and continuous variables

$$
\begin{aligned}
x_i &\geq 0 : \text{the quantity (volume) of ingredient } i \text{ used} \\
y &\geq 0 : \text{the total quantity of food produced.}
\end{aligned}
$$

The formulation is then expressed in the two following constraints: for each type of food constituent, the requirements must be satisfied by adding together the available ingredients

$$
\sum_{i=1}^{n} A_{fi} x_i - B_f y \leq 0 \quad , \quad f = 1, 2, \ldots, m,
$$

and because there is no loss of mass in the blending process, the final product comprises the sum of the ingredients by volume

$$
\sum_{i=1}^{n} x_i - y = 0.
$$

4.3 Transportation and Assignment Problems

4.3.1 The Transportation Problem

The transportation problem is the problem of sending quantities of identical goods from n different origins, such as factories, to m different destinations such as storage warehouses or customers. Let us use the indices

$$
\begin{aligned}
f &= 1, 2, \ldots, n \ : \text{the set of factories} \\
w &= 1, 2, \ldots, m \ : \text{the set of warehouses}
\end{aligned}
$$

Let the quantity of the goods required at each warehouse be D_w (the demands) and the quantity of the goods available at each factory be S_f (the supplies) and let C_{fw} be the cost of sending one unit of the good from factory f to warehouse w. The total cost of transporting goods is to be minimized. Each of D, S, and C is assumed to be non-negative. Initially we shall assume that the total supply and total demand exactly match, i.e.,

$$\sum_{f=1}^{n} S_f = \sum_{w=1}^{m} D_w .$$

The main decisions must be how much of the good is sent from a particular factory to a particular warehouse, so we introduce the decision variable $x_{fw} \geq 0$ to represent the amount to be sent from factory f to warehouse w.

The objective function

$$\min \quad \sum_{f=1}^{n} \sum_{w=1}^{m} C_{fw} x_{fw} \tag{4.3.1}$$

is built up by multiplying each unit delivery cost with the quantity of the good incurring that cost, and summing over all goods. The model has to obey the constraints

$$\sum_{f=1}^{n} x_{fw} = D_w \quad , \quad w = 1, 2, \ldots, m, \tag{4.3.2}$$

i.e., the total quantity sent from all factories to warehouse w must match its requirement D_w ; and

$$\sum_{w=1}^{m} x_{fw} = S_f \quad , \quad f = 1, 2, \ldots, n, \tag{4.3.3}$$

i.e., the total quantity sent to all warehouses from factory f must match its supply S_f.

Note that in (4.3.2) we may replace = by $\geq$, as we know it will be uneconomical for the optimal solution to supply too much, and in (4.3.3) we may replace = by $\leq$ as we know that in the optimal solution as much will be supplied as is demanded. These changes to inequalities rather than equations become more appropriate when the total supply and total demand are not equal. Then the problem becomes one of supplying as much as possible, at minimum cost, of what has been demanded, or not using all the supplies that are available.

The basic transportation problem was introduced by Kantorovich (1939,[322]), Hitchcock (1941,[269]), and Koopmans (1947,[340]) and solved by the Simplex algorithm by Dantzig (1951,[139]). It was later solved by the *stepping-stone algorithm* of Charnes and Cooper (1954,[113]), later improved on by Dennis (1958,[155]) and Reinfield & Vogel (1958,[460]). As with the assignment problem (see Sect. 4.3.3), the transportation problem does not need to be solved by LP methods, and the algorithm of Charnes & Cooper (1954,[113]) solves large problems easily. If the problem is posed as an LP problem and solved by LP software without recourse to ILP software, the optimal solution will again provide all variables with integer values provided D and S are integers. The reason for this is that the set of coefficients, which are all 0s or 1s, forms what is termed a totally unimodular coefficient matrix, which guarantees that an LP solution is an ILP solution. Unimodularity will be discussed further in Sect. 4.5.

Many variations on the basic structure of the transportation problem exist, as the following examples indicate:

1. The objective function may be maximize, for instance if the problem concerned maximizing the total satisfaction of groups of people who are being supplied with some service. In this type of problem, C_{fw} is no longer a cost but instead measures gain, contribution, or satisfaction.
2. By replacing = in the matching of the total supply to demand, and in (4.3.2) and (4.3.3) by using any possible combinations of $\leq, \geq$ or =, and by varying the objective function, Appa (1973,[27]) produced 54 variations on the basic structure of the transportation problem. Most of these have obvious practical application. All variations of the basic structure lead to totally unimodular coefficient matrices, as before.
3. Routes for delivery from one particular factory to one particular warehouse may be disallowed by associating a large cost with that route, when the problem involves the minimization of total cost, or by omitting the appropriate variable from the formulation. (For a maximization problem a zero gain can be associated with a particular route to discourage it. In order to ensure that a particular route is used maximally cost/gain should be set to encourage this route. Such artificial cost/gain values can later be removed when an optimal solution has been found and the value of that solution calculated after reinstating the true cost/gain values.)

Three problems that can be formulated as transportation problems, described by Appa (1973,[27]) but attributed by him to others, are now summarized:

1. Purchase/storage problem. Goods needed in specified amounts in each of a series of time periods at lowest total cost can be provided by purchase or from quantities obtained and stored earlier. The amount purchased and the amount stored, in each period, represent the sources of supply. The amount used directly and the amount put into storage, in each period, represent the demands. Storage is an additional cost and is added cumulatively to supplies of goods that are to be used, or could be used, in later periods. As goods obtained or stored in a later period cannot be supplied or put into store in an earlier period many routes have to be barred.
2. Catering/laundry problem. A caterer wishes to supply fresh napkins at cheapest total cost for use at functions over a series of days. Different numbers of napkins

are required on each day. Napkins can be purchased, at known cost, as required, but, in addition, soiled napkins can be laundered and provide a source of supply for later days. Napkins can either be laundered, at known cost, and returned within 48 h or within 24 h, at higher cost. In this problem there are three types of source of supply in each period — new napkins, quickly washed napkins, and slowly washed napkins, the last two options not always being available. The demand in each period has three dimensions — demand for direct use, demand for 24 h laundering for later use, and demand for 48 h laundering for later use. Clearly napkins cannot go backward in time, so certain routes will be barred. The problem is related to the previous one but has the new variation that the same napkins may possibly be used as both a source of supply and a demand many times over.

3. Rota problem. Personnel providing 24 h coverage can start shifts at points spaced equally throughout the day, e.g., at 4-h intervals. Shifts last an integer multiple of these intervals, e.g., 8 h. Different numbers of personnel are required at different times of the day. The problem is to use the minimum total number of personnel over a 24 h period to provide complete coverage. In each interval, the sources of supply will be the number of personnel starting in that interval and those available from an earlier interval. As in the other two problems time must not be allowed to run backward, so appropriate routes will be barred.

4.3.2 The Transshipment Problem

The transshipment problem may be regarded as a 2-stage transportation problem with goods being shipped from source to an intermediate destination before reaching the final destination. Let us assume that each item must be shipped from a factory to a distribution center and then from a distribution center to a warehouse. As with the transportation problem, this problem can be solved by a special purpose method and is a totally unimodular problem when supplies and demands are integers, so produces integer solutions when solved by LP methods.

A typical problem may be stated and formulated as follows. Let us extend the problem given by (4.3.1)–(4.3.3) by introducing an index

$$d = 1, 2, \ldots, r \text{ : the set of distribution centers} \tag{4.3.4}$$

Now we redefine our variables as

$$x_{fd} = \text{ quantity of the good sent from factory } f \text{ to distribution center } d \tag{4.3.5}$$

and we introduce the variables

$$y_{dw} \geq 0 \quad \begin{array}{l}\text{quantity of the good sent from distribution center } d\\ \text{to warehouse } w.\end{array} \tag{4.3.6}$$

We require the constraints

$$\sum_{d=1}^{r} x_{fd} = S_f \quad , \quad \forall f \tag{4.3.7}$$

$$\sum_{d=1}^{r} y_{dw} = D_w \quad , \quad \forall w \tag{4.3.8}$$

$$\sum_{f=1}^{n} x_{fd} = \sum_{w=1}^{m} y_{dw} \quad , \quad \forall d. \tag{4.3.9}$$

Let C_{fd} be the cost of sending one unit of the good from factory f to distribution point d and C'_{dw} be the cost of sending one unit of the good from distribution point d to warehouse w. The objective function is

$$\min \quad \sum_{f=1}^{n}\sum_{d=1}^{r} C_{fd}x_{fd} + \sum_{d=1}^{r}\sum_{w=1}^{m} C'_{dw}y_{dw}. \tag{4.3.10}$$

Figure 4.2 shows routes on a transportation network and on a transshipment network.

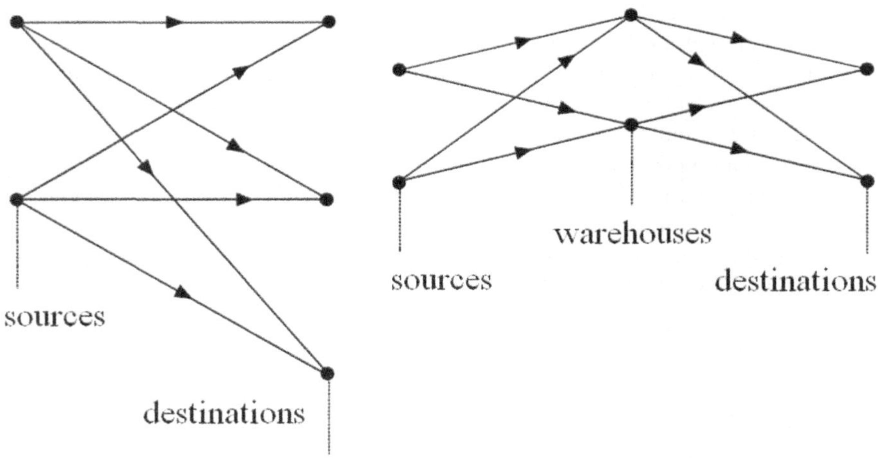

Fig. 4.2 Routes on transportation network (left) and transshipment network (right)

4.3.3 The Assignment Problem

The *assignment problem* is the problem of assigning a number of tasks to a number of persons to maximum advantage. In the problem the (tasks, persons) pair may be replaced by (objects, locations) and a variety of other pairs, and the order within the pair is not important. In addition, the notion of "maximum advantage" may in fact be to minimize cost or some other objectives. Thus the assignment problem has some similarities to the transportation problem but differs principally for the reason that the flows from source to destination are replaced by a single item or person. The formulation of a typical problem will now be given.

Let there be m tasks and n persons, $n \geq m$, indexed by

$$\begin{aligned} t &= 1, 2, \ldots, m \; : \text{the set of tasks} \\ p &= 1, 2, \ldots, n \; \; : \text{the set of persons.} \end{aligned} \qquad (4.3.11)$$

Let R_{tp} be the profit if task t is assigned to person p. The main decisions must refer to how the tasks are assigned to persons, so we introduce a binary decision variable, δ_{tp},

$$\delta_{tp} := \begin{cases} 1 \text{ , if task } t \text{ is assigned to person } p \\ 0 \text{ , otherwise.} \end{cases} \qquad (4.3.12)$$

If we want to relax the integrality condition we would incorporate these variables into an LP model by requiring

$$0 \leq \delta_{tp} \leq 1 \quad , \quad \forall \{tp\}. \qquad (4.3.13)$$

Then we require two types of constraints (a) to assign each task to a person and (b) to ensure each person is assigned at most one task. These constraints are modeled as follows:

$$\sum_{t=1}^{m} \delta_{tp} \leq 1 \quad , \quad \forall p \qquad (4.3.14)$$

ensuring that for person p the number of all the possible tasks assigned must be at most one. Conversely, the equations

$$\sum_{p=1}^{n} \delta_{tp} = 1 \quad , \quad \forall t \qquad (4.3.15)$$

ensure that for a task t the number of all the possible people to whom it is assigned must be exactly one. This constraint actually ensures that task t is assigned. Since we assumed that $n \geq m$, it is possible to do so.

The objective function is

$$\max \quad \sum_{t=1}^{m} \sum_{p=1}^{n} R_{tp}\delta_{tp}.$$

Clearly each δ_{tp} variable will be meaningless unless it takes on a value of 0 (the task is not assigned) or 1 (the task is assigned).

The assignment problem is a special case of the transportation problem where all supplies and demand are equal to one, and so can be solved as if it were an LP. The assignment problem was first solved by the *Hungarian algorithm* of Kuhn (1955,[345]), which was improved on by Jonker & Volgenat (1986,[293]) and Wright (1990,[589]). An alternative method for solving the problem is presented in Hung & Rom (1980,[276]). Further methods of solving the problem are discussed in Burkard & Derigs (1980,[105]) and computer codes are contained in Carpaneto et al. (1988,[109]). In fact, the assignment problem is normally solved by special purpose software, but clearly there is no harm, other than relative slowness, in using a general purpose LP solver.

More complex problems of allocation eventually lead to models that are not solvable by LP approaches. A typical example is a model to allocate school bus contracts described by Letchford (1996,[359]).

4.3.4 Transportation and Assignment Problems as Subproblems

When considering efficient problem solving, it is important to note that if an MILP model is formed from an assignment, transportation, or transshipment problem plus only a few other conditions, then it should solve rapidly and possibly produce integer solutions to appropriate variables without recourse to B&B. But the problem types tend not to occur in isolation as they are too simplistic, but occur in conjunction with other model features. They have been discussed in detail to develop principles of modeling rather than to illustrate complete practical models.

4.3.5 Matching Problems

The *perfect matching problem* is a variation of the assignment problem. It occurs when tasks have to be assigned to individuals, with two individuals being assigned to each task. The objective of the problem is to create the teams of two people assigned to each task in such a way as to maximize some gain, such as speed of completion of tasks or a monetary gain. The problem may be formulated as follows.

Suppose $2n$ people p must be assigned to n tasks, with exactly two people assigned to each task. Let $G_{pp'}$ be the gain if person p is assigned to the same task as person p' and assume $p < p'$. This restriction on the index combination avoids counting the same people assigned to one task twice because otherwise we would have $G_{pp'} = G_{p'p}$.

We introduce the following binary variables:

$$\delta_{pp'} := \begin{cases} 1 \text{ , if person } p \text{ is assigned to the same task as person } p' \\ 0 \text{ , otherwise} \end{cases}$$

for the following combination of indices

$$\begin{aligned} &p < p' \\ &p = 1, 2, \ldots, 2n - 1 \\ &p' = 2, 3, \ldots, 2n. \end{aligned}$$

The formulation is then to maximize the total gain provided by the assignment of people to tasks, given by

$$\max \quad \sum_{p=1}^{2n-1} \sum_{p'=p+1}^{2n} G_{pp'} \delta_{pp'}$$

subject to

$$\sum_{k<l} \delta_{kl} + \sum_{j>l} \delta_{lj} = 1 \quad , \quad l = 1, 2, \ldots, 2n - 1. \tag{4.3.16}$$

These constraints ensure that for each person l one other person, k if the person has a lower number than l in the numbering sequence or j if the person has a higher number than l in the numbering sequence, is assigned to the same task as l.

If the equality constraints (4.3.16) are replaced by $\leq$ constraints, the problem is described as the *matching problem*. An application of matching problems occurs embedded within a larger model in a vehicle and manpower scheduling model in Blais et al. (1990,[84]).

4.4 Network Flow Problems

A *network flow problem* is a problem that may be modeled as a *graph* consisting of *nodes* and *arcs* connecting pairs of nodes; the terms graph, nodes, and arc are explained below; sometimes, people also refer to nodes as *vertices*. Many problems can be modeled as network flow problems. Network problems are among the easiest models to solve and they are easy to describe with a graph or a picture.

Their basic concepts are arc or edges and nodes representing for instance railway lines and stations. In network flow problems the basic model can be thought of as comprising flows (of material, for instance) between points (e.g., flows from warehouse to customer). They can be solved by simple or by specialized Simplex algorithms. General network flow problems may be solved by special purpose codes as described in Bradley et al. (1977,[97]) or Mulvey (1978,[411]). Such codes solve problems many times more rapidly than the Simplex algorithm would on the equivalent LP problem.

4.4.1 Illustrating a Network Flow Problem

A node is defined to be a point on a graph and an arc to be a directed line segment joining two nodes. The arcs are considered to have a "direction," but it is not necessary for every pair of nodes to be directly connected. Material can flow along an arc and there is an associated unit cost for the flow. At all nodes material balance is preserved, i.e., the amount of material flowing in balances the amount of material flowing out. At some nodes, called *sources*, there is a specified flow in, and at some nodes, called *sinks*, there is a specified flow out. For example, in the transportation problem a supplier could represent a source and a customer could represent a sink. The network flow problem is then the problem of satisfying material flow requirements at each node such that the total cost of material flow is minimized. Figure 4.3 shows a typical small problem with 10 nodes and 15 directed arcs.

The problem would be formulated in LP terminology as

$$
\begin{aligned}
\min \quad & 2x_{13} + 2x_{14} + 3x_{24} + 5x_{25} + 2x_{34} \\
& +5x_{38} + 4x_{45} + 2x_{47} + 3x_{5,10} + 3x_{63} \\
& +5x_{68} + 2x_{69} + 3x_{76} + 2x_{79} + 5x_{7,10}
\end{aligned}
\tag{4.4.1}
$$

subject to material flow conservation at all the nodes, which is given, node by node, by

$$
\begin{array}{rrcl}
1: & x_{13} + x_{14} & = & 10 \\
2: & x_{24} + x_{25} & = & 10 \\
3: & x_{34} + x_{38} & = & x_{13} + x_{63} \\
4: & x_{45} + x_{47} & = & x_{14} + x_{24} + x_{34} \\
5: & x_{5,10} & = & x_{25} + x_{45} \\
6: & x_{63} + x_{68} + x_{69} & = & x_{76} \\
7: & x_{76} + x_{79} + x_{7,10} & = & x_{47} \\
8: & x_{38} + x_{68} & = & 5 \\
9: & x_{69} + x_{79} & = & 10 \\
10: & x_{7,10} + x_{5,10} & = & 5
\end{array}
\quad . \tag{4.4.2}
$$

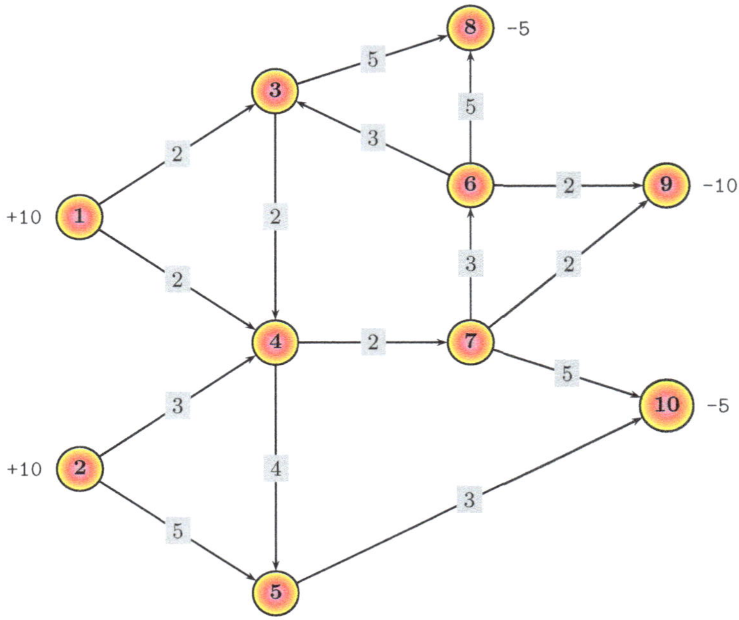

Fig. 4.3 Flows between nodes of a network. Costs are shown on the directed arcs connecting nodes (numbers in circles). Supplies are indicated to the left (+10,+10) as inflows to nodes 1 and 2, and demands to the right (−5, −10, −5) as outflows to nodes 8, 9, and 10

It is recommended that the reader now tries to solve the above problem (see Exercise 4.7.3).

4.4.2 The Structure and Importance of Network Flow Models

Network flow problems are important for another reason: Depending on the constraints and coefficients in the constraints, they form a totally unimodular matrix and then they belong to the class of *unimodular*[1] problems. These problems have the property that although they are apparently ILP or MILP problems they can in fact be solved without recourse to algorithms other than LP, i.e., when solved as LP problems they produce integer solutions. Typical network flow problems such as the *assignment* problem or the *transportation* problem are unimodular. These problems may occur as substructures in models; they are described in Sect. 4.3.

It will be useful to be aware of when models have the network flow structure or if they can be turned into network flow models. Apart from problems such as

[1] See Sect. 4.5 for further details.

the assignment, transportation, and transshipment, a wide class of LP problems are convertible to network flow problems. Work by Bixby and Cunningham (1980,[82]) and Baston *et al.* (1991,[51]) describes methods of determining if an LP problem may be converted into a network flow problem and provides methods to do so. It is also worthwhile to check whether a model is similar to a network flow model, perhaps with a small number of extra conditions. This will also be discussed further in Chap. 4.

4.4.3 Case Study: A Telephone Betting Scheduling Problem

Wilson & Willis (1983,[585]) provide an example of a telephone betting scheduling problem that is convertible to a network flow problem. Their problem and its formulation will now be discussed in more detail. They describe the essential problem as "How should telephone operators be assigned to shifts on each race day so that customer demand is met, the operating requirements are satisfied and the total cost to the Totalizer Agency Board (Victoria, Australia) is a minimum?" Let us provide a structured model formulation starting with the indices

$$\begin{aligned} s &= 1, 2, \ldots, m : \text{the set of shifts} \\ t &= 1, 2, \ldots, n : \text{the set of time periods.} \end{aligned} \tag{4.4.3}$$

The data are given by

$$\begin{aligned} A_{st} &: \text{1 if shift } s \text{ covers time period } t \text{, 0 otherwise} \\ D_t &: \text{operator demand for 15 min period } t \\ C_s &: \text{cost per operator of shift } s \\ M_t &: \text{maximum number of operators available for period } t \\ Q_{st} &: \text{1if shift } s \text{ starts or finishes at the beginning of period } t, \\ &\quad \text{0otherwise} \\ S_t &: \text{maximum number of staff permitted to commence} \\ &\quad \text{or end a shift in time period } t. \end{aligned} \tag{4.4.4}$$

We introduce integer variables, σ_s, denoting the number of operators on shift s. Now we are able to formulate the model starting with the objective function, which is to minimize the total cost over all shifts,

$$\min \sum_{s=1}^{m} C_s \sigma_s \tag{4.4.5}$$

subject to the three sets of constraints:

$$\sum_{s=1}^{m} A_{st} \sigma_s \geq D_t \quad , \quad t = 1, 2, \ldots, n \tag{4.4.6}$$

ensuring that there are sufficient operators available to cover a time period,

$$\sum_{s=1}^{m} A_{st}\sigma_s \le M_t \quad , \quad t = 1, 2, \ldots, n \tag{4.4.7}$$

guaranteeing that the number of operators used in a time period does not exceed the prescribed limit, and

$$\sum_{s=1}^{m} Q_{st}\sigma_t \le S_t \quad , \quad t = 1, 2, \ldots, n \tag{4.4.8}$$

ensuring that the number of staff commencing or ending a shift is within a maximum limit (otherwise there is congestion at hand over).

This formulation was quite slow to solve. As the constraint matrix comprises 0s and 1s, the authors suspected that a network model might be achievable and, in fact, established that it was. This reformulation is given below.

The concept of double indexed time periods (t, t') is now introduced. Such a time period runs from period t to period t', inclusively. Note that both indices now have the same domain, i.e., $t = 1, 2, \ldots, n$ and $t' = 1, 2, \ldots, n$. In terms of a network flow problem, t represents the source node and t' is the sink node. The data are now given as

$$\begin{aligned} D_{tt'} &: \text{minimum number of operators required in period } (t, t'), t \ne t' \\ C_{tt'} &: \text{cost of employing an operator in period } (t, t'), t \ne t' \\ M_{tt'} &: \text{maximum number of operators available in period } (t, t'), t \ne t'. \end{aligned} \tag{4.4.9}$$

Again we introduce integer variables, $\sigma_{tt'}$, but this time they count the number of operators assigned in period (t, t'), $t \ne t'$. In this formulation a forward arc $(t < t')$ represents the progression of time and a backward arc $(t > t')$ represents a shift (see Fig. 4.4).

The new model is

$$\min \quad \sum_{t=1}^{n} \sum_{\substack{t'=1 \\ t \ne t'}}^{n} C_{tt'}\sigma_{tt'} \tag{4.4.10}$$

subject to conservation of flow at source and sink

$$\sum_{t'=1}^{n} (\sigma_{0t'} - \sigma_{t'0}) + \sum_{t'=1}^{n} (\sigma_{nt'} - \sigma_{t'n}) = 0, \tag{4.4.11}$$

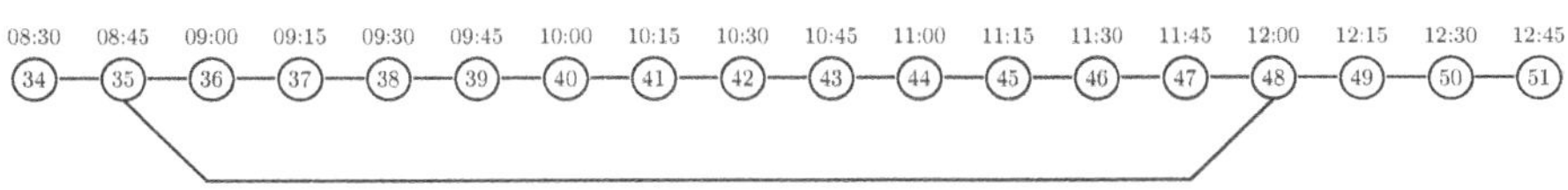

Fig. 4.4 A time expanded network. Backward arc (48, 35) represents a shift (from Wilson & Willis 1983,[585])

and conservation of flow at each node

$$\sum_{\substack{t'=1\\t\neq t'}}^{n}(\sigma_{tt'}-\sigma_{t't})=0 \quad , \quad t=2,3,\ldots,n-1, \tag{4.4.12}$$

and upper and lower limits on the number of operators allocated to any period

$$0\leq D_{tt'}\leq\sigma_{tt'}\leq M_{tt'}. \tag{4.4.13}$$

This formulation was solved rapidly by a network algorithm. This was found to be very useful for the client to allow many variations of the problem to be considered. The restriction that a shift must have a minimum size can be incorporated into the network algorithm, rather than by using constraints.

4.4.4 Other Applications of Network Modeling Technique

Further applications of the network modeling technique and the development of the approach appear in Glover et al. (1990,[225]). A complex network design problem is discussed in Sect. 10.6.

4.5 Unimodularity ⊖

A matrix A is called *totally unimodular* if the determinant of each square submatrix of A is ±1 or 0. For LP this will mean that if a problem is such that all the right-hand sides are integers and the coefficient matrix is unimodular, then the value given to the variable will be an integer. Hence, problems such as network problems (Sect. 4.4), assignment problems (Sect. 4.3.3), and transportation problems (Sect. 4.3.1) and many variants of these problems will have integer solutions once the problem has been solved as an LP. Thus there will be no need to specify the requirement that all the variables are integer in such problems when they are being solved by an LP algorithm as the requirement will be satisfied automatically. It should be noted that it is a non-trivial task to check a matrix for unimodularity, as can be seen below, and it is therefore useful to know which classes of problem will result in unimodular coefficient matrices. For further theoretical details on unimodular matrices and integral polyhedra the reader is referred to Nemhauser & Wolsey (1988,[421]).

For understanding the concept, let us consider a unimodular transportation matrix in a transportation problem with two sources and three destinations, and the constraint matrix A and submatrices S_3^1, S_3^2, and S_2:

$$\mathsf{A}=\begin{pmatrix}1\,1\,1\,0\,0\,0\\0\,0\,0\,1\,1\,1\\1\,0\,0\,1\,0\,0\\0\,1\,0\,0\,1\,0\\0\,0\,1\,0\,0\,1\end{pmatrix} \quad ; \quad \mathsf{S}_3^1=\begin{pmatrix}0\,1\,0\\0\,0\,1\\1\,0\,0\end{pmatrix}, \quad \mathsf{S}_3^2=\begin{pmatrix}1\,1\,1\\1\,0\,0\\0\,1\,0\end{pmatrix}, \quad \mathsf{S}_2=\begin{pmatrix}0\,1\\0\,0\end{pmatrix}.$$

It is required to check that each submatrix has determinant equal to $+1$, -1, or 0. Consider three such submatrices S_3^1, S_3^2, and S_2 arising from rows 3, 4, 5 and columns 3, 4, 5; rows 2, 3, 4 and columns 4, 5, 6; and rows 2, 4 and columns 3, 4 (resp.). The determinant of each submatrix is, in order, 1, 1, and 0. Gradually the whole set of submatrices can be checked and this becomes easier as patterns emerge, and we establish that the whole matrix is unimodular. Although this was a particular transportation problem, we find that larger transportation problems have a similar structure and the checking of submatrices can be done by induction from simpler patterns. Gradually it can be established that all transportation matrices are unimodular.

4.6 Summary and Recommended Bibliography

In this chapter we have investigated different types of problems that can be solved by an LP approach. We have seen that certain classes of problem will be easy to solve as approaches simpler than standard LP are applicable. We have seen how LP may be used to compare efficiency and how multiple objective problems may be structured with a view to solving them as straightforward LP problems. Thus the reader should now be familiar with the modeling and formulation of the following types of problems:

- Trimloss
- Food mix
- Transportation, assignment, and transshipment
- Network flow
- Problems with a unimodular structure

For network flow problems and algorithms to solve them we refer the reader to Bazaraa et al. (2010,[54]) and Jungnickel (2012,[296]).

4.7 Exercises

1. Solve the trimloss problem given in Sect. 4.1.1 using a modeling system and solver of your choice.
2. Solve the trimloss problem given in Sect. 4.1.2 using a modeling system and solver of your choice.
3. Solve the 10 node, 15 arc network flow problem given in Sect. 4.4.

Chapter 5
How Optimization Is Used in Practice: Case Studies in Linear Programming

This chapter contains several case studies in linear programming. The case studies arise from real-world problems now successfully being solved in industry. In the first case we look at optimization in the chemical industry. Then a blending problem is discussed that is apparently nonlinear, but can be converted to a linear problem. We then consider the techniques known as data envelopment analysis and goal programming, a special technique for multi-criteria optimization problems. Finally, we summarize and discuss some limitations of linear programming.

5.1 Optimizing the Production of a Chemical Reactor

A chemical company runs a reactor which can be operated in $M = 4$ different modes. Depending on the modes the reactor produces $P = 4$ different products p with given yield coefficients Y_{mp} listed in the recipe table below:

$m\backslash p$	1	2	3	4
1	61.0	12.1	9.5	4.3
2	61.0	17.0	9.0	7.6
3	61.0	8.0	6.0	3.2
4	61.0	10.0	6.0	2.3

Supplementary Information The online version contains supplementary material available at https://doi.org/10.1007/978-3-030-73237-0_5.

J. Kallrath, *Business Optimization Using Mathematical Programming*,
International Series in Operations Research & Management Science 307,
https://doi.org/10.1007/978-3-030-73237-0_5

The units in the table are tons/day. The production is linked in the following sense: if for instance $m = 1$ then the plant produces 9.5 tons/day of product 3, 12.1 tons/day of product 2, and so on.

For each of the products p there is a demand, D_p, given in tons per year in the following table:

p	1	2	3	4
D_p[tons]	20, 000	5800	4500	3000

Partial demand fulfillment is allowed, but it is not possible to sell more tons than specified by D_p. For each reactor mode m there is a net profit P_m per day summarized in the table [units are k£/day]

m	1	2	3	4
P_m[k£/day]	258	160	191	208

Note that net profit per day for each mode is calculated in advance based on the cost of operation in that mode, the recipe table listed above, and the specific costs[1] and selling prices for all the products. The question posed by the production planner is: On how many days should the reactor be operated in the different modes? The maximum operating time of the reactor is 330 days/year.

In order to answer this question, we introduce variables $d_m \geq 0$ denoting the number of days the reactor operates in mode m. Strictly speaking, d_m should be integral numbers but here we argue that it is sufficiently accurate to approximate it by real numbers. The production manager in the factory knows how to interpret, say, $d_2 = 118.1$. In addition it is worthwhile to introduce variables x_p describing the total amount of product p produced during a year. Thus, we have a total of 8 variables

$$d_m \geq 0 \quad , \quad m \in \{1, 2, \ldots, M\} \quad , \quad x_p \geq 0 \quad , \quad p \in \{1, 2, \ldots, P\}. \tag{5.1.1}$$

Let us first write down the restriction limiting reactor operating time:

$$\sum_{m=1}^{M} d_m \leq 330. \tag{5.1.2}$$

The total amount of product p produced during a year is

$$x_p = \sum_{m=1}^{M} Y_{mp} d_m \quad , \quad \forall p. \tag{5.1.3}$$

[1]The specific costs include the costs for raw material, energy, utilities, machines, and so on. Thus, our model is highly simplified.

Since the demand need not be satisfied exactly, it is sufficient to require to

$$x_p \leq D_p \quad , \quad \forall p. \tag{5.1.4}$$

Finally, we have the objective function

$$\max \quad z = \sum_{m=1}^{M} P_m d_m. \tag{5.1.5}$$

Note that in our model we could eliminate the variables x_p completely by combining (5.1.3) and (5.1.4). All the information needed could be derived from the d_m; they are the real decision variables. Thus (5.1.3) and (5.1.4) could be replaced by

$$\sum_{m=1}^{M} Y_{mp} d_m \leq D_p \quad , \quad \forall p, \tag{5.1.6}$$

but it may be more convenient also to have the x_p variables in the model because they tell us immediately how much is produced.

What can be learned from this example? The reader might be disappointed by this example and might consider this an insubstantial case. But the point is there are problems out in the world which are simple from a mathematical point of view but nevertheless may have a huge financial impact. Another point which is not so obvious is that the simple problem described in this section is the result of intensive discussion with the client. There are simplifications from the original problem but the problem presented reflects the structure of the business, and the model helps to improve the profitability of the business.

5.2 An Apparently Nonlinear Blending Problem

The case study discussed in this section is an example of how real-world problems appear, and how they can be solved, where the modeling aspects turns out to be crucial. In the beginning, when the customer presented the problem it was not even clear that in the end we would have an optimization problem. It sounded more likely that an expert system would be required, which would use rules to design feasible blends. This project also showed how important it is for modeling to involve several people with different backgrounds cooperating and being able to communicate with each other: the modeler capable of formulating a mathematical programming problem, the customer not deeply in touch with mathematics, but providing physical and chemical background and know-how for the project.

The objective of this project was the design and production of "recipes" for blending fluids at minimum cost, observing several product properties and requirements. Such properties were, for instance, viscosities, boiling points, and concentrations of alcohol. In addition, some logistic constraints had to be satisfied, e.g., all the content of some tanks of fluids had to be used, or that some materials were only available in limited amounts. The client's customers typically required L tons of blended liquid and specified some bounds for viscosity (η_-, η_+), boiling point (T_-^B, T_+^B), and alcohol concentration (C_-^A, C_+^A). We were further provided with the specific costs K_i [£/ton] to be paid for the components or materials $i \in \mathcal{I} := \{1, 2, \ldots, n\}$, and the physical and chemical properties of these materials.

5.2.1 *Formulating the Direct Problem*

In terms of good practice, it is often helpful to have the concept of *direct* and *inverse* problems in mind. If the values of all our decision variables are at hand, then we could derive all dependent information. The direct problem performs exactly this task: deriving all required information for a specified set of values of the decisions variables. People using *simulation* take the direct problem approach. Optimization problems are inverse problems. Optimization algorithms compute the values of the decision variables such that one of the derived pieces of information of the direct problem, the objective function, becomes optimal while all constraints are obeyed. In this sense, each optimization problem contains a direct problem. The optimization model is easily formulated provided we can identify the direct problem.

So let us ask: what is the direct problem in our customer's blending problem? Assuming we knew the design of the blending, could we derive all required information from it, or alternatively, what would be required to derive this information? How could we characterize the blending? One way to do this is to use the relative mass proportions [kg/kg], $x_i \geq 0$, of the components in the blended fluid. The n components have to observe the mass conservation

$$\sum_{i=1}^{n} x_i = 1. \tag{5.2.1}$$

A chemist would be more likely to characterize a fluid blend in terms of molecular weight fractions, w_i, i.e., mol/mol. The customer, a chemist himself, fortunately told us that these quantities x_i and w_i are coupled by the relations

$$w_i = \frac{x_i}{\mu_i} \Bigg/ \left(\sum_{k=1}^{n} \frac{x_k}{\mu_k} \right) \quad , \quad \forall i, \tag{5.2.2}$$

where μ_i denotes molecular weight. The sum of the w_i's is normalized as

$$\sum_{i=1}^{n} w_i = 1. \tag{5.2.3}$$

So, the $x_i \geq 0$ and $w_i \geq 0$ are our decisions variables, at least for the moment. Once we have characterized our blend, can we derive all relevant information? The answer is: only if we are able to compute the properties of the blend based on the properties of the components. Obviously, we need some physics and chemistry telling us how we could derive relevant blending properties. We ask the chemist for ideas or remember what we learned in high school. So, the next step of the modeling is to formulate relations which enable us to compute the properties of the mixture from the properties of the separate components. Viscosity blending follows the exponential relation

$$\eta = \prod_{i=1}^{n} \eta_i^{w_i}, \tag{5.2.4}$$

while alcohol blending is linear in mass fractions, i.e., the alcohol content C^A is

$$C^A = \sum_{i=1}^{n} C_i^A x_i. \tag{5.2.5}$$

The boiling point T^B of the blend is more difficult to compute because it can only be derived from the implicit condition

$$P(T^B) = 1, \tag{5.2.6}$$

which states that the vapor pressure P is equal to 1 atm at the boiling temperature T^B. At a given temperature T the vapor pressure $P(T)$ is a linear combination of the partial vapor pressures:

$$P(T) = \sum_{i=1}^{n} P_i(T) w_i, \tag{5.2.7}$$

where the partial vapor pressures $P_i(T)$ can be computed by the Antoine equation

$$P_i(T) = e^{A_i + B_i/(C_i+T)}, \tag{5.2.8}$$

where the constants A_i, B_i, and C_i are material parameters.

For formulating some logistic constraints on tank content or availability we also need to know the absolute amount of components in the blend, which can easily be computed from

$$X_i = Lx_i \quad , \quad \forall i, \tag{5.2.9}$$

where L is the amount [tons] of blended liquid required. Finally, the objective function to be minimized is

$$\min \quad Z = \sum_{i=1}^{n} (K_i L)\, x_i, \tag{5.2.10}$$

where K_i [£/ton] are the specific costs to be paid for the components.

5.2.2 Formulating the Inverse Problem

Having formulated the direct problem, we are able to formulate the inverse or optimization problem easily:

$$\min \quad Z = \sum_{i=1}^{n} (K_i L)\, x_i \tag{5.2.11}$$

subject to

$$\sum_{i=1}^{n} x_i = 1 \quad , \quad \sum_{i=1}^{n} w_i = 1, \tag{5.2.12}$$

and the bounds

$$\begin{array}{l} \eta_- \leq \eta \quad \leq \eta_+ \\ C_-^A \leq C^A \leq C_+^A \,, \\ T_-^B \leq T^B \leq T_+^B \end{array} \tag{5.2.13}$$

i.e., we have bounds on the viscosity, alcohol concentration, and boiling point of the blend. The system of constraints derived so far is completed by (5.2.1), and possibly the additional inequalities

$$x_i \geq X^{MIN} \quad , \quad i \in \mathscr{I}_1 \tag{5.2.14}$$

$$x_i \leq X^{MAX} \quad , \quad i \in \mathscr{I}_2 \tag{5.2.15}$$

or fixed bounds for some of the components

$$x_i = X_i^{FIX} \quad , \quad i \in \mathcal{J}_3 \tag{5.2.16}$$

representing the logistic requirements for appropriate subsets $\mathcal{J}_1, \mathcal{J}_2$, and $\mathcal{J}_3$ of $\mathcal{J}$.

5.2.3 Analyzing and Reformulating the Model

As a consequence of (5.2.4) and (5.2.7) which enter into (5.2.13), the problem appeared as a nonlinear, constrained optimization problem. However, a closer inspection shows that it can be transformed to a linear programming problem.

The objective function is linear in the mass fraction variables. The alcohol concentration is also linear in the variables x_i. Combining (5.2.5) and (5.2.13) yields

$$C_-^A \leq \sum_{i=1}^{n} C_i^A x_i \leq C_+^A. \tag{5.2.17}$$

Now we have to find some linear relation involving the viscosity. Remember that the logarithm is a (strictly) monotonically increasing function, i.e.,

$$x_1 < x_2 \Leftrightarrow \ln(x_1) < \ln(x_2). \tag{5.2.18}$$

For our purposes, it is sufficient to consider monotonically increasing functions $f(x)$, i.e., functions which have the property

$$x_1 \leq x_2 \Leftrightarrow f(x_1) \leq f(x_2). \tag{5.2.19}$$

The logarithm function has the property (5.2.19), so the viscosity condition and expression (5.2.4) can be transformed into a linear expression by taking the logarithm of both sides of (5.2.4) and (5.2.13) yielding

$$\ln(\eta_-) \leq \sum_{i=1}^{n} \ln(\eta_i) w_i \leq \ln(\eta_+). \tag{5.2.20}$$

Eventually, we found a linear relation, namely (5.2.20) involving the viscosity, and now we see why monotony is important. Our model will only guarantee that its solution satisfies (5.2.20). Exploiting (5.2.18) enables us to conclude safely from (5.2.20) that $\eta_- \leq \eta \leq \eta_+$. The boiling temperature of the mixture is bounded by the temperature limits T_-^B and T_+^B. Again, monotonicity can help us. Exploiting some physics, the bounds T_-^B and T_+^B on boiling temperature T^B imply that the vapor pressure at temperature T_-^B is smaller than 1 bar, $P(T_-^B) < 1$, and at T_+^B it is larger than 1 bar, $P(T_+^B) > 1$. To model the $>$ or $<$ condition, we introduce a small parameter $\varepsilon > 0$, say $\varepsilon \approx 10^{-6}$. Then we get

$$P(T_-^B) \leq 1 - \varepsilon < 1 < 1 + \varepsilon \leq P(T_+^B) \tag{5.2.21}$$

or, exploiting (5.2.7)

$$\sum_{i=1}^{n} P_i(T_-^B) w_i \leq 1 - \varepsilon \quad , \quad \sum_{i=1}^{n} P_i(T_+^B) w_i \geq 1 + \varepsilon . \tag{5.2.22}$$

The only remaining minor task is to eliminate the w_i and to formulate the model only in terms of the x_i. Let $\circ$ denote one of the relations $\{\leq, \geq, =\}$. A constraint of the form

$$\sum_{i=1}^{n} F_i w_i \circ F^* \tag{5.2.23}$$

can, by use of (5.2.2), be transformed to the equivalent form

$$\sum_{i=1}^{n} \frac{F_i - F^*}{\mu_i} x_i \circ 0 . \tag{5.2.24}$$

Now the model is a linear programming problem in the x_i appearing as

$$\min \quad Z = \sum_{i=1}^{n} (K_i T) x_i \tag{5.2.25}$$

subject to

$$x_i \geq 0 \quad , \quad \forall i \tag{5.2.26}$$

$$\sum_{i=1}^{n} x_i = 1 \tag{5.2.27}$$

$$\sum_{i=1}^{n} \frac{\ln(\eta_i) - \ln(\eta_-)}{\mu_i} x_i \geq 0 \quad , \quad \sum_{i=1}^{n} \frac{\ln(\eta_i) - \ln(\eta_+)}{\mu_i} x_i \leq 0 \tag{5.2.28}$$

$$\sum_{i=1}^{n} \frac{P_i(T_+^B) - (1 + \varepsilon)}{\mu_i} x_i \geq 0 \quad , \quad \sum_{i=1}^{n} \frac{P_i(T_-^B) - (1 - \varepsilon)}{\mu_i} x_i \leq 0 \tag{5.2.29}$$

$$C_-^A \leq \sum_{i=1}^{n} C_i^A x_i \leq C_+^A . \tag{5.2.30}$$

The system of constraints is completed by the logistic constraints

$$Mx_i \geq X^{MIN} \quad , \quad i \in \mathcal{J}_1 \tag{5.2.31}$$

$$Mx_i \leq X^{MAX} \quad , \quad i \in \mathcal{J}_2$$

or fixed bounds for some of the components

$$Mx_i = X_i^{FIX} \quad , \quad i \in \mathcal{J}_3. \tag{5.2.32}$$

Modeling can be improved by transforming the constraints (5.2.31)–(5.2.32) to bounds (which are treated more effectively in the Simplex algorithm). The bounds would be the inequality bounds (lower and upper bounds):

$$x_i \geq X^{MIN}/L \quad , \quad i \in \mathcal{J}_1 \tag{5.2.33}$$

$$x_i \leq X^{MAX}/L \quad , \quad i \in \mathcal{J}_2, \tag{5.2.34}$$

or the fixed bounds (for some of the components):

$$x_i = X_i^{FIX}/L \quad , \quad i \in \mathcal{J}_3. \tag{5.2.35}$$

Note that from matrix algebra we can derive the maximum number of blends we can expect to have in the blend. If $|\mathcal{J}_i|$ gives the number of products contained in the index set $\mathcal{J}_i$, then the number, n^P, of products in the blend is bounded by

$$n^P \leq m = 7 + |\mathcal{J}_1| + |\mathcal{J}_2| + |\mathcal{J}_3|, \tag{5.2.36}$$

where m is the total number of constraints in the problem. The reason is that we cannot have more basic (which implies non-zero) variables than constraints; see also Exercise 5 in Chapter 3. It is useful to know this because it enables us to tell the client in advance that all optimal solutions involve at most n^P of products in the blend.

5.3 Data Envelopment Analysis (DEA)

In this section a modeling approach to performance measurement will be introduced. The technique is called *Data Envelopment Analysis* (DEA) and has become popular for use in a variety of contexts. DEA provides a complete approach to performance appraisal and, in contrast to material in other sections in this chapter, is not simply a problem to be modeled using LP. The technique of DEA involves the use of LP to solve a set of inter-related problems to determine the relative efficiency of a group of *decision making units* (DMU) such as hospitals, universities, bank branches, retail outlets as in the example treated in Sect. 5.3.1, production plants of a larger company, libraries, local government offices or projects to invest in. The technique is based on ideas dating back to Farrell (1957,[177]) which were later developed into an LP approach by Charnes et al. (1978,[115]). The special feature of DEA is to measure performance in a way that tries to avoid subjectivity creeping into the evaluation of relative efficiency. The analysis will measure output(s) (e.g., turn over or quality levels) achieved from the input(s) — this can be costs, manpower required, raw materials and utilities used, etc. — provided and will compare the group of DMU by their strength in turning input into output. At the end of the analysis, the DEA technique will be able to say which units are (relatively) efficient and which are (relatively) inefficient.

In DEA first decision makers need to determine which input(s) need to be used in the decision making and assessment process. For example, a firm deciding between two projects to invest in might choose overall cost as the most important input; usually only one input is used. Following this, decision makers need to determine which factors have a large impact on the input (cost in this case). Then a set of outputs is chosen, perhaps 5–10 of these, e.g., number of customers served, sales value, and so on. The outputs are chosen which have principal effect on cost, if that were the chosen input, when it is transformed in the running of the business.

Once the set of inputs has been agreed, the analysis is no longer subjective. A mathematical model is used to determine sets of weights (weighting factors) to be attached to the input(s) and outputs of each DMU, which could be a department, section, or an outlet, to let it appear as efficient as possible. Thus each DMU has its own set of weights calculated for it to allow it to look its best relative to all other DMU. Clearly mathematical derivation of optimal efficiency avoids bias and establishes that DMU could not make themselves any more efficient by some other choice of weights (as these would not be optimal). Another feature of the modeling is that it does not matter in what units inputs are measured (provided there is consistency across all DMU) because DEA looks at trade-offs.

5.3.1 Example Illustrating DEA

An electrical retailer has outlets in Paris, London, and Bonn and identifies the major economic input at each branch to be the cost of running the branch for three major outputs with sales of type A, B, or C goods. Table 5.1 gives the values of these inputs and outputs in consistent units. Clearly it is difficult to say which branch is most efficient overall. Thus we shall construct a model for each outlet to determine its efficiency on a scale of zero to one, where "one" means completely efficient, relative to all other branches. First we consider the Paris model and formulate it as an LP problem.

Let $p_1, p_2, p_3 \geq 0$ represent weights to be attached to the three Paris sales values. These will be the variables whose value we wish to determine.

Table 5.1 Inputs (costs) and outputs (sales of type A, B, or C goods) used in the DEA of an electrical retailer's outlets Paris, London, and Bonn used in this section. As discussed in Sect. 5.3.3, the table also contains the additional outlet Amsterdam and the average of the original three outlets

	Paris	London	Bonn	Amsterdam	Average
Cost	110	150	200	180	153
Sales 'A'	50	50	100	67	67
Sales 'B'	50	100	90	80	80
Sales 'C'	50	40	60	50	50

We will regard the overall output from Paris as

$$50p_1 + 50p_2 + 50p_3. \tag{5.3.1}$$

We now construct the Paris model (with weight variables denoted by p) as

$$\max \frac{\text{output of Paris}}{\text{input of Paris}} = \max \frac{50p_1 + 50p_2 + 50p_3}{110} \tag{5.3.2}$$

maximizing the efficiency of the Paris outlet subject to the efficiency inequalities

$$\begin{aligned} (50p_1 + 50p_2 + 50p_3)/110 &\le 1 \\ (50p_1 + 100p_2 + 40p_3)/150 &\le 1 \\ (100p_1 + 90p_2 + 60p_3)/200 &\le 1 \end{aligned} \tag{5.3.3}$$

$$p_1 \ge \varepsilon > 0, \quad p_2 \ge \varepsilon > 0, \quad p_3 \ge \varepsilon > 0, \tag{5.3.4}$$

where ε is some small positive number avoiding that the variables just take the value zero. Note that the efficiency inequalities require us to measure output (nominator) and input (denominator) in the same units. This can be achieved by dimensionless weights or by allowing the weights to make the units of nominator and denominator compatible.

The origin of the objective function and constraints is as follows. The objective function is to maximize weighted output divided by input (the efficiency). Thus, the model is trying to find three weights that will show Paris off to best advantage as far as output per input is concerned. The idea behind this is to make the model non-subjective so that efficiency is chosen mathematically and the highest possible value for efficiency for Paris is selected. The Paris managers are essentially choosing, via the model, weight values for p_1, p_2, and p_3 so that they appear in the best light (and London and Bonn managers are able to do the same for their set of weights). This is in line with the premise "if you cannot look efficient under your own chosen weights, then you are not efficient." This avoids "wriggling" by managers who are now unable to say that the reason one outlet appears to be inefficient is because undue attention is being paid to some aspect. The solution to the LP problem is the maximum efficiency available. However, to avoid nonsensical results we have constraints on the choice of weights for Paris, as follows. We apply these weights to the outputs achieved at London and Bonn on their data and then ensure that their efficiency is not permitted to be larger than 1. In short: the efficiency for the best site is at most 1. Thus we have two constraints which construct the weighted output per input for London and Bonn, but using the weights from Paris, and constrain these quantities to be less than or equal to 1. By solving this model we obtain an objective function value (the efficiency) and three weights. The constraints of form $p \geq \varepsilon$ ensure that each output plays some part in the process.

Next we construct two more models beginning with the London model. Let l_1, $l_2, l_3 \geq 0$ be weights to be attached to the three London sales values. The model is

$$\max \quad \frac{50l_1 + 100l_2 + 40l_3}{150} \tag{5.3.5}$$

subject to

$$\begin{aligned} (50l_1 + 50l_2 + 50l_3)/110 &\leq 1 \\ (50l_1 + 100l_2 + 40l_3)/150 &\leq 1 \\ (100l_1 + 90l_2 + 60l_3)/200 &\leq 1 \end{aligned} \tag{5.3.6}$$

and

$$l_1 \geq \varepsilon > 0, \quad l_2 \geq \varepsilon > 0, \quad l_3 \geq \varepsilon > 0. \tag{5.3.7}$$

Finally, we have the Bonn model with weights $b_1, b_2, b_3 \geq 0$ to be attached to the three Bonn sales values. The model is

$$\max \quad \frac{100b_1 + 90b_2 + 60b_3}{200} \tag{5.3.8}$$

subject to

$$\begin{aligned} (50b_1 + 50b_2 + 50b_3)/110 &\leq 1 \\ (50b_1 + 100b_2 + 40b_3)/150 &\leq 1 \\ (100b_1 + 90b_2 + 60b_3)/200 &\leq 1 \end{aligned} \tag{5.3.9}$$

and

$$b_1 \geq \varepsilon > 0, \quad b_2 \geq \varepsilon > 0, \quad b_3 \geq \varepsilon > 0. \tag{5.3.10}$$

Thus we solve three independent but similarly structured LP problems to obtain efficiency values for each outlet and three weights that permit the efficiency values to be achieved.

5.3.2 *Efficiency*

It turns out that all three outlets have a relative efficiency of 1. It is recommended that the reader tries to establish this (see Exercise 5.2). This is perhaps unhelpful information but occurs because our example involved only a few outlets. Typically we might expect to have 5–10 outputs and 40–50 decision making units (outlets in the case of our example). If we have too many outputs we find that a large proportion of the decision making units turns out to be efficient (i.e., have relative efficiency of 1).

In fact it is usually appropriate in business terms to avoid having too many outputs for the following reason. Each output will represent a major dimension of the activity of the decision making unit. Research from the field of psychology has established that people find it difficult to handle more than about 5–6 concepts simultaneously. Thus to expect a manager to consider results with many dimensions involved may be unreasonable — it may destroy any intuitive "feel" for the problem. Thus, DEA will often deal with aggregate quantities as outputs and these should not be regarded as a gross simplification of reality but rather a nod toward practicality. The mathematical model could handle hundreds of outputs but the decision maker might be overwhelmed.

5.3.3 *Inefficiency*

If the calculated efficiency of a decision making unit turns out to be less than 1 (output $<$ input), then that unit is inefficient. The analysis will indicate that some linear combination of efficient units could undertake the same total activity as this unit (in terms of outputs) and do so for a smaller value of the input.

As an example, say there was another outlet at Amsterdam which had costs of 180 and sales of A, B, C of 67, 80, 50, respectively; see Table 5.1. This example has been contrived so that if we take the mean of Cost, Sales "A," Sales "B," and Sales "C," across the London and Bonn branches we get 460/3, 200/3, 240/3, and 150/3 respectively or 153, 67, 80, and 50. Hence we can see that this composite branch outperforms the Amsterdam branch by achieving the same output for an input of 153, i.e., 85% of the input of the Amsterdam branch.

When the analysis yields the above information the efficient units which comprise the composite branch are termed the *peer group* for Amsterdam. This information is helpful in convincing managers of the sense of the analysis. One more feature should be noted. DEA will only highlight information that could have been acquired in other ways. We do not need to use LP to show that Paris, London, and Bonn are better than Amsterdam once we know to look at them. However, the DEA approach pinpoints where to look and then common sense can be used to convince clients of the hard truths (under-performing at inefficient outlets) of our analysis. Alternatively, if the analysis is discredited in any way (e.g., disagreement over choice of outputs used) we have encouraged the start of a debate. These last few remarks are typical of much OR work — we use clever ideas to establish basic truths and then subsequently we can sometimes get to the heart of the problem without the clutter of technology.

5.3.4 More Than One Input

It may be desirable to include more than one input in the analysis. The inputs will be weighted in an analogous way to the outputs. Thus if we had a second input which had values of 50, 60, and 70 for the three cities Paris, London, and Bonn, respectively, all the divisors of 110, 150, 200 in the model would be replaced by, say

$$\begin{array}{l} 110w_1 + 50w_2 \\ 150w_1 + 60w_2\,. \\ 200w_1 + 70w_2 \end{array} \tag{5.3.11}$$

The Paris model would become

$$\max \quad \frac{50p_1 + 50p_2 + 50p_3}{110w_1 + 50w_2} \tag{5.3.12}$$

subject to

$$\begin{aligned} \frac{1}{110w_1+50w_2}(\,50p_1 + \ 50p_2 + \ 50p_3) &\le 1 \\ \frac{1}{150w_1+60w_2}(\,50p_1 + 100p_2 + \ 40p_3) &\le 1 \\ \frac{1}{200w_1+70w_2}(100p_1 + 90p_2 + \ 60p_3) &\le 1. \end{aligned} \tag{5.3.13}$$

After some cross-multiplication these constraints can be turned into conventional LP form. More than two inputs would require further weights and each of the above expressions would be extended. It is unlikely that we would require more than a small set of inputs (much fewer than the number of outputs) because of the conceptual problems referred to earlier.

5.3.5 Small Weights

In any analysis a unit may appear to be efficient if it is allowed (by the model) to give high weight to some outputs and low or nominal weight to others. This may be felt to lead to misleading results and over-reporting of efficiency when major dimensions of activity appear to be almost totally ignored. For example, it might be unreasonable for a high street store to claim to be 100% efficient by virtue of having a good staff canteen, which was one input, and ignoring all other inputs including those concerned with sales.

One way to build more realism into the analysis will be to insist on more appropriate values for the constant ε which was included in the model. In the relations (5.3.7) ε is a lower bound. A realistic lower bound in an analysis might be 50% of the mean value of an output or input. Thus we are saying that we can regard a weight as measuring (loosely speaking) marginal cost or marginal revenue in economic terms. It is therefore reasonable to force this value not to be unrealistically small. It is probably preferable when efficient units take on a "rounded profile," i.e., receiving some apparent contribution to their efficiency from most of their outputs. Placing lower bounds on weights will assist this. Note that for technical reasons if we put lower bounds on weights, then only one input can be used in the DEA model. Further discussion of the technicalities of DEA is contained in Thanassoulis et al. (1987,[543]) and Dyson & Thanassoulis (1988,[168]); see also the implemented example MCOL/GAMSLIB/*dea.gms* based on a tutorial by the same authors.

5.3.6 Applications of DEA

DEA has been used in measuring efficiency of bank and building society branches; cf. Ferrier & Lovell (1990,[179]), public houses; Athanassopoulos & Thanassoulis (1995,[39]), school bus services; cf. Sexton et al. (1994,[499]), public sector departments; cf. Smith & Mayston (1987,[514]), university departments; Jesson et al. (1987,[290]), and electricity distribution; Miliotis (1992,[399]). The technique is appropriate when an organization conducts similar activities at a variety of outlets, and thus financial services is an obvious candidate.

In the studies cited there is not always an obvious "saving" indicated by the DEA study. The benefits are likely to be longer term when a policy of rationalization or quality improvement is initiated as a result of deficiencies highlighted by the DEA study. The study by Sexton et al. (1994,[499]) did, however, indicate savings in US$millions. The technique is applicable in both large scale and relatively smaller scale enterprises.

5.3.7 A General Model for DEA

The usual DEA model with n decision making units, s outputs, and m inputs has indices

$$\begin{aligned} i &\in \{1, 2, \ldots, m\} &&: \text{the set of inputs} \\ j &\in \{1, 2, \ldots, n\} &&: \text{the set of decision making units} \\ r &\in \{1, 2, \ldots, s\} &&: \text{the set of outputs,} \end{aligned} \tag{5.3.14}$$

data

$$\begin{aligned} N_{ij} &: \text{the amount of input } i \text{ to unit } j \\ P_{rj} &: \text{the amount of output } r \text{ from unit } j, \end{aligned} \tag{5.3.15}$$

real variables

$$\begin{aligned} u_{rj} \geq 0 &: \text{the weight of output } r \text{ for unit } j \\ v_{ij} \geq 0 &: \text{the weight of input } i \text{ for unit } j, \end{aligned} \tag{5.3.16}$$

and so the model is stated as: maximize efficiency of unit k, i.e.,

$$\max \quad \left(\sum_{r=1}^{s} P_{rk} u_{rk}\right) \bigg/ \left(\sum_{i=1}^{m} N_{ik} v_{ik}\right) \tag{5.3.17}$$

subject to the efficiency of all units being no more than 1, i.e.,

$$\left(\sum_{r=1}^{s} P_{rj} u_{rk}\right) \bigg/ \left(\sum_{i=1}^{m} N_{ij} v_{ij}\right) \leq 1 \quad , \quad j = 1, 2, \ldots, n \tag{5.3.18}$$

and

$$u_{rj}, v_{ij} \geq \varepsilon \quad , \quad \forall \{rij\}. \tag{5.3.19}$$

Here k is the decision making unit under discussion and ε is a small constant.

This formulation has ratios in both the objective function and main constraints, which put it beyond LP. However, using the Charnes–Cooper transformation, this can easily be remedied as follows.

First we define a constraint

$$\sum_{i=1}^{m} N_{ik} v_{ik} = 1 \tag{5.3.20}$$

to restrict the denominator of the objective (5.3.17) to be equal to one.

Then the objective can be simply rewritten as

$$\max \quad \sum_{r=1}^{s} P_{rk} u_{rk}, \tag{5.3.21}$$

i.e., we are maximizing the numerator of the old objective function subject to the denominator being equal to one. We can think of the transformation as simply adjusting (scaling) the units in which the objective function is measured. In the main constraints (5.3.18) of the original formulation we can cross-multiply, bringing the denominator over to the right-hand side and giving

$$\sum_{r=1}^{s} P_{rj} u_{rk} \leq \sum_{i=1}^{m} N_{ij} v_{ij} \quad , \quad j = 1, 2, \ldots, n. \tag{5.3.22}$$

This operation is legitimate as the expression

$$\sum_{i=1}^{m} N_{ij} v_{ij} \quad , \quad j = 1, 2, \ldots, n \tag{5.3.23}$$

is non-zero.

A useful discussion on demystifying DEA is contained in Belton & Vickers (1993,[62]).

5.4 Vector Minimization and Goal Programming

One perceived limitation of LP and MILP is that models can only have a single objective function. We might feel this is unrealistic and it could be hard to articulate one objective. A chemical company might want to maximize the production output rate and simultaneously minimize the size of the chemical reactor to be built and maximize some parameters related to safety. The quality of the solution may be judged according to three criteria. This example shows that we may be interested in more than just maximizing profit; instead we may wish to optimize on profit, contribution, labor turnover, return on capital, return on stock, return per employee, and so on. Clearly some of these objectives may be contradictory, e.g., maximize profits versus minimize operating costs, or minimize risk versus maximize return on investment.

Optimization problems with more than one quantity to be maximized or minimized are called *multi-criteria optimization* or *vector optimization* problems. In Sect. 5.4.1 we explain and illustrate methods for solving them before considering an example in Sect. 5.4.2.

5.4.1 Solution Approaches for Multi-Criteria Optimization Problems

One approach to solving multi-criteria problems, i.e., *vector minimization problems*, is to express all objectives in terms of a common measure of goodness, e.g., money, but that is very often not possible. Consider again the reactor example above: it is not obvious how the objective related to the safety parameter could be converted into money. The problem really is how to compare different objectives on a common scale. When minimizing several objective functions simultaneously the concept of *Pareto optimal solutions* turns out to be useful. In a multi-criteria optimization problem a solution is said to be *Pareto optimal* if there exists no other solution that is at least as good according to every objective, and is strictly better according to at least one objective.

A special solution approach to multiple objective problems is to require that all the objectives should come close to some targets, e.g., the output rate should come as close as possible to 50 tons per hour and the reactor should fit in an area of 50 m by 50 m. The targets we set for a number of objectives are called *goals*. Our overall objective can then be regarded as to minimize the deviation overall of our goals from their target levels. The solutions derived are Pareto optimal.

Goal programming can be considered as an extension of linear programming in which targets are specified for a set of constraints. There are two basic approaches for goal programming: the *preemptive (lexicographic)* approach and the *Archimedean* approach.

In preemptive goal, programming goals are ordered according to importance and priorities. The goals at a certain priority level are considered to be infinitely[2] more important than the goals in the next lower level. Our reactor example above might have the ranking: reactor size, safety issues, and eventually production output rate. Preemptive goal programming is recommended if there is a ranking between incommensurate objectives available.

In the Archimedean approach, weights or penalties are applied for not achieving targets. The following example illustrates this. Let

$$2x + 3y \tag{5.4.1}$$

represent profit, and

$$y + 8z \tag{5.4.2}$$

represent return on capital in a simple LP model where there are a number of constraints involving the variables x, y, and z and other variables. In addition, let P be the desired level of profit and C the desired level of return on capital. P and C might be obtained from last year's figures plus some percentage to give targets for this year.

We now adjoin four non-negative variables $d_1, d_2, d_3, d_4 \geq 0$ as well as two new goal attainment constraints to our model

$$\begin{array}{llllll} 2x & +3y & & +d_1 - d_2 & & = P \text{ goal 1} \\ & y & +8z & & +d_3 - d_4 & = C \text{ goal 2.} \end{array} \tag{5.4.3}$$

[2]It would also be possible to define weights which express how much the ith objective is more important than the $(i+1)^{th}$ objective.

The objective function is to minimize deviation from target

$$\min \quad d_1 + d_2 + d_3 + d_4. \tag{5.4.4}$$

Any objective function attempted previously in the formulation would have to be expressed as a goal constraint. The problem is now an ordinary LP problem. (IP problems may be modified similarly.) Note the use of two d's in each constraint (with opposite signs) and the presence of all d's in the objective function (with the same sign). Note also how the d's perform a role to represent a free variable, namely the deviation from target. The technique for two goals can be extended to handle three or more goals.

One feature of goal programming is that every goal is treated as being equally important, and consequently an excess of 100 units in one goal would be compensated by a shortfall of 100 in another, or would be equivalent to excesses of 10 units in each of 10 goals. Neither of these sets of circumstances might be desirable, so two strategies may be introduced:

(a) Place upper limits on the values of d variables, e.g.,

$$d_1 \leq 10 \quad , \quad d_2 \leq 10 \quad , \quad d_3 \leq 5 \quad , \quad d_4 \leq 5, \tag{5.4.5}$$

which will keep deviation from a goal within reasonable bounds;

(b) In the objective function coefficients other than 1 may be used to indicate the relative importance of goals. For example, the objective

$$d_1 + d_2 + 10d_3 + 10d_4 \tag{5.4.6}$$

may be considered appropriate if one can reason that a unit deviation in return is ten times as important as a unit deviation in "profit."

Goal programming offers an alternative approach but should not be seen as without defects. The specific goal levels selected greatly determine the answer. Therefore, care is needed when selecting the targets. It is also important in which units the targets are measured. Therefore, some modelers still prefer to work with one objective function and then re-run the model with variations in the objective function coefficients.

Detailed treatment of goal programming appears in such books as Ignizio (1976,[282]) and Romero (1991,[471]) who introduce many variations on the basic idea.

A technique similar to goal programming may also be used to model *soft constraints*. Soft constraints are constraints which may be broken provided a penalty is paid. The way to do this is to let each constraint include a surplus variable which will also appear in the objective. For example, let x and y denote amounts to be produced of two quantities where the overall amount produced is limited to 100. Thus we have the constraint

$$x + y \leq 100. \tag{5.4.7}$$

If it is also possible to raise the limit of 100 to 110 provided a cost of \$20.00 is paid for each unit used above the limit, then this may be modeled as

$$x + y - s \leq 100 \tag{5.4.8}$$

$$s \leq 10 \tag{5.4.9}$$

and s will appear in the objective function (which is to minimize total cost) as

$$\ldots + 20s. \tag{5.4.10}$$

LP solvers may have special goal programming features. Goals can be constructed from either constraints or objective functions. If constraints are used, the goals are to minimize the violation of the constraints. These are met when the constraints are satisfied. In the preemptive case as many goals as possible are met in priority order. (It should be remembered that someone has to subjectively set weights to achieve this.) In the Archimedean case a weighted sum of penalties is minimized. If the goals are constructed from an objective function, then in the preemptive case a target for each objective function is calculated from the optimal value of that objective function (by percentage or absolute deviation). In the Archimedean case a multi-objective optimization problem is obtained, in which a weighted sum of the objective functions is minimized; here all goals are measured in the same unit.

5.4.2 A Case Study Involving Soft Constraints

A paper by Mitra et al. (1995,[409]) describes the use of a goal programming to model soft constraints to provide schedules for the coastguard operations of cutters. A simplified version of the problem appears below.

A number, N^C, of cutters have to be scheduled using a series of N^S possible schedules over a set of time periods, N^T, to satisfy a set, say, N^R, of schedule requirements. It is unlikely that all requirements could be satisfied, so what is wanted are schedules which satisfy requirements as best as they can. Thus scheduling constraints can be treated as soft constraints. It is desired to keep the schedules as close as possible to the requirements.

The problem may be formulated as follows, beginning with the indices

$i \in \{1, 2, \ldots, N^R\}$: schedule requirements that must be satisfied
$k \in \{1, 2, \ldots, N^C\}$: cutters
$l \in \{1, 2, \ldots, N^S\}$: possible schedules for cutter k
$t \in \{1, 2, \ldots, N^T\}$: time periods.

The data of the problem are given by

A_{itkl} : = 1 if possible cutter schedule l for cutter k contributes to schedule requirement i in period t, 0 otherwise
G_{it} : gain for each unit of over-achievement of schedule i in period t
P_{it} : penalty for each unit of under-achievement of schedule i in period t
R_{it} : schedule requirement for schedule i in period t.

We will need continuous variables

$o_{it} \geq 0$: measure of the amount of over-achievement of schedule requirement i in time period t
$u_{it} \geq 0$: measure of the amount of under-achievement of schedule requirement i in time period t.

(5.4.11)

and binary variables, $\delta_{kl} \in \{0, 1\}$,

$$\delta_{kl} := \begin{cases} 1 \text{ , if cutter } k \text{ is used in schedule } l \\ 0 \text{ , if otherwise} \end{cases} \quad , \quad \forall \{kl\}. \tag{5.4.12}$$

The formulation is then

$$\min \quad \sum_{i=1}^{N^R} \sum_{t=1}^{N^T} (P_{it} u_{it} + G_{it} o_{it}) \tag{5.4.13}$$

subject to

$$\sum_{k=1}^{N^C} \sum_{l=1}^{N_k^S} A_{itkl} \delta_{kl} + u_{it} - o_{it} = R_{it} \quad , \quad \forall \{it\}, \tag{5.4.14}$$

$$\sum_{l=1}^{N_k^S} \delta_{kl} = 1 \quad , \quad k \in \{1, 2, \ldots, N^C\}. \tag{5.4.15}$$

The principal constraint (5.4.14) requires that the contribution made to each schedule should be as close to the requirements as possible, and as it may not be possible to match these requirements exactly, relaxing variables are included to model the over- or under-achievement of the requirement. These variables then appear in the objective function multiplied by appropriate weights which will give priority to the achievement of certain requirements by weighting such requirements more heavily than others. When the objective function is minimized such measures of over- or under-achievement will be driven to zero where possible and a trade-off will take place among those achievements which cannot be met exactly. The final constraint (5.4.15) models the fact that each cutter must be assigned to some schedule. In some of the constraints R_{it} will represent a generous limit, while in others it will represent a tight limit.

When the model is solved schedules are provided. The mathematical formulation is not able to mimic all the rules that the schedulers will require. Once schedules are obtained they can then be further analyzed outside of the mathematical programming software system and some diagnostics used to partially modify schedules. This last statement is not meant to be read as an opt-out, but rather to reflect the fact that modeling all schedule requirements would lead to a large MILP problem which would be very difficult to solve; using the given model provides preliminary schedules of good quality which formerly had not been available to the schedulers who had been using various heuristic rules. Thus the mathematical programming model is providing a support system from which decisions may be negotiated.

5.4.3 A Case Study Exploiting a Hierarchy of Goals

Let us illustrate how lexicographic goal programming works by considering the following example with two variables x and y *s.t.* the inequality $42x + 13y \le 100$ as well as the trivial bounds $x \ge 0$ and $y \ge 0$. We are given three goals

name	criterion	type	A/P	Δ
goal 1 (OBJ1) – profit :	$5x + 2y - 20$	max	P	10
goal 2 (OBJ2 – waste):	$-3x + 15y - 48$	min	A	4
goal 3 (OBJ2 – bonus):	$1.5x + 21y - 3.8$	max	P	20

The multi-criteria LP or MILP problem is converted to a sequence of LP or MILP problems. The basic idea is to work down the list of goals according to the priority list given. Thus we start by maximizing the LP w.r.t. the first goal. Using *goalprog* from MCOL gives us the objective function value $z_1^* = -4.615385$. Using this value z_1^* enables us to convert goal 1 into the target constraint

$$5x + 2y - 20 \ge Z_1 = z_1^* - \frac{10}{100} \left| z_1^* \right| . \tag{5.4.16}$$

Note how we have constructed the target Z_1 for this goal (P indicates that we work percentage wise). In the example we have three goals with the optimization sense {max, min, max}. Two times we apply a percentage- wise relaxation, one time absolute. Inserting the value $z_1^* = -4.615385$ we get the target constraint

$$z_1^* = -4.615385 \quad \Rightarrow \quad 5x + 2y - 20 \geq -4.615385 - 0.1 \cdot |-4.615385| . \tag{5.4.17}$$

Now we minimize w.r.t. goal 2 adding (5.4.17) as an additional inequality. We obtain z_2^* and the next target constraint

$$z_2^* = 51.133603 \quad \Rightarrow \quad -3x + 15y - 48 \leq 51.133603 + 4. \tag{5.4.18}$$

Similar as the first goal, we have converted the second goal into a constraint (5.4.18) (here we allow an absolute deviation of 4) and maximize according to goal 3. Finally, we get $z_3^* = 141.943995$ and the solution $x = 0.238062$ and $y = 6.923186$. To be complete, we could also convert the third goal into a target constraint giving

$$1.5x + 21y - 3.8 \geq 141.943995 - 0.2 \cdot 141.943995 = 113.555196.$$

The summary of the computations and results ordered in rows are

x	y	z_1	Z_1	z_2	Z_2	z_3	Z_3
0.000000	7.692308	−4.615385	−5.076923				
0.315789	6.672065	−5.076923	−5.076923	51.133603	55.133603		
0.238062	6.923186	−4.963321	−5.076923	55.133603	55.133603	141.943995	113.555196

Note that lexicographic goal programming based on objective functions provides a useful technique to tackle multi-criteria optimization problems. However, we have to keep in mind that the sequence of the goals, i.e., the hierarchy of goals, influences the solution strongly. Therefore, the absolute or percentage deviations have to be chosen with care. The whole story of multi-criteria optimization is about the challenge of trade-offs between goals measured in possibly incompatible units, or units which are difficult to compare. In the example, it was just about costs and waste (lower costs imply more waste, less waste implies higher costs). In extreme cases, it could be a trade-offs between costs and lives as in the current COVID-19 situation; people do not like putting a cost on a life. This case has two conflicting goals which cannot be transformed to a common unit.

In addition to the lexicographic goal programming variant based on a hierarchy of objective functions, we could also use lexicographic ordered constraints, in which the overall goal is to minimize the violation of constraints. In the ideal case all constraints are fulfilled. Otherwise, we try to fulfill the constraints ordered by priorities as good as possible. Unfortunately, this also leads to some sorts of weights. We thus summarize that the absolute or percentage-wise deviations used in lexicographic goal programming based on objectives are much easier to interpret.

Finally, we mention that some MILP solvers, e.g., CPLEX or GUROBI, intrinsically support hierarchical solution approaches of multi-criteria optimization problems. For some implementations see MCOL/GAMSLIB/*epscm.gms* and *epscmmip.gms*.

5.5 Limitations of Linear Programming

As with other types of mathematical techniques used to solve real-world problems, we have to accept that linear programming has limitations. Some of these limitations might have come into the reader's mind while studying the case studies in this chapter. Below we discuss these limitations and provide ideas on what we can do about them.

5.5.1 *Single Objective*

The LP problem can only handle a single objective which may be unsatisfactory. It may also be difficult for users to agree on a single objective for the purposes of the model developer. One way round this is to use the technique of goal programming as in Sect. 5.4, but even that approach is not without restrictions. It may have to be accepted that the single objective is a compromise and that the results coming from the constraints will be the most useful information. Thus early results from a model with a compromise objective may be regarded as preliminary only. Within mathematical programming software systems the facility exists to re-run the problem using a series of different objective functions, either with small variations changing individual coefficients, or with more major differences between objectives. Results can then be compared, discussed, and argued over. This may then help users agree on a single objective function for the future.

5.5.2 *Assumption of Linearity*

In the constraints and objective function of an LP (and later also MILP) problem, the assumption has to be made that all modeling can be made linear. Exceptions are the use of SOS2 and the modeling of logical expressions (which are discrete but can be made linear). The assumption of linearity may be thought of as unnaturally restrictive. However, we have to keep in mind that many real-world problems are indeed linear, or are at least linear in great parts. The case studies discussed in Sect. 5.2 show that certain problems can even be transformed into linear problems. In addition, several possibilities exist to get round this difficulty. Firstly, SOS2 are helpful in modeling certain nonlinearities. Secondly, an LP or MILP model may be thought of as an approximation to a more realistic nonlinear model and may still provide useful results. As with arguments about multiple objectives, users can still benefit from a broad-brush approach to their problem and the linear model will be the subject of debate. Nonlinear models may be much more difficult or even impossible to solve, and the users have to accept a compromise. This argument may be hard to accept, but the modeler can appeal to practical considerations using the knowledge that LP and MILP technology is very advanced and may be useful in solving approximate versions of more complex problems than could never be solved using a nonlinear approach. The LP or MILP software will allow the user to solve many models rapidly and thereby build up a set of scenarios approximating the nonlinear model. Such an approach is feasible and may be very useful.

Nevertheless, if a problem is really nonlinear, then there exist methods [see Sect. 12.6] to tackle them. However, these methods require a strong mathematical background on the modeler's side.

5.5.3 Satisfaction of Constraints

This section holds for general optimization problems, not only LP or MILP. Optimization models are built on the assumption that all constraints must be strictly satisfied at optimality.[3] This means that, in a sense, all constraints are treated as being of equal importance. This may not be a totally realistic assumption as some constraints may represent restrictions, e.g., satisfying due dates, that a client might like to be satisfied, but if the model is being applied to several scenarios it might be unrealistic to expect every constraint to be satisfiable in every scenario. In a manner analogous to what is done in the model of the satisfiability problem [see Chap. 7], it may be necessary to introduce extra *violation variables* into certain constraints to represent exceptional shortfalls or surpluses. These will act similarly to slack variables, but a "cost" will be associated with such variables — the cost of not satisfying a constraint. Such constraints were referred to in Sect. 5.4 as *soft constraints*, while constraints which must not be broken were referred to as *hard constraints*. A collection of cases to be considered as scenarios based on different combinations of soft constraints will doubtless prove useful in convincing a client.

Ranging over right-hand side values may also provide a way of allowing constraints to be weakened as required. The model may be solved initially with larger than necessary right-hand side values and then smaller than necessary right-hand side values in constraints. Right-hand sides can be progressively tightened in subsequent runs of the model to test if the optimal solution remains feasible. This process will be particularly useful where an existing set of constraints is being expected to be satisfied in a new situation, e .g., a model from an old factory being applied to a new one. In such cases it might not be expected that all constraints could

[3]Of course, there are certain constraints, which *must* be fulfilled strictly. Constraints representing mass balances or other physical laws are examples of such constraints.

be satisfied, but the client wants to try to keep the model as similar to the old one as possible.

5.5.4 Structured Situations

Mathematical optimization models are assumed to be only applicable in highly structured situations. It would seem hard to design a model that would be useful for the organization of games involving prizes at a children's party! As many industrial or commercial environments are unstructured or volatile, mathematical programming would seem to be inappropriate. However, this view would be extreme. Structure can often be imposed on unstructured situations, and even if this is slightly artificial it will usually lead to a better understanding of the unstructured situation. It may also be possible to isolate parts of an unstructured system which are highly structured. By modeling these advantageously, the remaining efforts can be devoted to the unstructured parts.

5.5.5 Consistent and Available Data

Mathematical optimization models rely on data of consistent quality being readily available. This will not always be so. Indeed, collecting data is one of the most critical problems involved in solving large scale optimization problems. It may be glib to assume that we can obtain information on the costs of each of 20 stages of a production process. People we ask for information may not even be able to agree on a definition of cost let alone quantify aspects of it. Thus what we do will be a compromise. This may lead to a lack of user confidence in the model.

What we must do is to devise the best procedures we can for data collection and see that procedures are adhered to if the model is going to be used repeatedly on data which may change over time. New data must be collected under the same terms as the old, until a major overhaul of the model is undertaken. We must then make it clear to users of the model that results were obtained under certain assumptions and stress that results should still be helpful in providing decision support provided appropriate caution is maintained.

It is possible for model users to vary data by making use of *sensitivity*, *post-optimality*, and *ranging* analyses which may go a considerable way to getting round data deficiencies; see also the comments on *robust solutions* in Sect. 17.2.6; pp. 557. Finally, a technical approach for detecting data inconsistencies is provided in Sect. 14.1.2.3.

5.6 Summary

At the end of this chapter, the reader should be able to:

- Appreciate the formulation of more complex LP problems;
- Formulate certain more awkward LP problems such as blending problems;
- Perform efficiency modeling using DEA;
- Handle goal programming and problems requiring "soft" constraints;
- Be aware of ways of avoiding apparent nonlinearities in problems so that they may be formulated as LP models;
- Be aware of some of the limitations of linear programming and mathematical optimization in general.

5.7 Exercises

1. Lim is a company which manufactures two end-products, P1 and P2, for sale, using two scarce resources, R1 and R2, and two intermediate products, I1 and I2, which are only used in the manufacturing of the end-products. Lim is about to plan its manufacture for the next 4 months and wants to make as much profit as possible. It knows its manufacturing technology well, and has produced a table showing the requirements per unit for the end-products:

End-Product	R1	R2	I1	I2
P1	8	10	4	2
P2	6	12	3	3

so, for example, making 1 unit of end-product 1 requires 8 units of resource 1, 10 units of resource 2, 4 units of intermediate product 1, and 2 units of intermediate product 2. Lim has to make the intermediate products using the same scarce resources. Here is the requirements table.

Intermediate	R1	R2
I1	1.0	0.7
I2	0.6	1.2

Both intermediate and end-products can be stored from month to month. To store 1 unit of an end-product for one month costs £3, while to store a unit of intermediate product costs £1. The selling prices (in £) of the end-products vary month by month:

Product	Month 1	Month 2	Month 3	Month 4
P1	100	105	107	90
P2	90	100	110	115

No more than 30 units of any end-product can be sold in any month. The manufacturing costs /unit of end-products and intermediate products are (in £):

End-product	Month 1	Month 2	Month 3	Month 4
P1	40	42	55	35
P2	38	35	44	40

and

Intermediate	Month 1	Month 2	Month 3	Month 4
I1	6	6.2	5.3	4.8
I2	5.1	5.2	5.0	5.1

.

The availabilities of scarce resources vary widely from month to month:

Resource	Month 1	Month 2	Month 3	Month 4
R1	200	300	100	200
R2	250	400	50	240

.

Stocks currently stand at

	P1	P2	I1	I2
Level	40	38	60	50

and must be the same at the end of the 4th month.

(a) Formulate Lim's problem, identifying clearly the variables you use and the constraints that you construct.
(b) Construct a computer model of the problem that could be extended to deal with more end-products, intermediates, and time periods.
(c) Solve the problem and report upon your results.

2. Solve the three separate DEA problems (for Paris, London, and Bonn) given in Sect. 5.3.

Chapter 6
Modeling Structures Using Mixed Integer Programming

In this chapter we shall look at ways in which problems may be formulated using a series of devices mainly involving integer variables for counting indivisible entities (people, living animals, air planes, etc.), and particularly binary variables. Among other things, binary variables can be used to model

1. states,
2. logical conditions and logical expressions, and
3. special nonlinear terms and expressions.

In particular, we shall concentrate on the use of binary variables to model simple nonlinear features. Such features can be handled because binary variables allow us to model logical conditions. These approaches will then be illustrated by examples and later many of the approaches will appear in a series of short case studies.

6.1 Using Binary Variables to Model Logical Conditions

It is often convenient to express parts of a model using symbols from mathematical logic as a way of describing conditions within the model which must subsequently be converted into constraints in the model. The symbols are the connectives: disjunction *or* ($\vee$), conjunction *and* ($\wedge$), *not* or *negation* ($\neg$), *implication* (or *if*) ($\Longrightarrow$), and *equivalence* ($\Longleftrightarrow$). Many logical expressions, L_i, can be built from linking together statements using logical connectives, e.g.,

$$L_1 \vee L_2 \Rightarrow L_3. \tag{6.1.1}$$

Supplementary Information The online version contains supplementary material available at https://doi.org/10.1007/978-3-030-73237-0_6.

J. Kallrath, *Business Optimization Using Mathematical Programming*, International Series in Operations Research & Management Science 307, https://doi.org/10.1007/978-3-030-73237-0_6

To demonstrate the need for such concepts, consider the example of a two-stage chemical production plant producing pre-products and sales products. The production follows the rule: if any of the sales products S_1, S_2, and S_3 is produced, then at least one of the pre-products P_1 or P_2 must be produced. We could now introduce the *logical variables*[1] L_1^S, L_2^S, L_3^S, L_1^P, and L_2^P representing the decision that the product considered is produced. Then, our production rule is expressed as

$$(L_1^S \vee L_2^S \vee L_3^S) \Longrightarrow (L_1^P \vee L_2^P). \tag{6.1.2}$$

When dealing with logical expressions, it is helpful to use rules to transform logical terms into equivalent ones.[2] For instance, there are De Morgan's laws:

$$\neg(L_1 \vee L_2) \Leftrightarrow \neg L_1 \wedge \neg L_2 \quad , \quad \neg(L_1 \wedge L_2) \Leftrightarrow \neg L_1 \vee \neg L_2. \tag{6.1.3}$$

Further useful relations are the distributive laws

$$L_1 \vee (L_2 \wedge L_3) \quad \Leftrightarrow \quad (L_1 \vee L_2) \wedge (L_1 \vee L_3) \tag{6.1.4}$$

$$L_1 \wedge (L_2 \vee L_3) \quad \Leftrightarrow \quad (L_1 \wedge L_2) \vee (L_1 \wedge L_3) \tag{6.1.5}$$

and

$$(L_1 \vee L_2) \Longrightarrow L_3 \quad \Leftrightarrow \quad (L_1 \Longrightarrow L_3) \wedge (L_2 \Longrightarrow L_3). \tag{6.1.6}$$

6.1.1 General Integer Variables and Logical Conditions

If a machine is always set to exactly one of six modes numbered 1-6, then it might be expected that we could model this by introducing an integer variable α such that $0 \le \alpha \le 5$, and if $\alpha = i$ it is taken to mean that mode $i+1$ is selected. However, this rarely turns out to be practical because we usually also want to model some event like "if mode 3 or 5 is selected then ...". The implication must then take the form

$$\alpha = 2 \Longrightarrow \ldots, \quad \alpha \neq 2 \Longrightarrow \ldots, \quad \alpha = 4 \Longrightarrow \ldots, \quad \alpha \neq 4 \Longrightarrow \ldots \tag{6.1.7}$$

This turns out to be difficult, although not totally impossible, to model and requires the introduction of binary variables. Instead of a general integer variable it is better to introduce only a set of binary variables to model the machine modes as follows. Let us introduce the variables

$$\delta_i = \begin{cases} 1 \text{ , if the machine mode } i \text{ is selected} \\ 0 \text{ , otherwise} \end{cases} \quad , \quad i = 1, 2, \ldots, 6 \tag{6.1.8}$$

[1] Note that logical variables can only take the values *true* and *false*, or T and F, for short.

[2] We use the symbol $\Leftrightarrow$ to mean "is equivalent to."

and the constraint, to ensure that the machine is in exactly one mode,

$$\sum_{i=1}^{6} \delta_i = 1. \tag{6.1.9}$$

The implications can more easily be derived from

$$\delta_3 = 1 \Longrightarrow \ldots, \qquad \delta_3 = 0 \Longrightarrow \ldots \tag{6.1.10}$$

$$\delta_5 = 1 \Longrightarrow \ldots, \qquad \delta_5 = 0 \Longrightarrow \ldots \tag{6.1.11}$$

For instance, as we will see later, if we want to have the condition

$$\delta_3 = 1 \Longrightarrow \text{“extra raw materials are required”} \tag{6.1.12}$$

then this is easy to model. Thus, general integer variables are rarely used to represent *logical relations*, but rather they are used to represent actual quantities, e.g., number of supervisors on a shift. When the constraint (6.1.9) is introduced, the current setting of the state can be modeled by the variable α where

$$\alpha = \sum_{i=1}^{6} (i-1)\delta_i \tag{6.1.13}$$

and α is an integer variable with upper bound 5.

6.1.2 Transforming Logical into Arithmetical Expressions

In what follows, let us assume that any or all of n products can be produced. Let us further introduce logical variables L_p taking the value $true$ if we decide to produce product p (positive decision), and $false$ if we decide not to produce product p (negative decision). Production of individual products may or may not be interrelated. In addition, we will use binary variables $\delta_1, \delta_2, \ldots, \delta_n$ such that

$$\delta_p := \begin{cases} 1\text{ , if positive decision concerning } p \text{ is taken, i.e., } L_p = true \\ 0\text{ , if negative decision concerning } p \text{ is taken, i.e., } L_p = false. \end{cases} \tag{6.1.14}$$

Logical expressions or variables can have the two values *true* and *false*, and these correspond naturally to the values 1 and 0 that the binary variables can take. However, binary variables allow us to do arithmetic which is often an advantage. When we use a collection of binary variables, information concerning the collection can be handled easily in constraints by adding variables, e.g.,

$$\delta_1 + \delta_2 + \delta_3 + \delta_4 \tag{6.1.15}$$

will represent a quantity which can take a value between 0 and 4 and will represent the total number of positive decisions, i.e., the number of products being produced from the four that are possible.

In what follows we will give a translation between several logical and arithmetic expressions. If we introduce the binary variable δ to represent a logical decision (or variable) L, then the opposite of that decision $\neg L$ is given by

$$\delta' = 1 - \delta. \tag{6.1.16}$$

We can see this in the table

δ	$1-\delta$
1	0
0	1

(6.1.17)

Sometimes we might want to apply the *false* operator to a logical variable L. Such requirements can be expressed as

$$\delta = 0. \tag{6.1.18}$$

6.1.3 Logical Expressions with Two Arguments

Let us now concentrate on logical expressions involving two logical variables L_1 and L_2 which can assume the values *true* and *false*. Then we have the following transformations between logical operators and constraints

Logical operator	Logical expression	Constraint
or	$L_1 \vee L_2$	$\delta_1 + \delta_2 \geq 1$
and	$L_1 \wedge L_2$	$\delta_1 + \delta_2 = 2$
not	$\neg(L_1 \wedge L_2)$	$\delta_1 + \delta_2 \leq 1$
not	$\neg(L_1 \vee L_2)$	$\delta_1 = 0, \delta_2 = 0$
implication	$L_1 \Longrightarrow L_2$	$\delta_1 \leq \delta_2$
equivalence	$L_1 \Longleftrightarrow L_2$	$\delta_1 = \delta_2$.

(6.1.19)

We can check that a constraint logically performs the task required of it by considering all the possible values that the δ variables may take and the effect of these on the constraint. This is called constructing a *truth table*, an example of which for one constraint is shown below.

L_1	L_2	$L_1 \vee L_2$	δ_1	δ_2	$\delta_1 + \delta_2 \geq 1$
T	F	T	1	0	satisfied
T	T	T	1	1	satisfied
F	T	T	0	1	satisfied
F	F	F	0	0	violated.

(6.1.20)

The truth table enables us to establish that the logic is correctly modeled just as we saw in (6.1.17).

In the constraint representing the disjunction of L_1 and L_2, $L_1 \vee L_2$, meaning "$L_1 = T$ *or* $L_2 = T$", at least one of the two δ variables will be forced to take the value 1 and hence the inequality $\delta_1 + \delta_2 \geq 1$ implies that at least one logical variable takes the value *true*. Note that both variables could take the value 1, in which case both L_1 and L_2 are *true*. It was not explicit in the original requirements on the decisions that only one positive decision could be taken at a time. Thus, the possibility has been included that both decisions could be taken simultaneously. If we had wanted to model *exclusive or* or *exactly one of* in the above, then the equality constraint

$$\delta_1 + \delta_2 = 1 \tag{6.1.21}$$

would have been used.

This aspect of modeling logical conditions is a gentle reminder as to how important it is to obtain precise information from a client for whom a model is being produced. The word *or* may mean *at least one of* or may mean *exactly one of*. The differences between the logical conditions here may seem marginal when a model is being built initially, but may turn out to have major impact when a proposed solution is being evaluated — and it even affects feasibility.

The conjunction of L_1 and L_2, $L_1 \wedge L_2$ (meaning "$L_1 = T$ *and* $L_2 = T$") leads to

$$\delta_1 = 1 \tag{6.1.22}$$

and

$$\delta_2 = 1 \tag{6.1.23}$$

or, in short,

$$\delta_1 + \delta_2 = 2. \tag{6.1.24}$$

This may seem a somewhat pointless condition to include as it will clearly force δ_1 and δ_2 to take the values 1, and it might have been better to exclude them from being variables in the first place. However, it is included for the sake of completeness.

As can be seen from the development of $\wedge$ and $\vee$, the sum of several binary variables gives us a quantity which tells us how many positive decisions have been taken. Subsets of $\{1, 2, \ldots, n\}$ can be selected so that particular groups of decisions are constrained. For example,

$$\delta_3 + \delta_6 \tag{6.1.25}$$

provides an expression relating to two particular decisions.

The reverse conditions to $\vee$ and $\wedge$, $\neg(L_1 \vee L_2)$ and $\neg(L_1 \wedge L_2)$ can be modeled after applying De Morgan's laws (6.1.3). As $\neg(L_1 \wedge L_2) = \neg L_1 \vee \neg L_2$, then using the expression for *or* given in (6.1.19) we have

$$1 - \delta_1 + 1 - \delta_2 \geq 1, \tag{6.1.26}$$

which can be rearranged as

$$\delta_1 + \delta_2 \leq 1. \tag{6.1.27}$$

$\neg(L_1 \vee L_2) = \neg L_1 \wedge \neg L_2$ is handled in a corresponding way.

Using the relations now developed, we can build up complicated examples from easier ones. It is usually best to introduce a variable to represent $\neg L$, if such is required, e.g., δ'. Later the substitution $\delta' = 1 - \delta$ can be made in the constraints. For example, the statement that "if decision 1 is positive, then decision 2 is negative," i.e., $L_1 \Longrightarrow \neg L_2$, can be modeled as

$$\delta_1 \leq \delta_2', \tag{6.1.28}$$

where δ_2' corresponds to $\neg L_2$. Then (6.1.28) can be rearranged as

$$\delta_1 \leq 1 - \delta_2 \quad \text{or} \quad \delta_1 + \delta_2 \leq 1. \tag{6.1.29}$$

Notice that this expression is the same as that used for modeling $\neg(L_1 \wedge L_2)$ which is an equivalent expression for $\neg L_1 \vee \neg L_2$. That these should be equivalent logically can be seen as follows: $\delta_1 = 1$ requires $\delta_2 = 0$. However, if $\delta_2 = 1$, then it follows from the constraint (6.1.28) that δ_1 will be zero, suggesting that we are also modeling that "if decision 2 is taken then decision 1 cannot be taken." This is quite logical, because if decision 2 were taken then we must bar decision 1 from also being taken; otherwise, a contradiction results, namely that decision 2 could not be taken. We could also check this by examining the appropriate truth table.

6.1.4 Logical Expressions with More Than Two Arguments

The basic method used to transform logical expressions with more than two arguments into arithmetic ones is to apply the transformations such as (6.1.4) in order to connect logical expressions by $\wedge$. Each term connected by $\wedge$ then corresponds to a constraint.

The statement $L_1 \wedge (L_2 \vee L_3)$ is represented by the constraints

$$\delta_1 = 1 \quad , \quad \delta_2 + \delta_3 \geq 1. \tag{6.1.30}$$

To model the statement $L_1 \vee (L_2 \wedge L_3)$, we first apply (6.1.4) and get $(L_1 \vee L_2) \wedge (L_1 \vee L_3)$. That leads us immediately to

$$\delta_1 + \delta_2 \geq 1 \quad , \quad \delta_1 + \delta_3 \geq 1. \tag{6.1.31}$$

From these inequalities we could derive (by addition of inequalities) the inequality

$$2\delta_1 + \delta_2 + \delta_3 \geq 2, \tag{6.1.32}$$

which also provides an exact model for the original logical expression $L_1 \vee (L_2 \wedge L_3)$. This shows us that in general there is no unique mapping. However, note that (6.1.31) is "tighter" than (6.1.32), i.e., all integer solutions to (6.1.32) are valid in (6.1.31) and vice versa, but there may be fractional solutions to (6.1.31) that are not valid for (6.1.32). To see this consider the LP relaxation of both the cases (6.1.31) and (6.1.32). While $(\delta_1, \delta_2, \delta_3) = (0.8, 0.3, 0.1)$ fulfills (6.1.32) it violates the second inequality of (6.1.31). So we see that (6.1.31) cuts off more fractional combinations.

We are now in a position to formulate the logical expression (6.1.2) in the two-stage chemical production example. At first we apply (6.1.6) getting

$$(L_1^S \Longrightarrow L) \wedge (L_2^S \Longrightarrow L) \wedge (L_3^S \Longrightarrow L), \tag{6.1.33}$$

where we replace $L_1^P \vee L_2^P$ by L where $L \Longrightarrow L_1^P \vee L_2^P$. Let us associate a binary variable δ with L. Then, from table (6.1.19), we derive the constraints

$$\delta \leq \delta_1^P + \delta_2^P \tag{6.1.34}$$

and

$$\delta_1^S \leq \delta \quad , \quad \delta_2^S \leq \delta \quad , \quad \delta_3^S \leq \delta. \tag{6.1.35}$$

To conclude this section we list some useful relations involving three variables:

Relations	Constraint(s)
$L_1 \Longrightarrow (L_2 \wedge L_3)$	$\delta_1 \le \delta_2 \quad , \quad \delta_1 \le \delta_3$
$L_1 \Longrightarrow (L_2 \vee L_3)$	$\delta_1 \le \delta_2 + \delta_3$
$(L_1 \wedge L_2) \Longrightarrow L_3$	$\delta_1 + \delta_2 \le 1 + \delta_3$
$(L_1 \vee L_2) \Longrightarrow L_3$	$\delta_1 \le \delta_3 \quad , \quad \delta_2 \le \delta_3$
$L_1 \wedge (L_2 \vee L_3)$	$\delta_1 = 1 \quad , \quad \delta_2 + \delta_3 \ge 1$
$L_1 \vee (L_2 \wedge L_3)$	$\delta_1 + \delta_2 \ge 1 \quad , \quad \delta_1 + \delta_3 \ge 1.$

(6.1.36)

For convenience we also list some more general logical expressions involving n variables but still leading to one constraint. Let us introduce the convenient notation

$$\bigvee_{i=1}^{n} L_i := L_1 \vee L_2 \vee \ldots \vee L_n \quad , \quad \bigwedge_{i=1}^{n} L_i := L_1 \wedge L_2 \wedge \ldots \wedge L_n . \tag{6.1.37}$$

Some useful results which the reader might confirm in exercises are

$$\begin{array}{lll} \text{disjunction} & \bigvee_{i=1}^{n} L_i & : \sum_{i=1}^{n} \delta_i \ge 1 \\ \text{conjunction} & \bigwedge_{i=1}^{n} L_i & : \sum_{i=1}^{n} \delta_i = n \\ \text{implication} & \bigwedge_{i=1}^{k} L_i \Longrightarrow \bigvee_{i=k+1}^{n} L_i & : \sum_{i=k+1}^{n} \delta_i - \sum_{i=1}^{k} \delta_i \ge 1 - k \\ & \text{at least } k \text{ out of } n & : \sum_{i=1}^{n} \delta_i \ge k \\ & \text{exactly } k \text{ out of } n & : \sum_{i=1}^{n} \delta_i = k \\ & \text{at most } k \text{ out of } n & : \sum_{i=1}^{n} \delta_i \le k . \end{array} \tag{6.1.38}$$

Finally we generalize some relations containing the equivalence operator $\Longleftrightarrow$ and $n+1$ variables. To model the statement $L_{n+1} \Longleftrightarrow \bigvee_{i=1}^{n} L_i$ we use the constraints

$$\sum_{i=1}^{n} \delta_i \ge \delta_{n+1} \quad , \quad \delta_{n+1} \ge \delta_k \quad , \quad k = 1, \ldots, n. \tag{6.1.39}$$

The statement $L_{n+1} \Longleftrightarrow \bigwedge_{i=1}^{n} L_i$ is modeled by the constraints

$$-\sum_{i=1}^{n} \delta_i + \delta_{n+1} \ge 1 - n \quad , \quad \delta_{n+1} \le \delta_k \quad , \quad k = 1, \ldots, n. \tag{6.1.40}$$

Notice that when we model logic we organize what has to be modeled into the form

$$A \quad and \quad B \quad and \quad C \quad \ldots , \tag{6.1.41}$$

because we will model such an expression by a series of ILP constraints *all* of which must be satisfied. Thus, our model transforms to

$$\text{constraint } 'A' \quad and \quad \text{constraint } 'B' \quad and \quad \text{constraint } 'C' \quad \ldots.. \tag{6.1.42}$$

6.2 Logical Restrictions on Constraints

In order to handle logical restrictions placed on expressions which are themselves constraints it is helpful and necessary to find out which values the left-hand side of these constraints can take. Therefore, we first calculate the largest and smallest values the left-hand side of a constraint can take [see for instance Brearley et al. (1975,[98]) or McKinnon and Williams (1989,[390]) for further discussion]. Consider the following example:

Suppose we wish to place restrictions on the expression

$$A_1x_1 + A_2x_2 + \ldots + A_nx_n \circ B, \tag{6.2.1}$$

which should be rewritten as

$$A_1x_1 + A_2x_2 + \ldots + A_nx_n - B \circ 0, \tag{6.2.2}$$

where $\circ$ represents one of $=$, $\leq$, or $\geq$, and the variables $x_1, x_2, \ldots, x_n$ are continuous or integer, but each variable has a lower limit (is bounded below) and has an upper limit (is bounded above); in mathematical programming one usually calls these limits *bounds*. Let $L_1, L_2, \ldots, L_n$ represent the lower bounds of each variable (which will usually be zero but need not be, and must be finite) and $U_1, U_2, \ldots, U_n$ represent the upper bounds on each variable (which must be finite). Then the largest (smallest) value, $U(L)$, the expression (6.2.1) can take is

$$U = \sum_{i=1}^{n} A_iU_i - B \quad , \quad L = \sum_{i=1}^{n} A_iL_i - B, \tag{6.2.3}$$

unless any A_i is negative in which case the corresponding U_i should be replaced by L_i.

Now we can derive several logical restrictions on these constraints.

6.2.1 Bound Implications on Single Variables

The statements "if decision 1 is taken then variable x must be larger than some positive constant C" and "if decision 1 is not taken then variable x is zero" are modeled as

$$x \geq C\delta_1 \quad , \quad x \leq U\delta_1. \tag{6.2.4}$$

In the first statement we are assuming that $C > L$, otherwise the lower bound has no effect. Let us illustrate the upper bound implication considering the following example:

Example If production of *Xyrene* is started in a shift, there is a set-up cost of £1,000 irrespective of how much *Xyrene* is produced. The maximum quantity of *Xyrene* that can be produced in a shift is 500 tons. The (variable) cost of production of each tonne of *Xyrene* is £100. To model production cost C we introduce a decision variable δ such that

$$\delta = \begin{cases} 1\ , \text{if production of Xyrene is started in a shift} \\ 0\ , \text{otherwise,} \end{cases} \tag{6.2.5}$$

and a variable x to represent the amount of *Xyrene* produced, which may be zero. Then we require

$$x \leq 500\delta \quad , \quad C = 100x + 1000\delta.$$

6.2.2 Bound Implications on Constraints

The statements "if decision 1 is taken then $\sum_{i=1}^{n} A_i x_i \leq B$ must hold" is modeled as

$$\sum_{i=1}^{n} A_i x_i - B \leq U(1 - \delta_1). \tag{6.2.6}$$

Similarly, the statement "if decision 1 is taken then $\sum_{i=1}^{n} A_i x_i \geq B$ must hold" is modeled as

$$\sum_{i=1}^{n} A_i x_i - B \geq L(1 - \delta_1). \tag{6.2.7}$$

By contrast, the statement "if decision 1 is taken then $\sum_{i=1}^{n} A_i x_i = B$ must hold" is modeled by simultaneously requiring the constraints

$$\sum_{i=1}^{n} A_i x_i - B \leq U(1 - \delta_1) \quad , \quad \sum_{i=1}^{n} A_i x_i - B \geq L(1 - \delta_1).$$

Note that if $\delta_1 = 1$ we have

$$\sum_{i=1}^{n} A_i x_i - B \leq 0 \quad , \quad \sum_{i=1}^{n} A_i x_i - B \geq 0, \tag{6.2.8}$$

which is equivalent to $\sum_{i=1}^{n} A_i x_i = B$ as required. $\delta_1 = 0$ leads to

$$\sum_{i=1}^{n} A_i x_i - B \leq U \quad , \quad \sum_{i=1}^{n} A_i x_i - B \geq L, \tag{6.2.9}$$

which, according to the definition of L and U as upper and lower bounds, is always fulfilled. Following this argumentation and inspecting (6.2.6) and (6.2.7), we understand now, why in the beginning it was important to derive the lower and upper bounds.

Example A petroleum company is investigating a new region for likely drilling sites. The region is divided into six plots, each of which is equally likely to contain oil. Seismological information will provide the company with one of the following indications in the next month: the region is a good prospect, the region is a medium prospect, or the region is a poor prospect.

If the indication is that the region is a poor prospect, the company will take up drilling options on at least one and at most two of the plots; if the indication is that the region is a medium prospect, the company will take up drilling options on at least two and at most four of the plots, while if the indication is that the region is a good prospect, the company will take up drilling options on at least three and at most five of the plots.

The problem may be modeled as follows. Let δ_1, δ_2, and δ_3 decide on poor, medium, and good prospect, i.e.,

$$\delta_{1(2,3)} = \begin{cases} 1 \text{ , if the region is a poor (medium, good) prospect} \\ 0 \text{ , otherwise,} \end{cases} \tag{6.2.10}$$

and for $i = 1, 2, \ldots, 6$

$$\alpha_i = \begin{cases} 1 \text{ , if the option is taken up on plot } i \\ 0 \text{ , otherwise.} \end{cases} \tag{6.2.11}$$

The constraints are then

$$\delta_1 + \delta_2 + \delta_3 = 1 \tag{6.2.12}$$

saying that exactly one indication will be given. Let us now model that at least one drilling option is taken up if prospects are poor:

$$\sum_{i=1}^{6} \alpha_i \geq \delta_1. \tag{6.2.13}$$

We see that if the indication is poor then $\delta_1 = 1$, and at least one option is taken up; otherwise, the constraint is not binding as the minimum number of options taken up could be zero. The upper bound (at most two out of six) is modeled as

$$\sum_{i=1}^{6} \alpha_i - 2 \leq 4(1 - \delta_1), \tag{6.2.14}$$

which can be rewritten as

$$\sum_{i=1}^{6} \alpha_i \leq 6 - 4\delta_1. \tag{6.2.15}$$

If $\delta_1 = 1$, then the number of options taken up is at most two; otherwise, the constraint is not binding as the maximum number of options taken up could be six.

Modeling the other prospects proceeds similarly:

$$\sum_{i=1}^{6} \alpha_i \geq 2\delta_2. \tag{6.2.16}$$

If the indication is medium then $\delta_2 = 1$, and at least two options are taken up; otherwise, the constraint is not binding as the minimum number of options taken up could be zero.

$$\sum_{i=1}^{6} \alpha_i \leq 6 - 2\delta_2. \tag{6.2.17}$$

If $\delta_2 = 1$, then the number of options taken up is at most four; otherwise, the constraint is not binding as the maximum number of options taken up could be six.

$$\sum_{i=1}^{6} \alpha_i \geq 3\delta_3. \tag{6.2.18}$$

If the indication is good then $\delta_3 = 1$, and at least three options are taken up; otherwise, the constraint is not binding as the minimum number of options taken up could be zero.

$$\sum_{i=1}^{6} \alpha_i \leq 6 - \delta_3. \tag{6.2.19}$$

If $\delta_3 = 1$, then the number of options taken up is at most five; otherwise, the constraint is not binding as the maximum number of options taken up could be six.

6.2.3 Disjunctive Sets of Implications

The logical condition "if decision 3 is not taken then decisions 1 and 2 cannot be taken", could be modeled as

$$\delta_1 + \delta_2 \leq 2\delta_3. \tag{6.2.20}$$

However, it turns out that it is better to model the relationship as

$$\delta_1 \leq \delta_3 \quad , \quad \delta_2 \leq \delta_3. \tag{6.2.21}$$

The second formulation is such that more fractional values of the variables are excluded. This type of formulation is known as a *sharp formulation* which is more fully discussed in Jeroslow & Lowe (1984,[288] & 1985,[289]). Some indication of the strength of (6.2.21) compared to (6.2.20) can be seen from the fact that no integer solution in the variables δ_1, δ_2 is permitted by a non-integer solution for δ_3 in (6.2.21), but $\delta_3 = 0.5$ allows $\delta_1 = 1, \delta_2 = 0$ in (6.2.20). The idea for developing such formulations is to avoid the LP relaxation of an ILP formulation taking certain easily avoidable values for variables. Another more sophisticated example is the following.

If exactly one of a set of m constraints must hold, then this can be modeled by extending (6.2.6) above. Let the constraints be

$$\sum_{j=1}^{n} A_{ij}x_j \leq B_i \quad , \quad i = 1, 2, \ldots, m. \tag{6.2.22}$$

Then the requirement that exactly one must hold can be modeled by introducing the $\{0,1\}$ variables $\delta_1, \delta_2, \ldots, \delta_m$ and the constraints

$$\delta_1 + \delta_2 + \ldots + \delta_m = 1 \tag{6.2.23}$$

$$\sum_{j=1}^{n} A_{ij}x_j - B_i \leq U_i(1 - \delta_i) \quad , \quad i = 1, 2, \ldots, m, \tag{6.2.24}$$

where U_i is an upper bound on

$$\sum_{j=1}^{n} A_{ij}x_j - B_i. \tag{6.2.25}$$

There is an alternative way to formulate the above. The alternative formulation was developed by Jeroslow & Lowe (1984,[288]) and performs better on large problems because the LP relaxation of the constraints gives the convex hull of ILP solutions. As the formulation results in a larger model in terms of numbers of constraints and variables, the overhead associated with this when the model is solved implies that this formulation is not efficient for small problems. The alternative formulation requires the introduction of additional variables $x_{ij} \geq 0$ such that

$$x_j = \sum_{i=1}^{m} x_{ij} \quad , \quad j = 1, 2, \ldots, n. \tag{6.2.26}$$

Assuming

$$A_{ij} \geq 0, \quad i = 1, 2, \ldots, m, \quad j = 1, 2, \ldots, n, \tag{6.2.27}$$

the formulation then has the constraints

$$\sum_{j=1}^{n} A_{ij} x_{ij} - B_i \delta_i \leq 0 \quad , \quad i = 1, 2, \ldots, m \tag{6.2.28}$$

$$\sum_{i=1}^{m} \delta_i = 1, \tag{6.2.29}$$

where, as before, $\delta_1, \delta_2, \ldots, \delta_m$ are binary variables.

The constraint (6.2.29) requires that exactly one δ variable takes the value 1, δ_r say. Since we assumed that $A_{ij} \geq 0$ this then forces to zero all x_{ij} variables except x_{rj} $(j = 1, 2, \ldots, n)$. Let us inspect (6.2.28) in more detail:

$$\sum_{j=1}^{n} A_{ij} x_{ij} = 0 \quad , \quad \forall i \neq r \tag{6.2.30}$$

leading to $x_{ij} = 0$ for all $i \neq r$, and

$$\sum_{j=1}^{n} A_{rj} x_{rj} \leq B_r \tag{6.2.31}$$

ensuring that exactly one constraint of (6.2.22) holds, but with modified x variables. Note that due to $x_{ij} = 0$ for all $i \neq r$, (6.2.26) takes the simple form

$$x_j = x_{rj} \quad , \quad j = 1, 2, \ldots, n, \tag{6.2.32}$$

which enables us to replace x_{rj} in (6.2.31).

The topic briefly touched in this section finds it natural extension in disjunctive programming, a technique and discipline initiated by Egon Balas (2018,[44]) in the early 1970s, which has become a central tool for solving non-convex optimization problems like pure or mixed integer programs, through convexification (cutting plane) procedures combined with enumeration. It has played a major role in the revolution in the state of the art of integer programming that took place roughly during the period 1990-2010. Grossmann and Trespalacios (2013,[242]) describe systematic modeling of discrete-continuous optimization (both, MILP or MINLP) models through generalized disjunctive programming.

6.3 Modeling Non-Zero Variables

Let us consider the case in which we have a countable quantity σ which can take all integral values between 0 and 10 but for some reason not the value 4. Such a quantity σ could be represented if we first introduce a (continuous or integral) variable, ς, not restricted in sign with the bounds

$$-Z_1 \leq \varsigma \leq Z_2 \quad , \quad \varsigma \neq 0 \quad , \quad 0 < Z_1 = 4 \quad , \quad 0 < Z_2 = 6 \tag{6.3.1}$$

and then include the constraint

$$\sigma - 4 = \varsigma. \tag{6.3.2}$$

We can represent such non-zero variables in both continuous and integral cases by using a small non-negative parameter ε and two binary variables $\delta_1, \delta_2 \in \{0, 1\}$. Then we apply the ideas of Sect. 6.2.3 for the case $m = 2, n = 1$ to the disjunctive set of inequalities

$$\varsigma \leq -\varepsilon \quad \vee \quad \varsigma \geq \varepsilon \tag{6.3.3}$$

or, if we want to use only $\leq$ inequalities to apply (6.2.24)

$$\varsigma \leq -\varepsilon \quad \vee \quad -\varsigma \leq -\varepsilon. \tag{6.3.4}$$

If we apply (6.2.24) to this pair of inequalities we get the two inequalities

$$\varsigma + \Delta \leq U_1(1 - \delta_1) \quad , \quad -\varsigma + \Delta \leq U_2(1 - \delta_2) \tag{6.3.5}$$

with upper bounds $U_1 = Z_2 + \Delta$ and $U_2 = Z_1 + \Delta$. This leads to the constraints

$$\varsigma + (Z_2 + \Delta)\delta_1 \leq Z_2 \quad , \quad \varsigma - (Z_1 + \Delta)\delta_2 \geq -Z_1 \quad , \quad \delta_1 + \delta_2 = 1 \tag{6.3.6}$$

or the equivalent form,

$$\varsigma + \Delta\delta_1 \le \delta_2 Z_2 \quad , \quad \varsigma - \Delta\delta_2 \ge -\delta_1 Z_1 \quad , \quad \delta_1 + \delta_2 = 1. \tag{6.3.7}$$

If we inspect either (6.3.6) or (6.3.7) we see that the following implications hold:

$$\delta_1 = 1 \quad \Longrightarrow \delta_2 = 0 \quad , \quad -Z_1 \le \varsigma \le -\Delta \tag{6.3.8}$$

and

$$\delta_1 = 2 \quad \Longrightarrow \delta_1 = 0 \quad , \quad \Delta \le \varsigma \le Z_2, \tag{6.3.9}$$

which is exactly what we expected for our non-zero variable ς. Equations (6.3.6) or (6.3.7) are algebraically equivalent so they might lead to different performance in the B&B algorithm. Therefore, in applications one should try both formulations.

To give a practical example of how non-zero variables may enter a model, consider the situation in which a variable representing a temperature occurs. Let the temperatures be measured in degrees Celsius. The physical constraints in the model might be such that some device cannot operate if the temperature is close to or around the freezing point of water. In that case, we might represent the temperature as a non-zero variable and choose $\Delta = 1°\text{C}$.

6.4 Modeling Sets of All-Different Elements

A simple example of modeling the *all-different relation* is given by a group of salesmen, where each of them has to visit a different city. (We do not assume that every city has to be visited by a salesman; there may be more cities than salesmen. This problem can be modeled by the conventional approach using binary variables δ_{ij} expressing whether salesman i travels to city j and then adding the constraints

$$\sum_i \delta_{ij} \le 1 \quad , \quad \forall j, \tag{6.4.1}$$

which expresses that no city can be visited by more than one salesman, and

$$\sum_j \delta_{ij} = 1 \quad , \quad \forall i, \tag{6.4.2}$$

which ensures that each salesman visits exactly one city.

This case was easy because we could introduce a binary variable which implemented the *all-different relation* by the inequality (6.4.1).

A more complicated situation arises when we are facing a problem in which a set of continuous variables $x_i \ge 0$ is given together with the requirement that for each pair of indices $i \neq j$ we have to fulfill $x_i \neq x_j$. However, with the concept of non-zero variables introduced in Sect. 6.3, we can represent the *all-different relation* by introducing non-zero variables ς_{ij}, the constraints

$$\varsigma_{ij} = x_i - x_j \quad , \quad \forall i, j \mid i \neq j \tag{6.4.3}$$

and the constraints (6.3.6) or (6.3.7). Let us consider the following example to illustrate the use of modeling sets with all-different elements. A company wants to produce some electronic devices. These devices have, besides others, three variables representing frequencies f_1, f_2, and f_3, which need to be chosen in order to describe the functionality of the devices. For reasons physicists and electricians are well able to explain, low order resonances, say up to order 4, have to be avoided for the frequencies. That gives us the following relations between the continuous variables f_i :

$$f_i \neq nf_j \quad , \quad \forall i, j \mid i \neq j \quad , \quad n = 1, 2, 3, 4. \tag{6.4.4}$$

In order to describe this situation we would introduce the following non-zero variables ς_{ij} :

$$\varsigma_{ij} = f_i - nf_j \quad , \quad \forall i, j \mid i \neq j \quad , \quad n = 1, 2, 3, 4. \tag{6.4.5}$$

6.5 Modeling Absolute Value Terms ⊖

Sometimes in our model there appear expressions of the form $|x_1 - x_2|$. For instance, if on a machine the output rate of a certain product is called x_1 in the first time interval and x_2 in the second, then $|x_1 - x_2|$ represents the change of output rates between the time periods disregarding whether it is an increase or decrease of output rates, i.e.,

$$|x_1 - x_2| = \begin{cases} x_1 - x_2 \text{ , if } x_1 \geq x_2 \\ x_2 - x_1 \text{ , if } x_1 < x_2 \end{cases}. \tag{6.5.1}$$

While x_1 and x_2 are non-negative variables, the difference $x_1 - x_2$ could be positive or negative, but $|x_1 - x_2|$ is non-negative. We can interpret $|x_1 - x_2|$ as a nonlinear function, or alternatively, due to the presence of the if-statement, as a logical expression. Let us approach it here from a mathematical point of view. Let

$$l(\mathbf{x}) = l_1 x_1 + \ldots + l_k x_k \tag{6.5.2}$$

be any linear expression which appears as $|l(\mathbf{x})|$ in our model. In that case, we introduce two non-negative variables $a^+ \geq 0$ and $a^- \geq 0$. Then we can replace each instance of $|l(\mathbf{x})|$ with

$$|l(\mathbf{x})| = a^+ + a^- \tag{6.5.3}$$

and each instance of $l(\mathbf{x})$ with $a^+ - a^-$ provided that the constraint

$$l(\mathbf{x}) = a^+ - a^- \tag{6.5.4}$$

is appended. The expressions (6.5.3) will represent the absolute value $|l(\mathbf{x})|$ with the additional constraint, namely if at most one of the variables a^+ and a^- is non-zero, i.e.,

$$a^+ a^- = 0. \tag{6.5.5}$$

If we are explicitly interested in the absolute value term $|l(\mathbf{x})|$ we could introduce an additional non-negative variable, say $a \geq 0$, and the constraint

$$a = a^+ + a^-. \tag{6.5.6}$$

The appearance of $a^+ - a^-$ may remind us of the use of free variables introduced in Chap. 2. In one of the exercises at the end of Chap. 3 the reader has seen that at most one of the variables a^+ and a^- could take a positive value. However, in the present situation this is not the case. Since the expression $a^+ + a^-$ also occurs the columns associated with a^+ and a^- are linearly independent, i.e. both could become basic variables. The nonlinear condition (6.5.5) can either be modeled by special ordered sets of type 1 [see Sect. 6.8], or explicitly by introducing a binary variable δ, and the following constraints:

$$\begin{aligned} a^+ &\leq A\delta \\ a^- &\leq A(1-\delta), \end{aligned} \tag{6.5.7}$$

where A is an upper bound for $|l(\mathbf{x})|$. Note that $\delta = 0$ leads to $a^+ = 0$ while $\delta = 1$ leads to $a^- = 0$. So, in either case at least one of a^+ and a^- is zero as required.

There remains the question of how to choose the upper bound constant A. In terms of efficient MILP modeling we should try to choose bound A as small as possible [cf. Sect. 9.1.1.2] as that results in a tighter formulation. On the other hand bound A must not become too small as then we would not correctly model the absolute value term. A first, general, observation is the chain of inequalities

$$\begin{aligned} a^+ &\leq |l(\mathbf{x})| \leq |l_1| \cdot \max x_1 + \ldots + |l_k| \cdot \max x_k \\ a^- &\leq |l(\mathbf{x})| \leq |l_1| \cdot \max x_1 + \ldots + |l_k| \cdot \max x_k, \end{aligned} \tag{6.5.8}$$

which makes use of the fact that x_j are non-negative variables. A better upper bound can be derived if we distinguish the variables in $l(\mathbf{x})$ with positive and negative coefficients. Let $\mathscr{L}_1$ and $\mathscr{L}_2$ be the index sets of variables with positive and negative coefficients l_j, respectively. That gives us

$$|l(\mathbf{x})| \leq \max \left\{ \sum_{j \in \mathscr{L}_1} |l_j| \cdot \max x_j, \sum_{j \in \mathscr{L}_2} |l_j| \cdot \max x_j \right\}. \tag{6.5.9}$$

Let us illustrate the procedure with the following example

$$\max \quad Z = |x_1 - 2x_2| \tag{6.5.10}$$

subject to

$$x_1 \leq 3 \quad , \quad x_2 \leq 4. \tag{6.5.11}$$

Proceeding as described above we would get

$$\max \quad Z = a^+ + a^- \tag{6.5.12}$$

subject to

$$x_1 \leq 3 \quad , \quad x_2 \leq 4 \tag{6.5.13}$$

$$x_1 - 2x_2 = a^+ - a^- \tag{6.5.14}$$

and

$$\begin{aligned} a^+ &\leq \quad A\delta \\ a^- &\leq A(1-\delta). \end{aligned} \tag{6.5.15}$$

By applying (6.5.9) we obtain an appropriate value for A

$$A = \max \{|1| \cdot 3, |-2| \cdot 4\} = 8. \tag{6.5.16}$$

This example is contained in the model collection under the problem name *absval*. The result is $x_1 = 0$, $x_2 = 4$, $Z = 8$, $\delta = 0$, $a^+ = 0$, and $a^- = 8$. In that case our bound $A = 8$ was as tight as possible.

Table 6.1 For certain functions $f = f(x, y)$, this table provides the equivalent MILP formulations and the required inequalities. Note that $f = \min(x, y) = -\max(-x, -y)$, the minimum of two variables, is not modeled as using inequalities, but just related the maximum function

Function f	Description	Inequalities
$\max(x, y)$	Maximum of two variables	$f \geq x \quad \wedge \quad f \leq x + M_x\delta$ $f \geq y \quad \wedge \quad f \leq y + M_y(1-\delta)$
IF(δ, x, y)	Variable disjunction	$x - M_x(1-\delta) \leq f \leq x + M_x(1-\delta)$ $y - M_y\delta \leq f \leq y + M_y\delta$
$x\delta$	Product of two variables	$x - M(1-\delta) \leq f \leq x + M(1-\delta)$ $f \leq M\delta$
$xy = 0$	Complementarity	$-M_x(1-\delta) \leq x \leq M_x(1-\delta)$ $-M_y\delta \leq y \leq M_y\delta$

6.6 Nonlinear Terms and Equivalent MILP Formulations

In this section, we show how to represent or replace certain nonlinear or discontinuous functions by equivalent[3] MILP formulations. These formulations usually contain a binary variable δ, and a Big-M coefficient M appearing in carefully formulated inequalities. We summarize those equivalences for various functions $f = f(x, y)$ depending on two variables in Table 6.1.
The reader might be puzzled which values to assign to the Big-M coefficient M. The answer is: Large enough that we do not lose any feasible solution, and not larger than necessary, i.e., as small as possible. If x, for instance, represents the amount of production on machine, M could be the production capacity. If x is the amount of money to invest, a good choice of M could be the available budget. We will explain below why the Big-M coefficient should be chosen as small as possible.

For the functions and inequalities formulated it is sufficient to set them to the maximal values the variables x and y can assume in the problem at hand, i.e.,

$$M = \max\{\max x, \max y\}.$$

Let us for $f = f(x, y) = \max(x, y)$ show the equivalence in detail. If $y > x$, we have $f = \max(x, y) = y$. In that case, the binary variable can take only the value $\delta = 1$, resulting in the inequalities

$$f \geq x \quad \wedge \quad f \leq x + M \quad \wedge \quad f \geq y \quad \wedge \quad f \leq y.$$

[3]Sometimes, these equivalent MILP formulations are also called *linearization*. However, we avoid this term in this context, as we feel it is more appropriate to Taylor series expansions stopping after the linear term.

We observe that $f = y$ and that f is not restricted if $x < y$. If $x > y$, δ can only take the value $\delta = 0$ leading to the inequalities

$$f \geq x \quad \wedge \quad f \leq x \quad \wedge \quad f \geq y \quad \wedge \quad f \leq y + M$$

giving us $f = x$, as expected. In the case $x = y$ the binary variable δ is undetermined; it can take either the value 0 or 1. In both cases, we get $f = \max(x, y) = x = y$.

Likewise to $\max(x, y)$, we translate other bi-variate functions into linear inequalities involving binary, among them

$$f = f(x, y) = \mathrm{IF}(\delta, x, y) = \begin{cases} x, & \delta = 1 \\ y, & \delta = 0, \end{cases}$$

the product of a continuous variable x and a binary variable δ

$$f = f(x, y) = x\delta = \begin{cases} x, & \delta = 1 \\ 0, & \delta = 0, \end{cases}$$

and the complementarity of two arbitrary variables

$$xy = 0 \Longrightarrow \begin{cases} x = 0, & \delta = 1 \\ y = 0, & \delta = 0. \end{cases}$$

We leave it as an exercise to prove these equivalences for these functions.

There may also exist other equivalent MILP formulations for a function; we see this if we have a second look at $f = f(x, y) = x\delta$, the product of a continuous variables x and a binary variable δ. In Table 6.1 we had the two inequalities

$$x - M(1 - \delta) \leq f \leq x + M(1 - \delta).$$

We could also replace $x\delta$ by the three equivalent inequalities:

$$\begin{aligned} f &\leq M\delta, \\ f &\leq x, \\ f &\geq x - M(1 - \delta). \end{aligned}$$

Which equivalent formulation is better? The second is probably tighter, but it is not ideal yet. The reason is that MILP solvers perform extensive preprocessing, among them bound tightening. Therefore, a third formulation using an auxiliary variable u is usually even better:

$$\begin{aligned} f &\leq M\delta \\ u &\leq M(1 - \delta) \\ f &= x - u. \end{aligned}$$

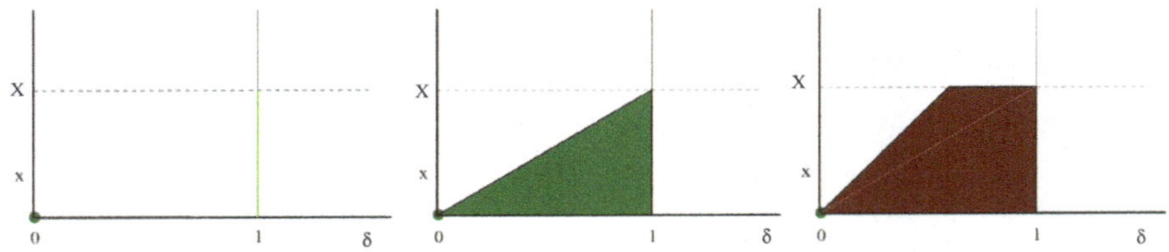

Fig. 6.1 Smallest Big-M coefficient. The figures from left to right show the integer feasible region, the convex hull, and the relaxed feasible region for $0 \leq \delta \leq 1$

Both inequalities are now of type $\leq$ (which have computational benefits for MILP solvers in this context) followed by a simple equality.

Let us now give a motivation why M has to be selected as small as possible. For a natural upper bound X on x, Fig. 6.1 shows from left to right the integer feasible region S

$$S := \left\{(x, \delta)|x \in \mathbb{R}_0^+ \wedge \delta \in \{0, 1\} \wedge x \leq M\delta \wedge x \leq X\right\}$$

with the point at (0,0) and the vertical line in the left figure, the convex hull S_{ch}

$$S_{\text{ch}} := \left\{(x, \delta)|x \in \mathbb{R}_0^+ \wedge \delta \in [0, 1] \wedge x \leq X\delta\right\},$$

i.e., the triangle in the figure in the middle, and the LP relaxation or relaxed feasible region S_{LP}, *resp.*,

$$S_{\text{LP}} := \left\{(x, \delta)|x \in \mathbb{R}_0^+ \wedge \delta \in [0, 1] \wedge x \leq M\delta \wedge x \leq X\right\}.$$

Note that for both, S_{ch} and S_{LP}, the binary variable δ has been relaxed to $0 \leq \delta \leq 1$. The convex hull S_{ch} is the smallest set containing all integer feasible points but has the binary variable relaxed. The three sets obey the chain of inequalities

$$S \subset S_{\text{ch}} \subseteq S_{\text{LP}}.$$

Now we can also better understand and express what we mean when we discuss the *goodness* or *tightness* of a formulation: The closer S_{LP} is to S_{ch}, the better the formulation. We also note that for $M = X$, the LP relaxation is identical to the convex hull. This illustrates why it is a good idea to select the Big-M coefficient as small as possible. However, we must not make the value for M too small. If we selected $M < X$, we would lose some integer feasible points.

Some commercial MILP solvers provide *indicator constraints* as an alternative to Big-M formulations for considering disjunctive constraints as in Sect. 6.2.3; cf. Belotti et al. (2016,[61]). Consider an optimization problem P (linear or nonlinear) with a disjunctive constraint $\mathsf{A}\mathbf{x} \leq \mathbf{b}$ and its associated binary indicator variable $\delta \in \{0, 1\}$. If $\delta = 1$ the disjunctive constraint $\mathsf{A}\mathbf{x} \leq \mathbf{b}$ has to be satisfied. On the other hand, if $\delta = 0$ the constraint may be violated. As the solver handles this relationship between the disjunctive constraint and the associated indicator variable internally, a seeming advantage is that the user does not have to think about the choice of M — for less experienced modelers this could be helpful. However, a consequence of not providing M is the loss of a tight connection of the disjunctive constraint to P. This causes a weakness when branching on δ. The potential advantage of the Big-M formulations

$$\mathsf{A}\mathbf{x} \le \mathbf{b} + M\delta$$

is — even in the worst case with badly selected values of M — an algebraic coupling through the domain relaxation (neglecting the integrality of δ), i.e., LP or NLP relaxation of P. We should also keep in mind that a solver can only base its operations on the algebra while a user might have intuition about the value of M.

If P is nonlinear, we could also model the disjunction using the complementarity condition

$$(\mathsf{A}\mathbf{x} - \mathbf{b})(1 - \delta) \le 0,$$

which unfortunately leads to a non-convex problem. Thus, it is better that we do not follow this approach.

As a summary we recommend: For reasonably sized values of M, i.e., M is not much larger than other coefficients in the model and the model does not suffer from the existence of M, use the Big-M approach. If M is very large compared[4] to all other coefficients (in that case, the LP relaxation is weak and numerical instabilities may arise), or it is even very difficult to construct the Big-M formulation within the model, it might be a good idea to resort to indicator constraints.

6.7 Modeling Products of Binary Variables

Certain nonlinear terms involving binary variables can also be treated by MILP models. The first example of such terms is the squares of binaries. Each time we encounter δ^2 we can just replace it by δ because $\delta^2 = \delta$ for 0 and 1, the only values binary variables can assume. This argumentation shows that we can also replace δ^k by δ. Things get more complicated if we want to include products of different binary variables such as $\delta_1\delta_2$. Nevertheless, it can be done at the cost of an extra binary

[4] It is not easy to define quantitatively what is meant by *very large compared to other coefficients* (e.g., is 1000 very large compared to 1?), without experience or knowledge of a solver's capabilities. Thus we prefer to give an example: If all values of input data and expected values of the variables are between 1 and 10, $M = 10^6$ is probably very large compared to other coefficients — it would be a good idea to try smaller values, e.g., $M = 100$, and see how the solver performs with that choice.

variable. Models including products of k binary variables $\delta_p = \prod_{i=1}^{k} \delta_i$, $\delta_i \in \{0, 1\}$ can be transformed directly into linear integer models according to

$$\delta_p \le \delta_i \quad , \quad \forall i \quad ; \quad -\delta_p + \sum_{i=1}^{k} \delta_i \le k - 1.$$

For $k = 2$, the term $\delta_1 \delta_2$ is thus replaced by the three inequalities

$$\delta_p \le \delta_1 \quad , \quad \delta_p \le \delta_2 \quad , \quad -\delta_p + \delta_1 + \delta_2 \le 1.$$

The first two of them guarantee that δ_p becomes zero if either of the binary variables in the product term is zero. The third one ensures that if all of them are one, δ_p becomes also one.

6.8 Special Ordered Sets

In Sect. 6.1.4 the modeling of a set of decisions where at most k decisions can be selected was introduced using the constraint

$$\delta_1 + \delta_2 + \ldots + \delta_n \le k. \tag{6.8.1}$$

If the decisions are to be mutually exclusive, then the modeling will require that only one decision can be selected, hence the appropriate constraint is

$$\delta_1 + \delta_2 + \ldots + \delta_n \le 1. \tag{6.8.2}$$

This is a commonly occurring condition. However, if the decisions represented by the δ variables are ordered in some way, for example if they represented different levels at which a facility could be opened, then it would be preferable to specify that the δ variables form a *special ordered set of type 1*. Special ordered sets of type 1 and type 2, or SOS1s and SOS2s for short, appear as special entities in the theory of integer programming problems and the software used to solve them. SOS1s were introduced by Beale & Tomlin (1970,[58]), and are further discussed by Hummeltenberg (1984,[275]) and Wilson (1990,[584]). They provide an efficient structure to handle mutual exclusivity constraints [see Sect. 6.8.1].

Another variant of special ordered sets, referred to as *type 2*, allows us to model piecewise linear functions, so that we may include certain nonlinear terms in otherwise linear models [see Sect. 6.8.2].

6.8.1 Special Ordered Sets of Type 1

A *special ordered set of type 1* (SOS1) is an ordered set of variables of which at most one may be non-zero. The variables can be any kind of variables.

In the most common instances of an SOS1 each variable will be a binary variable δ_i. The variation which insists that exactly one variable be non-zero is also common. This condition could be modeled in ILP terms by what is termed a *convexity row*

$$\sum_{i=1}^{n} \delta_i = 1. \tag{6.8.3}$$

Declaring the variables $\delta_1, \ldots, \delta_n$ as an SOS1 $\{\delta_1, \ldots, \delta_n\}$ and adding the convexity row (6.8.3) ensures that the variables $\delta_1, \ldots, \delta_n$ will only take the values 0 and 1. Note that is not necessary to declare the variables explicitly as binary variables.

We have, however, to keep in mind that the concept of special ordered sets has nothing to do with the convexity conditions. In Sect. 11.4 we will encounter an example in which there exists no such relation between the elements of SOS1.

As well as the convexity row, we have another definitional condition involved in the model because we have required that the variables be ordered. It is possible that variables will have a natural order, e.g., time order or size precedence order, or we may need to create one. These conditions will be modeled using what is called a *reference row* and its typical form is

$$x = \sum_{i=1}^{n} X_i \delta_i, \tag{6.8.4}$$

where the X_i are weights or some form of outputs associated with each decision δ_i, respectively. The reference row and the coefficients X_i establish some order between the variables δ_i.

Let us now consider a typical example illustrating the idea of SOS1. In a network design problem, a company wants to determine the optimal topology and capacities of pipelines connecting the nodes of the network, e.g., plants. The capacity of the pipelines can only be chosen from a finite set of distinct values $C_1, \ldots, C_n$. Let the capacities be ordered in the following sense: the larger the index, the larger the capacity, i.e.,

$$j > i \quad \Rightarrow \quad C_j > C_i. \tag{6.8.5}$$

Such a situation is illustrated in Fig. 6.2. To determine the correct capacity we introduce n binary variables δ_i such that

$$\delta_i := \begin{cases} 1\,, & \text{if size } C_i \text{ is selected for the pipeline} \\ 0\,, & \text{otherwise.} \end{cases} \tag{6.8.6}$$

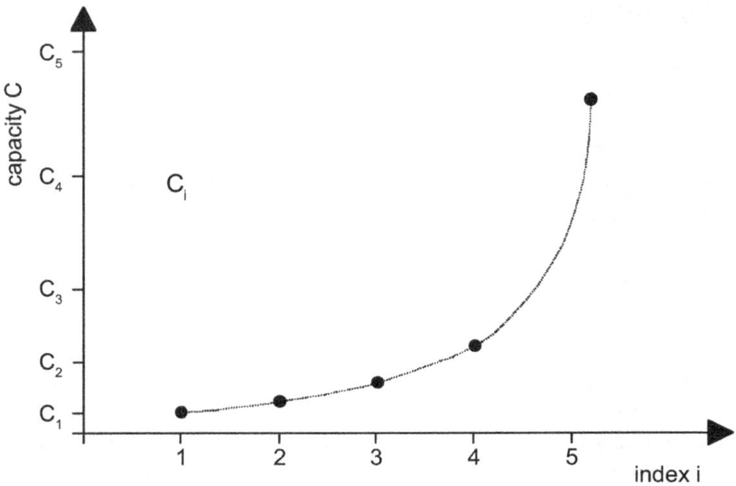

Fig. 6.2 Using SOS1 to select capacity size from a set of possible capacities, C_i

These binary variables will form an SOS1 and therefore it is not necessary to declare them as binary variables any more. Now we also see why the convexity constraint appears quite naturally: it is really a mutual exclusivity constraint. We need to ensure that exactly one capacity size is chosen, i.e.,

$$\sum_{i=1}^{n} \delta_i = 1. \tag{6.8.7}$$

The actual size, c, which is chosen can be computed from

$$c = \sum_{i=1}^{n} C_i \delta_i . \tag{6.8.8}$$

Equation (6.8.8) may perfectly serve as the reference row.

In Sect. 9.4.4 we will see that using an SOS1 solves a problem faster than using binary variables; the positive effect might be weaker for very strong commercial MILP solver. The reason is that the integer programming solution process, B&B, treats each special ordered set as one general integer variable with maximum value n, rather than n separate binary variables. It should be noted, however, that the concept of "order" which the reference row imposes is important and every collection of mutually exclusive decisions should not necessarily be modeled as an SOS1 unless the ordering concept is present in a natural way among the members of the set, e.g., the decisions refer to different time periods, which have a natural order, or to different sizes or levels of complexity. The benefits of rapid solution provided by the use of SOS1s may be lost if the sets have no natural order. Integer programming software normally facilitates the specification of sets of variables as SOS1s and uses a special purpose algorithm to aid problem solution when SOS1s are present. The specification of SOS1s — in a generic format — may look as

```
VARIABLES
  delta(n)
  c
CONSTRAINTS
  CONVEX: SUM(i = 1:n)         delta(i) = 1.0
  REFROW: SUM(i = 1:n) C(i)*delta(i) = c      $
SETS
  SET1: SUM(i = 1:n) delta(i) .S1. REFROW
```

In the above, the `SETS` section defines special ordered sets of variables in the model. Note that the δ_i need not to be declared as binary variables and that the `SUM` command in the `SET1` row is not really a summation but more a synonym for *union*.

6.8.2 Special Ordered Sets of Type 2

A *special ordered set of type 2* (SOS2) is an ordered set of variables usually denoted as $\{\lambda_1, \lambda_2, \ldots, \lambda_n\}$ of which at most two may take non-zero values. If two variables are non-zero, then they must be adjacent in the ordering.

Analogously with SOS1, the most common variation will be a set of variables $\lambda_1, \lambda_2, \ldots, \lambda_n$ with the convexity condition (6.8.3). An appropriate solution to the constraints might be $\lambda_2 = 0.75$, $\lambda_3 = 0.25$ (with all other λ variables zero) or, if only a single variable is non-zero, $\lambda_3 = 1$.

An SOS2 is unlikely to be used to model naturally occurring decisions. Instead, it is used to model certain nonlinear relationships required in models. These nonlinear relationships are replaced by a linearly interpolated one. Consider the example illustrated in Fig. 6.3.

Here y (cost) is related to x (production level) but the relationship cannot be modeled by a straight line as cost increases less sharply once production has reached higher levels. The relationship is modeled by a smooth curve. As LP and MILP only permit linear relationships between decision variables, we have to devise a way to incorporate such a curvilinear relationship into an existing LP or MILP model. If we consider the five straight line segments which have been drawn in the right part of Fig. 6.3, we could use these to model approximately the nonlinear function or curved relationship as they are a fairly close approximation to the actual behavior of the function (and we could improve the standard of approximation by using more than five segments). Switching to consider only the five segments, rather than the original function (left part of Fig. 6.3), we find that they can be modeled using SOS2.

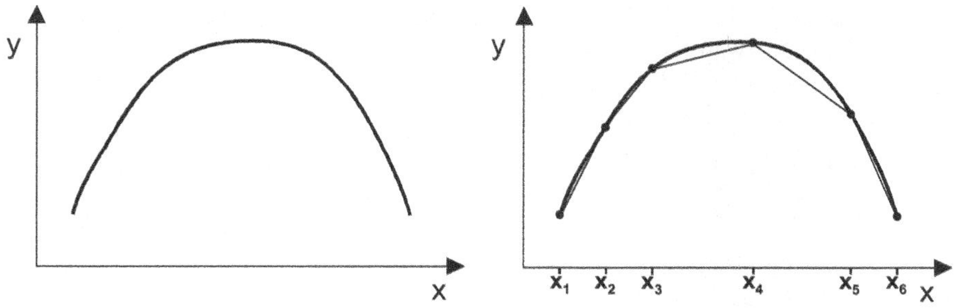

Fig. 6.3 Using SOS2 to model a nonlinear function. Left: The original nonlinear function. Right: Approximation of the nonlinear function by five segments

Each segment is clearly linear, but we need to have a way of switching from one segment to another at appropriate points, e.g., at (x_2, y_2). This is performed by the "If two variables are non-zero then they must be adjacent in the ordering" part of the definition of the SOS2. The segments in the right part of Fig. 6.3 are modeled as the SOS2 $\{\lambda_1, \lambda_2, \ldots, \lambda_6\}$ with the convexity row

$$\lambda_1 + \lambda_2 + \ldots + \lambda_6 = 1 \tag{6.8.9}$$

and the reference row

$$l = \lambda_1 + 2\lambda_2 + \ldots + 6\lambda_6 \tag{6.8.10}$$

establishing a monotonic relation in the problem. Cost will be defined by the equation

$$y = 100\lambda_1 + 300\lambda_2 + 450\lambda_3 + \ldots + 100\lambda_6. \tag{6.8.11}$$

Thus, a typical solution might yield

$$\lambda_2 = 0.75 \quad , \quad \lambda_3 = 0.25, \tag{6.8.12}$$

giving

$$l = 2 \cdot 0.75 + 3 \cdot 0.25 = 2.25 \tag{6.8.13}$$

and

$$y = 300 \cdot 0.75 + 450 \cdot 0.25 = 337.50. \tag{6.8.14}$$

Examining the graphs given in Fig. 6.3 shows that our solution is approximate, but a good approximation. Notice that the (invalid) solution $\lambda_1 = 0.5, \lambda_3 = 0.5$

gives a poor approximation, hence the need for non-zero variables to be adjacent. The approximation could have been improved by using more points in the SOS2, normally called grid points on the graph, but the more points used the slower the problem will be to solve. It should also be noted that the segments will "hug" the curve closely at certain parts and be further away at others depending on the curvature of the curve and the density of the grid points. Where the segments and curve diverge we can include more grid points, which need not all be equally spaced. It should be further noted that we can often approximate a curve from either side, which will either consistently underestimate or overestimate values. The modeler will need to decide whether it is better to underestimate or overestimate from the context of the application for which the model is being used. A rather tricky SOS2 application is given in Sect. 8.4.1. A piecewise linear approximation model to a transport model with a nonlinear objective is very instructive and implemented in MCOL/GAMSLIB/*trnspwlx.gms*.

Let us summarize how SOS2s provide a way of incorporating (approximately) certain nonlinear relationships within linear models using linear interpolation. Assume that we have an arbitrary nonlinear function $y = f(x)$ occurring somewhere in our model. Then we define a set of n grid points or breakpoints X_i (see Sect. 14.2.3 for how to construct optimal breakpoints systems), compute the corresponding function values $F_i = f(X_i)$, and define an SOS2 consisting of n variables λ_i. We substitute $f(x)$ with y in the model and add the constraints

$$x = \sum_{i=1}^{n} X_i \lambda_i \tag{6.8.15}$$

interpolating[5] in argument, and

$$y = \sum_{i=1}^{n} F_i \lambda_i \tag{6.8.16}$$

interpolating in function value, as well as the convexity constraint

$$\sum_{i=1}^{n} \lambda_i = 1. \tag{6.8.17}$$

Finally, (6.8.15) serves as the reference row to establish a monotonic relationship in our model. This is needed for the B&B algorithm. Equality (6.8.16) would also be a good reference row candidate if the coefficients F_i do establish order or monotonicity, e.g., if $f(x)$ is a monotonic function. It is always worth paying great

[5]In general *to interpolate* means to compute intermediate values of a quantity between a series of given values. But in our case, since the variables λ_i are elements in an SOS2, we interpolate between two adjacent values.

attention to the choice of the reference row. The stronger the monotonic relationship, the better.

There is another notion we would like to focus the reader's attention on: nonlinear functions occurring in the objective function can be easily and safely handled using SOS2. Great care is, however, necessary when applying it to equations containing nonlinear functions. Since the nonlinear functions are replaced by linear interpolants, the problem instance may become infeasible.

Alternative ways to model nonlinear functions exist, but they are not able to take advantage of LP or MILP software; rather, they use methods which are much slower and less reliable. As with SOS1s, software has been designed to support modeling using SOS2s.

In some LP/IP software systems special facilities are available for automatically modeling commonly occurring nonlinear relationships of a single variable such as x^2, x^3, e^x without the need to express the conditions more fully using SOS2s.[6]

Ordered sets are related to the earlier concept of *separable programming* developed by Miller (1963,[400]). When the function to be approximated is convex the straight line segments provide solutions that automatically satisfy the "non-zero variables must be adjacent" condition; the λ_i variables need not to be declared as a SOS2. When the convexity condition does not hold, SOS2s are required. The branching rules adopted by B&B integer programming methods to handle SOS2s are described in Chap. 9. Note that a general IP formulation could be used to model the relationships of SOS2s, but will not perform as well as a model using the tailor-made facilities. The branching rules of SOS2s are the key to this.

Other methods may also be used for modeling more complex relationships than those expressible as a smooth curve interrelating two decision variables y and x. Consider the relationship

$$y = xz \tag{6.8.18}$$

a simple product of two variables x and z.

If we use the logarithm function, denoted as log, we can express the simple product as

$$\log y = \log x + \log z, \tag{6.8.19}$$

which is a linear relationship in the logarithms. Thus, by introducing new variables to denote $\log x$, and $\log z$, and linking these to x and z by defining SOS2 connections between the graphs of x and $\log x$, and z and $\log z$, we can model values of $\log y$ which can later be reconverted to y values. However, if x and z vary over a fairly wide range, it is dangerous to have such variables and their logarithms all present in the same model as it may lead to unreliable results due to scaling difficulties [see

[6] `SCICONIC` is an example.

Sect. 9.2.2]. An alternative approach to handling products is described in the next subsection.

Finally, with a new continuous variable z and the inequality

$$z \geq g_i(x) := A_i(x - X_i) + F_i \quad , \quad \forall i = 1, \ldots, S-1, \tag{6.8.20}$$

we can show that linearly constrained optimization problems with piecewise linear, convex (concave) objective function as LP problems (only continuous variables) can be minimized (maximized). In (6.8.20), $g_i(x)$ describes the ith linear segment of the objective function $f(x)$. If $f(x)$ is to be minimized, it can be replaced by $\min z$. Similarly as the proof of the *Vertex Theorem* of LP, cf. Neumann & Morlock (1993,[422, Theorem 1.1.5, p.48]), the argument is, that if

$$z > g_i(x) \quad , \quad \forall i = 1, \ldots, S-1$$

is fulfilled, one can find an index i_*, so that

$$z_* = g_{i_*}(x) := A_{i_*}(x - X_i) + F_{i_*}$$

leads to a better solution (objective function value is on a segment) — here, convexity is required. As there is only a finite number of segments, on each segment the minimal objective function value is assumed. If one formulates the convex piecewise linear function $f(x)$ with SOS2 variables, the λ variables in the LP relaxation already automatically fulfill the SOS2 conditions.

To limit the increase of problem size due to number of breakpoints, there are two interesting approaches. Vielma & Nemhauser (2011,[564]) provide a formulation for $n = 1 + 2^k$ known breakpoints. Instead of n SOS2-variables, only k binary variables and $2k$ linear inequalities are required, i.e., the number of binary variables *and* the number of additional inequalities grow only with the logarithm of the number of breakpoints. Alternatively, Rebennack & Kallrath (2015,[455]) have developed a procedure to compute the global minimum of the number of breakpoints subject to a pre-given approximation accuracy of the function; see Sect. 14.2.3.

6.8.3 Linked Ordered Sets

Chains of *linked ordered sets* were first introduced by Beale & Forrest (1976,[57]) and are further developed in Beale & Daniel (1980,[56]). They provide a way to approximate product terms using a pair of linked SOS1s. Say we wish to model a product term of the form

$$z = x \cdot g(y), \tag{6.8.21}$$

where $g(y)$ is some known nonlinear function of a continuous variable y, and y can take only one of the n values

$$Y_1, Y_2, \ldots, Y_n. \tag{6.8.22}$$

Furthermore, x is a continuous variable with known (finite) upper and lower bounds such that

$$X_{\min} \leq x \leq X_{\max}. \tag{6.8.23}$$

All combinations of possible values of x, y, and z can be defined by introducing $2n$ non-negative variables

$$\lambda_{11}, \lambda_{12}, \ldots, \lambda_{1n} \tag{6.8.24}$$

$$\lambda_{21}, \lambda_{22}, \ldots, \lambda_{2n}, \tag{6.8.25}$$

and the constraints

$$\sum_{j=1}^{n} \lambda_{1j} + \sum_{j=1}^{n} \lambda_{2j} = 1 \tag{6.8.26}$$

$$y = \sum_{j=1}^{n} Y_j(\lambda_{1j} + \lambda_{2j}), \tag{6.8.27}$$

and the additional restriction that

$$\lambda_{ij} \neq 0 \Rightarrow \lambda_{rs} = 0 \quad \forall s \neq j, \quad \begin{matrix} i \in \{1,2\} \\ r \in \{1,2\} \end{matrix}, \quad j \in \{1, 2, \ldots, n\}. \tag{6.8.28}$$

Then x is given by

$$x = \sum_{j=1}^{n} X_{\min}\lambda_{1j} + \sum_{j=1}^{n} X_{\max}\lambda_{2j} \tag{6.8.29}$$

and z is given by

$$z = \sum_{j=1}^{n} X_{\min} \cdot g(Y_j) \cdot \lambda_{1j} + \sum_{j=1}^{n} X_{\max} \cdot g(Y_j) \cdot \lambda_{2j}. \tag{6.8.30}$$

The condition (6.8.28) will be handled by a modification of the branching conditions in the B&B algorithm [see Sect. 3.8.6]. Its operation is illustrated in (6.8.31).

Row \ Column	...	$j-2$	$j-1$	j	$j+1$	$j+2$	$j+3$
....							
i		.	.	.	.	.	.
					$\updownarrow$	$\updownarrow$	
$i+1$		.	.	.	.	.	.

(6.8.31)

The conditions (6.8.26), (6.8.27), and (6.8.28) ensure that y takes one of the n valid values, e.g., if $\lambda_{1k} + \lambda_{2k} = 1$, then y will take the value Y_k. The conditions (6.8.26) and (6.8.29) ensure that x takes a value lying on a line segment $X_{\min}$ and $X_{\max}$, because if $\lambda_{1k} + \lambda_{2k} = 1$ then x will take the value $\lambda_{1k} X_{\min} + \lambda_{2k} X_{\max}$, ensuring that all possible values of x are available. Finally the conditions (6.8.26), (6.8.30), and (6.8.28) ensure that a valid product term is calculated for z. This can be seen more clearly if the constant $g(Y_k)$ is factored out of the expression (all terms corresponding to $j \neq k$ are zero).

It should be noted that all expressions (6.8.26)–(6.8.30) are now linear, but (6.8.21) was nonlinear. The following example will illustrate the concept of linked ordered sets. Let $n = 4$, $X_{\min} = 3.4$, $X_{\max} = 8.5$, and let the values of y be 2.0, 2.5, 3.0, 3.8. Let the function of y be

$$g(y) = e^y + y^2. \tag{6.8.32}$$

The range of g consists of the four values 11.39, 18.43, 29.09, 59.14. Introduce the λ-variables along with the constraint

$$\lambda_{11} + \lambda_{12} + \lambda_{13} + \lambda_{14} + \lambda_{21} + \lambda_{22} + \lambda_{23} + \lambda_{24} = 1. \tag{6.8.33}$$

We then have

$$y = 2.0(\lambda_{11} + \lambda_{21}) + 2.5(\lambda_{12} + \lambda_{22}) + 3.0(\lambda_{13} + \lambda_{23}) + 3.8(\lambda_{14} + \lambda_{24}) \tag{6.8.34}$$

and

$$x = 3.4 \sum_{i=1}^{4} \lambda_{1i} + 8.5 \sum_{i=1}^{4} \lambda_{2i} \tag{6.8.35}$$

and

$$\begin{aligned} z = {} & 3.4(11.39\lambda_{11} + 18.43\lambda_{12} + 29.09\lambda_{13} + 59.14\lambda_{14}) \\ & + 8.5(11.39\lambda_{21} + 18.43\lambda_{22} + 29.09\lambda_{23} + 59.14\lambda_{24}). \end{aligned} \tag{6.8.36}$$

A typical solution might be $\lambda_{13} = 0.4, \lambda_{23} = 0.6$,with all other λ variables zero, which gives $y = 3$ in Eq. (6.8.34) and $x = 6.46$ in Eq. (6.8.35) and $z = 3.4(0.4)29.09 + 8.5(0.6)29.09 = 187.92$ in Eq. (6.8.36).

Note: If the variable y occurs in several product terms, we can define two sets of λ variables for each product term. Hence the section title is *linked ordered sets*.

6.8.4 Families of Special Ordered Sets

As was described in Sect. 6.8.3, a model may require several special ordered sets. Special ordered sets may also be used in families where there are links between each set. There are three likely purposes:

1. The sets model a precedence relationship. For example, special ordered sets may be used to model the times at which events may occur. Let us consider two events whose times t_1 and t_2 are given by

$$t_1 = \lambda_1 + 2\lambda_2 + \ldots + n\lambda_n \tag{6.8.37}$$

and

$$t_2 = \mu_1 + 2\mu_2 + \ldots + n\mu_n, \tag{6.8.38}$$

where $\lambda_1, \lambda_2, \ldots, \lambda_n$ and $\mu_1, \mu_2, \ldots, \mu_n$ are both special ordered sets of type 1. Then, if we also require that event 2 is at least one time unit later than event 1, we have the restriction

$$t_1 + 1 \leq t_2, \tag{6.8.39}$$

which thereby links two special ordered sets.

2. The sets model the prevention of overlap. As in the previous example two special ordered sets may be used to model the times at which two events may occur. If it is required that the two times differ by at least one unit, then we require the linking restriction

$$t_1 \neq t_2. \tag{6.8.40}$$

3. The sets model a link between each other, as in linked ordered sets.

In these three examples, the families may comprise SOS1 or SOS2 in cases (1) or (2), but only SOS2 in case (3). Additional inequalities may be added to formulations which use linked sets of special ordered sets in order to help B&B reach a solution. For a full discussion on sets of special ordered sets see Wilson (1990,[584]) and Sect. 6.9 for a discussion on improving formulations by the introduction of additional inequalities.

6.9 Improving Formulations by Adding Logical Inequalities

When an ILP or MILP model is solved using a MILP solver, the first step is to solve the LP problem which is obtained from the ILP/MILP problem by simply ignoring the conditions that the variables designated as integer variables must take integer values. The upper and lower bounds on such variables, e.g., an upper bound of 1 and a lower bound of zero for a binary variable, are retained in the model together with all the other constraints. The difference between the optimal value of the objective function of this LP problem and the optimal solution value of the ILP/MILP problem is called the *integrality gap*. It is important to keep this gap as small as possible as a small gap aids the B&B algorithms. The value of the gap will not be known in advance, but it is known from previous work on ILP models that certain methods of formulation consistently help decrease this gap. Such methods may involve adding additional constraints to the problem, which may appear counterproductive, but which is done to aid the B&B process with sound mathematical reasoning.

For example, suppose that we want to model the condition "product 1 can only be produced if products 2, 3, ..., n have been produced." The single constraint

$$(n-1)\delta_1 \le \delta_2 + \delta_3 + \ldots + \delta_n \tag{6.9.1}$$

is strictly adequate to model the condition, but is computationally inferior to including the $n-1$ constraints

$$\delta_1 \le \delta_i \quad , \quad i = 2, \ldots, n. \tag{6.9.2}$$

In the former case, in an LP relaxation δ_1 may still take the value $(n-2)/(n-1)$ if one of the remaining δ variables is zero. In contrast, in (6.9.2) δ_1 would have been forced to zero.

The modeling of logical implications also gives rise to circumstances where formulations may be not fully specified. In (6.1.36), the implication $\Longrightarrow$ was used to link one decision to another in the form $L_1 \Longrightarrow (L_2 \wedge L_3)$. This was modeled as

$$\delta_1 \le \delta_2 \quad , \quad \delta_1 \le \delta_3. \tag{6.9.3}$$

However, it is useful to note that it is still possible to produce product 2 *and* product 3, when product 1 is not produced. If we only want that to happen when product 1 is produced, then recasting the model of $\Longrightarrow$ as $\Longleftrightarrow$ may be appropriate. The model now requires the addition of the constraint

$$1 + \delta_1 \ge \delta_2 + \delta_3 \tag{6.9.4}$$

to complete the formulation. It could be that if we do not model $\Longleftrightarrow$ then there is the possibility of an error in our logic.

Thus as a general rule, it is useful in an implication ($\Longrightarrow$) to consider what the modeler will want to happen when the "implying" part does **not** hold and check whether what is really required is equivalence ($\Longleftrightarrow$). It is desirable to use equivalence ($\Longleftrightarrow$) wherever possible instead of implication ($\Longrightarrow$), and not just to rely on $\Longrightarrow$ and the objective function jointly giving the effect of $\Longleftrightarrow$ by discouraging any logical impossibilities. Thus, we can extend our earlier results by reconsidering the formulations summarized in the "implication" table (6.1.36). Here is the "equivalence" version of (6.1.36):

Relations	Constraint(s)		
$L_1 \Longleftrightarrow (L_2 \wedge L_3)$	$\delta_1 \le \delta_2$,	$\delta_1 \le \delta_3$,	$1 + \delta_1 \ge \delta_2 + \delta_3$
$L_1 \Longleftrightarrow (L_2 \vee L_3)$	$\delta_2 \le \delta_1$,	$\delta_3 \le \delta_1$,	$\delta_1 \le \delta_2 + \delta_3$
$(L_1 \wedge L_2) \Longleftrightarrow L_3$	$\delta_3 \le \delta_2$,	$\delta_3 \le \delta_1$,	$\delta_1 + \delta_2 \le 1 + \delta_3$
$(L_1 \vee L_2) \Longleftrightarrow L_3$	$\delta_1 \le \delta_3$,	$\delta_2 \le \delta_3$,	$\delta_1 + \delta_2 \ge \delta_3$.

(6.9.5)

The statement $\bigwedge_{i=1}^{k} L_i \Longrightarrow \bigvee_{i=k+1}^{n} L_i$ may make us consider checking if $\bigwedge_{i=1}^{k} L_i \Longleftrightarrow \bigvee_{i=k+1}^{n} L_i$ is what is required. If it is, it will be modeled as

$$\sum_{i=k+1}^{n} \delta_i - \sum_{i=1}^{k} \delta_i \ge 1 - k \quad ; \quad \delta_i \ge \delta_j \quad , \quad \begin{array}{l} i = 1, \ldots, k \\ j = k+1, \ldots, n. \end{array} \tag{6.9.6}$$

6.10 Summary

In this chapter, we have endeavored to show the types of statement that can be formulated using logical variables. Single statements and linked statements have been modeled using integer variables. The scope of the techniques involving special ordered sets has been considered and it has been shown how this extends the capability of LP and ILP. It is possible to use binary variables to perform the role that special ordered sets can take, but it is often better to use the higher level modeling objects provided by the sets. Toward the end of the chapter, we saw how formulations may be strengthened by adding constraints. Thus, the reader should be able to:

- model decisions using {0,1} variables;
- make use of the logical connectives *and*, *or*, and *not*;
- model logical restrictions involving two or more arguments;
- model logical restrictions on constraints;
- model special types of conditions including disjunctive constraints, non-zero variables, and all-different conditions;
- model nonlinearities using special ordered sets;
- appreciate the need to close the integrality gap by favoring certain types of formulations.

6.11 Exercises

1. Read carefully through Sect. 6.1.3 and transform the following statements into algebraic formulations.

   (a) $L_1 \vee \neg L_2$;
   (b) $\neg L_1 \Longrightarrow L_2$.

2. Read carefully through Sect. 6.1.4 and derive the constraints listed in table (6.1.36) representing the logical expressions involving three variables.
3. Model the two statements

   (a) $L_1 \vee L_2 \vee \ldots \vee L_k \vee \neg L_{k+1} \vee \neg L_{k+2} \vee \ldots \vee \neg L_n)$;
   (b) $\neg L_1 \wedge \neg L_2 \wedge \neg \ldots \wedge \neg L_k \Longrightarrow (L_{k+1} \vee L_{k+2} \vee \ldots \vee L_n)$.

4. In the traditional British game of Bickering, a team comprises four players. The British International Bickering squad has the four male players Al Aggressive, Bill Beastly, Leo Large, Nigel Nasty, and the four female players Edwina Eager, Patsy Putrid, Sharon Solid, and Tina Tall. Four of these players will be selected to form the mixed sex team. The rules of the game are such that

   (a) A team must contain at least one man and one woman;
   (b) If a team has at least two male players, then at least three players in the team must be right handed;
   (c) If a team has at least two female players, then at least three players in the team must be left handed.

   Leo, Nigel, Sharon, and Tina are the only right-handed players.
   Represent the logical conditions that must be satisfied by a team using ILP constraints.
5. Five married couples each decided to celebrate with a bottle of wine last week. From the information given, formulate and solve an ILP problem and hence put together each married couple and say which wine they drank and on which night. Each couple had wine on a different night. Five different wines were used.

   (a) Philip is married to Marie. They did not have wine on Wednesday night. Carl had wine on Wednesday night.
   (b) The Soave was not drunk on Friday night, nor was this wine drunk by Simon.
   (c) Simon and his wife had a bottle of wine the night after the couple who had the Spumante, but two nights after Margaret and her husband had wine.
   (d) Kathy did not have wine on Tuesday night, but she was the person who had the Chianti.
   (e) Olive and her husband, who is not Ray, enjoyed their wine on Friday.

   The men are Carl, Philip, Ray, Roland, and Simon. The women are Kathy, Margaret, Marie, Olive, and Vanessa. The wines are Chianti, Liebfrauenmilch, Riesling, Soave, and Spumante. The nights are Monday to Friday.

6. A company has a series of pipelines laid under the factory floor. The floor consists of a series of heavy square slabs. It is desired to inspect each pipeline and this can be done if the company lifts **one** of the slabs directly above each pipeline. The layout of slabs is as shown:

1	2	3	4	5	6	7	8	9
10	11	12	13	14	15	16	17	18
19	20	21	22	23	24	25	26	27
28	29	30	31	32	33	34	35	36

with pipelines lying under the five slab groups {2,3,4}, {7,16,17,18}, {10,11,12}, {13,14}, and {25,34,33,32}.
When a slab is lifted it suffers damage, as do all other slabs that touch it (unless they touch diagonally). The company wishes to minimize the total number of slabs damaged when all pipelines have been inspected.
Formulate and solve, using a modeling language and a MILP solver, this problem. (Note: this is quite a difficult problem.)
7. Refer back to the Lim's problem in Chap. 5.
Lim is now offered the opportunity to purchase I1 under the following conditions: In any month up to 400 units can be bought at a cost comprising a fixed component of £400, plus £1 per unit. So, for instance, if Lim bought I1 in all four months it would incur a cost of 4 times £400, plus £1 per unit. If it wishes, Lim can still manufacture I1 itself.

   (a) Extend your model to deal with the new opportunity.
   (b) Solve the problem you formulated.

8. A company wishes to build a corporate model of its operations. The following conditions must be met by the model:

   (a) A four year planning period will be considered in the model.
   (b) Up to five types of investment are available to the company in any year. The investments can be chosen to commence in any year and any of the investments may be repeated in any year. Thus up to twenty investments are available. These investments will be evaluated in terms of their net present value at the end of the four year planning horizon. The net present value of cash flows in future year n is defined as

$$\frac{\text{sum of cash inflows} - \text{sum of cash outflows}}{(1+r)^n}$$

   (c) The interest rate is 15% per annum, i.e., $r = 0.15$.
   (d) In each of the four years the net cash outflow must not exceed £250,000 (maximum amount available for investment).
   (e) At least one investment must be commenced in each year, to fulfill operating requirements.

The projects generate the following cash income/payment given in £1,000 units. (Negative entries represent outflows, i.e., loss.)

	Years	1	2	3	4
Project	1	−100	−50	150	150
	2	−100	−40	50	200
	3	40	−100	50	50
	4	−200	100	100	200
	5	−150	0	150	100

Formulate and solve the problem to maximize net present value at the horizon.

9. A manufacturing company produces three products, P_1, P_2, and P_3. Each product requires the use of certain machines in a specific order. The company has available the following machines: 2 Finishers, 3 Smoothers, 3 Sprays, 3 Drills, and 1 Beveller. (A product on a machine represents a "job.") The orders and times for each job are:

P_1	Machine:	Smooth	Spray	Drill	Finish	Spray
	Time:	25	18	27	19	5
P_2	Machine:	Drill	Bevel	Smooth	Finish	Spray
	Time:	8	17	19	6	12
P_3	Machine:	Bevel	Smooth	Finish	Spray	Drill
	Time:	16	8	19	12	2

Formulate and solve the problem as an ILP to minimize the overall time for the manufacture of one of each of the three products (given that each product must be machined in the sequence stated).

Chapter 7
Types of Mixed Integer Linear Programming Problems

In this chapter a number of standard ILP problems will be formulated. As in Chap. 4, we shall look at some straightforward problems that are easy to formulate and then consider harder ILP problems. For many of the problem types a case study is provided, and for some of these, a model relating to the case is supplied in MCOL.

7.1 Knapsack and Related Problems

7.1.1 The Knapsack Problem

To understand what knapsack problems are, consider burglar Bill planning — see Fig. 7.1 — to break into the home of a wealthy family. Bill has been watching the house for some time and knows that it contains many valuable items, some of which very much interest him. As Bill plans to perform the robbery completely on his own, the weights of the items are also important to him. Our well organized burglar has put together all the information collected intensively during the last 17 weeks. There are 8 single items of particular interest. The table lists the data related to these items: their values, V_i, in units of 1,000 USD and their weights, W_i, in kilograms:

i	1	2	3	4	5	6	7	8
V_i	15	100	90	60	40	15	10	1
W_i	2	20	20	30	40	30	60	10

.

Supplementary Information The online version contains supplementary material available at https://doi.org/10.1007/978-3-030-73237-0_7.

J. Kallrath, *Business Optimization Using Mathematical Programming*, International Series in Operations Research & Management Science 307, https://doi.org/10.1007/978-3-030-73237-0_7

Fig. 7.1 Knapsack problem supporting a burglar in selecting goods to steal. He wants to maximize the value of the goods to pack into the knapsack. Produced for this book by Diana Kallrath, Copyright ©2020

During the last 3 months Bill has done a tremendous amount of body-building, so expects to be able to carry (for a short while) a sack with a maximum weight of 102 kg. It becomes quite clear to him that if he can break in only once he cannot steal all the items. The weight of all items adds up to 212 kg. The question that has kept him busy for quite a while is which items he should steal in order to maximize the value of his booty. It so happens that our notorious burglar, in a previous robbery, acquired a lot of books, and in one of them he found out about the so-called knapsack problems. Here is what he has learned.

The *knapsack problem* is the problem of choosing how many of each of a set of items of given weights and values to place in a container (knapsack), which can accommodate a limited total weight of items, such that the total value of the items placed in the container is maximum. The problem and a way to solve it are described by Dantzig (1957,[141]). Clearly the major decisions are the quantity of each item to be included. The problem may be formulated as follows.

Let σ_i be the integer number of items of type i that are selected, W_i be the weight of an item of type i, V_i be the value of an item of type i ($i \in \{1, 2, \ldots, n\}$), and C be the capacity of the knapsack. The objective function is

$$\sum_{i=1}^{n} V_i \sigma_i, \tag{7.1.1}$$

which is to be maximized subject to the constraint that the total weight of included items is no more than the available capacity

$$\sum_{i=1}^{n} W_i \sigma_i \le C \quad , \quad \sigma_i \in \mathbb{N}_0. \tag{7.1.2}$$

This problem is clearly one of the simplest ILP problems to formulate, as it has only one constraint. Nevertheless, it is not trivial to solve, and as n is growing, the time required to solve the problem increases exponentially. For large problems, methods other than standard ILP are recommended. A full discussion is contained in the book devoted to the knapsack problem by Martello & Toth (1990,[387]). Applications of the problem are discussed in Salkin & de Kluyver (1975,[479]).

A common variation of the problem is to limit the number of items available of each type to 1. The problem is then termed the $\{0,1\}$-*Knapsack Problem* and σ_i is a binary variable. The problems are of further interest when they occur as subproblems within complex solution procedures for larger problems (see Sect. 7.1.2).

So with this knowledge our burglar is even capable of formulating an optimization problem and implementing it in a modeling language. His implementation is contained in the model collection under the model name *burglar*. Note that σ_i is written as $x(i)$ in the model.

If you solve the problem, you will find the following answer: $objective = 280$, $x(1) = 1$, $x(2) = 1$, $x(3)=1$, $x(4) = 1$, $x(5)=0$, $x(6) = 1$, $x(7) = 0$, $x(8) = 0$.

7.1.2 Case Study: Float Glass Manufacturing

The trimloss problem was introduced in Sect. 4.1.1. In the following case, the authors make use of knapsack problems for pattern selection within the trimloss problem.

Dyson & Gregory (1974,[167]) discuss the use of the trimloss (or cutting stock) problem formulation to handle a problem arising in float glass manufacturing scheduling. The following formulation requires the use of as little glass as possible so as to meet the list of orders. Let us use the indices

$$\begin{aligned} p &\in \{1, 2, \ldots, n\} \ : \text{for the pattern number} \\ o &\in \{1, 2, \ldots, m\} : \text{for the order number,} \end{aligned} \tag{7.1.3}$$

the data

$$\begin{aligned} N_o &: \text{the number of pieces of size } l_0 \text{ x } w_0 \text{ required,} \\ A_{op} &: \text{the number of pieces of size } l_0 \text{ x } w_0, \\ &\quad \text{produced by a single application of cutting pattern } p, \\ C_p &: \text{area of cutting pattern } p, \end{aligned} \tag{7.1.4}$$

and finally the integer variable

$$\sigma_p \geq 0 : \text{the number of times pattern } p \text{ is repeated.} \tag{7.1.5}$$

Then the formulation is

$$\min \quad \sum_{p=1}^{n} C_p \sigma_p \tag{7.1.6}$$

subject to

$$\sum_{p=1}^{n} A_{op} \sigma_p \geq N_o \quad , \quad \forall o \in \{1, 2, \ldots m\}. \tag{7.1.7}$$

It turns out that the total number of cutting patterns is about 10,000. Thus, a method is used to introduce pattern variables into the problem gradually and drop other pattern variables, when the problem was being solved by the optimization software. This technique is called *column generation* and makes use of solving intermediate knapsack problems to determine which patterns should be considered next.

7.1.3 The Generalized Assignment Problem

Continuing the analogy from the knapsack problem, let us assume that the burglar wishes to divide up the loot into separate containers and sell it (illegally!) to different persons. He wants to use three containers at most, each of which has a weight restriction of 90 kg, but the profit he will receive varies according to the sack in which an item is placed because the persons to whom each container will be sold, who are known to our burglar, will place a different value on each item. Clearly the burglar intends to sell the loot for the highest total profit and also wishes to sell all the items.

i	1	2	3	4	5	6	7	8
P_i [in k$] container 1	10	80	50	30	30	10	8	1
P_i [in k$] container 2	15	70	70	60	35	12	6	2
P_i [in k$] container 3	9	80	60	50	35	8	9	1
W_i [in kg]	2	20	20	30	40	30	60	10

The major decision that must be taken is to which container each item should be allocated. This problem is known as the *generalized assignment problem* (GAP), of which the above example forms a special case. In the more general case, the weights of items are also allowed to vary depending on which container they are assigned to. The GAP can be stated (cf. Martello & Toth (1990,[387])) as:

Given n items and m containers with indices

$$\begin{aligned} c &\in \{1, 2, \ldots, m\} : \text{the set of containers} \\ i &\in \{1, 2, \ldots, n\} : \text{the set of items} \end{aligned} \tag{7.1.8}$$

and data

$$\begin{aligned} &C_c \; : \text{capacity of container } c \\ &P_{ci} \; : \text{profit of item } i \text{ if assigned to container } c \quad , \\ &W_{ci} \; : \text{volume of item } i \text{ if assigned to container } c \end{aligned} \tag{7.1.9}$$

assign each item to exactly one container so as to maximize the total profit, without assigning to any container a total volume greater than its capacity. Note that the volume of an item can vary according to which container it is assigned to, as the quantity of protective wrapping required varies from container to container. The main decisions are thus to which containers particular items are assigned to. Accordingly we introduce binary variables, $\delta_{ci} \in \{0, 1\}$, such that

$$\delta_{ci} := \begin{cases} 1 \text{ , if item } i \text{ is assigned to container } c, \\ 0 \text{ , otherwise.} \end{cases} \tag{7.1.10}$$

The model formulation is

$$\max \quad \sum_{c=1}^{m} \sum_{i=1}^{n} P_{ci} \delta_{ci} \tag{7.1.11}$$

subject to the constraint that the total volume of a container is limited by capacity

$$\sum_{i=1}^{n} W_{ci} \delta_{ci} \leq C_c \quad , \quad \forall c \in \{1, 2, \ldots m\} \tag{7.1.12}$$

and the constraint that each item is assigned to exactly one container

$$\sum_{c=1}^{m} \delta_{ci} = 1 \quad , \quad \forall i \in \{1, 2, \ldots, n\}, \tag{7.1.13}$$

$$\delta_{ci} \in \{0, 1\} \quad , \quad c \in \{1, 2, \ldots, m\} \quad , \quad i \in \{1, 2, \ldots, n\}. \tag{7.1.14}$$

Note: In some versions of the problem, if $W_{c'i'} = 0$ for some c', i' in the above, then it is assumed that $\delta_{c'i'}$ is not a variable.

The burglar's GAP should now be solved as an exercise (see Sect. 7.10.). A further example of the GAP is solved in Sect. 14.1.3.3 by Lagrange relaxation.

7.1.4 *The Multiple Binary Knapsack Problem*

One further variation will be considered. The *multiple knapsack problem*, cf. Martello & Toth (1990,[387]), has similarities to the generalized assignment problem. In this problem essentially all conditions of the generalized assignment problem are present, except for the fact that not every item needs to be assigned to a container (or knapsack). Such a variation would be relevant if the size of each container was more restricted and the total size of all the containers was insufficient to contain all the items.

In formulation, the principal difference from the GAP given by (7.1.11–7.1.14) is that Eq. (7.1.13) is replaced by the inequalities

$$\sum_{c=1}^{m} \delta_{ci} \leq 1 \quad , \quad \forall i \in \{1, 2, \ldots, n\} \tag{7.1.15}$$

and W_{ci} is replaced by W'_i and P_{ci} by P'_c.

7.2 The Traveling Salesman Problem

In distribution it frequently arises that one or more vehicles or persons must make a circular trip around a number of locations, collecting or delivering goods at these locations. Finding the "best" route will be a feature of such problems. The *traveling salesman problem* (TSP) is the problem faced by a traveler wishing to devise one circuit to visit a series of cities starting from one city, visiting each other city *exactly once*, and returning to the original city. The distances from any city to any other city are known in advance, and the problem requires that the total distance traveled be minimized.

This problem seems deceptively simple at first. It appears to have similarities to the assignment problem as each city must be entered once and left once. However, the difficulty of the problem lies in obtaining a complete route that passes through *all* cities progressively. If there are 4 cities, a route that goes from city 1 to city 2 and back to city 1 again together with a route that goes from city 3 to city 4 and back to city 3 again has the virtue that each city is entered once and left once and all cities are visited but does not have the property that there is one circuit; there is no link between the pair of cities 1,2 and the pair of cities 3,4, so this solution would be regarded as invalid. In a valid solution the single circuit is called a tour and for invalid solutions circuits that encompass less than the full set of cities are called subtours. The TSP was first posed and solved by Dantzig et al. (1954,[145]) and further considered by Little et al. (1963,[369]). Considerable work on the problem has been generated and a good introduction to advanced work can be found in the book Lawler et al. (1985,[356]). The problem may be formulated as an ILP problem, but the formulation is not intuitive in the part that formulates conditions to avoid subtours.

Clearly the principal decision is the order in which the cities are visited, and each component of that order will be a route from a particular city to the next one in the order. Thus we introduce variables accordingly. Let us introduce binary variables $\delta_{ij} \in \{0, 1\}$ such that

$$\delta_{ij} := \begin{cases} 1 \text{ , if the tour goes directly from city } i \text{ to city } j \\ 0 \text{ , otherwise.} \end{cases} \tag{7.2.1}$$

Let the total number of cities be n and let C_{ij} be the distance from city i to city j ($i \in \{1, 2, \ldots n\}$, $j \in \{1, 2, \ldots, n\}$). The first group of constraints ensures that each city must be entered once:

$$\sum_{i=1 \wedge i \neq j}^{n} \delta_{ij} = 1 \quad , \quad j \in \{1, 2, \ldots, n\}. \tag{7.2.2}$$

Here the expression on the left is the total number of cities that lead directly to city j in the chosen route. Next, each city must be left once, i.e., the number of cities that are visited directly after any city must be 1

$$\sum_{j=1 \wedge i \neq j}^{n} \delta_{ij} = 1 \quad , \quad i \in \{1, 2, \ldots n\}. \tag{7.2.3}$$

Here the expression on the left is the total number of cities to which city i leads directly.

The objective function is

$$\min \sum_{i=1}^{n} \sum_{j=1 \wedge i \neq j}^{n} C_{ij}\delta_{ij}. \tag{7.2.4}$$

More complex constraints are now introduced to prevent subtours occurring. These also require the introduction of new variables. We introduce continuous variables y_i for $i > 1$. These variables can be interpreted as the sequence number in which each city is visited (assuming city 1 is visited first). Finally, we add the inequalities

$$y_i - y_j + (n-1)\delta_{ij} \leq n-2 \quad , \quad \forall \{ij \,|\, i > 1 \wedge j > 1 \wedge i \neq j\}, \tag{7.2.5}$$

which ensure that subtours do not occur; see Miller et al. (1960,[401]). These inequalities operate in the following way. Consider our earlier four-city example. The constraints will be

$$y_2 - y_3 + 3\delta_{23} \le 2, \quad y_3 - y_2 + 3\delta_{32} \le 2 \tag{7.2.6}$$

$$y_2 - y_4 + 3\delta_{24} \le 2, \quad y_4 - y_2 + 3\delta_{42} \le 2$$
$$y_3 - y_4 + 3\delta_{34} \le 2, \quad y_4 - y_3 + 3\delta_{43} \le 2.$$

In our invalid solution we have $\delta_{12}, \delta_{21}, \delta_{34}, \delta_{43} = 1$, and all other δ variables zero. Substituting in (7.2.6), we obtain

$$y_2 - y_3 \le 2, \quad y_3 - y_2 \le 2 \tag{7.2.7}$$

$$y_2 - y_4 \le 2, \quad y_4 - y_2 \le 2$$
$$y_3 - y_4 + 3 \le 2, \quad y_4 - y_3 + 3 \le 2.$$

If we add together the fifth and sixth constraints of (7.2.7), we get an inconsistency, *viz.* $6 \le 4$. In general, if $\delta_{ij} = 1$, then

$$y_i - y_j + n - 1 \le n - 2 \tag{7.2.8}$$

so

$$y_j \ge y_i + 1 \quad , \tag{7.2.9}$$

i.e., "one more" and if $\delta_{ij} = 0$,

$$y_i - y_j \le n - 2.$$

The biggest difference is $n - 2$, and y_i are all different, so they must take the values $2, 3, \ldots, n$. This will be impossible for a "solution" involving subtours. See Fig. 7.2 for valid and invalid solutions of a TSP with four cities.

The set of constraints given by (7.2.5) may be replaced by a set of constraints that have a slightly more realistic interpretation, as follows. As we wish to ensure subtours do not occur, if we partition the set of cities into two subsets, $\mathcal{N}_1$ and $\mathcal{N}_2$ (each subset containing at least two cities), then there must exist a link between the two subsets. For every valid partition we can repeat this argument. The constraints are then given by defining two disjoint sets, each containing at least two cities. So we introduce the sets $\mathcal{N}_1$ and $\mathcal{N}_2$ such that $\mathcal{N}_1 \cup \mathcal{N}_2 = \{1, 2, \ldots, n\}$, and the set of constraints

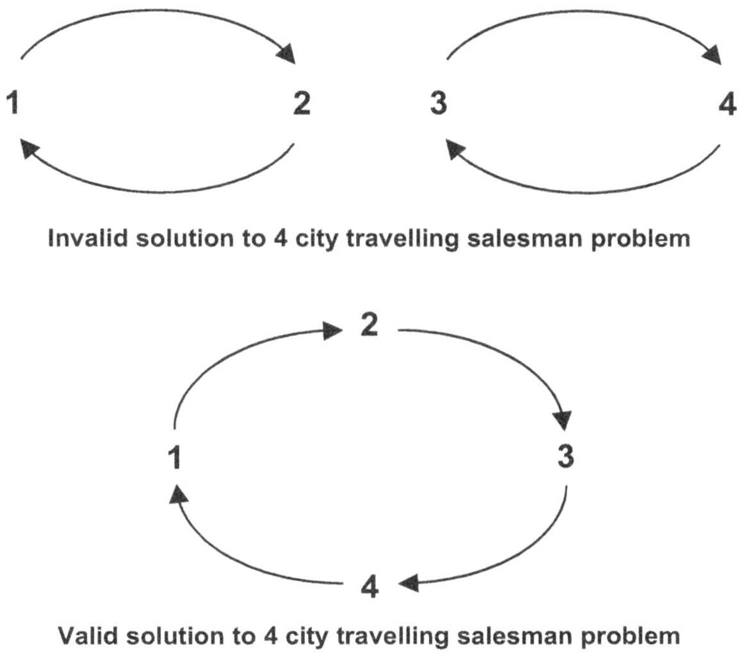

Fig. 7.2 Traveling salesman problem with four cities

$$\sum_{i \in \mathcal{N}_1,\ j \in \mathcal{N}_2} \delta_{ij} \geq 1. \tag{7.2.10}$$

For the case $n = 4$ the possible partitions are $\{\{1, 2\}, \{3, 4\}\}$, $\{\{1, 3\}, \{2, 4\}\}$, $\{\{1, 4\}, \{2, 3\}\}$, $\{\{2, 3\}, \{1, 4\}\}$, $\{\{2, 4\}, \{1, 3\}\}$, and $\{\{3, 4\}, \{1, 2\}\}$ and the constraints are

$$\begin{aligned}
&\delta_{13} + \delta_{14} + \delta_{23} + \delta_{24} \geq 1 \\
&\delta_{12} + \delta_{14} + \delta_{32} + \delta_{34} \geq 1 \\
&\delta_{12} + \delta_{13} + \delta_{42} + \delta_{43} \geq 1 \\
&\delta_{21} + \delta_{24} + \delta_{31} + \delta_{34} \geq 1 \\
&\delta_{21} + \delta_{23} + \delta_{41} + \delta_{43} \geq 1 \\
&\delta_{31} + \delta_{32} + \delta_{41} + \delta_{42} \geq 1.
\end{aligned} \tag{7.2.11}$$

The first and last of these constraints would be violated by the invalid solution described earlier ($\delta_{12} = 1, \delta_{21} = 1, \delta_{34} = 1, \delta_{43} = 1$).

Clearly the number of constraints becomes large as n rises. The formulation may seem tortuous, but it should be noted that even for a problem involving only 12 cities there are almost 40 million different routes to be considered. The TSP is hard to solve because even for a modest number of cities the number of variables and constraints required is large. In contrast to the assignment or transportation problems, the presence within a problem of elements that resemble the structure of the TSP will lead the modeler to expect that the problem will be hard to solve. However, by introducing special techniques larger problems can be solved, cf. Padberg & Rinaldi (1987,[432], 532 cities) or Applegate et al. (2007,[28], 85,900 cities). Note the technology increase between 1987 and 2007.

An application that is described in Vasquez-Maeques (1991,[563]) makes use of a model based on the TSP to allocate arrival slots to aircraft. Bohoris & Thomas (1995,[90]) describe a routing model that involves collection and depot staffing.

7.2.1 Postman Problems

The problem of finding a single route that covers each road in a network and returns to the starting point (such as might be required for a postal delivery route) while minimizing the total distance traveled is known as the *Chinese postman problem*, in tribute to its originator Kwan Mei-Ko (1962,[394]). It turns out that this problem can be solved without the use of LP or ILP techniques but by using an algorithm of Edmonds & Johnson (1973,[169]). However, if we place a limit on the number or length of roads that can be covered in a single route and, hence, force the complete network to be covered in several stages, we have a problem known as the *capacitated Chinese postman problem*, which is much harder to solve. Eglese (1994,[170]) describes an application of the capacitated Chinese postman problem to the scheduling of vehicles undertaking winter gritting of roads.

7.2.2 Vehicle Routing Problems

In the TSP one has to compute an optimal sequence of nodes that minimizes the costs of visiting all the nodes that make up a path, starting from a default source node and possibly returning to this node. In vehicle routing problems (VRPs, cf. Toth & Vigo (2002,[552]), Laporte et al. (2013,[352]), Toth & Vigo (2014,[553]) or Cacchiani et al. (2020,[107])) the task is to visit these nodes (depots, cities, locations of customers, etc.) using one or several fleets of vehicles. This essentially involves two major subtasks: A subset of nodes has to be allocated to a vehicle establishing a route or path, and over this subset a TSP has to be solved.

VRPs come in many flavors depending on the practical situation; cf. Irnich et al. (2014,[283]). Usually, they are subject to resource constraints (vehicle-specific travel times from one node to another node, weight and volume associated with the carrying load as, for instance, in Baldacci et al. (2010,[45])) and time-window constraints (time intervals for pickup and delivery as, for instance, in Desaulniers et al. (2001,[156])). Vehicles can serve only one route or several routes.

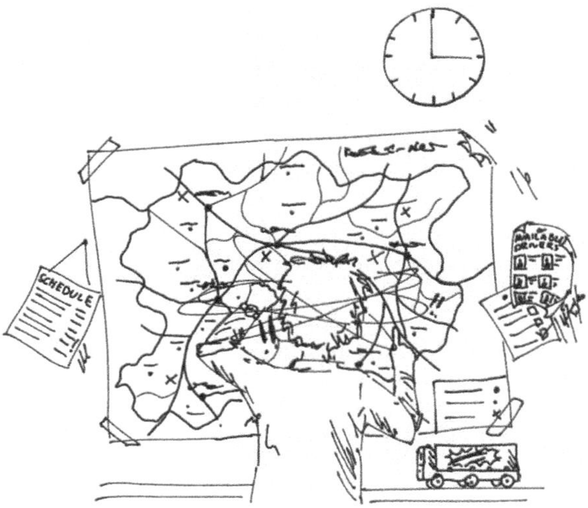

Fig. 7.3 Vehicle routing and dispatching. The dispatcher has a difficult life allocating driving destinations to vehicles, constructing tours, and allocating drivers to vehicles. Produced for this book by Diana Kallrath, Copyright ©2020

Exact algorithms as in Toth & Vigo (2002,[551]) or Mingozzi et al. (2013,[402]), or B&B as in Toth & Vigo (2002b,[550]) and B&C (cf. Wenger (2003,[570])) can work up to a few thousand nodes. Larger problem instances are usually solved by heuristics and metaheuristics.

While the model formulations are beyond what we want to host in this section, we want to stress that they support dispatchers in their difficult job — see Fig. 7.3 — allocating driving destinations to vehicles, constructing tours, and allocating drivers to vehicles — and meeting many constraints.

7.2.3 *Case Study: Heating Oil Delivery*

To illustrate real-world TSP and VRP problems, let us consider the heating oil delivery problem presented and formulated by Guéret et al. (2002,[245]). A fuel truck fills its tank — its capacity is C — at a refinery (the depot) and delivers heating oil to clients. The total number of kilometers driven must be minimized.

The depot and the client locations establish the set $\mathcal{S} = \{1, \ldots, S\}$ of sites. By definition, we identify site 1 as the depot. The locations of clients are then the subset $\mathcal{C} = \{2, \ldots, S\} \subset \mathcal{S}$. We introduce binary variables δ_{ij} that take the value 1 if site i immediately precedes site j in a tour, and 0 otherwise. Let D_{ij} be the distance between two sites i and j, and R_i the requested quantity ordered by client i with $R_1 = 0$.

The truck's finite capacity C puts the problem beyond a TSP problem. Therefore, we introduce variables q_i representing the total amount of oil delivered on the route to previous clients and client i. For example, if client 10 gets its oil delivered by a truck driving route $1-3-11-10-6-1$, then $q_{10}=R_3+R_{11}+R_{10}$. If the total demand $\sum_{i\in\mathscr{C}} R_i$ exceeds C, the truck has to do several tours, i.e., it drives back to the depot and refills its tank before serving the next client. Thus, a tour is a sequence of sites starting and ending with the depot.

With these data and variables δ_{ij} and q_i, we formulate the following model. The objective (7.2.12) of this problem is to minimize the total number of kilometers driven, i.e.,

$$\min \sum_{i\in\mathscr{S}} \sum_{j\in\mathscr{S}, i\neq j} D_{ij}\delta_{ij}. \tag{7.2.12}$$

Every client location has to be served once. This is expressed through the two assignment equalities

$$\sum_{i\in\mathscr{S}, i\neq j} \delta_{ij}=1 \quad , \quad \forall j\in\mathscr{C} \quad ; \quad \sum_{j\in\mathscr{S}, j\neq i} \delta_{ij}=1 \quad , \quad \forall i\in\mathscr{C} \tag{7.2.13}$$

that make the delivery enter and leave every client location exactly once.

The quantity q_i must be at least as large as the quantity ordered by client i and within the capacity limit C of the tankers

$$R_i \le q_i \le C \quad , \quad \forall i\in\mathscr{C}. \tag{7.2.14}$$

Furthermore, if client i is the first of a tour, then q_i is equal to the quantity ordered by this client. This constraint is expressed through the two sets of inequalities (7.2.14) and

$$q_i \le C+(R_i-C)\delta_{1i} \quad , \quad \forall i\in\mathscr{C}. \tag{7.2.15}$$

Indeed, if i is the first client of a tour, then $\delta_{1i}=1$ and, after simplification, (7.2.15) is equivalent to

$$q_i \le R_i \quad , \quad \forall i\in\mathscr{C}. \tag{7.2.16}$$

From (7.2.16) and (7.2.14) it follows that q_i is equal to the demand of client i. If i is not the first of a tour, $\delta_{1i}=0$ and (7.2.15) is equivalent to

$$q_i \le C, \tag{7.2.17}$$

which is redundant as it is already expressed in constraint (7.2.14). Let us now consider the case where i is not the first customer of the tour. Then q_i must equal the sum of quantities delivered between the depot and i inclusively. This means that if client j comes after client i in a tour, q_j must be equal to the quantity delivered on the tour from the depot to i, plus the quantity ordered by j. This relation is stated by

$$q_j \geq q_i + R_j - C + C\delta_{ij} + (C - R_j - R_i)\delta_{ji} \quad , \quad \forall\{(i,j) \in \mathscr{C} \times \mathscr{C} | i \neq j\}. \tag{7.2.18}$$

If j is the immediate successor of i in a tour, then $\delta_{ij} = 1$ and $\delta_{ji} = 0$, and (7.2.18) is equivalent to (7.2.19). Indeed, we have

$$q_j \geq q_i + R_j. \tag{7.2.19}$$

When j does not come immediately after i, constraint (7.2.18) remains valid. If j is the immediate predecessor of i, inequality (7.2.18) becomes

$$q_j \geq q_i - R_i. \tag{7.2.20}$$

This inequality means that the quantity delivered from the refinery up to j is not less than the quantity delivered between the depot and the successor i of j on the tour, a quantity that needs to be reduced by the delivery at i. If j is the immediate predecessor of i, then i is the immediate successor of j. By swapping the indices in (7.2.19), in addition to (7.2.20) we obtain the inequality

$$q_i \geq q_j + R_i. \tag{7.2.21}$$

The combination of inequalities (7.2.20) and (7.2.21) implies

$$q_i = q_j + R_i. \tag{7.2.22}$$

If i and j are not next to each other on a tour, we obtain

$$q_j \geq q_i + R_j - C. \tag{7.2.23}$$

As the terms on the right-hand side of the inequality sign are less than or equal to R_j, (7.2.23) is redundant as it is subsumed by the constraint (7.2.14). Note that the assignment of variables q_i to every site i guarantees that the capacity limits of the tankers are not exceeded while making any tour impossible that does not include the depot. Finally, the model is completed by

$$q_i \geq 0 \quad , \quad \forall i \in \mathscr{C}, \tag{7.2.24}$$

enforcing that the variables q_i are non-negative, and

$$\delta_{ij} \in \{0,1\} \quad , \quad \forall\{(i,j) \in \mathscr{S} \times \mathscr{S} | i \neq j\}, \tag{7.2.25}$$

expressing that δ_{ij} are binary variables.

MCOL contains *vehRoute.mos*, an implementation of this problem for some example data[1] with order demands $\mathbf{R} = 1000 \cdot (0, 14, 3, 6, 16, 15, 5)$ liters for six clients, distances (symmetric in this example) between sites

s/s	1	2	3	4	5	6	7
1	0	148	55	32	70	140	73
2	148	0	93	180	99	12	72
3	55	93	0	85	20	83	28
4	32	180	85	0	100	174	99
5	70	99	20	100	0	85	49
6	140	12	83	174	85	0	73
7	73	72	28	99	49	73	0

,

and the truck capacity $C = 39,000$ liters leading to the objective function value and distance driven of 497. During its first tour specified by $1 - 3 - 6 - 2 - 7 - 1$ the delivers are 37,000 liters to four clients: 2,3,6, and 7. 22,000 liters are delivered to clients 4 and 5 during the truck's second tour $1 - 4 - 5 - 1$. Note that tours are not explicit model objects; they are rather established in post-processing. Tour k is finished when the truck returns to the depot, and tour $k + 1$ begins when the truck leaves the depot again.

7.3 Facility Location Problems

7.3.1 The Uncapacitated Facility Location Problem

In distribution and logistics one problem that arises is that of locating a set of facilities, such as warehouses, each at one of a series of locations, so that clients may be supplied with goods from these locations in the most efficient way possible. In theory it may be possible to supply all customers from just one location, but this may be unwise because costs of distribution will be high to customers situated a long way from this one location. At the other extreme, it may be possible to set up many facilities so that each customer is supplied from a nearby location, but as costs will be associated with the opening up of each facility it is unlikely to be economic to use a large set of facilities. Thus the major decision to be taken involves jointly deciding how many facilities should be set up and at which locations. This problem is termed the *uncapacitated facility location problem.*

[1]The data *e4deliver.dat* and the Mosel implementation are from *e4deliver.mos* in FICO Xpress Optimization Examples Repository (https://examples.xpress.fico.com/example.pl?id=mosel_app_5_4).

In a typical instance of the problem, 20 locations might be considered as potential sites for renting warehouse space and 1,000 commercial customers are to be supplied with building materials such as bricks, cement, and timber from these locations. Sites on the outskirts of major industrial areas are favored locations because they are inexpensive, and have good road communication networks. As the materials are heavy, it is unwise to concentrate materials at only a few locations, but it is unclear how many locations should be used.

In other instances of the problem we may wish to collect items from the customers, rather than deliver to them. It may also be best not to supply/deliver all the requirements for a customer from one location, but rather to supply/deliver a fraction from each of several locations.

A general instance of the uncapacitated facility location problem may be formulated as follows. First we introduce indices:

$$\begin{aligned} i &\in \{1, 2, \ldots, m\} : \text{set of clients} \\ j &\in \{1, 2, \ldots, n\} : \text{set of locations} \end{aligned} \qquad (7.3.1)$$

and data:

$$\begin{aligned} C_j &: \text{the cost of locating a facility at location } j \\ H_{ij} &: \text{the cost of satisfying demand of client } i \text{ from location } j. \end{aligned} \qquad (7.3.2)$$

We introduce binary variables

$$\delta_j := \begin{cases} 1 \text{, if a facility is placed at location } j \\ 0 \text{, otherwise} \end{cases} \qquad (7.3.3)$$

and continuous variables $y_{ij} \geq 0$ measuring the fractional quantity of demand of client i that is satisfied from location j. The objective includes set-up costs and variables costs

$$\min \quad \sum_{j=1}^{n} C_j \delta_j + \sum_{j=1}^{n} \sum_{i=1}^{m} H_{ij} y_{ij}. \qquad (7.3.4)$$

The constraints are as follows. Each client must be completely supplied with no surplus, and hence, we have the constraint

$$\sum_{j=1}^{n} y_{ij} = 1 \quad , \quad i \in \{1, 2, \ldots, m\} \qquad (7.3.5)$$

and if $\delta_j = 0$, then $y_{ij} = 0$ as a client can only be supplied from a location if a facility is already located there, i.e.,

$$y_{ij} - \delta_j \leq 0 \quad , \quad i \in \{1, 2, \ldots, m\}; \quad j \in \{1, 2, \ldots, n\}. \qquad (7.3.6)$$

An application that makes use of a variation of the above model is used in Robinson et al. (1993,[462]) where the problem is one of the distributions using networks.

7.3.2 *The Capacitated Facility Location Problem*

In the uncapacitated facility location problem it was assumed that once a facility had been opened up at a particular location, then it could supply/deliver all the customers' requirements. This is perhaps unrealistic, and it might be expected that there would be further limitations at individual locations on amounts supplied/delivered, e.g., because of limitations on storage capacities at each site. When capacity limitations are introduced, the problem is termed the *capacitated facility location problem.*

The previous formulation in Sect. 7.3.1 may be adapted to provide a capacitated problem. y_{ij} is replaced by the binary variable γ_{ij} in the formulation, included in the objective function, and no longer do we permit fractional quantities of demand. The constraint (7.3.5) is replaced by

$$\sum_{j=1}^{n} \gamma_{ij} = D_i \quad , \quad i \in \{1, 2, \ldots, m\}, \tag{7.3.7}$$

where D_i is the demand of client i and (7.3.6) is replaced by

$$\sum_{i=1}^{m} \gamma_{ij} - U_j \delta_j \leq 0 \quad , \quad j \in \{1, 2, \ldots, n\}, \tag{7.3.8}$$

where U_j is the capacity of facility j.

An application that makes use of a facility location model similar to the above appears in Spencer et al. (1990,[517]). The application concerns the selection of sites for telemarketing centers. An application in the regional water industry described by Baker & Baia (1995,[41]) uses modifications of B&B routines to solve problems of capacitated location. The book *Location Science* by Laporte et al. (2015,[351]) is recommend for further reading.

7.4 Set Covering, Partitioning, and Packing

7.4.1 *The Set Covering Problem*

Let us consider a problem where up to eight depots ($j = 1, 2, \ldots, 8$) can be used to supply three sets ($i = 1, 2, 3$) of customers with specialized telecommunications

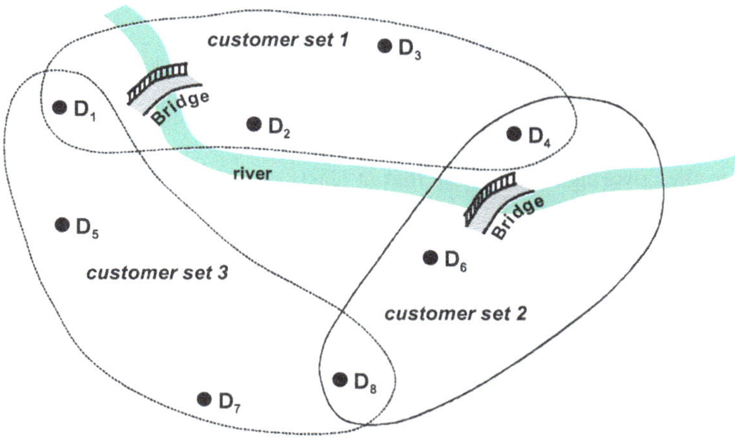

Fig. 7.4 A set covering problem. The river separates the customer set on the North from those on the South. Due to the river and bridges, the customer sets can be served by the following depots:

customer sets	*depots*
1	1, 2, 3, 4
2	4, 6, 8
3	1, 5, 7, 8

Note that depots 4 and 8 alone can cover all customer sets; same for (1,8) and (1,4)

equipment. A set-up cost is associated with each depot chosen to be used. Due to the river geography illustrated in Fig. 7.4, each set of customers can be supplied only by particular depots. Customers in set 1 are on the North side of the river and can receive deliveries from depots 1, 2, 3, and 4; depot 1 is just on the South side of the bridge leading to a short driving distance. The table below shows which depots may be used to supply each set of customers. The problem is to open sufficient depots, at minimum total cost, to ensure that all customers are supplied, i.e., are covered. A possible solution to this problem just stated would be to choose depots 4 and 8 (without reference to any costs), as it only involves two depots and it is clear that no single depot can supply all three customers.

A problem of the above type is referred to as a *set covering problem* (SCP), i.e., the problem of "covering" at minimal total cost, a series of m subsets of n elements by choosing at least one element from each subset. The problem may be stated as an ILP problem by introducing binary variables, δ_j, to indicate if a particular element j, in our example a depot, is chosen or not. Let the data be

$$
\begin{aligned}
C_j &: \text{the cost of choosing element } j \\
A_{ij} &:= \begin{cases} 1, & \text{if element } j \text{ occurs in subset } i \\ 0, & \text{otherwise.} \end{cases}
\end{aligned}
\qquad (7.4.1)
$$

The formulation is then

$$\min \quad \sum_{j=1}^{n} C_j\delta_j \tag{7.4.2}$$

subject to the covering requirement that at least one element must be chosen from each subset

$$\sum_{j=1}^{n} A_{ij}\delta_j \geq 1 \quad , \quad i \in \{1, 2, \ldots, m\}, \tag{7.4.3}$$

and the domain conditions

$$\delta_j \in \{0, 1\} \quad , \quad \forall j \in \{1, 2, ..n\}. \tag{7.4.4}$$

In this formulation each A_{ij} is an $m \times n$ matrix of binary entries. The main constraints of the problem ensure that each subset is covered by guaranteeing that at least one variable that occurs in it is non-zero.

Corresponding to Fig. 7.4, for the three-subset example we have

$$A_{ij} = \begin{pmatrix} 1\,1\,1\,1\,0\,0\,0\,0 \\ 0\,0\,0\,1\,0\,1\,0\,1 \\ 1\,0\,0\,0\,1\,0\,1\,1 \end{pmatrix}. \tag{7.4.5}$$

If we further assume that all cost coefficients are equal, i.e., $C_j = 1$ for all j, the problem becomes one of selecting the minimum number of covering depots (subsets) as opposed to a minimum-cost collection of covering depots. With these data, we get the detailed problem formulation

$$\min \quad \delta_1 + \delta_2 + \delta_3 + \delta_4 + \delta_5 + \delta_6 + \delta_7 + \delta_8 \tag{7.4.6}$$

subject to the cover inequalities

$$\begin{array}{llllllllll} \delta_1 & +\delta_2 & +\delta_3 & +\delta_4 & & & & & \geq 1 \\ & & & \delta_4 & & +\delta_6 & & +\delta_8 & \geq 1 \\ \delta_1 & & & & +\delta_5 & & +\delta_7 & +\delta_8 & \geq 1 \end{array} \tag{7.4.7}$$

and the domain conditions

$$\delta_j \in \{0, 1\} \quad , \quad \forall j \in \{1, 2, .., n\}. \tag{7.4.8}$$

An optimal solution to this problem is $\delta_4 = \delta_8 = 1$, with all other variables equal to zero and an objective function value of 2.

In a more abstract description, and somewhat different from our introduction above, the set covering problem is about a set $\mathscr{U}$, n subsets $\mathscr{S}_j \subseteq \mathscr{U}$, and the question whether for a natural number $k \le n$ a union of k or fewer subsets S_j exist, which corresponds to the set $\mathscr{U}$ (cover). Formulated as an optimization problem, one wants to determine a coverage with the smallest possible number of subsets S_j, or, if costs C_j are assigned to the subsets S_j, a coverage with lowest costs.

There are many practical applications of this problem, including assembly line balancing (Salveson (1955,[480]), locating emergency facilities (Toregas et al. (1971,[549])), and crew scheduling (Baker & Fisher (1981,[42])). For practical purposes, SCP instances with up to n variables and m constraints can be solved to optimality; approximation methods are often used for larger instances. With the increasing power of MILP solvers, n and m increase over time. The set covering problem has come into prominence of late because of its use to schedule airline crews. This application leads to large problems, but these can be solved, and the solutions obtained allow airlines to save sums equivalent to many millions of pounds by efficient policies of crew scheduling. Such problems typically have over 100,000 variables. Further discussion is contained in Ryan (1992,[475]). Because of the financial impact of this scheduling area, much research is under way to find fast methods for solving set covering problems.

7.4.2 The Set Partitioning Problem

The *set partitioning problem* (SPP) is a variation of the set covering problem where the inequalities of SCP are replaced by equations. Thus exactly one element from each subset must be chosen.

Let us use the situation of the previous section. The problem is now to select a minimum number of depots such that each customer may be supplied by exactly one of the selected depots. As before the set-up cost associated with each depot chosen to be used will be 1. Each set of customers can be supplied by particular depots only as given in Fig. 7.4. The requirement is to open exactly one depot for each set of customers at minimum total cost. For the three-subset example given, the formulation is

$$\min \quad \delta_1 + \delta_2 + \delta_3 + \delta_4 + \delta_5 + \delta_6 + \delta_7 + \delta_8 \tag{7.4.9}$$

subject to the partitioning equalities

$$\begin{aligned} \delta_1 + \delta_2 + \delta_3 + \delta_4 &= 1 \\ \delta_4 + \delta_6 + \delta_8 &= 1 \\ \delta_1 + \delta_5 + \delta_7 + \delta_8 &= 1 \end{aligned} \tag{7.4.10}$$

and domain conditions

$$\delta_j \in \{0, 1\} \quad , \quad \forall j \in \{1, 2, \ldots, 8\}. \tag{7.4.11}$$

An optimal solution to this problem is $\delta_2 = \delta_8 = 1$, with all other variables equal to zero. Note that the optimal solution of the previous section, $\delta_4 = \delta_8 = 1$, violates the second equality $\delta_4 + \delta_6 + \delta_8 = 1$.

The set partitioning problem has an application in airline crew scheduling, where each set represents flight legs that must be flown and each variable represents a feasible round-trip rotation for a crew. Each subset then represents a collection of feasible allocations of crews, exactly one of which must be chosen. Some applications are described in Arabeyre et al. (1969,[29]), Baker & Fisher (1981,[42]), and Gershkoff (1989,[214]).

7.4.3 *The Set Packing Problem*

The *set packing problem* is closely related to the set covering problem. In this problem we also have a series of subsets of a set of elements. Each element has a value, and we wish to choose at most one element from each subset to maximize the total value of the elements chosen.

Consider a problem in which three machines are available and each machine can make at most one of a restricted series of items in any period. Each item can be made on a single machine. As in the previous two sections, let the data be as in the table below.

Machine	Item Types
1	1, 2, 3, 4
2	4, 6, 8
3	1, 5, 7, 8

(7.4.12)

Let the values of all item types be 1, and then a formulation of the problem to manufacture as many different item types as possible is

$$\max \quad \delta_1 + \delta_2 + \delta_3 + \delta_4 + \delta_5 + \delta_6 + \delta_7 + \delta_8 \tag{7.4.13}$$

subject to the packing requirement that at most one item may be chosen from each subset of item types, i.e.,

$$\begin{aligned} \delta_1 + \delta_2 + \delta_3 + \delta_4 &\le 1 \\ \delta_4 + \delta_6 + \delta_8 &\le 1 \\ \delta_1 + \delta_5 + \delta_7 + \delta_8 &\le 1 \end{aligned} \tag{7.4.14}$$

and the domain conditions

$$\delta_j \in \{0, 1\} \quad , \quad \forall j \in \{1, 2, \ldots, 8\}. \tag{7.4.15}$$

An optimal solution, one of many, is $\delta_2 = \delta_6 = \delta_7 = 1$, with all other variables zero and the objective equal to 3.

7.4.4 Additional Applications

Set covering is involved in a model built to schedule vehicles and manpower, which is described in Blais et al. (1990,[84]). A set partitioning approach is used to pair crews for airlines in an application described in Anbil et al. (1991,[18]).

7.4.5 Case Study: Airline Management at Delta Air Lines

A case study will now be discussed that describes the use of extended covering and packing type models. The application is in airline management at Delta Air Lines and is described by Subramanian et al. (1994,[536]).

The problem faced by the company is that if an airline uses a too small aircraft, then it may lose potential customers, while if it uses a too large aircraft, it may be underutilized, which is wasteful and therefore expensive. The ideal is to have the right aircraft available in the right location at the right time. A model was developed to assign fleet types to flight legs. The model is organized to schedule a single day, which is typical, and forms part of a continuous cycle. In practice, this will be only partially true and adjustments are required, e.g., for weekends.

A simplified version of the model appearing in Subramanian et al. (1994,[536]) will now be given. Let us start by defining the indices

$$\begin{array}{ll} i \in \mathscr{C} & \text{, the set of cities} \\ k \in \mathscr{F} & \text{, the set of aircraft types} \\ l \in \mathscr{L} & \text{, the set of flight legs} \\ t \in \mathscr{T} & \text{, the set of time periods} \end{array} \tag{7.4.16}$$

and the subsets $\mathscr{L}_{dit} \subset \mathscr{L}$ and $\mathscr{L}_{oit} \subset \mathscr{L}$

$$\begin{array}{ll} \mathscr{L}_{dit} & \text{, the set of flight legs whose destination is city } i \text{ in period } t \\ \mathscr{L}_{oit} & \text{, the set of flight legs whose origin is city } i \text{ in period } t. \end{array} \tag{7.4.17}$$

Time periods are minutes and run across different days.

The relevant data are

$$
\begin{aligned}
C_{kl} &: \text{cost if an aircraft of type } k \text{ flies leg } l \\
S_k &: \text{the maximum number of available aircrafts of type } k.
\end{aligned}
\tag{7.4.18}
$$

If the aircraft cannot fly leg l, then we set $C_{kl} = \infty$. We need the binary variables , $\delta_{kl} \in \{0, 1\}$ defined as

$$
\delta_{kl} := \begin{cases} 1 \text{ , if an aircraft of type } k \text{ is assigned to leg } l \\ 0 \text{ , otherwise} \end{cases} \quad , \quad \begin{array}{l} \forall k \in \mathcal{F} \\ \forall l \in \mathcal{L}. \end{array}
\tag{7.4.19}
$$

Furthermore, we introduce the integer variables

$$
\begin{aligned}
\alpha_{kit} &\in \mathbb{N}_0 \text{ the number of aircrafts of type } k \text{ on the ground} \\
&\qquad \text{at city } i \text{ from time period } t \text{ to period time } t+1, \\
\mu_k &\in \mathbb{N}_0 \text{ the number of aircrafts of type } k \text{ that are used.}
\end{aligned}
\tag{7.4.20}
$$

The objective is to minimize total costs, i.e.,

$$
\sum_{k \in \mathcal{F}, l \in \mathcal{L}} C_{kl} \delta_{kl}
\tag{7.4.21}
$$

subject to

$$
\alpha_{k,i,t+1} = \alpha_{k,i,t-1} + \sum_{l \in \mathcal{L}_{dit}} \delta_{kl} - \sum_{l \in \mathcal{L}oit} \delta_{kl} \quad , \quad \begin{array}{l} k \in \mathcal{F} \\ i \in \mathcal{C} \\ t \in \mathcal{T}, \end{array}
\tag{7.4.22}
$$

i.e., the number of aircrafts on the ground at city i in time period $t+1$ is equal to the number of aircrafts on the ground at city i in time period $t-1$ plus all incoming aircrafts to city i in period t minus all departing aircrafts from city i in time period t. Next we model

$$
\sum_{k \in \mathcal{F}} \delta_{kl} = 1 \quad , \quad l \in \mathcal{L},
\tag{7.4.23}
$$

which states that exactly one aircraft is assigned to cover each leg. Furthermore, the number of aircrafts in use must balance, i.e.,

$$
\sum_{l \in \mathcal{L}} \delta_{kl} + \sum_{i \in \mathcal{C}, t \in \mathcal{T}} \alpha_{kit} - \mu_k = 0 \quad , \quad k \in \mathcal{F}.
\tag{7.4.24}
$$

Next, the number of aircrafts of type k that are used must not exceed the maximum.

$$
\mu_k \leq S_k.
\tag{7.4.25}
$$

Finally, if the destination of leg l_1 and the origin of leg l_2 are identical, then the constraint

$$\delta_{kl_1} - \delta_{kl_2} = 0 \quad , \quad k \in \mathscr{F} \tag{7.4.26}$$

must hold.

The constraints (7.4.26), as for many in the formulation, have indices that are specified in a general way, but such constraints are only valid for certain appropriate values of the indices and summations are only performed over appropriate values of indices. The notation has been intentionally simplified here to give the reader the "flavor" of the formulation. For full details, the reader must consult Subramanian et al. (1994,[536]).

Models were developed for Delta Air Lines consisting of tens of thousands of constraints and variables (Subramanian et al., 1994,[536]). Several different objective functions were also used and soft constraints were also incorporated. The model is able to perform the schedule changes rapidly; the planners can then consider the solutions. It was reported in Subramanian et al. (1994,[536]) that use of the model, called "coldstart," was expected to save Delta Air Lines $(US)300 million over a three year period.

7.5 Satisfiability

Consider the following problem. A child is presented with numbered bags of sweets from which she may choose as many or as few bags as she wishes to her own satisfaction, but with some restrictions. There are four bags of sweets and the child is told:

"You may not choose all three of the bags 1, 2 or 3."
"You may not choose all three of the bags 2, 3 or 4."

How should she choose?

Clearly in each of the two conditions one bag of sweets must not be chosen. If the child wishes to choose as many bags as possible, it would be advantageous to avoid choosing bag 2 (or 3), because then all the other three bags could be chosen.

The above is a simple case of the *satisfiability problem*. The satisfiability problem is the problem of finding a solution to a series of conditions, each defined using binary variables, each of which must be satisfied. Each condition comprises a subset of binary variables and to satisfy each condition at least one of its set of variables must be set to zero. Thus a typical problem would be to find a solution to the constraints

$$\delta_1 + \delta_2 + \delta_3 \leq 2 \quad , \quad \delta_2 + \delta_3 + \delta_4 \leq 2. \tag{7.5.1}$$

Clearly such a problem can easily be solved by

$$\delta_1 = 0 \quad , \quad \delta_2 = \delta_3 = 1 \quad , \quad \delta_4 = 0. \tag{7.5.2}$$

To make the problem more interesting, we introduce a further restriction that

$$\delta_1 = 1 - \delta_4 \tag{7.5.3}$$

and so this solution is not valid. The expression $1 - \delta_4$ is more typically written $\overline{\delta}$ or δ_4'.

The more typical form of satisfiability problem is to find a solution satisfying a set of constraints such as the following:

$$\begin{aligned} \delta_1 + \delta_2' + \delta_3 &\leq 2 \\ \delta_2 + \delta_3 + \delta_4 &\leq 2 \\ \delta_1 + \delta_2 + \delta_3' &\leq 2 \\ \delta_1 + \delta_4 &\leq 1. \end{aligned} \tag{7.5.4}$$

At the moment this problem has no objective function, nor is it known if a feasible solution exists (but $\delta_1 = \delta_4 = 0$, and any values for the other variables clearly provide a feasible solution). To make the problem into a more straightforward ILP problem we may introduce an extra binary variable, β_j, into each constraint and an objective function as follows:

$$\min \quad \beta_1 + \beta_2 + \beta_3 + \beta_4 \tag{7.5.5}$$

subject to

$$\begin{aligned} \delta_1 + \delta_2' + \delta_3 - \beta_1 &\leq 2 \\ \delta_2 + \delta_3 + \delta_4 - \beta_2 &\leq 2 \\ \delta_1 + \delta_2 + \delta_3' - \beta_3 &\leq 2 \\ \delta_1 + \delta_4 - \beta_4 &\leq 1. \end{aligned} \tag{7.5.6}$$

A value of 1 for any β variable ensures that the (modified) constraint in which it occurs is satisfied. The problem is now an ILP problem with all variables required to be $\{0,1\}$. For this problem, if the objective function has optimal value zero, then a feasible solution has been found to be the original satisfiability problem.

The general form of a constraint in a satisfiability problem is

$$\sum_{i \in \mathcal{I}} \gamma_i \leq |I| - 1, \quad \text{where} \quad \gamma_i = \delta_i, \text{ or } \gamma_i = 1 - \delta_i, \; \delta_i \in \{0, 1\} \quad , \quad i \in \mathcal{I}. \tag{7.5.7}$$

7.6 Bin Packing

7.6.1 *The Bin Packing Problem*

Consider our earlier encounter with the notorious burglar. After committing the crime it might have been true that the burglar was not faced with dividing the loot up into three containers, but rather he could decide how many containers would be appropriate and would then ship these off by sea to destinations beyond jurisdiction. Once that was decided, he would allocate each item to exactly one container. It is likely that his choice of how many containers to use will be governed by how much he can pack into each container. The problem that he faces is termed the *bin packing problem*.

The bin packing problem can be stated (cf. Martello & Toth (1990,[387])) as: given n items indexed by j and n identical bins indexed by i, with data

$$\begin{array}{ll} C & : \text{the capacity of each bin} \\ W_j & : \text{the weight of item } j, \end{array} \tag{7.6.1}$$

assign each item j to exactly one bin i so that the total weight of the items in each bin is less than or equal to C and the number of bins used is a minimum.

Here we need two sets of decision variables. We shall use the variables β to indicate if a particular bin is to be used or not and the variables δ to indicate if an item is assigned to a particular bin or not, i.e.,

$$\beta_i := \begin{cases} 1 \text{ , if bin } i \text{ is used} \\ 0 \text{ , otherwise} \end{cases} \quad , \quad \forall i \in \{1, 2, \ldots, n\} \tag{7.6.2}$$

and

$$\delta_{ij} := \begin{cases} 1 \text{ , if item } j \text{ is assigned to bin } i \\ 0 \text{ , otherwise} \end{cases} \quad , \quad \begin{array}{l} \forall i \in \{1, 2, \ldots, n\} \\ \forall j \in \{1, 2, \ldots, n\} \end{array} . \tag{7.6.3}$$

The formulation is then

$$\min \quad \sum_{i=1}^{n} \beta_i \tag{7.6.4}$$

subject to the requirement that the total weight packed into a bin must not exceed its capacity and is zero if the bin is not used

$$\sum_{j=1}^{n} W_j \delta_{ij} \le C \beta_i \quad , \quad i \in \{1, 2, \ldots, n\} \tag{7.6.5}$$

and that each item must be assigned to exactly one bin, i.e.,

$$\sum_{i=1}^{n} \delta_{ij} = 1 \quad , \quad j \in \{1, 2, \ldots, n\}, \tag{7.6.6}$$

where

$$0 \leq W_j \leq C \quad , \quad j \in \{1, 2, \ldots, n\} \quad , \quad C > 0. \tag{7.6.7}$$

7.6.2 The Capacitated Plant Location Problem

A further variation on problems involving the location of facilities and serving of customers will now be considered. It is a problem related to the bin packing problem, the *capacitated plant location problem*, that of deciding which plants to open to satisfy customers and assigning each customer to a sub-unit, $\mathcal{K}_j$ of a plant j. Cornuejols et al. (1977,[132]) describe an application of the problem to the banking sector. Leung & Magnanti (1989,[360]) provide a vertex packing formulation of the constraints of the problem. The formulation uses two sets of variables:

$$\beta_j := \begin{cases} 1 \text{ , if plant } j \text{ is open} \\ 0 \text{ , otherwise} \end{cases} \quad , \quad \forall j \in \mathcal{N} \tag{7.6.8}$$

and $\forall i \in \mathcal{M}$, $\forall j \in \mathcal{N}$, $\forall k \in \mathcal{K}_j$:

$$\delta_{ijk} := \begin{cases} 1 \text{ , if customer } i \text{ is assigned to sub-unit } k \text{ of plant } j \\ 0 \text{ , otherwise,} \end{cases} \tag{7.6.9}$$

where $\mathcal{K}_j = \{1, 2, \ldots, k_j\}$ is a dependent index set. The following constraints are present.

Each customer can be assigned to at most one sub-unit of one plant

$$\sum_{j \in N} \sum_{k \in \mathcal{K}_j} \delta_{ijk} \leq 1 \quad , \quad \forall i \in \mathcal{M}. \tag{7.6.10}$$

A customer can only be assigned to a sub-unit of a plant if the plant is open, i.e.,

$$\delta_{ijk} \leq \beta_j \quad i \in \mathcal{M}, \quad j \in \mathcal{N} \quad , \quad k \in \mathcal{K}_j. \tag{7.6.11}$$

Of course, an objective function will also be required, but we do not specify this here.

7.7 Clustering Problems

This section will discuss the capacitated clustering problem and the related p-median problem. These problems are similar to the facility location problems introduced earlier but are distinguished here to allow the reader to relate to other authors' use of the terminology. They will be presented in a somewhat abstract form to make the distinction from other similar problems more obvious.

7.7.1 *The Capacitated Clustering Problem*

The capacitated clustering problem (CCP) is characterized by (m, n, p) for $p \leq n$, $p \leq m$ with the following: Allocate n points uniquely to p out of m clusters so that the capacity of each cluster is not violated, and the points are allocated to maximize the homogeneity of points within the clusters while simultaneously trying to maximize the heterogeneity of the points between clusters (see for instance Mulvey & Beck (1984,[412]), Osman & Christofides (1994,[429])). Homogeneity of points within clusters is obtained by minimizing the sum of distances D_{ij} from each point i to the cluster center, j, to which it is allocated. Such problems often arise from positioning depots (the clusters) to supply groups of customers (the points), located at particular sites, in order to minimize the total transportation cost of supplying the customers.

We formulate the problem using the following indices:

$$\begin{aligned} i \in N &= \{1, \ldots, n\} \ , \text{ the set of points} \\ j \in M &= \{1, \ldots, m\} \ , \text{ the set of clusters} \end{aligned} \tag{7.7.1}$$

and data

$$\begin{aligned} C_j &: \text{the capacity of cluster } j \\ D_{ij} &: \text{the distance between point } i \text{ and the center of cluster } j \ . \\ W_{ij} &: \text{the weight of point } i \text{ if allocated to cluster } j \end{aligned} \tag{7.7.2}$$

It may be formulated using two sets of variables:

$$\beta_j := \begin{cases} 1 \ , \text{if cluster } j \text{ is used} \\ 0 \ , \text{otherwise} \end{cases} \quad , \quad \forall j \in \{1, 2, \ldots m\}, \tag{7.7.3}$$

and

$$\delta_{ij} := \begin{cases} 1 \ , \text{if point } i \text{ is assigned to cluster } j \\ 0 \ , \text{otherwise} \end{cases} \quad , \quad \begin{matrix} \forall i \in \{1, 2, \ldots, n\} \\ \forall j \in \{1, 2, \ldots, m\} \end{matrix} . \tag{7.7.4}$$

Then the objective function and constraints are given by

$$\min \quad \sum_{i=1}^{n}\sum_{j=1}^{m} D_{ij}\delta_{ij} \tag{7.7.5}$$

subject to the requirement that the weight of points assigned to a cluster must not exceed its capacity, i.e.,

$$\sum_{i=1}^{n} W_{ij}\delta_{ij} \leq C_j\beta_j \quad , \quad \forall j \in \{1, 2, \ldots, m\}, \tag{7.7.6}$$

and that each point must be assigned to exactly one cluster, i.e.,

$$\sum_{j=1}^{m} \delta_{ij} = 1 \quad , \quad \forall i \in \{1, 2, \ldots, n\}, \tag{7.7.7}$$

and the total number of clusters we can use is fixed

$$\sum_{j=1}^{m} \beta_j = p. \tag{7.7.8}$$

It is useful to add the following constraints as this will tighten the formulation:

$$\delta_{ij} \leq \beta_j \quad , \quad \forall i \in \{1, 2, \ldots, n\} \quad , \forall j \in \{1, 2, \ldots, m\}. \tag{7.7.9}$$

This problem has similarities to the bin packing problem.

7.7.2 The p-Median Problem

The p-median problem — cf. Christofides & Beasley (1982),[122]) or MCOL/GAMSLIB/*pmedian.gms* for an implementation — is the problem of locating p facilities (medians) on a network so as to minimize the total cost of allocating vertices (e.g., sales points) to these facilities. The facilities (medians) are themselves chosen from the set of vertices. This problem is clearly similar to CCP. The principal difference is that each of the p facilities will also be vertices. Let $\mathcal{N} = \{1, \ldots, n\}$ be the set of vertices. The problem can be formulated in terms of decision variables as $\delta_{ij} \in \{0, 1\}$, where

$$\delta_{ij} := \begin{cases} 1 \text{ , if vertex } i \text{ is allocated to vertex } j \\ 0 \text{ , otherwise} \end{cases} \quad , \quad \forall i, j \in \mathcal{N}. \tag{7.7.10}$$

Note that $\delta_{ij} = 1$ implies that j is a median vertex. D_{ij} is the cost (which may be determined by distance) of allocating vertex i to vertex j (when j is a median vertex).

The formulation is then

$$\min \quad \sum_{i=1}^{n}\sum_{j=1}^{n} D_{ij}\delta_{ij} \tag{7.7.11}$$

subject to

$$\sum_{\substack{i=1\\i\neq j}}^{n} \delta_{ij} \leq (n-1)\delta_{jj} \quad , \quad \forall j \in \mathcal{N}, \tag{7.7.12}$$

which ensures that nothing can be allocated to a vertex j unless that vertex is a median (this event is flagged by having $\delta_{jj} = 1$)

$$\sum_{j=1}^{n} \delta_{ij} = 1 \quad , \quad \forall i \in \mathcal{N}, \tag{7.7.13}$$

which ensures that each vertex i is allocated to some median j, and

$$\sum_{j=1}^{n} \delta_{jj} = p, \tag{7.7.14}$$

which ensures that exactly p vertices are medians.

7.8 Scheduling Problems

Scheduling problems arise where it is required to organize work, operations, tasks, or people into some time order. In many problems involving scarce resources it will be possible to determine if certain tasks can be performed, i.e., a feasible schedule exists, but then there is the associated problem of determining a working schedule that can be used in practice. Many different types of scheduling problem may be formulated and solved as ILP problems. Surveys of research in the area and applications appear in Lenstra et al. (1977,[358]), Potts & Van Wassenhove (1992,[445]), and Blazewicz et al. (1993,[85]). Three types of scheduling problems are introduced by means of examples; an additional one is investigated in more detail in a case study. It appears, however, that for solving scheduling problems, the most promising approaches using CP technique, as long as the objective function is simple, but for more complicated ones, e.g., for tardiness, or net present value (NPV) LP/MILP approaches seem preferable.

7.8.1 Example A: Scheduling Machine Operations

The following example illustrates a typical small scheduling problem. A factory produces two components by using time on three different machines. Each component is produced by six sequential operations. The times of these operations and the required machine are given in the table. Only one operation at a time may be carried out on a machine and the order of operations must be adhered to.

Operation	1	2	3	4	5	6
Machine required for component 1	1	2	1	2	1	1
Time for component 1	8	6	7	9	8	2
Machine required for component 2	3	2	3	2	3	3
Time for component 2	7	5	9	8	6	5

. (7.8.1)

The objective is to schedule with minimum total makespan, where makespan is defined as the time by which both components have been produced. Let t_i be the time at which component 1 commences operation i, s_i the time at which component 2 commences operation $i (i \in \{1, 2, 3, 4, 5, 6\})$, and e be the makespan. There are two types of constraints. The first are simple precedence constraints:

$$t_2 \geq t_1 + 8, \quad t_3 \geq t_2 + 6, \quad t_4 \geq t_3 + 7 \tag{7.8.2}$$

$$t_5 \geq t_4 + 9, \quad t_6 \geq t_5 + 8, \quad e \geq t_6 + 2$$

and

$$s_2 \geq s_1 + 7, \quad s_3 \geq s_2 + 5, \quad s_4 \geq s_3 + 9 \tag{7.8.3}$$

$$s_5 \geq s_4 + 8, \quad s_6 \geq s_5 + 6, \quad e \geq s_6 + 5.$$

e is the larger of two completion times *viz.* $t_6 + 2$ and $s_6 + 5$. The requirement is handled by two constraints, modeling that e is larger than each of the completion times. The effect is that e will be equal to the larger of the two completion times, as e is being minimized in the objective function.

Solving scheduling problems requires conflict resolution as different processes or jobs compete for scarce resources. As components 1 and 2 have to compete for the use of machine 2, one or other must take precedence to avoid conflict. The possible conflicts, denoting "component" by cpt and "operation" by op, are (cpt 1 op 2, cpt 2 op 2), (cpt 1 op 4, cpt 2 op 2), (cpt 1 op 2, cpt 2 op 4), (cpt 1 op 4, cpt 2 op 4). To correspond to each of these four possible conflicts, we introduce a binary variable $\delta_i \in \{0, 1\}, i = 1, 2, 3, 4$, such that

$$\delta_i := \begin{cases} 1 \text{ , if component 1 precedes component 2 at conflict } i \\ 0 \text{ , otherwise.} \end{cases} \tag{7.8.4}$$

We then require the second type of constraint that will model conflict resolution. A typical constraint requires us to model "either component 1 completes operation 2 before component 2 or component 2 completes operation 2 before component 1." This corresponds to

$$\text{either } s_2 \geq t_2 + 6 \text{ or } t_2 \geq s_2 + 5. \tag{7.8.5}$$

To model

$$\text{either } -t_2 + s_2 + 5 \leq 0 \text{ or } -s_2 + t_2 + 6 \leq 0, \tag{7.8.6}$$

we note that the maximum time all operations can take to produce both components consecutively is 80 units and introduce the constraints

$$-t_2 + s_2 + 5 \leq 80\delta_1 \text{ , } -s_2 + t_2 + 6 \leq 80(1 - \delta_1). \tag{7.8.7}$$

The complete set of constraints required to model conflict resolution is

$$t_2 \geq s_2 + 5 - 80\delta_1 \tag{7.8.8}$$

$$s_2 \geq t_2 + 6 - 80(1 - \delta_1) \tag{7.8.9}$$

$$t_2 \geq s_4 + 8 - 80\delta_2 \tag{7.8.10}$$

$$s_4 \geq t_2 + 6 - 80(1 - \delta_2) \tag{7.8.11}$$

$$t_4 \geq s_2 + 5 - 80\delta_3 \tag{7.8.12}$$

$$s_2 \geq t_4 + 9 - 80(1 - \delta_3) \tag{7.8.13}$$

$$t_4 \geq s_4 + 8 - 80\delta_4 \tag{7.8.14}$$

$$s_4 \geq t_4 + 9 - 80(1 - \delta_4). \tag{7.8.15}$$

The objective function is

$$\min \quad e. \tag{7.8.16}$$

This problem is typical of the smaller scheduling problems that can be formulated as ILP models. As problems grow in size with more machines and more components in conflict the problems become extremely difficult to solve as straightforward ILP models.

7.8.2 Example B: A Flowshop Problem

The flowshop problem is the problem of scheduling N jobs in a machine shop containing M machines, where each job has to be processed on every machine, and every job follows the same ordering of machines as it is processed. Once each job has been allocated a position in the schedule, no change is allowed and jobs cannot be subdivided. The objective of the problem is to minimize makespan.

A formulation of the flowshop problem is given in Wilson (1989,[583]) and Fig. 7.5 shows a typical schedule. This formulation is as follows, beginning with

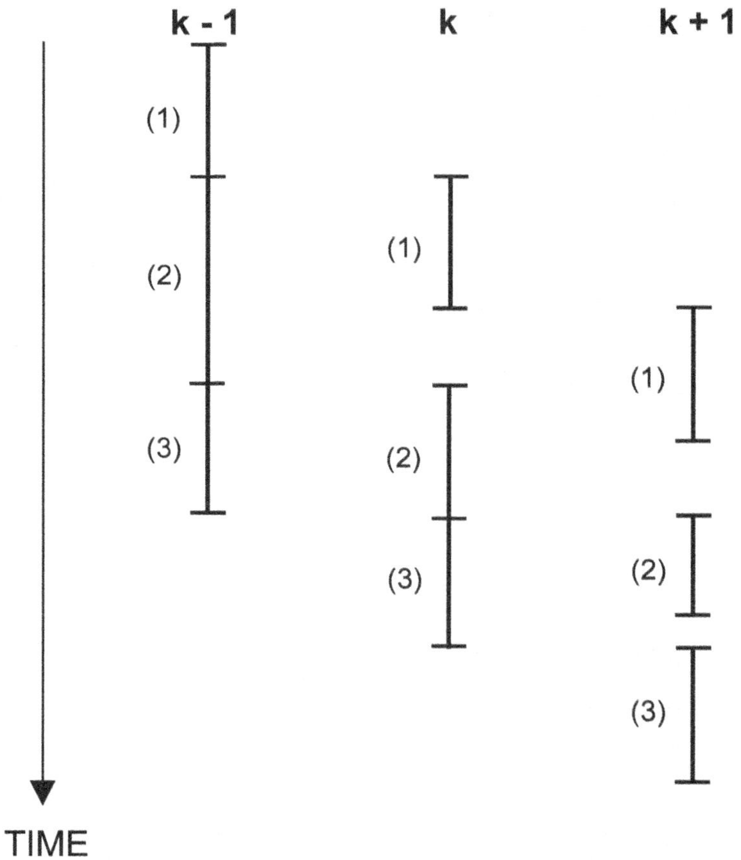

Fig. 7.5 Gantt chart showing the schedule

the indices:

$$
\begin{array}{lll}
i \in \mathcal{I} & = \{1, \ldots, N\} \ , & \text{the set of jobs} \\
j \in \mathcal{J} & = \{1, \ldots, N\} \ , & \text{the schedule positions for processing jobs} \\
k \in \mathcal{K} & = \{1, \ldots, M\} \ , & \text{the operation (machine) number.}
\end{array}
\tag{7.8.17}
$$

The data are the processing time P_{ik} of job i on machine k. We introduce binary variables $\delta_{ij} \in \{0, 1\}$ such that

$$
\delta_{ij} := \begin{cases} 1 \ , & \text{if job } i \text{ is processed in schedule position } j \\ 0 \ , & \text{otherwise,} \end{cases}
\tag{7.8.18}
$$

and continuous variables $s_{jk} \geq 0$ representing the time at which the job in schedule position j commences processing on machine k. Then we have the objective function

$$
\min \quad s_{NM} + \sum_{i=1}^{N} P_{iM} \delta_{iN}
\tag{7.8.19}
$$

subject to

$$
\sum_{i=1}^{N} \delta_{ij} = 1 \quad , \quad \forall j \in \mathcal{J} ,
\tag{7.8.20}
$$

$$
\sum_{j=1}^{N} \delta_{ij} = 1 \quad , \quad \forall i \in \mathcal{I} ,
\tag{7.8.21}
$$

i.e., jobs must be uniquely positioned for scheduling. The constraints

$$
s_{j+1,1} = s_{j1} + \sum_{i=1}^{N} P_{i1} \delta_{ij} \quad , \quad j = 1, 2, \ldots, N-1
\tag{7.8.22}
$$

$$
s_{11} = 0
\tag{7.8.23}
$$

and

$$
s_{1,k+1} = s_{1,k} + \sum_{i=1}^{N} P_{ik} \delta_{i1} \quad , \quad k = 1, 2, \ldots, M-1
\tag{7.8.24}
$$

ensure that there is no idle time on either machine 1 or on job 1. The constraint

$$s_{j,k+1} \geq s_{jk} + \sum_{i=1}^{N} P_{ik}\delta_{ij} \quad , \quad j = 2, 3, \ldots, N \quad , \quad k = 1, 2, \ldots, M - 1 \tag{7.8.25}$$

ensures that the commencement of job j on machine $(k + 1)$ is no earlier than its finish time on machine k. Finally, the constraint

$$s_{j+1,k} \geq s_{jk} + \sum_{i=1}^{N} P_{ik}\delta_{ij} \quad , \quad j = 1, 2, \ldots, N - 1 \quad , \quad k = 2, 3, \ldots, M \tag{7.8.26}$$

ensures that the commencement of job $j + 1$ on a machine must be no earlier than the finish time of job j on the same machine.

Some idea of the difficulty of this problem is indicated by the fact that it contains N^2 binary variables, and this will become cumbersome as N grows in size. An instructive example based on Guéret et al. (2002,[245]) is implemented in MCOL/GAMSLIB/*flowshop.gms*.

7.8.3 Example C: Scheduling Involving Job Switching

A further scheduling problem arises when *job switching* is involved, i.e., it is possible to commence one job and then switch to another before the first is complete. It may be formulated in the style of the TSP.

The problem is to determine the sequence of n jobs that must be completed in sequence, such that the completion time of the last job is minimized. Let $N = \{1, \ldots, n\}$ be an index set, and C_{ij} the cost of switching from production of job i to job j, and let

$$\delta_{ij} := \begin{cases} 1 \text{ , if job } i \text{ immediately precedes job } j \\ 0 \text{ , otherwise} \end{cases} \quad , \quad \forall i, j \in N, i \neq j. \tag{7.8.27}$$

Let $\mathscr{A}$ represent the set of admissible possibilities of i preceding j. Then the problem formulation is

$$\min \sum_{(i,j)\in\mathscr{A}} C_{ij}\delta_{ij} \tag{7.8.28}$$

subject to

$$\sum_{(i,j)\in\mathscr{A}} \delta_{ij} = n - 1 \tag{7.8.29}$$

as all jobs except the first in the sequence must have a predecessor

$$\sum_{i=1}^{n} \delta_{ij} \leq 1 \quad , \quad \forall j = 1, 2, \ldots, n, \tag{7.8.30}$$

as each job must be preceded by at most one other,

$$\sum_{j=1}^{n} \delta_{ij} \leq 1 \quad , \quad \forall i = 1, 2, \ldots, n, \tag{7.8.31}$$

as each job must have at most one successor, and

$$\sum_{(i,j)\in\mathscr{S}} \delta_{ij} \leq |W| - 1 \quad , \quad W \subset \{1, 2, \ldots, n\} \quad , \quad 2 \leq |W| \leq n - 1, \tag{7.8.32}$$

where $\mathscr{S}$ is the subset

$$\mathscr{S} := \{(i, j) \in \mathscr{A} \mid i \in W \ , \ j \in W\} \, . \tag{7.8.33}$$

The last set of constraints is included to avoid illegal subsequences of jobs occurring and is similar to what was done in Sect. 7.2.

7.8.4 Case Study: Bus Crew Scheduling

The Citti-Bus Company Ltd. wants to cover some varying demands for drivers by a set of crews who work 8 h shifts with a 1 h meal break. The day is divided into hourly periods. There are $N^S = 12$ possible shift types, all embedded in an 18 h day lasting from 06.00 to 24.00. Different shift types, s, start at different times. Shift 1 starts at 6am, shift 2 at 7 a.m., and eventually shift 12 at 5 p.m. In general: shift s starts at time $s + 5$ h. To save money we want to run the buses at minimum total cost.

To solve our problem by mathematical optimization let us check whether we understood the problem completely. We want to derive a schedule that allocates people to shifts. The problem is demand driven, if demand is high we need more personnel, while late in the night or early in the morning there is low demand. Those shifts covering the early morning hours or late night hours are the most costly ones as the table below indicates

shift	1	2	3	4-8	9	10	11	12
time	6-13	7-14	8-15	9-20	14-21	15-22	16-23	17-24 .
costs	100	90	80	50	60	70	80	100

So we expect solutions with less personnel on the less attractive shifts and more people on the mid-day shifts. When modeling real-world problems, we should try to have a clear idea of what economic driving force there exists in the model and what kind of results we anticipate. Sometimes deviations from our expectations may force us to think again about our problem and its formulation.

Let t be the index of work hours, $t = 1, 2, \ldots, T$. Since we have an 18 h work day, we set $T = 18$. So the time from 6am to 7am is referred to as the first work hour, i.e., $t = 1$. Using the time index, t, shift s starts when $t = s$. The data are

$$\begin{aligned} C_s &: \text{the cost of shift } s \\ R_t &: \text{the required manning level in work hour } t\text{, and} \\ W_h &: \text{the manning duties in each shift.} \end{aligned} \tag{7.8.34}$$

Note that in the requirement table work hour t in R_t refers to the absolute time interval $[t+5, t+6]$ and W_h is an indication table with index h ranging from 1 to 8. It is set to 1 if h is a working hour of a shift, and 0 otherwise. Let us have a look at an example: if someone works 3 h, then has a break, and then works 4 h, this table looks as follows:

h	1	2	3	4	5	6	7	8
W_h	1	1	1	0	1	1	1	1

Alternatively, we could also define an indication table B_{st} that indicates whether (if $B_{st} = 1$) shift s includes work hour t, or not ($B_{st} = 0$). At time t during the day shift s is then in its $(t - s + 1)^{st}$ hour of operation. The indication tables are thus related to each other by

$$B_{st} = \begin{cases} 0 & \text{, if } t < s \\ W_{t-s+1} & \text{, if } s \le t \le s+7 \\ 0 & \text{, if } t > s+7 \end{cases} \quad , \quad \forall \{st\}. \tag{7.8.35}$$

The basic decisions (or degrees of freedom) in our model are the numbers α_s of crews working on shift s. Thus we introduce integer variables $\alpha_s \in \{0, 1, \ldots, 9\}$ into our model. If we inspect the demand for drivers (table should be read as followsṡ: column 4 has the meaning “from 10 to 12 there is demand for 6 crews”)

t	6	7	8	10	12	14	16	17	19	20	23-24
R_t	3	9	10	6	7	6	8	10	8	3	2

specified in our small example model *buscrew*, it is unlikely that we would need more than 9 crews on a single shift. The reason is that for the data given we clearly need 3 crews on the first shift starting at 6am, and 2 crews on shifts 11 and 12. That covers already at least two crews that need to be present during peak times (8 to

10 a.m. and 5 to 7 p.m.). Thus, even 8 would be a safe upper bound for the number of crews per shift.

The objective function is formulated easily as

$$\min \quad \sum_{s=1}^{N^S} C_s \alpha_s . \tag{7.8.36}$$

It is subject to the constraint

$$\sum_{s=1}^{N^S} B_{st} \alpha_s \geq R_t \quad , \quad \forall t, \tag{7.8.37}$$

which is equivalent to

$$\sum_{s=\max\{1,t-7\}}^{\min\{t,T+1-7\}} W_{t-s+1} \alpha_s \geq R_t \quad , \quad \forall t. \tag{7.8.38}$$

7.9 Summary and Recommended Bibliography

In this chapter we have investigated how various standard ILP problems may be formulated. The models illustrate standard methods of formulation, in particular those involving binary variables. We have seen how these formulations may be used in practical cases by including the features not already present in the somewhat artificial model. Thus by studying this chapter the reader should be able to identify and formulate:

- Problems of the knapsack family
- Problems of the traveling salesman family
- Facility location problems
- Set covering, set partitioning, and set packing problems
- Satisfiability problems
- Problems of the bin packing family
- Clustering problems
- Scheduling problems

Readers interested in the *Traveling Salesman Problem* are referred to Applegate et al. (2007,[28]). Knapsack problems are well covered by Martello & Toth (1990,[387]).

7.10 Exercises

1. Formulate and solve a traveling salesperson problem over 8 towns where it is required that the salesperson starts at town 1 and performs a complete circuit of all towns such that the total distance traveled is minimized. The distance, in kilometers, from each town to each other town is given in the table below. (The distances are symmetric in this problem, i.e., the distance between two towns is the same irrespective of which direction the salesperson is traveling.)

town	1	2	3	4	5	6	7	8
1	-	185	270	315	124	123	99	180
2		-	215	219	187	154	210	190
3			-	210	220	238	198	150
4				-	218	134	157	149
5					-	168	143	129
6						-	198	201
7							-	180
8								-

2. Solve the following problem.
Three containers of size 26, 25, and 34 tons are available. It is required to allocate the following 8 loads each to a container. The loads are such that their cost of delivery, C, varies according to which container they are assigned and the weight, W, of each load also varies (due to the protective packaging needed) according to which container is used. Data on C are given in the following table.

Load	1	2	3	4	5	6	7	8
Container 1	27	12	12	16	24	31	41	13
Container 2	14	5	37	9	36	25	1	34
Container 3	34	34	20	9	19	19	3	34

Data on W are given in the following table.

Load	1	2	3	4	5	6	7	8
Container 1	21	13	9	5	7	15	5	24
Container 2	20	8	18	25	6	6	9	6
Container 3	16	16	18	24	11	11	16	18

Formulate and solve this problem to minimize the total cost of delivery.

3. Solve the problem given in Sect. 7.1.1.
4. Up to nine components may be attached to any of three circuit boards. Each component has a fixed weight and a fixed value if it is used. The total weight that

may be attached to each circuit board is restricted and is 80, 90, and 85 units, respectively, for the three boards. Each component may be attached to no more than one board.

Component	1	2	3	4	5	6	7	8	9
Weight	40	10	40	30	50	50	55	25	40
Value	80	20	60	40	60	60	65	25	30

Formulate and solve the problem of attaching components to boards so as to maximize the total value.

5. A company has 14,000 £ to invest. There are four possibilities denoted by i. With each investment we have associated costs, C_i, and yields, Y_i, in two years' time. The table entries are given in units of thousands of £ (k£).

i	1	2	3	4
C_i	5	7	4	3
Y_i	16	22	12	8

Formulate the company's problem of maximizing the yield in two years' time as an integer program.

6. Container loads consist of items chosen from the set of items numbered {1,2, ..., 20}. Each container load must include one specially marked item and any of the items may be chosen in advance. There are 10 containers and they must comprise the following items:

Container	Items
1	1, 8, 11, 17
2	2, 8, 9, 11, 12
3	3, 4, 5, 6, 15
4	7, 8, 12, 14, 16
5	6, 9, 12, 15, 20
6	10, 13, 18, 19
7	6, 8, 12, 15
8	12, 14, 16, 18
9	1, 5, 10, 20
10	7, 13, 17, 19

Odd numbered items cost £50.00 to mark in advance and even numbered items cost £100.00 to mark in advance (the cost is a total for *all* items of a particular type, and single items cannot be marked individually). Formulate and solve the problem of minimizing the total cost of advance marking of the special items.

7. Six jobs have to be processed on four machines. The table gives the processing times required by each job on each machine. Once a job has started on a machine,

it cannot be interrupted and all jobs must be machined in the order 1, 2, 3, 4. Only one job may be processed by a machine at any given time. Formulate and solve the problem of processing all jobs in minimum total time.

Job	1	2	3	4	5	6
Machine						
1	10	12	14	15	13	12
2	9	7	6	4	8	7
3	6	4	3	2	4	5
4	2	2	1	1	3	3

8. Group design problem: 24 participants should be divided into groups with a minimum of 3, and a maximum of 6 persons. After 8 permutations, each participant should have been at least once in a group with every other participant. Determine the number and size of the groups fulfilling this condition.

Chapter 8
Case Studies and Problem Formulations

In this chapter we shall consider applications of mathematical programming arising in various contexts. We shall examine the problem faced by the client, the model built, and the use to which it would be put. Problems will be ones that require financial modeling, modeling of the closing and opening of facilities, modeling across time periods, discounting over time, using opening and closing balances, and using hard and soft constraints in model development. Each case will demonstrate the need for a model of a particular type. A formulation of each model is provided in MCOL, so that the reader may solve the problem and investigate the results. In some cases it will be possible to formulate the problem in several contrasting ways. Some of the benefits of the alternative formulations will be discussed. In the second half of the chapter the focus will be on difficulties that commonly arise when we try to run models and solve problems. We discuss how to get around infeasibilities and illuminate certain aspects of sensitivity analysis.

8.1 A Depot Location Problem

There are six major towns in one area. The local fire-brigade service has decided to rationalize the location of fire departments (currently each town has a station) so that each town is within 20 min driving time of a fire department. Given the data for driving times (specified in minutes) in the "driving time" table below from each town t_1 to each town t_2,

Supplementary Information The online version contains supplementary material available at https://doi.org/10.1007/978-3-030-73237-0_8.

J. Kallrath, *Business Optimization Using Mathematical Programming*, International Series in Operations Research & Management Science 307, https://doi.org/10.1007/978-3-030-73237-0_8

$t_1 \backslash t_2$	1	2	3	4	5	6
1	0	15	25	35	35	25
2	15	0	30	40	25	15
3	25	30	0	20	30	25
4	35	40	20	0	20	30
5	35	25	35	20	0	19
6	25	15	25	30	19	0

the problem of deciding which fire departments remain open can be decided by using an integer program.

Let us define our variables. We need binary variables $\delta_t \in \{0, 1\}$ defined such that

$$\delta_t := \begin{cases} 1 \text{ , if a fire department is located in town } t \\ 0 \text{ , if not.} \end{cases} \tag{8.1.1}$$

The objective function is to minimize the number of fire departments:

$$\min \quad \delta_1 + \delta_2 + \delta_3 + \delta_4 + \delta_5 + \delta_6. \tag{8.1.2}$$

If we look at the data on driving times of fire engines, we are interested in cases when times do not exceed 20 min. Let us consider town 1 first. If we look at the second column in the table, we see that a fire department must be sited at town 1 or town 2 (or both) if town 1 is to be covered (driving times less than 20 min). Therefore we need a constraint

$$\text{Cover town 1:} \qquad \delta_1 + \delta_2 \geq 1. \tag{8.1.3}$$

Thus the problem is essentially a set covering problem where a "cover" is defined by the driving time being less than 20 minutes. By examining successive columns of the "driving time" table or the "covering" table below

$t_1 \backslash t_2$	1	2	3	4	5	6
1	1	1	0	0	0	0
2	1	1	0	0	0	1
3	0	0	1	1	0	0
4	0	0	1	1	1	0
5	0	0	0	1	1	1
6	0	1	0	0	1	1

we get the complete system of constraints

$$
\begin{array}{lcccccccl}
\text{Cover town 1:} & \delta_1 & +\delta_2 & & & & & \geq 1 \\
\text{Cover town 2:} & \delta_1 & +\delta_2 & & & & +\delta_6 & \geq 1 \\
\text{Cover town 3:} & & & \delta_3 & +\delta_4 & & & \geq 1 \\
\text{Cover town 4:} & & & \delta_3 & +\delta_4 & +\delta_5 & & \geq 1 \\
\text{Cover town 5:} & & & & \delta_4 & +\delta_5 & +\delta_6 & \geq 1 \\
\text{Cover town 6:} & & \delta_2 & & & +\delta_5 & +\delta_6 & \geq 1.
\end{array}
\tag{8.1.4}
$$

The constraints are obviously fulfilled if we put $\delta_2 = \delta_4 = 1$ and $\delta_1 = \delta_3 = \delta_5 = \delta_6 = 0$, i.e., it is sufficient to keep open two fire departments. This solution is also the optimal solution, and it is unique, i.e., there is no other solution with only two fire departments.

This problem, in a broader framework, is treated in *Location Science*, cf. [351].

8.2 Planning and Scheduling Across Time Periods

Often it is required that a model considers production across time periods. In each time period there is inventory available and production is made partly for use and partly for stock. Adjacent time intervals are connected by multi-period flow constraints. Restrictions that are applied include minimum and maximum inventory levels and satisfaction of customer demands. Goods held in inventory may also deteriorate. Production planning problems like the one discussed in Sect. 10.4 usually make use of modeling across time periods.

When building multi-period models, two important issues have to be discussed with the client: the length of the planning horizon and the resolution in time. A typical production planning model may cover a planning horizon of 1 year and the smallest unit in time may be a month. In scheduling problems very often the horizon is only 1 or 2 weeks and the resolution in time comes down to only a few hours or a day. The choice of both parameters has great influence on the size of the problem. The length of the planning horizon depends on the purpose the model is used for and also on the quality of the data entering the model. If the market demand forecast is not very accurate, it seems not to make much sense to plan too far into the future. The resolution in time depends on the degree of reality required and determines the accuracy of the model. It may also occur that the resolution in time is different for certain aspects contained in the model. In Sect. 10.4 it is shown how different time scales can be embedded in one model.

Hendry et al. (1996,[264]) describe an MILP model to schedule production across time periods in the copper industry. Three main production processes occur in the copper foundry: the melting process, the production of "logs" of copper, and the cutting of the logs into "billets" (smaller lengths). The process of converting the molten copper into one lot of logs is called a "drop" and there are about six standard billet lengths for each log diameter. Two main factors affect the cost of production: operational factors and the yield of copper obtained. Molten copper can be stored and production is considered over a two week cycle. The company is interested in minimization of production time.

8.2.1 Indices, Data, and Variables

The formulation is as follows. The model makes use of production requirements from an earlier stage, using another model, referred to as "stage one." As usual we start with the indices

$$\begin{aligned} m &\in \{1, \ldots, M\} \ , \ the \text{index of days} \\ d &\in \{1, \ldots, D\} \ , \ the \text{index of drop diameters} \\ t &\in \{1, \ldots, T\} \ , \ \text{index of periods (not used explicitly).} \end{aligned} \tag{8.2.1}$$

The data of the problem are described in the following list:

$$\begin{aligned}
W_m &: \text{the weighting assigned to the production of logs on day } m, \\
M_d &: \text{the maximum number of drops of diameter } d \text{ per day that} \\
&\quad \text{can be produced,} \\
N_d &: \text{the required number of drops of diameter d in period } t \\
&\quad \text{(output from "stage one" model),} \\
K_d &: \text{weight of copper needed to produce one drop of diameter } d, \\
v_1 &: \text{the inventory of molten copper held at the beginning of} \\
&\quad \text{the first day of the cycle,} \\
T_d &: \text{the time required to produce one drop of diameter } d, \\
O &: \text{the time to change over from the production of one log} \\
&\quad \text{diameter to another,} \\
C &: \text{the daily log production capacity in man-minutes,} \\
Y &: \text{the percentage of molten copper that can be used,} \\
M^S &: \text{the maximum tonnage of molten copper that can be stored,} \\
M^X &: \text{the maximum daily quantity of scrap copper that can be melted,} \\
M^N &: \text{the minimum daily quantity of scrap copper that can be melted.}
\end{aligned} \tag{8.2.2}$$

Continuous variables:

$$\begin{aligned}
c_m \geq 0 &: \text{the weight of scrap copper melted on day } m \\
v_m \geq 0 &: \text{the inventory of molten copper held at the beginning} \\
&\quad \text{of day } m \ (m \neq 1) \\
s_m \geq 0 &: \text{slack time at the log production process on day } m.
\end{aligned} \tag{8.2.3}$$

In order to count the number of drops and diameters we define the integer variables

$$\begin{aligned} \alpha_{md} &: \text{the number of drops of diameter } d \text{ produced on day } m \\ \beta_m &: \text{the number of different log diameters produced on day } m. \end{aligned} \tag{8.2.4}$$

Finally we introduce the binary variables

$$\gamma_m := \begin{cases} 1 \text{ , if it is decided to melt copper on day } m \\ 0 \text{ , otherwise} \end{cases} \tag{8.2.5}$$

and

$$\delta_{md} := \begin{cases} 1 \text{ , if logs of diameter } d \text{ are produced on day } m \\ 0 \text{ , otherwise.} \end{cases} \tag{8.2.6}$$

8.2.2 Objective Function

The objective function is

$$\min \sum_{m=1}^{M} \left[W_m \cdot \left(O\beta_m + \sum_{d=1}^{D} T_d \alpha_{md} \right) \right] , \tag{8.2.7}$$

which will minimize the total time, including change over time, required to produce the logs. The weighting given to each day is to ensure that the smallest number of days is used, and this is achieved by giving the smallest weighting to the first day and increasing the values over time.

8.2.3 Constraints

The following constraints were modeled. The first one,

$$\sum_{m=1}^{M} \alpha_{md} = N_d \quad , \quad \forall d \in D, \tag{8.2.8}$$

expresses that the total number of drops of each diameter produced in the period should equal the requirement. Then we formulate a typical inventory balance equation coupling adjacent time periods

$$\sum_{d=1}^{D} K_d \alpha_{md} + v_{m+1} = Y \cdot c_m + v_m \quad , \quad m \in \{1, \ldots, M-1\}. \tag{8.2.9}$$

The equations (8.2.9) guarantee that the inventory on day $(m + 1)$ plus the weight of copper used in production should equal the inventory on day m plus the yield from the melted copper. Some care is needed with the index m counting the days. Therefore, the equations are only considered for all but the last day. Sometimes, inventory balance constraints contain the stock level, v_m, at time m and time $m-1$. Inventory balance constraints are only complete if they connect appropriately to initial stock, v_1, and possibly, to final stock, v_{M+1}. If we do not connect appropriately to initial stock, we lose all stocks available at that time. Note that v_1 is not a variable but a constant.

The next constraint we have to consider is the inventory capacity, i.e., we have to guarantee

$$v_m \le M^S \quad , \quad \forall m \in \{1, 2, ..., M\} \tag{8.2.10}$$

that the inventory does not exceed its maximum capacity.

The weight of scrap copper melted on day m must be zero or lie between its minimum and maximum. In Hendry et al. (1996,[264]) this property is described by the inequalities

$$M^N - c_m \le M^X(1 - \gamma_m) \quad , \quad \forall m \in \{1, 2, ..., M\} \tag{8.2.11}$$

and

$$c_m \le M^X \gamma_m \quad , \quad \forall m \in \{1, 2, ..., M\}. \tag{8.2.12}$$

Note that the binary variable γ_m forces either

$$\gamma_m = 0 \Rightarrow c_m = 0 \tag{8.2.13}$$

or

$$\gamma_m = 1 \Rightarrow M^N \le c_m \le M^X. \tag{8.2.14}$$

We can replace the inequalities (8.2.11) and (8.2.12) by declaring c_m as a semi-continuous variable. As discussed in Sect. 9.4.5 this leads to a more efficient model.

The total time required to produce logs in day m plus change over time plus slack time equals the available time capacity, C, leading to the equation

$$\sum_{d=1}^{D} T_d \alpha_{md} + O\beta_m + s_m = C \quad , \quad \forall m \in \{1, 2, ..., M\}. \tag{8.2.15}$$

The number, β_m, of different log diameters produced on day m is given by the total number of decisions to produce logs of different types, i.e.,

$$\beta_m = \sum_{d=1}^{D} \delta_{md} \quad , \quad \forall m \in \{1, 2, ..., M\}. \tag{8.2.16}$$

The number, α_{md}, of drops of diameter d produced on day m must not exceed the maximum possible that is ensured by

$$\alpha_{md} \leq M_d \delta_{md} \quad , \quad \forall m \in \{1, 2, ..., M\} \quad , \quad \forall d \in D. \tag{8.2.17}$$

Finally, we wish to express that the number of drops of diameter d produced on day m must be non-zero if the decision to produce is taken, i.e.,

$$\alpha_{md} \geq \delta_{md} \quad , \quad \forall m \in \{1, 2, ..., M\} \quad , \quad \forall d \in D. \tag{8.2.18}$$

It was anticipated in Hendry et al. (1996,[264]) that the implementation of the model would lead to substantial improvements in performance for the company as results from the model-based optimization suggested a useful increase in copper yields.

8.3 Distribution Planning for a Brewery

In the UK, brewing companies brew two major different types of beer: ale and lager. This is also the case for our specific brewer who wanted to optimize the business using mathematical programming. Each type of beer comes in several varieties (liquid types), distinguished by taste, color, flavor, and strength and several other characteristics.

Most beers are sold in four major containers: cans, bottles, barrels, and "cellar tanks." When packaged in the latter form, the beer is delivered in bulk to a tank at the retail outlet (usually a public house) and then served to the customer straight from the tank. Not all liquid types go into each container, and in fact the actual number of commodities (i.e., liquid types in specific containers) is significantly smaller than the maximum number theoretically possible.

When a particular liquid is brewed at one of the breweries, it is taken to a packaging unit to be placed in a container. Sometimes this journey is very short if, for instance, the packaging unit is sited adjacent to the brewery and connected to it by pipeline. But since packaging units are expensive it is sometimes necessary to transport liquid in bulk from the brewery to a packaging unit of the correct type.

Each brewery belonging to the company has a separate limited capacity for lager and for ale, but the capacity within these two sub-categories may be freely allocated to any of the liquid types. A packaging unit can only put liquids into one type of container.

After liquids have been put into containers (and have become a commodity), they are then transported to customer zones (demand points). These are typically depots owned by the company, or the premises owned by wholesalers or supermarkets.

8.3.1 Dimensions, Indices, Data, and Variables

The parameters or dimensions of the model and the indices are summarized below

$b \in \{1, \ldots, N^B\}$, the set of breweries
$c \in \{1, \ldots, N^C\}$, the set of commodity types
$d \in \{1, \ldots, N^D\}$, the set of demand points
$l \in \{1, \ldots, N^L\}$, the set of liquid types, $N^L = N^{AL} + N^{LA}$
$p \in \{1, \ldots, N^P\}$, the set of packaging units
$t \in \{1, \ldots, N^T\}$, the set of containers (packing type),

where N^{AL} and N^{LA} denote the numbers of ales and lagers.

The company wished to consider the data as being driven by the commodities, that is, the combinations of liquid type in a particular container. A partial list might be

Commodity number	Liquid	Container
...	...	...
16	B23L	Barrels
17	B23L	Half pint bottles
18	G26A	Pint bottles
...	...	...

Data were collected and assembled into the following list:

C_{pb}^{TBP} the unit transport cost from brewery b to packaging unit p
C_{tdp}^{TPD} the unit transport cost of container type t from packaging unit p to demand point d
R_c^L the liquid required by commodity c
R_c^C the container type required by commodity c
C_{bl}^B the brewing capacity at brewery b for liquid l
C_p^P the total packaging capacity at p
D_{cd} the final demand for commodity c at demand point d
I_{lb}^{BL} 1if brewery b can brew liquid l
N_l^{BL} the number of breweries that can brew liquid l
B_b^U the maximum amount that can be brewed at b
B_b^L the minimum amount that can be brewed at b
P_p^U the maximum amount that can be packed at p
P_p^L the minimum amount that can be packed at p.

The continuous, non-negative decision variables needed for the problem are as follows:

b_{lb},	the amount of liquid l brewed at brewery b
s_{lp},	the throughput of liquid l through packaging unit p
l_{blp},	the amount of liquid l sent from brewery b to packaging unit p
p_{cp},	the amount of commodity c packed at packaging unit p
x_{pcd},	the quantity of commodity c sent from packaging unit p to demand point d.

Note that in order to reduce the number of variables in the model listed below the variables are only defined if certain logical conditions on data and indices are fulfilled. In addition we need the binary variables $\beta_{lb}, \delta_{lp} \in \{0, 1\}$ defined as

$$\beta_{lb} := \begin{cases} 1 \text{ , if any liquid } l \text{ is brewed at brewery } b \\ 0 \text{ , otherwise} \end{cases} \tag{8.3.1}$$

and

$$\delta_{lp} := \begin{cases} 1 \text{ , if any liquid } l \text{ is packaged at packaging unit } p \\ 0 \text{ , otherwise.} \end{cases} \tag{8.3.2}$$

Note that the β_{lb} variables are only defined if brewery b can brew liquid l and there is more than one brewery that can brew the liquid. The δ_{lp} are only defined if the capacity of the packaging unit p is non-zero.

Explicitly allowing for zero capacity at a packaging unit may seem peculiar, but one of the aims of the exercise was to consider the possible closure of one or more of the packaging units. Different scenarios could easily be considered by setting $C_p^P = 0$ for a packing unit that was to be closed.

Note that liquid 1 denotes *Ale*, while liquid 2 is *Lager*.

8.3.2 Objective Function

The objective function to be minimized was the total cost of transporting liquids from breweries to packaging units, plus the total cost of distributing commodities from packaging units to demand points.

$$\sum_{p=1}^{N^P} \sum_{b=1}^{N^B} \sum_{l=1}^{N^L} C_{pb}^{TBP} l_{blp} + \sum_{p=1}^{N^P} \sum_{c=1}^{N^C} \sum_{d=1}^{N^D} C_{\tau dp}^{TPD} x_{pcd} \quad , \quad \tau = R_c^C . \tag{8.3.3}$$

It would be easy to incorporate different brewery packaging costs for each brewery and packing unit, but these were not thought to be important in the initial stages of the study.

8.3.3 Constraints

The following constraints were operative.

Liquid balance at each brewery:

$$\sum_{p=1}^{N^P} l_{blp} = b_{lb} \quad , \quad \forall l \in \{1, 2, ..., N^L\} \quad , \quad \forall b \in \{1, 2, ..., N^B\}, \tag{8.3.4}$$

which ensures that for each liquid/brewery combination, all of it is sent to some packaging unit. The constraint

$$\sum_{c \text{ containing } l} p_{cp} = s_{lp} \quad , \quad \forall l \in \{1, 2, ..., N^L\} \quad , \quad \forall p \in \{1, 2, ..., N^P\} \tag{8.3.5}$$

defines the throughput of liquid l through packaging unit p as the total amount packaged of all commodities containing liquid l.

The total amount of ale brewed at a brewery is less than or equal to its ale capacity, and correspondingly for lager. Hence,

$$\sum_{p} l_{b,ale,p} \leq C^B_{b,ale} \quad , \quad \forall b \in \{1, 2, ..., N^B\} \tag{8.3.6}$$

$$\sum_{p} l_{b,lager,p} \leq C^B_{b,lager} \quad , \quad \forall b \in \{1, 2, ..., N^B\}. \tag{8.3.7}$$

Conservation of liquid going into the packaging units and being packed means that

$$\sum_{c \text{ requiring } l} p_{cp} = \sum_{b=1}^{N^B} l_{blp} \quad , \quad \forall l \in \{1, 2, ..., N^L\} \quad , \quad \forall p \in \{1, 2, ..., N^P\}. \tag{8.3.8}$$

The capacity limitation at the packaging unit means that

$$\sum_{c=1}^{N^C} p_{cp} \leq C^P_p \quad , \quad \forall p \in \{1, 2, ..., N^P\}. \tag{8.3.9}$$

Conservation of each commodity leaving each packaging unit gives us the equations

$$\sum_{d=1}^{N^D} x_{pcd} = p_{cp} \quad , \quad \forall p \in \{1, 2, ..., N^P\} \quad , \quad \forall c \in \{1, 2, ..., N^C\}. \tag{8.3.10}$$

Satisfying the demands for each commodity at each demand point exactly yields

$$\sum_{p=1}^{N^P} x_{pcd} = D_{cd} \quad , \quad \forall c \in \{1, 2, ..., N^C\} \quad , \quad \forall d \in \{1, 2, ..., N^D\}. \tag{8.3.11}$$

The throughput, s_{lp}, of liquid l through packaging unit p is zero if δ_{lp} is zero (forced by (8.3.12), otherwise (forced by (8.3.13)) it must be greater than the minimum packaging level for any commodity and less than the maximum packaging capacity

$$s_{lp} \le P_p^U \delta_{lp} \quad , \quad \forall l \in \{1, 2, ..., N^L\} \quad , \quad \forall p \in \{1, 2, ..., N^P\} \tag{8.3.12}$$

$$s_{lp} \ge P_p^L \delta_{lp} \quad , \quad \forall l \in \{1, 2, ..., N^L\} \quad , \quad \forall p \in \{1, 2, ..., N^P\}. \tag{8.3.13}$$

This is a common formulation device. As the δ_{lp} variables are binary (0 or 1), then if the binary variable is 0 the continuous variable is zero, while if it is 1 then the continuous variable is constrained to be less than the maximum value P_p^U and greater than the sensible minimum value P_p^L. (These types of constraints occur frequently enough for the semi-continuous variable construct to have been devised specially. For the sake of simplicity they are not used here. But see Sect. 9.4.5 for the benefits of such variables in B&B.)

Finally, we need constraints similar to (8.3.12) and (8.3.13) above, but applying to the brewing of liquids at breweries. These constraints only apply if there is more than one brewery that can brew a particular liquid.

$$b_{lb} \le B_l^U \beta_{lb} \quad \forall l \in \{1, 2, ..., N^L\}, b \in \{1, 2, ..., N^B\} \mid N_l^{BL} > 1 \tag{8.3.14}$$

$$b_{lb} \ge B_b^L \beta_{lb} \quad \forall l \in \{1, 2, ..., N^L\}, b \in \{1, 2, ..., N^B\} \mid N_b^{BL} > 1. \tag{8.3.15}$$

So if a β_{lb} is 0, the amount brewed is 0. But if a β_{lb} is 1, then the amount brewed must be at least the smallest sensible quantity to brew at the brewery but not more than the maximum amount that can be brewed by any brewery. We must also declare β_{lb} and δ_{lp} as binary variables; they can only take the value 0 or 1.

The data for most of the tables were held in ranges in a `Lotus 1-2-3` spreadsheet and imported by the `DISKDATA -l` command of `mp-model`. A useful approach is employed by the first `DISKDATA -l` command that fills up a table called `SIZES` from a spreadsheet range `SIZES`. The spreadsheet software is, in 2020, somewhat outdated. Currently, the majority uses Excel, but who knows what will be used in 15 years. Note that the `DISKDATA` command for reading text data is supported by `mp-model`'s successor `FICO Xpress Mosel`.

8.3.4 *Running the Model*

The brewery case study is contained in MCOL under the problem name *brewery*. Most of the data were assembled into the single `Lotus 1-2-3` spreadsheet file *brewdata.wk3*, but some were held in data files. Unfortunately, already after 20 years, this software seems not be used anymore — our data are lost. This should serve as a good warning and advise for the future. Only ASCII data have a reasonably long life time. Everything else depends strongly on market changes. The life time of software seems to be very limited — and even if the software still exists, backward compatibility is not guaranteed either.

8.4 Financial Modeling

In its proper meaning *financial optimization* is concerned with problems occurring in the financial services industry and is used by investment and brokerage houses. Most models in this area are related to *risk management* and *financial engineering*. An overview on these topics is found in Zenios (1993,[594]). This collection of contributions from different authors provides further references to financial optimization and is very much recommended to the reader interested in this field.

Financial optimization is relevant to investors in stock markets or fixed-income markets. It analyzes risks and leads to the constructions and maintenance of portfolios with specified risk–return characteristics. Many models are concerned with optimal portfolio systems and portfolio immunization strategies, and currency hedging strategies are another field of risk management.

The financial engineer designs and modifies financial instruments and tries to improve the marketability of such products and to meet the needs of investors.

The mathematics involved in financial modeling is usually very demanding. Most models are nonlinear and often lead to mixed integer nonlinear programming problems (see Sect. 12.6). Since in addition, the reader might not be familiar with the technical vocabulary involved in financial optimization, we decided not to incorporate material from this area into this book.

Instead, we just provide one example related to optimal purchase in the presence of tapered discounts (also a nonlinear problem) and another one concerned with yield management.

8.4.1 *Optimal Purchasing Strategies*

This problem has the following background: there are three suppliers of a good, and they have quoted various prices for various quantities of the good. We want to buy at the least total cost, yet not too much from any one supplier. Each supplier offers

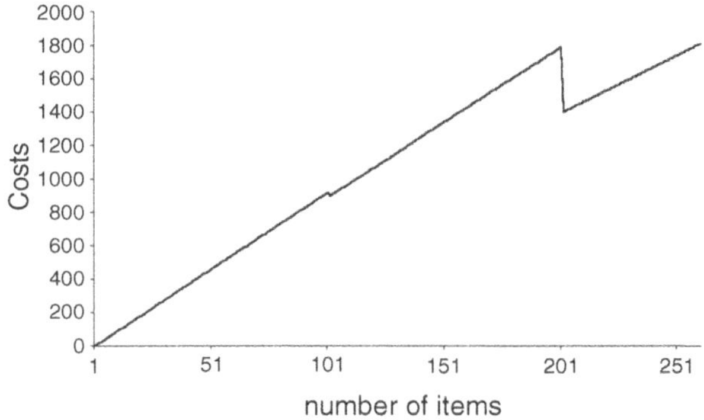

Fig. 8.1 Costs versus number of items up to 260

decreasing prices per item for increased lot sizes. For example we want to buy at most 40% from our known supplier Bob who offers us the following price ranges :

Ranges	0 – 100	101 – 200	201 – 1000
Prices	9.2	9	7
Breakpoint	100	200	1000

$$C_w(x) := \begin{cases} 9.2x\,, & 0 \le x \le 100 \\ 9.0x\,, & 100 \le x \le 200 \\ 7.0x\,, & 200 \le x \le 1000 \end{cases},$$

i.e., if we buy 150 items from Bob, the total cost is $150 \cdot 9 = 1350$. Figure 8.1 shows the cost as a function of the number of items for up to 260 items.

Let us treat the more general case of having N^S suppliers indexed by s and N^B breakpoints b, i.e., points at which the unit price changes. Discounts, which change in steps, are called tapered discounts and occur frequently in real-world problems.

We will provide two model formulations that describe the problem. The first attempt to model this problem is based on SOS2 introduced in Sect. 6.8.2, and the second one uses SOS1 introduced in Sect. 6.8.1 and is more elegant and efficient.

Let us start with the model using SOS2. Besides the set of breakpoints defined above, for numerical reasons this also requires a set of $N = 2N^B$ coarsened breakpoints (see below). The problem *purchase* supplied in MCOL shows how we could make use of size and dimensions of models specified in data files. While using the indices

$$\begin{aligned} b &\in \mathcal{B} = \{1, \ldots, N^B\}\ , \text{ the set of breakpoints} \\ k &\in \mathcal{K} = \{1, \ldots, N\}\ \ , \text{ the set of coarsened breakpoints} \\ s &\in \mathcal{S} = \{1, \ldots, N^S\}\ , \text{ the set of suppliers,} \end{aligned} \qquad (8.4.1)$$

our relevant data are

$$
\begin{array}{ll}
B_{sb}, & \text{the quantity of good at breakpoint } b \text{ of supplier } s \\
C_{sk}, & \text{total costs to supplier } s \text{ at breakpoint } k \\
\Delta, & \text{base factor for coarsening the grid} \\
\Delta_k, & \text{recursively defined coarsening factors at breakpoint } k \\
P_{sb}, & \text{price to supplier } s \text{ between breakpoints } b-1 \text{ and } b \\
P_{sb}^U, & \text{maximum percentages to supplier } s \text{ between } b-1 \text{ and } b \\
R, & \text{the total amount required} \\
X_{sk}, & \text{coarsened quantities at breakpoint } k \text{ to supplier } s.
\end{array}
\tag{8.4.2}
$$

This problem involves quantity discounts. The graph of cost against quantity purchased is discontinuous because of the unit costs, which are defined to be piecewise constant. So the situation we are facing here is certainly nonlinear, but it is a very special kind of nonlinearity. In Sect. 6.8.2 we saw how to use special ordered sets of type 2 for modeling nonlinear functions. Now we will show how we apply this concept to the current problem. To maintain sense in the special ordered set formulation, we must "fix" the discontinuity by stopping purchases just at the breakpoint. In order to do so, we introduce a single parameter Δ specifying the distance from the breakpoints and allowing us to derive several other data recursively. The coarsening factors are computed based on the parameter Δ, say $\Delta = 1.0$, according to the following recursive scheme:

$$
\Delta_1 = 0 \quad , \quad \Delta_2 = -\Delta \quad , \quad \Delta_k = -\Delta_{k-1} \quad , \quad k \in \{3, \ldots, N\}. \tag{8.4.3}
$$

Note that for $\Delta = 1.0$ this defines Δ_k recursively as $0, -1, 1, -1, \ldots, -1$. Based on Δ_k and the definitions:[1]

$$
k_1 = \left\lfloor \frac{k+1}{2} \right\rfloor \quad , \quad k_2 = \left\lfloor \frac{k}{2} \right\rfloor \tag{8.4.4}
$$

the breakpoints, let us call them *numerical breakpoints*, of the coarsened grid

$$
X_{s1} = 0 \quad , \quad X_{sk} = B_{sk_2} + \Delta_k \quad , \quad k \in \{2, \ldots, N\} \tag{8.4.5}
$$

can also be computed recursively. In the example given above the numerical breakpoints are as follows:

$$
\begin{array}{cccccc}
X_{s1} & X_{s2} & X_{s3} & X_{s4} & X_{s5} & X_{s6} \\
0 & 99 & 101 & 199 & 201 & 999
\end{array} .
$$

[1]The expression $\lfloor x \rfloor$ means return the largest integer number not exceeding x. If x is a positive number, then $\lfloor x \rfloor$ yields the integral part of x, i.e., $\lfloor 4.5 \rfloor = 4$.

Next we compute the total costs at the original breakpoints *approaching from the left and right*

$$C_{s1} = 0 \quad , \quad C_{sk} = P_{sk_1} \cdot B_{sk_2} \quad , \quad \forall k \in \{2, \dots, N\}. \tag{8.4.6}$$

To give an example: C_{s2} gives the cost of 920 at the breakpoint $B_{s1} = 100$ if we are approaching from the left, i.e., assuming a unit price of 9.2. C_{s3} gives the price at the same breakpoint if we approach from the right (assuming a unit price of 9.0). Note that while our basic data P_{sb} have the dimension *money/item*, C_{sk} represents the total cost and has the dimension *money*. If we evaluate the data for our supplier Bob, we get the following total costs at the breakpoints:

C_{s1}	C_{s2}	C_{s3}	C_{s4}	C_{s5}	C_{s6}
0	920	900	1800	1400	7000.

Our model will have the continuous variables $x_s \geq 0$ measuring the quantity to purchase from supplier s, and the variables λ_{ks} forming a special ordered set of type 2 as defined in Sect. 6.8.2 and obeying the convexity constraint

$$\sum_{k=1}^{N} \lambda_{ks} = 1 \quad , \quad \forall s \in \{1, 2, ..., N^S\}. \tag{8.4.7}$$

The decision variables x_s are related to the variables λ_{ks} by

$$x_s = \sum_{k=1}^{N} X_{sk} \cdot \lambda_{ks} \quad , \quad \forall s \in \mathscr{S}. \tag{8.4.8}$$

Note that Eq. (8.4.8) is also perfectly suited to serve as a reference row since the breakpoints X_{sk} provide a nice order. The rest of the model formulation is now straightforward. Let us begin with the objective function

$$\min \quad \sum_{s=1}^{N^S} \sum_{k=1}^{N} C_{sk} \lambda_{ks}. \tag{8.4.9}$$

Finally, we have to respect the minimum quantity, R, that must be bought

$$\sum_{s=1}^{N^S} x_s \geq R \tag{8.4.10}$$

and that no more than the maximum percentage is purchased from each supplier

$$x_s \leq P_s^U \cdot \frac{R}{100} \quad , \quad \forall s \in \mathscr{S}. \tag{8.4.11}$$

That concludes our model description. Some further details are found in the model file *purchase*.

A completely different approach to this problem is based on SOS1. In that approach, some δ_{sb} variables select which line segment we are in. The binary variables are defined such that

$$\delta_{sb} = \begin{cases} 1 \text{ , if } B_{sb-1} \leq y_{sb} \leq B_{sb} \\ 0 \text{ , otherwise} \end{cases} \quad , \quad \begin{matrix} \forall s \in \mathcal{S} \\ \forall b \in \mathcal{B} \end{matrix} , \tag{8.4.12}$$

where $B_{s0} = 0$, and the non-negative continuous variables y_{sb} relate to the variables x_s according to

$$x_s = \sum_{b=1}^{N^B} y_{sb} \quad , \quad \forall s \in \mathcal{S} . \tag{8.4.13}$$

The binary variables δ_b form an SOS1 with reference row

$$\sum_{b=1}^{N^B} B_{sb}\delta_{sb} \quad , \quad \forall s \in \mathcal{S} \tag{8.4.14}$$

and are subject to the convexity condition

$$\sum_{b=1}^{N^B} \delta_{sb} = 1 \quad , \quad \forall s \in \mathcal{S} . \tag{8.4.15}$$

The binary variables are used to ensure that at most one out of the N^B variables y_{sb} is non-zero. This property explains why (8.4.13) is valid and is enforced by the two inequalities

$$B_{sb-1}\delta_{sb} \leq y_{sb} \leq B_{sb}\delta_{sb} \quad , \quad \forall s \in \mathcal{S} \quad , \quad \forall b \in \mathcal{B} . \tag{8.4.16}$$

The model is completed by (8.4.10)–(8.4.11) and the objective function

$$\min \quad \sum_{s=1}^{N^S} \sum_{b=1}^{N^B} P_{sb} x_{sb} . \tag{8.4.17}$$

While the second model formulation is more elegant and easier to understand, it is also preferable from a mathematical point of view and gives a better representation of the real-world problem. The introduction of Δ imported some inaccuracy into the model. The second formulation is exact and it is a much more intuitive approach to the model.

Let us conclude with the remark that modeling is never unique. This case study shows very well how the same real-world problem can have rather different model formulations. Depending from where we start structuring the real-world problem and translate it into the mathematical language, we may obtain very different formulations.

8.4.2 A Yield Management Example

A container ship is traveling from Southampton to Singapore, stopping at Genoa, Port Said, Dubai, and Bombay to unload and take on cargo.

The marketing department of the company has estimated the revenue (yield) per thousand tons of cargo transported between each pair of ports and has also specified the maximum cargo that they think they can sell between each pair of ports.

We wish to find the best set of cargoes to accept, given the yields on each source–destination pair. The ship must never exceed its loading capacity of $C = 25$ ktons.

Let us start the model formulation by specifying the indices s for the port where we start and d for the destination port. We denote the number of ports by N. In this case we have $N = 6$. Our data are structured as

$$
\begin{aligned}
R^L_{sd}\ &, \text{ the minimum amount (in ktons) that must be} \\
&\quad \text{transported from } s \text{ to } d, \\
R^U_{sd}\ &, \text{ the maximum amount (in ktons) that can be} \\
&\quad \text{transported from } s \text{ to } d, \\
Y_{sd}\ &, \text{ the revenue per kton carried from port } s \text{ to port } d.
\end{aligned}
\tag{8.4.18}
$$

The tables below list these data for the cities Southampton (SH), Genoa, Port Said, Dubai, Bombay, and Singapore. We start with the yields Y_{sd} :

From/To	SH	Genoa	Port Said	Dubai	Bombay	Singapore
SH	.0	20.12	24.23	47.23	68.20	59.30
Genoa	.0	.0	14.80	24.98	52.45	50.12
Port Said	.0	.0	.0	39.50	60.65	48.23
Dubai	.0	.0	.0	.0	49.24	28.34
Bombay	.0	.0	.0	.0	.0	20.50
Singapore	.0	.0	.0	.0	.0	.0

then we have the minimum requirements R^L_{sd} :
and finally the maximum requirements R^U_{sd}:

From/To	SH	Genoa	Port Said	Dubai	Bombay	Singapore
SH	.0	1.500	1.000	5.000	1.800	4.000
Genoa	.0	.0	4.000	1.500	2.500	2.000
Port Said	.0	.0	.0	3.000	2.300	1.500
Dubai	.0	.0	.0	.0	1.200	1.500
Bombay	.0	.0	.0	.0	.0	1.800
Singapore	.0	.0	.0	.0	.0	.0

From/To	SH	Genoa	Port Said	Dubai	Bombay	Singapore
SH	.0	4.000	5.000	8.000	6.000	12.000
Genoa	.0	.0	10.000	4.000	8.000	6.000
Port Said	.0	.0	.0	8.000	5.000	8.000
Dubai	.0	.0	.0	.0	2.900	6.700
Bombay	.0	.0	.0	.0	.0	8.000
Singapore	.0	.0	.0	.0	.0	.0

.

The decision variables determine how much to carry from each port to another port later in the voyage. Let us call these continuous variables $c_{sd} \geq 0$, the number of ktons to carry from source port s to destination port d. The possible (s, d) pairs are

$(1, 2)$	$(1, 3)$	$(1, 4)$	$(1, 5)$	$(1, 6)$
	$(2, 3)$	$(2, 4)$	$(2, 5)$	$(2, 6)$
		$(3, 4)$	$(3, 5)$	$(3, 6)$
			$(4, 5)$	$(4, 6)$
				$(5, 6)$

so we introduce the variables c_{sd} only for these combinations. Note that we can easily represent these pairs by letting the first index s run from $s = 1, \ldots, N - 1$. The second index, d, then becomes a dependent index and runs only from $d = s + 1, \ldots, N$. This scheme is also applied to the objective function and the constraints.

The objective function to be maximized is the total yield, which is the sum over all (source, destination) pairs of the yield per kton carried times the number of ktons carried.

$$\max \sum_{s=1}^{N-1} \sum_{d=s+1}^{N} Y_{sd} c_{sd}. \tag{8.4.19}$$

At any time the ship cannot carry more than its capacity. Consider the load on board when a ship is just about to arrive at port p (p ranging from 2 to N). The amount on board is the sum of the amounts being carried from the ports before p to ports p and beyond. So

$$\sum_{s=1}^{p-1}\sum_{d=p}^{N} c_{sd} \le C \quad , \quad p \in \{2,\ldots,N\}. \tag{8.4.20}$$

We also have the minimum and maximum requirements

$$R_{sd}^{L} \le c_{sd} \le R_{sd}^{U} \quad , \quad s = 1,\ldots,N-1 \quad , \quad d = s+1,\ldots,N. \tag{8.4.21}$$

Since the constraints are just bounds on single variables, we better formulate them as bounds as this has numerical advantages. Note that the value the index d can take depends on which value s has been assigned. This model is stored under the problem name *yldmgmt*. A careful inspection of the model file shows how the non-trivial use of subscripts discussed above is implemented.

A further interesting example of yield management in the fishery industry is described by Glen (1995,[223]).

8.5 Post-Optimal Analysis

Post-optimal analysis starts when the optimizer finishes, displaying a message in the *log* such as "problem is infeasible," "problem is unbounded," or more desirably "optimal solution found." Unboundedness can usually be resolved relatively easy: in most cases a class of constraints has been forgotten. For nonlinear optimization problems resolving unboundedness can be more complicated. Similar as in linear optimization, a convex objective function to be maximized, or a concave objective function to be minimized, tends to increase to infinity or to decrease to minus infinity due to missing bounds on variables or missing constraints. The objective function $f(\mathbf{x})$ may have points $\mathbf{x}_p$ in the interior of the feasible region $\mathcal{S}$ with $\lim_{x\to\mathbf{x}_p} f(\mathbf{x}) = \pm\infty$. Due to missing bounds on variables or missing constraints, $\mathcal{S}$ may not be a compact set. In such cases, it may happen that for $x \to \pm\infty$, $f(\mathbf{x})$ only approaches the optimal value but never can reach it. Examples are $f(\mathbf{x}) = -1 + (x-1)/(x+1)$ or $f(\mathbf{x}) = e^{-x}$ with $\lim_{x\to\infty} f(\mathbf{x}) = 0$. These functions never assume the value zero.

Infeasibility is more serious and is discussed in further detail in Sect. 8.5.1. A finite optimal solution gives rise to the question "how stable is the solution?". Here, stability is related to how small changes in initial data affect the solution. Some optimizers have the ability to perform various post-optimal analyses, where the sensitivity of the optimal solution values to the assumed values of data can be assessed. These analyses usually go under the name *ranging*.

8.5.1 Getting Around Infeasibility

My problem is infeasible — what should I do? Diagnosing infeasibilities is probably the hardest part of modeling and optimization. Tracking down infeasibility in a large problem can be difficult and reinforces the need to experiment with modeling small instances of problems first to remove such inconsistencies. Novice users often ask "which constraint is wrong?" and are surprised when told that this is an ill-posed question. To see why, consider a small shop producing and selling bolts and screws. If the variables b and s denote the numbers of bolts and screws, we might have the constraints

$$\begin{aligned} b + \ \ s &\geq 6 \\ 2b + 3s &\leq 7. \end{aligned} \tag{8.5.1}$$

The first constraint might be a marketing constraint asking for a minimum demand to be satisfied, while the second one might refer to some labor availability. Neither of these constraints is wrong, they are just inconsistent when taken with the non-negativity constraints of the variables b and s.

Infeasibility can arise from any or all of the following causes:

a) Getting a constraint sense wrong.
b) Bad data.
c) The problem really is infeasible.
d) Over-enthusiastic modeling.

Cause a) occurs more often than one would like to admit — you have written $\geq$ when you meant $\leq$. Going over the model with a colleague tends to eliminate these errors (at the risk of making oneself look foolish!).

Simple data entry errors often contribute to cause b), with values like 1.00 being mistakenly typed as 100. Some defensive data checks are in order in all production models, but, of course, these are boring compared with the intellectual excitement of modeling and so get put off until the last moment. Displaying data graphically often helps to catch such errors and is a good reason for holding data in spreadsheets whose graphical facilities are advanced and easy to use.

If you are modeling a business situation that is currently in place (as opposed to doing a planning exercise or other theoretical work), then it is unlikely that a model, if correct, will be infeasible. The best way to see why the model is infeasible is to fix the decision variables to values you know work in practice and then see which constraints are violated.

Unfortunately, in a planning environment it is possible that the problem really is infeasible, and we have no good ideas as to feasible values for the decision variables, so the option in the previous paragraph is not open to us. The model is illogical in a sense. What can we do in this case? The model builder must rethink the model and examine each relationship carefully to find the source of the difficulty. A helpful discussion is contained in Greenberg (1993c,[236]). Also the notion of

Independent Infeasible Sets (IIS) has been developed to help the modeler.[2] They are not a panacea, but they do help in practice.

An IIS is a set of constraints with the property that if any one constraint is removed from the IIS then the remaining constraints are consistent. In a sense, an IIS is the smallest set of "nasty" constraints. Sometimes an IIS can have but a few members, and inspection of these gives a good guess as to where the problem is, but if an IIS has many members, the search for inconsistencies may still be hard work.

The last major cause (cause d) above) of infeasibilities that we shall discuss is what we might call a "lack of defensive modeling." While the model is logically correct, the data yield a model that is too tightly constrained to be feasible. Few constraints are, in practice, totally rigid. We might say that we only have 4 people available in a particular shift but in practice we might be able to get in (expensive) contractors or work (expensive) overtime. These constraints are, in some sense, "soft constraints." Some constraints, such as the number of hours in a day, or on physical processes, are, however, truly rigid.

The trick with soft constraints is to add a slack variable to the constraint, so that the original constraint can be violated, and add a penalty term into the objective function. Considering the second constraint in (8.5.1), we might introduce an amount, l, of extra labor that we acquire at a cost per hour of, say, L. The constraint becomes

$$2b + 3s \leq 7 + l. \tag{8.5.2}$$

The objective function

$$\max \quad p \tag{8.5.3}$$

(maximizing profit) changes into

$$\max \quad p - Ll. \tag{8.5.4}$$

It is likely that there is a bound, L^U, on the extra availability of labor, so we will have a bound

$$l \leq L^U. \tag{8.5.5}$$

With equality constraints we might allow positive or negative deviations from strict equality, each with a cost.

The difficulty with softening constraints is that realistic choices have to be made for the objective function entries of the slack variables. In the case of the labor slack we have presented above there is probably a market rate we could use, but this is not always the case. Then, if we choose too low a penalty cost our optimal solution

[2] An IIS facility is available, for instance, CPLEX or XPRESS-OPTIMIZER.

might involve having a non-zero slack even when the problem is otherwise feasible. There is no hard-and-fast rule here, and one has to look to see what options and economic resources are possible in practice. Further discussion on this is contained in Sect. 5.5.3.

8.5.2 Basic Concept of Ranging

LP solvers allow ranging on right-hand side coefficients and the objective function. Let us see what this means by considering a (profit) maximization problem and what happens to one particular variable, say variable x, which has an objective function coefficient P. First let us consider how the objective function changes as x is forced away from its optimal value x^0. In order to do so consider Fig. 8.2.

The objective function value will go down with a constant slope if we move x in either direction (more precisely, it will not go up) toward some maximum movement, at which point the role of degradation will increase. It is fairly straightforward, from the optimal solution of the LP, to calculate the slopes of the degradation, and the limits to which x can be moved before the slopes change. As indicated in the diagram, the slopes will generally not be equal for a movement up and a movement down. Not unreasonably, the slope in the direction of increasing x is called the *unit cost up*, and in the decreasing x direction the *unit cost down*. The values of x at which the slopes change are called *upper activity* and *lower activity*, respectively.

In a real industrial situation there may be a good reason why we would wish to set x away from its optimal value. For instance, the value used today might be only very slightly different from that used yesterday, and we would like to stick with a stable plan. Inspecting the unit costs up or down will immediately tell us whether our reluctance to change will have a small or a large effect on total profit.

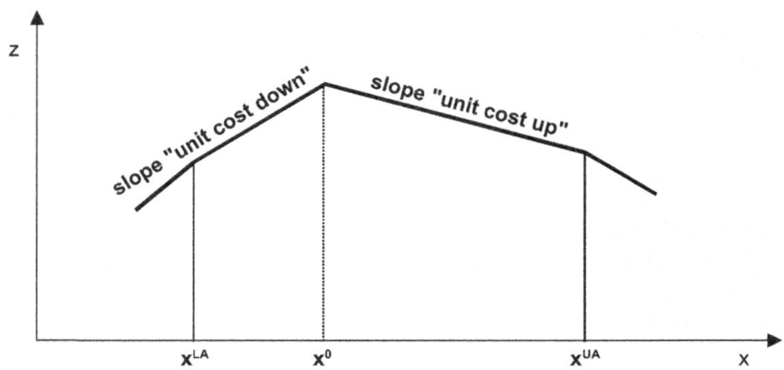

Fig. 8.2 Sensitivity analysis: objective versus optimal value

An alternative source of uncertainty might be the true value of P, the per unit profit of x. We might be able to see how the optimal amount of x indicated by the LP varies as P varies. Figure 8.3 illustrates this dependency. As P increases, the optimal amount of x to produce suddenly jumps to A, while as P decreases the optimal amount of x jumps down to B. The value of P at which x jumps to a higher level is called *upper profit*, and likewise the value of p at which x suddenly plummets is the *lower profit*. It can be shown that the new levels A and B are just the upper activity and lower activity we encountered when we were forcing x away from x^*.

At first sight, these discontinuities in the optimal solution of x look surprising — does not economically tell us that supply curves are smooth? The reason can be seen easily in a two-variable problem illustrated in Fig. 8.4.

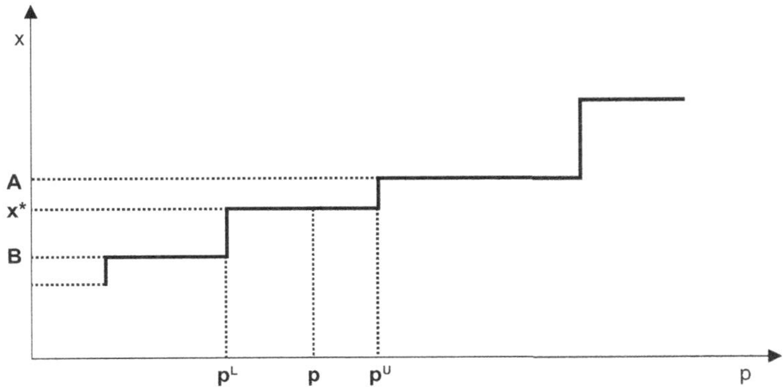

Fig. 8.3 Sensitivity analysis: optimal value versus unit profit

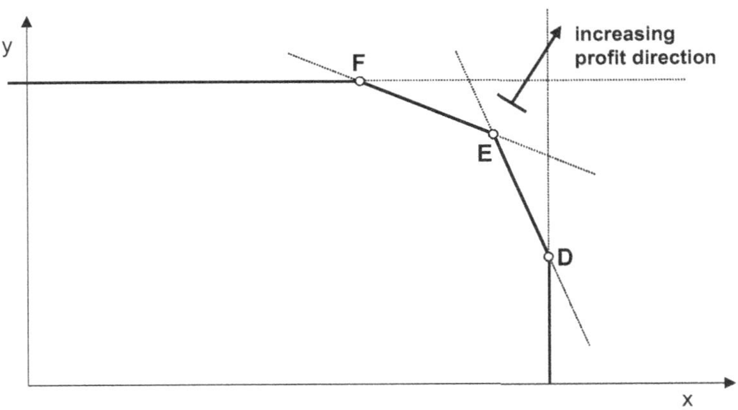

Fig. 8.4 Sensitivity analysis: slope of objective function

For a fixed per unit profit coefficient for y, as x's profit increases the optimal point will suddenly jump to D, while as it decreases it will suddenly jump to F. So a small change in P can lead to a large change in the best value of x.

8.5.3 Parametric Programming

Various other post-optimal analyses might be useful. For instance, we might like to see how the solution values change as the right-hand side of one or more constraints change, or as various profit coefficients change. So, we would like to trace out the solution to

$$\max \quad \mathbf{c}^T \mathbf{x} \tag{8.5.6}$$

subject to

$$\mathsf{A}\mathbf{x} \leq \mathbf{b} + \theta \boldsymbol{\beta} \quad , \quad \mathbf{x} \geq 0 \tag{8.5.7}$$

as θ varies from 0 to 1. Here β is a right-hand side change. Doing this is called *parametric programming*.

Similarly, we might like to see the solution of

$$\max \quad (\mathbf{c} + \theta \boldsymbol{\delta})^T \mathbf{x} \tag{8.5.8}$$

subject to

$$\mathsf{A}\mathbf{x} \leq \mathbf{b} \quad , \quad \mathbf{x} \geq 0, \tag{8.5.9}$$

as θ varies from 0 to 1. $\boldsymbol{\delta}$ is a vector of changes to the objective function. This is called *parametric programming on the cost row*.

If the right-hand side and objective function can change simultaneously, then we might have

$$\max \quad (\mathbf{c} + \theta \boldsymbol{\delta})^T x \tag{8.5.10}$$

subject to

$$\mathsf{A}\mathbf{x} \leq \mathbf{b} + \theta \boldsymbol{\beta} \quad , \quad \mathbf{x} \geq 0, \tag{8.5.11}$$

which is called *parametrics on the rim*.

Much harder to tackle is the case where coefficients in the constraint matrix change:

$$\max \quad \mathbf{c}^T \mathbf{x} \tag{8.5.12}$$

subject to

$$(\mathsf{A} + \theta\alpha)\mathbf{x} \leq \mathbf{b} \quad , \quad \mathbf{x} \geq 0. \tag{8.5.13}$$

We are not aware of any current optimization software that includes this facility.

Although the techniques of ranging and parametric programming are nice theoretically, they are of almost no use with large problems. The difficulty arises because *the* solution to almost all large LP problems is degenerate, i.e., there are many alternative solutions with the same objective function value. Minimal changes in the right-hand side or profit coefficients will then lead to a solution that is different in the decision variables, but nearly the same in the objective function value. Thus, the upper and lower activities will coincide with the current optimal value, and we will obtain little useful information.

When we do parametric programming on large problems, there are likely to be an enormous number of different solutions as we change θ. If we print out the different solutions each time we jump to a new vertex, then we will be drowned in output. The natural thing is to say that we want solutions for a set of θ, say $\theta \in \{0.1, 0.2, \ldots, 1.0\}$. But if that is what we want it is best to solve ten separate problems, which in the case of parametrics on the right-hand side for instance, give rise to problems P_i with $\beta_i = i/10$.

We would solve these problems for $i = 1, 2, \ldots, 10$ using the optimal basis for problem i to start the solution procedure for problem $i + 1$.

8.5.4 Sensitivity Analysis in MILP Problems

Finally we want to make some comments on sensitivity analysis in MILP problems. The concept of sensitivity is closely related to that of continuity. In continuous problems we expect that a "small" change in an input quantity will have "small" effects onto output quantities. However, when discussing the dependence of optimal x on its unit profit coefficient P, we saw that we can expect jumps. The situation is even worse in MILP problems. Consider the single constraint capacity problem

$$\max \quad 5\delta_1 + 100\delta_2 + 10\delta_3 \tag{8.5.14}$$

subject to

$$20\delta_1 + 100\delta_2 + 40\delta_3 \leq C \quad , \quad \delta_i \in \{0, 1\} \tag{8.5.15}$$

with capacity $C = 100$. The solution is obviously $\delta_1 = \delta_3 = 0$ and $\delta_2 = 1$ yielding an objective function value of 100. If C is slightly decreased for instance to $C = 100 - \varepsilon$ where ε is any positive real number, the solution changes to $\delta_1 = \delta_3 = 1$, $\delta_2 = 0$ and an objective function value of 15. In LP the objective function is a smooth function, but in ILP it is not.

Thus the consequence is that for MILP problems it is almost impossible to provide meaningful sensitivity analysis. We cannot expect better than this because of the lack of continuity of discrete problems. Work by Williams (1995,[581]) attempts to provide an explanation of sensitivity analysis and duality, where possible, for MILP problems, but the possibilities are limited.

8.6 Summary and Recommended Bibliography

At the end of this chapter the reader should now be familiar with:

- Several examples of models requiring the use of MILP
- The nature of modeling special features required in practical instances of depot location, distribution, optimal purchase, yield management problems, and scheduling across time periods
- The difficulty of detecting sources of infeasibility in models
- The potential of post-optimality and parametric operations on solved problems

It may also motivate the reader to have a deeper look into the literature related to certain fields of Operations Research, e.g., *Location Science* [351].

Chapter 9
User Control of the Optimization Process and Improving Efficiency

This chapter will provide information on how the solution time of an IP problem can be reduced significantly. This is important, as in contrast to ordinary LP problems, effective solution of IP problems depends critically upon good model formulation, the use of high-level branching constructs, and control of the B&B strategy. Good formulations are those whose LP feasible region is as "small as possible" not excluding any feasible MILP solution, or, to be precise, those whose LP relaxation has a feasible region which is close to the convex hull of the MILP problem's feasible set. In practice, this means, for example, that upper bounds should be as small as possible. Formulating models in this fashion is still largely the responsibility of the modeler, although work has been done on automatically reformulating mixed zero-one problems, cf. VanRoy & Wolsey (1987,[562]), leading to tighter formulations. Preprocessing can also improve the model formulation; cf. Achterberg et al. (2008,[4]), Gamrath et al. (2015,[206]), and Achterberg et al. (2020,[5]). The merit of good formulations is evident in practical applications such as Meyer (1969,[395]), Jeroslow & Lowe (1984,[288]), and Cheshire et al. (1984,[120]).

Most of the techniques described in the sections below are executed automatically by the (commercial) LP and MILP solvers available in 2020.

9.1 Preprocessing

Preprocessing methods introduce model changes to speed up the algorithms. The modifications are made, of course, in such a way that the feasible region of the MILP problem is not changed. They apply to both pure LP and MILP problems but they are much more important for MILP problems. There are also dual techniques that change the feasible region but ensure that at least one optimal solution remains in the feasible set. A selection of several preprocessing algorithms is found in Johnson

J. Kallrath, *Business Optimization Using Mathematical Programming*,
International Series in Operations Research & Management Science 307,
https://doi.org/10.1007/978-3-030-73237-0_9

et al. (1985,[292]). An early overview of simple and advanced preprocessing techniques is given by Savelsbergh (1994,[484]). The reader is also referred to the comprehensive survey of presolve[1] methods by Andersen & Andersen (1995,[20]). Achterberg et al. (2008,[4]) are a useful source for presolving, Gamrath et al. (2015,[206]) provide a more recent review on progress in presolving for mixed integer programming, and Achterberg et al. (2020,[5]) give recent insights into presolve reductions in mixed integer programming. Some common preprocessing methods are: presolve (arithmetic tests on constraints and variables, bound tightening), disaggregation of constraints, coefficient reduction, and clique and cover detection. In 2020, among the most important techniques in presolve are: probing, bound tightening, duplicate columns, dominated columns, duplicate rows, dominated rows, two-row reductions, implied integer detection, singleton rows, and substitutions (especially dual substitutions and those with two variables).

Many of these methods are implemented in commercial software but vendors are usually not very specific about which techniques they have implemented or methods used. For SCIP and the open-source solver CBC it is easier to find details about presolving; cf. Gamrath et al. (2015,[206]). In particular, during preprocessing constraints and variables are eliminated, variables and rows are driven to their upper and lower bounds, or an obvious basis is identified. Being aware of these methods may greatly improve the user's model building leading to more efficient models or may reduce the user's efforts if it is clear that the software already does certain operations automatically. For those reasons we demonstrate different preprocessing methods by simple examples.

9.1.1 Presolve

There are two important operations presolve performs: *arithmetic tests* and *bound tightening*.

9.1.1.1 Arithmetic Tests

Despite the good efforts of modelers, models are frequently built which contain unintended redundant features. For example, given upper bounds of 10, 8, 8 (respectively) and lower bounds of 2, 3, 5 (respectively) for x_1, x_2, and x_3 the

[1] The terms *preprocessing* and *presolve* are often used synonymously. Sometimes the term *presolve* is used for those procedures which try to reduce the problem size and to discover whether the problem is unbounded or infeasible. *Preprocessing* involves the presolving phase but includes all other techniques which try, for instance, to improve the MILP formulation. It might be interesting to point out here that transferring a solution back to the space of the optimal solution is called *postsolving* and is a non-trivial step in some cases.

constraint

$$2x_1 + x_2 - x_3 \leq 25 \tag{9.1.1}$$

is redundant as the left-hand side of the constraint can never exceed 23. To see this we can inspect the left-hand side of (9.1.1). The maximum value it can take is given by driving the variables to their upper and lower bounds yielding

$$\max(2x_1 + x_2 - x_3) = 2 \cdot 10 + 1 \cdot 8 - 1 \cdot 5 = 23. \tag{9.1.2}$$

Thus the inequality (9.1.1) can be removed from the problem.

To give another example of redundant constraints, let us assume $0 < b_1 < b_2$ and consider the three constraints

$$0 \leq x_1 \leq b_1 \quad , \quad 0 \leq x_2 \leq \infty \quad , \quad x_1 - x_2 \leq b_2. \tag{9.1.3}$$

Since $x_1 \leq b_1$ and x_2 is non-negative the maximum positive value the left-hand side of the last constraint could take is b_1 and thus the last constraint can never become binding and thus is redundant.

Further, in the problem

$$\max \quad 2x_1 + x_2 - x_3 \tag{9.1.4}$$

subject to

$$\begin{aligned} x_1 + x_2 + x_3 &\leq 100 \\ 2x_1 + 5x_2 + 3x_3 &\leq 200 \\ -3x_1 + x_2 + x_3 &\leq 150, \end{aligned} \tag{9.1.5}$$

where all variables have lower bounds of zero, the variable x_3 could be removed from the problem. It takes away from the availability of the three resources (9.1.5) and reduces the objective function. Thus in an optimal solution we have $x_3 = 0$.

The idea of checking a problem for redundant features is discussed in Brearley et al. (1975,[98]). Additional information appears in Tomlin & Welch (1983a,[548] & b,[547]). Although the small examples given seem unlikely to occur, in practice redundancy is likely to creep in when large and complex models are formulated as (a) redundancy may be masked within the problem, (b) a model may be made from a combination of models built by several modelers, perhaps at different locations, and although each submodel may contain no redundancies the combined model may do so, and (c) idleness in modelers, e.g., $\sum x_i = R$ and fix $R = 0$.

It is beneficial to remove redundancies if this can be done conveniently as it will aid model management, reduce computer storage requirements and speed up solution. A model with redundant features will not give "wrong" solutions, but it offers scope for improvement. Many LP software systems have facilities for removing redundancies prior to the optimization process starting, i.e., a number of these checks are made and constraints and variables adjusted, or indeed removed, accordingly. Solution values are organized so that the output restores any redundant feature of the model removed by presolve, e.g., shadow prices and reduced costs are calculated.

Arithmetic tests performed in presolving can also detect *infeasible problems*. The idea behind such tests is to transform the constraints such that a selected variable x_j appears alone on the left-hand side, i.e.,

$$x_j \circ B_i \pm \sum_{\forall k, k \neq j} A_{ik} x_k \quad , \quad \forall i, \tag{9.1.6}$$

where the symbol $\circ$ represents the relations $\leq$ or $\geq$. Let us apply this idea to the constraints

$$0 \leq x_1 \leq 1 \quad , \quad 7 \leq x_2 \leq 9 \quad , \quad x_1 + x_2 \leq 5. \tag{9.1.7}$$

Selecting x_2 the constraints system becomes

$$\begin{aligned} x_2 &\geq 7 \\ x_2 &\leq 9 \\ x_2 &\leq 5 - x_1 \leq 5 \end{aligned} \tag{9.1.8}$$

as x_1 is non-negative. The first and last constraints are obviously in conflict and thus our problem is infeasible.

9.1.1.2 Tightening Bounds

As we have already seen in Sect. 3.3.1 where we tightened the bound of the inequality (3.3.2) $x_1 + x_2 \leq 3.5$ yielding $x_1 + x_2 \leq 3$, in some integer problems tightening bounds eliminates the tree enumeration in the B&B algorithms completely. Tightening bounds is crucial — and probing builds on the same theory. In this section we will see how we can shrink the domain of a variable if we consider several constraints simultaneously.

Consider the constraints

$$L_i \leq \sum_j A_{ij} x_j \leq U_i \quad , \quad \forall i \tag{9.1.9}$$

and the bounds

$$X_j^- \leq x_j \leq X_j^+ \quad , \quad \forall j. \tag{9.1.10}$$

Then, iteratively, we could compute the extreme values

$$E_i = \sum_j A_{ij} x_j \quad , \qquad \forall i \tag{9.1.11}$$

of constraints by driving the variables to either their lower or upper bounds, i.e., each variable x_j is fixed to one of its finite (if any) bound values depending on the corresponding A_{ij} sign. For each constraint, selecting one variable at a time, E_i is updated in order to find an *implied* bound for the selected variable. If tighter, the *implied* bound replaces the actual bound. This procedure iterates over all constraints, and stops if (a) during a complete iteration no reduced bound is found, or (b) the number of iterations reaches a predefined maximum value. To demonstrate tightening of bounds consider the following bounds

$$\begin{aligned} 0 \le x_1 &\le \infty \\ 0 \le x_2 &\le \ 2 \\ 4 \le x_3 &\le 10 \end{aligned} \tag{9.1.12}$$

and let us analyze the constraint

$$4 \le 2x_1 + 4x_2 - 6x_3 \le 10. \tag{9.1.13}$$

A new lower bound for x_1 is derived from considering $4 \le 2x_1 + 4x_2 - 6x_3$, or equivalently from

$$x_1 \ge 2 - 2x_2 + 3x_3. \tag{9.1.14}$$

Inserting the appropriate extreme values for x_2 and x_3, i.e., $x_2 = X_2^+ = 2$ and $x_3 = X_3^- = 4$ we find $x_1 \ge 10$. The new upper bound of x_1 is found from considering $2x_1 + 4x_2 - 6x_3 \le 10$, or equivalently

$$x_1 \le 5 - 2x_2 + 3x_3 \tag{9.1.15}$$

with $x_2 = X_2^+ = 0$ and $x_3 = X_3^- = 10$ yielding $x_1 \le 35$. Thus the updated bounds for x_1 read

$$10 \le x_1 \le 35. \tag{9.1.16}$$

Understanding the concept of tightening bounds we can now also appreciate why it is important to model constraints of the form

$$y \le Y \cdot \delta \quad , \quad y \ge 0 \quad , \quad \delta \in \{0, 1\} \tag{9.1.17}$$

with the upper bound Y on y as small as possible. In the LP relaxation δ appears as $0 \le \delta \le 1$. Suppose, for example, that the largest feasible value of y is $Y' < Y$, and that the objective function to be minimized contains a term, say $C\delta$, representing some set-up cost associated with δ. If in the optimal solution of the LP relaxation we have $y = Y'$, and we have used (9.1.17), because of the presence of $C\delta$, we will obtain $\delta = Y'/Y < 1$ leading to two additional subproblems in the B&B algorithm. On the other hand, if we use the constraint

$$y \le Y' \cdot \delta \quad , \quad y \ge 0 \quad , \quad 0 \le \delta \le 1 \tag{9.1.18}$$

we get $\delta = 1$. Although in practice we might not always encounter the favorable case it illustrates the idea why we are advised to choose the upper bound on y as small as possible.

A message such as `LP relaxation tightened` produced by a MILP solver produced during optimization means presolve has been effective in integer aspects of the problem. Note that solvers include several preprocessing methods under presolve.

9.1.2 Disaggregation of Constraints

Disaggregation aims at finding some logical implications between binary and other variables. Such logical implications are introduced into the model by adding logical inequalities. They can be added if implications of the following kind are present: if a binary variable is zero or one, then a continuous variable is either at its lower bound (L) or upper bound (U). There are four possible implications and we list the inequalities they generate

$$\begin{array}{l} \delta = 0 \Rightarrow x = L \longrightarrow x - L \le (U - L)\delta \\ \delta = 0 \Rightarrow x = U \longrightarrow U - x \le (U - L)\delta \\ \delta = 1 \Rightarrow x = U \longrightarrow x - L \ge (U - L)\delta \\ \delta = 1 \Rightarrow x = L \longrightarrow U - x \ge (U - L)\delta . \end{array} \tag{9.1.19}$$

Consider the binary variable $\delta \in \{0, 1\}$ in the example

$$\begin{array}{rl} x_1 & \le 100\delta \\ -x_1 + x_2 \le & 20 \end{array} \quad \text{with} \quad \begin{array}{r} 0 \le x_1 \le 100 \\ 20 \le x_2 \le \ \ 80 \end{array} . \tag{9.1.20}$$

In order to see whether one of the implications listed in (9.1.19) is present let us see how the constraints look like when the binary variable takes the values 0 or 1:

$$\delta = 1 \Longrightarrow \begin{array}{rl} x_1 & \le 100 \\ -x_1 + x_2 \le & 20 \end{array} \tag{9.1.21}$$

and

$$\delta = 0 \Longrightarrow \begin{array}{l} x_1 \leq 0 \\ x_2 \leq 20 \end{array} \Longrightarrow \begin{array}{l} x_1 = 0 \\ x_2 = 20 \end{array}. \tag{9.1.22}$$

While $\delta = 1$ does not imply that x_1 or x_2 is at any of its bounds, $\delta = 0$ does so: both variables, x_1 and x_2, are at their lower bounds. Applying (9.1.19) for x_1 yields the logical inequality $x_1 \leq 0$ which does not help at all. But the logical inequality derived for x_2 reads

$$x_2 - 60 \cdot \delta \leq 20 \tag{9.1.23}$$

and improves our original model. To see this let us consider the original inequalities (9.1.20). If we combine both of them we get

$$x_2 \leq 20 + x_1 \leq 20 + 100\delta. \tag{9.1.24}$$

Our additional logical inequality (9.1.23) leads to the tighter inequality

$$x_2 \leq 20 + x_1 \leq 20 + 60\delta. \tag{9.1.25}$$

9.1.3 Coefficient Reduction

Coefficient reduction[2] is another preprocessing method which aims at tighter MILP formulations. The reasoning involved is closely related to the (0-1) knapsack constraints we got to know in Sect. 7.1.1. To see how and why coefficient reduction works consider the inequality

$$4\delta_1 + 3\delta_2 - 2\delta_3 \leq 6. \tag{9.1.26}$$

This inequality is not yet in the knapsack form because one of the coefficients is negative. In order to achieve that goal we introduce a new binary variable $\delta_i^{'} = 1-\delta_i$ which then has the coefficient $-A_{ij}$, i.e., a positive one. In our example this leads to the binary knapsack constraint

$$4\delta_1 + 3\delta_2 + 2\delta_3' \leq 8. \tag{9.1.27}$$

So in any case we can assume that all coefficients A_{ij} are positive, i.e., that the inequality appears as a knapsack constraint. For each constraint $\sum_j A_{ij}\delta_j \leq b_i$ (in

[2]In 2020, coefficient reduction still matters, but there are a number of more advanced techniques to tighten coefficients, such as those that consider other rows in the problem or cliques.

the example we consider only one) define the sum of all coefficients

$$S_i := \sum_j A_{ij}. \tag{9.1.28}$$

We can assume that $S_i > b_i$ as otherwise the constraint is not binding at all. Let A_{im} be the maximum coefficient in constraint i and let us rewrite the original constraint as

$$A_{im}\delta_m + \sum_{j \neq m} A_{ij}\delta_j \leq b_i. \tag{9.1.29}$$

We can also exclude the case $A_{im} > b_i$ as in that case we could put $\delta_m = 0$. The definition of S_i allows us to derive the following inequality:

$$\sum_{j \neq m} A_{ij}\delta_j \leq \min\{S_i - A_{im}, b_i\}. \tag{9.1.30}$$

In order to improve the constraints let us add the term $(S_i - b_i)\,\delta_m$ on both sides of the inequality and rearrange the inequality (9.1.29) a bit:

$$(S_i - b_i)\,\delta_m + \sum_{j \neq m} A_{ij}\delta_j \leq b_i + (S_i - b_i)\,\delta_m - A_{im}\delta_m \tag{9.1.31}$$

$$= (S - A_{im} - b_i)\delta_m + b_i. \tag{9.1.32}$$

Since we want to tighten the formulation let us derive the minimum value of the right-hand side. Under the assumption $S_i - A_{im} < b_i$ the minimum value, $S - A_{im}$, is achieved for $\delta_m = 1$ (note that the term in brackets is negative according to our assumption). Thus we end up with the inequality

$$(S_i - b_i)\,\delta_m + \sum_{j \neq m} A_{ij}\delta_j \leq S_i - A_{im}, \tag{9.1.33}$$

which replaces the original constraint $\sum_j A_{ij}\delta_j \leq b_i$. However, we have to show that we did not change the feasible region of the original MILP problem. If we inspect (9.1.33) for the case $\delta_m = 0$ we get (9.1.30) again; thus we are safe on this side. The case $\delta_m = 1$ in (9.1.29) gives

$$A_{im} + \sum_{j \neq m} A_{ij}\delta_j \leq b_i, \tag{9.1.34}$$

which corresponds to the original problem without losing something. Thus we have shown that (9.1.33) is a valid inequality. As we have $S_i - A_{im} < b_i$, and equivalently

$S_i - b_i < A_{im}$, we see that the coefficient of δ_m and that of the right-hand side of (9.1.33) are really reduced, compared to the original inequality.

In short, we can establish the following rule: if $S_i - A_{im} < b_i$ the coefficients A_{im} and b_i can be assigned new values

$$\begin{aligned} A_{im} &\longleftarrow S_i - b_i \\ b_i &\longleftarrow S_i - A_{im} \end{aligned} \tag{9.1.35}$$

replacing the original inequality. Then a new value of S_i is evaluated and the process iterates over all coefficients.

Let us demonstrate the procedure and apply it to the constraint

$$4\delta_1 + 3\delta_2 + 2\delta_3 \leq 8, \tag{9.1.36}$$

which yields $S = 9$. In this case we have $A_{11} = 4$ as the biggest coefficient, $S - A_{11} = 5 < 8 = b_1$ and thus the improved coefficients

$$\begin{aligned} A_{11} &\longleftarrow 1 = S_1 - b_1 \\ b_1 &\longleftarrow 5 = S_1 - A_{11}, \end{aligned} \tag{9.1.37}$$

which lead to the improved constraint

$$\delta_1 + 3\delta_2 + 2\delta_3 \leq 5. \tag{9.1.38}$$

The next coefficient to be replaced is A_{12} and eventually A_{13} so the final constraint is

$$\delta_1 + \delta_2 + \delta_3 \leq 2, \tag{9.1.39}$$

which claims that at most two of the binary variables can be different from zero. In terms of the knapsack interpretation this inequality ensures that at most two items can be chosen. If we inspect the original inequality (9.1.36) we can check this result and see again that we could choose indeed at most two items.

9.1.4 Clique Generation

Consider again n binary variables $\delta_j \in \{0, 1\}$ and a constraint

$$\sum_{j=1}^{n} A_j \delta_j \leq b, \tag{9.1.40}$$

which is again a knapsack constraint if all coefficients A_j are positive. As in Sect. 9.1.3 we assume that this is the case. There may be several constraints of such form, and we inspect them individually. We also assume that the coefficients A_j in the row of interest are sorted in non-increasing order, i.e., $A_1 \geq A_2 \geq \ldots \geq A_n$. A clique is defined as an inequality

$$\sum_{j \in S} \delta_j \leq 1 \quad , \quad S \subset \{1, \ldots, n\}. \tag{9.1.41}$$

The clique generation process inspects the two coefficients $A_j + A_{j+1}$ starting with $j = 1$. If $A_j + A_{j+1} > b$, then at most one of δ_j and δ_{j+1} can take the value 1 as otherwise (9.1.40) is violated. We continue incrementing j until $A_{j^*} + A_{j^*+1} \leq b$ for some value j^*. Then $S = \{1, \ldots, j^*\}$ is a clique because at most one of δ_{j_1}, δ_{j_2} with $j_1, j_2 \in S$ can be non-zero. Model formulation will benefit (tighter LP relaxation) from violated cliques so only these are added to the model constraints. Naturally, clique generation iterates over all the constraints.

Let us see how clique generation works on the constraint

$$1.8\delta_1 + 1.5\delta_2 + \delta_3 + \delta_4 \leq 2. \tag{9.1.42}$$

Working from left to right, the first pair that does not violate the constraint (9.1.42) is δ_3, δ_4 as $1 + 1 \leq 2$. So the clique generated and added to our model is

$$\delta_1 + \delta_2 + \delta_3 \leq 1. \tag{9.1.43}$$

Note that

$$\delta_1 + \delta_2 + \delta_4 \leq 1 \tag{9.1.44}$$

also holds as δ_3 and δ_4 appear alike in (9.1.42). Exercise 9.1 will provide further examples.

In 2020, cut generation as described above is not considered part of presolve but part of the root node processing. That said, cliques are also added in presolve if the lifted clique dominates original constraints.

9.1.5 *Cover Constraints*

As clique constraint, cover constraints usually also apply to 0-1 knapsack problems. Again we consider n binary variables $\delta_j \in \{0, 1\}$ and a constraint

$$\sum_{j=1}^{n} A_j \delta_j \leq b, \tag{9.1.45}$$

in which all coefficients A_j are positive but this time sorted in non-decreasing order in each constraint. The idea of covers is to find certain combinations of binary variables such that the sum of their associated coefficients exceeds the capacity of the knapsack. Let us now introduce a *cover* defined as the constraint

$$\sum_j \delta_j \le K. \tag{9.1.46}$$

If we find a subset $\mathscr{S} \subset \{1, \ldots, n\}$ of indices such that $\sum_{j \in \mathscr{S}} A_j > b$ but for any smaller subset $\mathscr{S}' \subset \mathscr{S}$, $\sum_{j \in \mathscr{S}'} A_j \le b$, then

$$\sum_{j \in \mathscr{S}} \delta_j \le |\mathscr{S}| - 1 \tag{9.1.47}$$

is a *minimal cover* (Crowder et al., 1983,[135]).

Based on this minimal cover we will derive a *lifted cover inequality* by considering the following knapsack problem:

$$M = \max \sum_{j \in \mathscr{S}} \delta_j \tag{9.1.48}$$

subject to

$$\sum_{j \in \mathscr{S}} A_j \delta_j \le b - A_k \quad , \quad \forall k \notin \mathscr{S}. \tag{9.1.49}$$

If $M > 0$, then δ_k is *lifted* in the minimal cover that now takes the form

$$\sum_{j \in \mathscr{S}} \delta_j + (|\mathscr{S}| - 1 - M)\, \delta_k \le |\mathscr{S}| - 1 \tag{9.1.50}$$

for if $\delta_k = 0$ this is the minimal cover constraint above, while if $\delta_k = 1$, then $\sum_{j \in \mathscr{S}} \delta_j \le M$ as M was selected with this property. The cover generation process iterates over all the constraints.

Let us demonstrate this approach using the constraint

$$9\delta_1 + 10\delta_2 + 10\delta_3 + 11\delta_4 \le 23. \tag{9.1.51}$$

The first task is to determine an appropriate set of indices $\mathscr{S}$. There are 4 minimal covers {1,2,3}, {2,3,4}, {1,2,4}, and {1,3,4} corresponding (9.1.51). Let us take the first one, $\mathscr{S} = \{1, 2, 3\}$, yielding the minimal cover

$$\delta_1 + \delta_2 + \delta_3 \le 2. \tag{9.1.52}$$

If we interpret the inequality (9.1.52) in terms of the knapsack problem it states that of the first three items at most two can be selected. The only index not in $\mathscr{S}$ is $i = 4$. So the lifting technique tries to add δ_4 to the minimal cover (9.1.52). In order to do so we have to solve the following knapsack problem

$$M = \max \quad \delta_1 + \delta_2 + \delta_3 \tag{9.1.53}$$

subject to

$$9\delta_1 + 10\delta_2 + 10\delta_3 \leq 23 - 11 = 12. \tag{9.1.54}$$

In this example it is easy to see that we have the maximum $M = 1$ because only one binary value can be different from zero. Thus δ_4 can join the minimal cover constraint with coefficient $2 - M$ yielding

$$\delta_1 + \delta_2 + \delta_3 + \delta_4 \leq 2. \tag{9.1.55}$$

The last thing we have to check is that the lifted cover (9.1.55) is violated if we would assign otherwise feasible values to the variables occurring in the cover. Indeed, $\delta_1 = \delta_2 = \delta_3 = \delta_4 = 1$ makes (9.1.55) infeasible.

The result (9.1.55) is not a surprise as (9.1.55) just ensures that at most two binary variables can be different from zero.

In the MILP solvers available in 2020, cover and knapsack cuts are not considered part of presolve anymore. They are rather part of the root node processing.

9.2 Efficient LP Solving

The use of software to solve LP problems rarely benefits from user intervention during the optimization process, provided the matrix is well scaled [see Sect. 9.2.2]. However, it is useful to make use of a *warm start* wherever possible. As with many things in live: Too many or too much may not be good. If one uses warm starts often during search one may not finish the search. Therefore, increase the number of B&B nodes in each run. Other reasons why warm starts are good: Pseudo-costs; promising branching variables have been identified.

9.2.1 Warm Starts

A solution saved from a previous related run may be advantageously used in the form of an advanced basis [see Sect. 3.8.1 for a discussion of basis]. That is what we call a *warm start*. Inside MILP solvers there is also something called *hot start*. If we already have a factorization of the basis such as when diving in the B&B tree, we can use that and save quite some time. If warm starts and hot starts are not possible the use of a *crash basis* is recommended. *Crash* is discussed in Maros & Mitra (1996,[386]). The idea is to force into the solution at an early stage all the variables that look promising, rather than to introduce them sequentially. The exploitation of *crash structures* can dramatically speed the solution of LP problems [cf. Gould & Reid (1989,[232])].

9.2.2 Scaling

The *scale* of the matrix of coefficients in an LP or MILP problem is the ratio between the coefficient with largest absolute value and the coefficient with smallest absolute value. A well scaled matrix is one where that ratio is relatively small. It is always worth striving for a well scaled matrix as this avoids certain difficulties that can be encountered during the optimization process, e.g., in the Simplex algorithm the division of a large coefficient by a small one might be required, creating a very large number, which leads to difficulties of computer representation of that number. A variety of scaling methods are available that can be imposed on the matrix once it is formed. However, it is advisable to try to avoid some foreseeable difficulties by the careful choice of units for variables to avoid dramatic differences, e.g., working in 1,000 units for salary data so that a salary is represented as 15.30 and working in fractional measurements of tolerances so that 0.0015 mm is represented as 1.5. Foolish choice of units might place the coefficients 15,300.0 and 0.0015 in the same model and lead to the quotient of these two quantities being created, i.e., 10,200,000.0, which is rather large. If several of such large numbers occur and we need to compute differences between them, a typical problem of large number arithmetic requires our attention: inaccurate results due to the cancelation of digits. Ideally, the non-zero matrix coefficients should range between 0.01 and 1,000 in absolute magnitude, and the optimal decision variables should not exceed 1,000. A sensible use of units of measurement is worthwhile: using grams and tons to measure similar quantities can cause problems. Similarly, measuring objectives such as corporate cash flow in single monetary units such as £, SFr, Yen or $, is unwise.

A discussion of types of scaling appears in Fulkerson & Wolfe (1962,[203]), Curtis & Reid (1972,[136]), and Tomlin (1975,[546]). Solvers offer a number of types of scaling. Its objective is to get the coefficients as close to 1 as possible, while not losing accuracy. The default scaling usually is row and column scaling by the maximum element method . Other choices are

1. row (divide each row by the largest absolute value element in that row),
2. column (divide each column by the largest absolute value element in that column),
3. maximin,
4. Curtis-Reid scaling (Curtis & Reid, 1972,[136]), and
5. geometric mean (replace "largest absolute value" in 1 or 2 by geometric mean).

It has to be said that the influence of scaling on solution times is very badly understood, but it still matters, in 2020. Unfortunately, it sometimes leads to the situation that after unscaling, the solution is a lot less accurate than the user might want.

9.3 Good Modeling Practice

Solution times can also greatly be influenced by observing a few rules which distinguish "bad" from "good" modeling. Good modeling leads to numerically efficient behavior of the solver. The "tricks of the trade" listed below demonstrate once more that good modeling requires a tight connection and deep understanding of both the model and how an algorithm is implemented in a commercial solver.

1. Structure of the objective function: collecting different cost contributions (transport, production, inventory) into specially introduced variables instead of putting them all into the objective function may increase the time needed to solve the problem. If the total cost to be minimized is the sum of several logically destined parts (for instance, transport, production, and inventory) there is often a tendency to specify the total cost by a constraint

$$c = c^T + c^P + c^D \tag{9.3.1}$$

   and then give separate equations for c^T, c^P, and c^D [refer to Sects. 10.4.2.14 and 10.4.3.4]. This tends not to be a good idea, as there are very few coefficients in the objective function, and so the early choices in the Simplex algorithm can be rather arbitrary. However, if presolve and/or *crash techniques* are used then the substitutions into the objective function will probably occur, and the objections given above rendered invalid. Actually, presolving nowadays identifies substitution variables and hence, modeling can be done quite intuitive and close to natural language.
2. Avoid zero right-hand side equations: constraints such as $\sum A_j x_j = 0$ should be avoided. The minimum ratio rule has difficulties picking out the best basic variables to be eliminated from the basis. Some modelers like to have all right-hand side coefficients equal to zero, and supply just bounds on variables. For instance, the constraint

$$x + 2y \le 6 \tag{9.3.2}$$

   can be written as

$$x + 2y - s = 0 \quad , \quad s \le 6. \tag{9.3.3}$$

   Problems formulated in this way can be degenerate, as the Simplex minimum ratio rule has many ties to consider. So, the general advice is to avoid constraints such as $\sum A_j x_j = 0$ if possible. (Although, after many iterations some non-zero right-hand side might show up in the transformed equations associated with $\sum A_j x_j = 0$.)

3. In Chap. 2, page 45, we saw that balance constraints appeared in two guises. Let x_1, x_2, x_3 indicate flows into a process and y_1 and y_2 represent flows out. A constraint is required to model the fact that flow must be conserved in the sense that total flow out cannot exceed total flow in. The constraint is

$$x_1 + x_2 + x_3 - y_1 - y_2 \geq 0. \tag{9.3.4}$$

A tighter form of (9.3.4) and flow conservation is of course

$$x_1 + x_2 + x_3 - y_1 - y_2 = 0, \tag{9.3.5}$$

which ensures that flow out and flow in are exactly equal to each other. Even if (9.3.5) is the correct formulation of flow conservation there is an advantage using (9.3.4). This replacement is valid under the following assumption: assume that y_1 and y_2 represent amount of products which are to be produced and sold and appear in a maximization problem with positive coefficients in the objective function. Let x_1, x_2, and x_3 be amounts of pre-products purchased or intermediate products produced in a production network. The numerical advantage for using (9.3.4) is that it gives more freedom to the solver and supports better analysis of infeasibilities in incorrectly formulated models. If the model is formulated correctly, then in accordance with the economic situation the solution should have the property $y_1 + y_2 = x_1 + x_2 + x_3$ because it pays to produce as much as possible. If that equation is not fulfilled the optimizer "destroys" worthwhile matter which is, at the very least, strange! We should reanalyze the model as something is economically not well understood or is incorrect.

4. When solving integer problems it is sometimes advantageous to introduce additional discrete variables because they enable the user to set priorities on variables and thus influence the variable selection procedure during the B&B process. This is illustrated below. But it is not just the advantage due to priorities but also the branching strategies in a wider sense as it allows us to branch on constraints. Consider the case that we introduce an additional integer variable α which we relate to some binary variables δ_i

$$\alpha = \sum_i \delta_i. \tag{9.3.6}$$

Let us now assume that the LP relaxation provides a fractional value $\alpha = 4.3$. Branching on α is then equivalent to branching on the inequalities

$$\sum_i \delta_i \leq 4 \quad \text{or} \quad \sum_i \delta_i \geq 5, \tag{9.3.7}$$

which can lead to a great improvement.

5. Using priorities on variables to branch on: consider the case where a chemical company wants to build a large chemical reactor in either Korea or Brazil. The reactor is either a batch reactor or a continuous reactor. One might introduce binary variables, δ_{lt}, such that

$$\delta_{lt} = \begin{cases} 1 \text{ , if a reactor of type } t \text{ is built at location } l \\ 0 \text{ , otherwise.} \end{cases} \tag{9.3.8}$$

However, let as ask what is the more important decision — the reactor type or the location? Because Korea and Brazil are rather dissimilar countries with different industrial infrastructure, markets, and qualities of workmanship the decision related to the country is the more important one. Thus, in order to control the branching process it would be a great advantage to introduce the binary variable

$$\alpha_l = \begin{cases} 1 \text{ , if a reactor is built at all at location } l \\ 0 \text{ , otherwise.} \end{cases} \tag{9.3.9}$$

This allows us to force the B&B algorithm to follow our reasoning: take the most important decision first, i.e., if it comes to variable selection choose α_l first. The variables are related by

$$\delta_{lt} \le \alpha_l \quad , \quad \forall \{lt\} \tag{9.3.10}$$

and

$$\alpha_l \le \sum_t \delta_{lt} \quad , \quad \forall l. \tag{9.3.11}$$

Note that if we decide that $\alpha_l = 0$, then for the binary variables δ_{lt} we have $\delta_{lt} = 0$ for all t. If we decide $\alpha_l = 1$, then we ask for

$$\sum_t \delta_{lt} \ge 1. \tag{9.3.12}$$

6. The relaxation of equations containing integer and continuous variables into inequalities can greatly improve the B&B algorithm. Let us illustrate this approach by the following example stemming from the chemical industry. Several units u can produce a set of products p in batches of size B_{up}. Thus we have the equations

$$B_{up}\beta_{upt} = x_{upt} \quad , \quad \forall u, p, t \mid B_{up} > 0, \tag{9.3.13}$$

which link the number of batches β_{upt} (typical values are between 0 and 20) with the total amounts x_{upt} produced in time period t. The production variables are further subject to capacity constraints, availability constraints, production recipe constraints, etc. and appear in the objective function. The B&B did not find any feasible integer solution within 4 h or within 20,000 nodes. First, let us try to understand why that is so. The economic driving forces in the model are the demands; typical values are a few hundred tons. The demands are satisfied by appropriate values of the production variables x_{upt}. Usually, in an optimal solution the variables x_{upt} take some continuous values. Thus, in the LP relaxation and in most nodes of the B&B tree β_{upt} will be fractional. Generating two subproblems, i.e., adding the bounds $\beta_{upt} \le \lfloor \beta_{upt} \rfloor$ or $\beta_{upt} \ge \lceil \beta_{upt} \rceil$, changed the values $x_{u'p't}$ on other units and for other products and produced new fractional values $\beta_{u'p't}$ in the solution of the subproblem. The solution to the problem is to give up the Eq. (9.3.13) and to replace them by the inequalities

$$B_{up}\beta_{upt} \ge x_{upt} \quad , \quad \forall u, p, t \mid B_{up} > 0 \tag{9.3.14}$$

and to advise the B&B algorithm to first investigate the node $\beta_{upt} \ge \lceil \beta_{upt} \rceil$. Now it is easy for the B&B algorithm to find feasible integer solution. If an LP relaxation produces a solution with fractional value of β_{upt} the subproblem has the same solution but with batch size $\lceil \beta_{upt} \rceil$. In order to avoid having arbitrary large value of β_{upt} these variables are slightly penalized in the objective (typical value of the objective function are above 10^8 £, the penalty term is of the order 10^3 £). Having in mind that the capacity constraints are applied to the production variables x_{upt}, one might argue that batch sizes which cover more than what is produced might violate the production capacity. This is true and therefore the inequality

$$B_{up}\beta_{upt} - x_{upt} \le \varepsilon B_{up} \quad , \quad \forall u, p, t \mid B_{up} > 0 \tag{9.3.15}$$

was added, reducing the violation to a few percent ($0.01 \le \varepsilon \le 0.1$) of the batch sizes involved. Now the B&B produced good solutions after 200 or 300 nodes in less than 5 min. Inspecting the solution showed that only four or five out of a few hundred of the original equations (9.3.13) were slightly violated. Furthermore, the specific violations coincided with the uncertainty in the capacities specified.

7. In the old days of integer optimization, an integer variable α had sometimes been represented by binary variables δ_k exploiting the expression

$$\alpha = \sum_{k=0}^{n} 2^k \delta_k .$$

This is not necessary anymore — and it is not recommended! However, in terms of presolving, MILP solvers usually perform better on binary variables than on integer variables. One may keep this in mind, if the modeling process really leaves a choice between binary and integer variables.

9.4 Choice of Branch in Integer Programming

Early in 1977, Beale (1977,[55]) gives a detailed description of branching methodologies. Linderoth & Savelsbergh (1999,[368]) provide a computational study of search strategies for mixed integer programming. In commercial software, several methods are normally provided for controlling the branching strategy; cf. Achterberg et al. (2005,[3]). These include user-specified priorities for variable selection, forced branching directions (up/down), search strategy (BFS/DFS), bounding pruning, pseudo-costs, and control of the cut-off used for pruning the tree. For problems with symmetry structure, *orbital branching* by Ostrowski et al. (2011,[430]) is very important. This branching method is based on computing groups of variables that are equivalent *w.r.t.* the symmetry remaining in the problem after branching, including symmetry that is not present at the root node. These groups of equivalent variables, called orbits, are used to create a valid partitioning of the feasible region that significantly reduces the effects of symmetry while still allowing a flexible branching rule.

If it is expected that branching on a variable will cause a large degradation in the objective function, that variable is defined to be important. It may be an advantage to give relatively important variables a high priority. They are then chosen for branching at an early stage with the objective of reducing the overall work: these branches require more effort than others and should therefore be done a few times at the top of the B&B tree rather than many times at the bottom of it. Pseudo-costs are unit rates of degradation used to estimate the effect of imposing integrality conditions; after the root node evaluation that after root node they need to be initialized which requires strong branching. If estimated integer solutions are consistently biased it may be advantageous to alter these. The ability to force the branching direction is probably of limited use in codes with a good default branching strategy.

9.4.1 Control of the Objective Function Cut-Off

Control of the objective cut-off is one area where the user can most easily contribute to the B&B search. If all variables appearing in the objective function are integer, it is often possible to derive a minimum separation of distinct integer feasible objective values, i.e., a number α such that any integer feasible objective value better than the current one is better by at least α. Moreover, it is often the case that users will be content with an integer feasible solution that they know to be within α of the optimal one, e.g., if the objective function values are expected to be of the order of millions of USD, it probably can be accepted, to set $\alpha = 100$, resulting in a maximal loss of the resolution 100 USD. The value α which is called `MIPADDCUTOFF` in `Xpress-Mosel` or `CHEAT` in GAMS can then be added to the value of the current best integer solution to give a sharper cut-off value. This may greatly speed the

B&B process in solving pure integer models and also those in which the integer component is small. In the latter case, the B&B tree often "fans out" and the algorithm would otherwise waste time searching for alternative integer solutions very close to the current one.

9.4.2 Branching Control

When solving MILP problems it is important for the user to provide as much information as possible that may contribute to the control of the process.[3] There are two main ways in which the B&B algorithm may be directed:

1. controlling which currently unsatisfied global entity (binary variable, integer variable, special ordered set member, partial integer, or semi-continuous variable, etc.) is chosen for branching, and the direction to branch first;
2. choosing which node to tackle and controlling how nodes are cut.

9.4.2.1 Entity Choice

There are a number of ways of controlling the entity choice. Priorities can be assigned to entities, so that those with the highest priority are chosen first. It is usually better to branch on variables representing major decisions rather than those representing consequences of such decisions. For example, as in example 5 in Sect. 9.3, a major decision may be the size of a facility and minor decisions based on it may be the location of components within that facility. If branching is performed first on variables given high priorities (i.e., major decisions) then these variables will be dealt with at the top of a B&B tree rather than at the bottom of the tree. As the effort required to optimize the LP problems created by branching on a major variable is usually high then it is better to do this only a few times at the top of the tree rather than many times at the bottom of the tree.

A good strategy is to go depth-first through the B&B tree to obtain a good integer solution as quickly as possible. This approach can help eliminate large parts of the tree and therefore speed up the integer search.

9.4.2.2 Choice of Branch or Node

Most integer programming solvers use a depth-first strategy until a first integer solution is found. If a node is cut-off or gives rise to an infeasible problem, then its brother (node at the same level on the other half of the branch) is explored next.

[3]This was still true in the 1990s. Nowadays, 2020, branching control, pseudo-costs, etc. are calculated once and updated automatically — there is nothing a user needs to "specify" anymore.

Where both descendants are fathomed (i.e., established that they are each in one of the three categories: infeasible; at a solution value poorer than the current best; or a better integer solution than that found so far), the node with the best estimate is chosen from all active nodes. Once an integer solution is found the Forrest-Hirst-Tomlin (1974,[196]) criterion is used. This chooses the node with the highest value of

$$\frac{e - z^{IP}}{z - e}, \tag{9.4.1}$$

where z is the objective value at the node, e is the estimate of the best integer solution that would follow from that node and z^{IP} is the current best integer solution. If a good estimate of the optimal integer solution is available in advance from, for example, the solution of a similar model, this may be used in place of z^{IP} before the first integer solution is found.

An estimate of the best integer solution that could follow from each node is made by using *pseudo-costs* [55] which give the unit rates of degradation of the reduced costs d_j for altering variable values. If these estimates are consistently biased it is advisable to alter the pseudo-costs to remedy this. Different up and down pseudo-costs can also be specified, where the up pseudo cost is the estimated degradation per unit movement associated with increasing a variable (e.g., 0.5 to 1) and the down pseudo cost is the corresponding estimate associated with decreasing a variable (e.g., 0.5 to 0). This is useful when binary variables are used to model start-up costs if the upper bound on the continuous variable, that is restricted by the binary variable, is fairly loose. In the LP relaxation they may take very small values and the solver is biased to branch down on them. Raising the down pseudo-costs rectifies this.

9.4.3 Priorities

There is a priority value associated with each global entity. The lower the number, the more likely the entity is to be selected for branching; the higher the number, the less likely. For instance, a binary variable representing whether a project should be funded might usefully be given a low priority number, whereas a variable representing some finer detail of the model may be given a high value. In `XPRESS-MP` the `DIRECTIVES` command is used to facilitate priority setting and branching choice. By default `XPRESS-MP` will explore the branch estimated to yield the best integer solution from each node irrespective of whether this forces the global entity up or down.

9.4.4 Branching on Special Ordered Sets

The most useful high-level branching constructs are special ordered sets [see Sect. 6.8] and semi-continuous variables discussed in Sect. 9.4.5. We saw previously that SOS1 are sets of variables in which, at most, one can be non-zero. They are particularly effective if the set members are ordered in some way. For example, they might represent the size of pipes to use in a pipeline, as introduced in Sect. 6.8.1. Instead of branching on each set member individually, bounds can be imposed on the size of pipe, and branching can take place on the entire set.

When the B&B algorithm operates on special ordered sets of types 1 or 2 the algorithm handles these entities in a special way. During the stages of the B&B algorithm we may find that the current LP relaxation solution at a node contains set variables at non-zero values in infeasible combinations. It is clearly desirable to exclude such combinations and this is done by branching. The approach used is to reduce the "distance" between the non-zeros by flagging subsets of variables in a method analogous to varying upper and lower bounds on integer variables to try to reduce the difference between bounds. If a set has elements $\lambda_1, \lambda_2, \ldots, \lambda_n$ and the LP relaxation solution suggests that more than one element is non-zero (or more than two in the case of SOS2, or two non-adjacent elements) then a pair of adjacent variables λ_r, λ_{r+1} is chosen to initiate the branching in the SOS1. It is actually a reference row value that is used to branch on:

either all variables λ_i with $X_i < R$ are set to zero
or all variables λ_i with $X_i \geq R$ are set to zero.

Let us use the example in Sect. 6.8.1 to illustrate the numerical advantage associated with SOS1. We had n different capacities C_i ordered according to increasing index [see Fig. 6.2]. To pick out the right capacity we introduced n binary variables δ_i such that

$$\delta_i := \begin{cases} 1 \text{ , if size } C_i \text{ is selected for the pipeline} \\ 0 \text{ , otherwise.} \end{cases} \tag{9.4.2}$$

These binary variables formed an SOS1 and therefore it was not necessary to declare them as binary variables any more. We ensured that exactly one capacity size was chosen, i.e.,

$$\sum_{i=1}^{n} \delta_i = 1. \tag{9.4.3}$$

The actual size, c, which is chosen was computed from

$$c = \sum_{i=1}^{n} C_i \delta_i . \tag{9.4.4}$$

Equation (9.4.4) served as the reference row.

In order to see what happens let us assume $n = 4, C_i = i$ and that the LP relaxation produced the following solution related to the set variables:

$$(\delta_1, \delta_2, \delta_3, \delta_4) = (0.47, 0.0, 0.0, 0.53) \tag{9.4.5}$$

and thus by evaluating the reference row (9.4.4)

$$c = 1 \cdot 0.47 + 4 \cdot 0.53 = 2.59. \tag{9.4.6}$$

Note that the pipe size 2.59 is what the model prefers to have, which means that we probably will have size 2 or size 3. Ordinary branching would now branch either on δ_1 or δ_4 because these are not integral. Let us assume for the moment that the variable selection will choose δ_1. The branching direction $\delta_1 \geq 1$ will give $c = 1$ which is far from the relaxation. The branching direction $\delta_1 \leq 0$ would probably make δ_2 or δ_3 fractional. If we branch on δ_4 the situation is similar. So progress is only poor. However, branching for SOS1 uses the value computed for c to separate the indices and to generate two subproblems. In this example with $c = 2.59$ it detects

$$C_2 \leq c \leq C_3, \tag{9.4.7}$$

which leads to the following two subproblems

$$\delta_1 + \delta_2 = 0 \quad \text{or} \quad \delta_3 + \delta_4 = 0. \tag{9.4.8}$$

We could also read this is branching on $c \geq C_3$ and $c \leq C_2$. In the first case there is a great chance that the next LP problem will yield $\delta_3 = 1$ while in the second case we will probably get $\delta_2 = 1$. Now we are able to understand why the order plays an important role in SOS1: The order is used to separate all indices in two sets of indices. The stronger the order, the better the separation works.

Most commercial software packages have SOS1 and SOS2 implemented so the branching described above will all happen automatically and no user intervention is required. Note, however, that the user can put branching priorities [see Sect. 9.4.3] on sets as well as individual entities.

9.4.5 Branching on Semi-Continuous and Partial Integer Variables

Semi-continuous variables are permitted to take the value zero or anything at least as large as unity and no larger than some upper bound U, i.e.,

$$\sigma = 0 \quad \vee \quad 1 \leq \sigma \leq U. \tag{9.4.9}$$

Branching can take place on these variables directly rather than indirectly on an associated binary variable. Consider the equivalent formulation of (9.4.9) using a binary variable δ and continuous variable x

$$\delta \leq x \leq U\delta. \tag{9.4.10}$$

It is easy to see that the implication table

$$\begin{aligned} \delta = 0 \Rightarrow &\quad x = 0 \\ \delta = 1 \Rightarrow &\ 1 \le x \le U \\ x = 0 \Rightarrow &\quad \delta = 0 \\ x = 1 \Rightarrow &\quad \delta = 1 \end{aligned} \tag{9.4.11}$$

holds which shows that (9.4.10) is indeed an equivalent formulation forcing x to behave as a semi-continuous variable.

Let us now investigate how this formulation and the binary variable δ behave while solving the LP relaxation. In most models δ would appear in the objective function with some coefficient representing cost while x enters positively representing a quantity being produced or sold. Therefore, δ will adjust itself to the smallest value possible, i.e.,

$$0 < \delta < 1 \quad , \quad \delta = \frac{x}{U}. \tag{9.4.12}$$

The consequence is that branching has to be applied to each binary variable representing a semi-continuous variable. Now consider the original relation (9.4.9). The LP solver would just see the relaxed inequalities

$$0 < \sigma < U. \tag{9.4.13}$$

The B&B logic then works as follows: if $\sigma \ge 1$ or $\sigma = 0$, then nothing is done. Only in the case $0 < \sigma < 1$ does branching become necessary. Thus, efficient implementations of semi-continuous variables leads to a smaller number of branches. Some tests performed at BASF involving about 2,500 semi-continuous variables and using (9.4.10) and (9.4.9) alternatively reduced the CPU times needed to solve a production planning problem by a factor 6.

A generalization of semi-continuous variables leads to partial integer variables already briefly mentioned on page 38. Partial integers might arise, for example, in the following context. A motor manufacturer wishes to allocate cars to its national distributors throughout Europe. Naturally, it has to send each national distributor an integral number of cars, but rounding a solution value of, say, 5,192.3 cars to Germany down to 5,192 is acceptable. However, rounding down the 16.4 cars allocated to Andorra might be less satisfactory. So the manufacturer would like to specify that values below, say, 50 must be integer, while those above 50 can be treated as real number and be rounded by hand. Partial integers allow the modeler to specify precisely this restriction.

9.5 Symmetry and Optimality

Symmetry is a problem when trying to close the gap between the upper and lower bounds, and thus proving global optimality. This is especially a problem when using deterministic global solver for solving non-convex NLP problem. A systematic treatment of symmetry in ILP and ways to reduce or to destroy symmetries are treated by Margot (2010,[385]). Here we want to focus only on techniques the modeler can apply in the sense of static constraints or inequalities for both linear and nonlinear optimization problems.

Symmetry is easiest understood in optimization problems with a geometric background, e.g., in computing minimal convex hulls as in Kallrath & Frey (2019,[315]), where one wants to reduce the symmetries: translational, rotational, and mirror symmetry. Translational symmetry can partially be reduced by fixing the coordinate center of the convex hull. Alternatively, one can fix a specific object to the origin of the coordinate system. Rotational symmetry can be destroyed by fixing two objects, for instance, to the positive x-axis. When placing congruent objects, a sequence metric can be enforced.

With commercial MILP solvers, these symmetry destroying equalities or inequalities are usually useful and help to close or at least to reduce the gap in shorter time. However, with deterministic global solvers it is always a trade-off. Without symmetry reducing techniques, i.e., with less constraints, they find better initial solutions in shorter time. Symmetry reducing techniques only pay out when one wants to close the gap, which is usually possible only for smaller NLP or MINLP problems.

9.6 Summary

In this chapter we have emphasized the need for user intervention in many instances of model solving. Although we might prefer to treat the solution process of LP or IP problems as a "black box," we find that this might not be appropriate in all cases. Certain models prove difficult and/or slow to solve and we find that attention to some details by the user may improve solution time tremendously. It can help a solver a lot if the user can provide an input feasible solution — this is also called MIP warmstart and is one of the key elements of the polylithic modeling and solution approached discussed in Chap. 14.

When we are considering models that need frequent resolving, it will be important to be able to perform these repeated runs in an efficient way. Thus the advantages to be gained from good use of presolve, scaling and branching choice (in IP) are important to the modeler and client. After reading this chapter the modeler should be aware of:

- the use of preprocessing techniques;
- the benefit from disaggregation of constraints;
- the technique of coefficient reduction in $\{0,1\}$ constraints;
- the identification of special constraints, such as cliques and covers;
- the benefits of well scaled coefficient matrices;
- some model features which should be avoided;

- the benefits of using SOS1 and SOS2 as well as semi-continuous variables;
- the need for good branching choices to be made in the B&B algorithm; and
- the positive impact of cutting planes and heuristics solvers offer.

9.7 Exercises

1. This exercise consists of two tasks:

   (i) Find all covers and cliques for the IP problem with binary variables $\delta_1, \ldots, \delta_5$

$$\max \quad 7\delta_1 + 2\delta_2 + 8\delta_3 + 4\delta_4 + 7\delta_5 \tag{9.7.1}$$

   subject to

$$\begin{aligned} 3\delta_1 + 4\delta_2 + 5\delta_3 + 8\delta_4 &\le 12 \\ 9\delta_1 + 5\delta_2 + 10\delta_4 + 7\delta_5 &\le 14 \\ 8\delta_1 + 5\delta_2 &\le 12 \qquad (9.7.2) \\ \delta_1, \delta_2, \delta_3, \delta_4, \delta_5 &\in \{0, 1\}. \end{aligned}$$

   (ii) Show if any of the covers or cliques you have found may be lifted.

2. Remove all redundant constraints and variables from the following LP problem and check if any bounds on the continuous variables, x_1, x_2, x_3, x_4, may be strengthened.

$$\max \quad 2x_1 + x_2 + 3x_3 + 5x_4 \tag{9.7.3}$$

   subject to

$$\begin{aligned} 2x_1 + 2x_2 + x_3 + x_4 &\le 20 \qquad (9.7.4) \\ 3x_2 + 2x_3 \qquad &\le 18 \\ 3x_1 + 4x_2 + 2x_3 + 5x_4 &\le 35 \\ x_1, x_4 &\le 8. \qquad (9.7.5) \end{aligned}$$

3. Show how the following constraint in binary variables, $\delta_1, \ldots, \delta_6$

$$10\delta_1 + 8\delta_2 + 8\delta_3 + 7\delta_4 + 6\delta_5 + 6\delta_6 \le 19 \tag{9.7.6}$$

   may be reduced to the form

$$3\delta_1 + 3\delta_2 + 3\delta_3 + 2\delta_4 + 2\delta_5 + 2\delta_6 \le 6. \tag{9.7.7}$$

Chapter 10
How Optimization Is Used in Practice: Case Studies in Integer Programming

This chapter contains several case studies with an industrial background which involve mixed integer programming techniques. We discuss real-world problems of increasing size and complexity. The first group of case studies considers a contract allocation problem, metal ingot production, and a project planning problem. This is followed by a more extensive scheduling problem in the carton industry. The next problem formulates and discusses a worldwide production planning problem in the chemical industry. Then, a complex scheduling problem with personnel resource constraints (again coming from the chemical industry) is presented. Finally, we see mathematical programming in use to optimize a telecommunication network.

10.1 What Can be Learned from Real-World Problems?

Many practical problems leading to MILP formulations need great care in model formulation. In contrast with ordinary LPs, effective solution depends critically upon good model formulation, the use of high-level branching constructs, and control of the B&B strategy. Good formulations are those whose LP relaxation is as close as possible to the MILP — or, to be precise, those whose LP relaxation has a feasible region which is close to the convex hull of the MILP's feasible sets. In practice, this means, for example, that upper bounds on variables should be as small as possible. Formulating models in this fashion is still largely the responsibility of the modeler who has to combine and to apply many of the theoretical concepts we learned in Chap. 9. The more modeling experience the modeler has the better.

Supplementary Information The online version contains supplementary material available at https://doi.org/10.1007/978-3-030-73237-0_10.

J. Kallrath, *Business Optimization Using Mathematical Programming*,
International Series in Operations Research & Management Science 307,
https://doi.org/10.1007/978-3-030-73237-0_10

Apart from own modeling it also helps taking advantage of how other modelers solved their problems. Thus, the real-world problems presented may serve to create a feeling for what is a good or what is a poor model. Further, an important point is that if an MILP model formulation causes great trouble, a reformulation could improve the situation significantly.

10.2 Three Instructive Solved Real-World Problems

The three case studies below originally appeared in Ashford & Daniel (1992,[37]) "Some Lessons in Solving Practical Integer Programs." They appear by kind permission of the publishers of the Journal of the Operational Research Society, Macmillan (Stockton Press, UK). All use real data and have recently been, or are currently being, used in practice, so their background has been disguised to some extent.

10.2.1 Contract Allocation

A public utility, which is divided into six regional districts, wishes to allocate ten power generation contracts to its regions as cheaply as possible. The cost per unit of power generated by each region for each contract is known. If part of a contract is allocated to a region, then it must be at least as big as a certain minimum size. For reliability reasons, no contract may be placed exclusively with only one district. Each district has a limited power generation capacity.

Letting r index the districts or regions and c the contracts, this problem may be formulated as an integer program as follows. Let the decision variables be $x_{rc} \geq 0$, the power generated by each region for each contract.

The problem data are:

$$\begin{aligned} A_r &: \text{the total power available from each district} \\ C_{rc} &: \text{the unit cost of generating power} \\ L_r &: \text{the minimum power level available from each} \\ &\quad \text{district } r \text{ for each contract it has to supply} \\ R_c &: \text{the total power required for each contract} \end{aligned} \tag{10.2.1}$$

The objective is to minimize the total cost:

$$\sum_r \sum_c C_{rc} x_{rc} \tag{10.2.2}$$

subject to the demand requirements of each contract:

$$\sum_r x_{rc} \geq R_c \quad , \quad \forall c, \tag{10.2.3}$$

the capacity limitations of each district:

$$\sum_c x_{rc} \leq A_r \quad , \quad \forall r, \tag{10.2.4}$$

and the non-negativity conditions:

$$x_{rc} \geq 0 \quad , \quad \forall \{rc\}. \tag{10.2.5}$$

Thus far it is a simple transportation problem. One way of ensuring that the minimum contract contributions are met is to introduce binary variables δ_{rc} forced to be unity if $x_{rc} > 0$ by constraints:

$$x_{rc} - M\delta_{rc} \leq 0 \quad , \quad \forall \{rc\}, \tag{10.2.6}$$

where M is some suitable large positive number. Since the availabilities ranged from 10 to 60 units, a suitable value for M might be 100. In this case the modeler originally used

$$x_{rc} - M\delta_{rc} \geq L_r - M \quad , \quad \forall \{rc\} \tag{10.2.7}$$

to ensure that the minimum contract contributions are met and

$$\sum_r \delta_{rc} \geq 2 \quad , \quad \forall c \tag{10.2.8}$$

to ensure that at least two districts fulfill each contract.

The application of this model was to a small problem with ten contracts and six districts. But this naive formulation is so bad that the B&B search remained unfinished after 5000 nodes.

It is more useful to replace M by a better (i.e., smaller) upper bound [see Sect. 9.1.1.2, and the discussion near (9.1.17)] on each x_{ij}, an obvious one being A_i. We might even do better by considering $\min\{A_r, R_c\}$. Thus, (10.2.6) becomes

$$x_{rc} - \min\{A_r, R_c\}\delta_{rc} \leq 0 \quad , \quad \forall \{rc\}. \tag{10.2.9}$$

Moreover, (10.2.7) is far better reformulated without upper bounds on the x_{ij} as:

$$x_{rc} - L_r\delta_{rc} \geq 0 \quad , \quad \forall \{rc\}. \tag{10.2.10}$$

With these modifications the B&B search is completed in only eight nodes.

However, it turns out that the key reason for the rapid completion of the search is constraint (10.2.8): that at least two districts are involved in each contract. Without it, the B&B search has much greater freedom in branching choice and although the minimum cost is found after only 24 nodes, it takes a further 1417 nodes (and about 20 s on a Pentium) to complete the search and prove optimality. Under these circumstances, the solution process can be greatly helped by using semi-continuous variables in the formulation, replacing the binary δ_{rc} variables, which enable better estimates of the value of the best integer solution (if any) which would follow from each integer infeasible node of the B&B tree. If σ_{rc} are the semi-continuous variables associated with x_{rc}, then (10.2.10) is replaced by:

$$x_{rc} - L_r \sigma_{rc} = 0 \quad , \quad \forall \{rc\}, \tag{10.2.11}$$

and constraints (10.2.6) are dropped. We would then only need to branch on σ_{rc} if $0 < x_{rc} < L_r$, whereas previously δ_{rc} could be fractional when x_{rc} is taking a legal value. Notice that we would still have required the binary variables δ_{rc} if requirement (10.2.8) were present.

The semi-continuous variable formulation is much easier to solve, the B&B search being completed in only 132 nodes (and 2 s on a Pentium).

So summarizing what we have learned: the problem shows the merits of least upper bounds and the efficiency of semi-continuous variables.

10.2.2 Metal Ingot Production

Metal is smelted in vessels of a fixed size and then cast into ingots of different weights. The problem is to determine how to split each vessel into ingots so as to satisfy given requirements for ingots of each weight with the smallest amount of wastage. Unwanted ingots produced have to be treated as waste again.

If i indexes the ingot type and c indexes the alternative splitting of a vessel or the combinations of ingots that may be cast from a single vessel, then we may define the problem data as follows:

$$\begin{aligned} &M_i \ : \text{weight of ingot of type } i \\ &N_{ic} : \text{the number of ingots of weight } i \text{ cast from combination } c \\ &R_i \ : \text{the requirement for ingots of each type or weight} \\ &W_c \ : \text{wastage (units of weight) from a vessel cast in combination } c \end{aligned} \tag{10.2.12}$$

The problem we have to model is very similar to the trimloss problems discussed in Sect. 4.1.2 (Case B). So we may benefit from that section. Denoting the number of vessels cast to each combination by α_c the problem may now be formulated: first we have to meet the requirements for each weight of ingot

$$\sum_c N_{ic} \alpha_c \geq R_c \quad , \quad \forall i, \tag{10.2.13}$$

while minimizing the wastage:

$$\sum_c W_c \alpha_c + \sum_i \sum_c M_i \left(N_{ic} - R_i \right). \tag{10.2.14}$$

The first term is the wastage associated with each combination and the second is that of unwanted ingots. Notice that after some algebra this is

$$\sum_c \left\{ W_c + \sum_i M_i N_{ic} \right\} \cdot \alpha_c - \sum_i M_i R_i . \tag{10.2.15}$$

The second term is a constant and can be ignored, while the expression within parentheses in the first is simply the total weight of each vessel, which does not depend on c at all. So the objective function may be taken to be just

$$\min \quad \sum_c \alpha_c . \tag{10.2.16}$$

One particular problem had 18 different weights of ingot and 1345 possible ways of casting each vessel. Unaided, the B&B search was unfinished after 8000 nodes. However, the structure of the objective (10.2.16) can be exploited by observing that once an integer feasible solution has been found any better solution must have an objective at least 1 unit smaller. Thus an addcut $\alpha = -0.999$ was specified. Now the B&B search took only 695 nodes to find the optimal solution and prove optimality.

Finally, let us summarize what we have learned from this real-world problem. Our problem of casting metal ingots is approached via the trimloss model with a very large number of integer variables. Specification of a suitable addcut proved the only way of curtailing the explosive growth of the B&B tree.

10.2.3 Project Planning

A number, N^P, of projects are to be scheduled over the coming few, say T, months. Each project requires varying numbers of personnel over its duration and yields a known return. For instance, project 1 uses up 3 people in the month it starts, 4 in the second month, and 2 in its last month. The company has a limited number of personnel available for the projects and wishes to schedule the projects so as to achieve the greatest benefit. The benefit from a project only starts to accrue when the project has been completed, and then it accrues at a rate of R_p per month for project p, up to the end of the time horizon. The scheduling is on a month by month basis, so projects may be considered to start at the beginning of one month and terminate at the end of one month. The problem is to decide when to start each project, subject to not using more people than are available in any month. We use the following indices

$$\begin{aligned} m &= 1, ..., M_p &&: \text{set of months within project} \\ p &= 1, ..., N^P &&: \text{set of projects} \\ t &= 1, ..., T &&: \text{set of months in the time horizon} \end{aligned} \tag{10.2.17}$$

The problem data may be defined by:

$$\begin{array}{ll} A_t & : \text{personnel available for all projects in month } t \\ D_p & : \text{duration of project } p \text{ in months} \\ P_{pm} & : \text{personnel required by project } p \text{ in month } m \text{ of its operation} \\ R_p & : \text{return per month when project } p \text{ finishes.} \end{array} \qquad (10.2.18)$$

The decisions may be represented by the binary variables δ_{pt},

$$\delta_{pt} := \begin{cases} 1 \text{ , if project } p \text{ starts in month } t \\ 0 \text{ , otherwise} \end{cases} \qquad (10.2.19)$$

and integer variables $s_p = 0, 1, 2, \ldots$ representing the start of project p. From (10.2.23) it follows that all s_p will take integral values automatically. So they need not be declared as integer variables.

To reflect the fact that benefit from a project accrues at its completion, the company decided to credit each project started by its return, R_p, multiplied by the number of months from when it was completed to the end of the horizon. If the project is started in period t it finishes in month $t + D_p - 1$, so we get the benefit R_p for $T - D_p - t + 1$ months. To set up the objective function we must consider all the possible projects, and all starting months that let the project finish before the end of the planning period. For the project to complete it must start no later than month $T - D_p$. Thus, the total benefit to the company is

$$\max \sum_{p=1}^{N^P} \sum_{t=1}^{T-D_p+1} \left[R_p (T - D_p - t + 1) \right] \delta_{pt}. \qquad (10.2.20)$$

A project can only be done at most once, so

$$\sum_{t=1}^{T} \delta_{pt} \leq 1 \quad , \quad \forall p. \qquad (10.2.21)$$

The number of personnel required does not exceed those available, so

$$\sum_{p=1}^{N^P} \sum_{u=\max\{1,t-D_p+1\}}^{t} P_{p,t-u+1} \delta_{pu} \leq A_t \quad , \quad \forall t. \qquad (10.2.22)$$

Notice that in month t, project p will require $P_{p,t-u+1}$ people if it has begun in month u. Furthermore, it will not require any personnel if it was begun more than $t - D_p$ months ago. The start month of a project is given by

$$s_p = \sum_{t=1}^{T-D_p+1} t\delta_{pt} \quad , \quad \forall p. \qquad (10.2.23)$$

If the variables are declared to be binary by

$$\delta_{pt} \in \{0, 1\} \quad , \quad \forall p \quad , \quad t = 1, ..., T - D_p + 1, \tag{10.2.24}$$

the problem can now be solved as an integer program. MCOL provides a problem named *projschd* which contains a small example of our scheduling problem.

In this particular application, the company's planning period was initially 50 months and there were 17 projects to schedule. This gave a problem with 615 binary variables, which was solved after 742 nodes of the B&B search. However, the problem may be formulated with SOS1 rather than binary variables. It is convenient to define $\delta_{p,T+1}$ to be a binary variable taking the value unity if and only if project p is not begun at all. Then

$$S_p = \{\delta_{p1}, \delta_{p2}, ..., \delta_{p,T-d_j+1}, \delta_{p,T+1}\} \tag{10.2.25}$$

forms an SOS1 whose members are naturally ordered by month. This may be specified by the single non-computational reference row:

$$\sum_{p=1}^{N^P} \sum_{t=1}^{T-D_p+1} t\delta_{pt} + (T+1)\delta_{p,T+1}. \tag{10.2.26}$$

It remains to add the term $\delta_{p,T+1}$ to the convexity row (10.2.21) above, which becomes an equality. The binary conditions can now be dropped.

The optimal solution was now found and proven in only 187 nodes of the B&B search. The company wished to explore how the project schedule would change if 5 years were available to complete the projects, so the model was re-run with a time horizon of 60. In this case there were 785 δ_{pt} variables. The binary variable formulation was solved after 8152 nodes of the B&B search, whereas the SOS1 formulation was solved in only 1398 nodes. Note that although the ordering between the set members is weak, the SOS1 formulation is still very effective on a large example of the model.

10.2.4 Conclusions

Three practical integer programming models have been discussed with a view to elucidating what makes formulations good, illustrating how poor formulations can be improved and suggesting appropriate branching strategies. Each model is of fairly modest size, but none is particularly easy to solve. Some lessons that may be drawn from this exercise are as follows. Poor formulations such as the initial contract allocation model are catastrophic in the sense that they may render a solution, for practical purposes, impossible. Restrictions on the choice of integer solution, such as the lower bound on the number of districts supplying contracts in the first model, and a slightly lower number of time periods in the project planning model, are a great help in solving the whole integer program. High-level branching constructs, such as semi-continuous variables and special ordered sets, should always be considered. In the contracting problem, the solution time was reduced by a factor of ten by using semi-continuous variables, and an even more dramatic reduction in solution time was achieved by using Sl sets in the project planning models. Indeed, the larger project planning model, in 1997, could not be solved without SOS1. However, the ordering between set members must be fairly strong for them to be effective: they should not be used for every set of mutually exclusive decisions.

Integer programs do require more care in both their formulation and solution than linear programs of a similar size. But this is no reason not to attempt using them and, even if exploratory runs fail to find the optimum quickly, it is possible that some further work will yield an improved formulation or solution strategy that makes the models effective.

10.3 A Case Study in Production Scheduling

The following case study describes an example of problem formulation requiring the modeling of awkward logic and the use of special ordered sets. It is taken from Ashford & Daniel (1991,[36]) "Practical Aspects of Mathematical Programming" which originally appeared in Operational Research Tutorial Papers 1991. The case study appears by kind permission of its publishers, the Operational Research Society, Birmingham, England.

A carton manufacturer produces a variety of cartons for the storage of liquid products by processing board through machines, known as converters, which cut, score, and print the customer's design on it, making carton "blanks" which are subsequently folded and sealed on machines known as sealers. The converters are the bottleneck of the process, there being significant slack in other parts of the operation. Customers place orders weekly, so the company has the task of scheduling a week's production each Monday with a view to:

Minimizing the number of changes of machine configuration.

Each time a different size or shape of carton is produced, the machine cutter heads need re-alignment. This setting-up process may take up to 4 shifts, during which time the machine cannot produce.

Minimizing the duration of the ink changes in the converters.

There are constraints on the total number of inks being used at any time on each machine and also on the number of inks used printing adjacent cartons. Light to dark ink changes are more rapid than dark to light ones. Ink changes take between 15 min and 1 h.

Minimizing the number of overtime shifts worked.

The workforce is highly unionized and overtime is expensive.

Minimizing shortfalls in meeting customer orders.

The scheduling task was divided into three stages:

- Dividing the week's orders into groups of common size and shape and scheduling cartons on the basis of their groups to minimize the time spent changing machine configurations.
- Scheduling the orders within groups to reduce the ink change-over time.
- Scheduling orders through the sealers.

Orders for the same size and shape of carton are aggregated into groups and were scheduled by an integer programming model, balancing the cost of overtime working and configuration changes against those of tardy and short delivery. The ink and sealer scheduling was done heuristically, given the group schedules. A data management shell gave the user access to order and production data and displayed generated schedules. It manages the other modules of the system and was written in COBOL. The group scheduling MILP was generated by the `XPRESS-MP` model builder and optimized by the `XPRESS-MP` optimizer. The ink and sealer heuristics were written as macros in `LOTUS 123`.

The group scheduling model was formulated as follows. Each machine can produce up to 4 groups simultaneously, and there are a limited number of such configurations in which it may be run. There are up to 16 production shifts each week including those worked on overtime. Machine configurations remain unchanged throughout each shift.

The indices used in the model are:

$$\begin{array}{lll} m & = 1, 2, \dots, M & : \text{set of machines} \\ c & = 1, 2, \dots, C_m & : \text{set of configurations} \\ g & = 1, 2, \dots, G & : \text{set of groups of orders} \\ t & = 1, 2, \dots, T & : \text{set of shifts.} \end{array} \tag{10.3.1}$$

Note $c = 1$ means setting up and $t = 1$ means the last shift of the preceding week. The data are then:

$$\begin{array}{ll} S_t & : \text{the setting up cost (per machine, per shift)} \\ P_g & : \text{the penalty cost of not meeting an order (per carton)} \\ D_g & : \text{the demand for all items in each group} \\ O_t & \ \text{the overtime cost (per machine, per shift)} \\ H_{mcc'} & : \text{the number of shifts required to change from configuration } c \\ & \ \text{to } c' \text{ on machine } m \\ R_{mcg} & : \text{the production rate (number of cartons of group } g \text{ produced} \\ & \ \text{on machine } m \text{ in configuration } c \text{ per shift)} \\ I_m & : \text{the initial configuration of machine } m \\ OT & : \text{the set of overtime shifts} \\ SU & : \text{the set of shifts in which setting up is permitted.} \end{array} \tag{10.3.2}$$

The decision variables are: configuration, δ_{mct}

$$\delta_{mct} := \begin{cases} 1\ , \text{if machine } m \text{ is in configuration } c \text{ in shift } t \\ 0\ , \text{otherwise,} \end{cases} \tag{10.3.3}$$

and overtime production, σ_{mct}

$$\sigma_{mct} := \begin{cases} 1\ , \text{if } m \text{ produces in configuration } c \text{ in overtime shift } t \\ 0\ , \text{otherwise,} \end{cases} \tag{10.3.4}$$

and, finally, the shortfall, z_g, of order group g (in cartons).

Machines can only be idle in a producing configuration if left unattended in an overtime shift. So the production in a non-overtime shift is given by δ_{mct}.

Multi-shift machine set ups are, by custom and practice, never split across unworked overtime shifts. Most configuration changes can be completed in one or two shifts. Moreover, the production manager would only allow setting up in 10 mid-week shifts. Usually machine configurations would be changed in overtime shifts worked at the week-end.

The model minimizes the sum of the setting up, shortfall, and overtime costs:

$$\min \quad \sum_{m=1}^{M}\sum_{t=2}^{T} S_t\delta_{m1t} + \sum_{g=1}^{G} P_g z_g + \sum_{m=1}^{M}\sum_{c=2}^{C_m}\sum_{t\in OT} O_t\sigma_{mct} \tag{10.3.5}$$

subject to the following constraints:

The overtime production can only be in the current configuration for each machine, so:

$$\sigma_{mct} \le \delta_{mct} \quad , \quad \forall m \quad , \quad c = 2, 3, ..., C_m \quad , \quad t \in OT. \tag{10.3.6}$$

No machine, M, can be in two distinct producing configurations (say c' and c), u shifts apart, if it takes u or more shifts to change from one configuration to the other, i.e., $H_{mc'c} \ge u$. So:

$$\sum_{c'=2\wedge c'\neq c\wedge H_{mc'c}\ge u}^{C_m} \delta_{mc't} + \delta_{mc(t+u)} \le 1, \quad \begin{array}{l} \forall m, \quad c = 2, 3, ..., C_m \\ u = 1, 2, 3, 4, \quad t = 1, 2, ..., T-u. \end{array} \tag{10.3.7}$$

The weekly demand, less the shortfall in production cannot exceed the number of cartons produced, so:

$$\sum_{m=1}^{M}\sum_{c=2}^{C_m}\left\{\sum_{t=2\wedge t\notin OT}^{T} R_{mcg}\delta_{mct} + \sum_{t=2\wedge t\in OT}^{T} R_{mcg}\sigma_{mct}\right\} + z_g \ge D_g \quad , \quad \forall g. \tag{10.3.8}$$

The model must begin in the configuration of each machine at the end of the previous week, so:

$$\delta_{mct} := \begin{cases} 1 \text{ , if } c = I_m \\ 0 \text{ , otherwise} \end{cases} \quad , \quad \forall m. \tag{10.3.9}$$

Of course, each machine must be in exactly one configuration (including setting up) in each shift, so:

$$\sum_{c=1}^{C_m} \delta_{mct} = 1 \quad , \quad \forall m \quad , \quad t = 2, 3, ..., T \tag{10.3.10}$$

and

$$\sum_{c=1}^{C_m} \sigma_{mct} = 1 \quad , \quad \forall m \quad , \quad t \in OT. \tag{10.3.11}$$

The configuration variables are binary, so:

$$\delta_{mct} \in \{0, 1\}, \quad \forall m \quad ; \quad c = 1, 2, ..., C_m; \quad t = 2, 3, ..., T \tag{10.3.12}$$

and

$$\sigma_{mct} \in \{0, 1\}, \quad \forall m \quad ; \quad c = 1, 2, ..., C_m; \quad t \in OT. \tag{10.3.13}$$

Notice that there is no idle configuration in this formulation. This was unnecessary because there is no penalty for producing more cartons from any group than required. Simple post-processing of the solution removed any redundant production.

As each machine must be in exactly one configuration in each shift, it is possible to replace the binary variables by SOS1 variables:

$$\{\delta_{mct}, c = 1, 2, ..., C_m\} \quad , \quad \forall m \quad , \quad t = 2, 3, ..., T \tag{10.3.14}$$

and

$$\{\sigma_{mct}, c = 1, 2, ..., C_m\} \quad , \quad \forall m \quad , \quad t \in OT. \tag{10.3.15}$$

There is, however, no natural ordering between the set members, except that the setting-up configuration in which nothing is produced might be considered first. Thus a reference row of the form:

$$\sum_{m=1}^{M} \sum_{t=2}^{T} \sum_{c=1}^{C_m} c\delta_{mct} + \sum_{m=1}^{M} \sum_{t \in OT}^{T} \sum_{c=1}^{C_m} c\sigma_{mct} \tag{10.3.16}$$

could be used.

This formulation was specified to the `XPRESS-MP` model builder and is stored under the problem name *carton* in MCOL.

The model provided is in `XPRESS-MP` form, but it bears such a close resemblance to the algebraic formulation that its meaning should be relatively transparent. However, there are a number of aspects of this application which merit discussion.

Notice the separation between structure and data. This model is run and re-run unchanged but with different data files (specified after the `DISKDATA` keyword). This is convenient from an implementation standpoint and also supports model validation and testing. The separation could be made more complete by reading from file the index maxima, overtime shifts, and shifts in which set-ups are allowed.

The advantages of using a modeling language are well illustrated by this case. For instance, (10.3.7) is relatively easy to write down in algebra, but it would be unpleasant to write a C or FORTRAN program to generate the constraints in MPS notation.

The specification is self-documenting: model specification statements follow the analyst's algebra and comments can be inserted anywhere (after the "!" symbol). This facilitates model maintenance.

Some data processing can be done with the model. The `REQ` array identifies configurations that may be required and this is used to restrict the configuration variables generated. The "|" symbol may be read as "given that" and the variable, constraint, or term is only generated if the following logical expression is true. Together with the use of the factory's operational restrictions on permissible set-up shifts this constrains the choice available to the optimizer, greatly reducing the solution times. This is a characteristic of many MILP applications which may only be solvable at all when such operational restrictions are taken into account.

The actual application used slightly different (confidential) data from those given and had additional complicating constraints, such as mid-week due dates. Initially, there were 5 machines, 19 shifts, and 31 possible configurations for the machines (including setting up). A typical run had 310 binary variables in total and the B&B search was completed in only 150 nodes and 18 s on a 33 MHz 486 PC. However, this is highly data dependent and some sets of orders could take substantially longer to schedule, particularly after the acquisition of a new converter and introduction of new product groups. The SOS1 formulation requires less storage overheads during the B&B, so gave some savings in solution time, but the ordering between the set members was not strong enough to make this substantial on all runs. An addcut value, $\alpha = 0.1\%$ of the objective value, proved very effective in reducing solution time when scheduling large sets of orders.

The group scheduling model always solved within a few minutes. A complete run of the system typically takes about half an hour with the majority of the time spent in the ink and sealer scheduling heuristics. Turnaround time was considered entirely satisfactory by the client, who reckoned that the quality of schedules was superior to those generated manually. Moreover, the system is re-run whenever the week's orders change, which usually happens by the second day. Manual schedules took a day to produce and re-scheduling mid-week was virtually impossible. The relatively rapid response offered by the system was a crucial factor in its successful implementation.

Finally, let us refer to two scheduling problems connected to the pulp and paper industry. The first one, solved by Westerlund et al. (2007,[577]), is from one of the largest pulp and paper companies in Finland (Metsä Tissue group and their factory in Mänttä). The factory produces tissue paper from recycled paper as well as virgin mass (i.e., from new birch, spruce and pine fibers) if necessary. When scheduling the whole production, the tissue products (toilet paper, kitchen paper, etc.) to be produced with the tissue paper machines are initially scheduled for the coming weeks, given the incoming orders for the products. In short-term (daily) scheduling one then needs to decide how to utilize the available raw materials (the recycled mass and virgin fiber) in most cost efficient way. The different scheduling horizons are because the raw material must be prepared in a separate process and recycled mass degrades because of aging in only a few days and can thus not be stored in storages for longer time without the use of costly chemicals. The second scheduling problem is from the paper converting industry Walki Wisa in Pietarsaari (Finland). Here, a trimloss problem has been solved separately but the long and short-term scheduling problems for the utilization of all machines in the factory are given and have been solved by Roslöf (2002,[473]).

10.4 Optimal Worldwide Production Plans ⊖

10.4.1 Brief Description of the Problem

We start with a rather general description of the model properties. The model describes a multi-site production network including a variety of different products which can be produced in a plant (one plant at each site) depending on the mode chosen for the plant to operate in. All products are subject to sales requirements of customers according to predefined market demands considering several different market scenarios.

There are production sites located in different regions. It is possible to transport each product from each production facility to any other site, subject to predefined minimal transportation rates. Some of the production sites can manufacture several products, others cannot.

Within the framework of this production network the objective function (contribution margin) included selling prices and the following costs: variable production costs, change-over costs, site- and product-specific inventory costs, additional inventory costs (associated with the rental of additional inventory capacity), transportation costs, and finally costs for external purchase. Since all balancing is performed on the base of German Marks (DM), exchange rates are incorporated into the model. The optimizer should compute a production plan for running all plants to maximize contribution margin, and plans for inventory status, shipping and sales.

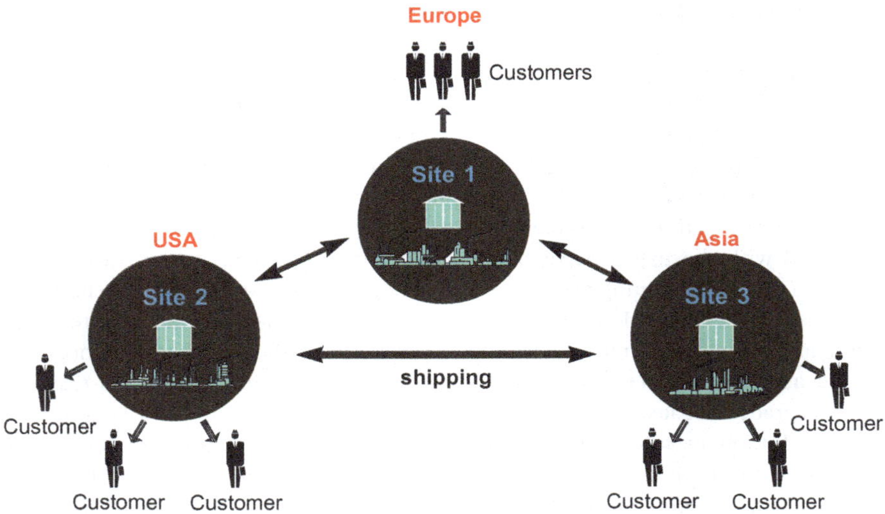

Fig. 10.1 Production network with three sites

The original work (Kallrath 1995,[300]) resulted in a production planning model for three plants in Germany, USA, and Japan [see Fig. 10.1]. Each of the plants can produce products with the quality of production being the same at all production facilities. Capacities are site-dependent, change-over times, minimal production charges and minimal plant utilization rates are site and product dependent. Production is only possible if a minimum plant utilization rate is achieved, and the production plan is only accepted if minimum production charges are respected. The model describes a scenario with product change-over times dependent on production site and minimum requirements for the quantity of product to be produced in the course of a year or a month, limited storage capacity, limited transportation capacities, minimum transport quantities, non-zero transportation times. Market demands to be expected over the year can be included in the form of demand quantities and associated probabilities for different, e.g., pessimistic and optimistic scenarios.

In the next subsection we present a model describing a production network with three sites and three products covering a year's planning horizon; in Sect. 10.4.5 a more general case is described. For reasons of confidentiality it is not possible to provide real data for this case study.

10.4.2 *Mathematical Formulation of the Model*

10.4.2.1 General Framework

Throughout this model description the following set of indices will be used:

$$
\begin{array}{lll}
i &= 1, \ldots, N_A & : \text{production sites (or markets)} \\
j &= 1, \ldots, N_P & : \text{products} \\
k &= 1, \ldots, N_K & : \text{production slices within the planning horizon} \\
m &= 1, \ldots, M_i & : \text{number of production modes at site } i \\
r &= 1, \ldots, R & : \text{production slices within a commercial} \\
 & & \text{time interval horizon} \\
t &= 1, \ldots, N_T & : \text{commercial time intervals} \\
w &= 1, \ldots, N_W & : \text{different market demand scenarios.}
\end{array}
\qquad (10.4.1)
$$

At first we will consider a scenario similar to the original work (Kallrath, 1995,[300]) with $N_A = 3$ production sites or plants i. Each of them can operate in $M_i = 3$ different modes producing $N_P = 3$ products j. In that special case we have a one-to-one correspondence between modes and products, i.e., $m = j$. As the case study in Sect. 5.1 shows, in the chemical industry we have usually the case that a plant might produce several products in a given mode. This feature leads to mode changes. We exploit the fact that we have only three modes to describe the mode changes.[1]

Each production site is also a sales point and is responsible for delivering the products to customers associated with that sales region. In this model we do not distinguish between production sites and sales points. However, in more general scenarios with, for instance, more sales points than production sites, one should keep sites and sales points logically separated from each other.

10.4.2.2 Time Discretization

We divide the entire planning horizon (1 year in our case) into $N_K = 12$ discrete production slices (1 month each).

Moreover, regarding delivery or sale, a commercial time scale of 12 periods (months) with $N_T = N_K / R$ is chosen. Usually, the production schedule has a finer resolution than the commercial plans for sales and shipping. Using two time scales, the resolution is chosen adequately for the purpose of both production planners and marketing people. If k refers to a production slice in time interval t, then the function

$$k(t, r) := R(t - 1) + r \qquad (10.4.2)$$

[1] In Sect. 10.4.5 we present a reformulated version of the model which is capable of handling an arbitrary number of modes and changes between them.

gives the absolute number $k(t,r)$ of that production slice referenced by t and r and connects both time scales. For abbreviation, if we want to cover the whole planning horizon, we use k rather than $k(t,r)$. Note that $N_K = k(N_T, R)$. If the production time scale and the commercial time scale are identical, we have $R = 1$, and $k = k(t,r) := t$. This feature can also be used to describe a production plan covering 12 months which, for production, considers weeks during the first 3 months and considers months for the rest of the planning horizon. In that case we would use R_t instead of R with

$$R_t = \begin{cases} 4 \,, & 1 \le t \le 3 \\ 1 \,, & 4 \le t \le 12. \end{cases} \tag{10.4.3}$$

10.4.2.3 Including Several Market Demand Scenarios

The index w is used to indicate different market demand scenarios. Each market demand scenario has a probability of W_w; the sum of them is

$$\sum_{w=1}^{N_W} W_w = 1. \tag{10.4.4}$$

The stochastic features will be associated only with the commercial quantities, i.e., selling, storing, purchasing, and shipping; the variables describing these aspects will have the index w. All variables related to production (state variables, mode changing variables, etc.) are not affected.

10.4.2.4 The Variables

At first we introduce the non-negative (continuous) *production variables*

$$\begin{array}{ll} p_{ij}^{TOT} & : \text{aggregated production (in tons) of product } j \text{ at site } i \\ m_{ijk} & : \text{production (in tons) of product } j \text{ at site } i \text{ in period } k \\ \kappa_{ik} & : \text{semi-continuous variable related to minimum utilization.} \end{array} \tag{10.4.5}$$

The worldwide production network and its current state is characterized by the *state variables* $\delta_{imk} \in \{0,1\}$,

$$\delta_{imk} := \begin{cases} 1 \,, & \text{if plant } i \text{ is in mode } m \text{ at the end of period } k \\ 0 \,, & \text{otherwise.} \end{cases} \tag{10.4.6}$$

These variables carry the information of the mode or status of the plant.

In order to describe mode changes we introduce the mode changing binary variables

$$\xi_{ikm_1m_2} = \begin{cases} 1 \text{ , if } \delta_{im_1k-1} = \delta_{im_2k} = 1 \\ 0 \text{ , otherwise,} \end{cases} \tag{10.4.7}$$

and the binary variable χ_{ik}

$$\chi_{ik} := \begin{cases} 1, \text{ if a mode-changed occurred at site } i \text{ in period } k \\ 0, \text{ otherwise} \end{cases} \tag{10.4.8}$$

indicating whether a mode change occurred at site i in period k or not.

To follow the stock we use the non-negative continuous *stock variables*

$$\begin{aligned} s_{ijtw} &: \text{total stock (in tons) of product } j \text{ at site } i \\ s^A_{ijtw} &: \text{rented stock capacity (in tons) of product } j \text{ at site } i. \end{aligned} \tag{10.4.9}$$

For sales and transport we introduce the non-negative continuous variables

$$p_{ijtsw} : \text{amount of product } j \text{ shipped from site } i \text{ to site } s, \tag{10.4.10}$$

which for $s \neq i$ indicate the quantity of product j shipped at time t from plant i to site s, and for $s = i$ represent the quantity sold at site i. In addition we have the non-negative semi-continuous *transport variables*

$$\sigma_{ijtsw} : \text{(dimensionless) amount of product } j \text{ shipped from site } i \text{ to } s. \tag{10.4.11}$$

Finally, we need some non-negative variables describing the purchase of products from external competitors

$$p^E_{sjt} : \text{external purchase (in tons) of product } j \text{ in period } t. \tag{10.4.12}$$

Note that most of the variables are not needed for all combinations of indices but rather for a few combinations indicated by some logical qualifiers; therefore we refer to these variables as sparse variables. XPRESS-MP keeps track automatically of sparse variables appearing in constraints. To keep things simple we do not give the logical qualifiers in the model description.

10.4.2.5 The State of the Production Network

The worldwide production network and its current state is characterized by the *state variables* $\delta_{imk} \in \{0, 1\}$,

$$\delta_{imk} := \begin{cases} 1 \text{ , if plant } i \text{ is in mode } m \text{ at the end of period } k \\ 0 \text{ , otherwise} \end{cases} \quad ; \quad \begin{array}{l} \forall i \\ \forall m \\ \forall k. \end{array} \qquad (10.4.13)$$

These variables are used to guarantee that at the end of time interval k the plant at site i is in a unique mode. This is achieved by the equations

$$\sum_{m=1}^{M_i} \delta_{imk} = 1 \;\; ; \quad \begin{array}{l} \forall i \\ \forall k. \end{array} \qquad (10.4.14)$$

Due to the presence of Eq. (10.4.14) it is sufficient to request the binary character of the δ_{imk} explicitly only for $N_A N_K \sum_{m=1}^{M_i}(M_i - 1)$ of them.

Note that some initial data $\Delta_{im} = \delta_{im0}$ have to be provided to define the known status of plant i before we start planning. Of course, these initial data must satisfy the condition

$$\sum_{m=1}^{M_i} \Delta_{im} = 1 \quad , \quad \forall i. \qquad (10.4.15)$$

10.4.2.6 Exploiting Fixed Setup Plans

Sometimes the state of all plants may be given in advance, and one may want to fix the states of all plants to the states known from another optimization run. Therefore, the model provides the option to use the states of all plants according to the bounds

$$\delta_{imk} = \Delta_{imk}^F, \qquad (10.4.16)$$

where Δ_{imk}^F give the state of all plants during the whole period T_P and I^Δ is a switch specifying whether a fixed plan is to be used ($I^\Delta = 1$) or not ($I^\Delta = 1$). This switch, the default is $I^\Delta = 1$, allows fixing the setup while all other variables can be optimized with respect to the fixed modes.

10.4.2.7 Keeping Track of Mode Changes

It is one of the most fundamental assumptions in this model that we can have at most one mode change per period. If the state variables take the values $\delta_{im_1k-1} = \delta_{im_2k} = 1$ we have a mode change from mode m_1 to m_2 in time interval k.

Mode changes are tracked by the binary variables

$$\xi_{ikm_1m_2} = \begin{cases} 1 \text{ , if } \delta_{im_1k-1} = \delta_{im_2k} = 1 \\ 0 \text{ , otherwise} \end{cases} \quad ; \quad \begin{array}{l} \forall i \\ \forall m \\ \forall k. \end{array} \tag{10.4.17}$$

This variable is unity if at the end of period $k-1$ the plant is in mode m_1 and at the end of period k it is in mode m_2. So the variables $\xi_{ikm_1m_2}$, $m_1 \neq m_2$, tell us whether a mode change is taking place from mode m_1 to mode m_2 during time interval k. The $M_i \times M_i$ matrix ξ_{ik} (for fixed i and k) either is identical to zero or there is at most one element which is unity. The state variables δ and the set-up change variables ξ are connected by

$$\xi_{ikm_1m_2} \geq \delta_{im_1k-1} + \delta_{im_2k} - 1 \quad ; \quad \forall i \;\; \forall k \;\; \forall m_1 \;\; \forall m_2 \,|m_2 \neq m_1. \tag{10.4.18}$$

The conditions (10.4.14) and (10.4.18), together with the definition (10.4.43) of the change-over costs, and the maximum property of the solution ensure that the $\xi_{ikm_1m_2}$ automatically assume only the values 0 or 1; the saving in computing time by comparison with the explicit declaration as binary variables is considerable. To understand this property consider two subsequent time intervals $k-1$ and k. For a particular binary variable $\xi_{ikm_1m_2}$ we can analyze four different cases summarized in the table below leading to four different inequalities:

δ_{im_1k-1}	δ_{im_2k}	$\xi_{ikm_1m_2} \geq$
0	0	−1
1	0	0
0	1	0
1	1	1

The case $\delta_{im_1k-1} = \delta_{im_2k} = 1$ represents a set-up change from mode m_1 to m_2 in time interval k. This situation is illustrated in Fig. 10.2.

If for a particular mode m the binary decision variables take the value zero in consecutive time intervals, i.e., $\delta_{imk-1} = \delta_{imk} = 0$, no production of a product j associated with that mode m is possible in time interval k.

The total number of change-overs during the entire production period at site i can be limited in a simple way by

$$\sum_{k=1}^{N_K} \sum_{m_1=1}^{N_P} \sum_{\substack{m_2=1 \\ m_2 \neq m_1}}^{N_P} \xi_{ikm_1m_2} \leq U_i \; ; \quad \forall i. \tag{10.4.19}$$

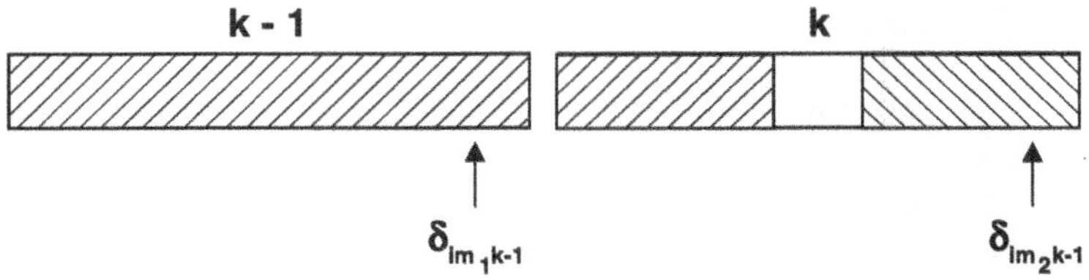

Fig. 10.2 Illustration of a set-up change. The white gap in the right part of the figure represents a set-up change from mode m_1 to mode m_2. The size of the gap indicates the time needed for the set-up change relative to the length of time interval k

10.4.2.8 Coupling Modes and Production

Let us now couple the production to the modes. This could be described by an indication matrix I_{imj}

$$I_{imj} := \begin{cases} 1 \text{ , if at site } i \text{ mode } m \text{ allows production of } j \\ 0 \text{ , otherwise.} \end{cases} \tag{10.4.20}$$

Note that this approach allows us to model that a certain product can be produced in different modes. We now formulate the following relation

$$m_{ijk} \leq \sum_{m=1|I_{imj}=1}^{M_i} C_{ik}\delta_{imk-1} + C_{ik}\delta_{imk} \quad , \quad \forall\{ijk\} \tag{10.4.21}$$

with site- and time-dependent logistic capacities C'_{ik} [explained below] specified in tons per time interval, e.g., tons per month. In our special scenario in which products and modes are uniquely coupled that equation just reduces to

$$m_{ijk} \leq C_{ik}\delta_{i,m=j,k-1} + C_{ik}\delta_{i,m=j,k} \quad , \quad \forall\{ijk\}. \tag{10.4.22}$$

The inequalities (10.4.22) lead either to $m_{ijk} = 0$, $m_{ijk} \leq C_{ik}$ or $m_{ijk} \leq 2C_{ik}$. Only the first case is relevant and restricts m_{ijk}. The second inequality becomes valid if a mode change occurs but is replaced by the stronger inequality (10.4.25). The third case is also dominated by (10.4.25). The logistic capacities are either derived from the theoretical annual capacities C_i explained below according to

$$C_{ik} := \min\left\{\frac{C_i}{N_K}, C'_{ik}\right\}, \tag{10.4.23}$$

or they are derived from the capacities C'_{ik} specified *ad hoc* for the production time intervals considering, e.g., plant shutdown or maintenance work. The annual production capacities C_i (tons per year) of plant i give the amount of production if a plant were producing continuously for a whole year. If no change-over occurs in a specific time interval r, production is possible at full capacity C_{ir}. Otherwise, the theoretical production capacity C_i is modified in this interval to a reduced capacity $C^R_{im_1,m_2k}$ according to

$$C^R_{ikm_1m_2} = C_{ik} - \frac{\Delta_{im_1m_2}}{T_P} C_i \quad , \quad \forall i \ \forall k \ \forall m_1, m_2 \,|m_2 \neq m_1 . \tag{10.4.24}$$

Note that the mode change-over times $\Delta_{im_1m_2}$ [in days] are used to reduce the capacity when change-overs occur. Typically, values vary between 2 and 8 days. By definition, we put $\Delta_{imm} = 0$.

It is, of course, also not valid for the sum of the products produced during a time interval r in which a change-over takes place to be greater than the modified capacity, i.e.,

$$\sum_{j=1}^{N_P} m_{ijk} \leq C_{ir} + \sum_{m_1=1}^{M_i} \sum_{\substack{m_2=1 \\ m_2 \neq m_1}}^{M_i} \left[C^R_{ikm_1m_2} - C_{ir} \right] \xi_{ikm_1m_2} \quad ; \quad \forall \{ik\}. \tag{10.4.25}$$

If no change-over takes place in a given time interval, the conditions (10.4.22), (10.4.24), and (10.4.25) reduce to $m_{ijk} \leq C_{ik}$. In the opposite case, for a unique pair (m_1, m_2), (10.4.25) reduces to $\sum_{j=1}^{N_P} m_{ijk} \leq C^R_{ikm_1m_2}$, which furthermore requires $m_{im_1k} + m_{im_2k} \leq C_{ikm_1m_2}$.

10.4.2.9 Minimum Production Requirements

Finally, minimum production requirements, $\mathscr{C}_{ij}$, still have to be taken into account over the entire production schedule, i.e.,

$$\sum_{t=1}^{N_T} \sum_{r=1}^{R} m_{ijk(t,r)} \geq \mathscr{C}_{ij} \quad ; \quad \forall \{ij\}. \tag{10.4.26}$$

These conditions ($N_S N_P N_K$ coefficients) are replaced by the equivalent system of equations with $(1 + N_K)$ coefficients

$$p^{TOT}_{ij} := \sum_{t=1}^{N_T} \sum_{r=1}^{R} m_{ijk(t,r)} \ ; \quad \forall \{ij\}, \tag{10.4.27}$$

and bounds

$$p_{ij}^{TOT} \geq \mathscr{C}_{ij} \ ; \ \ \forall \{ij\}. \tag{10.4.28}$$

Moreover, the total production at a site must not fall below given global production limits $\mathscr{C}_i$, i.e.,

$$\sum_{j=1}^{N_P} p_{ij}^{TOT} \geq \mathscr{C}_i \ ; \ \ \forall i. \tag{10.4.29}$$

Finally, monthly minimum requirements $\mathscr{C}_{ijt}$

$$\sum_{r=1}^{R} m_{ijk(t,r)} \geq \mathscr{C}_{ijt} \ ; \ \ \forall \{ijt\} \tag{10.4.30}$$

remain to be fulfilled.

10.4.2.10 Modeling Stock Balances and Inventories

The stock balance equations are a special type of *flow balance constraints* [see Sect. 2.3] and include the variables s_{ijtw} which describe the stock of product j at site i at the end of time interval t. These constraints read

$$s_{ijtw} = s_{ijt-1w} + p_{ijtw}^{E} + \sum_{r=1}^{R} m_{ijk(t,r)} - \left[\sum_{s=1}^{N_S} p_{ijtsw} - \sum_{\substack{s=1 \\ s \neq i}}^{N_S} p_{sjd(t,s,i)iw} \right] \ , \ \begin{array}{c} \forall i \\ \forall j \\ \forall t \\ \forall w. \end{array} \tag{10.4.31}$$

This relation takes account of production, transfer to other sites, procurement of goods from other sites, the initial stock $s_{ij0} = S_{ij}^0$, quantities p_{ijtw}^E purchased from external sources, which can likewise be kept in store, and the direct delivery or the sale. In the last sum the index s marks additions to stock which already take account of the transportation time, i.e., $d(t, i_1, i_2) = t - T_{i_1 i_2}$, $T_{i_1 i_2}$ denoting the transportation time (in integer number of periods) from site i_1 to site i_2. Naturally, only those terms with positive index $d(t, i_1, i_2) > 0$ are taken into consideration. Note that because $T_{i_1 i_2}$ is used in the index it is required that $T_{i_1 i_2}$ is measured in units of the period and that it is integral.

The next step is now to ensure that a certain minimum stock is guaranteed. If S_{ij}^- denotes the minimum stock for product j at plant i and S_{ij}^0 the initial stocks, the above requirement implies that

$$s_{ijtw} \geq S_{ij}^- \quad ; \quad \forall\{ijtw\}. \tag{10.4.32}$$

For logistic reasons there is a further requirement for the last time interval that a prescribed final stock $\mathscr{M}_{ij}$ is achieved, i.e.,

$$s_{ijN_Tw} \geq \mathscr{M}_{ij} \quad ; \quad \forall\{ijw\}. \tag{10.4.33}$$

It must also be ensured that the local inventory capacity S_{ij} is not exceeded, i.e.,

$$s_{ijtw} \leq S_{ij} \quad ; \quad \forall i \ \forall j \ \forall t \ \forall w. \tag{10.4.34}$$

It seems reasonable to introduce additional variables s_{ijtw}^A which represent the amount of product in additional storage, which can be provided, for example, by leasing, and thus do not lead to (10.4.34) but to the softer condition

$$s_{ijtw} \leq S_{ij} + s_{ijtw}^A \quad ; \quad \forall\{ijtw\}, \tag{10.4.35}$$

where the capacity of the additional store itself may be bounded by

$$s_{ijtw}^A \leq S_{ij}^A \quad ; \quad \forall\{ijtw\}. \tag{10.4.36}$$

10.4.2.11 Modeling Transport

Transport is only possible if the amount to be shipped is within the bounds P^- (and P^+). The condition

$$p_{ijtsw} = 0 \ \vee \ P^- \leq p_{ijtsw} \leq P^+ \ , \ z \neq i \quad ; \quad \forall\{ijtsw\} \tag{10.4.37}$$

is formulated using the semi-continuous variables σ_{ijtsw} which satisfy the condition

$$\sigma_{ijtsw} = 0 \ \vee \ 1 \leq \sigma_{ijtsw} \leq \sigma^+ \ , \ z \neq i \ ; \ \forall\{ijtsw\}. \tag{10.4.38}$$

Note that these variables are scaled to unity and related to the original variables according to

$$p_{ijtsw} = P^- \sigma_{ijtsw} \quad , \quad \sigma^+ = P^+/P^-, \ z \neq i \ ; \ \forall\{ijtsw\}. \tag{10.4.39}$$

The distinction between the variables p_{ijtsw} and σ_{ijtsw} is not necessary from an algebraic point of view. We could replace all occurrences of p_{ijtsw} by $P^{-}\sigma_{ijtsw}$. However, both approaches have different effects on scaling and thus on numerical performance. In this case study it was much better to have both variables p_{ijtsw} and σ_{ijtsw} in the model.

10.4.2.12 External Purchase

The quantities purchased from other suppliers are subject to the bounds

$$p_{ijtw}^{E} \leq P_{ijt}^{E} \cdot \mathscr{B}_{ij} \quad ; \quad \forall\{ijtw\}. \tag{10.4.40}$$

Here, the P_{ijt}^{E} are upper bounds of external purchase in tons, while the $\mathscr{B}_{ij} \in \{0, 1\}$ specify whether externally purchased product is permissible at all. External purchase becomes very important if there are conditions added to the model that demand has to be satisfied.

10.4.2.13 Modeling Sales and Demands

Certainly no more must be delivered than is required by the demand, i.e.,

$$p_{ijtiw} \leq D_{ijtw} \quad ; \quad \forall\{ijtw\}. \tag{10.4.41}$$

The constants D_{ijtw} indicate how much of product j at time t is demanded with a specific probability at site i.

Sometimes we might want to model that demand is satisfied, i.e.,

$$p_{ijtiw} = D_{ijtw} \quad ; \quad \forall\{ijtw\}. \tag{10.4.42}$$

In that case we should make sure that external purchase is possible as otherwise we might produce an infeasible problem due to lack of own production capacity.

10.4.2.14 Defining the Objective Function

The objective function is now to include the change-over costs, inventory costs, transportation costs, and production costs. First, with the plant-specific and product-specific change-over costs $C_{im_1m_2}^{U}$ we have the change-over costs

$$c^{U} := \sum_{i=1}^{N_A} \sum_{t=1}^{N_T} \sum_{r=1}^{R} \sum_{m_1=1}^{M_i} \sum_{\substack{m_2=1 \\ m_2 \neq m_1}}^{Mi} C_{im_1m_2}^{U} \xi_{ik(t,r)m_1m_2}. \tag{10.4.43}$$

Using the constant inventory costs C_{ij}^{I} for product j at plant i for one time interval, the inventory costs c^{I} are given as

$$c^{I} := \sum_{i=1}^{N_A} \sum_{j=1}^{N_P} \sum_{w=1}^{N_D} \left(W_w C_{ij}^{I} \sum_{t=1}^{N_T} s_{ijtw} \right) . \tag{10.4.44}$$

Apart from these inventory costs, we also consider the additional costs c^{M} for a possible leased store with specific leased inventory costs C_{ij}^{M}.

$$c^{M} := \sum_{i=1}^{N_A} \sum_{j=1}^{N_P} \sum_{w=1}^{N_D} \left(W_w \left(C_{ij}^{M} - C_{ij}^{I} \right) \sum_{t=1}^{N_T} s_{ijtw}^{A} \right) . \tag{10.4.45}$$

These costs C_{ij}^{M} are surplus costs for the additional inventory capacity. Part of the costs is already covered by the inventory cost C_{ij}^{I} because the definition of s_{ijtw} in (10.4.31) includes both types of inventory, i.e., part of the cost related to s_{ijtw}^{A} is already included in c^{L}.

With the individual costs C_{is}^{T} the entire transportation costs are given by

$$c^{T} := \sum_{i=1}^{N_A} \sum_{j=1}^{N_P} \sum_{t=1}^{N_T} \sum_{\substack{s=1 \\ z \neq i}}^{N_S} \sum_{w=1}^{N_D} W_w C_{is}^{T} p_{ijtsw} . \tag{10.4.46}$$

Finally, it remains to take account of the production cost c^{P}:

$$c^{P} := \sum_{i=1}^{N_A} \sum_{j=1}^{N_P} \sum_{t=1}^{N_T} \sum_{r=1}^{R} C_{ij}^{P} m_{ijk(t,r)} \tag{10.4.47}$$

and the costs c^{B}

$$c^{B} = \sum_{i=1}^{N_A} \sum_{j=1}^{N_P} \sum_{t=1}^{N_T} \sum_{w=1}^{N_D} W_w C_{ijt}^{E} p_{ijtw}^{E} , \tag{10.4.48}$$

which arise from external purchase at purchasing prices C_{ijt}^{E}. The income e is computed from the sales

$$e = \sum_{i=1}^{N_A} \sum_{j=1}^{N_P} \sum_{t=1}^{N_T} \sum_{w=1}^{N_D} W_w E_{ijt} p_{ijtiw} . \tag{10.4.49}$$

This results in the objective function to be maximized as the contribution

$$Z(\delta, \xi, m, p, s, s^A, p^E) = e - (c^U + c^I + c^L + c^T + c^P + c^B). \qquad (10.4.50)$$

10.4.3 Remarks on the Model Formulation

Let us now focus on a few points in the model formulation. Most of these remarks are related to techniques and tricks which might be appreciated by practitioners.

10.4.3.1 Including Minimum Utilization Rates

During the modeling and validation phase it became necessary to add an additional requirement which guarantees a minimum utilization rate U_i^R for each plant. As is seen in Fig. 10.3 there are times in the production plan when the plant produces only a tiny amount. Such solutions are not very likely to be accepted. Therefore, the client asked for the inclusion of minimum utilization rates. That would force the plant to stop production completely during some time intervals if not enough is produced. Fixing this problem leads to another class of semi-continuous variables κ_{ik} which force the solution to represent only scenarios with zero-production or a utilization rate with more than $100 \cdot U_i^R$ %, i.e.,

$$U_i^R C_{ik} \kappa_{ik} = \sum_{j=1}^{N_P} m_{ijk} \quad , \quad \forall \{ik\}. \qquad (10.4.51)$$

Since κ_{ik} is semi-continuous, i.e., $\kappa_{ik} = 0$ or $1 \leq \kappa_{ik} \leq \kappa^+$, Eq. (10.4.51) actually requires that

$$\sum_{j=1}^{N_P} m_{ijk} = 0 \quad \text{or} \quad \sum_{j=1}^{N_P} m_{ijk} \geq U_i^R C_{ik} \quad , \quad \forall \{ik\}. \qquad (10.4.52)$$

To reduce the number of variables in the model, κ_{ik} is only defined for those indices (i, k), for which C_{ik} is positive.

10.4.3.2 Exploiting Sparsity

Commercial LP-solvers use a *revised Simplex algorithm*, which operates very efficiently on sparse matrices. In particular, it can be helpful to keep the ratio of the number of rows (constraints) and the number of columns (variables) and also the *density*, i.e., the ratio of the number of non-zero coefficients to the total number

Fig. 10.3 Production plan. The units of measure are missing for reasons of confidentiality

of coefficients, as low as possible. For this reason, (10.4.19) is replaced by the two constraints

$$\chi_{ik} := \sum_{m_1=1}^{M_i} \sum_{m_2=1}^{M_i} \xi_{ikm_1m_2} \quad , \quad \forall\{ik\} \tag{10.4.53}$$

and

$$\sum_{k=1}^{N_K} \chi_{ik} \leq U_i \quad , \quad \forall i. \tag{10.4.54}$$

The system (10.4.19) consists of one row and $N_A N_K (N_P - 1) N_P$ columns[2] for the variables $\xi_{ikm_1m_2}$ has $N_A N_K (N_P - 1) N_P$ coefficients, i.e., the density is unity. In the case of (10.4.53,10.4.54) one obtains $N_A N_K + N_A$ rows and $N_A N_K (N_P - 1) N_P + N_A N_K$ columns, but

$$N_A N_K + N_A N_K (N_P - 1) N_P + N_A N_K \tag{10.4.55}$$

coefficients. From that we derive the density

$$\begin{aligned} \rho &= \frac{\text{no. of non-zero coefficients}}{(\text{no. of rows}) \cdot (\text{no. of columns})} \\ &= \frac{N_A N_K [(N_P - 1) N_P + 2]}{(N_A N_K)^2 (1 + \frac{1}{N_K}) [(N_P - 1) N_P + 1]} = \frac{96}{91} \cdot \frac{1}{N_S N_K} . \end{aligned} \tag{10.4.56}$$

By introducing $N_A N_K$ additional constraints and $N_A N_K$ new variables we achieve a (local) density which is reduced by a factor of $N_A N_K$, i.e., in our case for $N_A = 3$ and $N_K = 12$ a factor of 36.

Let us conclude with the remark that besides scaling both the size and the density of the problem affect the performance of the solver. It is subject to testing and model reformulations to find out which model formulation is the most efficient. In this case study using the larger but sparser system produced faster solutions.

10.4.3.3 Avoiding Zero Right-Hand Side Equations

For numerical reasons discussed below while building a model the analyst should try to avoid equations of the form

$$\sum_{j=1}^{n} A_{ij} x_j = 0, \tag{10.4.57}$$

[2]Here we used the special case information that $M_i = N_P$.

or should include them only with great care. The problem which might occur including many of these equations is related to the "minimum ratio rule" which governs the elimination of non-basic variables in the Simplex algorithm. From Sect. 3.2.2 we remember

$$x_r := \frac{b_i}{A_{rs}} = \min \left\{ \frac{b_i}{A_{is}} \middle| A_{is} > 0 \right\}, \tag{10.4.58}$$

where that basic variable x_i with value b_i leading to the smallest value b_i/A_{is} is eliminated and x_r is the value of the new basic variable. Note that b_i corresponds to the right-hand side of an equality constraint. Therefore, if several zero right-hand side equalities are present the algorithm might have difficulties selecting the basic variable to be eliminated.

Nevertheless, the advice of avoiding such structures should always be subject to careful testing of the numerics. The relations (10.4.53) improved the performance of the Simplex algorithm. This example also shows how such equations may enter a model: they are used to define quantities later needed. However, we may keep in mind that during preprocessing, defined variables often get replaced again; these are usually storage overhead and also they make hide constraints in which other variables would be present (and these are original variables with non-zero coefficient in objective function). Exceptions are if they help for branching!

10.4.3.4 The Structure of the Objective Function

A frequent structural feature of objective functions in production planning problems is that it is built up from several components representing different costs or income terms. The Eqs. (10.4.43)–(10.4.49) define these components. This approach has three disadvantages leading to a weaker performance of the Simplex based LP-solver:

- it introduces additional zero right-hand side equalities,
- it introduces unbalanced scaling, and
- it makes "pricing" more difficult (choice of new basic variables).

The first point is not a major problem because only a few equations are added. The second one is a more serious problem. Assume that all the other variables and their coefficients are well balanced, i.e., all variables have values of the order of one and their coefficients take values of the order of 1000. Now, using defining equalities and the short form of the objective function introduces variables which have much larger values but the objective function has unit coefficient in all entries. The third point may be even worse. Remember from Sect. 3.2.2 that the reduced costs d_j are computed according to the formula

$$d_j = c_j - \mathbf{c}_B^T \mathscr{B}^{-1} \mathbf{A}_j, \tag{10.4.59}$$

where c_j denotes the coefficient in the objective function associated with non-basic variables x_j under examination, $\mathbf{c}_B$ collects all coefficients in the objective function associated with the basic variables, $\mathscr{B}$ is the basic matrix, and $\mathbf{A}_j$ refers to the present columns associated with x_j. Although reduced cost can be different from zero when the variables do not appear in the objective function, in this case study it helped to bring all variables into the objective function $Z = Z(\delta, \xi, m, p, s, s^A, p^E)$

$$
\begin{aligned}
Z = & \sum_{i=1}^{N_A}\sum_{j=1}^{N_P}\sum_{t=1}^{N_T}\sum_{w=1}^{N_D} W_w E_{ijt} p_{ijtiw} - \sum_{i=1}^{N_A}\sum_{j=1}^{N_P}\sum_{t=1}^{N_T}\sum_{\substack{s=1\\ z\neq i}}^{N_S}\sum_{w=1}^{N_D} W_w C_{is}^T p_{ijtsw} \\
& - \sum_{i=1}^{N_A}\sum_{t=1}^{N_T}\sum_{r=1}^{R}\sum_{m_1=1}^{M_i}\sum_{\substack{m_2=1\\ m_2\neq m_1}}^{N_P} C_{im_1m_2}^U \xi_{ik(t,r)m_1m_2} - \sum_{i=1}^{N_A}\sum_{j=1}^{N_P}\sum_{w=1}^{N_D}\left(W_w C_{ij}^I \sum_{t=1}^{N_T} s_{ijtw}\right) \\
& - \sum_{i=1}^{N_A}\sum_{j=1}^{N_P}\sum_{w=1}^{N_D}\left(W_w \left(C_{ij}^M - C_{ij}^I\right)\sum_{t=1}^{N_T} s_{ijtw}^A\right) - \sum_{i=1}^{N_A}\sum_{j=1}^{N_P}\sum_{t=1}^{N_T}\sum_{r=1}^{R} C_{ij}^P m_{ijk(t,r)} \\
& - \sum_{i=1}^{N_A}\sum_{j=1}^{N_P}\sum_{t=1}^{N_T}\sum_{r=1}^{R} C_{ij}^P m_{ijk(t,r)} - \sum_{i=1}^{N_A}\sum_{j=1}^{N_P}\sum_{t=1}^{N_T}\sum_{w=1}^{N_D} W_w C_{ijt}^E p_{ijtw}^E .
\end{aligned}
$$

10.4.4 Model Performance

The model leads to a mixed integer linear programming problem (MILP) with 72 binary, 252 semi-continuous variables and 1071 continuous variables, and 974 non-trivial linear conditions. The structure of the model was matched to the algorithmic fundamentals of the software early in the formulation stage leading to a density of 0.33% and about 4500 non-zero matrix elements. This facilitated the optimum use of the numerical problem solving properties of the XPRESS-MP package. The first feasible mixed integer solution is accepted as the solution. This heuristic is justified because our trials indicate that its associated contribution margin deviates by only a few percent from that of the continuous problem (well within the error associated with the input data) and because it eliminates the need for the time consuming complete search for the absolute optimal solution via a B&B algorithm. The first feasible integer solution is usually found within 40 s using XPRESS-MP on a Pentium. In some cases, we could even prove optimality in minutes. But proving optimality turned out to be very data dependent. In some cases we could not prove it within days.

10.4.5 Reformulations of the Model

Although the client was quite happy that the problem could be solved very fast, there remains the open question of whether the solution found is the optimal solution. Additional functionalities or wishes are to use the production planning model to plan for more than 3 products, to relax the mode-product relations and to include site- and mode-dependent capacities. Note that our model made explicit use of the fact that we had only 3 products in all constraints connecting the mode variables δ and the set-up change variables ξ.

Thus, we need to reformulate the model first to meet our client's reality, and second to decrease the integrality gap. In many cases reformulations of the model help to get smaller integrality gaps. In some cases reformulations enable us to find valid inequalities which can be used to reduce the gap.

10.4.5.1 Estimating the Quality of the Solution

If we cannot prove optimality (in real-world problems we always have to expect that to happen) we can discuss the quality of solutions by inspecting the upper and lower bounds derived by the B&B algorithm [see Fig. 3.9 and discussion of bounds on page 130]. In a maximization problem the upper bound, z^U, is provided by the LP relaxations while the lower bound, z^L, corresponds to the best integer solution found. So we have the bounds

$$z^L \leq z^* \leq z^U \tag{10.4.60}$$

on the objective function value z^* of the (unknown) optimal solution. In a maximization problem the difference $z^U - z^*$ is called the *integrality gap*. If the search is terminated before z^* has been computed, the difference $z^U - z^L$ is used as an upper bound on the integrality gap.

Assuming that both z^L and z^U are positive the quality of our solution can also be expressed by the relative expression

$$p := 100 \cdot \frac{z^U - z^L}{z^U}, \tag{10.4.61}$$

which expresses that the difference between the best solution found and the (unknown) optimal solution is at most $p\%$. If that measure does not satisfy our client we have to try to improve the bounds or to prove optimality.

While the lower bound z^L increases if we allow the algorithm to seek for further integer feasible solutions the upper bound z^U decreases very slowly during the computations. The upper bound can be decreased faster if it is possible to find effective cuts and to apply B&C techniques.

10.4.5.2 Including Mode-Dependent Capacities

To include more than 3 products and to generalize the mode-product relation we modify the model as follows. At first, we change the data related to capacity. Instead of using C_i and C_{ik} we introduce the following data

D	: days per period
H_{ik}	: number of days available for production and change-over on site i in period k
$H^R_{ikm_1m_2}$	: number of days available for production at site i in period k if a change-over from mode m_1 to m_2 occurs in period k
R^P_{imj}	: production rates in tons/day, the amount of product j which could be produced in mode m in one day at site i.

Notice that H_{ik} depends on k which gives us the opportunity to model temporary shutdowns (maintenance, test runs, etc.). The production rates R^P_{imj} will also be used to indicate whether product j can be produced in mode m at site i at all. If $R^P_{imj} = 0$ this is not possible. It is possible that a certain product can be produced in several modes.

We can relate the new data tables, H_{ik} and $H^R_{ikm_1m_2}$ to our original data as follows

$$H_{ik} = \frac{C_{ik}}{C_i} D \qquad (10.4.62)$$

and

$$H^R_{ikm_1m_2} = \max\left\{0, H_{ik} - \Delta_{im_1m_2}\right\} . \qquad (10.4.63)$$

Note that we do not require that H_{ik} or $H^R_{ikm_1m_2}$ is an integer quantity, and that

$$\Delta_{imm} = 0 \quad , \quad \forall i \quad , \quad m = 1, \ldots, M_i . \qquad (10.4.64)$$

It is also possible to model the fact that months have different number of days. The table H_{ik} can be used to keep track of that.

The use of these data leads to some model changes affecting (10.4.21), (10.4.22), (10.4.24), and (10.4.25). However, as the reformulation of the change-overs is also changed we postpone the discussion of the modified relations to Sect. 10.4.5.3.

10.4.5.3 Modes, Change-Overs and Production

In order to include more products (and also modes) and to improve the model formulation with respect to the integrality gap we modify the meaning of two types of variables already used:

$$m_{imk} \geq 0 \text{ number of days plant } i \text{ in period } k \text{ is in mode } m$$

$$\xi_{ikm_1m_2} = \begin{cases} 1 \text{ , if } \delta_{im_1k-1} = \delta_{im_2k} = 1 \\ 0 \text{ , otherwise.} \end{cases} \tag{10.4.65}$$

Note the difference: m_{imk} now does not give the amount of tons, but just the number of days (fractional days are allowed) in which the plant is in mode m during period k. The variables m_{imk} can easily be related to a new variable, p^T_{ijk},

$$p^T_{ijk} \geq 0 \quad , \quad \text{tons of product } j \text{ produced at site } i \text{ in period } k. \tag{10.4.66}$$

Thus mode, number of days available per mode, and tons of products are connected by

$$p^T_{ijk} \leq \sum_{m=1|R^P_{imj}>0}^{M_i} R^P_{imj} m_{imk} \quad , \quad \forall \{ijk\}. \tag{10.4.67}$$

Note that this relation is an inequality. Even if the plant is in a mode in which a certain product could be produced there is no need to produce at full capacity.

The variable $\xi_{ikm_1m_2}$ has almost the same meaning as before. It is unity if at the end of period $k-1$ the plant is in mode m_1 and at the end of period k it is in mode m_2. While so far we interpreted $\xi_{ikm_1m_2}$ as a variable only describing whether a change-over occurs or not it can now also be used to check whether production continues. If the plant is in mode m both at the end of period $k-1$ and k, then we have $\xi_{ikmm} = 1$.

In Sect. 9.3 we mentioned that it is sometimes advantageous to introduce additional discrete variables. We now introduce additional binary variables with the following meaning

$$\alpha_{imk} := \begin{cases} 1, \text{ if plant } i \text{ is in mode } m \text{ for some time in period } k \\ 0, \text{ otherwise} \end{cases} \tag{10.4.68}$$

$$\beta_{imk} := \begin{cases} 1, \text{ if mode } m \text{ is started at site } i \text{ in period } k \\ 0, \text{ otherwise} \end{cases} \tag{10.4.69}$$

and finally

$$\gamma_{imk} := \begin{cases} 1, \text{ if mode } m \text{ is terminated at site } i \text{ in period } k \\ 0, \text{ otherwise.} \end{cases} \tag{10.4.70}$$

These binary variables are related to others by the constraints

$$\beta_{imk} = \sum_{m_1 \neq m} \xi_{ikm_1m} \quad , \quad \forall i, \quad m = 1, \ldots, M_i, \quad \forall k \tag{10.4.71}$$

and

$$\gamma_{imk} = \sum_{m_1 \neq m} \xi_{ikmm_1} \quad , \quad \forall i, \quad m = 1, \ldots, M_i, \quad \forall k$$

To express whether mode m is used at all in period k we have

$$\alpha_{imk} = \delta_{imk-1} + \delta_{imk} - \xi_{ikmm} \quad , \quad \forall i, \quad m = 1, \ldots, M_i \quad , \quad k = 2, \ldots, N^T \tag{10.4.72}$$

and

$$\alpha_{im1} = \Delta_{im} + \delta_{im1} - \xi_{i1mm} \quad , \quad \forall i, \quad m = 1, \ldots, M_i, \tag{10.4.73}$$

where $\Delta_{im} = \delta_{im0}$ denotes the known status of plant i before we start planning. Note that it is not necessary to declare ξ, β, and γ as binary variables if we have declared δ and α as binary variables.

Finally we have the following block of constraints:

$$\gamma_{imk} = \delta_{imk-1} - \xi_{ikmm} \quad , \quad \forall i, \quad m = 1, \ldots, M_i \quad , \quad k = 2, \ldots, N^T \tag{10.4.74}$$

or

$$\gamma_{im1} = \Delta_{im} - \xi_{i1mm} \quad , \quad \forall i, \quad m = 1, \ldots, M_i \tag{10.4.75}$$

and

$$\beta_{imk} = \delta_{imk} - \xi_{ikmm} \quad , \quad \forall i, \quad m = 1, \ldots, M_i \quad , \quad k = 2, \ldots, N^T. \tag{10.4.76}$$

The constraints above describe the change-over activities. For production planning problems under mode constraint (i.e., a single production mode for each period and change-over variables representing the change of mode over time), a tight linear programming formulation of the change-over activities is obtained by representing the evolution of the mode over time as a unit flow problem [see Karmarkar & Schrage (1985,[326]) and Wolsey (1989,[587])].

Note that (10.4.14) is not required any longer. This follows from (10.4.15) and inspection of (10.4.73)

$$\sum_{m=1}^{M_i} \alpha_{im1} = 1 + \sum_{m=1}^{M_i} \delta_{im1} - \sum_{m=1}^{M_i} \xi_{i1mm} \quad , \quad \forall i, \quad m = 1, \ldots, M_i \tag{10.4.77}$$

If in the first period a set-up change takes place, then $\sum_{m=1}^{M_i} \xi_{i1mm} = 0$ but $\sum_{m=1}^{M_i} \alpha_{im1} = 2$ as production is possible in exactly two modes. Thus we have

$\sum_{m=1}^{M_i} \delta_{im1} = 1$. If no set-up change occurred then production was possible in only one mode which gives $\sum_{m=1}^{M_i} \alpha_{im1} = \sum_{m=1}^{M_i} \xi_{i1mm} = 1$ which again leads to $\sum_{m=1}^{M_i} \delta_{im1} = 1$.

10.4.5.4 Reformulated Capacity Constraints

Let us now start to formulate the capacity constraints as a function of the change-over variables. While in the first model formulation capacity constraints restricted the amount of tons which could be produced, capacity constraints are now used to restrict the number of days the plant is in a certain mode. We first note that

$$m_{imk} < H^R_{immk} \cdot \alpha_{imk} \quad , \quad \forall i, \quad m = 1, \dots, M_i \quad , \quad \forall k \qquad (10.4.78)$$

is a valid upper bound. Remember that α_{imk} carries the information whether mode m is used in period k or not. If the plant is never in mode m during period k, then $\alpha_{imk} = 0$ and the plant indeed spends zero days in mode m. If mode m is chosen, i.e., $\alpha_{imk} = 1$, the inequality reduces to $m_{imk} < H_{ik}$ due to (10.4.64). Fractional values of α_{imk} during the LP relaxation or within the tree reduce the time the plant can be in mode m leading to smaller amounts of the products being produced in this mode.

Next, we want to compute the available capacities subject to change-overs. These constraints read

$$m_{imk} \le \sum_{m=1}^{M_i} H^R_{imm_2k} \cdot \xi_{ikm_2m} + \sum_{\substack{m_2=1 \\ m_2 \ne m}}^{M_i} H^R_{im_2mk} \cdot \xi_{ikmm_2} \quad , \quad \begin{matrix} \forall i \\ m = 1, \dots, M_i \\ \forall k \end{matrix} \qquad (10.4.79)$$

and

$$\sum_{m_1=1 | H_{ik} \ne 0}^{M_i} m_{imk} \le H_{ik} - \sum_{\substack{m_1=1 \\ H_{ik} \ne 0}}^{M_i} \sum_{\substack{m_2=1 \\ m_2 \ne m_1}}^{M_i} \Delta_{im_1m_2} \cdot \xi_{ikm_1m_2} \quad , \quad \begin{matrix} \forall i \\ \forall k. \end{matrix} \qquad (10.4.80)$$

Constraint (10.4.79) defines the upper limit on the number of days of production of product j at site i in period k as a function of the change-over variables. Days for mode m become available in period k only if either the site status switches *to* mode m in period k [first term in the right-hand side of (10.4.79)] or the site status switches *from* mode m in period k (i.e., has status j at the end of period $k-1$). When the mode of the site is m at the end of period $k-1$ *and* k (i.e., $\xi_{ikmm} = 1$ and no change-over occurs in period k), the value of the available capacity H^R_{ikmm} is counted once in the first terms. In any integer solution, by constraint (10.4.79), at most two modes have a positive upper bound on available days in each period.

Constraint (10.4.80) defines the global available capacity in period k to be equal to the number of days available minus the number of days used for change-over. The second term can only be applied if $\Delta_{im_1m_2} \leq H_{ik}$. Otherwise, we should fix $\xi_{ikm_1m_2} = 0$ because there is not enough capacity in period k to perform the change over from mode m_1 to mode m_2.

10.4.5.5 Some Remarks on the Reformulation

What is the benefit of the reformulation? First, it became necessary to meet the client's reality. That was not possible with the old formulation. The new formulation is very general and is open to further generalizations. Second, the reformulated model leads to a smaller integrality gap. Third, when applying XPRESS-MP's B&C facilities some general types of cuts could be detected which were not present before.

10.4.6 What Can be Learned from This Case Study?

This case study showed clearly that a model is an object subject to many changes, modifications, and generalizations. The client involved in building and validating the model understood what was going on and saw other features in the real-world problem which could be incorporated easily while others required substantial modifications to the current model formulation. During this process the model tended to increase in complexity and also in the number of variables and constraints. The advantage was clearly that each model was based on an "easier" model which was well understood and validated and not the subject of infeasibilities which may often occur in the first phase of a project. In the end, the client had a model which was significantly improved in terms of the integrality gap and solution time. The modeler has used model formulations which moved away from the *natural* variables which were obvious in the first phase. In order to prove optimality, both B&B and B&C algorithms had to be used.

10.5 A Complex Scheduling Problem ⊖

This case study discusses a complex precedence and resource constrained scheduling problem.[3] In our case the scarce resource is personnel. It shows how an analyst could approach such a problem and how to put some structure in it. Several submodels will be identified which are already very demanding in themselves.

[3]See also Patterson et al. (1989,[438]) for a general discussion of such problems.

The whole model is far from being solvable by a pure MILP approach, but it shows something which is also very important for working on real-world problems: knowing the limits of certain optimization techniques and, despite those limits, being able to bring the client's and analyst's worlds together and to provide an acceptable solution.

10.5.1 Description of the Problem

The client[4] uses a set of machines, employs a number of workers, and receives orders from customers. Each order demands a certain amount of a product, which can be produced on the client's machines. The machines are operated and supervised by the workers. Orders are often split up into several identical jobs, which are necessary in order to produce the required amount of the product, because frequently orders demand more product than the machine capacity. A job for a given order is processed on a machine according to a procedure. A procedure defines and describes the structure of a job. It consists of a sequence of tasks defining how to produce some amount of a product. The amount of product produced by a job depends on the capacity of the machine. The tasks have to be carried out in a predetermined order. Each task has a demand for labor and a certain duration, defined by the detailed personnel profile. The workers are allocated to the different tasks in order to keep the jobs running. Allocation of the workers has to comply with working regulation rules, e.g., taking breaks, washing, equally spread labor among the workers, limits on labor intensive work, over-occupation rate and time. The mathematical formulation of this scheduling problem includes an assignment model, a sequencing model, and a time-indexed formulation in order to incorporate the detailed personnel profile. The objective is to optimize this production system, i.e., to minimize the makespan and/or to minimize the (variation in the) number of workers.

10.5.2 Structuring the Problem

Obviously we are facing a complex model. Therefore, it is advisable to structure the model carefully. We will do this by identifying the most important objects and their attributes. Note that our description already provides a classification into *main* and *special features*. This is important because clients often ask for many details which turn out as *nice to have* but not totally important.

---

[4]The term client is used to refer to one who asks for support from a mathematical consultant. The term customer is used to denote one who purchases goods from the client.

10.5.2.1 Orders, Procedures, Tasks, and Jobs

Main Features An order is assigned to a single machine and once the first job of an order has started, the order cannot be preempted. Once a machine is chosen it requires a specific procedure that must be used. A procedure contains a sequence of several tasks, thus determining the structure of a job. Orders are the demands of the customers for some specific product. An order basically consists of an order number, a machine number, a priority, an amount of product, an earliest start time, and a due date. Using the due date a latest start time for an order can be derived. The interval of possible start times may vary depending on whether some pre-product has to be produced or not (see below). Depending on the quantity of product ordered either one job has to run or several identical batch jobs have to run producing the product. How many jobs are needed is calculated by dividing the amount demanded by the capacity of the machine. The due dates are not rigid, as in some cases partial satisfaction of orders is allowed, i.e., 80% of the jobs have to be finished by the due date, while for the other 20% of the jobs some additional time is available. Each order or job has a priority. This priority determines their relevance to the production system and mainly controls the appropriate sequencing of the orders. There are limited buffer times possible between all the tasks of a job, i.e., the start of a task can be shifted over a certain time window after the completion time of the previous task. Thus, for a certain time its labor requirement may be used for more urgent tasks, but there is a point when a task has to start. No new jobs are started in the last hour of a shift, because the workers are entitled to go for a washing break in the last hour of a shift.

Special Features (Precedence Relations) Some products are produced according to multi-stage recipes connecting different pre-products and the finished product, thus, connecting different machines. Therefore, jobs may be coupled by amounts and time. Time coupling means that some tasks of jobs in different orders are synchronized. Furthermore, sometimes it may occur that the last task of one job (e.g., getting out material from a pre-product job) corresponds to the first or even the second task (loading that material on the appropriate machine) of the following dependent job, i.e., these tasks start at the same time, are executed simultaneously, and the personnel is shared by both tasks. Some jobs can only start if a specified amount of a pre-product (produced in a job of another order) is available. The amount of pre-product needed is determined by taking into consideration the stocks of this pre-product. Thus, the job producing it does not have to run as many times as calculated, i.e., the number of jobs needed to run can be reduced. And, it is often necessary to start the following dependent job even earlier than just after the last task of the preceding job. This is because there are no explicit set-up times, as they have been integrated into the procedures as tasks. The same holds for reconstruction times. Some procedures not only produce a main product but also deliver by-products which are often recycled using another procedure, i.e., at the end of some job it may be necessary to start two different new jobs: one job processing the main product toward the finished product and one job processing (recycling) the by-products. Often the recycling process does not have to start immediately after the end of the preceding job. Like the tasks it has a limited buffer, but it must have started by the end of this time slice.

10.5.2.2 Labor, Shifts, Workers and Their Relations

Main Features The machines are operated 24 h a day. The workers work in two 12-h shifts, a day shift starting at six in the morning and a night shift starting at six in the evening. The number of workers on each shift is at most 15. There are four different groups of workers A, B, C, D working in a 4-shift mode, e.g., shift A starts off as the day shift, the next day it becomes the night shift and then there are two days off. This results in the following sequence of shifts:

Day 1	Day 2	Day 3	Day 4	Day 5	...
A,B	C,A	D,C	B,D	A,B	...

For several reasons a worker may be absent, e.g., breaks, meetings, courses, holidays, sickness, etc. The available manpower per shift is adjusted accordingly. The workers are assigned to the machines according to the labor requirement of the different tasks. The labor requirement denotes whether a worker has to be present during the whole duration of a task or only from time to time to supervise the task. The former are labor intensive tasks which require one or more workers and the latter are normal tasks which require a fraction of a worker, i.e., a worker can operate and supervise more than one task at the same time. Only a certain percentage of the number of workers is allowed to execute labor intensive tasks at each time in order to have capacity for normal tasks. Workers have a long break and a (short) washing break during the shift. The long break has to be taken between 12.00 noon and 2.00 pm for the day shift and 12.00 midnight and 2.00 am for the night shift. The workers of a shift have to go in three groups, one group after the other. The washing break has to take place in the last hour of the shift. As a prerequisite, no new jobs are started in the last hour of a shift.

Special Features workers have different skills: some can operate every machine, others most of them, while again others can only operate a few machines. Workers are given a priority for each machine which expresses their qualification in operating this machine. The labor intensive work load has to be spread equally over the workers of a shift. They can be asked to work more than 100%, but this over-occupation rate is limited to a certain percentage and there is a limit in hours on how long the over-occupation can last, too. However, they do not have to work longer than their 12-h shift. Assigning workers to tasks with fractional labor requirements on machines of the same group is preferred, because that would reduce their distances for moving between machines.

10.5.2.3 Machines

Main Features The total number of machines is approximately 70, but only a fraction of the machines is in use simultaneously. The other machines are not active and not available to the production system. Furthermore, not all the active machines are available, too. They are either in use or idle, because some are assigned exclusion times and, thus, are temporarily not available. However, they are accessible again after this time ends. Exclusion times can be processing jobs that cannot be preempted, large repairs, normal maintenance, rebuilding, and so on. There are different machines that can produce the same product using the same procedure. The only difference is the capacity of the machines involved. There are different machines using different procedures but producing the same product, and there are machines where each can process different procedures producing different products. Hence, there is a choice between different machines for an order and competition between orders for running on a specific machine. This competition is a general one, i.e., once an order and a machine have been linked all jobs of this order run on this machine according to the respective procedure. Each machine has a different capacity determining how much product can be produced with one job.

Special Features The machines are divided into four groups representing the different physical floors in the multi-storey plant. Once one machine of a group processes a job it requires at least one worker to stay with this group of machines permanently.

10.5.2.4 Services

Special Features The workers have to fulfill some duties that do not require the use of machines. These services follow the same structure as the production system. They are defined by a procedure and if they have to be executed the production system receives an order for this service, which is then introduced into the system like any other order.

10.5.2.5 Objectives

The client in general receives orders in advance for up to 6 months, but certain products have such a high demand that up to 12 months of orders are known in advance. On the operational side it is necessary to know what orders, i.e., jobs and tasks, are currently in process or will be processed in the near future (2 weeks). Because of these requirements two different planning horizons are considered. Long-term planning is a time span of 1 year divided into slots of 12 h, i.e. , one slot represents one shift, giving 730 time slices. Short-term planning has a time window of 2 weeks and a resolution of 1 h, giving 336 time slices. Long-term planning will work on average labor demand for the orders and no single tasks are considered. This plan is meant to help with the management of the negotiation of due dates for new customer orders arriving at the system, of the availability of labor, and of the absence of workers, i.e., courses, training, holidays, long-term illness. It is expected to be used once a month. Short-term planning seeks to improve the planning of the operational side with a horizon of 2 weeks into the past and future (2 and 12 days, respectively) providing the client with an improved scheduling plan. This planning is based on the detailed data, using an hourly time slot system. It is meant to help react to sudden changes during the operation of the machines, e.g., accidents, urgent orders, break downs, etc. The short-term planning system is expected to be used several times a day.

10.5.3 Mathematical Formulation of the Problem

10.5.3.1 General Framework

Our problem includes an assignment model and a sequencing model. The resource constraints, i.e., the detailed personnel profile, can be included using a time-indexed formulation. These substructures motivate us to define a sequence of mathematical models with increasing complexity:

Model E := Time-indexed formulation with preassigned job-to-machine relations and given sequence on each machine
Model D := Model E including buffer times between tasks
Model C := Model D including the sequencing problems for preassigned job-to-machine relations
Model B := Model C including the assignment problem
Model A := Model B including all special features of the problem

Obviously, we have the inclusion: $E \subset D \subset C \subset B \subset A$. In this book we will only give a mixed integer formulation of Model E. Problem E on its own is already an extremely hard problem. Some of the ingredients of this model have already been encountered in Sect. 10.2.3.

10.5.3.2 Time Discretization

The smallest time interval to be considered is 1 h. Tasks which last fractions of an hour are rounded to the next upper or lower integer.

10.5.3.3 Indices

The following set of indices will be used

$$
\begin{aligned}
i &= 1,\ldots,N^O &&: \text{set of orders}\\
j &= 1,\ldots,N_i^J &&: \text{set of jobs in order } i\\
k &= 1,\ldots,N_i^K &&: \text{set of tasks within the jobs of order } i\\
m &= 1,\ldots,N^M &&: \text{set of machines}\\
t &= 1,\ldots,N^T &&: \text{set of time slices.}
\end{aligned}
\qquad (10.5.1)
$$

10.5.3.4 Data

Here are the data defining an instance of the problem

C_{jt} : costs if job j starts in t
D_i : due date for order i
L_t : labor capacity, i.e., the number of workers available in period t. This value may vary between 4 and 15.
L_{jt} : labor requirement, i.e., the number of personnel required by job j in the tth period of the processing time. Typical values are 1/3, 1/2, (2/3), 1, (2), (3). The numbers in brackets are less frequent.
L_{ikm} : labor requirement, i.e., the number of personnel required by task k of each job of order i on machine m. Typical values are 1/3, 1/2, (2/3), 1, (2), (3). The numbers in brackets are less frequent.
N^P : constant labor capacity during the whole planning horizon. This value may vary between 4 and 15.
P_j : processing time of job j
$P_{j_1 j_2}$: minimum separation between the start times of job j_1 and j_2
$R_{j_1 j_2}$: is defined in such a way, that $P_{j_1 j_2} + R_{j_1 j_2}$ is the maximum
$T_j^{L(U)}$: earliest (latest) possible start time of job j.

(10.5.2)

10.5.3.5 Main Decision Variables

We will use the binary variables

$$
\delta_{jt} := \begin{cases} 1\,, & \text{if job } j \text{ starts in period } t\\ 0\,, & \text{otherwise} \end{cases} \qquad (10.5.3)
$$

and

$$\alpha_{jt} := \begin{cases} 1 \text{ , if job } j \text{ starts in or before period } t \\ 0 \text{ , otherwise.} \end{cases} \tag{10.5.4}$$

10.5.3.6 Other Variables

Furthermore we will use some variables which are easily derived from the basic variables.

$$\begin{aligned} \mu \ &\in \mathbb{N} : \text{integer variable denoting the makespan or completion} \\ &\qquad \text{time of all jobs resp. orders} \\ \sigma_j \ &\in \mathbb{N} : \text{integer variable denoting the starting time of job } j \end{aligned} \tag{10.5.5}$$

10.5.3.7 Auxiliary Sets

For convenience we introduce the following definitions and sets

$$\begin{aligned} \mathcal{N} &:= \{1, 2, \ldots, N^J\} \\ \mathcal{T} &:= \{1, 2, \ldots, N^T\} \\ \mathcal{T}_j &:= \{t | T_j^L \le t \le T_j^U\}. \end{aligned} \tag{10.5.6}$$

10.5.4 *Time-Indexed Formulations*

Let us suppose there is one order i preassigned per machine, and each order consists of N_i^J identical jobs with processing times P_j. Since we are not interested in solving the assignment or sequencing problem we need no longer be concerned with orders and jobs. So we only use jobs, but keep in mind that some attributes which now are related to jobs are really associated with orders.

In order to minimize the makespan μ we define one artificial job called makespan j_*, which has to wait until all other jobs are finished and has processing time 1. Although we want to minimize the makespan, for mathematical reasons it might help to introduce some costs C_{jt} if job j starts in t and to consider an objective function which minimizes the sum of all starting costs. Such an objective function occurs if the labor constraints are dualized. If we need exact agreement with the problem description we can put the C_{jt} to zero and consider only the artificial makespan job with cost $C_{j_*t} = t - 1$.

So the model considered involves a discrete-time horizon $\mathcal{T}$, a set $\mathcal{N}$ of jobs, starting cost functions C_{jt} defined on $\mathcal{T}$ and $\mathcal{N}$ representing the cost incurred by job j if it starts at time t, an acyclic precedence digraph $\mathcal{D} = (\mathcal{N}, A)$, and three values P_j, P_{lj} and $R_{lj} \geq 0$ for all $(l, j) \in A$ with $P_j \leq P_{lj}$. P_j is the processing time of job j, P_{lj} is the minimum separation between the start times of l and j, respectively, whereas $P_{lj} + R_{lj}$ is the maximum separation.

In addition there are labor constraints: once started each job i requires a different amount L_{ju} of a limited resource in each of the $u = 1, \ldots, P_j$ periods of its processing time, while L_t is the total amount of resource available in period t.

10.5.4.1 The Delta Formulation

There is a "natural" time-indexed formulation for this problem that we first met in Sect. 10.2.3. Again, we use binary variables

$$\delta_{jt} := \begin{cases} 1 \text{ , if job } j \text{ starts in period } t \\ 0 \text{ , otherwise} \end{cases} \quad , \quad \forall j \quad , \quad \forall t \in \mathcal{T}_j . \tag{10.5.7}$$

Note that

$$\sigma_j = \sum_{t \in \mathcal{T}_j} t \delta_{jt} \quad , \quad \forall j \in \mathcal{N} \tag{10.5.8}$$

is the start time of job j. We now obtain the formulation :

$$\min \quad \sum_{j \in N} \sum_{t \in \mathcal{T}_j} C_{jt} \delta_{jt} \tag{10.5.9}$$

$$\sum_{t \in \tau} \delta_{jt} = 1 \quad , \quad \forall j \in \mathcal{N} \tag{10.5.10}$$

$$\sigma_j - \sigma_l \equiv \sum_{t \in \tau} t \, \delta_{jt} - \sum_{t \in \tau} t \, \delta_{lt} \geq P_{lj} \quad , \quad \forall \{lj\} \in A \tag{10.5.11}$$

$$\sigma_j - \sigma_l \equiv \sum_{t \in \tau} t \, \delta_{jt} - \sum_{t \in \tau} t \, \delta_{lt} \leq P_{lj} + B_{lj} \quad , \quad \forall \{lj\} \in A . \tag{10.5.12}$$

The labor constraint reads

$$\sum_{j \in \mathcal{N}} \sum_{s=\max\{1, t-P_j+1\}}^{t} L_{j,t-s+1} \, \delta_{js} \leq L_t \quad , \quad \forall t \in \mathcal{T} . \tag{10.5.13}$$

The makespan, μ, expressed in the δ_{jt} variables is given by

$$\mu = \sum_{t \in \mathcal{T}_j} (t-1) \delta_{j_* t} , \tag{10.5.14}$$

and thus the objective function minimizing the makespan follows as

$$\min \sum_{t \in \mathcal{T}_j} (t-1)\delta_{j_* t}. \tag{10.5.15}$$

Several variants of this model have been proposed in the literature, see for instance Pritsker et al. (1969,[448]), Talbot & Patterson (1978,[538]), and Christofides et al. (1987,[123]). In the special case that $L_{ju} = 1$ for all j, u and $L_t = 1$ for all t, it reduces to a single-machine scheduling problem. The time-indexed formulation of this problem, but without the precedence and buffer constraints (10.5.11) and (10.5.12), has recently been studied from a polyhedral point of view by Sousa & Wolsey (1992,[515]) and van den Akker (1994,[558]). Tavares (1995,[539]) contains a recent review of different approaches to project scheduling.

The constraints (10.5.11) and (10.5.12) can be disaggregated giving:

$$\sum_{s=1}^{t} \delta_{is} \geq \sum_{s=1}^{t+P_{lj}} \delta_{js} \quad , \quad \forall \{ij\} \in A \tag{10.5.16}$$

and

$$\sum_{s=1}^{t} \delta_{is} \leq \sum_{s=1}^{t+P_{lj}+R_{lj}} \delta_{js} \quad , \quad \forall \{ij\} \in A \tag{10.5.17}$$

10.5.4.2 The Alpha Formulation

Making a standard variable change for time-indexed formulations by introducing for each job j and each period $t \in \mathcal{T}$, we define

$$\alpha_{jt} = \begin{cases} 1 \text{ , if job } j \text{ starts in or before period } t \\ 0 \text{ , otherwise} \end{cases} \tag{10.5.18}$$

Thus, $\delta_{jt} = \alpha_{jt} = 0$ for all j and all $t < T_j^L$, where T_j^L is the earliest start time of j, $\delta_{jT_j^L} = \alpha_{jT_j^L}$ for all j, and $\delta_{jt} = \alpha_{jt} - \alpha_{j,t-1}$ for all j and $t = T_j^L + 1, \ldots, N^T$. To support the understanding of the variables α_{jt} we give the following relation

$$\alpha_{jt} = \begin{cases} 1 \text{ , if } t \geq \sigma_j \\ 0 \text{ , if } t < \sigma_j , \end{cases} \tag{10.5.19}$$

in which σ_j denotes the starting time of job j.

Formally, variables α_{jt} need to satisfy the following constraints

$$\alpha_{jN^T} = 1 \quad , \quad \forall j \in \mathcal{N} \tag{10.5.20}$$

$$\alpha_{jt} \leq \alpha_{j,t+1} \quad , \quad \forall j \in \mathcal{N} \quad , \quad t = T_j^L, \ldots, N^T - 1 \tag{10.5.21}$$

$$\alpha_{lt} \geq \alpha_{j,t+P_{lj}} \quad , \quad \forall \{ljt\} \tag{10.5.22}$$

$$\alpha_{lt} \leq \alpha_{j,t+P_{lj}+R_{lj}} \quad , \quad \forall \{ljt\}. \tag{10.5.23}$$

Now the labor constraint reads

$$\sum_{j \in N} \sum_{s=t-P_j+1}^{t} L_{j,t-s+1} (\alpha_{js} - \alpha_{j,s-1}) \leq L_t \quad , \quad \forall t \in \mathcal{T} \tag{10.5.24}$$

$$\alpha_{jt} \geq 0. \tag{10.5.25}$$

The makespan, μ, expressed in the α_{jt} variables is given by

$$\mu = N^T - \sum_{t \in \mathcal{T}_j} \alpha_{j_* t} \tag{10.5.26}$$

and, thus the objective function minimizing the makespan has the following form

$$\min \left(N^T - \sum_{t \in \mathcal{T}_j} \alpha_{j_* t} \right) , \tag{10.5.27}$$

which is a special form of (10.5.9) with most of the values $C_{jt} = 0$ and those for $j = j_*$ are put to $C_{j_* t} = -1$. The earlier the makespan job j_*, i.e., the earlier all other jobs are finished, the larger is the sum $\sum_{t \in \mathcal{T}_j} \alpha_{j_* t}$ (according to the definition of the α_{jt} variables). Instead of maximizing this sum we minimize the term in (10.5.27). The constant term N^T is not really important, of course, but it supports the interpretation of the term as the makespan.

Now we have two formulations and we might ask ourselves which one is better. How good is the δ or α formulation? The answer: it can be shown that the polyhedron (10.5.20)–(10.5.23) and (10.5.25) is integer. In other words: the linear programming relaxations of the δ and α formulations are integer if no labor constraints are present or active.

For any acyclic precedence digraph, this result can be derived from Flin et al. (1982,[188]) where it is essentially shown that the optimization problem defined on the polyhedron (10.5.20-10.5.23) and (10.5.25) is the dual of some minimum-cost network flow problem. As a consequence we note that the problem without the labor constraints is easily solvable as a linear program or as a network flow problem. In the presence of the additional labor constraints (10.5.13) or (10.5.24), the given formulation is not necessarily integral any more. As it contains the knapsack constraints (10.5.13) or (10.5.24) plus the generalized upper bound constraints (10.5.10), the knapsack cuts exploiting the GUB structure derived in Wolsey (1990,[588]) could be used to tighten the above formulation.

10.5.5 Numerical Experiments

In order to approach the full model we will first present some simple scenarios with only a few orders and not too many jobs and tasks. We consider two groups of scenarios of different complexity. Later, we will apply the model to a client's prototype.

10.5.5.1 Description of Small Scenarios

This problem considers only about 15 to 18 jobs which last only up to 4 h and belong to at most 4 orders. The scenarios assume that $N^P = 3$ or $N^P = 4$ workers are available and have to process N^O orders i each consisting of a number of N_i^J jobs. All jobs of the same order (jobs are counted $1, \ldots, j, \ldots, N^J$; in the table below, jobs from j_i^F and j_i^L belong to order i) are characterized by a unique personnel profile. These data are summarized in the following tables:

i	j_i^F	j_i^L	P_j	N_i^J	M_i^O	D_i
1	1	2	4	2	10	8
2	3	4	3	2	8	6
3	5	7	2	3	9	6
4	8	10	3	3	12	9
5	11	13	4	3	18	12
$\sum$					57	41

,

$i\backslash t$	1	2	3	4	M_i^J
1	1	1	1	2	5
2	2	1	1		4
3	1	2			3
4	1	2	1		4
5	1	2	1	2	6

,

where M_i^J specifies the total amount of personnel (in hours) required by each job in order i. This allows us to compute the total number of personnel M^O (in hours) required by order i. The following relations hold

$$M_i^O := N_i^J M_i^J . \tag{10.5.28}$$

Furthermore, the table gives the value of the total duration D_i of an order if labor is not an active restriction, i.e.,

$$D_i := P_j N_i^J . \tag{10.5.29}$$

The objective is to minimize the makespan. An "artificial" last job is introduced that follows all the others, and the objective is to start this job as soon as possible.

The first example considers the first three orders of the tables. The makespan is $\mu = 10$, i.e., all jobs are finished within 10 time units. Although no order-to-machine assignment was given in advance, it turns out that each order runs on a separate machine, and no machine is used by more than one order. The starting times are summarized in the table

$i\backslash j$	1	2	3
1	3	7	
2	1	5	
3	1	3	7

. (10.5.30)

The schedule associated with these starting times is

$i\backslash t$	1	2	3	4	5	6	7	8	9	10	11	12
1			1	1	1	2	1	1	1	2		
2	2	1	1		2	1	1					
3	1	2	1	2			1	2				
P	3	3	3	3	3	3	3	3	1	2		

.

For the second example the table below gives the starting times obtained for a scenario of 4 orders each connected to one of 4 machines. Each order is assigned to one machine, and each machine is used by at most one job. The minimum makespan μ or completion time for three workers available is $\mu = 14$.

$i\backslash j$	1	2	3
1	7	11	
2	1	9	
3	3	5	7
4	1	4	11

.

The schedule associated with this starting time is

$i\backslash t$	1	2	3	4	5	6	7	8	9	10	11	12	13	14	15
1							1	1	1	2	1	1	1	2	
2	2	1	1						2	1	1				
3			1	2	1	2	1	2							
4	1	2	1	1	2	1					1	2	1		
P	3	3	3	3	3	3	2	3	3	3	3	3	2	2	

.

In addition, the case $N^P = 4$ has been solved. The optimal solution $\mu = 11$ and its schedule are

$i\backslash j$	1	2	3
1	1	5	
2	1	8	
3	3	6	10
4	1	5	9

,

$i\backslash t$	1	2	3	4	5	6	7	8	9	10	11	12
1	1	1	1	2	1	1	1	2				
2	2	1	1					2	1	1		
3			1	2		1	2			1	2	
4	1	2	1		1	2	1		1	2	1	
P	4	4	4	4	2	4	4	4	2	4	3	

.

10.5.5.2 A Client's Prototype

This prototype has been set by a BASF customer. It covers 840 time slices and considers $N^J = 88$ jobs leading to 17,000 binary variables. The subject of this investigation is planning a 5-week period in hours ($N^T = 840$ h) with optimal labor occupation rates. $N^O = 10$ orders have to be scheduled on $N^M = 9$ machines. Each order runs on one pre-selected machine, orders 6 and 7 use the same machine. The sequence should be decided by the optimizer. The necessary amount of each order is produced not at once but by $N_{c(k)}$ parts ($N_{c(k)} \approx 3 - 30$) called *jobs*. These jobs each consist of 1 up to 6 steps or *tasks* k, characterized by a specific labor requirement L_{jk} and different durations P_{jk}. As all jobs of an order are assumed to be identical, the order of jobs within an order is fixed to avoid symmetries. The jobs are counted with numbers $1, \ldots, N^J$. A due date for the last job (job with number $N_{l(k)}$) of each order (= delivery time of order) is given (Table 10.1). There are L_t shift workers present ($L_t = 2 - 10$) at time t. In this prototype it is assumed that the

Table 10.1 This table lists for all orders i the given due dates D_i and the attributes $A_{ik} - B_{ik} : M_{ik}$ of each task, i.e., start time — finish time from start of the corresponding task: labor requirement in units of $1/6$ worker

i	D_i	$A_{ik} - B_{ik} : M_{ik}$				
1	264	1–2: 12	3: 6	4–9:2	10–11:6	12–24:3	25–28:12
2	762	1: 12	2–7: 2	8:6			
3	432	1–2: 6	3–19: 2	20–21:6	22–23:9		
4	552	1–2: 2	3–9: 2	10–11:6	12–24:2	25–26:12	
5	624	1–2: 6	3–9: 2	10–12:6	13–24:2		
6	744	1–12: 6					
7	840	1–2: 6	3–9: 3	10–12:6			
8	840	1–2: 12	3:6	4–24:2	25–27:6	28–34:3	35–37:12
9	744	1–2: 12	3–19: 3	20–24:6	25–29:3	30–31:6	
10	840	1–5: 6	6 :12	7–15:2	16–21:6		

number L_t of present personnel is constant during the whole planning period, i.e., $L_t = N^P$ for all t. Times for breaks and variations because of holidays or illness are not taken into consideration, nor is it the aim to assign workers to machines or jobs. The objective of the optimization is minimizing the schedule length (makespan μ).

The production of orders 2–6 and 9–10 is connected, i.e., the output of one order serves as a pre-product for the following one. The remaining orders are not restricted by their pre-products. The production of one job j_{i_1} of order i_1 requires $p = p(i_1, i_2)$ jobs j_{i_2} of order i_2. We denote this by the notation $(i_1 \leftarrow p \cdot i_2)$. In this model the following prescriptions represent the connections between the different orders:

$$\{(3 \leftarrow 2.5 \cdot 2), (4 \leftarrow 3 \cdot 3), (5 \leftarrow 1 \cdot 4), (6 \leftarrow 1 \cdot 5), (10 \leftarrow 0.3 \cdot 9)\}. \qquad (10.5.31)$$

From theses prescriptions and from the given orders the vector of jobs to produce (7, 30, 12, 4, 4, 4, 10, 4, 3, 10) is computed, i.e., in total $N^J = 88$ jobs. The directed graphs displayed in Fig. 10.4 show the precedence relations between the jobs of all orders. The sequence of the orders 6 and 7 (jobs 58–61 and 62–71) is not known beforehand.

The orders and jobs are completely characterized by the data stored in MCOL.

From the duration of the jobs of an order, the given deadlines and the links between several jobs, restricted start time intervals can be computed similar to the test for local consistency before starting the B&B procedure with constraint programming techniques [see Sect. 10.5.6]. By this the intervals between earliest and latest start time for most jobs are getting smaller. For orders not connected to the production of others, the latest start time is given by the duration of a job and the due date of the order. If the production of an order is linked to others, then apart

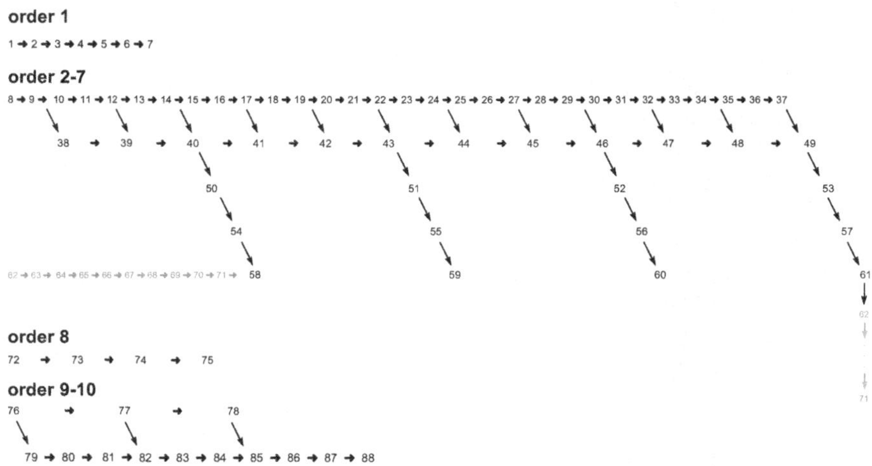

Fig. 10.4 Precedence relations between jobs

from the due date of the final product, the due dates for the pre-products (e.g., for safety reasons) lead in most cases to additional restrictions.

The prototype presented in this section contains another feature which leads to a sequencing problem: orders 6 and 7 are processed on the same machine. One order should only start if the other has been completed. The *or*-relation ("order 6 is produced before order 7" *or* "order 7 is produced before order 6") might be expressed by an additional binary variable η indicating whether order 6 is produced before order 7, i.e., whether the last job of order 6 ends before the first job of order 7, or vice versa. For the sake of simplicity and because we only want to formulate *Model E* we do not use this approach. Instead of that we solve the problem without the sequencing component and check afterward whether the sequencing conditions are fulfilled. If not we solve two subproblems forcing the order 6 to be produced before order 7 or vice versa.

Results As already mentioned in Sect. 10.5.4 the LP relaxation of the given formulation is integral without labor constraints only if besides (10.5.24), the precedence constraints are disaggregated, namely only if the constraints (10.5.16) and (10.5.17) are added to the formulation and if a final fictitious job is added to model the makespan objective. Our experience shows that this result also holds if the labor constraints do not become active. Unfortunately, this is only the case for relatively trivial cases $N^P \geq 15$. For the more interesting case $N^P = 4$, just computing the LP relaxation requires already about 2 h CPU time on an IBM RS/6000 workstation. Table 10.2 contains the results of an investigation on the influence of the number L_t of available personnel on the total makespan of all given orders. These results were achieved by using *constraint programming techniques* [257] further discussed in Sect. 10.5.6. The result for $N^P = 4$ is shown in Fig. 10.5 as Gantt charts and occupation diagram.

In order to evaluate the quality of our solution we will derive some bounds on the optimal solution value. If we relax the labor constraint, i.e., assume that enough personnel is available then we can proceed as follows: based on precedence constraints between the jobs we derive $\mu \geq 362$ as a lower bound on the makespan. For values $N^P \leq 2.82$ this bound can be improved by dividing the sum of required man hours by the number of present personnel. For $N^P = 2$ the bound $\mu \geq 512$

Table 10.2 This table presents the best results found with respect to objective function z_1. It contains the computation times Δ_t ($N^P = 3, 3\frac{1}{3}, 4\frac{1}{2}$ on IBM RS-6000, $N^P = 4$ with 486 PC (66 MHz, 32 MB main memory)), the makespan t_e resp. The sum of completion times Z with given number of personnel N^P and occupation rates U in percent. p rates the quality of the solution found in percent according to the definition (10.4.61); 0 indicates proven optimality

L_t	$\Delta_t[h]$	t_e	$t_e[h]$	$G[\%]$
3	1:00:48	$22^d 17^h$	545	50.6
$3\frac{1}{3}$	0:30:07	$19^d 11^h$	467	29.0
4	0:24:21	$15^d 20^h$	380	6.4
$4\frac{1}{2}$	0:43:32	$15^d 02^h$	362	0.0

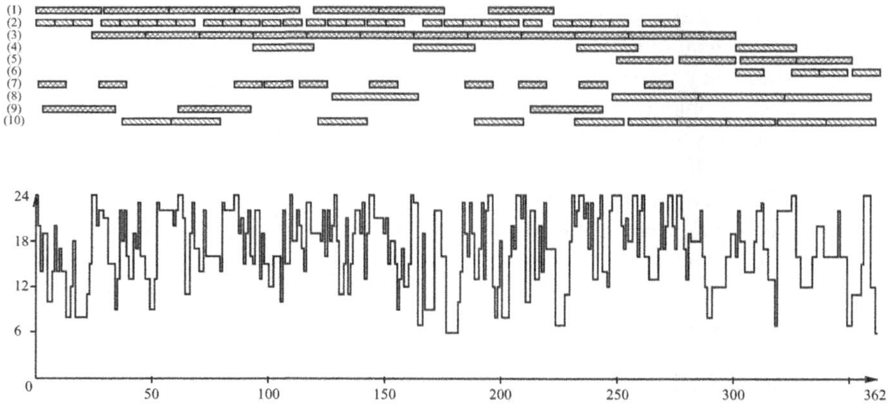

Fig. 10.5 Gantt chart and personnel occupation diagram

follows. The jobs of orders 1, 2, 8 and 9 all start with personnel requirement 2, so that if $N^P = 2$ workers in total are available two jobs of these orders cannot be processed at the same time. In addition orders 4 and 10 contain tasks with personnel requirement 2, so only the task before these work intensive tasks may be executed at the same time like other jobs of orders 1, 2, 4, 8, 9 or 10. Therefore, we derive $\mu \geq 604$ if $N^P < 2\frac{1}{6}$.

So far, the MILP approach using the time-indexed formulation described in Sect. 10.5.4 was not very efficient. Only for $N^P \geq 6$, we were able to derive the results summarized in the table[5] below:

N^P	9	8	7	6	5	4	3
t_c	362	362	362	362	362	362	≤ 408
n^*	27	29	165	346	277		
Time [sec]	66	91	2746	48249	76635		

.

Note the drastic increase in the CPU time required to prove optimality when N^P decreases. For $N^P \geq 5$ the optimality of the solution has been proven. For $N^P = 4$ no MILP approach was able to produce an integer feasible solution because there are too many nodes in the B&B tree. The problem becomes much harder when the labor constraints become more restrictive, i.e., for smaller values of N^P.

[5]The values for $N^P \leq 4$ have been computed by Colombani & Heipcke (1997,[127]) in less than 2 min using the constraint programming software package `SchedEns`.

10.5.6 What Can be Learned from This Case Study?

In the actual formulation of the problem only a few types of binary variables have been used. Even that caused great problems. Much care is needed in order to save the model from an inflation of binary variables. So in the long run, if end users keep on insisting on increasing the size of their problems (the length of the planning horizon, time resolution, etc.) and there is no huge breakthrough in mathematics which would move MILP into the class of polynomial problems [see remarks on page 555], we are going to have to give up trying to solve MILP problems of a certain size.

In that case the last resort is to use heuristics, e.g., simulated annealing, tabu search, or as discussed below, *constraint programming*. Heipcke (1995,[257]) applied constraint programming techniques to solve this scheduling problem. Constraint programming techniques originated in artificial intelligence research. They can be used to represent complex solution spaces in a natural manner by local relations between variables, generally called constraints. There are no restrictions or limitations on the structure of these relations. Local propagation techniques support the search for optimal solutions in discrete problems. They can significantly reduce the size of the solution space to be inspected. That approach could produce solutions, and when coupled with the LP relaxation allowed us to estimate the quality of the solution. For a full discussion and results achieved for that problem using an object-oriented CP tool the reader is referred to Heipcke et al. (1995,[257]). Further work by Colombani & Heipcke (1997,[127]) using the constraint programming package `SchedEns` succeeded in solving problems with a few hundred jobs and 2000 tasks in less than 20 min.

Another promising attempt (private communication with R. Möhring) is the implementation of a B&B approach for the BASF scheduling problem that is based on work by Bartusch et al. (1988,[50]). This approach permits temporal constraints in the form of arbitrary, context-sensitive time lags between starting and/or completion times of activities and time-dependent resource requirements and availabilities in the form of piece-wise constant step functions. The B&B algorithm constructs feasible schedules along the time axis, starting at time 0 and, at every decision time t, branches into feasible decisions (about which set of jobs to start at t) with respect to the resource constraints, while always observing the temporal constraints. It thus always maintains a set of best possible feasible partial schedules for the different subproblems created by these decisions and seeks to either complete them into a full schedule, or to discard a branch because the best value that can be obtained by this branch exceeds the best objective value known so far.

This approach is combined with heuristics to generate good feasible solutions in order to start with good upper bounds, different strategies for exploring the subproblems created (best-first-search, depth-first-search, user-defined search), and several methods for generating good lower bounds for best possible value of the subproblems.

In addition, it uses a powerful domination rule which permits discarding a subproblem if the currently "active" jobs in the partial schedule have been encountered already in a different branch with a better lower bound. This domination rule is based on an efficient search tree implementation of all sets of active jobs encountered so far. For three different scenarios with four different personnel resources, each was solved to optimality or proven to be infeasible within 90 min. The reason for the success is that this approach uses a specialized B&B which is expected to do better than a general B&B in MILP.

What can we conclude from the case study in this section? Are scheduling problems too complex to be tackled by mathematical optimization? The answer is clearly: no! But we should be prepared for the fact that solving scheduling problems is very demanding in know-how, and may require more than standard software. There is no straightforward solution to this class of problems and each problem is a challenge.

10.6 Telecommunication Service Network ⊖

In this study, the client wants to determine the cost-minimal configuration of a corporation's telecommunications service network operating nationwide based upon given demands for voice and data transmission. The whole problem involves two models for optimizing the Virtual Private Network and toll-free number service costs based on given voice traffic demands. In the case study we focus on the optimization of private line service cost based upon given data traffic demands.

The mathematical formulation of this subproblem basically corresponds to that of a capacitated network design problem, where the sites of the corporation are the nodes and the possible private line connections are the edges of the network. More precisely, the problem is modeled as a special type of minimum-cost (capacitated) multi-commodity network flow problem, in which the capacities and the corresponding cost of the edges may take on certain discrete possible values, while the requirements on the flows (demands of data transmission) through the edges have continuous values. In addition each of the commodities (demands) has to be routed via a single path through the network to be designed. The corresponding MILP model basically just involves binary variables for the determination of the edge capacities and for the decisions on the paths through the network.

In Sect. 10.6.1, we give an overview of the problem avoiding any mathematics. Section 10.6.2 introduces the mathematical model formulation for the data-network subproblem. Section 10.6.3 deals with improvements and reformulations of this model.

10.6.1 Description of the Model

10.6.1.1 Technical Aspects of Private Lines

A private line is an analog or digital physical line between any two so-called Points of Presence (POPs) of the carrier, which is rented exclusively to one customer for voice and/or data communications. It is not used by any other customer of the telecommunications carrier. Each carrier has its own POPs which are nodes of its nationwide network. In order to be able to utilize private line services the customer has to lease so-called access lines, which connect a corporation's site to the closest POP of the chosen carrier.

Today, *private line services* are mainly used for data traffic, although voice traffic is also possible on those lines. Customers have to pay utilization-independent monthly recurring charges which are based upon the distance covered by the line and upon the bandwidth of the line. The bandwidth or transmission rate refers to the information-carrying capacity of a line. It is generally specified in bits per second (bps), or kilobits per second (Kbps), respectively. Bandwidth is provided by the telecom carriers in certain discrete values, usually in integral multiples of 56 or 64 Kbps, depending on the carrier. (Basically there is no difference between 56 Kbps multiples and 64 Kbps multiples, because in the case of 64 Kbps multiples the difference of 8 Kbps is used for network management information. The performance of both is the same.) The following types of private lines are available on the market:

- Digital Service 0 (DS0) or Digital Data Service (DDS) provides digital transmission at a speed of 56 Kbps.
- Digital Service 1 (DS1) or Terrestrial 1.544 Mbps-line Service (T1) supports transmission of voice and data over digital lines at 1.544 Mbps. Each DS1 represents 24 DS0 channels at 56 Kbps.
- Fractional T1 Service (FT1) supports voice and data transmission over digital lines in multiples of 56 Kbps up to 768 Kbps.

DS0, FT1, and DS1 lines can be split up into an appropriate number of single channels according to the customer's demand.

In general an *access line* is also a private line, but one which is leased for the distance from the customer's site to the carrier's POP.

10.6.1.2 Tariff Structure of Private Line Services

There are two different types of tariffs for private line services:

- non-recurring installation costs,
- periodically recurring charges (monthly recurring charges for access lines or private POP-to-POP lines).

Under certain circumstances (e.g., access lines) part or all of the installation charges are waived by the communications carrier.

The monthly recurring charges for POP-to-POP private lines depend on:

- chosen carrier,
- distance (according to the distance rate band, state),
- bandwidth,
- terms of contract with the carrier.

Additionally leased access lines from the sites to the carrier's POP are necessary for both end nodes of a private line. Obviously, the recurring charges for private lines and leased access lines are independent of the transmitted volume.

The dependence of the tariff for private lines on the *bandwidth* is described by a step function: the tariff increases nonlinearly with the number of leased 56 Kbps-channels [see Fig. 10.6]. There is a connection-dependent break-even point, at which the cost of a number of single 56 Kbps-channels (or the cost of a fractional T1 connection) becomes equal to that of a DS1 or T1-line. This break-even is found at about 6–10 single 56 Kbps-channels depending on the considered connection. The cost of a number of T1 lines increases linearly with the number of T1s.

The dependence of the tariff on the *distance* is determined by pricing plans of the carriers: carriers have defined their own service-dependent distance rate bands. The distance is usually determined by the mileage calculated from the horizontal and vertical geographical co-ordinates of the customer sites involved. Some carriers

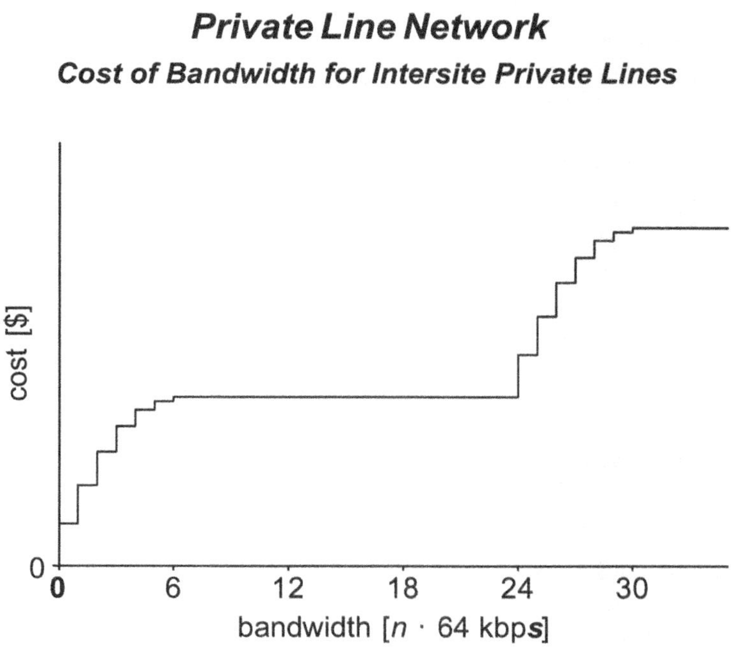

Fig. 10.6 Cost of bandwidth for POP-to-POP private lines

also distinguish between interstate and intrastate rates: interstate applies to service between two different states of the US, intrastate refers to service within a single state. The private line discounts depend on:

- chosen carrier,
- monthly volume (measured in US $) for all private lines,
- terms of contract.

10.6.1.3 Demands on Private Line Services

It is relatively complicated to investigate the corporation's demand for data transmission because it is technically difficult to measure the existing data transmissions on each of about 200 existing lines within a sufficiently large typical time period. The identification of original source and final destination of all those transmissions turned out to be especially difficult.

Therefore it is assumed that all existing bandwidths configured for each logical site-to-site data connection are really needed. Further investigations on how to measure data traffic are being conducted.

10.6.1.4 Private Line Network Optimization

The objective of this optimization is to (re-)design the structure of the corporation's private line network such that the given traffic demands between the sites of the corporation are routed via cost-optimal paths through the network. Certain predefined sites — so-called hub sites — with the appropriate knowledge and equipment, should be used for routing purposes.

10.6.2 Mathematical Model Formulation

The model discussed here covers the usage of private lines for data transmission. We will merely consider one carrier at a time, and we will subsequently compare the carrier-dependent optimization results. In practice it might be preferable to consider two or three carriers simultaneously and to distribute the total service volume in order to some emergency backup lines. Based on the financial situation, the simplification seems to be sensible, as higher total service volumes are assigned to a single carrier so that higher overall volume discounts could be expected. Other preconditions, simplifications, and restrictions, that apply for this model, are the following:

- consideration of the total cost of a single time period (1 month);
- cost of private POP-to-POP lines as a step function of the number of 56Kbps trunks;

- no consideration of access costs;
- no variation of the term of the contract: all charges based upon a 3 years' term of contract.

The model addresses the design of the corporation's private line network where the demands of data transmissions are given as bandwidths required between any two sites of the corporation. The objective is to find for all those logical connections optimal routings via certain hub sites with minimum overall network cost.

This network design problem is modeled as a minimum-cost capacitated multi-commodity network flow problem. The model involves binary flow variables, integer and SOS1 of design variables, and real cost variables. The set of constraints consists of

- flow conservation constraints (Sect. 10.6.2.2),
- edge capacity constraints (Sect. 10.6.2.3),
- additional constraints (Sect. 10.6.2.4).

10.6.2.1 General Foundations

Given a set $\mathcal{S}$ of N^S corporation sites with certain data traffic demands between the sites, the objective of this model is to design an optimal private line network — using a number of N^H predefined sites as hub sites for routing purposes — in order to minimize the total data traffic cost of the corporation.

The problem is modeled as a *minimum-cost capacitated multi-commodity network flow problem*, where

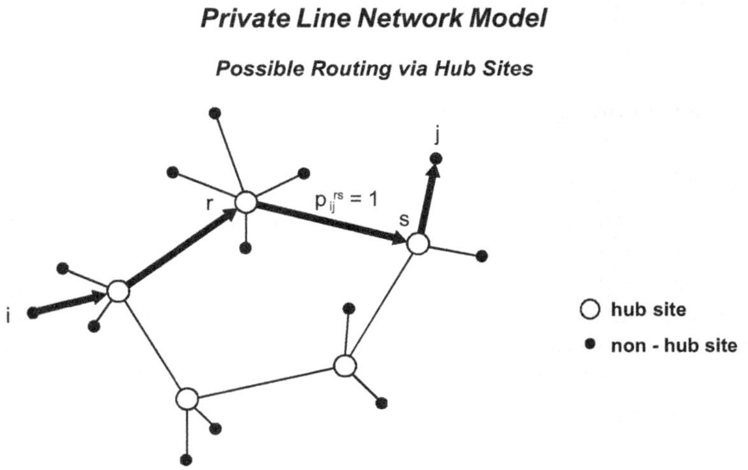

Fig. 10.7 A possible routing via hub sites for a demand D_{ij}

- the sites form the set of *nodes* of the network,
- the hub sites are a *subset of the nodes set*,
- the directed physical connections between any two sites are the *directed edges* of the network,
- the paths through the network are *sequences of directed edges*,
- the total bandwidth installed on an edge is the *capacity* of the edge, and
- the logical demands on bandwidth between any two sites of the corporation are the *commodities*.

The objective is to install capacity on the edges and find a feasible (simultaneous) routing of the demands on the resulting capacitated network, so that the overall capacity cost is minimized. The *capacities* of the edges can be purchased in integral multiples of two modularities, namely 56 Kbps (DS0, FT1 $= n \cdot 56$ Kbps, $1 < n \leq 12$) and $24 \cdot 56$ Kbps (T1).

The *capacity costs* of an edge include distance-, bandwidth-, and possibly state-dependent charges for using the private line services of the telecommunications carrier. While we assume that the cost of T1 lines is a linear function of the number of T1s, that is

$$c(n \cdot \mathrm{T1}) = n \cdot c(\mathrm{T1}) \tag{10.6.1}$$

we take into account that the cost of fractional T1 channels increases nonlinearly with the number of channels:

$$c(n \cdot 56\,\mathrm{Kbps}) < n \cdot c(56\,\mathrm{Kbps}) \tag{10.6.2}$$

and

$$c((n+1) \cdot 56\,\mathrm{Kbps}) - c(n \cdot 56\,\mathrm{Kbps}) < c(n \cdot 56\,\mathrm{Kbps}) - c((n-1) \cdot 56\,\mathrm{Kbps}). \tag{10.6.3}$$

Therefore we specify for each edge between any two sites r and s a grid of fractional T1 channels $(k-1) \cdot 56$ Kbps, $k \in \{1, .., N_{rs}^K\}$, where N_{rs}^K corresponds to the break-even point, at which a T1 line becomes less expensive than $N_{rs}^K \cdot 56$ Kbps channels, and let the optimizer decide on the optimal grid point.

For simplification reasons we assume that the *traffic demands* D_{ij} between any sites i and j of the corporation's network are given as — not necessarily integral — multiples of 56 Kbps. We associate commodities $[i, j]$ with these demands D_{ij} and denote the total number of demands $D_{ij} > 0$ by $N^D = \left|\{[i, j] \mid i, j \in \mathcal{S} \wedge D_{ij} > 0\}\right|$. A *feasible routing* of a commodity $[i, j]$ in the capacitated network has the following characteristics [see Fig. 10.7 and Brockmüller et al. (1997,[99]).]:

- All of D_{ij} are routed on a *single* directed path $\mathcal{P}_{ij}$. (A distribution of a demand on more than one path is not possible, as, e.g., two paths with a data transmission rate of 112 Kbps do not add up to one connection with a transmission rate of 224 Kbps.)

- If D_{ij} is routed on $\mathscr{P}_{ij} = \{i, r_1, r_2, \ldots, r_N, j\}$, then the nodes r_h, $h \in \{1, \ldots, N\}$, are hub sites. That is, traffic from node i to node j can either be sent directly, or via a number of hub nodes.

- For any undirected edge $\{r, s\}$ the total traffic routed using the edge (in either direction) is at most equal to the selected capacity of the edge.

The model description is simplified if we use the following sets. Let $\mathscr{S}$ denote the set of all nodes of the network ($|\ \mathscr{S}\ | = N^S$) and let $\mathscr{H} \subseteq \mathscr{S}$ denote the set of hub nodes ($|\ \mathscr{H}\ | = N^H$). The set $\mathscr{A}$ of *feasible edges* of the problem is given by

$$\mathscr{A} = \mathscr{A}^H \cup \mathscr{A}^I \cup \mathscr{A}^O \cup \mathscr{A}^D, \tag{10.6.4}$$

where the sets describe the edges between different sites

$$\begin{array}{lll} \mathscr{A}^H & := \{(r, s) \mid r, s \in \mathscr{H}\} & \text{between hub sites,} \\ \mathscr{A}^I & := \{(r, s) \mid r \in \mathscr{H}, s \in \mathscr{S} \setminus \mathscr{H}\} & \text{from a hub site to a non-hub site,} \\ \mathscr{A}^O & := \{(r, s) \mid r \in \mathscr{S} \setminus \mathscr{H}, s \in \mathscr{H}\} & \text{from a non-hub site to a hub site,} \\ \mathscr{A}^D & := \{(r, s) \mid r, s \in \mathscr{S}, D_{rs} > 0\} & \text{between sites } r\&s \text{with } D_{rs} > 0. \end{array} \tag{10.6.5}$$

Let (r, s) be a feasible edge. Then the set $\mathscr{K}(r, s)$ of *commodities that can be routed via* (r, s) is given by

$$\mathscr{K}(r, s) = \begin{cases} \{[i, j] \mid i, j \in \mathscr{S}, D^{ij} > 0\} & \text{if } (r, s) \in \mathscr{A}^H \\ \{[i, j] \mid i = r, j \in \mathscr{S}, D_{ij} > 0\} & \text{if } (r, s) \in \mathscr{A}^O \\ \{[i, j] \mid i \in \mathscr{S}, j = s, D_{ij} > 0\} & \text{if } (r, s) \in \mathscr{A}^I \\ \{[i, j] \mid i = r, j = s, D_{ij} > 0\} & \text{if } (r, s) \in \mathscr{A}^D \setminus \mathscr{A}^H. \end{cases} \tag{10.6.6}$$

Let $\mathscr{E}$ be the set of undirected edges obtained from $\mathscr{A}$:

$$\mathscr{E} = \{\{r, s\} \mid (r, s) \in \mathscr{A}\}. \tag{10.6.7}$$

The following variables will be used throughout this model:

π_{ij}^{rs} $\in \{0, 1\}$, equal to 1 if directed edge (r, s) appears in routing of D^{ij}, $[i, j] \in \mathscr{K}(r, s)$, and equal to 0 otherwise (*flow variables*);

γ_{rs} $\in \mathbb{N}_0$, capacity variable controlling the number of T1s installed on $\{r, s\} \in \mathscr{E}$ (*integer T1 design variables*);

β_{rsk} $\in \{0, 1\}$, SOS1 of capacity control, equal to 1 if a bandwidth of $(k - 1) \cdot 56$Kbps channels is installed on $\{r, s\} \in \mathscr{E}$, and equal to 0 otherwise (*SOS1 of FT1 design variables*);

c_{rs}^P [\$], cost of total bandwidth capacity installed on $\{r, s\} \in \mathscr{E}$.

10.6.2.2 Flow Conservation Constraints

Let the binary variable π_{ij}^{rs} indicate whether the directed edge (r, s) appears in the routing of commodity $[i, j]$ ($\pi_{ij}^{rs} = 1$), or not ($\pi_{ij}^{rs} = 0$). These variables are defined for feasible combinations of edges and commodities only, that is to say for all $[i, j], (r, s)$ with $[i, j] \in \mathscr{K}(r, s)$. We want to determine the single routing $\mathscr{R}_{ij}$ through the network for demand D_{ij} via the following set of directed edges:

$$\mathscr{R}_{ij} = \left\{(r, s) \in \mathscr{A} \,\middle|\, [i, j] \in \mathscr{K}(r, s) \wedge \pi_{ij}^{rs} = 1\right\} \quad , \quad \forall [i, j] \,\middle|\, D_{ij} > 0. \tag{10.6.8}$$

This set defines a feasible routing if the set includes

- one edge starting at i,
- one edge terminating in j (not necessarily different from the first edge), and if
- for each edge in the set terminating in a node $q \neq j$ there is another edge in the set starting in node q:

$$\forall (r, q) \in \mathscr{R}_{ij}, q \neq j : \qquad \exists s \in \mathscr{S}, (q, s) \in \mathscr{R}_{ij}. \tag{10.6.9}$$

This is ensured by the following *flow conservation constraints*: *source of a flow* is node i, there is either a direct private line from i to j for there is a flow from i to one and only one hub site $\neq j$:

$$\sum_{\substack{s=1 \\ [i,j]\in\mathscr{K}(i,s)}}^{N^S} \pi_{ij}^{is} = \pi_{ij}^{ij} + \sum_{\substack{s=1 \\ s\in\mathscr{H}\setminus\{i,j\}}}^{N^S} \pi_{ij}^{is} = 1 \quad , \quad \forall [i, j] \,\middle|\, D_{ij} > 0 \, . \tag{10.6.10}$$

Sink of a flow is node j: there is either a direct private line from i to j or there is a flow from one and only one hub site $\neq i$ to j:

$$\sum_{\substack{r=1 \\ [i,j]\in\mathscr{K}(r,j)}}^{N^S} \pi_{ij}^{rj} = \pi_{ij}^{ij} + \sum_{\substack{r=1 \\ r\in\mathscr{H}\setminus\{i,j\}}}^{N^S} \pi_{ij}^{rj} = 1 \quad , \quad \forall [i, j] \,\middle|\, D_{ij} > 0 \, . \tag{10.6.11}$$

Flow via a hub site q is conserved, if q is not a starting or terminating node: the total flow into q equals the total flow out of q for all possible routings of commodities $[i, j]$, $D_{ij} > 0$:

$$\sum_{\substack{r=1 \\ [i,j]\in\mathscr{K}(r,q)}}^{N^S} \pi_{ij}^{rq} = \sum_{\substack{s=1 \\ [i,j]\in\mathscr{K}(q,s)}}^{N^S} \pi_{ij}^{qs} \quad , \quad \forall [i, j] \,\middle|\, D_{ij} > 0 \;\; \forall q \in \mathscr{H} \setminus \{i, j\}. \tag{10.6.12}$$

10.6.2.3 Edge Capacity Constraints

Let $\gamma_{rs} \in \mathbb{N}_0$ be an integer variable counting the number of T1s installed on an undirected edge $\{r, s\} \in \mathscr{E}$, and let $\beta_{rsk} \in \{0, 1\}$ be variables of an SOS1. β_{rsk} is equal to 1 if a fractional T1 with a bandwidth of $(k-1) \cdot 56$ Kbps channels is installed on $\{r, s\}$, and equal to 0 otherwise. The sum of T1s plus fractional T1 channels determines the total capacity of the edge. The total flow via edge $\{r, s\} \in \mathscr{E}$ (in either direction) may not exceed the total capacity of the edge, which is guaranteed by requiring

$$\sum_{\substack{i,j=1\\ [i,j]\in\mathscr{K}(r,s)}}^{N^S} D_{ij}\pi_{ij}^{rs} + \sum_{\substack{i,j=1\\ [i,j]\in\mathscr{K}(s,r)}}^{N^S} D_{ij}\pi_{ij}^{sr} \le \sum_{k=1}^{N_{rs}^K} (k-1)\beta_{rsk} + 24\gamma_{rs} \quad , \quad \forall\{r, s\} \in \mathscr{E}. \tag{10.6.13}$$

The left-hand side of (10.6.13) is the total flow via $\{r, s\}$ in either direction measured in 56 Kbps channels, while the right-hand side is the total capacity installed on $\{r, s\}$, also measured in 56 Kbps channels.

In addition we have to state the usual convexity constraints for the design variable β_{rsk} controlling the number of selected FT1 channels, that is to say

$$\sum_{k=1}^{N_{rs}^K} \beta_{rsk} = 1 \quad , \quad \forall\{r, s\} \in \mathscr{E}, \tag{10.6.14}$$

which means that there is a definite FT1 bandwidth for each connection $\{r, s\} \in \mathscr{E}$. Note that the FT1 bandwidth is possibly equal to 0 which is the case if $\beta_{rs(k=1)} = 1$. As reference rows for the β_{rsk} variables we use the capacity constraints (10.6.13).

10.6.2.4 Additional Constraints

In this section we describe some further constraints which arise in the design of the corporation's private line network. Since these constraints do not critically influence the structure of the model, we have not yet incorporated these constraints in the prototype models.

Based on technical reasons there may be a restriction on the maximum number of nodes used for routing a commodity $[i, j]$ through the network. If at most M_{ij} nodes are allowed for routing $[i, j]$, that means that at most $M_{ij} + 1$ edges are allowed on the routing path $\mathscr{R}_{ij}$ for $[i, j]$. Since the flow variables π_{ij}^{rs} indicate whether edges appear on the path for $[i, j]$ or not, one can count the total number of edges on the path by adding up all the π_{ij}^{rs} variables for the commodity $[i, j]$. Hence the *maximum node constraints* read as follows:

$$\sum_{\substack{r,s=1 \\ [i,j]\in\mathscr{K}(r,s)}}^{N^S} \left(\pi_{ij}^{rs} + \pi_{ij}^{sr}\right) \leq M_{ij} + 1 \quad , \quad \forall [i, j] \| D_{ij} > 0. \tag{10.6.15}$$

Based upon availability or just in order to exclude certain (feasible) edges from being used for routing purposes, one may want to specify limits for the maximum capacity of an edge $\{r, s\} \in \mathscr{E}$. If at most a bandwidth of $K_{rs} \cdot 56$ Kbps is allowed for $\{r, s\}$, then this is ensured by the following *maximum edge capacity constraints*:

$$\sum_{k=1}^{N_{rs}^K} (k-1)\beta_{rsk} + 24\gamma_{rs} \leq K_{rs} \qquad \forall \{r, s\} \in \mathscr{E}. \tag{10.6.16}$$

Some logical site-to-site connections $[i, j]$ require more than one data transmission bandwidth for different applications. These different demands may, but need not necessarily, be routed via different paths through the network. These additional requirements just raise the number of commodities which have to be routed through the network. The extension of the model described above is straightforward: One just has to use the suitable substitutions

$$[i, j] \longrightarrow [i, j, d] \quad , \quad \pi_{ij}^{rs} \longrightarrow \pi_{ijd}^{rs}, \tag{10.6.17}$$

where the index d counts the different demands per connection $[i, j]$.

Some vital site-to-site traffic demands D_{ij} require backups in case of a failure in the regular routing path. We have to differentiate between two cases:

- The backup path(s) should additionally be available at all times at the same capacity as the regular path.
- The backup path(s) should provide a certain backup capacity which could be made available in cases of emergency by closing the edges on the path(s) for all other communications.

The first case is easy to handle. Let Λ_{ijd} denote the required number of backup paths for demand D_{ij}. Then the modified source flow constraints

$$\pi_{ijd}^{ijd} + \sum_{\substack{s=1 \\ s\in\mathscr{H}\setminus\{i,j\}}}^{N^S} \pi_{ijd}^{is} = 1 + \Lambda_{ijd} \quad , \quad \forall [i, j, d] \left| D_{ij} > 0 \right. \tag{10.6.18}$$

and the modified sink flow constraints

$$\pi_{ijd}^{ijd} + \sum_{\substack{r=1 \\ r \in \mathscr{H} \setminus \{i,j\}}}^{N^S} \pi_{ijd}^{rj} = 1 + \Lambda_{ijd} \quad , \quad \forall [i,j,d] \,|D_{ij} > 0 \tag{10.6.19}$$

ensure that there are Λ_{ijd} backup paths available. The additional constraints

$$\sum_{\substack{r=1 \\ [i,j,d] \in \mathscr{K}(r,q)}}^{N^S} \pi_{ijd}^{rq} \leq 1 \quad , \quad \forall q \in \mathscr{H} \setminus \{i,j\} \tag{10.6.20}$$

make sure that each hub site is used for at most one routing path of $[i, j, d]$, i.e., i and j are the only nodes common to the regular path and the backup path(s).

The second case of backups is more difficult to handle. In this case one has to introduce additional backup flow variables Λ_{ijd} which fulfil flow conservation constraints corresponding to (10.6.10), (10.6.11) and (10.6.12), where Λ_{ijd} replaces the "1" on the right-hand sides of (10.6.10) and (10.6.11). The additional constraints

$$\sum_{\substack{r=1 \\ [i,j,d] \in \mathscr{K}(r,q)}}^{N^S} \pi_{ijd}^{rq} + \sum_{\substack{r=1 \\ [i,j,d] \in \mathscr{K}(r,q)}}^{N^S} \Lambda_{ijd} \leq 1 \quad , \quad \forall q \in \mathscr{H} \setminus \{i,j\} \tag{10.6.21}$$

make sure that each hub site is used for at most one routing path of $[i, j, d]$. Furthermore the (huge number of) additional capacity constraints

$$\sum_{k=1}^{N_{rs}^K} (k-1)\beta_{rsk} + 24\gamma_{rs} \geq B_{ijd} \left[\Lambda_{ijd} + \Lambda_{ijd} \right] \quad , \quad \begin{array}{l} \forall \{r,s\} \in \mathscr{E} \\ \forall [i,j,d] \,| B_{ijd} > 0 \end{array} \tag{10.6.22}$$

ensure that there is at least enough capacity available (on the edges of the backup path) to cover the backup bandwidth B_{ijd}.

10.6.2.5 Objective Function of the Model

The objective is to minimize the total POP-POP private line cost. If C_{rsk}^{FT} is the cost of a fractional T1 line with a bandwidth of $(k-1) \cdot 56$ Kbps between r and $s (s > r)$, and C_{rs}^{T1} is the cost of a T1 line between r and s, then the *edge capacity cost* of $\{r, s\}$ adds up to

$$c_{rs}^P = \sum_{k=1}^{N_{rs}^K} C_{rsk}^{FT} \beta_{rsk} + C_{rs}^{T1} \gamma_{rs} \quad , \quad \forall \{r,s\} \in \mathscr{E}. \tag{10.6.23}$$

Therefore the overall cost to be minimized for the network of POP-to-POP private lines is given by

$$Z = \sum_{\substack{r,s=1 \\ \{r,s\}\in\mathcal{E}}}^{N^S} \sum_{k=1}^{N_{rs}^K} C_{rsk}^{FT}\, \beta_{rsk} + \sum_{\substack{r,s=1 \\ \{r,s\}\in\mathcal{E}}}^{N^S} C_{rs}^{T1}\, \gamma_{rs}. \tag{10.6.24}$$

Note that $C_{rs(k=1)}^{FT} = 0$.

10.6.2.6 Estimation of Problem Size

The size of the problem depends on the following dimensions:

$$\begin{aligned} N^S &: \text{ number of corporation's sites,} \\ N^H &: \text{ number of hub sites,} \\ N^D &: \text{ number of demands } D_{ij} > 0, \\ N^K &: \text{ number of fractional T1 grid points,} \end{aligned}$$

and on the number, $|\mathcal{E}|$, of feasible undirected edges in the network. This number is limited by

$$|\mathcal{E}| \leq \frac{N^H(N^H-1)}{2} + \frac{N^H(N^S-N^H)}{2} + N^D, \tag{10.6.25}$$

where the first term of the sum counts the edges between hub sites, the second counts the edges between non-hub sites and hub sites, and the last is a rough upper limit of the edges between non-hub sites with demands.

The total number of columns results from the following selection of variables:

$$\begin{array}{llll} \beta_{rsk}: & |\mathcal{E}| \cdot N^K & \text{SOS1,} & \text{fractional T1 capacity variable} \\ \gamma_{rs}: & |\mathcal{E}| & \text{integer,} & \text{T1 capacity variable} \\ \pi_{ij}^{rs}: & |\mathcal{E}| \cdot N^D & \text{binary,} & \text{flow variables} \\ c_{rs}^P: & |\mathcal{E}| & \text{real,} & \text{auxiliary cost variables,} \end{array} \tag{10.6.26}$$

which means that we have to expect about 5000 SOS1 members, 400 integers, 35,000 binary, and 400 continuous variables for the real problem.

The total number of rows results from the following system of constraints:

$$\begin{array}{lll} (10.6.10): & N^D & \text{flows into the network,} \\ (10.6.11): & N^D & \text{flows out of the network,} \\ (10.6.12): & \mid \mathscr{H} \mid \cdot N^D & \text{flow conservation at hubs,} \\ (10.6.13): & \mid \mathscr{E} \mid & \text{capacity constraints,} \\ (10.6.14): & \mid \mathscr{E} \mid & \text{convexity for SOS1 } \beta, \end{array} \quad (10.6.27)$$

which means that we have to expect about 2000 rows for the real problem.

10.6.2.7 Computational Needs

Several scenarios with different numbers of sites (s), hubs (h), and demands (d) were solved on a RS6000/340H workstation using XPRESS-MP 8.09 without the presolve capability being activated. The problem sizes (expressed by the number of variables, n, and number of constraints, m), the number of binary variables, n^B, the number of variables in special ordered sets, n^S, and the density, ρ, of the problems are shown in the table below:

Scenario	n	m	n^B	n^S	ρ
(12 s, 5 h, 16 days)	2066	231	288	48/1728	1.27
(15 s, 7 h, 18 days)	3821	369	690	77/3054	0.80
(26 s, 6 h, 31 d)	6148	616	891	135/5122	0.48
(20 s, 8 h, 24 d)	6259	581	1256	124/4879	0.51

These scenarios had the following solution statistics:

Scenario	LP	IP	LP-gap	Time
(12 s, 5 h, 16 days)	6856	12072	43.21%	26 min
(15 s, 7 h, 18 days)	8479	13854	38.80%	14h 12min
(26 s, 6 h, 31 days)	9453	%	51.01%	??
(20 s, 8 h, 24 days)	9690	%	43.90%	??

The initial model which was used in these test runs did not involve the integer variables γ_{rs}. Just the S1 variables β_{rsk} were used to determine the edge capacities. Therefore there are no integer variables in these scenarios, and the number of S1 members was much higher than could be expected from the problem size estimations above. Nevertheless the large integrality gaps indicated that there was a strong need to improve the model formulation.

10.6.3 Analysis and Reformulations of the Models

This section points out the basic structure of the model and describes the reformulations and improvements which should help to speed up the solution process. Valid inequalities and some problem-specific cuts are given.

10.6.3.1 Basic Structure of the Model

As pointed out before, the basic structure of the model is most similar to the structure of a general minimum-cost capacitated multi-commodity network flow problem involving flow conservation constraints like (10.6.12), edge capacity constraints like (10.6.13), edge capacity cost like (10.6.23), and an objective function being the sum of the edge cost to be minimized. The network flow model described here has the following specialities:

- no distribution of flows on different paths possible (=> π binary),
- only two different discrete capacity batches available (=> β, γ integer),
- nonlinear cost function for one of the capacity batch sizes (=> β S1),
- two different types of node [=> additional flow constraints (10.6.10), (10.6.11) for non-hub sites].

10.6.3.2 Some Valid Inequalities: Edge Capacity Cuts

We expected to get useful valid inequalities by investigating relationships between the flow variables π_{ij}^{rs} and the design variables γ_{rs} and β_{rsk}. A first set of valid inequalities is obtained by reducing the capacity constraints (10.6.13) for total flows through an edge to constraints for single flows through the edge. The inequalities

$$D_{ij} \cdot \left(\pi_{ij}^{rs} + \pi_{ij}^{sr}\right) \le \sum_{k=1}^{N_{rs}^{K}} (k-1)\beta_{rsk} + 24\gamma_{rs} \quad , \quad \begin{array}{l} \forall \{r,s\} \in \mathscr{E} \\ \forall [i,j] \left| D_{ij} > 0 \right. \end{array} \tag{10.6.28}$$

express that the single flow for demand D_{ij} [instead of all flows $\sum D_{ij}$ as in (10.6.13)] is less than or equal to the total capacity in $\{r, s\}$ and gives the valid inequalities (*edge capacity cuts*)

$$\Delta_{ij} \cdot \left(\pi_{ij}^{rs} + \pi_{ij}^{sr}\right) \le \sum_{\substack{k=1 \\ (k-1)\ge G_{ij}}}^{N_{rs}^{K}} \beta_{rsk} + \gamma_{rs} \quad , \quad \begin{array}{l} \forall \{r,s\} \in \mathscr{E} \\ \forall [i,j] \left| D_{ij} > 0, \right. \end{array} \tag{10.6.29}$$

where $\Delta_{ij} := \left\lceil \frac{D_{ij}}{24} \right\rceil$ is the up-rounded integer associated with $D_{ij}/24$ and $G_{ij} := D_{ij} - 24(\Delta_{ij} - 1)$ is the fractional part of $D_{ij}/24$. These cuts (10.6.29) have only binary and integer variables, and all coefficients are integral.

Let us first try to give an intuitive interpretation of (10.6.29). Remember that γ_{rs} is an integer number measuring the number of T1 lines and that each T1 line consists of at most 2456 Kbps channels. The variables β_{rsk} measure the bandwidth produced by these channels. They are members of SOS1, and for each index pair $\{rs\}$ they are one for exactly one k. The original inequality (10.6.28) is expressed in terms of channels while the cuts are based on the number of T1s. Let us consider an example in which the demand is $D_{ij} = 30$. It is not likely that an optimal solution will chose more capacity then necessary. Thus, the demand is covered by two T1 lines, i.e., $\gamma_{rs} = 2$, and six more channels, i.e., $\beta_{rs7} = 1$. Let us now consider (10.6.29). Δ_{ij} measures the demand in units of the channels but it is up-rounded. Consider also the integer value, $\Delta'_{ij} = \Delta_{ij} - 1$ of the demand in units of channels. In our example we get $\Delta_{ij} = 3$. The demand $\Delta'_{ij} = 2$ is covered by two T1 lines, i.e., $\gamma_{rs} = 2$. How do we get the additional six more channels? (10.6.29) does not know about the channels anymore. But the first term on the right-hand side of (10.6.29) has already excluded those indices k which lead to less than 6 channels. On the other hand, in an optimal solution the variable β_{rsk} with smallest index k is chosen.

How can one find such cuts? The first step is to divide (10.6.28) by 24 yielding

$$\frac{D_{ij}}{24} \cdot \left(\pi_{ij}^{rs} + \pi_{ij}^{sr}\right) \le \sum_{k=1}^{N_{rs}^K} \frac{k-1}{24}\beta_{rsk} + \gamma_{rs} \quad , \quad \begin{array}{l} \forall \{r,s\} \in \mathcal{E} \\ \forall [i,j] \,|\, D_{ij} > 0 \end{array} \qquad (10.6.30)$$

and observing that the inequalities (10.6.30) contain only binary or integer variables but still have some fractional coefficients. Down-rounding and up-rounding[6] in inequalities with integer variables are common techniques. If we inspect (10.6.30) we see that the first right-hand side term needs to cover the fractional part, 0.25, of the demand $D_{ij}/24 = 2.25$ while γ_{rs} takes care of the integer part, 2. What sort of solutions can we expect in an LP relaxation which involves (10.6.30)? Typically, some of the β_{rsk} are different from zero in LP relaxations, e.g., $\beta_{rs5} = \beta_{rs9} = 0.5$ which also gives 6 channels. The cuts (10.6.29) can never have β_{rs5} being different from zero (they just do not appear in the sum), and also in LP relaxations there is a tendency to select the β_{rsk} with index k as small as possible. So, this paragraph gives an idea how the cuts are derived and why they lead to a tighter formulation. The smaller the demand the more efficient is this formulation.

After we have some intuitive understanding concerning the foundations of the cuts, we are now able to show formally that they are valid inequalities: if the demand

[6]Take the following example: The inequality $3.2\alpha_1 + 3.9\alpha_2 \le 8$ with integer variables α_1 and α_2 has the valid inequality $4\alpha_1 + 4\alpha_2 \le 8$ which is equivalent to $\alpha_1 + \alpha_2 \le 2$. If we draw the feasible regions associated with the original constraint and $\alpha_1 + \alpha_2 \le 2$ we see that the latter gives a smaller feasible region and thus is the tighter constraint.

$[ij]$ flows over the edge $\{rs\}$ then $\pi_{ij}^{rs} + \pi_{ij}^{sr} = 1$ and two cases are possible. The first case is $\gamma_{rs} \geq \Delta_{ij}$ (that is the case when $G_{ij} > N_{rs}^K$). The second case is $\gamma_{rs} = \Delta_{ij} - 1$ (that is the case when $G_{ij} \leq N_{rs}^K$) which gives $\sum_{k-1 \geq G_{ij}} \beta_{rsk} = 1$, i.e., $\beta_{rsk} = 0$ for $k < G_{ij} + 1$. So we have shown that the inequalities (10.6.29) are valid.

There are a lot of (in some cases redundant) inequalities, namely a number of $|\mathscr{E}| \cdot N^D$. They ideally should be used as dynamic cuts in the B&C process of the optimization run. However, just adding all these cuts as static, additional inequalities to the model yielded much better LP relaxation values.

10.6.3.3 Some Improvements to the Model Formulation

The *total demand* D_i^T originating or terminating at a site i, measured in multiples of 56 Kbps, is calculated by

$$D_i^T = \sum_{\substack{j=2 \\ j>i}}^{N^S} D_{ij} + \sum_{\substack{j=1 \\ j<i}}^{N^S-1} D_{ji} \quad , \quad \forall i. \tag{10.6.31}$$

The number of β_{rsk} variables indicating the number of selected 56 Kbps channels for $\{r, s\}$ is limited by the total demand (measured in multiples of 56 Kbps) if either r or s is not a hub site. This maximum value N_{rs}^{K*} is given by

$$\begin{aligned} N_{rs}^{K*} &:= \min\left\{N_{rs}^K, D_r^T + 1\right\} \quad , \quad \forall r \in \mathscr{S} \setminus \mathscr{H} \quad \forall s \\ N_{rs}^{K*} &:= \min\left\{N_{rs}^K, D_s^T + 1\right\} \quad , \quad \forall r \; \forall s \in \mathscr{S} \setminus \mathscr{H}. \end{aligned} \tag{10.6.32}$$

We use N_{rs}^{K*} instead of N_{rs}^K in order to reduce the number of variables β_{rsk}. Furthermore, it is possible to fix some of the γ_{rs} variables in cases where the total demand at a non-hub site is less than the maximum number N_{rs}^{K*} of β_{rsk} which is related to the break-even point where an additional number of single 56 Kbps channels would become more expensive than a T1 line. In these cases the following bounds apply

$$\begin{aligned} \gamma_{rs} &= 0 \text{ if } D_r^T < N_{rs}^{K*} \; \forall r \in \mathscr{S} \setminus \mathscr{H} \quad \forall s \\ \gamma_{rs} &= 0 \text{ if } D_s^T < N_{rs}^{K*} \; \forall r \quad \forall s \in \mathscr{S} \setminus \mathscr{H} \\ \gamma_{rs} &= 0 \text{ if } D_{rs} < N_{rs}^{K*} \; \forall r, s \in \mathscr{S} \setminus \mathscr{H}. \end{aligned} \tag{10.6.33}$$

The integration of these two improvements into the model formulation led to a slightly better LP relaxation value and to a remarkable reduction of the total solution time, e.g., 30% for the (20 s, 8 h, 2 days) scenario.

10.6.3.4 A Surrogate Problem with a Simplified Cost Function

We now simplify the problem by using a linearization of the cost function for the FT1 channels. In this case the decision on the number of 56 Kbps channels may be modeled via integer variables ω_{rs} rather than the S1 variables β_{rsk}. The relationship between the ω_{rs} and β_{rsk} variables is defined by

$$\omega_{rs} := \sum_{k=1}^{N_{rs}^{K*}} (k-1) \cdot \beta_{rsk} \quad \in \mathbb{N}_0 \qquad \forall \{r,s\} \in \mathscr{E}. \tag{10.6.34}$$

The capacity of an edge $\{r, s\}$ (again measured in multiples of 56 Kbps) is given by

$$\omega_{rs} + 24\gamma_{rs} \tag{10.6.35}$$

and the cost of a single 56 Kbps channel is approximated by

$$C_{rs}^{56} := \frac{C_{rs(k=N_{rs}^{K*})}^{FT1}}{N_{rs}^{K*} - 1} \tag{10.6.36}$$

such that the total *approximated* capacity costs of edge $\{r, s\}$ are calculated by

$$\tilde{c}_{rs}^{P} = C_{rs}^{56}\,\omega_{rs} + C_{rs}^{T1}\,\gamma_{rs}\,. \tag{10.6.37}$$

Simple edge capacity cuts analogous to (10.6.29) for this surrogate model are obtained by considering the reduced capacity constraints for single flows through an edge. The inequalities

$$\omega_{rs} + 24\gamma_{rs} \geq D_{ij}\left(\pi_{ij}^{rs} + \pi_{ij}^{sr}\right) \quad , \quad \forall i \, \Big| \, D_i^T > 0 \quad , \ \forall \{r,s\} \in \mathscr{E} \tag{10.6.38}$$

give the valid inequalities

$$\begin{array}{l} \gamma_{rs} \geq \Delta_{ij}\left(\pi_{ij}^{rs} + \pi_{ij}^{sr}\right) \text{ and } \omega_{rs} \geq G_{ij}\left[\Delta_{ij}\left(\pi_{ij}^{rs} + \pi_{ij}^{sr}\right) - \gamma_{rs}\right] \\ \left(\forall [i,j] \, \middle| \, D_{ij} > 0 \ \wedge G_{ij} \leq N_{rs}^{K*} \quad \forall \{r,s\} \in \mathscr{E}\right), \end{array} \tag{10.6.39}$$

where $\Delta_{ij} = \left\lceil \frac{D_{ij}}{24} \right\rceil$ and $G_{ij} = D_{ij} - 24(\Delta_{ij} - 1)$ are the same as defined for (10.6.29).

Simple node flow cuts form an additional set of valid inequalities. They are of the same type as (10.6.39) but looking at all demands at a site i and all outgoing and incoming flows. In this case the capacity constraints for the sum of all the edges adjacent to i

$$\sum_{\substack{\{r,s\}\in\mathscr{E} \\ r=i\vee s=i}} (\omega_{rs} + 24\gamma_{rs}) \geq D_i^T \quad , \quad \forall i \,\middle|\, D_i^T > 0 \tag{10.6.40}$$

give the valid inequalities

$$\sum_{\substack{\{r,s\}\in\mathscr{E} \\ r=i\vee s=i}}^{N^S} \omega_{rs} \geq G_i^T \cdot \left(\Delta_i^T - \sum_{\substack{\{r,s\}\in\mathscr{E} \\ r=i\vee s=i}}^{N^S} \gamma_{rs} \right) \quad , \quad \forall i \,\middle|\, D_i^T > 0, \tag{10.6.41}$$

where $\Delta_i^T := \left\lceil \frac{D_i^T}{24} \right\rceil$ and $G_i^T := D_i^T - 24(\Delta_i^T - 1)$. Note that compared to (10.6.39) the set of node flow inequalities is much smaller. The reasoning for these cuts is similar as in Sect. 10.6.3.2.

This surrogate model involving the approximate cost function, which

- incorporates integers ω_{rs} instead of binary variables β_{rsk} which establish SOS1,
- includes the edge capacity cuts (10.6.39) as ModelCuts,[7] but merely for those demands required between the end nodes of the considered edge (that is for all $[r,s]$, $D_{rs} > 0$, $\{r,s\} \in \mathscr{E}$), and
- includes all possible node flow cuts (10.6.41) as ModelCuts,

was tested with the (20 s, 8 h, 24 days) prototype scenario. The optimizer just needed 2–5% of the time spent to solve the original problem with the first model formulations. And the cost approximation was fairly good for the considered scenario (8% deviation).

10.6.3.5 More Valid Inequalities: Node Flow Cuts

Since the simple node flow cuts appear to speed up the solution process, we consider the corresponding inequalities for the original problem with nonlinear FT1 cost function. Looking at all demands at a site i and all outgoing and incoming flows the capacity constraints for the sum of all the edges adjacent to i in this case read

$$\sum_{\substack{\{r,s\}\in\mathscr{E} \\ r=i\vee s=i}} \left(\sum_{k=1}^{N_{rs}^{K*}} (k-1)\beta_{rsk} + 24\gamma_{rs} \right) \geq D_i^T \quad , \quad \forall i \,\middle|\, D_i^T > 0 \tag{10.6.42}$$

[7] ModelCuts are problem-specific cuts, i.e., valid inequalities that will cut-off unwanted fractional values of binary or integer variables and that are otherwise redundant constraints. They are added directly to the model formulation. In contrast, in B&C cuts are added dynamically in the tree to cut-off unwanted fractional variables.

and give the valid inequalities (*node flow cuts*)

$$\sum_{\substack{\{r,s\}\in\mathcal{E}\\ r=i\vee s=i}} \sum_{k=1}^{N_{rs}^{K*}} (k-1)\beta_{rsk} \geq G_i^T \left(\Delta_i^T - \sum_{\substack{\{r,s\}\in\mathcal{E}\\ r=i\vee s=i}} \gamma_{rs} \right) \quad , \quad \forall i \left| D_i^T > 0, \right. \tag{10.6.43}$$

where $\Delta_i^T = \left\lceil \frac{D_i^T}{24} \right\rceil$ and $G_i^T = D_i^T - 24(\Delta_i^T - 1)$ are the same as for (10.6.41).

10.6.3.6 Some Remarks on Performance

For small problems solutions were obtained in a reasonable amount of time, when the demands fulfilled the following essential condition:

$$D_{ij} \in \mathbb{N}_0 \quad , \quad \forall [i,j] \left| D_{ij} > 0. \right. \tag{10.6.44}$$

Otherwise the solution times became huge, even for small problem sizes. Since demand is a vague quantity anyway, we gave the recommendation to the customer to round it to the nearest integer in order to speed up the solution. Problems larger than (12 s, 5 h, 16 days) were only tackled with integral demands. So far, it is possible to prove optimality for a scenario of (28 s, 8 h, 32 days) involving 2088 global entities.

10.7 Synchronization of Batch and Continuous Processes

This case study is based on Blackburn et al. (2014,[83]) and serves to illustrate the usage of continuous-time formulations. So far, we have used discrete-time formulations to describe production planning problems. Scheduling problems with detailed time resolution needs, however, often lead to very large problems when using time-discrete formulations.

In Janak et al. (2006,[285]) a large-scale scheduling problem faced in a very complex BASF plant is described and solved by using a continuous-time formulation, in which time is allocated to event points. The problem involves all of the structural elements from the above-stated list except for the sixth one ("multi-component flow and nonlinear blending"). Here we describe only a small part of the plant and only the major aspects and challenges. Raw materials are fed into up to three batch reactors and undergo a batch operation. Each time, the full batch size is used to produce an intermediate which is sent to two buffer tanks before being processed in a continuous unit as displayed in Fig. 10.8. Since certain products run faster on the continuous unit compared to their batch processing time on the reactors, the reactors are allowed to start prior to the start of the continuous unit to prepare the campaign

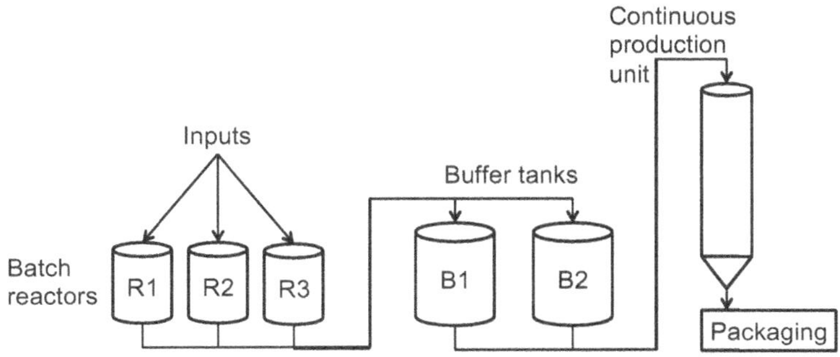

Fig. 10.8 Combination of batch and continuous production process

and pre-fill the tanks. The challenge is to synchronize the batch and continuous production by exploiting the two buffer tanks to best meet the demand quantities and due dates of the individual products.

The time horizon for this scheduling problem is 4–8 weeks. In this time frame, about 10 to 20 different products need to be produced. These products fall into two categories. Products of type 1 are subject to tracing production batches for legal purposes. This implies keeping track of each batch produced on one of the reactors, keeping the tanks separate, and emptying each tank completely before recharging it with a new batch. Products of type 2 are not subject to tracing, which technically implies that we can combine the two buffer tanks to one large buffer reservoir.

Apart from constraints such as the one above, multiple optimization criteria have to be taken into account. Of highest priority is that, ideally, the continuous unit should not be interrupted. Moreover, the second reactor should be used at a minimal number of batches because it should rather serve other products in the larger network topology. As the demand of the products does not exactly fit to what the three batch reactors with their individual batch sizes can produce in any combination, we also try to minimize both underproduction and overproduction. Minimizing the gaps between the batches on the individual reactors is another goal to be pursued.

In order to solve this complex multi-criteria optimization problem, we apply a two phase approach: In Phase 1 we make the simplifying assumption that reactors are *always* available and can produce ahead to pre-fill the buffer tanks. Under this assumption we minimize the total interruption time of the campaigns on the continuous unit. Furthermore, we derive various lower bounds on certain quantities (indicated by the arrow below) by applying the following hierarchy of objectives in a lexicographic goal programming approach; cf. Schniederjans (1995,[490]):

1. minimize overproduction (demand vs. reactor batch sizes) → lower bound on unavoidable overproduction,
2. minimize underproduction → lower bound on unavoidable underproduction,
3. minimize interruption on the continuous unit → lower bound on unavoidable interruptions,

4. minimize time for pre-filling the tanks $\rightarrow$ ensures closest connection of the current campaigns and the associated reactor batches to the previous campaign, and
5. minimize the number of charges on reactor 2 $\rightarrow$ lower bound on the minimal number of batches on reactor 2.

The results of Phase 1 either provide useful information for the planner, such as the unavoidable over- and underproduction under ideal conditions, or they are used in Phase 2 to improve the numerical behavior of the problem and facilitate the determination of the optimal solution.

In Phase 2 we account for possible reactor blockades or the unavailability of reactors. Such blockades might require a later start of a campaign on the continuous unit in order to avoid interruptions. Ideally we deviate from the campaign start determined in Phase 1 as little as possible. Therefore, we minimize the time shift by taking into account the lower bound on the unavoidable interruptions from Phase 1. Since we are also interested in dense production schedules on the batch reactors we minimize the finishing time of the reactor finishing last. Finally, we minimize the number of batches on reactor 2 and overproduction under these new conditions. Overall, the hierarchy of objectives in Phase 2 can be summarized as:

1. minimize time shifts of the campaigns,
2. minimize finishing time of the reactor finishing last,
3. minimize number of batches on reactor 2, and
4. minimize overproduction.

The above-described approach produces a feasible schedule within 10 min which is optimal with respect to the outlined hierarchy of individual objective functions. Figure 10.9 shows a typical Gantt chart with the three batch reactors, the continuous plant and the two tanks aggregated to one large tank. The gray shaded area on the continuous plant C represents cleaning time. The gray-blue shaded areas on the reactors starting at time 0 indicate time intervals at which the reactors are not yet available. The pink color represents a blockade. The red product campaign is characterized by consecutive batches without gaps, while the yellow product batches show gaps. The tank diagram at the bottom nicely illustrates that tanks are prepared prior to the start of the continuous plant.

This model, implemented in `GAMS`, is embedded into `SAP APO` and has been successfully in use since 2006. The reasons for its success include a strong management support as well as a talented planner having a good understanding of all the details on the production floor.

Planners appreciate the tool a lot because it brings strong reliability into their daily work and helps to reduce their time spent on generating feasible schedules. Since the generated schedules reflect reality to a very high degree, the planner can rely on the fact that the tanks do not overflow or the continuous process does not have to be interrupted because a tank runs out of material to be processed. The explicit determination of unavoidable over- or underproduction is another benefit. It provides valuable information for joint meetings of the production planner, the

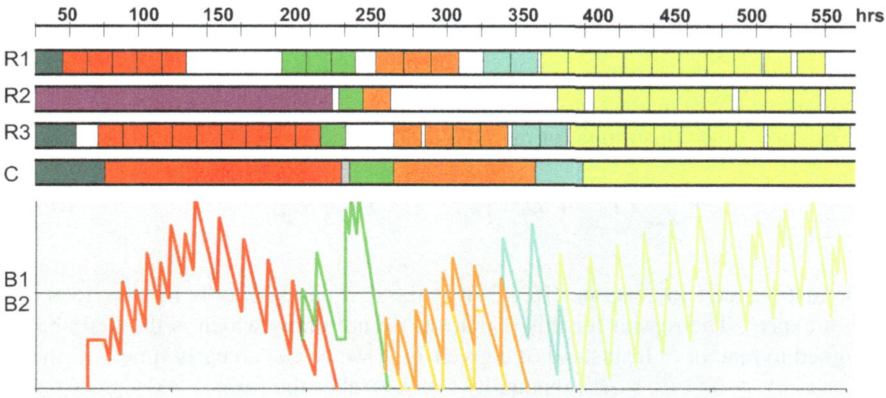

Fig. 10.9 Combination of batch and continuous production process — Sample optimization result. The lower part shows the combined tank level of B1 and B2. Once one tank is empty near the end of a campaign, it can be used to prepare for the next campaign (colored overlaps)

supply chain and marketing groups about the feasibility of the original (high-level) plans created by those groups. Last but not least, the financial benefit of about 1-2 million USD per year due to smoother operations and a better utilization of the production assets is highly appreciated.

10.7.1 Time Sequencing Constraints

Time sequencing constraints are needed to establish a sequence of charging points. We require that the time associated with event point c cannot be earlier than the time for the previous event point plus the filling time associated with the previous tank charge, i.e.,

$$t_c - t_{c-1} \geq d_c^{\mathrm{F}} \quad ; \quad \forall\{s\} \quad , \quad \forall\{c \in \mathscr{C}_p | c > 1\}. \tag{10.7.1}$$

As we have tank-independent filling times, we use

$$t_c - t_{c-1} \geq \sum_{r \in \mathscr{R}} D_r^{\mathrm{F}} \delta_{c-1,r}^{\mathrm{R}} \quad ; \quad \forall\{c \in \mathscr{C}_p | c > 1\}. \tag{10.7.2}$$

10.7.2 Reactor Availability Constraints

Reactors are defined to be available from an earliest time, T_r^{E}. In the rolling campaign mode we could also use a variable t_r^{E} which determines the optimal availability point. A reactor is blocked from the time it starts a reaction task until the end of charging to a tank. Thus, the total time the reactor is blocked is the sum of the times for the reaction task plus the filling task, i.e., $B_{rp}^{\mathrm{D}} = D_{rp}^{\mathrm{RT}} + D_{rp}^{\mathrm{CT}}$; as previously, we will neglect the index p.

The early time constraints for pre-given early times T_r^{EF} read

$$t_c \geq \sum_{r\in\mathscr{R}} \left(T_r^{\mathrm{EF}} + D_r^{\mathrm{RT}}\right) \delta_{cr}^{\mathrm{R}} \quad ; \quad \forall\{c \in \mathscr{C}_p'\}. \tag{10.7.3}$$

Note that we have to generate (10.7.3) for all $c \in \mathscr{C}_p'$ and not only for $c = 1$, as one might expect. The reason for this is that we do not know which is the first charge assigned to reactor r. If, instead of a given T_r^{EF}, we model an early time $-t_r^{\mathrm{E}}$ in the past, i.e., $t_r^{\mathrm{E}} \geq 0 \leq t_r^{\mathrm{E}} \leq B_r^{\mathrm{D}}$, inequality (10.7.3) takes the form

$$t_c \geq \sum_{r\in\mathscr{R}} D_r^{\mathrm{RT}} \delta_{cr}^{\mathrm{R}} - \sum_{r\in\mathscr{R}} t_r^{\mathrm{E}} \delta_{cr}^{\mathrm{R}} = \sum_{r\in\mathscr{R}} D_r^{\mathrm{RT}} \delta_{cr}^{\mathrm{R}} - \sum_{r\in\mathscr{R}} t_{cr}^{\delta} \quad ; \quad \forall\{c \in \mathscr{C}_p'\}, \tag{10.7.4}$$

where the nonlinear product terms $t_r^{\mathrm{E}} \delta_{cr}^{\mathrm{R}}$ have been replaced by $t_{cr}^{\delta} = t_r^{\mathrm{E}} \delta_{cr}^{\mathrm{R}}$. The continuous variables t_{cr}^{δ} are calculated by

$$t_{cr}^{\delta} \leq t_r^{\mathrm{E}} \quad ; \quad \forall\{c \in \mathscr{C}_p'\} \quad , \quad \forall r \tag{10.7.5}$$

$$t_{cr}^{\delta} \leq B_r^{\mathrm{D}} \delta_{cr}^{\mathrm{R}} \quad ; \quad \forall\{c \in \mathscr{C}_p'\} \quad , \quad \forall r \tag{10.7.6}$$

and

$$t_{cr}^{\delta} \geq t_r^{\mathrm{E}} - B_r^{\mathrm{D}}(1 - \delta_{cr}^{\mathrm{R}}) \quad ; \quad \forall\{c \in \mathscr{C}_p'\} \quad , \quad \forall r. \tag{10.7.7}$$

In order to reduce the usage of t_r^{E}, the sum $\sum_r t_r^{\mathrm{E}}$ is minimized in an additional goal programming step.

To model the batch reaction times $B_r^{\mathrm{D}} = D_r^{\mathrm{RT}} + D_r^{\mathrm{FT}}$ we proceed as follows. If c_1 denotes any charge preceding c, i.e., $c_1 < c$, we obtain the following relationship between the times associated with c_1 and c:

$$t_c - t_{c_1} \geq \sum_{r\in\mathscr{R}} B_r^{\mathrm{D}} \rho_{cc_1r} \quad ; \quad \forall\{c \in \mathscr{C}_p | c > 1\}, \tag{10.7.8}$$

where $\rho_{cc_1r} = 1$ if reactor r produces both charge c_1 and c. The number of inequalities (10.7.8) increases quadratically in N_p^{C}. It might be sufficient to use the smaller number of constraints

$$t_c - t_{c_1} \geq \sum_{r\in\mathscr{R}} B_r^{\mathrm{D}} \rho_{cc_1r} \quad ; \quad \forall\{c, c_1 \in \mathscr{C}_p' | \max(1, c - K_{\max}) \leq c_1 < c\}, \tag{10.7.9}$$

instead of (10.7.8) where the control parameter $K_{\max}$ might be set, for example, to, $K_{\max} = 3$; for $K_{\max} = N_p^{\mathrm{C}} - 1$ (10.7.9) and (10.7.8) are equivalent. For $K_{\max} = 3$ (10.7.9) is generated for $c_1 = c-1, c-2$, and $c-3$. Note that the model may lead to a wrong result, i.e., $t_c - t_{c_1} < B_r^{\mathrm{D}}$, when

$$\delta_{cr}^{\mathrm{R}} = \delta_{c-K_{\max},r}^{\mathrm{R}} = 1 \quad , \quad \delta_{c_1 r}^{\mathrm{R}} = 0 \quad ; \quad \forall \{c_1 \in \mathscr{C}_p' | c - K_{\max}) \le c_1 < c-1\}.$$

This case applies to a situation in which a particular reactor is only rarely used. It is still not very likely that (10.7.8) is violated because the other reactors need to fulfill (10.7.9) for small differences $c - c_1$ as well. Either way this needs to be checked in the post optimal analysis.

The binary variables $\rho_{cc_1r} := \delta_{cr}^{\mathrm{R}} \delta_{c_1 r}^{\mathrm{R}}$ are subject to the following constraints.

$$\rho_{cc_1r} \le \delta_{cr}^{\mathrm{R}} \quad ; \quad \forall \{c, c_1 \in \mathscr{C}_p', r | \max(1, c - K_{\max}) \le c_1 < c\}, \tag{10.7.10}$$

$$\rho_{cc_1r} \le \delta_{c_1 r}^{\mathrm{R}} \quad ; \quad \forall \{c, c_1 \in \mathscr{C}_p', r | \max(1, c - K_{\max}) \le c_1 < c\}, \tag{10.7.11}$$

and

$$\rho_{cc_1r} \ge \delta_{cr}^{\mathrm{R}} + \delta_{c_1 r}^{\mathrm{R}} - 1 \quad ; \quad \forall \{c, c_1 \in \mathscr{C}_p' | \max(1, c - K_{\max}) \le c_1 < c\} \quad , \quad \forall \{r\}. \tag{10.7.12}$$

10.7.3 Exploiting Free Reactor Time — Delaying Campaign Starts

The optimal early starting times t_r^{F} in campaign k are matched to the available free reactor times D_r^{fr} by the inequality

$$B_r^{\mathrm{D}} \alpha_{1r}^0 + t_r^{\mathrm{F}} = D_r^{\mathrm{fr}} + r^{\mathrm{F}} \quad ; \quad \forall r, \tag{10.7.13}$$

where r^{F} is a relaxation variable. Note that formulating (10.7.13) as an equality may be too restrictive as it enforces that the available free time D_r^{fr} is used for pre-filling. It might be better to use

$$B_r^{\mathrm{D}} \alpha_{1r}^0 + t_r^{\mathrm{F}} \le D_r^{\mathrm{fr}} + r^{\mathrm{F}} \quad ; \quad \forall r, \tag{10.7.14}$$

which states that we can use pre-filling and early time only up to the limit $D_r^{\mathrm{fr}} + r^{\mathrm{F}}$. Note that r^{F} does not depend on reactor r which is consistent with the interpretation that it represents the delayed start of a campaign.

10.7.4 Restricting the Latest Time a Reactor Is Available

In order to prepare for campaign $k+1$ with known optimal early times and tank pre-filling we implement the constraint

$$t_c + D_r^{\mathrm{CT}} \le \left(D_r^{\mathrm{stop}} + r_r^{\mathrm{S}}\right) + M_{\mathrm{k}}^{\mathrm{stop}}(1-\delta_{cr}) \quad ; \quad \forall r, \tag{10.7.15}$$

with the preferred termination time D_r^{stop} after which reactor r is not used any more for campaign k. The inequality (10.7.15) is activated if $\delta_{cr} = 1$; otherwise it is redundant due to the usage of an appropriate value of $M_{\mathrm{k}}^{\mathrm{stop}}$. The big-M value $M_{\mathrm{k}}^{\mathrm{stop}}$ can be chosen as the estimated campaign duration D_k^*

$$M_k^{\mathrm{stop}} := D_k^* := D_k / R_k^{\mathrm{S}} \quad ; \quad \forall k. \tag{10.7.16}$$

Alternatively, we could use the k and r dependent tighter big-M value

$$M_{rk}^{\mathrm{stop}} := D_k^* - D_r^{\mathrm{stop}} - B_r \quad ; \quad \forall \{rk\}. \tag{10.7.17}$$

The relaxation variables r_r^{S} in (10.7.15) are used to avoid infeasibilities in case the termination time cannot be met. In an auxiliary model we minimize $\max_r r_r^{\mathrm{S}}$, i.e., the maximum relaxation time by adding the inequalities

$$r_{\max}^{\mathrm{S}} \ge r_r^{\mathrm{S}} \quad ; \quad \forall r \tag{10.7.18}$$

and minimizing $r_{\max}^{\mathrm{S}}$. The overall effect is that we try to connect the adjacent campaigns k and $k+1$ best possible: in campaign k we try to match the termination times to the ideal free reactors times known for campaign $k+1$, while in campaign $k+1$ we match the early times and pre-filling to the available times derived in campaign k from $D_r^{\mathrm{stop}} + r_r^{\mathrm{S}}$.

10.8 Summary and Recommended Bibliography

In this chapter we have considered several typical real-world problems (completely solved to the customer's satisfaction), two larger problems originating in chemical industry, and finally, the optimization of a telecommunication service network. While the first of the larger ones could be successfully implemented and solved by paying attention to critical mathematical issues and finally applying B&C, the second one still awaits solution. For solving the telecommunication problem B&C is needed again. The examples show that it is very important to get the data structure right and to work on the mathematical formulation of the problem. Reformulations of the problems are usually necessary to solve larger real-world problems. This point is very important for the new practitioner to keep in mind. Although computer facilities are improving in power, the benefits which can be gained from efficient model building and clever reformulations outperform the hardware aspects by a long way. Thus by studying this chapter the reader should be able to:

- model more complex problems;
- appreciate the scope and difficulty of problems in practice;
- understand the gains to be made in the solution of practical problems by formulating them in a sensible manner and by controlling the search for optimality;
- formulate own allocation problems.

Further reading on allocation problems: Westerlund et al. (2007,[578]) have developed a generic model for n-dimensional allocation. This model covers scheduling problems (1-dimensional allocation), trimloss problems (2-dimensional allocation), packing problems (3-dimensional allocation), joint packing and scheduling problems (4-dimensional allocation) and more generally, allocation problems in any number of dimensions.

10.9 Exercises

1. ParkBench Ltd. (PB) manufactures park benches. It has three primary inputs, beechwood, steel, and labor. Steel and labor have a stable price over the four quarters of the year, but beechwood's price varies by quarter. PB produces two sorts of benches, the Superior (S) and the Executive (E). The requirements for the inputs are given in the table below (e.g., one Superior bench requires 14.2 units of beechwood, 1.83 units of steel and 1 unit of labor).

	Beechwood	Steel	Labor
Superior	14.2	1.83	1
Executive	22.4	1.94	1.26

The selling prices of the benches vary by season (quarters Q1, Q2, Q3, Q4 in this case).

£	Q1	Q2	Q3	Q4
Superior	850	890	860	750
Executive	1050	1200	900	810

At any time of the year steel costs £100 per unit and labor costs $90 per unit. The price [cost/unit] of beechwood, however, is much more volatile. Skilled labor is in short supply, and its availability is given as well. Price and labor availability are given by:

Quarter	$Q1$	$Q2$	$Q3$	$Q4$
Cost/unit [£/unit]	28	34	22	18
Skilled labor availability	48	42	23	46

It is possible to store up to 10 benches in total from one quarter to another at a cost of £ 82 per bench per quarter. At the start of Quarter 1 PB has 2 Superior and 4 Executive benches in stock, and it must have the same numbers in stock at the end of Quarter 4. Demand is, of course, seasonal. The Sales Department's estimates of maximum demands by quarter are:

Units	$Q1$	$Q2$	$Q3$	$Q4$
Superior	10	20	30	10
Executive	8	25	30	5

(a) Formulate and solve PB's problem, clearly stating any assumptions you make. In particular, assume that a plan that gives fractions of benches is acceptable.
(b) Present your solution in a short report, explaining why it is plausible. In particular, interpret the reduced costs and dual values.

2. Refer back to the previous exercise: PB's beechwood supplier now offers them a deal. If in any quarter they buy more than a certain number of units they can have the rest of their purchases that quarter at a lower price per unit. In other words, if they buy a quantity below the breakpoint then they pay Price 1 (the normal price given above), but each unit they buy above the breakpoint only costs them Price 2. Here are the data:

	Q1	Q2	Q3	Q4
Price 1	28	34	22	18
Breakpoint	300	500	350	250
Price 2	20	25	20	15

Reformulate your problem, and find the new policy. Note: Free versions of the modeling systems have restrictions on the number of constraints, variables, and nodes in the B&B search. This problem, if formulated in a natural way, should easily fit into this a free version. If your formulation does not fit into the software you should try to reformulate it.

Chapter 11
Beyond LP and MILP Problems ⊖

This chapter mentions several optimization problems which go beyond linear and mixed integer linear optimization. The focus is rather on motivation. Therefore, it is not intended to cover these topics in complete depth, but the reader should at least be aware that modeling real-world problems is not restricted to linear models. In fractional programming we show how to transform the problem to linear programming, and successive linear programming as a special solution technique of nonlinear optimization. Next, we briefly discuss stochastic optimization. For quadratic programming, which is again a special case of nonlinear optimization, we provide an equivalent formulation based on special ordered sets. Nonlinear optimization is covered in more detail in the next chapter followed by separate chapters on deterministic global optimization in practice and polylithic modeling and solution approaches.

Nonlinearities may occur in the objective function, in the constraints, or in both. They may appear as nonlinear functions such as x^n, $\sqrt{x}$, e^x, or $\sin x$. Sometimes, usually in refinery problems or problems in the food industry, we encounter nonlinear terms of the form $A(x)x$, which we call *nonlinearity in the coefficients*,[1] as instead of Ax with coefficient A, the coefficient is a function involving the variable x as an argument. This type of nonlinearity is very suitable for sequential linear programming, sometimes also called successive linear programming or recursion.

Although this is not the topic of this book, a model might contain differential equations in the constraints or integral expressions in the objective function. They could be included by appropriate discretization techniques and might bring additional nonlinearities into the model.

Supplementary Information The online version contains supplementary material available at https://doi.org/10.1007/978-3-030-73237-0_11.

[1]Of course, they are not really *coefficients* but rather functions multiplied onto the variable.

J. Kallrath, *Business Optimization Using Mathematical Programming*, International Series in Operations Research & Management Science 307, https://doi.org/10.1007/978-3-030-73237-0_11

11.1 Fractional Programming *

The minimization problem LFPP

$$\min \frac{\mathbf{p}^{\mathrm{T}}\mathbf{x}+u}{\mathbf{q}^{\mathrm{T}}\mathbf{x}+v} \tag{11.1.1}$$

subject to

$$\mathsf{A}\mathbf{x} \leq \mathbf{b}, \quad \mathbf{x} \geq 0 \tag{11.1.2}$$

is called a *linear fractional programming problem*. If the feasible set $\mathscr{X} \subseteq \mathbb{R}^n$ of (11.1.2) is nonempty and $\mathbf{q}^{\mathrm{T}}\mathbf{x}+v > 0$ for all $\mathbf{x} \in \mathscr{X}$, LFPP is equivalent to the LP problem

$$\min \mathbf{p}^{\mathrm{T}}\mathbf{y} + uw$$

subject to

$$\begin{aligned} \mathsf{A}\mathbf{y} - \mathbf{b}w &\leq 0 \\ \mathbf{q}^{\mathrm{T}}\mathbf{y} + vw &= 1, \quad \mathbf{y} \geq 0, \quad w \geq 0. \end{aligned}$$

This equivalence is based on the transformation

$$w = \frac{1}{\mathbf{q}^{\mathrm{T}}\mathbf{x}+v}, \qquad \mathbf{y} = w\mathbf{x}$$

relating the scalar variable $w > 0$ to the denominator of (11.1.1). The connection

$$\mathbf{y} = w\mathbf{x} \tag{11.1.3}$$

between the original variables $\mathbf{x}$ and the new variables $\mathbf{y}$ enables us to transform the solution from the equivalent LP back to the original problem. To illustrate this, consider the example

$$\min \frac{x_1+1}{x_2+2}$$

subject to

$$x_1 + x_2 \leq 1, \quad x_1, x_2 \geq 0.$$

The equivalent LP model

$$\min y_1 + w$$

subject to

$$y_1 + y_2 - w \leq 0, \quad y_2 + 2w = 1, \quad x_1, x_2, w \geq 0$$

has the solution

$$y_1 = 0, \quad y_2 = \frac{1}{3}, \quad w = \frac{1}{3}$$

computed using *fractionalProg* in MCOL. Exploiting (11.1.3), we compute the solution of the original problem

$$x_1 = 0, \quad x_2 = 1, \quad z = \frac{1}{3}.$$

In this optimal solution, the nominator is as small as possible, while the denominator is as large as possible. Note that this is sufficient for the global optimum but not necessary.

If we want to minimize sums of ratios

$$\min \sum_{i=1}^{p} \frac{n_i(x)}{d_i(x)},$$

there is no general transformation as above. As this problem has many local minima, we have to resort to global optimization techniques described in Chap. 13. If the nominator functions $n_i(x)$ are convex and the denominator functions $d_i(x)$ are concave, the global minimum can be computed by the Branch & Bound algorithm developed by Benson (2001,[71]). This paper also contains references to practical applications in multi-stage stochastic shipping problems, government contracting problems, and bond portfolio optimization problems.

11.2 Recursion or Successive Linear Programming

Recursion better known as *Successive Linear Programming* (SLP) is a technique whereby LP may be used to solve certain nonlinear problems. Some coefficients in an LP problem are defined to be functions of the optimal values of LP variables. When an LP problem has been solved, the coefficients are re-evaluated and the LP is solved again. Under some assumptions this method may converge to a local[2] (though not necessarily the global[3]) optimum. Those problems most amenable to SLP are LPs with only a few nonlinear terms.

In the first subsection we will give an example illustrating the idea of recursion. The second one discusses the pooling problem and provides some deeper mathematical background on recursion and SLP.

11.2.1 An Example

This small recursion example is simplified from a real-problem concerning the disposal of toxic sludge. Sludge has several nasty components, e.g., cadmium. There are various sources of toxic sludge, which are taken to intermediate processing areas, where they are inevitably subject to some blending, and then finally are deposed at dumping grounds (end points). Formally, sludge including components c originates at source i and has to go via processing area j to final dumping ground k. The maximum tonnage of nasty component c that can be dumped at k is M_{ck}. Sludge arriving at processing area j is blended, so its components cannot be kept separately. The indices of the problem are

$$\begin{aligned} c &= 1, 2, \ldots, C &&: \text{components} \\ i &= 1, 2, \ldots, I &&: \text{sources} \\ j &= 1, 2, \ldots, J &&: \text{processing areas} \\ k &= 1, 2, \ldots, K &&: \text{destinations (end points).} \end{aligned} \tag{11.2.1}$$

The data are

$$\begin{aligned} &A_i &&: \text{availability at } i \\ &C_{jk} &&: \text{per ton disposal cost from } j \text{ to } k \\ &F_{ci} &&: \text{fraction of } c \text{ in material from } i \\ &\lambda_{jc} &&: \text{assumed fraction of } c \text{ at } j \text{ (recursed)} \\ &M_{ck} &&: \text{maximum tons disposal of } c \text{ at } k. \end{aligned} \tag{11.2.2}$$

Finally, we introduce the variables

$$\begin{aligned} q_{cj} &\geq 0 : \text{throughput of component } c \text{ at } j \text{ in tons} \\ t_j &\geq 0 : \text{quantity of component } c \text{ passing through } j \\ x_{ij} &\geq 0 : \text{amount [tons] sent from } i \text{ to } j \\ y_{jk} &\geq 0 : \text{amount [tons] sent from } j \text{ to } k. \end{aligned} \tag{11.2.3}$$

---

[2]A solution is called a *local optimum* if there exists no better feasible solution in its "vicinity."

[3]A solution is a *global optimum* if there exists no better feasible solution.

Then the problem formulation starts with the disposal costs to be minimized

$$\min \quad \sum_{j=1}^{J}\sum_{k=1}^{K} C_{jk}y_{jk}. \tag{11.2.4}$$

We want to get rid of all sludge at i, i.e.,

$$\sum_{j=1}^{J} x_{ij} = A_i, \quad \forall i. \tag{11.2.5}$$

Then let us consider the flow into j

$$\sum_{i=1}^{I} x_{ij} = t_j, \quad \forall j, \tag{11.2.6}$$

and the flow out of j

$$\sum_{k=1}^{K} y_{jk} = t_j, \quad \forall j. \tag{11.2.7}$$

Next, we consider the quantity q_{cj} of component c passing through j:

$$q_{cj} = \sum_{i=1}^{I} F_{ci}x_{ij}. \tag{11.2.8}$$

Now we have to ensure that not too much c is present at k, i.e., there is an upper limit of tons of c which can be disposed of at k. Modeling this feature is more complicated. Component c comes in with each flow y_{jk} from processing area j to end point k. If we knew the concentration, λ_{cj}, or the fraction of c contained in the stream originating at treatment place j, then the constraint just would read

$$\sum_{j=1}^{J} \lambda_{cj}y_{jk} \le M_{ck}, \quad \forall \{ck\}. \tag{11.2.9}$$

The problem is, of course, that we do not know the fraction λ_{cj}. If we accept nonlinear constraints, the only problem is how this quantity is related to other variables. The essence of this formulation is that: the proportion of blended sludge at processing area j, i.e., the toxic component c, is equal to the total quantity of toxic component arriving at the treatment works divided by the total quantity t_j of sludge arriving there, i.e.,

$$\lambda_{cj} := \frac{\sum_{i=1}^{I} F_{ci} x_{ij}}{t_j} = \frac{q_{cj}}{t_j}, \tag{11.2.10}$$

or equivalently

$$t_j \lambda_{cj} = q_{cj}. \tag{11.2.11}$$

Both formulations are nonlinear equations. The second formulation has one numerical advantage: we avoid possible divisions by zero.

If we want to avoid nonlinear constraints, i.e., we want to have a linear model, we will have some difficulties. However, the following observation might help. Once we have all information concerning the flow rates arriving at j we can compute the fractions λ_{cj}. So let us assume for a moment we know the values of λ_{cj}, or (which is what happens in real life) we guess some reasonable values for the λ_{cj}, solve the LP, and then re-calculate their values from the LP solution according to (11.2.10). This is the key idea of recursion. Formally, λ_{cj} enters the expression as any other constant data do. Once x_{ij} (or q_{cj}) and t_j are known from a solved LP problem we apply (11.2.10) and compute λ_{cj}. With the updated values λ_{cj} we solve a new LP problem. The recursion process continues until some form of convergence occurs; see Appendix C.7 for a formal definition of convergence. Note that we cannot guarantee convergence to a global optimum — and may in fact converge only to a local optimum or not at all. So, the λ_{jc} are unknowns which are recursed to q_{cj}/t_j. XPRESS-OPTIMIZER has a recursion facility which lets the user specify coefficients in the matrix as functions of the optimal values of variables. For instance, a coefficient a_{ij} can be specified as being set equal to the value of x_k/x_l. When the coefficients to be recursed have been defined, XPRESS-OPTIMIZER can be instructed to iterate until convergence has been established.

Now let us summarize: we have learned that recursion is a rather simple method to solve certain nonlinear problems. There is no guarantee that this procedure converges. A more sophisticated method which looks very similar is *successive linear programming*. This algorithm guarantees convergence by putting bounds on the difference between the estimate and the next solution value, while recursion does not necessarily do so.

11.2.2 The Pooling Problem

The pooling problem occurs frequently as a subproblem in chemical process, petroleum and food industry and within several types of network flow problems, e.g., wastewater networks, crude oil refinery planning, in which stream rates and concentrations are needed — and are conserved. Typical applications occur if liquid, non-reacting chemicals having different concentrations of contaminants need to be blended and pooled. The pooling problem leads to a special nonlinear optimization problem with a structure which allows us to apply successive linear programming successfully. To understand the problem, let us consider a pool of a liquid substance, such as oil. There are several pipes bringing in oil at different flow rates and different concentrations of a certain component that we are interested in. Pipes leaving the pool carry oil at different flow rates but with the same concentration of the component of interest. An early and more detailed discussion of the pooling problem itself is contained in Fieldhouse (1993,[182]) and Main (1993,[381]). A more recent survey is by Misener & Floudas (2009,[404]). Here we just concentrate on some original formulations and elementary mathematical solution approaches to solving the pooling problem.

Let us try to generalize the pooling problem and to improve our mathematical understanding of it — this will lead us to *sequential linear programming* (SLP) and *distributive recursion* (DR). We will see below that both methods are mathematically equivalent but DR has some numerical advantages over SLP. In modeling a petrochemical network of plants one often faces the problem of pooling n feed streams i of different quality. The feed streams have unknown inflow rates x_i. The quality[4] of feed stream i may be characterized by the contents or concentration C_i of aromatic compounds in that stream, for instance. The pool yields a product of the average aromatic compound concentration c,

$$c = \sum_{i=1}^{n} C_i x_i \big/ x \quad \text{or} \quad cx = \sum_{i=1}^{n} C_i x_i , \tag{11.2.12}$$

where x is the total pool volume

$$x = \sum_{i=1}^{n} x_i , \tag{11.2.13}$$

or the total inflow rate. Note that the quantities x, x_i, and c are unknown variables while the concentrations C_i may be known in advance (of course, they might also be unknown variables, but let us consider the simpler case). As shown in Fig. 11.1a, streams leaving the pool have concentration c. This is the important feature: output streams from a pool have the same concentration. If such a stream is fed into a process unit as illustrated in Fig. 11.1b, then, for instance, the total amount or volume of aromatic compounds in this unit could be limited by an upper bound A^+.

[4]In refinery industry the expression *quality* or *quality constraints* is commonly used. The quality itself can be measured in dimensionless quantities such as concentration (*tons/ton*) or in appropriate physical units (take viscosity as an example).

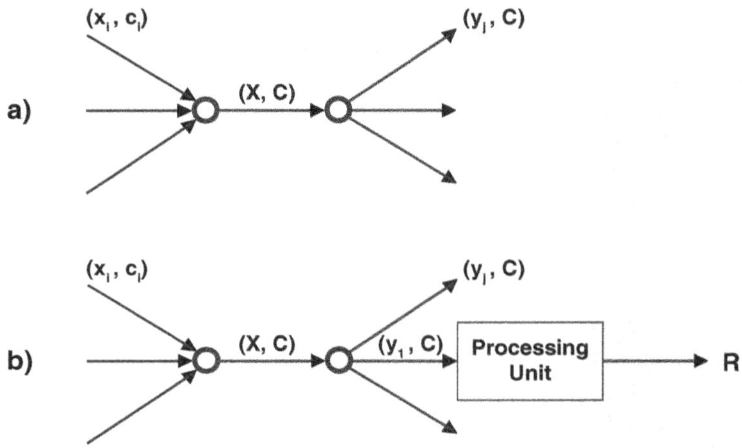

Fig. 11.1 The pooling problem and a process unit fed by a pool

In this case a quality constraint

$$cy_1 \leq A^+ \tag{11.2.14}$$

needs to be added, where y_1 is the outgoing flow rate of the pool and ingoing flow rate w.r.t. the process unit. Both (11.2.12) and (11.2.14) contain products of variables and are thus nonlinear constraints.

If only few nonlinear equations are present, special techniques such as SLP or DR are applied.[5] In SLP, the nonlinear product terms cx and cy_1 are replaced by their Taylor series approximations. Letting c^0 and x^0 be the current values of c and x, the first order approximation of cx is

$$cx \cong c^0x^0 + x^0(c - c^0) + c^0(x - x^0) = c^0x + x^0\Delta c. \tag{11.2.15}$$

Similarly, the first order approximation of cy_1 is

$$cy_1 \cong c^0y_1^0 + y_1^0(c - c^0) + c^0(y_1 - y_1^0) = c^0y + y_1^0\Delta c. \tag{11.2.16}$$

The right-hand side of (11.2.16) is linear and has the unknowns x and Δc,

$$\Delta c := c - c^0. \tag{11.2.17}$$

[5]In principle the methods work with an arbitrary number of variables — but if problems get larger, convergence problems and the significance of good initial guesses increase.

The pool concentration c acts as a nonlinear variable, i.e., it appears in a nonlinear term and its change Δc is determined by solving the LP. This leads to a new value of c, determined by (11.2.17), which is used to initiate the next iteration, replacing c^0. Note that x acts as a special linear variable with nonconstant coefficient c_0 changing in each iteration. Most SLP implementations include bounds on all Δc variables of the form

$$-S \leq \Delta c \leq S,$$

where S is a bound on the step size Δc imposed to ensure that the Taylor series approximations are sufficiently accurate. SLP and DR have different approaches towards the presence of these bounds and the logic for varying them. Using the rules described in Zhang et al. (1985,[595]), convergence of SLP to a local optimum can be proven for any differentiable nonlinear constraint or objective function. Nonlinear optimization theory tells us, because our problem is non-convex, that there is no mathematical technique available, which is based *only* on evaluating function values or values of derivatives of a local optimum x_*, to find out whether x_* is a global optimum. However, in Sect. 12.7 we will learn about *deterministic global optimization* algorithms which are able to prove global optimality.

Note that Δc plays the role of the recursed variable introduced in Sect. 11.2. Instead of using the unknown variables Δc it is possible to use another variable defined as

$$e = x^0 \Delta c. \tag{11.2.18}$$

This leads us to DR. Using this variable the above relations take the form

$$x = \sum_{i=1}^{n} x_i \tag{11.2.19}$$

$$\sum_{i=1}^{n} C_i x_i = c^0 x + e, \tag{11.2.20}$$

and

$$c^0 y_1 + \frac{y_1^0}{x^0} e \leq A^+. \tag{11.2.21}$$

This approach is called *distributive recursion* with the error variable e and the distribution factor

$$\frac{y_1^0}{x^0} \leq 1. \tag{11.2.22}$$

From a numerical point of view, the advantage of this approach is that the LP-matrix has much better scaling properties. Note that the new variable e only occurs with coefficients less than or equal to unity. The next iteration is, of course, initialized with

$$c^0 \leftarrow c^0 + \frac{e}{x^0}. \tag{11.2.23}$$

Furthermore, it should be noted that e has the units of concentration $\times$ amount (volume), i.e., amount or volume, and that e is a free or unconstrained variable, i.e.,

$$-\infty \leq e \leq \infty. \tag{11.2.24}$$

As well as SLP, DR applies a damping strategy on to this free variable and requests

$$-E^- \leq e \leq E^+. \tag{11.2.25}$$

If the matrix generator does not support such variables they can easily be produced by introducing two non-negative variables e^+ and e^- and replacing the variable e by the expression $e^+ - e^-$ at all occurrences of e. Since e^+ and e^- have the same column coefficient wherever they occur they are linearly dependent and therefore cannot enter the basis simultaneously, i.e., at most one of them can take a non-negative value.

Since e usually appears in flow balance equations, convergence properties are observed to be better than in the SLP approach when putting further restrictions on the choice of e.

We have already briefly mentioned that in nonlinear problems in general, and in particular in the pooling problem, we have to distinguish between local and global optima. In practice it is observed that the pooling problem usually has several local optima. Depending on the initial guesses the solver finds different local optima. Thus, solving models involving pooling problems requires great care and deep understanding of the underlying real-world problems. Let us give at least one bit of general advice: one should certainly try to avoid setting initial guess variables to zero. Once a good solution is found one should keep the values of the recursed variables and use them as initial values in further computations.

More recent approaches in refinery modeling exploit continuous-time formulations, special techniques for dealing with bilinear terms, and global optimization techniques; cf. Jia & Ierapetritou (2003,[291]), Misener & Floudas (2009,[404]), Xiao & Floudas (2016,[591]), or Castillo et al.. (2017,[110]). A strong impact on refinery modeling is also that MILP solvers now provide extended features to support bilinear terms.

11.3 Optimization Under Uncertainty*

In most part of this book we have developed and discussed deterministic models assuming that the data are deterministic, although, in practice, input data may be uncertain. In Sect. 3.4.3 we have described how LP problems can be handled when variations in coefficients took place within the problem. If it turns out that some coefficients in a mathematical programming problem are random or subject to fluctuations, but governed by some probability distribution, the problem is often approached by *stochastic programming*, a commonly used modeling and solution technique to deal with *optimization under uncertainty*. Therefore, to start with, let us first have a broader look at optimization under uncertainty.

11.3.1 Motivation and Overview

In most applications and in many software packages input data are assumed to be exact and deterministic. This assumption leads to deterministic models. If some of the input data are subject to uncertainty we enter the realm of *optimization under uncertainty*, where we find optimization problems in which at least some of the input data are seen as random data, or in which some constraints even only hold with a given probability. Optimization under uncertainty can be useful in, for instance,

1. Energy industry: stream flows in water units (inflows in hydro-thermal models), spot market raw material and energy prices, load profiles (demand), generator outages, or delivery of solar and wind energy.
2. Production planning: uncertain demands, or output variations or plants.
3. Finance: uncertainty in prices or returns of the financial instruments, or the volatility of interest rates and currency exchange rates.

The nature of uncertainty can have its roots in physical random processes, market uncertainties, or major unexpected events. Among the reasons for uncertainties we find:

1. Physical or technical parameters only known to a certain degree of accuracy. Safe intervals can usually be specified for such input parameters.
2. Process uncertainties, e.g., stochastic fluctuations in a feed stream to a reactor, or processing times subject to uncertainties.
3. Demand uncertainties occur in many situations: supply chain planning (cf. Gupta & Maranas, 2003,[247]), investment planning (cf. Chakraborty et al. 2004,[112]), or strategic design optimization problems, cf. Kallrath (2002,[303]), involving uncertain demand and price over a long planning horizon of 10 to 20 years.

Optimization under uncertainty applied to real-world problems needs to cope with conceptual issues such as identifying the stochastic nature of uncertain processes or uncertain data entering the optimization model, dealing with feasibility and optimality, or interpreting the results properly. In the late 1990s, stochastic programming was mostly used in energy industry and finance but less frequently outside these sectors. Nowadays, we see other areas of use, including in healthcare (e.g., scheduling) and in public health (e.g., pandemic control and logistics for key resources such as ventilators) in water resources (beyond hydroelectric scheduling) in mining (e.g., with uncertain ore prices and ore grade).

A road block to wider adoption is two-fold: Unlike deterministic optimization, there is not a single standard model. There are chance constraints, two-stage models, multi-stage models, etc. And, unlike `AMPL`, `GAMS`, `Mosel`, etc., for mathematical programming, there is not a single easy interface that is widely accepted for optimization under uncertainty. So, one needs greater expertise on the modeling side and one has a bigger hurdle on the software side. That is changing but is still an issue, especially, when it comes to industrial strength software.

The first step to modeling real-world problems involving uncertain input data is to carefully analyze the nature of the uncertainty. Zimmermann (2000,[598]) gives a good overview of what one has to take into consideration. It is crucial that the assumptions, which are the basis of the various solution approaches, are checked. Technically, statistics including time series analysis has been around for a long time and can be used to fit stochastic processes to time series of data. Sahinidis (2004,[477]) has provided an excellent overview of techniques used to solve optimization problems under uncertainty. Kallrath (2008,[308]) has presented two pricing problems under uncertainty in chemical process industry and some additional discussion points. Below we list and comment on some techniques which already have been or may be used in real-world projects.

1. *Sensitivity analysis* is conceptually difficult in the context of MIP and is, from a mathematical point of view, not a serious approach to solving optimization problems under uncertainty (Wallace (2000,[569])). However, this approach can be successfully applied to an optimization model embedded into a Monte Carlo simulation to establish how strong the objective function values depend on fluctuations of some input data. If the problem can quickly be solved to optimality one could proceed as follow: The problem is solved for a set of input data generated randomly around nominal values. The distribution of a few thousand input values is mapped to a distribution of objective function values.
2. *Stochastic Programming* (SP), and in particular *multi-stage stochastic models*, also called *recourse models*, have been used for quite a long time (cf. Dantzig (1955,[140]), Kall (1976,[298]), or Birge (1997,[78])). In SP, models contain the probability information of the stochastic uncertainty. An important assumption is that the probability distributions do not depend on the decision variables in most cases. While stochastic MILP is an active field of research (cf. the surveys of Klein-Haneveld and van der Vlerk (1999,[335]) or Sen & Higle (1999,[498]), and, more recently, in the *Handbook of Stochastic Programming*, Ruszczyński & Shapiro (2003,[474]), Schultz (2003,[495]), or Andrade et al. (2005,[24])) industrial strength software has yet to enter the stage.

3. *Chance constrained programming* (CCP) deals with probabilistically constrained programming problems and dates back to Charnes & Cooper (1959,[114]); for more recent references cf. Prekopa (1995,[446]), Gupta et al. (2000,[249]), Gupta & Maranas (2003,[247]), Orçun et al. (1996,[427]), Arellano-Garcia et al. (2003,[31], 2004,[32]).
4. *Fuzzy set modeling* supporting uncertainties which fall into the class of vague information and which are expressible as linguistic expressions. Fuzzy set theory in the context of LP problems has been used, for instance, by Zimmermann (1987,[597]; 1991,[596]) and Rommelfanger (1993,[472]).
5. *Robust optimization* — cf. Ben-Tal et al. (2000,[63]), Bertsimas & Sim (2003,[75]), Beyer & Sendhoff (2007,[76]), or Gregory et al. (2011,[239]) — is an approach to dealing with optimization under uncertainty if the uncertainty does not have a stochastic background and/or that information on the underlying distribution is not or hardly available (which is, unfortunately, often the case in real-world optimization problems). Ideally, the solutions of robust optimization should be robust against deviations from the nominal values of the input data. A robust optimization approach for planning and scheduling under uncertainty has been developed by Floudas and his co-workers; cf. Lin et al. (2004,[367]) and Janak et al. (2007,[286]). They have developed a complete theoretical framework for *general* MILP problems. Scheduling and planning problems are a sub-class of the problems the theory can address. They have also solved large scale scheduling applications by robust optimization approaches. While in SP the number of variables increases drastically with the number of scenarios, in this robust optimization approach by Floudas and his co-workers the number of variables and constraints approximately only doubles (for interval uncertainties). An argument sometimes used against robust optimization is that it is too restrictive and pessimistic. This argument becomes invalid because the approach presented by Lin et al. (2004,[367]) allows to specify the probability at which probabilistic constraints have to be fulfilled. One advantage of this robust optimization approach is that for uniform distributions, the type and the complexity of the problem do not change — MILP problems remain MILP problems. Another advantage is that the number of variables and constraints increases only moderately. Having talked about SP, where the distribution is assumed to be known, and having talked about robust optimization, where the uncertain parameter takes a worst case value in an uncertainty set, it is worthwhile to consider *distributionally robust optimization* (DRO), which has recently emerged as a very attractive middle ground. In the data-driven variant of DRO nature selects a worst case probability distribution that does not deviate too much from probability distributions that are consistent with the available data. The survey paper by Rahimian & Mehrotra (2019,[449]) may serve as a good starting point. Robust optimization can also be seen as bilevel programming (see Sect. 14.1.3.4). This allows to use the cutting-set methods by Mutapcic and Boyd (2009,[417]), an interesting approach though technically very demanding.
6. Decisions based on *Markov processes* (cf. Meyn 2002,[396]) and/or the control of *time-discrete stochastic processes* allow for decision-dependent probability

distributions but typically require stronger assumptions on the stochasticity. Cheng et al. (2003,[117]) have given an excellent illustration of such techniques applied to design and planning under uncertainty. An important distinction between Markov decision processes (MDPs) and multi-stage stochastic programs is primarily in: tractable MDPs have finite (small sets) of available actions at each stage while stochastic programs exploit convexity and mathematical programming structures to allow for more complicated action spaces.

It is strongly recommended that if some data, e.g., demand forecasts in planning models or production data in scheduling, are subject to uncertainties, one should consider whether the assumption that planning and modeling is exclusively based on deterministic data can be discarded and uncertainty can be modeled. If probability distributions for the uncertain input data can be provided, SP, CCP, and DRO are the means of choice.

Despite the broad range of solution approaches one important modeling issue is usually not found in publications about solving real-world problems: A rigorous discussion why the selected approach is appropriate to be applied in the real-world problem of interest. This should involve a discussion of the nature of uncertainty: Is it a random process? Is it a Brownian motion? Is there some hidden or unknown causality in the process? In the context of energy design planning, costs for raw materials oil, gas and nuclear, transport and sales prices and the amount of demand are not only subject to complicated market dynamics (usually not covered by the model), but they are also to random events such as disasters like earthquakes, volcano eruptions, pandemic, or sometimes political instabilities and wars. Mathematically, this leads to the representation $y = g(x_i, u_j)$ where we neither know the causal parameters x_i and the stochastic components u_j nor the functional relationship g. Are expected value approaches helpful if we do not know the probability distribution function or if we cannot repeat the decision as often as we need to? In energy industry one could argue that the power demand over a day is well-known after a few years of collecting historical data and only subject to small stochastic scattering. However, with growing market penetration of renewable energy sources, both solar and wind, with the latter having greater short-term uncertainty on the energy availability side, the overall uncertainty increases significantly. In moderate times without strong fluctuations in economy, in financial service industry, the value of stocks can be approximated by Brownian motions. In times of turbulence the dynamics are completely different, and most extrapolation techniques fail. Even in the best case, in which we can compute the optimal expected value, can we also provide the variance of the expected value and is it a small one? If not, the result is rather useless. All this is not to say one should not use a certain approach. But we stress to provide good evidence why the selected approach is appropriate for the problem at hand. And, unfortunately, there are cases, e.g., strategic design decisions of 20 years, in which we probably cannot do much more than quantifying the sensitivity and looking for robust solutions.

11.3.2 Stochastic Programming

Stochastic programming (SP) finds many applications in energy industry and financial optimization in which uncertainty sometimes is even the dominating factor: uncertainty in prices or returns of the financial instruments, the volatility of interest rates or currency exchange rates, and others. Stochastic programming is a very complex topic and we only focus on its connection with recursion and the value of the stochastic extension. For a more detailed treatment the reader is referred to Kall & Wallace (1994,[299]). Some problems are difficult to solve while some may be solved approximately. Recursion provides a way to solve some stochastic programs, e.g., to transform an LP problem with random coefficients into a solvable form (Dantzig et al., 1981,[146]). We may formulate it as a stochastic two-stage model multi-stage, with or without recourse, which may then be solved by recursion. Unfortunately, in practice the probability distribution is often unknown, and only the expected values of random quantities are available.

To illustrate the idea of SP and a two-stage recourse problem, consider the case of fixing the production of a chemical plant with uncertain demands and demand-dependent selling prices for the next 6 months. The production amounts represent the stage-1 decision. As soon as the production decision has been made based on an assumed demand, and the actual demand becomes available, a stage-2 decision is necessary. If demand is lower than anticipated, the excess production might be sold at a lower price. If the demand turns out to be higher, an extra production (driving the plant above its limit for a short period) might be considered to be sold at a higher price. To illustrate the idea of two-stage recourse problems in more quantitative details and to get familiar with the nomenclature we discuss and provide an implementation of the newsvendor problem in the following section.

11.3.2.1 Example: The Newsvendor Problem

The newsvendor problem has been first stated in the eighteenth century, when a bank needed to determine the level of cash reserves it required to cover the demand from its customers. It has been widely used in economic literature, e.g., to analyze supply chains in fashion and seasonal product industries. Since the 1950s, the newsvendor problem has been extensively studied in operations research and extended to model a variety of real-life problems; cf. Choi (2012,[121]).

The simplest and most elementary version of the newsvendor problem is an optimal stocking problem in which a newsvendor needs to decide how many newspapers to order for future uncertain demand given that the newspapers becomes obsolete at the end of the day, i.e., how many newspapers x should the newsvendor purchase early in the morning from a newspaper company at a price of C per newspaper; see Fig. 11.2. Customers demand D (unknown at purchase time) has to be satisfied; otherwise a penalty $Q > C$ occurs per uncovered newspaper. There is an upper bound X^+ on the number of newspapers the vendor can purchase (e.g., due

Fig. 11.2 Newsvendor problem. At stage 1 (left), the newsvendor has to decide how many newspapers to purchase early in the morning (demand is not known). At stage 2 (right), the newsvendor learns about the real demand (of course, demand somewhat depends on his ability to find customers). Produced for this book by Diana Kallrath, Copyright ©2020

to carrying limits). The objective function is to minimize the total cost (purchasing costs + penalty costs). This example may appear very simple and artificial, but the equivalent situation, for instance, in energy industry is very realistic:

$$\begin{array}{ll} \text{newspapers} & \Leftrightarrow \text{energy produced or purchased} \\ \text{demand} & \Leftrightarrow \text{uncertain load (demand) of energy} \\ \text{upper bound } X^+ & \Leftrightarrow \text{energy production/availability limits} \end{array} \quad .$$

A good starting point to approach this problem is to consider the deterministic version of the problem with certain demand D formulated as the LP:

$$\begin{aligned} \min z &= Cx + Qy \\ s.t. \quad & 0 \le x \le X^+ \\ & x + y \ge D, \quad y \ge 0, \end{aligned} \tag{11.3.1}$$

where the variables x and y represent purchased newspapers and uncovered demand. Strictly speaking, this is an MILP problem, but we neglect the integrality of x and y. This deterministic version of the problem has an analytic solution: The optimal solution for known D is: purchase

$$x = \min\{D, X^+\} \tag{11.3.2}$$

newspapers. But how many should be purchased when D is uncertain?

In presence of uncertain input data one can get an overview and a feeling for the sensitivity of the solution *w.r.t.* the uncertain input data by solving the deterministic problem for several scenarios. Therefore, let us consider the following data

$$C = 1.5,\ Q = 2 \text{ and } X^+ = 50$$

for the scenarios s_1, s_2, s_3, s_4, and s_5 with demand data

$$(D_1, D_2, D_3, D_4, D_5) = (13, 23, 25, 31, 33).$$

We assume that each scenario has a probability 20% for its realization. Now we proceed as follows: We compute the optimal solution (11.3.2) for each scenario separately, i.e., we will put $D = D_i$ for $i = 1, 2, 3, 4,,$ and 5. Next, we insert each scenario solution (the purchase value x) into the other scenarios to explore the effect of having the wrong demand. This way we obtain the scenario solution T_{ij} (*wait-and-see approach* or WS solution):

	s_1	s_2	s_3	s_4	s_5	average
s_1	13; 19.5	13; 39.5	13; 43.5	13; 55.5	13; 59.5	13; 43.5
s_2	23; 34.5	23; 34.5	23; 38.5	23; 50.5	23; 54.5	23; 42.5
s_3	25; 37.5	25; 37.5	25; 37.5	25; 49.5	25; 53.5	25; 43.1
s_4	31; 46.5	31; 46.5	31; 46.5	31; 46.5	31; 50.5	31; 47.3
s_5	33; 49.5	33; 49.5	33; 49.5	33; 49.5	33; 49.5	33; 49.5
EV	25; 37.5	25; 37.5	25; 37.5	25; 49.5	25; 53.5	25; 43.1
SS	23; 34.5	23; 34.5	23; 38.5	23; 50.5	23; 54.5	23; 42.5

Note that T_{ij} gives $(x_i; z_i)$ with optimal purchase $x_i = D_i$ of the deterministic problem for $D = D_i$ and the corresponding objective function value $z_i = 1.5x_i$, i.e., if scenario s_i is realized for known $D = D_i$. For $i \neq j$, T_{ij} gives $(x_i; z_{ij})$ the objective function value obtained for the deterministic problem for $D = D_j$ while fixing $x = x_i$. For $i < j$ $(i > j)$, the newsvendor has bought too few (too many) newspapers.

Now let us consider the solutions of these scenarios and certain averages. The scenarios s_i with demands (13,23,25,31,33) have average demand

$$D_* = \frac{13 + 23 + 25 + 31 + 33}{5} = 25.$$

In the scenario solution table we can consider the average cost per row (last column). These averages are also called *expected values*. Then we solve each scenario with the purchase fixed to the average ('expected value') demand, i.e., $x = D_*$ displayed in the EV [Expected Value] row of the table.

From the deterministic model (11.3.1), we approach the two-stage stochastic solution by duplicating the demand inequality for each scenario s:

$$\min z = Cx + \sum_s p_s Q y_s$$

$$s.t. \quad 0 \leq x \leq X^+$$

$$x + y_s \geq D_s, \quad y_s \geq 0 \quad ; \quad \forall s. \tag{11.3.3}$$

Note that the recourse costs $\sum_s p_s Q y_s$ associated with the second-stage variables y_s are evaluated with the probability p_s of the scenarios. The solution of problem (11.3.3) is called *here-and-now* solution and is interpreted as follows: We have to *decide now how many newspapers to buy* but we *incorporate the scenario-based expectation into the decision*. The stochastic approach is superior to just considering individual scenarios or averages as it explicitly considers the structure of the problem in the model. The stochastic solution, or policy, is to purchase $x_{\mathrm{ss}} = 23$ newspapers (the objective function value is the recourse value $\hat{R} = z_{\mathrm{ss}} = 42.5$). We compare this to the individual scenarios and the averages, i.e., we solve each scenario with its purchase fixed to the stochastic solution $x = x_{\mathrm{ss}} = 23$. As a result, recourse corrects the first-stage variable (purchase x).

For the newsvendor example we can now quantify the value $\hat{V}$ (or VSS) of the stochastic solution:

$$\hat{V} = \hat{E} - \hat{R}. \tag{11.3.4}$$

The *expected result* $\hat{E}$ of using the EV solution is $\hat{E} = 43.1$; see *average value* in the EV row of the table. The solution of the expected value problem, obtained by replacing all uncertain input data with their expected values, is used to fix the first-stage variables when computing the WS solutions. For the newsvendor problem we obtain the special relation

$$\hat{V} = \text{averageEVsolution - averageSPsolution}.$$

Now we can interpret the value of the stochastic solution: $\hat{V} = 0.6$ quantifies the potential for improvement if we switch from the deterministic to the stochastic model. The larger the $\hat{V}$ normalized to the objective function, the more it makes sense to switch to the stochastic model. If $\hat{V}$ is small, it is not worthwhile to resort to SP.

Finally, let us focus on the *expected value of perfect information* [EVPI], which is determined by comparing the objective function values of the following two problems: The here-and-now or recourse solution ($\hat{R} = 42.5$) obtained by solving the stochastic optimization problem, and the wait-and-see solution ($\hat{W} = 37.5$) resulting from solving each scenario individually and computing the expected values of the objective functions. For the newsvendor problem we get $EVPI = 5.0$ with the following interpretation: The EVPI (5.0, in this case) quantifies the (average) improvement if we knew which scenario will become reality. It is the price for gaining access to perfect information. The larger the EVPI (normalized to the objective function), the more sensible it is to invest in better forecast. While the difference $\hat{R} - \hat{W}$ measures the EVPI, the difference $\hat{E} - \hat{R}$ is $\hat{V}$. The quantities are useful for answering the question whether it is worthwhile to resort to SP. $\hat{V}$ requires the computation of the stochastic solution $\hat{R}$ (HN=here-and-now, also called RP=recourse problem). The relevance of the stochastic approach is, for minimization problems, better expressed by the chain of inequalities (11.3.4), $\hat{W} \leq \hat{R} \leq \hat{E}$. The stochastic solution is bounded by the values $\hat{W}$ (of the wait-and-see solution) and $\hat{E}$. If $\hat{E} - \hat{W}$ is small there is no need for a stochastic extension. If $\hat{E} - \hat{W}$ is large, SP can make a significant difference. We have to keep in mind that these conclusions are only safe if the scenarios represent the probability distribution of uncertain input data well.

11.3.2.2 Scenario-Based Stochastic Optimization

Among the techniques to solve optimization problems under uncertainty is scenario-based stochastic optimization [cf. Dantzig (1955,[140]), Kall (1976,[298]), Kall & Wallace (2004,[299]), Schultz (1995,[494]), Birge (1997,[78]), Carøe & Schultz (1999,[108]), Klein-Haneveld & van der Vlerk (1999,[335]), the *Handbook of Stochastic Programming* [474], Schultz (2003,[495]), Andrade (2005,[24]), Sen (2004,[497]), or Mitra et al. (2004,[410])]. For scenario-based chance constrained programming, cf. Prekopa (1995,[446]), Gupta et al. (2000,[249]), Gupta & Maranas (2003,[247]), Orçun et al. (1996,[427]), Arellano-Garcia et al. (2003,[31], 2004,[32]). These approaches have in common that they construct a deterministic equivalent problem of the original model and its uncertainties. Valente et al. (2009,[557]) discuss this also with respect to embedding SP and scenario trees in AMLs.

In the context of energy optimization, one of the largest application fields of SP, generating reasonable scenarios representing the uncertainties, is a formidable task. The newsvendor problem had one period (no time-dependence) but two stages. Stages are characterized as follows: at the beginning of a stage new information about some uncertain events becomes available. At the end of a stage we see recourse decisions, i.e., adjustments (with stage 1 being an exception). Note that stages are not necessarily connected to periods. The overall scheme for stages is: Make a decision at stage $t = 1$, observe uncertain input data resolved at stage $t + 1$ and make adjustments, proceed for all stages t until $t = T$. In production planning problems we may have many time periods (months), but only two stages ($T = 2$), perhaps four ($T = 4$, after 1 month, 3 months, and 6 months). In the energy sector, we typically have 8760 time periods (1 year horizon at a resolution of 1 hour) but only 52 stages (weeks). Decisions use only previously observed events as a source of information; for future events we have only expectations (this is called *non-anticipativity* of the stochastic process).

For a production set of several thousands of demands over the planning horizon, it is not evident how to create scenarios covering the evolution of demands, sales prices as well as the costs of energy and raw materials. Many input data such as inflows in hydro-thermal models might be correlated or anti-correlated, others may be completely independent; the same holds for raw materials. These relations, if known, are usually built into the scenarios. Constructing appropriate scenarios is an art; cf. Di Domenica et al. (2007,[159]). Scenarios should, on the one hand, cover the expected situations reasonably, and should, on the other hand, not be too similar to avoid large the numbers of scenarios. To summarize the overall procedure: Solving stochastic programs is always based on *deterministic equivalents*: In Step 1 we approximate the stochastic process ξ and its *continuous* probability distribution by a *discrete* probability distribution. In Step 2 we derive scenarios from the discrete distributions. The deterministic equivalent includes all scenarios and stages; the number of scenarios increases exponentially. Therefore, to keep the problem size limited, we should define good stages, construct reasonable scenarios, and ask ourselves *how many scenarios are necessary*. If scenarios differ only slightly, can we reduce the number of scenarios? This leads us to *scenario reduction*, for which Henrion et al. (2008,[265]; 2009,[266]) have provided algorithms and software to reduce scenario trees by combining similar and near-by scenarios.

11.3.2.3 Terminology and Technical Preliminaries

Stochastic programming separates into distribution problems (expected value and wait-and-see) and recourse problems (distribution-based and scenario-based), both allowing the stochastic measures EVPI and VSS, as well as chance constrained programming. Let us connect the basic definition with more formal mathematical concepts.

The *expected value (EV) solution* is defined as the solution of the Expected Value Problem, which is the deterministic problem obtained by replacing all uncertain input data by their expected values, i.e., determining the expected values of all uncertain input data (exploiting the probability distribution), and solving the deterministic problem using the expected values of the uncertain input data. The EV solution is used to compute the EEV.

The *wait-and-see (WS)* solution results from solving the deterministic problem for all individual scenarios s. This way we obtain a set of solutions (not one implementable solution), the distribution of the optimal objective function values over all scenarios, and the wait-and-see value $\hat{W}$ defined as the expected value of this distribution. The implicit assumption of the WS approach is that the decision maker does not need to decide now and can always *wait* until uncertainty turns into certainty (*see*). The WS solution gives us some feeling of how the optimal solution varies as a function of the scenarios, i.e., it seems similar to some kind of sensitivity analysis, but it does not support any decision directly.

At this stage, we provide a few technical definitions used in probability theory related to the concept of the expected value of a random variable subject to a probability distribution.

The expected value $\mathbb{E}[x]$ of a continuous real-valued random variable $x = x(\omega)$ is the integral *w.r.t.* to a probability measure

$$\mathbb{E}[x] := \int_{\Omega} x(\omega)\mathrm{d}p(\omega)$$

with probability space Ω, ω representing the random process, and a probability measure $\mathrm{d}p(\omega)$, e.g., $f(x)\mathrm{d}x$. In the special case of a probability density function which is Riemann-integrable we have

$$\mathbb{E}[x] := \int_{\mathbb{R}} xf(x)dx,$$

i.e., $\mathrm{d}p(\omega) = f(x)\mathrm{d}x$ with a probability density function $f(x)$.

The expected value $\mathbb{E}[x]$ of a discrete real-valued random variable x_i, $i \in \mathscr{I}$, with associated probabilities $p_i = p(x = x_i)$ over a countable index set $\mathscr{I}$, is the sum

$$\mathbb{E}[x] := \sum_{i \in \mathscr{I}} p_i x_i .$$

Note that we have a non-zero probability p_i for x_i.

11.3.2.4 Practical Usage and Policies

Some words on the practical use and interpretation of SP: Multi-stage scenario-based solutions give us an approximation to the solution (rather a policy) of the stochastic problem with the following properties:

1. Recourse has a backward correction effect of future uncertainties on the must-decide-now decision at stage 1 (for which we have deterministic information).
2. At each node of the scenario tree we get the optimal values of the decisions for certain realizations of the stochastic input data.

There are two ways to use two-stage SP for real-world decision problems:

1. Use only the stage-1 decision and apply a rolling-horizon approach when reaching the time connected to stage 2.
2. Use the stage-1 decision, wait until the uncertain input data becomes certain during stage 2 and look up the corresponding stage-2 decision variables.

11.3.2.5 The Value of the Stochastic Extension

Imagine a company which has a deterministic model and solution approach in place. They are aware that some input data are uncertain and also have some ideas about various scenarios how these uncertain data might evolve. Now they are wondering whether it is worthwhile to enhance the deterministic model by rigid methods, e.g., to exploit scenario-based stochastic optimization, to cover the uncertainty. Is it worth investing several months or even years to build up the know-how and to modify and extend the model? What financial benefit can be expected when including stochasticity into the model? Is it realistic to expect a several-percent increase of profit or a decrease of cost?

These questions are not easy to answer. As safe answer would be: "It depends." Let us try to work out what exactly it depends on and how we can find a satisfying answer. Although practitioners often start straight ahead with scenarios, in the beginning it is a safer to start working out what the underlying stochastic process is. This may result in scenarios. If the stochastic process is a fully random process, e.g., Brownian motion, scenarios can be difficult to construct and Monte Carlo or sampling methods are preferable.

In any case, sensitivity analysis is a good starting point. Varying some of the uncertain input parameters or evaluating different scenarios over a certain timespan shows how stable or sensitive the solution and its objective function value is *w.r.t.* to changes in the uncertain input data.

For two-stage recourse problems have introduced a concept to evaluate the potential of the stochastic extension: The *value of the stochastic solution* (VSS or $\hat{V}$), which is a measure of whether SP can improve decision making, and the *expected value of perfect information* (EVPI) measuring whether or not it is reasonable to pay for obtaining perfect information of the future. The computation of $\hat{V}$ and EVPI usually exploits the constructed scenarios.

For minimization models, we have already seen in the newsvendor example that the inequalities $\hat{W} \leq \hat{R} \leq \hat{E}$ hold. The value $\hat{E}$ (or EEV) denotes the expected result of using the solution of the deterministic model EV (all random input data are replaced by their expected values). The quantity $\hat{W}$ (or WS), known as the *wait-and-see solution* value, denotes the expected value of the optimal solution for each scenario. The value $\hat{R}$ (or RP for *recourse problem*), also known as the *here-and-now solution*, denotes the optimal solution value of the *recourse problem*. The difference EVPI $= \hat{R} - \hat{W}$ denotes the expected value of perfect information and compares the here-and-now and wait-and-see approaches. A small EVPI indicates a low additional profit when reaching perfect information. The difference $\hat{V} = \hat{E} - \hat{R}$ denotes the value of the stochastic solution and compares the here-and-now and expected values approaches. A small $\hat{V}$ means that the approximation of the stochastic program by the program with expected values instead of random variables is a good one. Calculating these two measures is relatively simple for two-stage models and is illustrated in *newsvendor.gms* contained in the book library MCOL; see Sect. 11.3.2.1. The computation of $\hat{V}$ requires that we know the value of $\hat{E}$ and proceeds as follows:

1. Solve the expected value problem EV (replace all uncertain data with their expected values),
2. Fix the first-stage solution for each scenario (WS model) at the optimal solution obtained by solving the EV problem in Step 1,
3. Solve the resulting problem for each scenario, and
4. Calculate the expected value of the objective function over the set of scenarios for these modified WS problems.

Note that the computation of $\hat{V}$ requires the solution of the stochastic problem. Therefore, we are less interested in $\hat{V}$ itself (we will get its value only after we have solved the stochastic problem) but rather focus on inequality (11.3.4), which gives us the bounds $\hat{W}$ and $\hat{E}$ on the stochastic solution.

Transferring these concepts to the multi-stage case is not straightforward and problematic. There exist a few possible approaches, one of which is presented below and originates from Escudero et al. (2007,[175]). In their paper, they generalize the definition of bounds for the optimal values of the objective function for various deterministic equivalent models in multi-stage stochastic linear programs. In particular, they introduce a chain of expected values when fixing the value of the decision variables at the optimal value in the related average scenario model, EV. The final value of the chain happens to be the expected value of using the expected value solution, $\hat{E}$, in two-stage models. Differences between the values in this chain indicate the need to solve the stochastic model, RP. In each stage, they allow us to compute the value of the stochastic solution, $\hat{V}$, and to check how good the approximation of the stochastic program by the deterministic one is up to the stage where the expected values are used instead of the random variables. The proposed extension of the bounds is primarily useful for avoiding to obtain the RP value when the average based solution is good enough, as it also happens for the two-stage problem.

Various difficulties occur in the multi-stage case and the generalization of the two measures described above for this case. In particular, it is not as obvious which variables must be fixed in the WS models. There are different approaches in literature. The easiest one is to only fix the first-stage variable at the optimal solution as in the two-stage case. The variables of the following stages are free to take on different values for different scenarios. The problem with this approach is that it can happen that the first-stage solution in the EV problem performs better than the solution of the RP one. The reason is that in the multi-stage case RP contains the non-anticipativity requirement which is ignored when the WS models are solved. Let us consider a multi-stage model where constraints only relate two consecutive stages. We denote the optimal solution value of this problem as $\hat{R}$. We deal with the uncertainty in the stochastic parameters via scenario trees. We now define the EV problem, where the uncertain parameters are replaced by their expected values (overlined quantities):

$$\min c_1 x_1 + \bar{c}_2 \bar{x}_2 + \ldots \bar{c}_T \bar{x}_T$$

$$s.t. \quad W_1 x_1 = h_1$$

$$\bar{T}_{t-1}x_{t-1} + \bar{W}_t x_t = \bar{h}_t, \quad t = 2, \ldots, T$$
$$x_t \geq 0, \quad t = 1, \ldots, T. \tag{11.3.5}$$

Let $\bar{x}_t^*$ be the optimal solution of problem (11.3.5). Let the expected result in t using the expected value solution, denoted by $\hat{E}_t$; $t = 2, \ldots, T$, be the optimal value of the RP model, where the decision variables $x_1, \ldots, x_{t-1}$ until stage $t-1$ are fixed at the optimal values obtained from the solution of the average scenario problem (11.3.5):

$$\hat{E}_t := \begin{cases} \text{RP model} \\ s.t.\ x_\tau^s = \bar{x}_\tau^*, \quad \tau = 1, \ldots, t-1, \quad s = 1, \ldots, S. \end{cases} \tag{11.3.6}$$

We use RP in the definition of $\hat{E}_t$ for clarification and simplification purposes to formulate it. However, the computation of $\hat{E}_t$ does not require the complexity which is required for obtaining RP (except, obviously, for obtaining $\hat{E}_1$ that coincides with the value RP). Since we fix the 0–1 variables until stage $t-1$, the model for obtaining $\hat{E}_t$ can be decomposed (if there are no continuous variables until stage $t-1$) in G_t independent submodels (with much smaller dimensions than the ones for RP). For $\hat{E}_T$ in particular we have S scenario-based submodels, the dimension of which is the same as the dimension of the mean value problem EV, so that $\hat{E}_T$ is the weighted sum of the optimal solution value of these submodels.

Extending the definition (11.3.6) to $t = 1$ and defining $\hat{E}_1 = \hat{R}$ generates the sequence of expected values $\hat{E}_1, \hat{E}_2, \ldots, \hat{E}_T$, for which the inequalities

$$\hat{E}_{t+1} \leq \hat{E}_t, \quad t = 1, \ldots, T-1 \tag{11.3.7}$$

can be proven (in maximization problems).

We define the value of the stochastic solution in t, denoted by $\hat{V}_t$, as

$$\hat{V}_t = \hat{E}_t - \hat{R}, \quad t = 1, \ldots, T. \tag{11.3.8}$$

Useful results (for maximization problems) are the inequalities

$$0 \leq \hat{V}_t \leq \hat{V}_{t+1}, \quad t = 1, \ldots, T-1 \tag{11.3.9}$$

and

$$\hat{V}_t \leq \bar{E} - \hat{E}_t, \quad t = 1, \ldots, T, \tag{11.3.10}$$

where $\bar{E}$ denotes the objective function value of the EV problem.

The sequence (11.3.9) of non-negative values describes the cost of ignoring uncertainty until stage t when making decisions in multi-stage models. The expected value of the solution that provides the average scenario problem, $\hat{E}$, defined for two-stage problems, is equal to the value of $\hat{E}_T$ in multi-stage models. $\hat{E}_T$ allows us to calculate the maximum cost that we would be prepared to pay to ignore uncertainty at all stages or in the complete time horizon. *In particular, Escudero et al. (2007,[175]) introduce a chain of expected values when fixing the value of the decision variables at the optimal value in the related average scenario model, EV. The final value of the chain happens to be the expected value of using the expected value solution, $\hat{E}$, in two-stage models. Differences between the values in this chain indicate the need to solve the stochastic model, RP. At each stage, they allow us to compute the value of the stochastic solution, VSS, and to check how good the approximation of the stochastic program by the deterministic one is up to the stage when the expected values are used instead of the random variables.*

Let us summarize the computational procedure: To obtain $\hat{E}_t$, one needs to solve the EV (just one problem) and save the optimal solution values. Then one fixes all scenario variables up to stage $t-1$. There are no non-anticipativity constraints, and one can decompose the full problem into smaller submodels and compute the expected value solution. It is the same situation as the one in the traditional $\hat{E}$, for the two-stage case. In *Portfolio.gms* this approach is labeled by EEV(t).

The difficulty, however, is that models or problems, in which many variables have been fixed, are often infeasible. Thus, if all problems $\hat{E}_2, \dots, \hat{E}_t$ are infeasible, we will not obtain any information about the quality of the approach above. A possible alternative that does not fix the variables which generate infeasibility in the model calculates an estimation of $\hat{E}_t$.

The intermediate values, $\hat{E}_t$, and then $\hat{V}_t$, give us information about a suitable choice of the number of stages in the model for $t = 2, \dots, T$. Similar successive values $\hat{E}_t \approx \hat{E}_{t+1}$, and then $\hat{V}_t \approx \hat{V}_{t+1}$, would indicate that the deterministic problem until stage t is a good approximation to the stochastic problem, and therefore it would not be necessary to define later stages.

A sufficient condition for $\hat{E}_t = \hat{E}_{t+1}$, and then $\hat{V}_t = \hat{V}_{t+1}$, is the independence of $z_{t+1}(s)$ for scenario s. This means that the optimal values at stage $t+1$ are insensitive to the value of the random elements. In such situations, finding the optimal solution for one particular $\xi^{(s)}$ ($\xi^{(\bar{s})}$, for example), would yield the same result, and it is unnecessary to define a further stage.

In the second approach, labeled EEVhat(t) in *Portfolio.gms*, we compute $E\wedge EV_t$, the feasible expected value in t using the solution of the average scenario model, the optimal value of the RP model, where the decision variables until stage $t-1$ are fixed at zero if they are fixed at zero in the optimal solution of the average scenario model (11.3.5). That is,

$$E\wedge EV_t := \begin{cases} \text{RP model} \\ s.t.\ s_\tau^* \le \bar{x}_\tau^* M_\tau, \quad \tau = 1, \dots, t-1, \end{cases} \tag{11.3.11}$$

where $M_1, \dots, M_{t-1}$ are sufficiently large constants.

The approach described in the last paragraph provides us with a feasible policy for our multi-stage model but is quite unrealistic.

Therefore Escudero et al. (2007,[175]) have developed a dynamic approach, labeled EDEV(t) in *Portfolio.gms*, which gives tighter bounds, works tightly with the tree structure of the multi-stage problem and therefore uses the notations described earlier. To obtain the $EDEV_t$ values, one needs to solve one EV_g problem per scenario group ($g \in \mathcal{G}$). All of them (EV_g) are deterministic models where the parameter estimations are updated and based on average values only for the scenarios that belong to group g, i.e., model EV_g is solved for each scenario group at each stage of the problem and the estimates are updated. The expected result of using these dynamic solutions of the model based on average scenarios is obtained immediately from the solution of these models. The computational steps are summarized below:

Step 1: $t = 1$. Solve problem (11.3.5), i.e., the Expected Value problem EV_g for scenario group $g = 1$ (this is the traditional EV problem). Save the optimal solution value Z^1_{EV} and the optimal solution values $\bar{x}^*_1$ of the first-stage variables.

Step 2: Repeat for $t = 1, \ldots, T-1$:
Define the set of $|\mathcal{G}_t|$ average scenario problems EV_g for the scenario subtrees corresponding to each group of scenarios at the next stage $g \in \mathcal{G}_t$, where the random parameters of subsequent stages are estimated by their expected values. All variables of the previous stages are fixed at the optimal solution values obtained in the chain $EV_{a(g)}$; $g \in \mathcal{G}$; $\tau = 1, \ldots, t-1$. When solving the $|\mathcal{G}_t|$ problems, store the optimal objective values Z^g_{EV} and the optimal values of the variables of the current stage as $\bar{x}^{*,g}_t$ for $g \in \mathcal{G}_t$. In detail: At this stage $t = 2$, one has $|\mathcal{G}_2|$ scenario groups. For each $g \in \mathcal{G}_2$, one obtains a subtree. One needs to solve $|\mathcal{G}_2|$ Expected Value problems, each of them is the corresponding EV problem for the subtree with root node 1 and node g in stage 2. In all of these problems, the first-stage variables are fixed to the values obtained in Step 1. For each EV_g problem one saves the optimal solution value Z^g_{EV} and the optimal solution values for the second-stage variables. At stage $t = 3$, one has $|\mathcal{G}_3|$ scenario groups. For each $g \in \mathcal{G}_3$, one obtains a subtree again. Now one needs to solve $|\mathcal{G}_3|$ Expected Value problems, each of them is the corresponding EV problem for the subtree with root node 1 and node g at stage 3. In all of these problems, the variables are fixed for stages 1 and 2 to the values obtained in Step 1 and 2 ($t = 2$). Again, for each EV_g problem, one saves the optimal solution value Z^g_{EV} and the optimal solution values for stage 3. This continues for all stages except the last one (see below).

Step 3: Define the set of $|\mathcal{G}_t| = S$ average scenario problems EV_g, $g \in \mathcal{G}_t$ for the scenario subtrees corresponding to each group of scenarios at the last stage, where all the variables of the previous stages are fixed, except for the last one, at the optimal solution values obtained in the chain of models $EV_{a(g)}$; $g \in \mathcal{G}_\tau$; $\tau = 1, \ldots, t-1$. So, at the last stage, $t = T$, we have $\mathcal{G}_T = S$ and need to solve $|\mathcal{G}_T|$ EV_g problems. Each of them is a related scenario problem with variables of the last stage only.

With the procedure given in Steps 1 to 3, the expected value solutions are updated. Let us denote the expected result in t of using the dynamic solution of the average scenario by $EDEV_t$, $t = 1, \ldots, T$, as the expected value of the optimal values of the EV_g, $g \in \mathcal{G}_t$ problems, i.e.,

$$EDEV_t = \sum_{g \in \mathcal{G}_t} p_g Z_{EV}^g, \quad t = 1, \ldots, T, \tag{11.3.12}$$

where p_g is the probability of scenario group g computed as

$$p_g = \sum_{s \in \mathcal{S}_g} p_s. \tag{11.3.13}$$

We define the dynamic value of the stochastic solution $\hat{V}^D$ as

$$\hat{V}^D = EDEV_T - \hat{R}. \tag{11.3.14}$$

We need to solve $|\mathcal{G}_1| + |\mathcal{G}_2| + \ldots + |\mathcal{G}_T| = |\mathcal{G}|$ models to obtain $\hat{V}^D$, but all submodels have small dimensions. We can, similarly to the expected result in t using the expected value solution described above, define the *dynamic value of the stochastic solution*, $\hat{V}_t^D$, for each stage t. For the dynamic versions of $\hat{E}$ and VSS there exist corresponding inequalities to (11.3.7), (11.3.9) and (11.3.10).

So, in each update step at stage t one gets a new chain of inequalities as

$$\hat{W}_t \leq \hat{R} \leq EDEV_t \leq \hat{E}_t. \tag{11.3.15}$$

In particular, at each stage, the list of computational steps to decide whether the RP model needs to be solved or not looks similar to that given for two-stage models.

11.3.3 Recommended Literature

Here we provide to the reader a selected list of books and articles on stochastic optimization.

1. The textbook *Introduction to Stochastic Programming* by provides a thorough introduction to SP with detailed examples and including some material on integer SP — a book strongly recommended.
2. *Lectures on Stochastic Programming: Modeling and Theory* by A. Shapiro et al. (2009,[500]) has a good introductory chapter on stochastic programming models, gives a strong focus on two-stage and multi-stage problems as well as chance constrained programming, and is a useful resource on the topic of statistical inference (sample average techniques used, for instance, by `DECIS` solver embedded in `GAMS`).
3. *The Value of the Stochastic Solution in Multi-stage Problems* by Escudero et al. (2007,[175]). An illuminating treatment of the stochastic key parameter EEV and its extension to multi-stage models.

4. An interesting SP approach to *planning of offshore gas field developments under uncertainty in reserves* is by Goel & Grossmann (2004,[226]).

11.4 Quadratic Programming

Quadratic programming is a special case of nonlinear programming. In this section we do not cover the theory or efficient algorithms for solving such problems. For detailed background on quadratic programming we refer the reader to, for instance, to Gill et al. (1981,[217]) or Furini et al. (2019,[204]). We provide a simple trick to solve quadratic programming problem using special ordered sets and just an MILP solver.

Let us consider the quadratic programming problem (QP):[6]

$$\min \frac{1}{2}\mathbf{x}^{\mathrm{T}}\mathsf{G}\mathbf{x} + \mathbf{g}^{\mathrm{T}}\mathbf{x} \tag{11.4.1}$$

subject to

$$\mathsf{A}^{\mathrm{T}}\mathbf{x} \geq \mathbf{b}, \quad \mathbf{x} \geq \mathbf{0}. \tag{11.4.2}$$

Introducing Lagrange multipliers $\boldsymbol{\lambda}$ for the constraints $\mathsf{A}^{\mathrm{T}}\mathbf{x} \geq \mathbf{b}$, and $\boldsymbol{\pi}$ for the non-negativity bounds $\mathbf{x} \geq \mathbf{0}$ leads us to the Lagrangian function

$$L(\mathbf{x}, \boldsymbol{\lambda}, \boldsymbol{\pi}) = \frac{1}{2}\mathbf{x}^{\mathrm{T}}\mathsf{G}\mathbf{x} + \mathbf{g}^{\mathrm{T}}\mathbf{x} - \boldsymbol{\lambda}^{\mathrm{T}}(\mathsf{A}^{\mathrm{T}}\mathbf{x} - \mathbf{b}) - \boldsymbol{\pi}^{\mathrm{T}}\mathbf{x}. \tag{11.4.3}$$

Defining slack variables $\mathbf{r} := \mathsf{A}^{\mathrm{T}}\mathbf{x} - \mathbf{b}$, the Karush–Kuhn–Tucker (KKT) conditions follow as:

$$\boldsymbol{\pi} - \mathsf{G}\mathbf{x} + \mathsf{A}\boldsymbol{\lambda} = \mathbf{g} \tag{11.4.4}$$

$$\mathbf{r} - \mathsf{A}^{\mathrm{T}}\mathbf{x} = -\mathbf{b} \tag{11.4.5}$$

$$\boldsymbol{\pi}, \mathbf{r}, \mathbf{x}, \boldsymbol{\lambda} \geq \mathbf{0}, \quad \boldsymbol{\pi}^{\mathrm{T}}\mathbf{x} = 0, \quad \mathbf{r}^{\mathrm{T}}\boldsymbol{\lambda} = \mathbf{0}, \tag{11.4.6}$$

[6] Historically, the term *quadratic programming* (QP) refers to a quadratic objective function and linear constraints. Nowadays, sometimes, the more general problem with a quadratic objective function and quadratic constraints is also referred to as quadratic programming. Sometimes it is named QPQC. If the QP or QPQC involves integer variables, one also sees the naming MIQCQP. Thus, the advice is: Always check carefully for the definition when authors refer to quadratic programming.

where $\boldsymbol{\pi}^{\mathrm{T}}\mathbf{x} = 0$ and $\mathbf{r}^{\mathrm{T}}\boldsymbol{\lambda} = \mathbf{0}$ are the complementary slackness conditions which state that for each $\mathbf{x}$, the variable itself or the associated $\boldsymbol{\pi}$ are zero (and the same for $\mathbf{r}$ and its associated $\boldsymbol{\lambda}$). A solution of a QP must satisfy the KKT conditions which in this formulation provide necessary conditions for local minima. If G is positive semi-definite, then a local minimum is a global minimum.

It is easy to extend this to the case where there are upper bounds on the $\mathbf{x}$ variables. For instance, $\mathbf{x} \leq \mathbf{U}$ can be modeled as

$$-\mathbf{x} \geq -\mathbf{U} \qquad (11.4.7)$$

requiring us to introduce slack variables $\mathbf{s} := \mathbf{U} - \mathbf{x}$ and a set of multipliers, say $\boldsymbol{\sigma}$, for these constraints. The complementary slackness conditions then read as

$$\boldsymbol{\sigma}^{\mathrm{T}}\mathbf{s} = 0, \qquad (11.4.8)$$

and (11.4.4) is extended to

$$\boldsymbol{\pi} - \boldsymbol{\sigma} - \mathsf{G}\mathbf{x} + \mathsf{A}\boldsymbol{\lambda} = \mathbf{g}. \qquad (11.4.9)$$

Another extension is that equality constraints have free variable $\boldsymbol{\lambda}$ associated with them.

Note that the KKT conditions establish a nonlinear system of equations. However, the nonlinearity arises only in the complementary slackness conditions $\boldsymbol{\pi}^{\mathrm{T}}\mathbf{x} = 0$ and $\mathbf{r}^{\mathrm{T}}\boldsymbol{\lambda} = 0$. The complementary slackness conditions can be modeled as special ordered sets of type 1.

Therefore, using MILP software such as CPLEX or XPRESS-OPTIMIZER supporting SOS1, the quadratic programming problem is solved by deriving its KKT conditions and solving them as a LP problem modeling the complementary slackness conditions by SOS1. If S1 are unavailable in the MILP solver at hand, the complementary conditions can also be modeled with additional binary variables. Let X and P be upper limits on $\mathbf{x}$ and $\boldsymbol{\pi}$, and δ an additional (vector) binary variable. Then the condition $\boldsymbol{\pi}^{\mathrm{T}}\mathbf{x} = 0$ can be replaced equivalently by

$$\mathbf{x} \leq X\boldsymbol{\delta}, \quad \boldsymbol{\pi} \leq P(1 - \boldsymbol{\delta}), \qquad (11.4.10)$$

which enforces that either $\mathbf{x}$ or $\boldsymbol{\pi}$ is zero. The inequalities (11.4.10) do not exclude the case that both of them are zero.

Note that this MILP problem has no real objective function anymore. The original objective function (11.4.1) is only contained in the KKT conditions. As we have to provide an objective function to the MILP solver, we provide some dummy objective function.

Here is a small example demonstrating how the method works.

$$\min \quad 3x^2 + 2y^2 + z^2 + xy + \frac{1}{2}zx + xy - \frac{2}{5}zy + \frac{1}{2}xz - \frac{2}{5}yz \qquad (11.4.11)$$

subject to

$$\begin{array}{r} x + y + z \geq 1.00 \\ 1.3x + 1.2y + 1.08z \geq 1.12 \end{array}, \quad \begin{array}{l} 0 \leq x \leq 0.75 \\ 0 \leq y \leq 0.75 \\ 0 \leq z \leq 0.75 \end{array}. \tag{11.4.12}$$

Let us express this problem using the structure of the general QP. Observing

$$\mathbf{x}^{\mathrm{T}}\mathsf{G}\mathbf{x} = \sum_i \sum_j G_{ij} x_i x_j$$

and $x_1 = x$, $x_2 = y$, and $x_3 = z$, we derive

$$\mathsf{G} := \begin{pmatrix} 6.0 & 2.0 & 1.0 \\ 2.0 & 4.0 & -0.8 \\ 1.0 & -0.8 & 2.0 \end{pmatrix}, \quad \mathbf{g} := \begin{pmatrix} 0 \\ 0 \\ 0 \end{pmatrix}$$

and

$$\mathsf{A}^{\mathrm{T}} := \begin{pmatrix} 1.0 & 1.0 & 1.00 \\ 1.3 & 1.2 & 1.08 \end{pmatrix}, \quad \mathbf{b} := \begin{pmatrix} 1.00 \\ 1.12 \end{pmatrix}, \quad \mathbf{U} := \begin{pmatrix} 0.75 \\ 0.75 \\ 0.75 \end{pmatrix}.$$

The model formulation below uses SOS1 as discussed above.

```
! To minimize 1/2x'Gx + g'x        (' indicates transpose)
!  subject to
!    A'x >= b , x>= 0 , x<= U      [ A is m*n , G is n*n ]

! Lagrangian L(x,lambda,sigma,pi)  =
!  1/2x'Gx +g'x -lambda'(A'x-b)-sigma'(U-x)-pi'x

! Kuhn Tucker conditions imply
!  pi-sigma-Gx+A*lambda = g
!      i.e.:     pi + A*lambda = g + sigma + Gx
!  A'x-r   = b
!  x+s = U
!
!  pi'*x=0; s'*sigma=0; r'*lambda=0 (complementary slackness)
!  lambda is a free variable if A'x=b,
!                          and the corresponding r doesn't exist
!  sigma(j) and s(j) do not exist if U(j) infinite

LET N=3                    ! Dimension of x (three variables)
LET M=2                    ! Number of linear constraints
LET HUGE = 1.0e10          ! Upper bounds > HUGE are assumed not
                           ! to exist
```

```
TABLES
 G    (N,N)
 B    (M)
 AT   (M,N)
 g    (N)
 U    (N)
 IFEQ(M)                 ! 1 if row i is equality, else 0

DATA                     ! entries not specified are zero
 G(1,1)   = 6 ,  2   ,  1
 G(2,1)   = 2 ,  4   , -0.8
 G(3,1)   = 1 , -0.8 ,  2
 B(1)     = 1 , 1.12
 AT(1,1)  = 1   ,  1   , 1
 AT(2,1)  = 1.3 ,  1.2 , 1.08
 g(1)     = 0    , 0    , 0
 U(1)     = 0.75 , 0.75, 0.75
 IFEQ(1)  = 0 ,  0

VARIABLES
 x  (N)
 pi (N)
 lambda(M)
 r(i=1:M | IFEQ(i).ne.1)
 s     (j=1:N | U(j) < HUGE)
 sigma(j=1:N | U(j) < HUGE)

CONSTRAINTS

! Reference row
 dummy:                                         &
  -SUM(j=1:N) x (j)                             &
  +SUM(j=1:N) pi(j)                             &
  -SUM(i=1:M | IFEQ(i).eq.0) r(i)               &
  +SUM(i=1:M | IFEQ(i).eq.0) lambda(i)          &
  -SUM(j=1:N | U(i) < HUGE ) s     (j)          &
  +SUM(j=1:N | U(i) < HUGE ) sigma(j)      $

 PI(j=1:N):  pi(j) + SUM(i=1:M) AT(i,j) * lambda(j)       &
          = g(j) + sigma(j) + SUM(jc=1:N) G(jc,j) * x(j)
 R(i=1:M):  -r(i) + SUM(j=1:N) AT(i,j) * x(j)  = B(i)
 UPx(j=1:N | U(j) < HUGE):     x(j) + s(j) = U(j)

BOUNDS
 lambda(i=1:M | IFEQ(i).eq.1) .FR. ! Free for = constraints

SETS      ! For complementary slackness
 SSpi(j=1:N): pi(j) + x(j)                    .S1. dummy
 SSl(i=1:M | IFEQ(i).eq.0): r(i) + lambda(i)  .S1. dummy
 SSs(j=1:N | U(j) < HUGE) : s(j) + sigma (j)  .S1. dummy

END
```

Applying CPLEX or XPRESS-OPTIMIZER to the problem stored as *quadrat* leads to the results

$$\begin{array}{lll} x = x_1 = 0 & \pi_1 = 0.4 & \\ y = x_2 = 0.368421\,, & \pi_2 = 0.0\,, & \begin{array}{l}\lambda_1 = 0.968421\\ \lambda_2 = 0\end{array}\,, \\ z = x_3 = 0.631579 & \pi_3 = 0.0 & \end{array} \quad \begin{array}{l} \sigma_1 = 0 \\ \sigma_2 = 0\,. \\ \sigma_3 = 0 \end{array}$$

This solution is the global minimum, as all eigenvalues (7.2734, 3.629 6, 1.0970) of G are positive and G is thus positive definite.

11.5 Summary and Recommended Bibliography

In this chapter we have briefly investigated some extensions of mathematical programming to handle nonlinear problems. We have mentioned how probabilistic models can be tackled using recursion or stochastic programming and how nonlinear objective functions (e.g., quadratic) can be handled. We have also seen how IP problems might behave when nonlinearity is introduced. Discussion has been kept brief as the types of problems dealt with in this chapter are much less common than those in earlier chapters and because the solution of such problems is awkward and hard to guarantee.

Readers interested in the *Pooling Problem* are referred to Tawarmalani & Sahinidis (2002,[541]), Audet et al. (2004,[40]), Misener & Floudas (2009,[404]).

11.6 Exercises

1. Minimize $|x - y| + 4\,|y - 3|$ s.t. $x + 2y \le 10$ and $x \ge 0$, $y \ge 0$ using the approach described in Sect. 6.5 exploiting an MILP solver. Hint: Resort to Sect. 6.5.
2. Minimize $|x - y| + 4\,|y - 3|$ s.t. $x + 2y \le 10$ and $x \ge 0$, $y \ge 0$ using the approximation $|u| \approx \sqrt{u^2 + \delta^2}$ exploiting an NLP solver; δ is small number, e.g., $\delta = 0.0125$.
3. Minimize $\max(e^{0.5x} + 7 - 0.2x, 9 - x^2 + 3x - 2)$ over the interval $x \in [-5, 3]$ using the approximation $\max(u, v) \approx u + v + \sqrt{(u - v)^2 + \delta^2}$ exploiting an NLP solver; δ is small number, e.g., $\delta = 0.0125$.
4. Maximize $\min(e^{0.5x} + 7 - 0.2x, 9 - x^2 + 3x - 2)$ over the interval $x \in [-5, 3]$ using the approximation $\min(u, v) \approx u + v - \sqrt{(u - v)^2 + \delta^2}$ exploiting an NLP solver; δ is small number, e.g., $\delta = 0.0125$.

Chapter 12
Mathematical Solution Techniques — The Nonlinear World

This chapter provides some of the mathematical and algorithmic backgrounds to solve NLP and MINLP problems to local or global optimality. Covering nonlinear, continuous, or mixed integer optimization in great depth is beyond the scope of this book. Therefore only some essential aspects and ideas are introduced and some basics are presented. Readers with further interest are referred to Gill et al. (1981,[217]), Spelluci (1993,[516]), Burer & Letchford (2012,[102]) for a survey on non-convex MINLP, Belotti et al. (2013,[60]) on MINLP, and Boukouvala et al. (2016,[94]) for advances on global optimization. Special techniques for NLP problems, often used in oil or food industry, such as recursion or sequential linear programming and distributive recursion, have already been covered in Sect. 11.2.

12.1 Unconstrained Optimization

Be $f : X \subseteq \mathbb{R}^n \rightarrow \mathbb{R}, \mathbf{x} \rightarrow f(\mathbf{x})$ a scalar- and real-valued function which can be evaluated on X; this is the case, for instance, if f is continuous on X. The problem

$$\min_{\mathbf{x}\in X}\{f(\mathbf{x})\} \quad \Longleftrightarrow \quad f_* := \min\{f(\mathbf{x}) \mid \mathbf{x} \in X\}, \tag{12.1.1}$$

is called *unconstrained optimization problem*. A vector $\mathbf{x}_*$ is called a *local minimum*, *local minimizer*, or *minimizing point* of $f(\mathbf{x})$ and can be formally expressed by

$$\mathbf{x}_* = \arg\min\{\min_{\mathbf{x}\in U\subset X} f(\mathbf{x})\} := \{\mathbf{x} \mid f(\mathbf{x}) \leq f(\mathbf{x}'),\ \forall\, \mathbf{x}' \in U \subset X\}. \tag{12.1.2}$$

Supplementary Information The online version contains supplementary material available at https://doi.org/10.1007/978-3-030-73237-0_12.

J. Kallrath, *Business Optimization Using Mathematical Programming*,
International Series in Operations Research & Management Science 307,
https://doi.org/10.1007/978-3-030-73237-0_12

Maximization problems can be reduced, exploiting the relationship

$$\max_{\mathbf{x} \in X} \{f(\mathbf{x})\} = -\min_{\mathbf{x} \in X}\{-f(\mathbf{x})\},$$

to minimization problems; therefore, depending on the situation, it is sometimes preferable to deal with a maximization or a minimization problem. A typical difficulty in nonlinear, non-convex optimization is that it is usually only possible to compute local optima, but it is very difficult to prove that such a local optimum is a global optimum. In short, a global optimum is the best solution of all feasible points while a local optimum $\mathbf{x}_*$ is only the best solution in a neighborhood or environment of $\mathbf{x}_*$. Formally, a local minimum is defined as: In a given minimization problem, a point $\mathbf{x}_* \in X$ is a *local minimum w.r.t. a neighborhood* $U_{\mathbf{x}_*}$ around $\mathbf{x}_*$ (or just *local minimum*), if

$$f(\mathbf{x}_*) \leq f(\mathbf{x}), \quad \forall\, \mathbf{x} \in U_{\mathbf{x}_*}.$$

If $f_* = f(\mathbf{x}_*) \leq f(\mathbf{x})$ holds for all $\mathbf{x} \in X$, $\mathbf{x}_*$ is called a *global minimum* of f.

The first solution approaches were derivative-free methods (DFM), i.e., search procedures that merely evaluated function values $f(\mathbf{x})$. One of the most commonly used methods is the (downhill) Simplex method, also known as Nelder–Mead method, [Spendley et al. (1962,[518]), Nelder & Mead (1965,[419])] not to be confused with the Simplex algorithm in Chap. 3. Another approach is the *alternating variables method.* In each iteration k, only variable x_k is varied to reduce $f(\mathbf{x})$; all other variables remain fixed at their values. Both methods are easy to implement and suitable to solve problems with non-smooth functions. Although they usually show a rather slow convergence behavior, in some situations where derivative-based methods (DBM) prove to be problematic, they can be very useful. This is especially true when DBM are numerically sensitive to the initial values or when one needs to distinguish between local and global extrema. In such cases, DFM can be used to obtain initial values for DBM although both also depend on initial values.

DBM used in local solvers require derivatives of first, and often second order; so let us always assume that $f(\mathbf{x})$ has the required smoothness properties, i.e., that the derivatives are continuous to the degree required. In detail, we need:

- the gradient of the scalar, real-valued differentiable function $f(\mathbf{x})$ of the vector $\mathbf{x}$; $f : X = \mathbb{R}^n \rightarrow \mathbb{R}$, $\mathbf{x} \rightarrow f(\mathbf{x})$

$$\nabla f(\mathbf{x}) := \left(\frac{\partial}{\partial x_1} f(\mathbf{x}), \ldots, \frac{\partial}{\partial x_n} f(\mathbf{x})\right) \in \mathbb{R}^n, \tag{12.1.3}$$

- the Hessian Matrix[1] H of the function $f(\mathbf{x})$

$$\mathsf{H}(\mathbf{x}) \equiv \nabla^2 f(\mathbf{x}) := \frac{\partial}{\partial x_i}\left(\frac{\partial}{\partial x_j} f(\mathbf{x})\right) = \left(\frac{\partial^2}{\partial x_i \partial x_j} f(\mathbf{x})\right) \in \mathcal{M}(n,n), \tag{12.1.4}$$

- the Jacobian[2] J

$$\mathsf{J}(\mathbf{x}) \equiv \nabla \mathbf{g}(\mathbf{x}) := (\nabla g_1(\mathbf{x}), \ldots, \nabla g_m(\mathbf{x})) = \left(\frac{\partial}{\partial x_j} g_i(\mathbf{x})\right) \in \mathcal{M}(m,n), \tag{12.1.5}$$

i.e., the gradient of the vector-valued function $\mathbf{g}(\mathbf{x}) = (g_1(\mathbf{x}), \ldots, g_m(\mathbf{x}))^{\mathrm{T}}$. The Hessian matrix of the scalar function $f(\mathbf{x})$ is the Jacobian matrix of the gradient $\nabla f(\mathbf{x})$. As in one dimension, the minimizer $\mathbf{x}_*$ has to fulfill the necessary condition

$$\nabla f(\mathbf{x}_*) = \mathbf{0} \quad , \quad \nabla f(\mathbf{x}_*) \in \mathbb{R}^n. \tag{12.1.6}$$

Using a Taylor series expansion of $f(\mathbf{x})$ around the point $\mathbf{x}_*$, the following theorem (sufficient condition) can be proven:

Theorem 12.1 *For $\mathbf{x}_*$ to be a locally minimizing point, it is sufficient that (12.1.6) is fulfilled and that the Hessian matrix $\mathsf{H}_* := \mathsf{H}(\mathbf{x}*)$ is positive definite , i.e.,*

$$\mathbf{s}^{\mathrm{T}}\mathsf{H}_*\mathbf{s} > 0 \quad , \quad \forall \mathbf{s} \neq 0 \quad , \quad \mathbf{s} \in \mathbb{R}^n.$$

Here, $\mathbf{s}$ denotes a non-zero vector. A basic approach to computing a numerical solution of the minimization problem (12.1.1) is the *line search algorithm.* Based on a known solution $\mathbf{x}_k$ in iteration k, the next value $\mathbf{x}_{k+1}$ in iteration $k+1$ is calculated using the following steps:

- determination of a search direction $\mathbf{s}_k$;
- solve the line search subproblem, i.e., determine the minimum of the auxiliary function[3] $f(\mathbf{x}_k + \alpha_k \mathbf{s}_k)$ using a *line search*, where $\alpha_k > 0$ is a matching damping factor; and
- computation of the solution $\mathbf{x}_{k+1} := \mathbf{x}_k + \alpha_k \mathbf{s}_k$ for the next iteration step.

Algorithms for the solution of (12.1.1) differ in the way the search direction $\mathbf{s}_k$ is calculated. *Damped procedures* use a damping factor α, $0 < \alpha \leq 1$. Most of them

[1]Named after the German mathematician Ludwig Otto Hesse (1811–1874). $\mathcal{M}(m,n)$ denotes the set of all matrices with m rows and n columns.

[2]Named after the German mathematician Carl Gustav Jacob Jacobi (1804–1851).

[3]Usually, $f(\mathbf{x}_k + \alpha \mathbf{s}_k)$ is not exactly minimized w.r.t. α. One possible heuristic is to evaluate f for $\alpha_m = 2^{-m}$ for $m = 0, 1, 2, \ldots,$ and to stop the line search when $f(\mathbf{x}_k + \alpha_m \mathbf{s}_k) \leq f(\mathbf{x}_k)$.

use a line search procedure to compute α_k in iteration k. *Undamped procedures* set $\alpha_k = 1$. If the exact minimum of $f(\mathbf{x}_k + \alpha_k \mathbf{s}_k)$ is computed, then the gradient $\nabla f(\mathbf{x}_{k+1})$ in the new point $\mathbf{x}_{k+1}$ is perpendicular to the search direction $\mathbf{s}_k$, i.e., $\mathbf{s}_k^{\mathrm{T}} \nabla f(\mathbf{x}_{k+1}) = 0$. This follows from the necessary condition

$$0 = \frac{\mathrm{d}}{\mathrm{d}\alpha_k} f(\mathbf{x}_k + \alpha_k \mathbf{s}_k) = [\nabla f(\mathbf{x}_k + \alpha_k \mathbf{s}_k)]^{\mathrm{T}} \mathbf{s}_k .$$

There are various classic optimization procedures to solve (12.1.1) based on this fundamental concept. *Descent method* is a line search method, in which the search direction $\mathbf{s}_k$ satisfies

$$\nabla f(\mathbf{x}_k)^{\mathrm{T}} \mathbf{s}_k < 0. \tag{12.1.7}$$

The *method of the steepest descent* uses $\mathbf{s}_k = -\nabla f(\mathbf{x}_k)$; obviously, condition (12.1.7) is satisfied for this choice of $\mathbf{s}_k$. The gradient can be calculated analytically or numerically approximated by finite differences. Typically, one uses asymmetric differences involving derivatives of degree two $f''(\mathbf{x})$

$$\nabla_i f(\mathbf{x}) = \frac{\partial f}{\partial x_i}(\mathbf{x}) \approx \frac{f(\mathbf{x} + h\mathbf{e}_i) - f(\mathbf{x})}{h} + f''(\mathbf{x})h$$

or the symmetrical differences involving derivatives of degree three $f'''(\mathbf{x})$

$$\nabla_i f(\mathbf{x}) = \frac{\partial f}{\partial x_i}(\mathbf{x}) \approx \frac{f(\mathbf{x} + \frac{1}{2}h\mathbf{e}_i) - f(\mathbf{x} - \frac{1}{2}h\mathbf{e}_i)}{h} + \tfrac{1}{24} f'''(\mathbf{x})h^2,$$

where $\mathbf{e}_i$ is the unit vector along the i-th coordinate axis. From the numerical point of view, symmetric differences are preferable to asymmetric differences, as they are more accurate due to a smaller approximation error[4] which is of order h^2.

A different way of calculating the search direction $\mathbf{s}_k$ is to use a second order Taylor series expansion of $f(\mathbf{x})$ around $\mathbf{x}_k$. This corresponds to a quadratic approximation of the function $f(\mathbf{x})$, i.e., $\mathbf{x} = \mathbf{x}_k + \mathbf{s}_k$, $f(\mathbf{x}) = f(\mathbf{x}_k + \mathbf{s}_k)$ and

$$f(\mathbf{x}_k + \mathbf{s}_k) \approx f(\mathbf{x}_k) + \nabla f(\mathbf{x}_k)^{\mathrm{T}} \mathbf{s}_k + \tfrac{1}{2} \mathbf{s}_k^{\mathrm{T}} \mathsf{H}_k \mathbf{s}_k . \tag{12.1.8}$$

If both the gradient $\nabla f(\mathbf{x})$ and the Hessian H are analytically given, (12.1.8) enables us to derive the classic *Newton procedure*. Applying the necessary con-

[4] This raises the question of how to choose the size of the increments h. As illustrated by Press et al. (1992,[447]), the optimal choice of this size depends on the curvature, i.e., the second derivative of $g : \mathbb{R} \to \mathbb{R}$, $u \to g(u)$. Since the second derivation is, however, unknown in most cases, we refer the reader to Press et al. (1992,[447, p.180]) and their heuristic approximation $h \approx 2\varepsilon_g^{1/3} u$, where ε_g is the relative precision with which $g(u)$ is calculated. For functions that are not too complicated this corresponds approximately to machine accuracy, i.e., $\varepsilon_g \approx \varepsilon_m$.

ditions (12.1.6) on (12.1.8), we obtain

$$\mathsf{H}_k \mathbf{s}_k = -\nabla f(\mathbf{x}_k), \tag{12.1.9}$$

which allows us to compute $\mathbf{s}_k$ according to

$$\mathbf{s}_k = -\mathsf{H}_k^{-1} \nabla f(\mathbf{x}_k). \tag{12.1.10}$$

In most practical problems, the Hessian H is not available. However, if $\nabla f(\mathbf{x})$ is analytically known, or can be numerically expressed using asymmetric or symmetric finite differences with sufficient accuracy, the *Quasi-Newton method* can be applied; cf. Werner (1992,[571, Section 7.3]). There are two types of quasi-Newton methods. Either H_k in (12.1.10) is replaced by the finite differences-based symmetric approximation $\bar{\mathsf{H}}_k$

$$\tfrac{1}{2}\left(\bar{\mathsf{H}}_k + \bar{\mathsf{H}}_k^{\mathrm{T}}\right),$$

or the inverse matrix H_k^{-1} is replaced by an approximated symmetric, positive-definite matrix $\tilde{\mathsf{H}}_k$; the latter is numerically more efficient. The initial matrix $\tilde{\mathsf{H}}_0$ can be any positive-definite matrix. In the absence of initial values, usually the unit matrix $\mathbb{1}$ is used to initialize $\tilde{\mathsf{H}}_0$, or if specific information is available, a diagonal matrix. These methods are sometimes also called *variable metric methods*, cf. Press et al. (1992,[447, p.418–422]). The efficiency of this Quasi-Newton method depends on the quality of the update procedure, in which $\tilde{\mathsf{H}}_{k+1}$ is calculated from $\tilde{\mathsf{H}}_k$; some update formulas operate directly on the inverse matrix H_k^{-1} and generate H_{k+1}^{-1}.

12.2 Constrained Optimization — Foundations and Theorems

Similarly to linear programming, *constrained optimization* is also called *nonlinear programming* (NLP) for historical reasons. An NLP problem with n variables, n_2 equations, and n_3 inequalities is defined as **Problem NLP**

$$\begin{aligned} \text{Minimize:}\quad & f(\mathbf{x}), && \mathbf{x} \in \mathbb{R}^n, \\ \textit{subject to}\quad & \mathbf{F}_2(\mathbf{x}) = 0, && \mathbf{F}_2 : \mathbb{R}^n \to \mathbb{R}^{n_2}, \\ & \mathbf{F}_3(\mathbf{x}) \geq 0, && \mathbf{F}_3 : \mathbb{R}^n \to \mathbb{R}^{n_3}. \end{aligned} \tag{12.2.1}$$

The functions $f(\mathbf{x})$, $\mathbf{F}_2(\mathbf{x})$, and $\mathbf{F}_3(\mathbf{x})$ are here assumed as continuously differentiable over the whole vector space $\mathbb{R}^n$. The vector inequality $\mathbf{F}_3(\mathbf{x}) \geq 0$ is the short form of the n_3 inequalities $F_{3k}(\mathbf{x}) \geq 0$, $1 \leq k \leq n_3$. The set of all feasible points

$$\mathscr{S} := \{\mathbf{x} \mid \mathbf{F}_2(\mathbf{x}) = 0 \wedge \mathbf{F}_3(\mathbf{x}) \geq 0\} \tag{12.2.2}$$

is called *feasible region* or *feasible set* $\mathcal{S}$. The set of active constraints in the point $\mathbf{x}$ is determined by the index set

$$\mathcal{I}(\mathbf{x}) := \{i \mid F_{3i}(\mathbf{x}) = 0, \quad i = 1, \ldots, n_3\},$$

which is sometimes also called the set of active inequalities. In the early 1950s, Kuhn & Tucker (1951,[346]) extended the theory of Lagrange multipliers, which until then had been used to solve equality-constrained optimization problems, to problems containing now both equalities and inequalities. The Kuhn–Tucker Theory is based on the following definition of the Lagrangian function:

$$L(\mathbf{x}, \boldsymbol{\lambda}, \boldsymbol{\mu}) := f(\mathbf{x}) - \boldsymbol{\lambda}^{\mathrm{T}}\mathbf{F}_2(\mathbf{x}) - \boldsymbol{\mu}^{\mathrm{T}}\mathbf{F3}(\mathbf{x}), \tag{12.2.3}$$

which contains the objective function $f(\mathbf{x})$ with the constraints $\mathbf{F}_2(\mathbf{x})$ and $\mathbf{F}_3(\mathbf{x})$. The vector variables $\boldsymbol{\lambda} \in \mathbb{R}^{n_2}$ and $\boldsymbol{\mu} \in \mathbb{R}^{n_3}$ are called *Lagrange multipliers*; they are added as additional variables to the original NLP problem and its variables.

Below we summarize some theorems and results of NLP theory. The necessary conditions for the existence of a local optimum proven by Kuhn & Tucker (1951,[346]) are presented using the Jacobians J_2 and J_3 of $\mathbf{F}_2$ and $\mathbf{F}_3$; cf. Ravindran et al. (1987,[453]):

Theorem 12.2 *If* $\mathbf{x}_*$ *is a solution of the NLP problem and if the functions* $f(\mathbf{x})$, $\mathbf{F}_2(\mathbf{x})$, *and* $\mathbf{F}_3(\mathbf{x})$ *are continuous-differentiable and fulfill some regularity conditions, then vectors* μ_* *and* λ_* *exist such that* $\mathbf{x}_*$, μ_* *and* λ_* *satisfy the following conditions:*

$$\mathbf{F}_2(\mathbf{x}) = 0 \quad , \tag{12.2.4}$$

$$\mathbf{F}_3(\mathbf{x}) \geq 0 \quad ,$$

$$\boldsymbol{\mu}^{\mathrm{T}}\mathbf{F}_3(x) = 0 \quad ,$$

$$\boldsymbol{\mu} \geq 0 \quad ,$$

$$\nabla f(\mathbf{x}) - \boldsymbol{\lambda}^{\mathrm{T}}\mathrm{J}_2(\mathbf{x}) - \boldsymbol{\mu}^{\mathrm{T}}\mathrm{J}_3(\mathbf{x}) = 0 \quad , \tag{12.2.5}$$

cf. Collatz & Wetterling (1971,[126]), or Fletcher (1987,[187]). Equations (12.2.4)–(12.2.5) are also called in short *(Karush*[5]*–)Kuhn–Tucker (KKT) Conditions*, or simply first order conditions. A point $(\mathbf{x}_*, \mu)$ that satisfies these conditions is called *Karush–Kuhn–Tucker point*, or *KKT point* in short.

In addition, the functions $f(\mathbf{x})$, $\mathbf{F}_2(\mathbf{x})$, and $\mathbf{F}_3(\mathbf{x})$ need to fulfill certain *regularity conditions* or *constraint qualifications* specified by Kuhn and Tucker to exclude certain irregular cases. Alternative forms of the regularity conditions are discussed,

---

[5] It was only later detected that Karush (1939,[329]) had already proven the same result in his 1939 master's thesis at the University of Chicago. In his review article Kuhn (1976,[344]) gave a historical overview of inequality-constrained optimization.

for instance, by Bomze & Grossmann (1993,[91]) or Gill et al. (1981,[217]). A very general formulation can be found at Bock (1987,[87]).

Let $\mathcal{I}(\mathbf{x}')$ be the set of active inequalities in point $\mathbf{x}'$. Let $\tilde{\mathbf{F}}_3$ be the function consisting of all functions F_{3i} for which $i \in \mathcal{I}(\mathbf{x}')$; $\tilde{\mathsf{J}}_3$ denotes the associated Jacobian. Furthermore, $\mathbf{u}^{\mathrm{T}} := (\mathbf{F}_2^{\mathrm{T}}, \tilde{\mathbf{F}}_3^{\mathrm{T}}) : \mathbb{R}^n \rightarrow \mathbb{R}N$ and $N := n_2 + \mid \mathcal{I} \mid$. Let $L(\mathbf{x}, \boldsymbol{\mu}, \boldsymbol{\lambda})$ be the *Lagrangian function* (12.2.3) of **NLP**. Finally, let $\mathsf{J}^N = \mathsf{J}^N(\mathbf{x}) = \partial\mathbf{u}/\partial\mathbf{x}$ be the Jacobian of $\mathbf{u}$ associated with the equations and active inequalities. A feasible point $\mathbf{x}'$ is called *regular*, if $\mathrm{rank}(\mathsf{J}^N(\mathbf{x}')) = N$; cf. Bock (1987,[87, p.48]) and Appendix C.5 for a definition of the *rank* of a matrix. Now, we state Theorem 12.3:

Theorem 12.3 *Let $\mathbf{x}_* \in \mathbb{R}^n$ be a regular point and a local minimization point of Problem **NLP** (12.2.1). Then there exist vectors $\boldsymbol{\mu}_*$ and $\boldsymbol{\lambda}_*$ so that $\mathbf{x}_*$, $\boldsymbol{\mu}_*$, and $\boldsymbol{\lambda}_*$ fulfill the KKT conditions [Eqs. (12.2.4)–(12.2.5)].*

It should be noted that the difference between Theorem 12.2 and 12.3 is in the assumptions, i.e., in the regularity conditions. If we further define the set of directions

$$\mathcal{T}(\mathbf{x}_*) := \left\{ \mathbf{p} \neq 0 \;\middle|\; \begin{array}{ll} \mathsf{J}_2(\mathbf{x}_*)\mathbf{p} = 0, & \\ \tilde{\mathsf{J}}_3(\mathbf{x}_*)\mathbf{p} \geq 0, & \tilde{\mu}_{i*}\tilde{\mathsf{J}}_3(\mathbf{x}_*)\mathbf{p} = 0, \;\; \forall\, i \in \mathcal{I}(\mathbf{x}_*) \end{array} \right\},$$

and the Hessian

$$\mathsf{H}(\mathbf{x}_*, \boldsymbol{\mu}_*, \boldsymbol{\lambda}_*) := \frac{\partial^2}{\partial \mathbf{x}^2} L(\mathbf{x}_*, \boldsymbol{\mu}_*, \boldsymbol{\lambda}_*),$$

by extending the assumptions in Theorem 12.2, the following results can be derived:

Theorem 12.4 (Second Order Necessary Conditions) *If the functions $f(\mathbf{x})$, $\mathbf{F}_2(\mathbf{x})$, and $\mathbf{F}_3(\mathbf{x})$ are twice continuous-differentiable, the second order necessary conditions for a KKT point to be $(\mathbf{x}_*, \boldsymbol{\mu}_*, \boldsymbol{\lambda}_*)$ a local optimum are:*

$$\mathbf{p}^{\mathrm{T}}\mathsf{H}(\mathbf{x}_*, \boldsymbol{\mu}_*, \boldsymbol{\lambda}_*)\,\mathbf{p} \geq 0, \quad \forall\, \mathbf{p} \in \mathcal{T}(\mathbf{x}_*). \tag{12.2.6}$$

The interpretation of this theorem is that the Hessian of the Lagrangian function for all directions $\mathbf{p} \in \mathcal{T}(\mathbf{x}_*)$ is positive definite.

Theorem 12.5 (Second Order Sufficient Conditions) *Let $(\mathbf{x}_*, \boldsymbol{\mu}_*, \boldsymbol{\lambda}_*)$ be a KKT point of Problem **NLP** (12.2.1). For all directions $\mathbf{p} \in \mathcal{T}(\mathbf{x}_*)$, let the Hessian of the Lagrangian function be positive definite, i.e.,*

$$\mathbf{p}^{\mathrm{T}}\mathsf{H}(\mathbf{x}_*, \boldsymbol{\mu}_*, \boldsymbol{\lambda}_*)\,\mathbf{p} > 0, \quad \forall\, \mathbf{p} \in \mathcal{T}(\mathbf{x}_*). \tag{12.2.7}$$

*Then $\mathbf{x}_*$ is a strict local minimum of Problem **NLP**.*

Fletcher (1987,[187]) provides a proof of this theorem, further discussion of the second order conditions, and even a less restrictive formulation of the regularity conditions, based on a local characterization of the linearized constraints.

For a special class of problems, namely convex problems, the following theorem is proven [cf. Kuhn & Tucker (1951,[346]), Collatz & Wetterling (1971,[126]), or Bomze & Grossmann (1993,[91])]:

Theorem 12.6 (Kuhn–Tucker Theorem; Sufficient Condition) . *Given the nonlinear Optimization problem* ***NLP*** *with a convex objective function* $f(\mathbf{x})$*, linear equalities* $\mathbf{F}_2(\mathbf{x})$*, and linear and concave inequalities* $\mathbf{F}_3(\mathbf{x})$*. If there exist a solution* $(\mathbf{x}_*, \boldsymbol{\mu}_*, \boldsymbol{\lambda}_*)$ *that satisfies the KKT conditions (12.2.4) to (12.2.5), then* $\mathbf{x}_*$ *is an optimal solution of* ***NLP****.*

If the functions $f(\mathbf{x})$, $\mathbf{F}_2(\mathbf{x})$, and $\mathbf{F}_3(\mathbf{x})$ satisfy the assumptions[6] of Theorem 12.6, **NLP** is called a convex optimization problem. For convex optimization problems, local optimality implies global optimality; cf. [436].

Algorithms for the solution of (12.2.1) are, for instance, found in Gill et al. (1981,[217]) or Fletcher (1987,[187]). Most of them rely in some way on linearization techniques. Inequalities are taken into account by active-set strategies. Among the most powerful nonlinear optimization algorithms are the *Generalized Reduced Gradient method* described in Sect. 12.3, *Sequential Quadratic Programming* explained in Sect. 12.4, and *Interior-points methods* for problems with many inequalities; cf. Bazaraa et al. (1993,[53]) or Wright (1996,[590]).

12.3 Reduced Gradient Methods

Reduced gradient methods are a natural extension of the Simplex algorithm for nonlinear optimization problems. In each iteration step, the active inequalities are used to separate the variables into independent (free) and dependent ones. This is followed by minimizing the objective function *w.r.t.* to the free variables. This procedure was originally developed by Abadie & Carpenter (1969,[2]); recent developments are in Abadie (1978,[1]), Lasdon et al. (1978,[354]) and Lasdon & Goods (1978,[353]) as well as Gill et al. (1981,[217, Section 6.3]) or Spelluci (1993,[516, pp. 361]). It is used in some commercial software packages such as MINOS [cf. Murtagh & Saunders (1978,[415]; 1982,[416])] or CONOPT, developed by Drud (1994,[164]). These solvers in turn are available in the modeling language GAMS [101], which has the consequence that they are very frequently used for solving practical problems that lead to large and sparse NLP problems. To get some

[6] These assumptions guarantee that both the feasible region and the objective function are convex.

insight into the reduced gradient method, consider the optimization problem

$$\min_{\mathbf{x},\mathbf{y}} \left\{ f(\mathbf{x}) + \mathbf{c}^{\mathrm{T}}\mathbf{x} + \mathbf{d}^{\mathrm{T}}\mathbf{y} \;\middle|\; \begin{matrix} \mathbf{h}(\mathbf{x}) + \mathsf{A}_1\mathbf{y} \circ \mathbf{b}_1 \\ \mathsf{A}_2\mathbf{x}+\mathsf{A}_3\mathbf{y} \circ \mathbf{b}_2 \end{matrix}, \mathbf{L} \leq \mathbf{x}, \mathbf{y} \leq \mathbf{U} \right\} \tag{12.3.1}$$

with $\mathbf{x} \in \mathbb{R}^{n_x}$, $\mathbf{y} \in \mathbb{R}^{n_y}$, $\mathbf{g} : \mathbb{R}^{n_x} \rightarrow \mathbb{R}^{m_1}$, $\mathsf{A}_1 \in \mathcal{M}_{m_1,n_y}$, $\mathsf{A}_2 \in \mathcal{M}_{m_2,n_x}$, $\mathsf{A}_3 \in \mathcal{M}_{m_2,n_y}$ and $\mathbf{b}_i \in \mathbb{R}^{m_i}$; $\circ$ stands for one of the relations $\leq$, $=$ or $\geq$ in the individual components of the vector constraints. Note that the variables $\mathbf{y}$ are only used in the linear terms; they are therefore called *linear variables*. Accordingly, the terms *nonlinear variables* as well as *linear* and *nonlinear constraints* are used in this section.

If $\mathbf{h}(\mathbf{x}) = 0$, i.e., only the objective function contains nonlinear terms, the *reduced gradient method* described, for instance, in Murtagh & Saunders (1978,[415]) can be combined with a quasi-Newton algorithm, cf. P. Wolfe (The reduced-gradient method. unpublished manuscript, RAND Corporation, 1962). As in linear programming, we introduce a slack variable $\mathbf{u} \geq \mathbf{0}$ to write the inequality $\mathsf{A}\mathbf{x} \leq \mathbf{b}$ as equality $\mathsf{A}\mathbf{x} + \mathbf{u} = \mathbf{b}$. Then we apply a zero-point transformation $\mathbf{s} = \mathbf{u} - \mathbf{b}$ to get $\mathsf{A}\mathbf{x} + \mathbf{s} = \mathbf{0}$, which allows us to use basic and non-basic variables $\mathbf{x}^{\mathrm{B}}$ and $\mathbf{x}^{\mathrm{N}}$ as in LP to obtain the representation $\mathsf{B}\mathbf{x}^{\mathrm{B}}+\mathsf{N}\mathbf{x}^{\mathrm{N}} = \mathbf{0}$. For problems with nonlinear objective function using $n^s \leq n^x$ with so-called superbasic variables $\mathbf{x}^{\mathrm{S}}$ the decomposition

$$\mathsf{B}\mathbf{x}^{\mathrm{B}}+\mathsf{S}\mathbf{x}^{\mathrm{S}}+\mathsf{N}\mathbf{x}^{\mathrm{N}} = \mathbf{0}$$

is constructed. For non-degenerate optimal solutions both the variables $\mathbf{x}^{\mathrm{S}}$ and the variables $\mathbf{x}^{\mathrm{B}}$ take values between their lower and upper bounds, while the non-basic variables $\mathbf{x}^{\mathrm{N}}$ as in LP are fixed to their bounds. However, in the reduced gradient method both $\mathbf{x}^{\mathrm{S}}$ and $\mathbf{x}^{\mathrm{N}}$ (as in LP) are seen as independent variables used to minimize the objective function value and the sum of the infeasibilities, while $\mathbf{x}^{\mathrm{B}}$ is still used to satisfy the linear constraints. If you notice that with the current number n^s no improvement can be achieved, some of the non-basic variables were selected and kept as superbasic variable, i.e., the value of n^s will be increased. If, on the other hand, a basic or superbasic variable takes a value at one of its bounds, it will be kept as a non-basic variable and n^s is reduced by one.

To determine the superbasic variables we introduce the matrix

$$\mathsf{Z} = \left(-\mathsf{B}^{-1}\mathsf{S}\ \mathbb{1}\ 0\right)^{\mathrm{T}} \tag{12.3.2}$$

composed by the blocks $\mathsf{B}^{-1}\mathsf{S}$, $\mathbb{1}$ and 0-matrix. Note that although Z is defined like this in (12.3.2), it is not calculated directly in this way. Instead, one exploits the LU decomposition of the basic matrix B, i.e., the product representation $\mathsf{B} = \mathsf{L}\mathsf{U}$, where the matrices L and U are lower and upper triangular matrices, and computes $\mathsf{Z}\mathbf{q}$ and $\mathsf{Z}^{\mathrm{T}}\mathbf{g}$ as solutions of linear systems of equations involving B and B^{T}. The superbasic variables are calculated using a quasi-Newton procedure; a search direction results from

$$\mathsf{R}^{\mathrm{T}}\mathsf{R}\mathbf{q} = -\mathsf{Z}^{\mathrm{T}}\mathbf{g}, \quad \mathbf{g} := \nabla_{\mathbf{x}} f(\mathbf{x}),$$

where $\mathsf{Z}^{\mathrm{T}}\mathbf{g}$ is the reduced gradient and R is an upper triangular matrix. This matrix R can be calculated by various update procedures using

$$\mathsf{R}^{\mathrm{T}}\mathsf{R} \approx \mathsf{Z}^{\mathrm{T}}\mathsf{H}\mathsf{Z}$$

to approximate the Hessian H, i.e., the second derivatives of $f(\mathbf{x})$. As soon as $\mathbf{q}$ is calculated, $\mathbf{p} = \mathsf{Z}\mathbf{q}$ follows as the search direction for all variables. This is followed by a line search to solve the one-dimensional problem

$$\min_{\alpha} \left\{ f(\mathbf{x}+\alpha\mathbf{p}) \;\middle|\; 0 \le \alpha \le \alpha^{+} \right\},$$

where α^{+} is derived from the limits of the variable. As in LP, the dual values $\boldsymbol{\pi}$ or shadow prices can be computed by solving a linear system of equations:

$$\mathbf{g}_{\mathsf{B}} - \mathsf{B}^{\mathrm{T}}\boldsymbol{\pi} = 0,$$

where $\mathbf{g}_{\mathsf{B}}$ is the gradient of the objective function. The corresponding quantity for the superbasic variable is

$$\mathsf{Z}^{\mathrm{T}}\mathbf{g} := \mathbf{g}_{\mathsf{S}} - \mathbf{s}^{\mathrm{T}}\boldsymbol{\pi} = 0 \quad ;$$

in the optimal solution point $\mathbf{x}_*$ holds $\mathsf{Z}^{\mathrm{T}}\mathbf{g} = \mathbf{0}$.

If $\mathbf{h}(\mathbf{x}) \neq 0$, then, as implemented in GAMS/MINOS, we can use the projective Lagrangian algorithm invented by Robinson (1972,[461]) and described in Murtagh & Saunders (1982,[416]). In this iterative procedure the constraints are linearized; the sequence of problems with nonlinear objective function, but linear constraints, is solved using the reduced gradient method.

At the beginning of iteration k, $\mathbf{x}_k$ is an estimated value of the nonlinear variables and $\boldsymbol{\lambda}_k$ is an estimated value of the Lagrange multipliers (or dual values) associated with the nonlinear constraints. The linearizations result in

$$\mathbf{h}^{\mathrm{L}}(\mathbf{x}, \mathbf{x}_k) = \mathbf{h}(\mathbf{x}_k) + \mathsf{J}(\mathbf{x}_k)(\mathbf{x} - \mathbf{x}_k)$$

or in short form $\mathbf{h}^{\mathrm{L}} = \mathbf{h}_k + \mathsf{J}_k(\mathbf{x} - \mathbf{x}_k)$ where $\mathsf{J}_k = \mathsf{J}(\mathbf{x}_k)$ the Jacobian at $\mathbf{x}_k$; the i-th row of J_k is the gradient vector of the i-th nonlinear second order condition. This results in the minimization problem

$$\min_{\mathbf{x},\mathbf{y}} \left\{ f(\mathbf{x}) + \mathbf{c}^{\mathrm{T}}\mathbf{x} + \mathbf{d}^{\mathrm{T}}\mathbf{y} - \boldsymbol{\lambda}_k \mathrm{T}(\mathbf{x} - \mathbf{x}_k) + \frac{1}{2}\rho(\mathbf{h} - \mathbf{h}^{\mathrm{L}})^{\mathrm{T}}(\mathbf{h} - \mathbf{h}^{\mathrm{L}}) \right\} \qquad (12.3.3)$$

subject to the constraints

$$\begin{matrix} \mathbf{h}^{\mathrm{L}} + \mathsf{A}_1\mathbf{y} \circ \mathbf{b}_1 \\ \mathsf{A}_2\mathbf{x}+\mathsf{A}_3\mathbf{y} \circ \mathbf{b}_2 \end{matrix}, \quad \mathbf{L} \le \mathbf{x}, \mathbf{y} \le \mathbf{U}$$

with a suitable scalar parameter ρ in the quadratic penalty function

$$\frac{1}{2}\rho(\mathbf{h} - \mathbf{h}^{\mathrm{L}})^{\mathrm{T}}(\mathbf{h} - \mathbf{h}^{\mathrm{L}}).$$

As on page 431 we use the slack variables $\mathbf{s}_1$ and $\mathbf{s}_2$ where the right sides $\mathbf{b}_1$ and $\mathbf{b}_2$ find their way into bounds of $\mathbf{s}_1$ and $\mathbf{s}_2$. This results in the linear system of equations

$$\begin{pmatrix} \mathsf{J}_k & \mathsf{A}_1 \\ \mathsf{A}_2 & \mathsf{A}_3 \end{pmatrix} \begin{pmatrix} \mathbf{x} \\ \mathbf{y} \end{pmatrix} + \begin{pmatrix} \mathbb{1} & 0 \\ 0 & \mathbb{1} \end{pmatrix} \begin{pmatrix} \mathbf{s}_1 \\ \mathbf{s}_2 \end{pmatrix} = \begin{pmatrix} \mathsf{J}_k\mathbf{x}_k - \mathbf{g}_k \\ \mathbf{0} \end{pmatrix}$$

and the reduced gradient procedure from page 431 can be applied.

A disadvantage of the presented reduced gradient method is the linearization, which places it in the group of sequential linear methods; these procedures linearize the constraints and use a composed Lagrangian function as an objective function in which penalty terms occur.

Sequential linear methods iterate along the tangent direction during the line search. On the other hand, the Generalized Reduced Gradient method developed by Drud (1994,[164]) used in his solver CONOPT moves along the current hyper-surface induced by the given constraints, which is not unusual for this class. In each iteration of the line search, the basic variables are modified in such a way that they lead to a feasible solution. The line search in the GRG procedure itself is numerically more expensive, but this pays off if the number of (outer) iterations is small. The GRG iterative procedure in CONOPT is described shortly as follows:

1. Initializing the procedure.
2. Determining a feasible point.
3. Calculation of the Jacobian associated with the constraints J.
4. Selection of the set of basic variables $\mathbf{x}^{\mathrm{B}}$ so that the sub-matrix of J, established by the columns associated with these basic variables $\mathbf{x}^{\mathrm{B}}$, is regular; followed by a factorization, i.e., LU decomposition of the basis matrix B. The other variables are kept as non-basic variables $\mathbf{x}^{\mathrm{N}}$.
5. Calculation of the Lagrange multipliers $\mathsf{B}^{\mathrm{T}}\boldsymbol{\pi} = \mathbf{g}$.
6. Calculation of the reduced gradient $\mathbf{r} := \mathbf{g} - \mathsf{B}^{\mathrm{T}}\boldsymbol{\pi}$.
7. Abort if the value of $\mathbf{r}$ projected on the barriers is sufficiently small; the current point is considered an optimal solution. Otherwise, proceed to step 8.
8. Determination of a search direction $\mathbf{d}$ for the non-basic variables; the calculation is based on the knowledge of $\mathbf{r}$ and allows the calculation of the tangent direction.

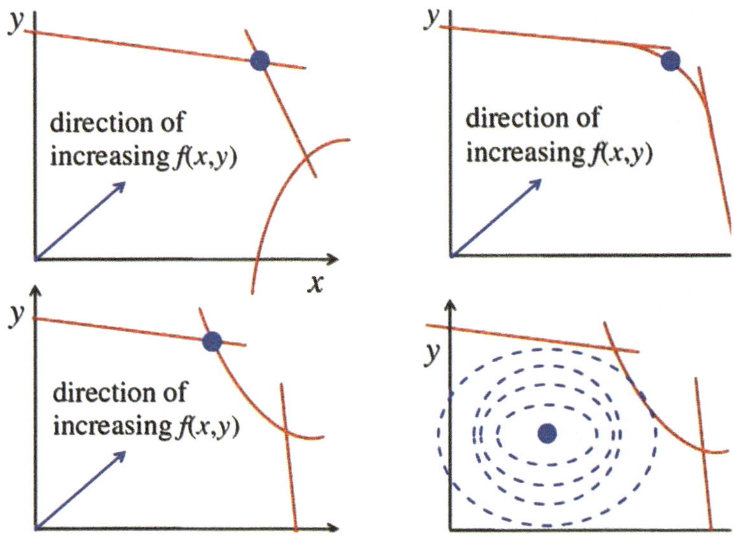

Fig. 12.1 NLP problems have different solution structures. The blue dot indicates the solution point. Upper left: A vertex solution of two intersecting linear constraints. Upper right: A solution on the curvilinear constraint. Lower left: A vertex solution at an intersection point of a linear and a curvilinear constraint. Lower right: A locally optimal interior solution

9. Perform a line search along direction **d**. At each step, the basic variables $\mathbf{x}^{\mathrm{B}}$ are modified so that $\mathbf{h}(\mathbf{x}^{\mathrm{B}}, \mathbf{x}^{\mathrm{N}}) = \mathbf{b}$ is fulfilled, where the factorized representation of B is used in a Pseudo-Newton method.
10. Continue with step 3.

In NLP problems various situations can occur as displayed in Fig. 12.1: (1) The solution is a locally optimal corner or vertex solution. In this case, there are no superbasic variables; the solution is dominated by the constraints. (2) The solution is a locally optimal point on a curvilinear constraint, i.e., it is on the boundary of the feasible region, but it is not a vertex solution. (3) The solution is a locally optimal interior solution. The largest component of the reduced gradient **r** is less than some optimality tolerance.

12.4 Sequential Quadratic Programming

Sequential quadratic programming (SQP) is a very efficient optimization method [520] for solving problem (12.2.1). A widely used SQP-based solver is SNOPT developed by Gill et al. (1997,[219]). The basic idea of the procedure is to solve (12.2.1) by solving a sequence of quadratic optimization problems. The subproblem in iteration k appears as

$$\min_{\Delta\mathbf{x}} \left\{ \frac{1}{2}\Delta\mathbf{x}^{\mathrm{T}}\mathsf{Q}_k\Delta\mathbf{x} + \nabla f(\mathbf{x}_k)^{\mathrm{T}}\Delta\mathbf{x} \right\}, \quad \Delta\mathbf{x} \in \mathbb{R}^n \tag{12.4.4}$$

$$\mathsf{J}_2(\mathbf{x}_k)^{\mathrm{T}}\Delta\mathbf{x} + \mathbf{F}_2(\mathbf{x}_k) = 0, \quad \mathsf{J}_3(\mathbf{x}_k)^{\mathrm{T}}\Delta\mathbf{x} + \mathbf{F}_3(\mathbf{x}_k) \geq 0,$$

where index k denotes the quantities known at the beginning of iteration k, and $\Delta\mathbf{x}$ is the correction vector to be determined. To solve this subproblem [cf. Gill et al. (1981, Section 6.5.3)], the constraints are linearized and the Taylor series of the objective function is truncated after the quadratic term. There is no need to consider the constant term $f(\mathbf{x}_k)$ any longer. The necessary conditions of the Lagrangian function corresponding to (12.4.4) are

$$\mathsf{Q}_k\Delta\mathbf{x} + \nabla f(\mathbf{x}_k) - \mathsf{J}_2(\mathbf{x}_k)\tilde{\lambda}_{k+1} - \mathsf{J}_3(\mathbf{x}_k)\tilde{\mu}_{k+1} = \mathbf{0}.$$

If $\boldsymbol{\lambda}_k$ denotes the vector of the Lagrange multipliers (for simplicity's sake there is no distinction between the Lagrange multipliers $\boldsymbol{\lambda}$ and $\boldsymbol{\mu}$ for equations and inequalities), known at the beginning of iteration k, and if $\Delta\mathbf{x}_k$, $\tilde{\boldsymbol{\lambda}}_k$, and $\tilde{\boldsymbol{\mu}}_k$ are the solutions of (12.4.4) of iteration k, the next iteration $k+1$ follows as

$$\begin{pmatrix} \mathbf{x}_{k+1} \\ \boldsymbol{\lambda}_{k+1} \\ \boldsymbol{\mu}_{k+1} \end{pmatrix} = \begin{pmatrix} \mathbf{x}_k \\ \boldsymbol{\lambda}_k \\ \boldsymbol{\mu}_k \end{pmatrix} + \alpha_k \begin{pmatrix} \Delta\mathbf{x}_k \\ \Delta\boldsymbol{\lambda}_k \\ \Delta\boldsymbol{\mu}_k \end{pmatrix}, \quad \begin{pmatrix} \Delta\boldsymbol{\lambda}_k = \tilde{\boldsymbol{\lambda}}_k - \boldsymbol{\lambda}_k \\ \Delta\boldsymbol{\mu}_k = \tilde{\boldsymbol{\mu}}_k - \boldsymbol{\mu}_k \end{pmatrix},$$

where α_k is an appropriate damping factor.

For the solution of the quadratic subproblems, the reader is referred to Gill et al. (1981, Section 5.3.2) and Fletcher (1987,[187], Chapter 10).

12.5 Interior-Point Methods

Interior-Point Methods (IPMs) are algorithms for computing local solutions of nonlinear constrained optimization problems, implemented, for instance, in IPOPT by Wächter & Biegler (2006,[568]), a solver freely available and also embedded in GAMS. This solver expects equality constraints and bounds on variables. Logarithmic barrier functions take care of the bounds. Using the necessary and sufficient conditions for the existence of local optima (KKT conditions from Sect. 12.2), the optimization problem is reduced to solving a nonlinear system of equations, for instance, by Newton's method. In LP, IPMs are particularly suitable for large, sparse matrices or those that are almost degenerate; see also Appendix 3.8.5.

12.6 Mixed Integer Nonlinear Programming

In Sect. 11.2.2 we formulated a pooling problem and learned that it is a special case of nonlinear programming. Let us now assume that in addition to the constraints (11.2.12) and (11.2.14) on page 397 some processing units can only be operated in specific, i.e., discrete modes. Then we have a mixed integer nonlinear optimization problem. These problems are far more difficult than the ones we have looked at so far. Nevertheless, there are algorithms capable of solving such problems. An early overview of methods designed for solving mixed integer nonlinear problems is given in Leyffer (1993,[361]). Readers interested in these algorithms are also referred to Floudas (1995,[189]) or to the reviews by Grossmann (2002,[240]) or Trespalacios & Grossmann (2014,[555]). Recommended is also the retrospective on optimization by Biegler & Grossmann (2004,[77]) including a tree of classes of optimization problems with a strong focus on process systems engineering. Finally, a recent review and comparison of solvers for convex MINLP is by Kronqvist et al. (2019,[343]).

12.6.1 Definition of an MINLP Problem

In Sect. 2.6.2 we have already provided a formal definition of MINLP problems. We repeat it here for convenience. For $\mathbf{x}^{\mathrm{T}} = (x_1, \ldots, x_{n_c})$ and $\mathbf{y}^{\mathrm{T}} = (y_1, \ldots, y_{n_d})$, objective function $f(\mathbf{x}, \mathbf{y})$ and constraints $g(\mathbf{x}, \mathbf{y})$ and $h(\mathbf{x}, \mathbf{y})$ an optimization problem

$$\min \left\{ f(\mathbf{x}, \mathbf{y}) \left| \begin{array}{l} \mathbf{g}(\mathbf{x}, \mathbf{y}) = 0 \\ \mathbf{h}(\mathbf{x}, \mathbf{y}) \geq 0 \end{array} \right. , \quad \begin{array}{l} \mathbf{x} \in X \subseteq \mathbb{R}^{n_c} \\ \mathbf{y} \in U \subseteq \mathbb{Z}^{n_d} \end{array} \right\} \tag{12.6.1}$$

is called a *mixed integer nonlinear programming problem*, if the domain U is discrete, e.g., $U = \mathbb{N}_0 = \{0, 1, 2, 3, \ldots\}$ and at least one of the functions $f(\mathbf{x}, \mathbf{y})$, $\mathbf{g}(\mathbf{x}, \mathbf{y})$, and $\mathbf{h}(\mathbf{x}, \mathbf{y})$ is nonlinear.

The continuous variables in (12.6.1) could, for instance, describe the states (temperature, pressure, etc.), flow rates, or design parameters of plants or chemical reactors. The discrete variables, often binary variables, may be used to describe the topology of a process network or to represent the existence or non-existence of plants. The good news is that mostly, binary variables appear only in linear terms and rarely in nonlinear ones. Consider, for example, the following pure integer nonlinear problem:

$$\min_{y_1, y_2} \left\{ 3y_1 + 2y_2^2 \left| \begin{array}{c} y_1^4 - y_2 - 15 = 0 \\ y_1 + y_2 - 3 \geq 0 \end{array} \right. , \quad y_1, y_2 \in \mathbb{N}_0 \right\}$$

A feasible solution is $y_1 = 3$ and $y_2 = 66$. The unique optimal solution is $\mathbf{y}^* = (y_1, y_2)^* = (2, 1)$ and $f(\mathbf{y}^*) = 8$.

12.6.2 Some General Comments on MINLP

MINLP problems such as (12.6.1) are very difficult optimization problems. They belong to the class $\mathcal{NP}$-complete problems.[7]

We have learned that MILP problems are combinatorial optimization problems for which the B&B algorithm based on LP relaxation proves to be sufficiently efficient. A similar statement is also valid for quadratic programming (QP) problems. LP and QP problems are special cases of nonlinear programming (NLP) problems. Usually, NLP problems cannot be solved in a finite number of steps but only iteratively. Nevertheless, solving NLP problems is usually easier than solving MILP problems. The reason for this, a fact well-known to numerical analysts, is that many NLP problems can be solved locally using sequential quadratic programming (a similar technique to sequential linear programming described in Sect. 11.2.2) and have convergence rates of second order.[8] This property allows us to determine local solutions quickly. The matter becomes more complicated for strongly non-convex NLP problems with many local extrema. One way is to use deterministic global optimization techniques as discussed in Sect. 12.7. Alternatively, metaheuristics, for instance, simulated annealing and genetic algorithms are commonly used for this kind of problem (but they are not able to prove global optimality).

Unfortunately, MINLP problems combine all the difficulties of both its subclasses: MILP and NLP. Even worse, in addition they have properties absent in NLP or MILP. While for convex NLP problems a local minimum is identical to the global minimum, we find that this result does not hold for MINLP problems.

It is difficult to solve (12.6.1) in its general form. Therefore, in the 1980s and 1990s only special instances of (12.6.1) had been investigated. A significant assumption or requirement is convexity. A set of points is called *convex* if for any given two points belonging to the set the straight line connecting the points also completely belongs to the set. More formally, the set $M \subseteq \mathbb{R}^n$ is called *convex* if and only if $\mathbf{x}_1 \in M$ and $\mathbf{x}_2 \in M$ imply: $\lambda_1\mathbf{x}_1 + \lambda_2\mathbf{x}_2 \in M$ for all $\lambda_{1,2} \in [0, 1]$ and

---

[7] No algorithm is known which can solve any $\mathcal{NP}$-complete problem in polynomial time (the solve time is bounded by a polynomial function of the problem size). It is thought that if such an algorithm were found, it would also be able to solve other $\mathcal{NP}$-complete problems in polynomial time, e.g., the resource constrained scheduling problem presented in Sect. 10.5. See Nemhauser & Wolsey (1988) for further material on exact definition and explanation on class $\mathcal{NP}$.

[8] Second order convergence rate implies that we double the number of accurate digits after the decimal point in each iteration. As derivatives, especially, the Hessian, are subject to numerical errors; in practice one is usually content to prove and achieve superlinear convergence.

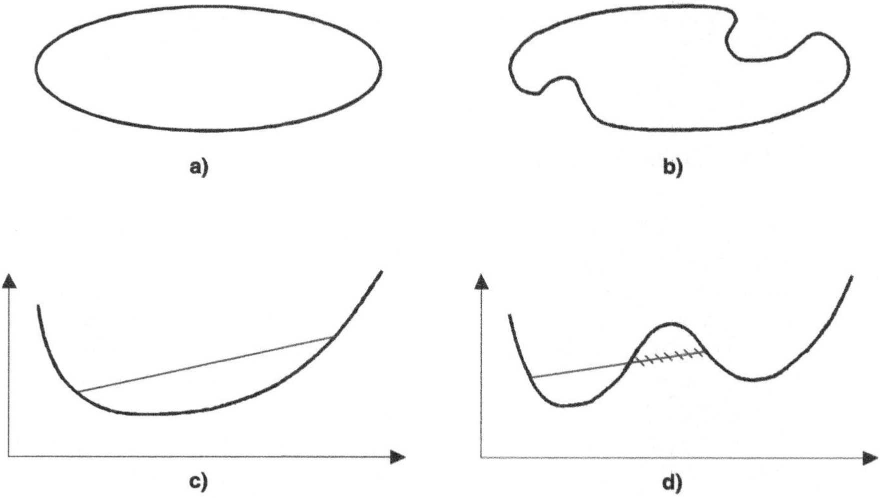

Fig. 12.2 Convex and non-convex sets and functions. (**a**) Convex set. (**b**) Non-convex set. (**c**) Convex function. (**d**) Non-convex function

$\lambda_1 + \lambda_2 = 1$. A function $f : X \subseteq \mathbb{R}^n \rightarrow \mathbb{R}$ is called *convex* if and only if

$$\lambda_1 f(\mathbf{x}_1) + \lambda_2 f(\mathbf{x}_2) \geq f(\lambda_1 \mathbf{x}_1 + \lambda_2 \mathbf{x}_2), \quad \forall \lambda_1, \lambda_2 \geq 0 \mid \lambda_1 + \lambda_2 = 1, \qquad (12.6.2)$$

which implies that the line connecting two points $f(\mathbf{x}_1)$ and $f(\mathbf{x}_2)$ of the graph of f never lies "below" the graph [see Fig. 12.2c], and that the tangent at a point $(\mathbf{x}_0, f(\mathbf{x}_0))$ of the graph always lies "below" the graph , i.e.,

$$f(\mathbf{x}) \geq f(\mathbf{x}_0) + f'(\mathbf{x}_0)(\mathbf{x} - \mathbf{x}_0), \quad \forall \mathbf{x}. \qquad (12.6.3)$$

Solution algorithms in discrete optimization belong to two classes: *deterministic* and *heuristic methods*. All deterministic methods use implicit enumeration and tree search rules, which try to avoid analyzing sub-trees. B&B, for instance, uses rules to fathom branches of the tree to avoid exploring them without ruling out the optimal solution. Deterministic methods can decide whether a given solution is optimal or not. Heuristic methods lack this feature. Unfortunately, so far, efficient exact methods for solving large-scale non-convex MINLP problems still wait to be invented.

A simple deterministic method is to list all combinations U of discrete variables $\mathbf{y}_i$. Each binary vector $\mathbf{y}_i$ generates an NLP in the continuous variable vector $\mathbf{x}$. If we solve this NLP problem, it is either infeasible or yields a solution, i.e., a pair $(\mathbf{x}_i, z_i = f(\mathbf{x}_i, \mathbf{y}_i))$. After having solved all NLP problems, we choose the pair with smallest z_i (let us refer to it using the index i^*). Thus, the solution is given by the triple $(\mathbf{x}^* = \mathbf{x}_{i^*}, \mathbf{y} = \mathbf{y}_{i^*}, z_i^* = f(\mathbf{x}_{i^*}, \mathbf{y}_{i^*}))$. This method is sometimes referred to as an exhaustive search, and, of course, only works if U has a limited number of elements and if the NLP subproblems allow us to determine their global minima. Although the convexity assumption is fulfilled for convex (continuous) problems, the method is of no practical use because of the prohibitively high numerical effort.

12.6.3 Deterministic Methods for Solving MINLP Problems

Deterministic methods for solving (convex) MINLP problems — cf. Nowak (2005,[423]) or Kronqvist (2018,[343]) for good reviews — in the 1980s fall into three classes:

1. Branch & Bound (Gupta & Ravindran, 1985,[248]),
2. Generalized Benders Decomposition (Geoffrion, 1972,[212]),
3. Outer Approximation (Duran & Grossmann, 1986,[165]), and
4. Convexification coupled to B&B entered the stage in the 1990s, followed somewhat later by B&C.

The B&B algorithm for MINLP problems by Gupta & Ravindran (1985,[248]) is based on the same ideas as the B&B algorithm for solving MILP problems. The first step is to solve the problem generated by relaxing the integrality condition on the variables. If a solution of the relaxed problem fulfills all integrality conditions, it is also a solution of the original problem. Otherwise, in a minimization problem the relaxed problem provides a lower bound (of course only if the global minimum can be determined) and the search tree is built up. A feasible integer solution provides an upper bound. A major drawback of the B&B algorithm applied to MINLP problems is that nodes deeper in the tree cannot benefit so greatly from information available at previous nodes as is the case in B&B algorithms for solving MILP problems exploiting warm starts using the dual Simplex algorithm.

The Generalized Benders Decomposition (GBD) method divides the variables into two sets: complicating and non-complicating variables. In MINLP models, the class of complicating variables is made up by the discrete (usually binary) variables of the problem at hand. Then the algorithm generates a sequence of NLP subproblems (produced by fixing the binary variables $\mathbf{y}^k$) and solves the so-called MILP Master problems in the space of the complicating variables. The NLP subproblems yield upper bounds for the original problem while the MILP Master problems yield additional combinations of binary variables $\mathbf{y}^k$ for subsequent NLP subproblems. Under convexity assumptions, the Master problems generate a sequence of lower bounds increasing monotonically. The algorithm terminates if lower and upper bounds equal or cross each other.

Outer Approximation (Duran & Grossmann, 1986,[165]) also consists of a sequence of NLP subproblems (produced by fixing the binary variables $\mathbf{y}^k$) generated by MILP Master problems. The significant difference lies in the definition of the master problems. Algorithms based on Outer Approximation (OA) describe the feasible region as the intersection of an infinite collection of sets with a simpler structure, e.g., polyhedra. In Outer Approximation, Master problems are generated by "outer approximations" (linearizations, or Taylor series expansions) of the nonlinear constraints in *those* points which are the optimal solutions of the NLP subproblems; that is, a finite collection of sets. The key idea of the algorithm by Duran & Grossmann (1986,[165]) is to solve the MINLP with a much smaller set of points, i.e. tangential planes. In convex MINLP problems, a superset of the feasible region is established. Thus, the OA Master problems (MILP problem in both discrete and continuous variables) produce a sequence of lower bounds monotonically increasing. The termination criterion is the same as above.

While the GBD Master problems have fewer variables and constraints, the OA algorithm provides tighter bounds and needs fewer iterations for convergence. Both GBD and OA algorithms have heuristic extensions for non-convex MINLP. In many instances they are even capable of proving optimality.

12.6.4 Algorithms and Software for Solving Non-convex MINLP Problems

DICOPT by Duran & Grossmann (1986,[165]) and Viswanathan & Grossmann (1990,[567]) was, in the 1980s, the only commercial software available for solving MINLP problem (12.6.1) of realistic size. It uses *Outer Approximation* with some extensions for non-convex problems.

One of the few and early MINLP algorithms and programs becoming available in the 1990s is αECP; cf. Westerlund et al. (2018,[579]) for a recent paper on this approach. The αECP method is an extension of Kelley's cutting plane method which was originally given for convex NLP problems; Kelley (1960,[331]). In Westerlund & Pettersson (1995,[574]) the method was extended to convex MINLP problems and in Westerlund et al. (1998,[576]) it was further extended to MINLP problems with pseudo-convex constraints. The method was further extended in Westerlund & Pörn (2002,[573]) and the actual version of the method converges to the global optimal solution for non-convex MINLP problems having a pseudo-convex objective function and pseudo-convex inequality constraints.

Another software package for mixed integer nonlinear optimization is `MINOPT` by Schweiger et al. (1996,[496]), developed at the Department of Chemical Engineering of Princeton University. This package can even handle MINLP problems with differential constraints or mixed integer optimal control problems.

The period from 2000 to 2020 shows many streams of activities covering computational complexity, convexification, decomposition, finding feasible solutions and deterministic approaches to computing global solutions of MINLPs leading, for instance, to the solvers ANTIGONE, BARON, COUENNE, LINDO, and DECOA, a parallel decomposition-based MINLP solver implemented in `Python` and `Pyomo`; cf. Burer & Letchford (2012,[102]), Bonami et al. (2012,[92]) for algorithms and software for convex MINLP, Belotti et al. (2013,[60]), Kilinç & Sahinidis (2017,[333]), Sahinidis (2019,[478]) and further references therein,

Nowak (2019,[424]), Muts et al. (2020,[418]) — and Sect. 12.7. While Gurobi's MILP solver is known as a strong MILP solver, it is worthwhile to mention Gurobi's efforts toward solving non-convex mixed integer quadratically constrained problems being part of their solver.

An impressive MINLP application is from one of the factories of Danisco in Finland (today owned by DuPont), where a compound named Betaine (a food ingredient) is separated from molasses (a liquid solution from sugar beets). Betaine is used for different purposes, but among all used as a nutrient in animal foods for farming. The mathematical problem in this application is very exciting since it resulted in an extremely complex MINLP model where the separation system has been formulated by Emet & Westerlund (2008,[173]) as a two point boundary value problem of a system of partial differential equations included as constraints in the MINLP model. More on discretized differential equations as constraints in NLP or MINLP problems is found in Andrei (2013,[25]).

12.7 Global Optimization — Mathematical Background

The global minimum of problem (2.6.10) can only be calculated with methods of global optimization, if we do not request convexity; cf. [190–192, 271–273, 330, 363, 406, 540, 542] — the literature reviews by Floudas et al. (2005,[194]) and of Misener & Floudas (2012,[406]) are particularly recommended. In the case of deterministic methods — and only those are to be considered — an upper and lower limit z^o and z^u for the objective function is determined for a given ε-bound, so that finally $z^u \leq f(\mathbf{x}, \mathbf{y}) \leq z^o$ and $z^o - z^u \leq \varepsilon$ apply. The upper bound z^o is the objective function value $z^o = f(\mathbf{x}, \mathbf{y})$ of a feasible point $(\mathbf{x}, \mathbf{y})$ or, better still, of a local minimum $(\mathbf{x}_*, \mathbf{y}_*)$ obtained by a solution method for determining stationary points in a continuous nonlinear problem or heuristic. While obtaining an upper bound is relatively simple, the determination of the lower bound z^u is more difficult and the following methods are suitable, e.g.,

- *interval methods* [330] or [452], which are not discussed here,
- *convex relaxation* using convex underestimator (see Fig. 12.3) and convexified sets combined with a B&B process, and also
- *piecewise linear underestimators* (MILP formulation).

A convex function f_u is called *convex underestimator* for a given function f on a set $\mathcal{S}$, if $f_u(\mathbf{x}) \leq f(\mathbf{x})$ for all $\mathbf{x} \in \mathcal{S}$. As an example, the non-convex function $f(x_1, x_2) = x_1 x_2$ is considered on the interval $[X_1^-, X_1^+] \times [X_2^-, X_2^+]$; it consists of only one bilinear term and can be linearly convex underestimated by the variable y and the following four inequalities:

$$
\begin{aligned}
X_1^- x_2 + X_2^- x_1 - X_1^- X_2^- &\leq y \leq -X_1^+ x_2 - X_2^- x_1 + X_1^- X_2^+ \\
X_1^+ x_2 + X_2^+ x_1 - X_1^+ X_2^+ &\leq y \leq -X_1^- x_2 - X_2^+ x_1 + X_1^- X_2^+;
\end{aligned}
$$

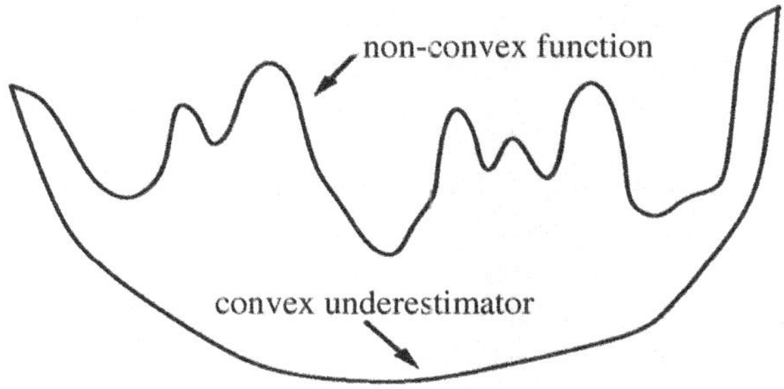

Fig. 12.3 Non-convex function and a convex underestimator. The convex underestimator displayed is not necessarily the best convex underestimator. The best convex underestimator possible is the convex hull function of the given non-convex function

cf. Al-Khayyal & Falk (1983,[16]), Al-Khayyal (1990,[15]) and McCormick (1976,[389]). The error between the f and this convex underestimator y is bounded by

$$\max_{x_1,x_2 \in [X_1^-,X_1^+]\times[X_2^-,X_2^+]} (x_1x_2 - y) = \frac{1}{4}\left(X_1^+ - X_1^-\right)\left(X_2^+ - X_2^-\right).$$

For rational terms x_1/x_2, which have no sign change in the interval $[X_1^-, X_1^+] \times [X_2^-, X_2^+]$, we obtain the linear inequalities

$$\begin{array}{ll} y \geq X_1^-/x_2 + x_1/X_2^+ - X_1^-/X_2^+ & \text{, if } X_1^- \geq 0 \\ y \geq x_1/X_2^+ + X_1^- x_2/X_2^- X_2^+ + X_1^-/X_2^- & \text{, if } X_1^- < 0 \\ y \geq X_1^+/x_2 + x_1/X_2^- - X_1^-/X_2^+ & \text{, if } X_1^+ \geq 0 \\ y \geq x_1/X_2^- - X_1^+ x_2/X_2^- X_2^+ + X_1^-/X_2^- & \text{, if } X_1^- < 0. \end{array}$$

Finally, we provide the best linear convex underestimator for univariate concave functions f(x), i.e., functions with $f''(x) < 0$ on the interval $[X^-, X^+]$,

$$f(x) \geq f(X^-) + \frac{f(X^+) - f(X^-)}{X^+ - X^-}\left(x - X^-\right).$$

Generally, twice continuous-differentiable non-convex functions $f(\mathbf{x})$ can be derived for $\mathbf{x} \in \mathbb{R}^n$ in the interval $[\mathbf{X}^-, \mathbf{X}^+]$, as in (1994,[384]) by

$$F(\mathbf{x}) := f(\mathbf{x}) - \sum_{i=1}^{n} \alpha_i \left(X_i^+ - x_i\right)\left(x_i - X_i^-\right)$$

with $F(\mathbf{x}) \leq f(\mathbf{x})$ and

$$\alpha_i \geq -\frac{1}{2} \min_{i,\mathbf{x}\in[\mathbf{X}^-,\mathbf{X}^+]} \lambda_i \left\{\mathsf{H}(\mathbf{x})\right\} \geq 0, \qquad (12.7.4)$$

where $\mathsf{H}(\mathbf{x})$ is the Hessian matrix of $f(\mathbf{x})$ and $\lambda_i\,\{\mathsf{H}(\mathbf{x})\}$ is the ith eigenvalue of $\mathsf{H}(\mathbf{x})$. Note that the smallest eigenvalue $\mathsf{H}(\mathbf{x})$ assumed on the interval $[\mathbf{X}^-, \mathbf{X}^+]$ must be calculated in each case. The maximum distance is given by

$$d_{\max} := \max_{\mathbf{x}\in[\mathbf{X}^-,\mathbf{X}^+]} (f(\mathbf{x}) - F(\mathbf{x})) = \frac{1}{4}\sum_{i=1}^{n} \alpha_i \left(X_i^+ - X_i^-\right)^2 . \qquad (12.7.5)$$

$F(\mathbf{x})$ is convex on the interval $[\mathbf{X}^-, \mathbf{X}^+]$ if and only if $\mathsf{H}_F(\mathbf{x})+2\Delta$ is positive definite for all $\mathbf{x} \in [\mathbf{X}^-, \mathbf{X}^+]$. Here, $\mathsf{H}_F(\mathbf{x})$ is the Hessian of F and $\Delta := diag\{\alpha_i\}$ is called diagonal shift matrix. The method outlined here is called the *non-uniform diagonal shift method* [7, 8] and [6, Sections 3.4 and 3.5]. If instead of the α_i-values, a global α, i.e., $\Delta := \alpha\mathbb{1}$ is used, one speaks of a *uniform diagonal shift procedure* — in this case, $d_{\max}$ is proportional to α.

The eigenvalues can be calculated exactly or with methods of *interval arithmetic* — cf. [452] or [330]. This step is important for the efficiency of the procedure; the α_i should be as small as possible after (12.7.4) in order to make $d_{\max}$ as small and the underestimators as accurate as possible. For the example

$$\min_{-1\le x_1\le 2,\,-1\le x_2\le 1} f(x_1, x_2), \qquad f(x_1, x_2) := \cos x_1 \sin x_2 - \frac{x_1}{x_2^2}$$

the Hessian follows — in the general case, e.g., with an automatic differentiation algorithm — as

$$\mathsf{H}_f(\mathbf{x}) := \begin{pmatrix} -\cos x_1 \sin x_2 & -\sin x_1 \cos x_2 + \frac{2x_2}{\left(x_2^2+1\right)^2} \\ -\sin x_1 \cos x_2 + \frac{2x_2}{\left(x_2^2+1\right)^2} & -\cos x_1 \sin x_2 + \frac{2x_1\left(x_2^2+1\right)^2 - 8x_1x_2^2\left(x_2^2+1\right)^2}{\left(x_2^2+1\right)^4} \end{pmatrix} .$$

For the application of interval arithmetic, it is important and necessary that $\mathsf{H}_f(\mathbf{x})$ can be calculated analytically. If one evaluates this matrix under consideration of the indicated bounds of the variables x_1 and x_2, then the interval estimation follows

$$\mathsf{H}_f(\mathbf{x}) \subseteq [\mathsf{H}_f] = \begin{pmatrix} [-0.84148, +0.84148] & [-3.00000, +2.84148] \\ [-3.00000, +2.84148] & [-40.84148, +32.84148] \end{pmatrix} .$$

However, these elements can be further refined by determining the global minimum or maximum for each element over the permissible range. This way we get the optimal interval-Hessian

$$[\mathsf{H}_f]_* = \begin{pmatrix} [-0.84148, +0.84148] & [-1.52288, +1.38086] \\ [-1.52288, +1.38086] & [-2.00608, +4.00181] \end{pmatrix} .$$

The following results are relevant here. The exact smallest eigenvalue for $(x_1, x_2) \in [-1, 2] \times [-1, 1]$ is $\lambda^e_{\min}(\mathsf{H}_f) = -2.3934$; this can be approximated, for example, by calculating the eigenvalues of the matrix $\mathsf{H}_f(\mathbf{x})$ for a very fine grid. The smallest eigenvalue of interval matrices can be calculated, e.g., with the method of Hertz [268]; the basic idea is to construct a finite subset of real matrices from the interval-Hessian so that the smallest eigenvalue over all these matrices of the subset is equal to the smallest eigenvalue of the interval-matrix. We get $\lambda^e_{\min}([\mathsf{H}_f]) = -41.0652$ and $\lambda^e_{\min}([\mathsf{H}_f]_*) = -3.0817$. Thus, $\alpha = 1.1967$ after (12.7.4) is sufficient to guarantee the convexity of the underestimator

$$F(\mathbf{x}) := f(\mathbf{x}) - \alpha\,(2 - x_1)\,(x_1 + 1) - \alpha\,(1 - x_2)\,(x_2 + 1)\,.$$

A convex set M_c is called *convexification* of a given set M if $M_c \supseteq M$. The set S implicitly defined by the inequalities[9] $\mathbf{g}(\mathbf{x}, \mathbf{y}) \geq 0$ is spatially divided into subsets S^i in a B&B method. The convex minimization problem $\min_{(\mathbf{x},\mathbf{y}) \in S^i_c} f^i_c(\mathbf{x}, \mathbf{y})$ is solved by convexification S^i_c. It is solved by means of a procedure similar to Outer Approximation described above providing the solution z^i_c from which the lower bound $z^u := \min_i z^i_c$ follows. By further refinement of the division and thus better adaptation of the convex underestimators $f^i_c(\mathbf{x}, \mathbf{y})$ to the division, a monotonously growing sequence of lower bounds z^u is created. The refinement is continued until $z^o - z^u \leq \varepsilon$ is valid.

It can be shown that this process class converges toward the global minimum up to every ε-bound — but only in theory, as the unavoidable rounding errors that occur during the practical implementation of a numerical algorithm on a computer are not taken into account; when solving practical problems, one should therefore not request accuracies, for instance, smaller than $\varepsilon < 10^{-6}$. The fact that the influence of rounding errors can be considerable and have serious consequences, and can even lead to completely false statements, has often been demonstrated on the basis of practically relevant problems. With regard to rounding errors, it can be useful in global optimization to check the results using methods of interval arithmetic.

The solver GLOMIQO (Global Mixed Integer Quadratic Optimizer), and its successor ANTIGONE, developed by Misener & Floudas (2012,[406]) is particularly suitable for solving non-convex NLP and MINLP problems that are only quadratic in constraints and objective function. Here the problem is reformulated first; if possible, variables in equalities are eliminated. For example

$$Q_{ij}x_ix_j + A_kx_k = B$$

under the assumption $A_k \neq 0$ the variable x_k is replaced by

$$x_k = \frac{Q_{ij}x_ix_j - B}{A_k}.$$

[9] Equations $\mathbf{h}(\mathbf{x}, \mathbf{y}) = 0$ are replaced by the inequality pairs $-\delta \leq \mathbf{h}(\mathbf{x}, \mathbf{y}) \leq \delta$ with any given $\delta > 0$. However, this procedure should only be used if there is no other way; it is not very efficient.

If lower and upper limits X_k^- and X_k^+ are known for x_k, the inequalities

$$X_k^- \le \frac{Q_{ij}x_i x_j - B}{A_k} \le X_k^+$$

are added to the problem.

Bilinear terms $x_i x_j$ are replaced by the introduction of an additional variable

$$y_{ij} := x_i + x_j$$

leading to

$$x_i x_j = \frac{1}{2}\left(y_{ij}^2 - x_i^2 - x_j^2\right). \tag{12.7.6}$$

Although the right-hand side of (12.7.6) now contains three additional quadratic terms, to our advantage each of them only contain one variable. Two special features of GLOMIQO, and its successor ANTIGONE, are the generation of *RLT conditions* and a graph-theoretical analysis to identify convex substructures which have a positive effect in connection with convex underestimators. RLT stands for *reformulation-linearization techniques*; the RLT conditions are dynamically added to the problem if required. GLOMIQO, for example, considers all variables x_i that occur nonlinearly anywhere in the model and multiplies them by all linear equalities with row vector $\mathbf{A}_m$ and column vector $\mathbf{x}$

$$\mathbf{A}_m\mathbf{x} = \sum_j A_{mj}x_j = b_m,$$

containing only continuous variables x_j, i.e.,

$$\sum_j \left(A_{mj}x_j - b_m\right) x_i = \sum_j A_{mj}x_j x_i - b_m x_i = 0, \quad \forall\{im\}.$$

The inequalities

$$b_m^- \le \mathbf{A}_m\mathbf{x} \le b_m^+$$

are treated similarly and lead to the additional inequalities

$$\begin{aligned}
&\left(\mathbf{A}_m\mathbf{x} - b_m^+\right)\left(x_i - X_i^-\right) \le 0, \quad \left(\mathbf{A}_m\mathbf{x} - b_m^+\right)\left(X_i^+ - x_i\right) \le 0, \quad \forall\{im\},\\
&\left(b_m^- - \mathbf{A}_m\mathbf{x}\right)\left(x_i - X_i^-\right) \le 0, \quad \left(b_m^- - \mathbf{A}_m\mathbf{x}\right)\left(X_i^+ - x_i\right) \le 0, \quad \forall\{im\},
\end{aligned}$$

which are not all generated a priori, but only when required within B&B.

12.8 Summary and Recommended Bibliography

In this chapter, we have covered different ways of solving NLP and MINLP problems and have investigated certain aspects of software which will solve such problems. Thus the reader should now be familiar with:

- A standard form of NLP problems;
- The use of first and second order necessary and sufficient conditions (KKT conditions) and some very general principles of NLP and MINLP; and
- How the general reduced gradient method works.

A good *Introduction to Nonlinear and Global Optimization* is by Eligius M. T. Hendrix & Boglárka G.-Tóth (2010,[263]). Methods to solve MINLP problems are well covered by Nowak (2005,[423]). For further reading on nonlinear optimization applications we recommend Andrei (2013,[25]).

12.9 Exercises

1. Use a local NLP solver to solve a) max $f(x)$ with $f(x) = (10 - x)\sin^9(2\pi x)$ for $x \in [0, 10]$. b) Solve a) under the constraint $x + f(x) \le 4$. Think about appropriate initial values for x.
2. Use a local NLP solver to minimize $f(x, y) = x^2 + y^2$ subject to $x^2 - 16y \le 0$ and $1.6 - y(x - 1.6) \le 0$ and $-7x^2 + 39x - 2y \le 42$. Think about appropriate initial values for x and y.
3. Let $t \in [0, 2\pi]$. Use a local NLP solver to minimize

$$
\begin{aligned}
f(t) = & \sum_{k=-1}^{+1} \frac{A_k}{\left\{3r_0^2 - 4r_0^2\cos\theta - 2r_0^2\left[\sin^2\theta\cos\left(t + \frac{2}{3}\pi k\right) - \cos^2\theta\right]\right\}^6} \\
& - \sum_{k=-1}^{+1} \frac{B_k}{\left\{3r_0^2 - 4r_0^2\cos\theta - 2r_0^2\left[\sin^2\theta\cos\left(t + \frac{2}{3}\pi k\right) - \cos^2\theta\right]\right\}^3}
\end{aligned}
$$

with $r_0 = 0.154, \theta = 109°.5$, and

$$
\begin{aligned}
A_{-1} &= 588600 & B_{-1} &= 1079.1 \\
A_0 &= 600800 & B_0 &= 1071.5\,. \\
A_{+1} &= 481300 & B_{+1} &= 1064.6
\end{aligned}
$$

Hint: Construct your own multi-start algorithm with initial values for t in $[0, \pi/2]$, $[\pi/2, \pi]$, $[\pi, 3\pi/2]$, and $[3\pi/2, 2\pi]$.

Chapter 13
Global Optimization in Practice

Global optimization techniques, cf. Horst & Pardalos (1995,[271]), Floudas (2000,[190]), Floudas & Gounaris (2009,[191]), or Misener & Floudas (2012,[406]), are suitable for solving non-convex NLP or MINLP problems. When in this book we refer to global optimization, we mean *deterministic global optimization*, where a globally optimal objective function value can be computed up to a given $\varepsilon > 0$. If ε approaches the machine accuracy, fundamental questions arise which are better answered in the discipline *Reliable Computing*; we therefore limit ourselves to $\varepsilon > 10^{-6}$, which is sufficient for most practical problems. In practice, the computational effort growth exponentially with the number of nonlinear variables has a limiting effect. Since 2002, several commercially available solvers have been developed, including BARON [215], LINDO [491] and GLOMIQO [406] and ANTIGONE, which are systematically described in the review article by Misener & Floudas (2012,[406]).

Petroleum industry tries to solve planning and scheduling problems , which contain pooling problems in their core, using deterministic global optimization. The production of natural gas is also related to this. Questions in the environment of water systems or food production often also contain pooling problems and therefore require global optimization techniques. Further application examples can be found in Floudas et al. (1999,[193]), Floudas (2000,[190]), or Kallrath (2004,[305]). It is also worth mentioning that with the methods of global optimization as in Maranas & Floudas (1994,[384]) it is possible to determine all zeros of a system of nonlinear equations. This is applied, e.g., in the determination of all stationary states of chemical reactors; Maranas & Floudas (1994,[384]) apply it to the determination of possible energy levels of molecules, Ratschek & Rokne (1993,[451]) to circuit

Supplementary Information The online version contains supplementary material available at https://doi.org/10.1007/978-3-030-73237-0_13.

J. Kallrath, *Business Optimization Using Mathematical Programming*,
International Series in Operations Research & Management Science 307,
https://doi.org/10.1007/978-3-030-73237-0_13

design problems. Applications to parameter estimation and robotics can be found in Jaulin et al. (2001,[287]). Misener & Floudas (2012,[406]) also include applications from the field of *Computational Geometry*.

In this chapter, we have commented on several applications originating from cutting and packing, a field for which the book by Scheithauer (2018,[488]) provides a very good overview and introduction. For smaller cutting or packing problems it is possible to prove global optimality.

13.1 Global Optimization Applied to Real-World Problems

There are many good reasons why deterministic global optimization is or should be used when solving non-convex nonlinear (continuous and mixed integer) unconstrained or constrained real-world optimization problems. The arguments are:

1. determination of a global minimum or maximum of the objective function f over the feasible region,
2. determination of lower and upper bounds on the global minimum or maximum,
3. determination of a collection of good quality local solutions in a local vicinity of the global solution,
4. determination of all isolated solution points of the set of equality and inequality constraints , and
5. guaranteed proof that a constrained nonlinear problem is feasible or infeasible.

Point 1 is of course the most important if we are solving objective functions the values of which are of the order of billions of dollars; even a tenth of a percent is a lot of money. Point 2 provides us with a quality estimation if we cannot prove global optimality, but at least we have guaranteed bounds quantifying the maximal difference to the theoretical global extremum. Point 3 is often relevant for practical purposes and the robustness of the solution. Point 4 is not so obvious, and possibly more often used in science: sometimes it is useful to know all roots of an equation system. Only exact methods — deterministic global optimization belongs to this class — can prove that a problem is infeasible; this is Point 5.

An important practical issue when solving complicated problems is to find feasible points at all. Local solvers may just produce the message *problem is infeasible*, which only means that it is *locally infeasible*. The global solvers mentioned above have multi-start techniques and local search strategies embedded which initialize a local NLP solver. That way, they have a much better chance to come up with at least a feasible point.

Closing the gap is still a challenge for most global solvers. And there is an important trade-off between determining feasible points, on the one hand, and closing the gap, on the other hand. For only determining feasible points the problem should be as simple as possible and contain as few nonlinear terms as possible. For closing the gap, in many cases one needs extra constraints destroying symmetry and degeneracy — which in turn makes it more difficult to find feasible points.

After the advent of global solvers in the early 2000s, now solvers can deal with larger problems with more variables than in 2002, have embedded better techniques and use reformulation and transformations, cf. Skjäl et al. (2012,[513]), Skjäl & Westerlund (2014,[512]) or Lundell (2018,[377]), to close the gap, but they are by far not at the same level as MILP solvers presently. Some global solvers still lack trigonometric functions. Robustness and stability are issues to be improved. So, there is a long way to go to meet industrial standards. But it is worthwhile to try global solvers for the problem at hand — and, in some operative industrial software they are used, for instance, in pulp and paper industry.

13.2 A Trimloss Problem in Paper Industry

Cutting and packing industry significantly[1] contributes to the GDP of several countries. As in Sect. 4.1, we are faced with the task of cutting paper products of various sizes from a large paper roll (master roll) in order to satisfy customer demands. This time, however, not only the choice of patterns and their multiplicity but also the patterns themselves should be the degree of freedom of optimization. This question, which is dealt with in papers by Westerlund et al. (1994,[575]) and Harjunkoski (1997,[253]), leads to a nonlinear, non-convex integer optimization problem.

The master roll has width W_{mr} and a length assumed to be infinite. Each order or product paper roll i, $i \in \mathscr{I} = \{1, \ldots, N\}$, is identified by its width b_i. It is assumed that all product rolls have the same infinite length. In general, cutting an entire order from the master roll without trimloss is impossible. A combination of optimal waste pattern minimizes the waste. In order to identify the best scheme, a maximum number of different patterns N^P is suggested, where a pattern is defined by the positioning of the knives. Each pattern contains one or more pieces of order widths W_i, $i \in \mathscr{I} = \{1, \ldots, N\}$, and can be repeated several times throughout the scheme to meet the demand for D_i pieces of order width W_i. The selection of each pattern is described by a binary variable δ_p, $p \in \mathscr{P} = \{1, \ldots, N^{\mathrm{P}}\}$, the number of repetitions of the pattern p by the integer variable μ_p. The number of products of size i in the pattern p is characterized by the integer variable α_{ip}. Later, we want to arrange the patterns so that the more frequently used ones are listed first. The patterns are still subject to some special constraints. For example, each pattern must have a minimum used total width of F. The number of knives in use is limited by

[1]The pulp and paper industry plays an important role worldwide. There are in the order of 3000 paper mills worldwide, producing a total of 394 million tons of paper and paperboard, in 2010. Europe (including Russia) has approximately 900 paper mills, while Germany has about 180. The largest producer in the world is the Finnish UPM group with an annual tonnage of 12.7 million tons, followed by Stora Enso with 11.8 million tons and by International Paper with 9.7 million tons per year. Santos & Almada (2012,[482]) report that in Portugal the pulp and paper industry contributes over 4% of the GDP and 5% of the active employees.

K, which means that there cannot be more than K order widths in a pattern. This results in the following minimization problem

$$z = \min_{\mu_p, \delta_p, \alpha_{ip}} \sum_{p \in \mathcal{P}} \left(C_p^{\mathrm{R}} \mu_p + C_p^{\mathrm{P}} \delta_p \right)$$

with roll costs C_p^{R} to be paid for each type p roll used, and fixed costs C_p^{P} to be paid when pattern p is used. The following constraints must be observed: Fulfillment of demand

$$\sum_{p \in \mathcal{P}} \mu_p \alpha_{ip} = D_i \quad , \quad \forall i. \tag{13.2.1}$$

Consideration of the lower and upper width when selecting patterns

$$F \delta_p \leq \sum_{p \in \mathcal{P}} W_i \alpha_{ip} \leq W_{\mathrm{mr}} \delta_p \quad , \quad p \in \mathcal{P}. \tag{13.2.2}$$

Limitation of the cutting knives and thus the number of order widths in a pattern

$$\delta_p \leq \sum_{i \in \mathcal{I}} \alpha_{ip} \leq K \delta_p \quad , \quad p \in \mathcal{P}.$$

Relating the pattern selection variable δ_p to the pattern multiplicity μ_p

$$\delta_p \leq \mu_p \leq P \delta_p \quad , \quad p \in \mathcal{P},$$

where P is an upper limit on the usage of pattern p. A special cutting condition which is derived from demand

$$\sum_{p \in \mathcal{P}} \mu_p \geq \max \left\{ \left\lceil \frac{\sum_{i \in \mathcal{I}} D_i}{K} \right\rceil , \left\lceil \frac{\sum_{i \in \mathcal{I}} W_i D_i}{W_{\mathrm{mr}}} \right\rceil \right\}$$

can give us some indication on how to select P.

It is important for proving optimality to destroy degeneracy, i.e., if pattern p is selected, then — in order to avoid defining patterns which are not used at all — pattern $p-1$ must also be selected

$$\delta_p \leq \delta_{p-1} \quad , \quad p \in \mathcal{P}.$$

Pattern $p-1$ should be used more often than pattern p

$$\mu_p \leq \mu_{p-1} \quad , \quad p \in \mathcal{P},$$

and finally the integrality conditions

$$\delta_p \in \{0, 1\} \quad , \quad p \in \mathscr{P},$$

$$\mu_p \in \{0, M_p\} \cap \mathbb{N}_0 \quad , \quad p \in \mathscr{P}$$

and

$$\alpha_{ip} \in \{0, N^K\} \cap \mathbb{N}_0 \quad , \quad i \in \mathscr{I} \quad , \quad p \in \mathscr{P}.$$

Setting the cost coefficients to

$$C_p^{\mathrm{P}} = 1 \quad , \quad C_p^{\mathrm{R}} = 0 \quad , \quad p \in \mathscr{P}$$

gives the minimal number of patterns, while

$$C_p^{\mathrm{P}} = 0 \quad , \quad C_p^{\mathrm{R}} = 1 \quad , \quad p \in \mathscr{P}$$

minimizes the number of rolls. Mixed costs or cost such as

$$C_p^{\mathrm{P}} = 1 \quad , \quad C_p^{\mathrm{R}} = 0.1p \quad , \quad p \in \mathscr{P}$$

are possible as well. The implementation *TrimMINLP.gms* has been set up for $W_{\mathrm{mr}} = 2360$, $F = 2225$, $K = 12$ and

$$D = (6, 6, 9, 9, 12, 15)$$
$$W = (300, 280, 265, 240, 225, 208).$$

For minimizing the number of patterns we set $N^{\mathrm{P}} = 4$ and obtain a proven minimum of two patterns ($\mu_1 = \mu_2 = 3$) with the global solver ANTIGONE, BARON, COUENNE, and LINDO within seconds. Six rolls are also the result when we minimize the number of rolls (for the setting $N^{\mathrm{P}} = 6$), but in that case we get four patterns ($\mu_1 = \mu_2 = 2$, $\mu_3 = \mu_4 = 1$). Using the mixed costs

$$C_p^{\mathrm{P}} = 1 \quad , \quad C_p^{\mathrm{R}} = 2 \quad , \quad p \in \mathscr{P}$$

leads again to $\mu_1 = \mu_2 = 3$, i.e., two patterns and six rolls.

For up to ten order widths, this MINLP model works well and the gap is closed within minutes. Above ten orders and also somewhat depending on the demand spectrum, solution times increase strongly. However, as we learn in Sect. 14.1.3.2, there are more efficient ways — for instance, column generation techniques — for solving special cases of the problem above than resorting to solving it as a MINLP problem. However, real-world situations may ask for features which destroy the structure of decomposition techniques such as column generation. Typical situations or complicating features are minimizing the number of patterns (that is why we provide this MINLP formulation) resulting in fewer knife changing operations, exact demand fulfillment as in (13.2.1), the restriction that only patterns with a certain minimal filling width F (equivalently, maximal trimloss) as above in (13.2.2) are considered feasible, or the requirement that orders should be covered by a minimal number of patterns.

Let us also refer to another trimloss problem in paper converting industry solved by Westerlund & Isaksson (1998,[572]). The production in this application is much lower but also in this case the entire production in the factory is considered. The production capacity in this industrial application is 100,000 metric tons of converted paper per year (the turnover of the factory being about 100 million euro) but also in this case the MINLP application is sufficiently large, calculated it in the number of MINLP problems solved on a yearly basis, which is about 3500. The trimloss (mainly on laminated paper) was about 4.5% of the production in late 1990s and decreased to about 1.7% (calculated on a yearly basis) after the system was taken into use at the Walki Wisa factory in Pietarsaari, Finland.

A similar technique has been used by Karelathi et al. (2011,[324]) to solve a large scale production planning in the stainless steel industry. The problem involves the cutting of all steel products in the Outokumpu Stainless Steel mill in Tornio, Finland. The factory is one of the largest stainless steel mills in the World, with a yearly production of about 2 million metric tons of stainless steel. The number of trimloss problems solved in the factory is huge as well. The total number of MINLP problems solved per year is about 150,000 and the total number of variables optimized on a yearly basis is about 150 million in this largescale industrial application. The system solving the trimloss problems has been in daily use since beginning of the century (started in 2000). Thus, about 3 million MINLP (trimloss) problems have been solved, including about 3,000,000,000 variables in this giant industrial application, since the system was taking in use.

13.3 Cutting and Packing Involving Convex Objects

The bottoms of large tanks, such as those used in breweries, are bent from a circular base and convexly deformed. They are made of high quality stainless steel and the circles are cut from rectangular metal plates [coils], which are subject, for example, to production restrictions not exceeding $L = S_1^{\mathrm{P}} = 8\mathrm{m}$ and $B = S_2^{\mathrm{P}} = 4\mathrm{m}$ in width. Required circles exceeding these dimensions are divided into convex polygons and then welded together. Since stainless steel is very expensive, the circles or polygons have to be cut from minimal area coils; special models and solutions, which guarantee the non-intersection of convex polygons by means of separating hyperplanes, have been developed by Kallrath (2009,[310]). Alternatively, previously produced rectangles can be stored in inventory, i.e., their dimensions are known. In Rebennack et al. (2009,[456]), this task is solved by a column numbering method in which non-convex NLP problems are solved globally.

Paint to be applied in large quantities to walls is often stored in buckets of elliptical basic shape so that the paint rollers can be as large as possible and the weight of the full paint buckets does not become too great. For the transport of such buckets it can be useful to combine large and small buckets on truck loading areas in order to achieve the best possible area utilization. Since ellipses also approximate well to other geometric figures by overlapping, packing or cutting problems with ellipses or later in three dimensions with ellipsoids are already very interesting. Kallrath & Rebennack (2014,[318]) have developed a separating line approach for this purpose, which has ultimately led to a closed algebraic formulation of the problem and allows to place a given set of ellipses on rectangles with minimum area.

Consider the case that all objects $i \in \mathscr{I}$ are to be placed into a design rectangle $\mathscr{R}$ within a target rectangle of specified size, i.e., $x_d^{\mathrm{P}} \leq S_d^{\mathrm{P}}$, $d \in \{1, 2\}$; the area

$$a = x_1^{\mathrm{P}} x_2^{\mathrm{P}} \tag{13.3.3}$$

of this design rectangle should become minimal.

The modeling process involves representing the objects, ensuring that their interior does not overlap with interior of other objects, and modeling the objective function. The placement of the objects usually allows to *translate* and *rotate* them.

Using appropriate inequalities, we must ensure that no objects overlap (they are allowed to touch) or exceed the boundaries of $\mathscr{R}$. Formulating these conditions for circles alone is simple, as outlined in Sect. 13.3.1.1, but for keeping circles, polygons and ellipses apart, it becomes very challenging to construct separating lines or separating planes in 3D (for instance, when packing ellipsoids as in Kallrath (2017,[313])).

13.3.1 Modeling the Cutting Constraints

We briefly describe the conditions for circles and polygons. For ellipses we refer the reader to Kallrath & Rebennack (2014,[318]).

13.3.1.1 Cutting Constraints for Circles

If our objects are only circles with radii R_i, the non-overlap conditions are quite simple. If $\mathbf{x}_i$ and $\mathbf{x}_{i'}$ denote the centers of the circles i and i', then non-overlap is ensured by

$$(\mathbf{x}_i - \mathbf{x}_{i'})^2 \geq (R_i + R_{i'})^2 \quad , \quad \forall \{ (i, i') \mid i < i' \}. \tag{13.3.4}$$

For n circles we obtain $n(n-1)/2$ inequalities of the type (13.3.4).

In this context (circle cutting or packing), it is worthwhile to briefly refer to Kepler's conjecture: For packing congruent circles or spheres in unrestricted 2D and 3D geometry, the highest packing density δ (fraction of area or volume covered by the objects) is realized for face-centered cubic and hexagonal close packing arrangements with $\delta_{2\mathrm{D}} = \delta_{3\mathrm{D}} = \pi/\sqrt{18}$. About 300 years after Kepler, the 2D case was proven by the Norwegian mathematician Axel Thue (1909,[544]). The much more difficult 3D case was proven, a century after Thue, by Hales (2005,[251]) — interestingly enough, in his proof he has used LP and B&B!

Unlike the situation in Kepler's conjecture, in our case we have non-congruent circles and a rectangular design container with free length and width. The inequalities

$$x_{id} \geq R_i \quad ; \quad \forall\{i,d\} \tag{13.3.5}$$

and

$$x_{id} + R_i \leq x_d^{\mathrm{P}} \leq S_d^{\mathrm{P}} \quad ; \quad \forall\{i,d\} \tag{13.3.6}$$

guarantee that no circle exceeds the boundaries of the rectangle.

For up to ten circles, feasible solutions are found within seconds — global optimality is proven in less than an hour.

13.3.1.2 Cutting Conditions for Polygons

A polygon p is characterized by its K_p vertices $V_{p1}, \ldots, V_{pK_p}$, or by its coordinates $\mathbf{X}_{pk}$, $k = 1, \ldots, K_p$. It suffices to consider convex polygons. Interesting applications of non-convex polygons in the textile industry can be reduced to convex polygons as non-convex polygons can always be built up by convex polygons. Polygons are implicitly and completely described by their centers, the direction from the center to the vertices and distances to the centers as well as by the orientation; see Fig. 13.1. The center $\mathbf{X}_p^0$ of the original, Unmoved, and non-rotated polygon is defined by

$$\mathbf{X}_p^0 = \frac{1}{K_p}\sum_{k=1}^{K_p} \mathbf{X}_{pk} \quad ; \quad \forall\{p\}. \tag{13.3.7}$$

The polygons can now be placed in any position, using the vector $\mathbf{x}_q^0$

$$\mathbf{x}_p^0 = \frac{1}{K_p}\sum_{k=1}^{K_p} \mathbf{x}_{pk} \quad ; \quad \forall\{p\} \tag{13.3.8}$$

as the center. Preservation of shape and orientation is determined by the equation

$$\mathbf{x}_{pk} = \mathbf{x}_p^0 + \begin{pmatrix} \cos\alpha_p & \sin\alpha_p \\ -\sin\alpha_p & \cos\alpha_p \end{pmatrix} \left(\mathbf{X}_{pk} - \mathbf{X}_p^0\right) \quad ; \quad \forall\{pk\}. \tag{13.3.9}$$

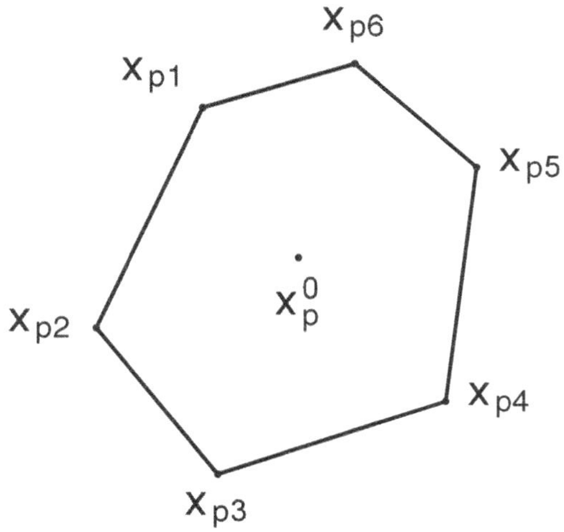

Fig. 13.1 Representation of polygons exploiting the vertices and the center. Reprinted by permission from Springer Nature, Journal of Global Optimization, Kallrath (2009,[310]), Fig. 1

If one wants to avoid trigonometric terms, instead of the rotation angle α_p, we can also use $v_p := \cos\alpha_p$ and $w_p := \sin\alpha_p$ as free variables with bounds $-1 \leq v_{\alpha_p} \leq +1$ and $-1 \leq w_{\alpha_p} \leq +1$, where v_p and w_p are further connected by the trigonometric identity

$$v_p^2 + w_p^2 = 1 \quad ; \quad \forall\{p\}. \tag{13.3.10}$$

For polygons with a symmetry axis we only need to consider angles α_p in the interval from 0° to 180°, i.e., $-1 \leq w_{\alpha_p} \leq 1$. Further symmetry properties can be used for regular polygons.

The condition that a convex polygon completely lies in $\mathcal{R}$ is reduced to the condition that all vertices lie inside $\mathcal{R}$, i.e.,

$$X_{pkd} \leq x_d^{\mathrm{P}} \leq S_{\max,d} \quad ; \quad \forall\{pkd\}. \tag{13.3.11}$$

In order to ensure that two polygons p and p' do not overlap, we have to proceed somewhat differently than with circles, where non-overlap was easily enforced by a distance condition between the centers. With any convex objects in the plane, including polygons and later ellipses, we can take advantage of the fact that every straight line that does not run through the interior of the object has the property that the object lies on one side of the straight line. For every two convex objects that are free of overlap except for points of contact, at least one straight line (in higher dimensions, at least one separating hyperplane) can be found, where both objects lie on different sides of the straight line.

For polygons it follows from the convexity of both polygons that all vertices of p and p' lie on the separating line; see Fig. 13.2. Let p and p' be polygons with vertex sets K_p and $K_{p'}$; the sets K_p and $K_{p'}$ can have a different cardinality. The straight

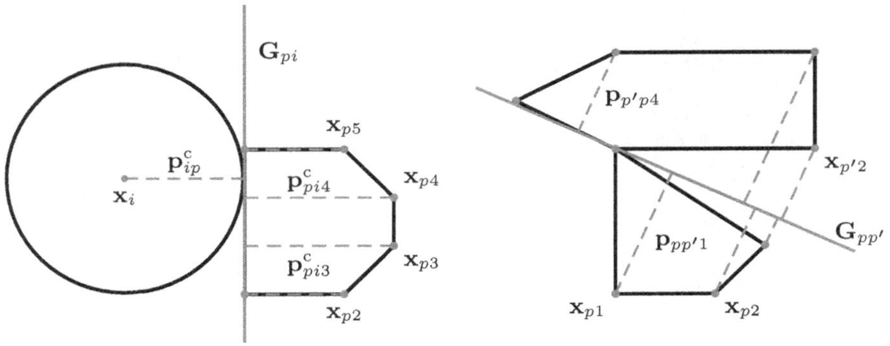

Fig. 13.2 Lines separating polygons and circles (**a**) as well as lines separating polygons (**b**), ensuring that they do not overlap. Reprinted by permission from Springer Nature, Journal of Global Optimization, Kallrath (2009,[310]), Fig. 2

lines $\mathbf{G}_{pp'}$ which separate the polygons p and p' contain the variables $\mathbf{g}_{pp'}$, $\mathbf{m}_{pp'}$, and $\lambda_{pp'}$ for each polygon combination pp' and have the form

$$\mathbf{G}_{pp'} := G_{pp'}(\lambda) = \mathbf{g}_{pp'} + \mathbf{m}_{pp'}\lambda_{pp'} \quad ; \quad \forall\{p, p'|p' > p\}, \tag{13.3.12}$$

where $\lambda \in \mathbb{R}$ parameterizes the straight line. The direction vector $\mathbf{m}_{pp'}$ is normalized to 1, i.e.,

$$\mathbf{m}_{pp'}^2 = 1 \quad ; \quad \forall\{p, p'|p' > p\}. \tag{13.3.13}$$

In general, condition (13.3.13) is numerically problematic because it reinforces the non-convex nature of the problem. In this case, however, (13.3.13) is helpful because we can easily normalize the normal vector $\mathbf{n}_{pp}$ with respect to $\mathbf{G}_{pp'}$ by

$$\left(n_{pp'1}, n_{pp'2}\right)^{\mathrm{T}} = \left(m_{pp'2}, -m_{pp'1}\right)^{\mathrm{T}} \quad ; \quad \forall\{p, p'|p' > p\}. \tag{13.3.14}$$

The K_p connection vectors $\mathbf{p}_{pp'k}$ from $\mathbf{G}_{pp'}$ to the corner point V_{pk} of polygon p are represented by

$$\mathbf{p}_{pp'k} = \mathbf{x}_{pk} - \left(\mathbf{g}_{pp'} + \mathbf{m}_{pp'd}\lambda_{pp'k}\right) \quad ; \quad \forall\{p, p', k|p' > p \wedge k \le K_p\}, \tag{13.3.15}$$

while for the $K_{p'}$ vertices $V_{p'k}$ of the polygon p' the connection vectors are calculated according to

$$\mathbf{p}_{p'pk} = \mathbf{x}_{p'k} - \left(\mathbf{g}_{pp'} + \mathbf{m}_{pp'd}\lambda_{p'pk}\right) \quad ; \quad \forall\{p, p', k|p' > p \wedge k \le K_{p'}\}. \tag{13.3.16}$$

The auxiliary variables $\lambda_{pp'k}$ and $\lambda_{p'pk}$ are needed to calculate the footpoints $\mathbf{p}_{pp'k}$ and $\mathbf{p}_{p'pk}$. The two polygons p and p' are separated by the conditions of parallelism

$$\mathbf{p}_{pp'k} = \Delta_{pp'k}\mathbf{n}_{pp'} \quad ; \quad \forall\{p, p', k|p' > p \wedge k \leq K_p\}, \tag{13.3.17}$$

and anti-parallelism

$$\mathbf{p}_{p'pk} = -\Delta_{p'pk}\mathbf{n}_{pp'} \quad ; \quad \forall\{p, p', k|p' > p \wedge k \leq K_{p'}\}, \tag{13.3.18}$$

where the variables $\Delta_{pp'k}$ and $\Delta_{p'pk}$ measure the distances of the vertices from the straight line.

To ensure that polygons do not intersect with circles, in (13.3.13) to (13.3.18) we replace polygon p' by circle i and make the following changes: The variables $\Delta_{p'pk}$ in (13.3.18) are fixed to the radii R_i, i.e., the straight line is tangent to the circle, Δ^{c}_{pik} corresponds to $\Delta_{pp'k}$ in (13.3.17), which now has the form

$$\mathbf{p}^{c}_{pik} = \Delta^{c}_{pik}\mathbf{n}_{pi} \quad ; \quad \forall\{p, i, k|k \leq K_p\}, \tag{13.3.19}$$

and (13.3.18) becomes

$$\mathbf{p}^{c}_{ip} = -R(i)\mathbf{n}_{ip} \quad ; \quad \forall\{ip\}. \tag{13.3.20}$$

Feasible configurations are found within minutes — global optimality is proven only for a few polygons.

13.3.2 Problem Structure and Symmetry

The cutting problems formulated above present themselves as NLP problems with the following non-convex aspects:

1. Firstly, the bilinear objective function (13.3.3). If the length or width of $\mathscr{R}$ is fixed (strip packing), this is reduced to a linear objective function.
2. The condition of non-overlap leads to a geometric situation which obviously results in a non-convex region. To understand this, imagine $\mathscr{R}$ with an already fixed object i_f. The allowed range of the remaining objects $i \in \mathscr{I}\backslash\{i_f\}$ concerning i_f is $\mathscr{R}$ without the range covered by i_f. If the objects are only circles, then (13.3.4) is the relevant inequality with only quadratic and bilinear terms. Also in the case of polygons and ellipses, the problem formulation does not require more than quadratic equalities or inequalities. The normalization equations (13.3.13) are particularly problematic.

Since all non-convex terms are quadratic in nature, it is not surprising that the solvers BARON, GLOMIQO, or ANTIGONE perform particularly well in this problem class and is the only global solver capable of closing the gap between the upper and lower bound for most problems, i.e., really calculating the global optimum. Apart from non-convexity, the presented problems are still difficult because of their hidden combinatorial character. So far, combinatorial problems have only been encountered in connection with mixed integer problems. The combinatorial structure of the traveling salesman problem results from the possible permutations for the order of the cities to be visited. The situation is similar here, which can be illustrated as follows. Suppose you allocate each object i with the pearls of a pearl necklace, which is then placed through the rectangle or meanders through $\mathscr{R}$. By the path of this pearl chain takes through $\mathscr{R}$, the objects can take many relative positions to each other, i.e., object j can be positioned to the left, below, right, or above object i. The objects can be placed in the same position as the pearl chain. Polygons and ellipses also have an orientation. Viewed in this way, the degrees of freedom consist of the order of the objects along the chain, the positions on the chain, and the path the chain takes through $\mathscr{R}$. The degrees of freedom are the same as the degrees of freedom of the objects along the chain. Strictly speaking, the order could be treated by complete enumeration but is only possible for small instances. The positions on the chain, represented by the centers of the objects, as well as the orientation are mapped in the NLP model. The path of the chain through $\mathscr{R}$ is a difficult problem in itself. The difficulty of the problem partly results from symmetry. Symmetry can be reduced a little by placing the center of an arbitrary object in the first quadrant of $\mathscr{R}$. For example, if we choose circle i_*, it means

$$x_{i_*d} \leq \frac{1}{2} x_d^{\mathrm{P}} \quad ; \quad \forall \{d\}. \tag{13.3.21}$$

For a polygon p_*, the inequality

$$x_{p_*d}^0 \leq \frac{1}{2} x_d^{\mathrm{P}} \quad ; \quad \forall \{d\} \tag{13.3.22}$$

has a positive effect on the computing time. Symmetry degeneration by the presence of congruent objects, i.e., those that are identical in shape and area, can be reduced by assigning them to the same congruence class I^{co}. For objects i and i' in the same congruence class we apply the ordering inequalities

$$x_{i1} + 5x_{i2} \leq x_{i'1} + 5x_{i'2} \quad ; \quad \forall \{(i, i') | i < i' \wedge I_i^{\mathrm{co}} = I_{i'}^{\mathrm{co}}\}. \tag{13.3.23}$$

The choice of coefficients is rather arbitrary. Another degree of degeneracy that can cause problems for a global solver is associated with free objects. These are mostly smaller objects that can move between the larger ones without touching them and without changing the value of the objective function, i.e., the area a of the rectangle. This is unproblematic for cutting problems, but rather undesirable for packing problems. One way of dealing with this problem is to including the center coordinates in the objective function to be minimized; this results in the selected objects being placed as far to the left as possible at the left bottom of the coordinate origin.

13.3.3 Some Results

Kallrath (2009,[310]) contains results for circles and polygons. Optimality could be proven for up to 10 circles, for small numbers of polygons as well. The results obtained by Misener & Floudas (2012,[406]) with GLOMIQO calculate optimal solutions in much shorter time. This can also be seen for smaller problems with ellipses.

13.4 Summary and Recommended Bibliography

In this chapter we have considered various non-convex, nonlinear optimization problems and formulations requiring global optimization techniques. Thus the reader should now be familiar with:

- the flavor of applications in the global optimization; and
- limits of solver used in deterministic global optimization.

The *Handbook of Test Problems of Local and Global Optimization* by Floudas et al. (1999,[193]) is still a valuable source to get familiar with typical problems in the field.

13.5 Exercises

1. Use a global NLP solver to solve Exercise 1 in Sect. 12.9.
2. Use a global NLP solver to solve Exercise 2 in Sect. 12.9.
3. Use a global NLP solver to solve Exercise 3 in Sect. 12.9.

Chapter 14
Polylithic Modeling and Solution Approaches

This chapter deals with polylithic modeling and solution approaches. Such approaches allow to considerably extending the set of solvable practical problems both in their quality (structure) and in their size (the number of variables and constraints). These approaches are illustrated by problems from paper industry, which were solved with the help of polylithic modeling and solution approaches. In detail, roll minimization based on column generation, simultaneous minimization of waste, and the number of used patterns, as well as format production, are treated.

14.1 Polylithic Modeling and Solution Approaches (PMSAs)

Many practical mixed integer optimization problems are difficult to solve. Instead of tackling these difficult problems directly as monolithic problems, they can also be solved equivalently as sequences of models. This situation leads to polylithic modeling and solution approaches including, for example,

1. *column generation* [cf. Lübbecke & Desrosiers (2005,[376]) or Sect. 2.6.3],
2. *Branch and Price*, an extension of column generation to mixed integer problems [cf. Sect. 3.3.4 or Barnhart et al. (1998,[49])],
3. *Benders Decomposition* — see Sect. 14.1.3.2,
4. Evaluation of auxiliary problems to derive tighter bounds for the original problem, which in turn leads to an easier solution of the original problem,

Supplementary Information The online version contains supplementary material available at https://doi.org/10.1007/978-3-030-73237-0_14.

J. Kallrath, *Business Optimization Using Mathematical Programming*, International Series in Operations Research & Management Science 307, https://doi.org/10.1007/978-3-030-73237-0_14

5. *hybrid methods* in which constructive heuristics are used together with exact optimization algorithms, or
6. *lexicographic goal programming* for solving multi-criteria optimization problems (see Sect. 5.4).

To the advanced reader, the field of *MatHeuristics* may appear closely related to PMSAs. Maniezzo et al. (2009,[383]) provide a good introductory reference to MatHeuristics. MatHeuristics refers to a broad class of hybrid algorithms in which exact approaches and metaheuristics are combined. So, obviously, there is a strong focus on or a connection to metaheuristics. PMSAs are broader — as we will see below — and, in some sense, even include MatHeuristics.

One might feel inclined to view PMSAs from two main angles: modeling (4,5,6) and solving techniques (1,2,3). Solving seems to be more generic as the solving techniques usually have an advanced theoretical basis and can be applied to many problems. But a deeper look shows that they exploit or require a lot of structure and a slight change in the model (additional or different constraints), and they do not work — thus, they are not strictly generic. Polylithic modeling is, definitely, less generic — and results in custom or tailor-made solution approaches. They follow the principal *anything goes*: submodels need not be exact models; they may be simplifications. Alternatively, the master is an approximation, solved by a tailored solution technique, and only submodels are solved to optimality. In practice, modeling and solving are interconnected and require deep knowledge and experience from the modeler. Therefore, the two viewing angles are possible but somewhat artificial.

14.1.1 Idea and Foundations of Polylithic Solution Approaches

Based on the Greek term *monolithos* (monolithic, a stone consisting of only one block), Kallrath (2009,[309]; 2011,[311]) has used the corresponding term *polylithic* for modeling and solution approaches in which mixed integer or non-convex nonlinear optimization problems are solved with the help of customized methods using several linked models or algorithms.

14.1.1.1 Monolithic Models and Solution Approaches

A monolithic model simply consists of a model with data, variables and constraints. It is solved by *one* call to a solution algorithm solving an LP, MILP, NLP, or MINLP problems — for the problems we have presented so far in this book, this use of a general purpose solver is the standard. This keeps the structure of the model and its solution relatively simple and clear. When using an algebraic modeling language, the possible constraints on the variables are set only once. The model output and solver results flow directly into an attractive, well-structured output report or a visualized presentation of results.

Fig. 14.1 Polylithic versus monolithic modeling. In contrast to monolithic methods (left), polylithic models consist of a set of models (right) linked to one another with regard to their inputs and outputs. Produced for this book by Diana Kallrath, Copyright ©2020

14.1.1.2 Polylithic Modeling and Solution Approaches

In contrast to monolithic methods, polylithic models consist of a set of models linked to one another with regard to their inputs and outputs, i.e., model $m + 1$ can use results of the first m models; see Fig. 14.1. This can be used to initialize certain variables or to put tighter bounds on variables. Examples of polylithic solutions are decomposition methods, such as the column generation method [cf. Gilmore & Gomory (1961,[220]) or Lübbecke & Desrosiers (2005,[376])], Branch and Price [cf. Barnhart et al. (1998,[49])], or hybrid methods [cf. Pochet &Wolsey (2006,[441])], in which constructive heuristics and/or local search and enhancement procedures are combined with exact MIP algorithms to calculate allowable points or good lower and upper bounds. In a natural way, tailor-made algorithms are obtained. However, this requires a great care and attention in programming with regard to the following points:

1. Bounds and initial values of variables must be tracked and updated very carefully.
2. The results of discrete variables are usually not integer but deviate slightly from them, e.g., 0.9999987 instead of 1. However, as these result values are still used, it not only makes sense but is also often necessary to round them. However, this can lead to infeasibilities. Caution is advised.
3. Variables and constraints in different models should be declared as local objects; otherwise, misunderstandings should be avoided by choosing names.
4. The maintenance effort for polylithic methods is considerably higher than for monolithic ones; the extension of additional functionalities in model and solution approach can be complex.

PMSAs have a wide range of applications; they significantly increase the amount of practical problems that we can solve in a reasonable amount of time. We find them in problem-specific preprocessing, in general mathematical algorithms and in structured primal[1] heuristics and hybrid methods. Customized polylithic solution approaches with thousands or millions of solution calls represent a particular challenge for algebraic modeling languages. Warm or hot starts, which can be used to prevent the model from having to be translated and compiled over and over again, become very important.

PMSAs are recommended when the problem cannot be solved directly — they represent a powerful but last resort. They can also be seen as a kind of bridge technology that can be replaced if significant improvements are made in the area of solution algorithms.

14.1.2 Problem-Specific Preprocessing

In Sect. 9.1, several preprocessing methods have already been presented, and it has been pointed out that commercial software vendors use these methods in their solvers but often do not reveal exactly which methods and how they have been implemented. In addition to this lack of transparency, another argument for the development of one's own preprocessing techniques and the exploitation of problem-specific structures and properties is that problem owners usually know more about their problems than the software can automatically identify.

14.1.2.1 Dynamic Reduction of Big-M Coefficients

In Chap. 9, Big-M coefficients (in short, Big-Ms) were used at various points. Using an illustrative production planning example, the significance of minimal Big-Ms in the activation or deactivation of inequalities such as

$$x \leq X + M(1 - \delta)$$

with a continuous variable x and a binary variable δ is shown in *dynBigM* (MCOL). In this implementation, M is reduced iteratively in a sequence of models, cf. Kallrath (2009,[309]). The number of nodes in a Branch-and-Bound algorithm can thus be significantly reduced. In general, this method is effective if the presolving techniques built into the commercial solver do not lead to a model tightening.

To illustrate the approach, let us consider a simple production planning example in which it is not obvious how one should choose the value M explicitly and *a priori*. Apart from this, we want to use only a simple MILP solver not exploiting

[1] Here, the term *primal* refers to the primary problem, i.e., the optimization problem in its original form. Primary heuristics and procedures provide only primal allowable points. Only by the addition of dual information, an optimality proof is possible.

any presolving techniques. We illustrate the idea of dynamic improvement of Big-M formulations by considering a small production example with 7 products $i = 1, \ldots, 7$ of sales prices $Y_i = 1 + i$ in \$/tons and individual machine production capacities C_i in tons/day. We consider only one specific day. Therefore, we neglect the index for different days. The total demand to be satisfied per day is D tons, i.e.,

$$\sum_i x_i = D, \tag{14.1.1}$$

where the variable x_i denotes the amount in tons of product i produced on that day. Production is only possible if product i has been selected, i.e.,

$$x_i \leq M_i \delta_i \quad , \quad \forall i, \tag{14.1.2}$$

where M_i are sufficiently large numbers. An obvious choice is to set $M_i = C_i$.

Another complication is that the amounts produced for selected products should not differ by more than Δ tons. If δ_i denotes the selection binary variable, we can formulate this condition by

$$\left|x_{i_1} - x_{i_2}\right| \leq \Delta + N_{i_1 i_2}(2 - \delta_{i_1} - \delta_{i_2}) \quad , \quad \forall\{(i_1, i_2)|i_2 \neq i_1\}, \tag{14.1.3}$$

where $N_{i_1 i_2}$ are sufficiently large numbers. Actually, the coefficients $N_{i_1 i_2}$ also represent Big-M values; a tight choice is

$$N_{i_1 i_2} = \max\{M_{i_1}, M_{i_2}\} - \Delta. \tag{14.1.4}$$

Note that inequality (14.1.3) only becomes active if both products i_1 and i_2 have been selected. The absolute value term in (14.1.3) is replaced by two inequalities

$$x_{i_1} - x_{i_2} \leq \Delta + N_{i_1 i_2}(2 - \delta_{i_1} - \delta_{i_2}) \quad , \quad \forall\{(i_1, i_2)|i_2 \neq i_1\}, \tag{14.1.5}$$

and

$$x_{i_2} - x_{i_1} \leq \Delta + N_{i_1 i_2}(2 - \delta_{i_1} - \delta_{i_2}) \quad , \quad \forall\{(i_1, i_2)|i_2 \neq i_1\}. \tag{14.1.6}$$

If product i is selected, P_i people need to be allocated to its production. As there are only $P_{\max}$ people available, we obtain the knapsack constraint

$$\sum_i P_i \delta_i \leq P_{\max}. \tag{14.1.7}$$

The idea of dynamically improving Big-M, i.e., to reduce the value M_i, is to solve a sequence of auxiliary problems generated by fixing $\delta_i = 1$ and maximizing x_i; we call this maximum X_i. This value allows us to update M_i from $M_i^{(1)} = C_i$ to $M_i^{(2)} = X_i$. The method is effective if $X_i < C_i$. We could try a second round: the next step could be to tighten $N_{i_1 i_2}$ and then go back to M_i.

As a numerical example, we use sales prices $Y_i = 1 + i$, $\Delta = 20$ and

$$C = [100, 80, 90, 60, 100, 110, 100], \quad P = [11, 6, 6, 5, 5, 4, 1],$$
$$D = 120, \quad P_{\max} = 19. \tag{14.1.8}$$

The sales prices just serve to provide some reasonable objective function. In this specific case, our procedure relies on setting all the binary variables to 1. In the general case, this may not even result in a feasible problem. Therefore, one has to keep in mind that such approaches are very problem-specific and are always tailor-made. Exploiting the problem structure, the use of such techniques can be very effective as we show below.

To demonstrate the effect of dynamically improving the Big-M coefficients, we avoid the strong commercial MILP solvers as their presolving strategies are very strong but rather use the public domain MILP solver `CoinGLPK` by Makhorin (2009,[382]) and set $M_i = C_i$ and $N_{i_1 i_2} = \max_i C_i = 110$. We obtain the optimal solution after 40 nodes. The first round of dynamic improving $M_i^{(1)} = C_i$ yields

$$M_i^{(2)} = [70, 70, 70, 60, 70, 70, 70].$$

Using these improved values, $M_i^{(2)}$ results in the optimal solution in just 21 nodes. A further round on M_i does not lead to any improvement. Tightening $N_{i_1 i_2}$ has not reduced the number of iterations. We have also applied CPLEX to this problem. CPLEX solves the problem to optimality in the root node for both $M_i^{(1)} = C_i$ and $M_i^{(2)}$; however, we can see a difference in the Simplex iteration count of 16 for the original formulation, and of 2 for the tightened formulation.

To summarize the dynamic model improvement such as Big-M formulation or models containing integer variables, the method is effective when the presolving techniques of the commercial solver are unable to tighten the formulation, but only a sequence of tailor-made auxiliary problems can do so. The approach can be easily implemented in any modeling language.

14.1.2.2 Bound Tightening for Integer Variables

With auxiliary problems and PMSAs, tightened upper limits can be calculated for integer variables. An example is the interval reduction of the number of batches in Lin et al. (2005,[365], Table 4), in which a reactor portfolio problem is solved. The objective function of the auxiliary problem was to maximize the number of batches. In the context of integer variables, this is highly recommended as it can significantly improve the lower bound of the original minimization problem. This becomes clear when you consider the difference in the upper bound of an integer variable $\alpha \leq 3$ instead of $\alpha \leq 3.4$.

14.1.2.3 Data Consistency Checks

In Sect. 2.9, we focused on data collection, while in Sect. 5.5.5, we have discussed the problem of consistent and obtainable data when modeling real-world problems. In complex applications, it is not easy to ensure or check the consistency of the input data, as problems usually only become apparent when several constraints need to be fulfilled at the same time. Automatic feasibility checks are therefore essential in production planning and scheduling, both in supply network planning and in energy industry. In process industry, for example, such tests in scheduling models could be simple tests for identifying inconsistent batch sizes. Useful are also auxiliary models for checking the consistency of state task networks and topology, compatibility of production facilities, warehousing, and demand. Finally, we can exploit advanced procedures such as single job analysis to accurately solve the scheduling problem for each job, identifying unavoidable underproduction or delays.

In particular, unavoidable underproduction or delays should be rigorously addressed using the multi-criteria optimization (first, minimizing violations and then solving the original problem) outlined in Sect. 5.4 rather than penalizing only underproduction or delays. As discussed by Kallrath & Maindl (2006,[316], pp. 344 f.), penalties with coefficients that cannot be interpreted economically lead to numerical problems and complicate the interpretation of the integrality gap.

14.1.3 Mathematical Algorithms

Here, we consider general mathematical algorithms, namely, decomposition methods, Branch and Bound, Branch and Cut, dynamic programming (DP), and goal programming (GP) for multi-criteria optimization problems.

14.1.3.1 Branch-and-Bound and Branch-and-Cut Methodologies

Most commercial solvers use Branch&Bound (B&B) to solve MIP or non-convex NLP or MINLP problems, usually branching on variables. There are special branching strategies for semi-continuous variables (SCVs) or special ordered sets. Since not all solvers support these structures, Kalvelagen (2003,[321]) has coded the branching method for SCVs into GAMS to solve an MINLP problem with additional SCVs. Problem-specific branching on other structures such as constraints can rarely be found in commercial solvers — another area for PMSAs.

Automatic addition of inequalities can be found in B&C, Outer Approximation, and goal programming as discussed in Sect. 5.4. Useful resources are Karuppiah & Grossmann (2008,[328]) for global optimization techniques of non-convex MINLPs with structures suitable for decomposition, Rebennack et al. (2011,[459]) for a detailed tutorial on B&C methods for solving the stable set problem (SSP), and Rebennack et al. (2011,[458]) for progress in graph reduction methods within B&C

methods for the SSP. Sometimes, other problem-specific cuts are available that can be added to the problem either statically or dynamically. The dynamic case is, of course, the more interesting one and is illustrated by several examples in the model libraries of AMLs such as `GAMS` (e.g., MCOL/GAMSLIB/*tsp5.gms*) or `Mosel`. For example, the tour elimination constraints are dynamically added to the core model of the traveling salesman problem (TSP) model. While this technique only works for small TSPs, it shows how dynamic addition of inequalities can be applied to other problems. For the TSP, one can also learn about including cut generation and callbacks for B&C.

14.1.3.2 Decomposition Methods

Decomposition techniques break down the problem into smaller problems that can be solved sequentially or in parallel. There are not only standardized techniques such as Dantzig–Wolfe decomposition (a special column generation method) to solve LP problems with special structures or Benders Decomposition [cf. Benders (1962,[64]) or Floudas (1995,[189], Chap. 6)] but also structure-utilizing, customized methods. Nowak (2005,[423]) gives a good overview on relaxation and decomposition methods for MINLP. Note that `SAS/OR` contains a decomposition solver, based on the PhD thesis of Galati (2009,[205]), for LP and ILP. SAS is the first commercial modeling and optimization software company offering this functionality saving the effort of developing one's own decomposition algorithms. This may lead to progress comparable to the situation with B&C thirty or forty years ago when MILP solvers did not yet have generic B&C technology embedded.

An alternative to decomposition is an aggregation approach (cf. [372, 373] and the references therein), where the original large problem is (iteratively) substituted for a smaller aggregated problem. If the original problem has a special structure, iterative aggregation can be combined with the decomposition approach, cf. [370].

Benders Decomposition (BD)

This technique was developed by Benders (1962,[64]) to break down MILP problems into a sequence of master problems (MP) and subproblems. The MPs are IP problems with discrete variables, while the subproblems are LPs. The MP gives an integer solution to the subproblem, which either proves the optimality of this solution with respect to the original problem or otherwise returns a permissibility or optimality inequality to the MP. BD has also been applied to very large LPs — preferably in the world of stochastic optimization, where BD is also known as the L-shaped method; cf. Birge & Louveaux (2000,[79]). If BD is applied to NLPs it is a *generalized BD*. If BD is also applied to the subproblems, one speaks of a *nested BD*.

Column Enumeration (CE) and Column Generation (CG)

The term *column* is derived from the language environment of linear programming. Since the 1950s, *columns* have stood for variables and *rows* for constraints. In the Simplex algorithm, non-basic variables are examined as candidates for possible basic variables in the pricing step. In dynamic variants of the Simplex algorithm, these non-basic variables are generated dynamically.

In the context of CE and CG, the term *columns* has a broader meaning and stands for specific objects that occur in a model; cf. Sect. 2.6.3. In cutting stock problems, columns usually stand for patterns, in network flow problems for permissible paths through a network and in vehicle routing problems for subsets of orders that can be assigned to a vehicle or a feasible tour, whereby feasible tours as in Desrochers et al. (1992,[157]) are generated by solving the shortest path problems. Heinz et al. (2012,[256]) have solved steel mill slab design problems and compare various methods to CG.

The basic idea of CG is to split an optimization problem $\mathscr{P}$ into a master problem $\mathscr{P}_{\mathrm{M}}$ and a subproblem $\mathscr{P}_{\mathrm{U}}$. The decomposition usually has a natural interpretation. Nonlinear problems $\mathscr{P}$ can be completely reduced to solving LPs, as in the case of trimloss minimization. Decompositions have to be carefully constructed so that both problems $\mathscr{P}_{\mathrm{M}}$ and $\mathscr{P}_{\mathrm{U}}$ can be solved in short time. The most famous example is probably the Gilmore and Gomory (1961,[220]; 1963,[221]) column generation method for calculating the minimum number of rolls to meet the demand for smaller order rolls of given width. The monolithic version of this problem leads to an MINLP problem with a very large number of variables.

CG provides an optimal solution to the LP problems, and it can be used to solve the LP relaxation of MILP/ILP problems. The idea of CG is to generate not all possible columns but the favorable ones only. This can be achieved by solving the pricing problem. At each iteration, solving the pricing problem generates a new column by optimizing the reduced cost, and therefore the generated column would be the favorable one. Then, the restricted master problem selects a combination of columns to create a feasible solution. It is also possible to solve certain types of subproblems (i.e., generate columns) using heuristics which reduce the solution time significantly.

In simple cases, it is possible to create all columns a priori (CE). If this is not possible, starting from an initial set of columns, new columns are dynamically added to the problem; good review articles for dynamic CG are, for example, Barnhart et al. (1998,[49]) or Lübbecke & Desrosiers (2005,[376]).

CE can be regarded as a special variant of CG that is applicable when only a small number of columns exist or when the consideration of a small number of columns is sufficient. This is the case, for example, in practical trimloss problems, as usually only patterns with a small strip loss are accepted; this already excludes most patterns. CE naturally leads to a selection problem of existing columns; this selection problem is also called partitioning model. Schrage (2006,[492], Sect. 11.7) gives illustrative examples from the application areas: decomposition, matching, overlapping, partitioning, and packing problems. Rebennack et al. (2009,[456]) have developed a general framework for decomposing a given set of objects into subsets and subsequently assigning them to existing objects and applied it to a trimloss problem in the metal industry. Despite the limitation of the number of columns, CE offers some important advantages. These include the easy applicability to MIP problems, the lack of need to solve a pricing problem to create new columns, and the low implementation effort.

Column Generation and Branch-and-Price Methodology

CE can be regarded as a special variant of CG. Generally speaking, we can use CG for solving well-structured MILP problems with several hundreds of thousands or millions of variables, i.e., columns. These problems lead to very large LP problems when the integer conditions are relaxed. If the LP problem contains so many variables (columns) that it cannot be solved with a direct LP solver (revised Simplex, interior-point method), the so-called *master problem* (MP) starts with a small subset of variables and becomes the *restricted MP* (RMP). After solving the RMP, new variables are identified in a *pricing problem* (PP). This step corresponds to the identification of non-basic variables that are included in the basis of the Simplex procedure and led to the designation column generation. The RMP is now solved with the new set and number of variables. The procedure terminates when no new variables can be identified using the pricing problem. The simplest variant of the CG procedure can be found in the Dantzig–Wolfe decomposition (Dantzig & Wolfe 1960,[144]). Note that CG would not provide an integer solution. However, at the last step, one can solve the restricted master problem as an integer problem to find an integer solution. However, this would be a heuristic and would not guarantee optimality. CG can guarantee optimality only for the LP models.

Gilmore and Gomory (1961) were the first to apply the idea of CG to an IP problem: the trimloss problem in paper industry. In this case, the PP is a knapsack problem, i.e., also an IP problem — here, in the knapsack problem, the new columns are created. This problem is unique in the sense that the gaps created by the RMP — here patterns — are sufficient to create the optimal integer solution to the problem. This is generally not the case as mentioned above. If both problems, $\mathscr{P}_{\mathrm{M}}$ and $\mathscr{P}_{\mathrm{U}}$, contain integer variables, the CG is embedded in a B&B procedure; this is therefore called *Branch and Price* (B&P); cf. Barnhart et al. (1998,[49]) and Savelsbergh (2001,[486]). In a nutshell, B&P is IP with CG. During the branching process, new columns are created — another reason for the term Branch and Price. B&P (often also linked to Branch and Cut) is tailor-made implementations with decomposition structures. Despite the considerable implementation effort, B&P is one of the most powerful techniques for solving MIP problems with CG. A list of successful applications in various areas can be found in Kallrath (2008,[307]).

14.1.3.3 Lagrange Relaxation

Lagrange relaxation (LR) — cf. Geoffrion (1974,[213]), Fisher (1985,[186]), Martin (1999,[388]), or Guignard (2003,[246]) — is not so much a mathematical method to determine feasible points, but rather an important method for the improvement of the lower bounds of a minimization problem, using dual information. The Branch-and-Bound method used in Sect. 3.8.6 for solving MILP problems is initially based on a relaxation of the LP problem; the relaxation results from neglecting the integrality condition of the discrete variable. For combinatorial problems such as the traveling salesman problem, however, this domain relaxation is very weak. The LR performs much better, as we will demonstrate later with the *generalized assignment problem* introduced in Sect. 7.1.3. LR is a widely used technique, especially in discrete optimization. The basic idea here is to neglect problematic restrictions and to include them in the objective function multiplied by a Lagrange multiplier to be determined instead. If — for fixed Lagrange multipliers — the resulting optimization problem is easier to solve, there is good hope that the lower bound can be efficiently improved by iteratively determining the Lagrange multipliers using a subgradient method. Let us consider the (mixed) integer linear program P:

$$\min_{\mathbf{x}} z = \mathbf{c}^{\mathrm{T}}\mathbf{x}$$

$$\mathsf{A}\mathbf{x} \geq \mathbf{b}$$

$$\mathsf{B}\mathbf{x} \geq \mathbf{d}$$

$$\mathbf{x} \geq 0 \quad , \quad x_j \in \mathbb{N}_0 \quad \forall j \in J.$$

First of all, it must always be decided which constraint is to be dualized. If we assume that the problem

$$\min_{\mathbf{x}} z = \mathbf{c}^{\mathrm{T}}\mathbf{x}$$

$$\mathsf{B}\mathbf{x} \geq \mathbf{d}$$

$$\mathbf{x} \geq 0 \quad , \quad x_j \in \mathbb{N}_0 \quad \forall j \in J$$

is relatively easy to solve, it is a good idea to relax $\mathsf{A}\mathbf{x} \geq \mathbf{b}$; that is what we want to prepare in the following. The set Γ

$$\Gamma := \{\mathbf{x} | \mathsf{B}\mathbf{x} \geq \mathbf{d}, x_j \in \mathbb{N}_0 \quad \forall j \in J\}$$

is the set of all integer feasible points of the inequality $\mathsf{B}\mathbf{x} \geq \mathbf{d}$. Exploiting Γ, we can also represent P as

$$\min_{\mathbf{x}} z = \mathbf{c}^{\mathrm{T}}\mathbf{x}$$

$$\mathsf{A}\mathbf{x} \geq \mathbf{b} \quad , \quad \mathbf{x} \in \Gamma.$$

If P has an optimal solution and conv(Γ) denotes the relaxation of Γ or the convex hull of Γ, which is obtained by replacing the discrete point set Γ with its convex hull, problem D

$$\begin{aligned} L(\mathbf{u}) &= \min\left\{\mathbf{c}^{\mathrm{T}}\mathbf{x} - \mathbf{u}^{\mathrm{T}}(\mathsf{A}\mathbf{x} - \mathbf{b})\,|\mathbf{x} \in \Gamma\right\} \\ &= \min\left\{\mathbf{u}^{\mathrm{T}}\mathbf{b} + \mathbf{c}^{\mathrm{T}}\mathbf{x} - \mathbf{u}^{\mathrm{T}}\mathsf{A}\mathbf{x}|\mathbf{x} \in \Gamma\right\} \\ &= \min\left\{\mathbf{b}^{\mathrm{T}}\mathbf{u} + \left(\mathbf{c} - \mathsf{A}^{\mathrm{T}}\mathbf{u}\right)^{\mathrm{T}}\mathbf{x}|\mathbf{x} \in \Gamma\right\} \end{aligned}$$

satisfies the equality

$$\min\{\mathbf{c}^{\mathrm{T}}\mathbf{x}|\mathsf{A}\mathbf{x} \geq \mathbf{b}, \mathbf{x} \in \mathrm{conv}(\Gamma)\} = \max\{L(\mathbf{u})|\mathbf{u} \geq 0\} = \max D,$$

i.e., the optimal solution of the relaxed problem can be obtained from the optimal solution of the dual problem associated with problem P. If $\bar{\Gamma}$ designates the LP relaxation of Γ, i.e., if we neglect the integrality condition of **x** and relax(P) the LP relaxation of P, then due to the relaxation properties, the following applies

$$\Gamma \subseteq \mathrm{conv}(\Gamma) \subseteq \bar{\Gamma}$$

resulting in the chain of inequalities

$$\min[\mathrm{relax}(P)] \leq \min[\mathrm{conv}(P)] = \max D \leq \min P,$$

i.e., for every non-negative **u**, the LR gives us a lower bound of P, which is also deduced directly from the structure of $L(\mathbf{u})$, because for every feasible point **x** and non-negative **u**, a positive term $\mathbf{u}^{\mathrm{T}}(\mathsf{A}\mathbf{x} - \mathbf{b})$ is subtracted from the objective function $\mathbf{c}^{\mathrm{T}}\mathbf{x}$. In this sense, problem D to be solved is a relaxation of P. In the case $\min[\mathrm{relax}(P)] = \min P$, LP and Lagrange relaxation agree; in other cases, the LR is usually stronger than the LP relaxation. $L(\mathbf{u})$ is a piecewise linear and concave function; thus, it can be differentiated *w.r.t.* **u** up to a finite set of discrete points $\bar{\mathbf{u}}$. It is the lack of differentiability at a finite number of points that leads to the concept of the subgradient. A real vector $\boldsymbol{\gamma} \in \mathbb{R}^m$ is a subgradient of $L(\mathbf{u})$ in a point $\bar{\mathbf{u}}$ exactly when

$$L(\mathbf{u}) \leq L(\bar{\mathbf{u}}) + (\mathbf{u} - \bar{\mathbf{u}})^{\mathrm{T}}\boldsymbol{\gamma}.$$

If $L(\mathbf{u})$ is differentiable *w.r.t.* to **u**, $\boldsymbol{\gamma}$ corresponds to the direction vector of the tangent line. In the non-differentiable case, $L(\bar{\mathbf{u}}) + (\mathbf{u} - \bar{\mathbf{u}})^{\mathrm{T}}\boldsymbol{\gamma}$ corresponds to a straight line that matches $L(\mathbf{u})$ in, but otherwise overestimates $L(\mathbf{u})$; there are infinitely many of them. The set of all subgradients $\boldsymbol{\gamma}$ in one point is also called subdifferential $\mathscr{G}(\bar{\mathbf{u}})$ of $L(\mathbf{u})$ in $\bar{\mathbf{u}}$. In the differentiable case, the subdifferential contains only one subgradient: the gradient $\nabla L = \partial L/\partial \mathbf{u}$. With the concept of the subgradient, the optimum of the non-differentiable function $L(\mathbf{u})$ can be calculated exploiting the subgradient method described in the following section.

Subgradient Method

The subgradient method extends the method of steepest ascent to piecewise linear convex or concave functions such as $L(\mathbf{u})$. Although a step toward a subgradient does not necessarily lead to an improvement of $L(\mathbf{u})$, it makes sense to follow the direction of a subgradient and proceed iteratively in the subgradient procedure as follows:

Step 0: Solution of the LP relaxation, whereby the constraints to be relaxed must be included in the model. The dual values of the constraints to be relaxed follow from the solution of this problem. In addition, a primary solution is required from which an upper bound can be calculated.

Step 1: An iteration counter k is initialized to $k = 0$. A vector $\mathbf{u}^0 > 0$ must be known; here, for example, the dual variables associated with the dualized constraints can be used from LP relaxation. Finally, an accuracy limit $\epsilon > 0$ is defined.

Step 2: Calculate the subgradient $\boldsymbol{\gamma}^k = \mathbf{g}(\mathbf{x}^k) = \mathbf{b} - \mathsf{A}\mathbf{x}^k$, where $\mathbf{g}(\mathbf{x}^k)$ in the differentiable case denotes the gradient of $L(\mathbf{u})$ in $\mathbf{u}^k$, and $\mathbf{x}^k \in \Gamma(\mathbf{u}^k)$, i.e., for fixed $\mathbf{u} = \mathbf{u}^k$, one has to calculate the value of the Lagrangian function $L(\mathbf{u}^k)$ and $\mathbf{x}^k$ as the optimal solution of the dual problem

$$L^k = L(\mathbf{u} = \mathbf{u}^k) = \min_{\mathbf{x}} \left\{ \mathbf{b}^{\mathrm{T}}\mathbf{u} + \left(\mathbf{c} - \mathsf{A}^{\mathrm{T}}\mathbf{u}\right)^{\mathrm{T}} \mathbf{x} \right\}$$

$$\mathsf{B}\mathbf{x} \geq \mathbf{d}, x_j \in \mathbb{N}_0 \quad \forall k \in k.$$

Step 3: Set $\mathbf{u}^{k+1} := \max\{0, \mathbf{u}^k + t_k \boldsymbol{\gamma}^k\}$ with a positive (scalar) increment t_k suitable for control.

Step 4: If $\Delta^k := \left\| \mathbf{u}^{k+1} - \mathbf{u}^k \right\| < \varepsilon$, termination occurs, or otherwise j is incremented by one and we continue with Step 2.

The control of the step size t_j — cf. Bazaraa & Sherali (1981,[52]) — is heuristic in nature and requires a problem-specific adaptation but follows the significant theoretical result of Poljak (1967,[442]) that the sequence $\{L^k\}$ converges against the limit $L(\bar{\mathbf{u}})$, if the sequence $\{t_k\}$ converges toward zero and for the sum of the sequence members holds

$$\sum_{k=0}^{\infty} t_k = \infty.$$

This divergence ultimately guarantees that the step sizes do not become too small. Held et al. (1974,[262]) recommend

$$t_k := \theta_k \frac{\left[L_{\mathrm{U}} - L^k\right]}{\left\| \boldsymbol{\gamma}_k \right\|^2}$$

with $0 < \varepsilon \leq \theta_k \leq 2$, where L_U is an upper limit for $L(\mathbf{u})$. As a consequence of the weak duality theorem, any primary solution can be chosen for $L(\mathbf{u})$. However, the solution of the primary problem to determine $L(\mathbf{u})$ forces us to determine at least one integer permissible solution; in some cases, a constructive heuristic can help here. At the initialization of $\mathbf{u}^0$, the recourse to the primary solution of LP relaxation was sufficient. Initially, you can set θ_0 to $\theta_0 = 2$ and halve the value of θ_k if L^k does not increase during several iterations.

Example: Generalized Assignment Problem

As an example and similar as in Sect. 7.1.3, we consider a set of tasks $i \in I$ that should be assigned to different machines $j \in J$, where each task i is assigned to only one machine j. The assignment $i \rightarrow j$ causes costs c_{ij} and uses machine capacity C_{ij}. If we are interested in a cost-minimal solution, this can be calculated by the *generalized assignment problem*

$$\min \sum_i \sum_j c_{ij} x_{ij}$$

$$\sum_j x_{ij} = 1 \quad , \quad \forall i \tag{14.1.9}$$

$$\sum_j C_{ij} x_{ij} \leq d_j \quad , \quad \forall j \tag{14.1.10}$$

$$x_{ij} \in \{0, 1\} \quad , \quad \forall \{ij\}.$$

If we think about using LR, we first have to be clear about which restrictions should be relaxed in the Lagrange sense. This is where (14.1.9) or (14.1.10) comes into question; we opt for the assignment equation (14.1.9), as this is more difficult than the knapsack inequality (14.1.10), which is one of the easier ones in the class of MILP problems — in general, equations that contain discrete variables are usually candidates for difficulties and unpleasant surprises. This choice to relax (14.1.10) in the sense of Lagrange becomes all the more meaningful, the more a larger number of tasks are confronted with a smaller number of machines. Let us now — as in Martin (1999,[388]) — consider three tasks and two machines with capacities of 13 and 11 as well as costs and capacity utilization for the example

$$c_{ij} := \begin{pmatrix} 9 & 2 \\ 1 & 2 \\ 3 & 8 \end{pmatrix} \quad , \quad C_{ij} := \begin{pmatrix} 6 & 8 \\ 7 & 5 \\ 9 & 6 \end{pmatrix} .$$

To describe the problem in a vector framework, we define

$$\mathsf{A} := \begin{pmatrix} 1\,1\,0\,0\,0\,0 \\ 0\,0\,1\,1\,0\,0 \\ 0\,0\,0\,0\,1\,1 \end{pmatrix} \quad , \quad \mathbf{b} := \begin{pmatrix} 1 \\ 1 \\ 1 \end{pmatrix},$$

and the vectors

$$\mathbf{c}^{\mathrm{T}} := (c_{11}, c_{12}, c_{21}, c_{22}, c_{31}, c_{32})$$
$$\mathbf{x} := (x_{11}, x_{12}, x_{21}, x_{22}, x_{31}, x_{32})^{\mathrm{T}}.$$

This allows us to write the allocation equation as

$$\mathsf{A}\mathbf{x} = \mathbf{b}.$$

In **Step 0**, we solve the relaxed LP problem, i.e.,

$$\min \sum_i \sum_j c_{ij} x_{ij} \tag{14.1.11}$$

$$\sum_j x_{ij} = 1 \quad , \quad \forall i \tag{14.1.12}$$

$$\sum_j C_{ij} x_{ij} \le d_j \quad , \quad \forall j \tag{14.1.13}$$

$$\mathbf{x} \ge 0$$

and get $z_{\mathrm{LP}} = 6.428$ as result; the dual values (2, 2, 4.286) associated with (14.1.12) are useful as well. For this small problem, we can also easily calculate the optimal integer solution; we get $z_{\mathrm{IP}} = 18$ — so the LP relaxation returns only a very weak lower bound of about 36%.

Now let us see whether the LR can help us. First, we initialize $u^0 := (2, 2, 4.286)$. We obtain an upper bound L_{U} from the obvious solution $x_{11} = x_{11} = x_{32} = 1$ and $x_{ij} = 0$ for all other combinations of i and j, i.e., $\sum_i \sum_j c_{ij} x_{ij} = 19$. Let us remember again how the problem $L(u^k)$

$$\min \sum_i \sum_j c_{ij} x_{ij} + \sum_i u^j \sum_j (1 - x_{ij}) \tag{14.1.14}$$

$$6x_{11} + 7x_{21} + 9x_{31} \le 13 \tag{14.1.15}$$

$$8x_{12} + 5x_{22} + 6x_{32} \le 11 \tag{14.1.16}$$

$$x_{11}, x_{12}, x_{21}, x_{22}, x_{31}, x_{32} \in \{0, 1\}$$

is structured and start the procedure as

{**Input**}
An upper bound L_U
An initial value $u^0 \geq 0$
{**Initialization**}
$\theta_0 := 2$
{**Subgradient iterations**}
for $k := 0, 1, \ldots$ **do**
$\boldsymbol{\gamma}^k := \mathbf{g}(x_j) = \mathbf{b} - \mathsf{A}\mathbf{x}^k$ {subgradient of $L(\mathbf{u}^k)$}
$t^k := \theta_k \left[L_U - L(\mathbf{u}^k)\right] \left\|\boldsymbol{\gamma}^k\right\|^{-2}$ {step size}
$u^{k+1} := \max\{0, \mathbf{u}^k + t_k \boldsymbol{\gamma}^k\}$
if $\left\|\mathbf{u}^{k+1} - \mathbf{u}^k\right\| < \varepsilon$ **then**
Stop
end if
if no improvement in more than K iterations **then**
$\theta_{k+1} := \theta_k/2$
else
$\theta_{k+1} := \theta_k$
end if
$k := k + 1$
end for

The following table generated using *lagRel.gms*[2] in MCOL (see also MCOL/GAMS LIB/*gapmin.gms*) summarizes the results.

k	$L(\mathbf{u}^k)$	θ^k	$\|\gamma_k\|$	t^k	Δ^k	u_1^k	u_2^k	u_3^k
1	7.000	2.0000	2.00	12.000	12.000	14.0000	14.0000	4.2857
2	2.286	2.0000	2.00	16.714	16.714	0.0000	14.0000	21.0000
3	−8.000	1.0000	2.00	13.500	13.500	13.5000	14.0000	7.5000
4	5.500	1.0000	2.00	6.750	6.750	13.5000	7.2500	14.2500
5	12.250	1.0000	1.00	6.750	6.750	13.5000	14.0000	14.2500
6	6.000	1.0000	1.00	13.000	13.000	13.5000	1.0000	14.2500
7	6.000	0.5000	1.00	6.500	6.500	13.5000	7.5000	14.2500
8	12.250	0.5000	2.00	1.688	1.688	15.1875	7.5000	12.5625
9	9.375	0.2500	2.00	1.203	1.203	13.9844	7.5000	13.7656
10	11.781	0.2500	2.00	0.902	0.902	13.0820	7.5000	14.6680
11	11.414	0.1250	2.00	0.474	0.474	13.5562	7.5000	14.1938
12	12.362	0.1250	2.00	0.415	0.415	13.9710	7.5000	13.7790
13	11.808	0.1250	2.00	0.450	0.450	13.5215	7.5000	14.2285

[2] This GAMS file and part of the description in this section are by Erwin Kalvelagen (www.amsterdamoptimization.com), who kindly gave his permission to make this file part of MCOL.

14	12.293	0.0625	2.00	0.210	0.210	13.7311	7.5000	14.0189
15	12.288	0.0625	2.00	0.210	0.210	13.5213	7.5000	14.2287
16	12.293	0.0313	2.00	0.105	0.105	13.6261	7.5000	14.1239
17	12.498	0.0313	2.00	0.102	0.102	13.5245	7.5000	14.2255
18	12.299	0.0313	2.00	0.105	0.105	13.6292	7.5000	14.1208
19	12.492	0.0156	2.00	0.051	0.051	13.5784	7.5000	14.1716
20	12.407	0.0156	2.00	0.052	0.052	13.6299	7.5000	14.1201
21	12.490	0.0078	2.00	0.025	0.025	13.6045	7.5000	14.1455
22	12.459	0.0078	2.00	0.026	0.026	13.6300	7.5000	14.1200
23	12.490	0.0039	2.00	0.013	0.013	13.6173	7.5000	14.1327
24	12.485	0.0039	2.00	0.013	0.013	13.6300	7.5000	14.1200
25	12.490	0.0020	2.00	0.006	0.006	13.6237	7.5000	14.1263
26	12.497	0.0020	2.00	0.006	0.006	13.6300	7.5000	14.1200
27	12.490	0.0010	2.00	0.003	0.003	13.6269	7.5000	14.1231
28	12.496	0.0010	2.00	0.003	0.003	13.6237	7.5000	14.1263
29	12.497	0.0005	2.00	0.002	0.002	13.6253	7.5000	14.1247
30	12.499	0.0005	2.00	0.002	0.002	13.6237	7.5000	14.1263
31	12.497	0.0005	2.00	0.002	0.002	13.6253	7.5000	14.1247
32	12.499	0.0002	2.00	0.001	0.001	13.6245	7.5000	14.1255
33	12.499	0.0002	2.00	0.001	0.001	13.6253	7.5000	14.1247
34	12.499	0.0001	2.00	0.000	0.000	13.6249	7.5000	14.1251
35	12.500	0.0001	2.00	0.000	0.000	13.6253	7.5000	14.1247
36	12.499	0.0001	2.00	0.000	0.000	13.6249	7.5000	14.1251
37	12.500	0.0001	2.00	0.000	0.000	13.6251	7.5000	14.1249

As the subgradients γ^k are calculated according to

$$\gamma_i^k := 1 - \sum_j x_{ij}^k,$$

in this example, we only obtain the value 1 or 2. The convergence in increment t^k and in the values of u^k is easy to see. Column t^k shows that every time $K = 2$ consecutive values L^k did not lead to an improvement, the value of θ was halved.

The LR provides us with bound 12.5 — this is almost 70%. Such improvements over the LP relaxation also encourage the LR to be integrated directly into B&B processes for solving MILP problems. LR bounds can be further strengthening taking into account a double-decomposition structure of the problem, cf. [371, 374].

Variations for Nonlinear Problems

In principle, the LR can also be applied to nonlinear problems. For the cutting problems discussed in Sect. 13.3 and especially for the placement of circles in

Sect. 13.3.1.1, we want to dualize the conditions

$$\left(\mathbf{x}_i - \mathbf{x}_j\right)^2 \geq \left(R_i + R_j\right)^2 \quad , \quad \forall\{(i,j) \mid i < j\}.$$

For non-convex problems, however, considerable duality gaps are to be expected, i.e., the lower bounds calculated from the LR are weak. Thus, in the non-convex case, procedures based on exact penalty functions are preferable; cf. Pillo et al. (2012,[440]). In exact penalty methods, the penalty constant is chosen so that the optimal solution to the problem containing the penalty term matches that of the original.

14.1.3.4 Bilevel Programming

Bilevel Programming (BLP) problems are mathematical programming problems that contain an optimization problem in the constraints; cf. Bracken & McGill (1973,[96]). Alternatively, one might say, it is an optimization problem constrained by another optimization problem.

Bard (1998,[48]) is a good starting point for practical BLP with a focus on algorithms and applications (economics, resource allocation, transportation network design). Many new real-world problems in the area of energy networks can only be efficiently solved as mixed integer bilevel programs. Among them are the natural gas cash-out problem treated by Dempe et al. (2011,[153]), the deregulated electricity market equilibrium problem, biofuel problems, and a problem of designing coupled energy carrier networks. BLP models for describing migration processes are also in the list of the most popular new topics of bilevel programming, as well as allocation, information protection, and cybersecurity problems. In the notation by Floudas et al. (1999,[193, Chap. 9, pp. 205]), a general form of BLP is given by

$$\min_{\mathbf{x},\mathbf{y}} F(\mathbf{x},\mathbf{y})$$

subject to

$$\mathbf{G}(\mathbf{x},\mathbf{y}) \leq \mathbf{0}$$

$$\mathbf{H}(\mathbf{x},\mathbf{y}) = \mathbf{0}$$

$$\mathbf{y} \in \arg\min_{\mathbf{z}} f(\mathbf{x},\mathbf{z})$$

subject to

$$\mathbf{g}(\mathbf{x},\mathbf{z}) \leq \mathbf{0}$$

$$\mathbf{h}(\mathbf{x},\mathbf{z}) = \mathbf{0}$$

$$\mathbf{z} \in \mathbb{R}^{n2}$$

$$\mathbf{x} \in \mathbb{R}^{n1} \quad , \quad \mathbf{y} \in \mathbb{R}^{n2}.$$

F and $\mathbf{G}$ and $\mathbf{H}$ are the outer (planner's or *leader*'s view in Stackelberg games) problem objective function and inequality and equality constraints, while f and $\mathbf{g}$ and $\mathbf{h}$ are the inner (behavioral or *follower*'s view in Stackelberg games) problem objective function and inequality and equality constraints, respectively. While the outer problem has $\mathbf{x}$ and $\mathbf{y}$ as decision variables, the inner problem has only $\mathbf{y}$ as a decision variable, i.e., the inner problem contains $\mathbf{x}$ as an input parameter. This structure can also be extended to multilevel programming, cf. the special three-level optimization problem treated by Dempe et al. (2020,[154]).

The levels cannot be changed. A high-level decision maker is able to influence the decisions made at lower levels, without having complete control over their actions. The objective function of one department of an organization is determined, in part, by variables controlled by other departments operating at higher or lower levels. Let us illustrate this by some typical real-world applications.

1. In economic planning at the regional or national level, a typical situation is given by the leader, i.e., the government controlling policy variables (e.g., tax rates or import quotas) and maximizing employment or minimizing the use of a resource. The follower is the industry to be regulated; industry wants to maximize, for instance, the net income subject to economic and governmental constraints, cf. Migdalas et al. (1998,[398]) and several papers therein.
2. Determining price support levels for biofuel crops, cf. Bard (1998,[48, Chap. 12]): the leader is again the government determining the level of tax credits for each biofuel product; the leader wants to minimize total outlays. The follower, who wants to minimize costs, is the petro-chemical industry.
3. Resource allocation in a decentralized company, cf. Aiyoshi & Shimizu (1981,[10]) or Bard (1988,[47]): the leader is the central resource supplier, who allocates products to manufacturers and wants to maximize profit of the company as a whole. the followers are the production facilities at different locations; they determine their own production mix and output to achieve maximal performance of their own unit.
4. Transportation system network design, cf. Bard (1998,[48, Chap. 10]): the leader is the central planner controlling investment costs, e.g., which links (roads or railroad tracks) to improve; the objective is to influence users' preferences to minimize total costs. The follower is the individual users. Their route selection determines the traffic flows and therefore the operational costs. The users try to minimize the cost of their routes.
5. Chemical and biological engineering: the leader is the overall process design with minimal operating costs. The follower minimizes the Gibbs free energy, along with constraints involving mass and energy balance; cf. the original work by Clark & Westerberg (1990,[125]) and Ferris et al. (2009,[180]) for more examples.

BLP can be understood as a logical extension of mathematical programming or as a generalization of a particular problem in game theory (Stackelberg games). Therefore, literature is extensive and diverse with many contributions found in economy, cf. Mirrlees (1999,[403]) — an example from this paper is coded and provided as MCOL\GAMS\EMPLIB*mirrless.12*. More coded GAMS examples in the same location named *fld9*.** and *ccmg*.** are from Floudas et al. (1999,[193, Chap. 9, pp. 205]) and Conejo et al. (2006,[131]).

Medium- and large-scale BLP problems are difficult to solve. Enumeration, penalty methods, and reformulation techniques, cf. Dempe & Kue (2017,[152]), are frequently used for solving BLP problems. Especially, linear BLP problems can be solved by vertex enumeration. In reformulation techniques, the bilevel problem is transformed into a single level one by replacing the inner problems with its KKT optimality conditions. BLP problems are also difficult from a theoretical point of view; cf. Dempe (2002,[151]). They do not need to have a solution. Even restricting all outer and inner functions to be continuous and bounded does not guarantee the existence of a solution.

14.1.4 Primal Heuristics

Primal heuristics allow us to calculate good primal solutions as an approximation for the optimal solution or starting values for further use in improvement procedures. We distinguish between structured primal heuristics and hybrid methods. The former follow a certain modeling or algorithmic procedure and can be used quite comprehensively. Hybrid methods combine, e.g., exact MIP algorithms with constructive heuristics or improvement methods — also called *metaheuristics* — like *simulated annealing*, *genetic algorithms*, or *taboo search*.

In its simplest form, local search methods can improve the initial solution determined by a constructive heuristic. In solving the MIP problem, the basic idea of primal heuristics is to repeatedly solve a subproblem on a smaller subset of discrete variables, usually binary variables. The binary variables can be chosen randomly or controlled by a metaheuristic. The selected binary variables are then subject to MIP optimization, while the others are fixed to their already known values. The neighborhood of a discrete solution can be chosen very versatilely. For a vector of binary variables, any other vector can be called a neighbor if these vectors differ only in a maximum of three components. In combination with MIP techniques, such neighborhoods lead to another polylithic method called local branching; cf. Fischetti & Lodi (2003,[185]). In this context, we also refer to the RINS heuristics implemented in CPLEX or XPRESS (cf. Danna et al. 2005,[138]) and to the *feasibility pump* proposed by Fischetti & Glover (2005,[184]), which can also be regarded as polylithic.

If possible, a bound for the objective function should be derived in addition to a primal feasible point. While the primal heuristic provides an upper bound for a minimization problem, in favorable cases a lower bound of the original problem can be calculated if a feasible point is known. In other cases, a lower bound can be calculated by solving an auxiliary problem, where the auxiliary problem should be a relaxation of the original problem, making it easier to solve. In Sect. 14.1.5 we show that primal heuristics allow the calculation of upper *and* lower bounds in favorable cases, if the problem or the objective function of the problem has a basic structure that minimizes the number of active objects and increases the problem size in the number of objects.

14.1.4.1 Structured Primal Heuristics

Structured primal heuristics follow certain rules for computing feasible points of mixed integer or non-convex, nonlinear optimization problems.

LP-Guided Dives, Relax-and-Fix

Primal feasible solutions can be computed by systematically dealing with the critical discrete variables, mostly binary variables. These include, for example, the *dive-and-fix* (DF) and *relax-and-fix* (RF) methods, sometimes also called fix-and-relax (FR), as well as *moving window technique* or *sliding window technique*, cf. Van Dinter et al. (2013,[162]). In DF, one solves the LP relaxation of an MIP problem followed by fixing a subset of the fractional variables to appropriate bounds. Near-integer-fix (NIF) is a variant of DF where the fixed variables can be fixed to the next integer point.

While NIF and DF can easily lead to infeasible problems, the situation with RF and FR is somewhat more advantageous. FR is a progressive process by exploiting decomposition structures into products, activities, time, or geometry. Using FR, the author has solved small 6×6 instances of the Eternity II problem (cf. Benoist & Burreau 2008,[70], or Muñoz et al. 2009,[414]) as MILP. This problem is a highly combinatorial, NP-complete edge-matching problem with little structure but $256! \cdot 4^{256}$ possible combinations, most of which are probably infeasible.

FR requires that we are able to decompose the discrete variables δ of an original problem $\mathscr{P}$ into R disjoints or weakly overlapping subsets $\mathscr{S}_r$, $r \in \mathscr{R} := \{1, \ldots, R\}$ with, for instance, a time-forward structure. An example of this is rolling production planning, in which early time periods can be more important than later ones. Based on these decompositions $\mathscr{S}_r$, R MIP problems $\mathscr{P}_r$ are solved, which eventually result in a composite feasible point of $\mathscr{P}$. In problem $\mathscr{P}_r$, $2 \le r \le R$, the values of $\delta \in \mathscr{S}_{r-1}$ are fixed to their optimum values from $\mathscr{P}_{r-1}$; integrality is required only for $\delta \in \mathscr{S}_r$. Since $\mathscr{P}_1$ represents a relaxation of $\mathscr{P}$, in a minimization problem, the objective function value of $\mathscr{P}_1$ gives a lower bound of the objective function value of $\mathscr{P}$. At the transition from $r-1$ to r, it may happen that $\mathscr{P}_r$ becomes infeasible. In most cases, however, it is sufficient to install a small overlap between $\mathscr{S}_{r-1}$ and $\mathscr{S}_r$ to ensure feasibility, in which additional free discrete variables at the end of partition $\mathscr{S}_{r-1}$ establish a connection to partition $\mathscr{S}_r$. Even in cases without a feasibility problem, this small overlap increases the quality of the point created by FR (δ_R, x_R), which is also a feasible point of the original problem. When using CPLEX or XPRESS-OPTIMIZER, the MILP solver can build directly on the feasible point. In favorite cases, the integrality gap between the upper and lower bounds can then be reduced or even closed.

FR works particularly well in energy industry. A pump storage plant may serve as an example. With an hourly discretization over one year and the binary decision variables (use of the highly pumped reservoir water or energy generation by a thermal power plant), we can easily arrive at tens of thousands of binary variables as well as at similarly large numbers of constraints (balancing and coupling of reservoir levels in adjacent time periods). The aim is to minimize start-up costs for pumps and thermal power plants. Since late and early periods are hardly coupled due to the lack of storage possibilities, a few partitions, each of which may, for example, comprise three months, are sufficient to generate good feasible starting points and to calculate the optimal solution of the original and complete problem in two minutes downstream with the MILP solver's mipstart functionality.

Linear Approximations for NLP Problems

Beale & Forrest (1976,[57]) have applied the idea of piecewise linear approximations to the calculation of global minima of non-convex, nonlinear functions. They have introduced structured sets named *special ordered sets of type 2* (in short, SOS2) with special branching strategies described in Sect. 6.8. Since then, numerous contributions to SOS2 have appeared, including Farias et al. (2000,[148]; 2008,[149]), with discontinuous, separable, piecewise linear functions, Vielma et al. (2009,[565]), with a development of a uniform approach and extensions to mixed integer models for non-separable piecewise linear functions, Misener & Floudas (2010,[405]), with piecewise linear formulations for two- and three-dimensional functions, or Rebennack & Kallrath (2015,[454, 455]) for computing minimal numbers of breakpoints and error-controlled piecewise linear approximations.

Homotopy Sequences of Models

Homotopy methods can be useful for computing initial feasible points from a sequence of relaxed models. For example, in a scheduling problem $\mathscr{P}$, due dates can be relaxed, resulting in the relaxed model $\mathscr{R}$. The optimal solution or any feasible point of $\mathscr{R}$ is also a feasible point of $\mathscr{P}$ if the due dates in $\mathscr{P}$ are modeled using slack variables. In an auxiliary model, the sum of these slack variables can be minimized.

Rolling horizon approaches using RF described above can also be understood as a homotopy. This is, however, a very simple case, as it only works on variables and their domain specification or bounds. More powerful, but more difficult to implement, are homotopy methods and are based on model sequences in which the models really differ by constraints and features.

In the design study described in Kallrath (1999,[301]), the costs for a site-wide production network are minimized. These include costs for raw materials, investment costs, and variable costs for reprocessing plants as well as a penalty cost term for residual impurities occurring at a certain point of the production network.

This MINLP problem, which is difficult due to its structure and size, is solved in a sequence of submodels of different complexity implemented in the modeling language GAMS. In this homotopy, model $m+1$ is initialized using the results from model m. For example, a simple linear model provides first results for calculating the concentrations c_{jk}^{in} occurring in the pooling problem and for initializing two simplified nonlinear models. The integer conditions of the binary variables were first relaxed and only considered in the last step. In summary, the following sequence of submodels of increasing complexity has been used:

problem type		*problem ingredients or results*
LP	yields	approximation of concentrations
NLP 1	includes	pooling, local cleaning units (LCUs)
NLP 2	yields	relaxed semi-continuous flows
RMINLP 1	includes	connections and free pools
RMINLP 2	includes	local LCUs
MINLP	is the	full model with all binary variables.

14.1.4.2 Hybrid Methods

In hybrid methods, exact optimization algorithms are combined with constructive heuristics or improvement procedures (local search procedures or metaheuristics like Simulated Annealing, Genetic Algorithms, or Tabu Search) combined to find feasible points of the optimization problem.

In constructive heuristics, the structure of the optimization problem is exploited to create a primal feasible point. This point could be further improved by the improvement procedures described above. This means that initial values can be assigned to discrete values in MIP problems, which is, for instance, very useful when using the solver's *mipstart* feature. In favorable situations, hybrid techniques can also be used to compute lower bounds for minimization problems; an example is the exhaustion heuristics in Kallrath et al. (2014,[319]) used to minimize the number of patterns in a cutting stock problem. Like the structured primary heuristics in Sect. 14.1.4.1, hybrid procedures are problem-specific; but they are tailor-made solution techniques, which are not easily transferable to other problems. The development of hybrid procedures is therefore the *art* to generate good feasible points with as little computation time as possible. In this context, *everything is allowed*.

14.1.5 Proving Optimality Using PMSAs

In the previous section, we have learned about using PMSAs as primal heuristics, i.e., to obtain feasible points. Special classes of problems that allows using primal heuristics to calculate lower bounds are MILP or MINLP problems that become smaller, i.e., have fewer variables and constraints when the objective function approaches the optimum. Examples are:

1. *Pattern minimization in cutting stock problems*: Kallrath et al. (2014,[319]) have used a PMSA to solve a cutting stock problem to optimality, i.e., closing the gap between the upper and lower bounds of an MILP problem. In that example, the number of patterns was to be minimized. The upper bound was given by the best integer feasible found. The key ideas were variables and an index set $\mathcal{J}$ of patterns from which appropriate patterns could be constructed. Originally, the problem was very large, as the number of pattern was identical to the number of order widths. However, as the objective function was to minimize the number of patterns, with each feasible solution found, a smaller upper bound was found, and the index space of patterns could be reduced. Finally, in this iterative procedure, due to the reduced index space, the problem became so small that it was possible to solve it to optimality.
2. *Calculation of optimal breakpoint systems*. For continuous piecewise linear approximations of nonlinear functions by SOS2 variables, Rebennack & Kallrath (2015,[455]) have calculated minimum breakpoint systems with different methods, where the approximator, over- or underestimator is sufficient for a given accuracy δ. The minimum breakpoint system cannot always be determined directly by solving an MINLP problem. In cases where direct calculation is not possible, a hybrid method first generates a permissible, but not necessarily minimum breakpoint system, which satisfies the δ accuracy. However, an upper bound for the number of breakpoints is now known, which reduces the cardinality of the set number $\mathcal{B}$ of possible breakpoints in the direct MINLP model and makes it possible to solve the MINLP problem.

So, the generic idea is: Consider an MILP, NLP, or MINLP problem over an index space $\mathcal{I}_k$ in iteration k. The objective function has the property that from its minimized values, a reduced index space of the original problem can be derived, leading to smaller problem with fewer variables and constraints. If during the iterations of the polylithic procedure, the problem becomes so small that optimality can be proven, we are done and have obtained feasibility and optimality. In these cases of objective functions leading to reduced problem size, PMSAs are both primal heuristics and dual techniques leading to improved lower bounds and, in favorite cases, to closed gaps. Examples of objective functions are minimizing the number of

1. patterns in cutting stock problems,
2. breakpoints to obtain piecewise linear approximations of nonlinear functions,
3. people working on shifts, and
4. rail cars, airplanes, or rental cars in a fleet.

The objective function is to minimize the number of elements in an index space involved in the optimization problem. The problem in iteration k, $k = 0, 1, 2, \ldots$ then involves a binary variable δ_i, $i \in \mathscr{I}_k$, indicating, for instance, whether index element i representing a pattern, breakpoint, person, or car has been activated or selected. The objective function is simply

$$\min \sum_{i \in \mathscr{I}_k} \delta_i .$$

Note that the index elements have to be ordered and allow to accessing its adjacent index, e.g., p01, p02, ... , p99, as this allows us to exploit index shrinking by

$$\delta_i \geq \delta_{i+1} \quad , \quad i \in \mathscr{I}_k, \tag{14.1.17}$$

and to define a sequence of index sets for which holds $|\mathscr{I}_{k+1}| \leq |\mathscr{I}_k|$ due to

$$\mathscr{I}_{k+1} = \{i | \delta_i \in \mathscr{I}_k \wedge \delta_i = 1 \text{ in iteration } k\}.$$

If they are not ordered, for instance, in product portfolio minimization, (14.1.17) cannot be enforced and the index space shrinking approach does not work.

14.2 PMSAs Applied to Real-World Problems

PMSAs are very problem-specific. In this section, we outline a few ideas for cutting stock and packing problems and illustrate an evolutionary approach used to solve a simple supply chain problem. We also provide a strategy to compute optimal breakpoints for approximating nonlinear functions.

14.2.1 Cutting Stock and Packing

Cutting stock and packing problems allow many PMSAs among them complete enumeration discussed in Sect. 14.2.1.1, various incremental, swapping, exchange or tour-reversing approaches in Sect. 14.2.1.2, or solving irregular strip packing problems by hybridizing and LP by Gomes & Oliveira (2006,[229]).

14.2.1.1 Complete Enumeration

The cutting stock problem formulated as an MINLP problem in Sect. 13.2 requires a lot of computing time when the number of orders increases above 10. Exact solutions — up to a certain problem size — can still be obtained using complete enumeration (CE). In CE we generate all existing patterns $p \in \mathcal{P}$ a priori, obtain the order multiplicity of order i within pattern p, and then solve the master problem. The master problem is the minimization problem

$$z = \min_{\mu_p, \delta_p, \alpha_{ip}} \sum_{p \in \mathcal{P}} \left(C_p^{\mathrm{R}} \mu_p + C_p^{\mathrm{P}} \delta_p \right)$$

with variable pattern costs C_p^{R} to be paid for each master roll cut to pattern p and fixed costs C_p^{P} to be paid when pattern p is used at all. The only variables are the integer variables μ_p (pattern multiplicity) and the binary variables δ_p indicating whether pattern p has been selected in the optimal solution.

The constraints are to fulfill demands

$$\sum_{p \in \mathcal{P}} \mu_p \alpha_{ip} = D_i \quad , \quad \forall i, \tag{14.2.18}$$

and to relate the pattern selection variable δ_p to the pattern multiplicity μ_p

$$\delta_p \le \mu_p \le P \delta_p \quad , \quad p \in \mathcal{P},$$

where P is an upper limit on the usage of pattern p.

CE has the advantage over column generation that it does not depend on problem structure as much as far as the inner structure of the pattern is concerned. The limits of practical usefulness of CE are on the one hand the number of patterns to be stored, and on the other hand the size of the master problem which can be solved by the available MILP solver. The implementation *TrimCE.gms* contains a procedural part establishing all feasible patterns inspired by Karelathi (2002,[323]).

If the master problem must be solved in no more than two minutes and becomes too large for the available MILP solver (300 to 400 patterns for `CoinCBC`, and 5,000 to 6,000 patterns for CPLEX) to stay within the target time limit, we resort to the multi-core decomposition ideas of parallel PMSAs in Kallrath et al. (2020,[314]). Let us assume we have 2000 patterns and want to use the free MILP solver COINCBC on a machine with 16 cores. Then, we randomly select 400 patterns and solve six master problems (with six different pattern sets) in parallel. This constellation even allows us to exploit two cores for each master problem. This approach, although global optimality is not guaranteed, has been used successfully on a real-world problem in paper industry. In most cases, it actually gets the global minimum — this has been verified by running certain instances using CPLEX.

Another example of column enumeration for a cutting stock problem is found in Rebennack et al. (2009,[456]). In this example, columns are subsets of disks to be cut from a set of existing rectangular plates (coils in steel industry) of different size. Exploiting these columns, the MINLP problem is decomposed into non-convex NLP problems. Finally, an MILP master problem has to be solved.

14.2.1.2 Incremental, Swapping, and Tour-Reversing Approaches

If the problem size of cutting or packing problems becomes too large, one may use incremental algorithms, i.e., exploiting additive placements of objects or subsets of objects until all objects have been placed, to construct at least feasible solutions. Especially, bin packing or strip packing problems — the width of a rectangle is known, and the length is to be minimized as in Kallrath (2009,[310]), or Kallrath & Rebennack (2014,[318]) — are very suitable for using incremental algorithms; the problem of computing a minimal-radius circular container hosting a given set of disks is another example (see Ryu et al. (2020,[476])). In such cases, an initial set of randomly selected objects is placed followed by small number of additional objects to be placed. The problem size is limited by partitioning the sets of objects into two subsets, $\mathscr{S}_{\mathrm{p}}$ and $\mathscr{S}_{\mathrm{f}}$. $\mathscr{S}_{\mathrm{p}}$ contains objects we want to keep fixed and not subject to the non-overlapping constraints; this is a subset of the objects already placed. $\mathscr{S}_{\mathrm{f}}$ consists of objects free to be placed and subject to non-overlapping constraints, i.e., new objects to be added and objects near the boundary between the already placed objects free to be relocated and the new ones. This approach has similarities to relax-and-fix in Sect. 14.1.4.1 and is also used for ellipsoid packing by Kallrath (2017,[313]).

Another approach is the 2-exchange heuristic by Gomes & Oliveira (2002,[228]) in the context of nesting problems or the approach by Kallrath & Frey (2019,[315]) for packing circles into perimeter-minimizing convex hulls is to swap objects and rearrange them. However, this is not easy because maintaining feasibility can be a serious problem. For the same problem, Kallrath & Frey (2019,[315]) have also implemented a tour approach that covers the combinatorial part of the outer disks establishing the perimeter of the convex hull.

14.2.2 Evolutionary Approach

Evolutionary algorithms (EAs) are generic population-based metaheuristics inspired by biological evolution, such as reproduction, mutation, recombination, and selection. Candidate solutions to the optimization problem play the role of individuals in a population, and the *fitness function* determines the quality of the solutions. Evolution of the population then takes place after the iterated application of reproduction, mutation, recombination, and selection. To illustrate this approach in the context of mathematical programming and PMSAs, we use the Supply Management Problem with Lower-Bounded Demand (SMPLD) by Chauhan et al. (2004,[116]). A set of providers supplies products of one type to a set of consumers, the quantity of products that can be delivered lies between given minimum and maximum values, and the costs proposed by each provider are linear functions of the quantity being delivered. The SMPLD consists in finding the quantity of products which will be supplied by each provider to each consumer minimizing the total cost (production and set-up costs):

Satisfy demand D_j for a set of customers $j \in \mathcal{J}$ by production x_{ij} from plants $i \in \mathcal{I}$ with capacity C_i. The non-negative continuous production variables x_{ij} are subject to the semi-continuous condition

$$x_{ij} = 0 \vee L_i \leq x_{ij} \leq C_i . \tag{14.2.19}$$

The corresponding MILP model with binary variables δ_{ij} and x_{ij} is given by the following relations, starting with the objective function:

$$\sum_j P_{ij} x_{ij} + S_{ij} \delta_{ij} , \tag{14.2.20}$$

where P_{ij} and S_{ij} are the specific production and set-up costs.

The capacity constraint is

$$\sum_j x_{ij} \leq C_i \quad , \quad \forall i , \tag{14.2.21}$$

and the demand constraint reads

$$\sum_i x_{ij} \geq D_j \quad , \quad \forall j . \tag{14.2.22}$$

As the objective function involves the binary variables δ_{ij}, we formulate the semi-continuous constraints (14.2.19) using the binary variables

$$L_i \delta_{ij} \leq x_{ij} \leq C_i \delta_{ij} \quad , \quad \forall \{ij\} .$$

This MILP problem, for the data provided in the implementation in *ea-smpld.gms*, has minimal cost of 732 (just mentioned to compare it to the EA solution); the solution is obtained in milliseconds as it is only a small problem.

Let us now describe how to solve this MILP problem using an evolutionary approach. In real-world problems, we might face the problem that some binary or integer variables make it impossible to solve the problem in reasonable time; sequencing problems fall into this class. The EA controls the binary variables δ_{ij}. For each fixation, the resulting LP is solved – and the fixed coefficients δ_{ij} are updated. Note that from the optimization model's point of view, δ_{ij} are not variables anymore but fixed coefficients leading to either $x_{ij} = 0$ or $L_i \leq x_{ij} \leq C_i$. Constraints (14.2.21) and (14.2.22) are relaxed as fixation can lead to infeasibilities. Therefore, we introduce the relaxation variables $t_i \geq 0$ and $s_j \geq 0$ and obtain the modified constraints

$$\sum_j x_{ij} \leq C_i + t_i \quad , \quad \forall i , \tag{14.2.23}$$

and

$$\sum_i x_{ij} \geq D_j + s_j \quad , \quad \forall j. \tag{14.2.24}$$

In the EA framework, the objective function (14.2.20) becomes the fitness function

$$\sum_j P_{ij} x_{ij} + S_{ij} \delta_{ij} + V \tag{14.2.25}$$

with the violation term

$$V := \sum_i P_i t_i + \sum_j P_j s_j. \tag{14.2.26}$$

To get the EA running, in Step 0 we assign initial values δ_{ij}. Using a random generator with 50% probability, we set $\delta_{ij} = 0$ and with 50% probability $\delta_{ij} = 1$.

Step 1: For given δ_{ij} (and thus $x_{ij} = 0$ or $L_i \leq x_{ij} \leq C_i$), we solve the resulting LP. If all relaxation variables vanish, we have a feasible point and an upper bound on both the fitness *and* the objective functions; we store δ_{ij} and x_{ij}. Otherwise, if some relaxation variables have non-zero values, we only have an upper bound on the fitness function.

Step 2: We apply a mutation operator to δ_{ij}. For instance, with a 10% mutation probability, we replace δ_{ij} by $1-\delta_{ij}$ (the other δ_{ij} coefficients remain unchanged) and move to Step 1 if the total number of iterations has not been reached.

This approach is implemented in *ea-smpld.gms* — this file contains further implementation details and the data. Note that as this EA implementation is a heuristics, it does not provide a lower bound. Therefore, we can stop, for instance, when we have found the first feasible solution (all relaxation variables take value zero), or after a certain number of iterations. In this example, the EA finds a solution with objective function value of 732, the optimal MILP value, after 80 iterations.

EA can be extended to problems with more discrete variables. In a scheduling problem, the EA could take care of the sequencing variables, while the remaining MILP part is solved by an MILP solver for the sequencing variables fixed. In the sense of a PMSA, EAs can be useful to generate feasible initial points, but it may be slow depending on the number of discrete variables. Note that if the number of discrete variables is small, it may be worthwhile to think about complete enumeration. Depending on the problem structure and the size of the feasible space, it may take a while before a feasible solution is obtained. An EA can produce a significant advantage if it reduces an MINLP to an MILP problem.

14.2.3 Optimal Breakpoints

In Sect. 6.8.2, we have learned how SOS2 for given sets of breakpoints X_b can be used to interpolate univariate nonlinear functions $f(x)$ by piecewise linear functions $\ell(x)$; both should be functions on $\mathbb{D} := [X_-, X_+] \subset \mathbb{R}$. The size of the resulting MILP problem depends crucially on the number of breakpoints, and one can expect that the MILP problem has significantly more variables than the original MINLP or NLP problem. We want to compute such breakpoint systems aiming for small numbers of breakpoints and distribute them so that an approximation accuracy δ is achieved. Rebennack & Kallrath (2015,[455], RK2015, hereafter) have shown that if f has finitely many points of discontinuity, then one can construct a piecewise linear δ-approximator function $\ell : \mathbb{D} \rightarrow \mathbb{R}$ for f with finitely many support areas. However, one may not be able to impose continuity on the function ℓ. There are various tasks related to the construction of optimal breakpoint systems:

1. Compute the *proven* minimal number of breakpoints required to piecewise linearly and continuously approximate any continuous function over a compactum (the methodology also works if the function has finitely many discontinuities) so that an approximation accuracy δ is achieved.
2. For a given number B of breakpoints, compute the tightest possible piecewise linear and continuous approximator; tightest in the sense of minimizing the largest deviation between the approximator $\ell(x)$ and the function $f(x)$.
3. For a given $\delta > 0$, compute piecewise linear and continuous δ-under- and δ-overestimators.

Note that we do not only talk about interpolation any longer but also about approximators, under- and overestimators. For univariate functions, the concept of under- and overestimators leads to piecewise linear δ-tubes around the function f and becomes very important when we want to replace nonlinear function terms in constraints. We can then use δ-under- and δ-overestimators for inner and outer approximations of the feasible region. Especially, if nonlinear terms occur in difficult constraints, and even more complicated, in equality constraints, we rather use linear under- and overestimators. Note that 2δ-underestimators for a continuous function f with a minimal number of breakpoints can be calculated by shifting a δ-approximator for f with a minimal number of breakpoints up by value δ.

As in Sect. 6.8.2, we consider a univariate function $f(x)$. For a set of breakpoints x_b, $b \in \mathcal{B}$, which are now unknown and subject to

$$X_- \leq x_b \leq x_{b+1} \leq X_+,$$

we have the equalities for convexity and relating the breakpoints to the argument x

$$\sum_{b \in \mathcal{B}} \lambda_b = 1 \quad , \quad x = \sum_{b \in \mathcal{B}} x_b \lambda_b, \tag{14.2.27}$$

and the function value

$$y = \ell(x) = \sum_{b \in \mathscr{B}} \phi_b \lambda_b \quad ; \quad \phi_b := \phi(x_b) = f(x_b) + s_b \quad , \quad \forall b \in \mathscr{B} \tag{14.2.28}$$

with shift variables s_b measuring the deviation of $\ell(x)$ from the function values $f(x)$ at the breakpoints; in pure interpolation we have $s_b = 0$. Note that instead of the function values $F_b = f(x_b)$ in (14.2.28), we use $\phi(x_b)$ to allow for approximation and not only interpolation.

Unlike in Sect. 6.8.2, the breakpoints b now have unknown positions x_b — they become the fundamental variables of our optimization problem; in addition, we have the s_b variables. With binary variables χ_b indicating whether breakpoint b is used, we want to minimize the number $B = |\mathscr{B}|$ of breakpoints,

$$\min \sum_{b \in \mathscr{B}} \chi_b,$$

and compute their positions x_b, satisfying the accuracy inequality

$$|\ell(x) - f(x)| \leq \delta \quad , \quad \forall x \in \mathbb{D}. \tag{14.2.29}$$

Inequality (14.2.29) is a continuum requirement, i.e., x in (14.2.29) is not a decision variable and can vary in the interval $\mathbb{D}$, and it turns the problem into the class of semi-infinite programming (finite number of variables, but infinitely many constraints). With a few more constraints coupling χ_b to the usage or existence of breakpoint b or x_b, *respectively*, RK2015 call this problem "OBSC" for *O*ptimal *B*breakpoint *S*ystem using a *C*ontinuum approach for x.

To solve this difficult semi-infinite MINLP problem and to obtain a computationally tractable mathematical program, one discretizes the continuum constraints (14.2.29) into I finite constraints of the form

$$|\ell(x_i) - f(x_i)| \leq \delta \quad , \quad \forall i \in \mathbb{I} := \{1, \ldots, I\}, \tag{14.2.30}$$

for appropriately selected grid points x_i. Applying this approach to *each* of the B breakpoints x_b in formulation OBSC leads to a *D*iscretized *O*ptimal *B*reakpoint *S*ystem (OBSD in RK2015). The size (in terms of number of variables and constraints) of OBSD depends strongly on the numbers B and I. The MINLP problem OBSD is generally too large and difficult to solve. Only for the modest numbers of breakpoints and not too many discretization points, there is a chance of solving these problems to global optimality. Alternatively, one can solve the optimal distribution of a fixed number B of breakpoints for the discretized continuum constraint

$$|\ell(x_i) - f(x_i)| \leq \mu \quad , \quad \forall i \in \mathbb{I}$$

and minimize μ followed by checking whether $\mu \leq \delta$.

RK2015 have used the idea of formulation OBSD and discretize each interval (x_{b-1}, x_b) into I equidistant grid points. By forcing the usage of exactly $B-1$ breakpoints (no need for counting $x_0 = X_-$ nor $x_B = X_+$ as breakpoints), the binary variables χ_b are eliminated, and one obtains the NLP problem FBSD:

$$\mu^* = \min \ \mu \tag{14.2.31}$$

$$\text{s.t.} \quad x_b - x_{b-1} \geq \frac{1}{M} \quad , \quad \forall b \tag{14.2.32}$$

$$l_{bi} = \phi(x_{b-1}) + \frac{\phi(x_b) - \phi(x_{b-1})}{x_b - x_{b-1}}(x_{bi} - x_{b-1}) \quad , \quad \forall\{bi\}$$

$$x_{bi} = x_{b-1} + \frac{i}{I+1}(x_b - x_{b-1}) \quad , \quad \forall\{bi\} \tag{14.2.33}$$

$$|l_{bi} - f(x_{bi})| \leq \mu \quad , \quad \forall\{bi\} \tag{14.2.34}$$

$$x_{bi} \in [X_-, X_+], \quad -\infty \leq l_{bi} \leq +\infty \text{ (free)} \quad , \quad \forall\{bi\}$$

$$\mu \geq 0 \quad ; \quad s_b \in [-\delta, +\delta], \quad x_b \in [X_-, X_+] \quad , \quad \forall\{b\} \tag{14.2.35}$$

At the breakpoints, the function deviation is bounded by δ saving the discretization points x_{bi} at the breakpoints. The solution of the FBSD minimization problem provides a breakpoint system x_b, the shift variables s_b, and the minimal value μ^* as functions of B and I, i.e., $\mu^* = \mu^*(B, I)$ and $x_b = x_b(B, I)$.

The simultaneous computation of all breakpoints by solving one optimization model is computationally very expensive. Thus, RK2015 have proposed PMSAs, especially, forward and backward schemes. In the forward scheme, one moves from $x_0 = X_-$ to breakpoint x_1 as far as possible away from x_1 fulfilling the accuracy requirement on $[x_0, x_1]$ and then successively from x_{b-1} to the next breakpoint $x_b \leq X_+$ covering the whole interval.

Finally, we present an example. Motivated by the fact that many global solvers do not support trigonometric functions, we want to represent $\sin x$ by a piecewise linear approximator using five breakpoints. Solving the model, for instance, using the implementation *optGrid.gms* provided in MCOL, we find — after 18 min of computing time using the solver LINDO–that $\sin x$ can be approximated to an accuracy of $\delta = 0.00582257548536694$ using the breakpoints $(b_0, b_1, b_2, b_3, b_4)$ on the interval $[0, \pi/2]$:

b	0	1	2	3	4
X_b	0	0.570343493004343	0.939440621802004	1.26316079512676	$\pi/2$

.

The range $[0, \pi/2]$ with the five breakpoints above can be extended to $[0, 2\pi]$ using 16 breakpoints by

$$x_b = \begin{cases} X_b & , \quad b = 0, \ldots, 4 \\ \pi/2 - X_{8-b} \ , & \quad b = 5, \ldots, 8 \\ -x_{16-b} & , \quad b = 9, \ldots, 16. \end{cases}$$

Finally, extensions to higher dimensions have been analyzed by Rebennack & Kallrath (2015,[454]). Two-dimensional functions $f(x_1, x_2)$, for instance, can be approximated or interpolated by exploiting triangulations.

14.3 Summary and Recommended Bibliography

In this chapter, we have presented polylithic modeling and solution approaches (PMSAs) for solving difficult or large MILP, NLP, and MINLP problems. Thus, the reader should now be familiar with

- the main benefits of PMSAs: problem-specific preprocessing and construction of solution algorithms;
- the importance of computing initial values and providing initial approximations to obtain primal feasible points of these MILP, NLP, and MINLP problems;
- proving optimality using PMSAs; and
- typical situations in which PMSAs can be useful.

In the literature, the field of *MatHeuristics* is closely related to PMSAs. A good introductory reference is by Maniezzo et al. (2009,[383]). A good review of mathematical models of irregular packing problems is by Leao et al. (2019,[357]).

14.4 Exercises

1. Solve the farmer's problem in Sects. 2.6.1 and 3.3.1 by *column enumeration*, i.e., generate all feasible combinations of calves and pigs, evaluate the objective function as on page 86, and pick the best combination as the optimal solution.
2. Implement the *column generation* (CG) method by Gilmore & Gomory in the AML of your choice for solving a trimloss problem similar to the one in Sect. 13.2. Exploit the following background material:
   The CG method by Gilmore & Gomory (1961,[220]) is based on a feasible initialization (feasible patterns, no underfulfillment of demand) and the iteration loop (master problem, subproblem = pricing problem). The master problem (MP) is given by

$$\min \sum_p \mu_p \quad s.t. \quad \sum_i N_{ip}\mu_p \geq D_i \quad , \quad \forall i \quad ; \quad \mu_p \in \{0, 1, 2, 3, \ldots\} \quad , \quad \forall p, \tag{14.4.1}$$

   where $D_i = (8, 16, 12, 7, 14, 16)$ is the requested number of rolls, and N_{ip} specifies how often order width i is contained in pattern p. During the iterations, the integrality conditions are replaced by the non-negativity conditions $\mu_p \geq 0$. The MP yields μ_p and the values of the dual variables or shadow prices P_i, which are required as input to the pricing problem (PP) generating new patterns. The PP is a knapsack problem with objective function

$$\min \left(1 - \sum_i P_i \alpha_i\right) \quad s.t. \quad \sum_i W_i \alpha_i \leq W \quad , \quad \begin{matrix} \alpha_i \in \{0, 1, 2, 3, \ldots\} \\ W_i = (33, 36, 38, 43, 49, 53) \end{matrix} \quad ; \quad \forall i, \tag{14.4.2}$$

with master roll width $W = 220$ and order widths W_i. The objective function in (14.4.2) corresponds to the reduced costs $\bar{c}_j := c_j - \pi^{\mathrm{T}} A_j = c_j - \sum_i \pi_i A_{ij}$ of the MP; we recognize the non-basic variables indexed by j, column vector $\mathbf{A}_j := (\alpha_1, \ldots, \alpha_I)^{\mathrm{T}}$ representing a new pattern leading to updated N_{ip}, and the dual variables $\boldsymbol{\pi}$ associated with the demand inequality. In (14.4.1), all objective function coefficients are 1 and $\pi_i = P_i$. As the MP is a minimization problem, iterations terminate when $\bar{c}_j = 0$, or for numerical reasons, when $\bar{c}_j \geq -\varepsilon$, where ε is a small, non-negative number, e.g., $\varepsilon = 10^{-6}$. After the iterations are done, an MP is solved as an MILP problem for the patterns generated and $\mu_p \in \{0, 1, 2, 3, \ldots\}$.

Hints: Work out a feasible initialization, implement the relaxed master problem, the pricing problem, the iteration loop, and finally the MILP master problem.

3. Extend the column generation method implemented in Exercise 2 by considering that only K knives are available. Hint: Start by thinking where this condition enters the game (MP or PP?). Run numerical experiments for $K = 3$ and $K = 2$.

Chapter 15
Cutting and Packing Beyond and Within Mathematical Programming

This chapter, based on material provided and written by Prof. Dr. Yuriy Stoyan & Prof. Dr. Tatiana Romanova,[1] is devoted to the phi-function technique used for mathematical modeling of cutting and packing (C&P) problems. Phi-functions are constructed here for some 2D and 3D geometric objects. Phi-functions can be described by quite simple formulas. A general solution strategy using phi-functions is outlined. Conceptually, the Phi-function approach exploits NLP and MINLP and can be understood as cutting and packing beyond and within Mathematical Programming. It also exploits polylithic modeling and solution techniques.

15.1 Introduction

The cutting and packing (C&P) problem is a part of computational geometry that has rich industrial applications in garment industry, sheet metal cutting, furniture making, shoe manufacturing, etc. The common task in these areas is to cut a certain set of figures of specified shapes and sizes from a given sheet (or strip) of material (textile, wood, metal, etc.), cf. the tutorial by Bennell & Oliveira (2008,[66]) and references therein. To minimize waste one wants to cut figures as close to each other as possible; in other words one needs to design a layout as close to the optimum as possible before the actual cutting.

Supplementary Information The online version contains supplementary material available at https://doi.org/10.1007/978-3-030-73237-0_15.

[1]The National Academy of Sciences of Ukraine, Institute of Mechanical Engineering Problems, Department of Mathematical Modeling and Optimal Design, Kharkiv, Ukraine & Kharkiv National University of Radioelectronics, Department of Applied Mathematics.

J. Kallrath, *Business Optimization Using Mathematical Programming*,
International Series in Operations Research & Management Science 307,
https://doi.org/10.1007/978-3-030-73237-0_15

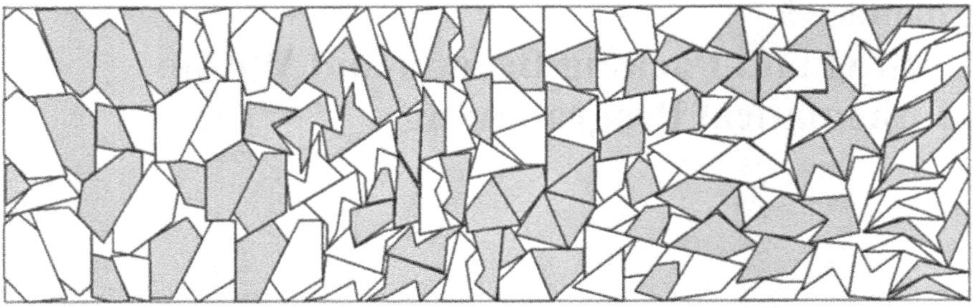

Fig. 15.1 Strip packing polygons into a rectangle of given width and infinite length. This is Fig. 7 from Taylor & Francis: Journal of the Operational Research Society, Vol. 67, Stoyan et al. (2016), Cutting and Packing Problems for Irregular Objects with Continuous Rotations: Mathematical Modelling and Nonlinear Optimization. Fig. 7 [530]. Copyright ©2016 The Operational Research Society, reprinted by permission of Taylor & Francis Ltd, www.tandfonline.com on behalf of The Operational Research Society

This is a mathematical problem which can be formalized as follows: given a strip of fixed width W and infinite length, say $S = \{x \geq 0,\ 0 \leq y \leq W\}$, cut out n given figures from the rectangle $\{0 \leq x \leq L,\ 0 \leq y \leq W\} \subset S$ without overlaps, so that L takes its minimum value, see Fig. 15.1.

In other applications, one needs to arrange a given set of objects within a certain area (say, shipment on a deck of a freight car or electronic components on a panel), and again one wants to minimize the use of space or maximize the number of objects.

Clearly these two types of problems — cutting and packing — are mathematically equivalent; they are known as the C&P problem (it is also called nesting problem, marker making, stock cutting, containment problem, etc.). In some cases, it involves additional restrictions on the minimal or maximal distance between certain objects or from the objects to the border of the container. The tutorial by Bennell et al. (2008,[66]) summarizes previous studies of C&P problems and its history.

Many other applications involve 3D geometry: packing pills into a bottle, placing crates and barrels into a cargo compartment, 3D laser cutting, modeling of granular media and liquids, and radiosurgery treatment planning, and 3D printing (just to name a few). Thus, the C&P problem naturally extends to three dimensions. Figure 15.2 illustrates a 3D packing problem — objects of various shape and dimensions are packed into a convex container in order to minimize its dimensions.

The problem is NP-hard, and as a result solution methodologies predominantly utilize heuristics; most existing methods of cutting and packing are restricted to objects of certain shapes and type and impose various limitations on their layout. The most popular and most frequently cited tool in modern literature on the C&P problem is the so-called *No-Fit Polygon*; it is designed to work only for polygonal objects without rotation and one of the objects that can be translated; cf. Bennell & Oliveira (2008,[66]). A detailed review of tools and mathematical models for irregular packing problems is provided in Leao et al. (2019,[357]). The book by Scheithauer (2018,[488]) introduces the fundamental knowledge for dealing with

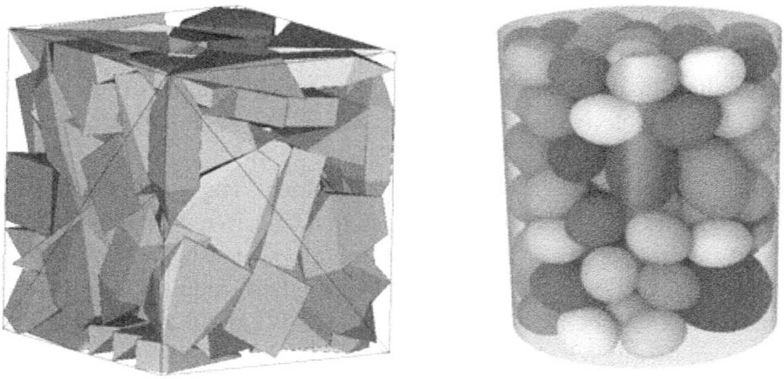

Fig. 15.2 3D packing examples: Local optimal placement of polyhedra (left, Fig. 12a in Romanova et al. 2018,[463]) and 100 ellipsoids (right, Fig. 10 in Romanova et al. 2020,[468])

C&P problems and presents modeling and solution approaches. However, not all advances are published in books or academic journals because many commercial companies closely guard their products.

The goal of this chapter is studying the cutting and packing problem in a formal mathematical manner. In this study, objects of very general shape (called phi-objects) are considered, and their layouts are characterized by means of special functions (called phi-functions) whose construction involves a certain degree of flexibility. The concepts of the phi-object and the phi-function have their roots in topology, but the phi-functions turn out to be highly convenient for practical solutions of C&P problems. In particular, since the construction of phi-functions is flexible, the advantage of this fact is taken to develop more efficient algorithms. The concept of *quasi phi-functions* extends the phi-function domain by introducing auxiliary variables. The use of quasi phi-functions, as an alternative to phi-functions, allows for simplified descriptions of non-overlapping constraints, at the expense of a larger number of variables. The quasi phi-functions concept is introduced in Stoyan et al. (2016,[531]). It is motivated by the idea proposed by Kallrath (2009,[310]) of using separation lines to model non-overlapping constraints for circles and convex polygons. The phi- and quasi phi-functions have been successfully used to model a variety of packing, cutting, and covering problems [118, 433, 433–435, 464, 465, 467, 526, 527, 535], including applications in space engineering [463, 522, 528, 529], 3D printing [466], terminal debugging [469, 470], and material sciences [166]. The principal goal of this chapter is to present the basic foundations of the phi-function technique and demonstrate practical benefits of this tool [68, 69, 118, 119, 521, 523–525, 530–532].

This chapter is organized as follows: in Sect. 15.2 phi-objects are introduced, in Sect. 15.3 phi-functions are defined and an overview of their properties is given, in Sect. 15.4 a reduction of the C&P problem to a constrained optimization problem using the phi-functions is provided, and in Sect. 15.5 various approaches to its solution are described. Illustrative examples are presented in Sect. 15.6.

15.2 Phi-objects

The first goal is to describe a general mathematical model for the cutting and packing (C&P) problem that should adequately represent virtually all existing applications.

The basic task is to place a set of geometric objects (later on, simply *objects*) T_i, $i \in \{1, 2, \ldots, n\} = I_n$, into a container T_0 so that certain restrictions on the location of the objects are met and a given objective function (measuring the "quality" of the placement) reaches its extreme value. We specify these requirements below.

This basic task can be also rephrased differently: given a (large) object T_0, we need to cut a set of smaller objects $\{T_1, \ldots, T_n\}$ from it. These objects are 2D or 3D geometric figures, i.e., subsets of $\mathbb{R}^2$ or $\mathbb{R}^3$. Generalization to any dimension is not pursued here.

The multiplicity of shapes of T_i and T_0, as well as the variety of restrictions and forms of the objective function, generates a wide spectrum of realizations of this basic problem. A unified approach to all such applications is presented with the ultimate goal of designing efficient algorithms for solving the C&P problem.

15.2.1 Phi-objects

A class of objects for the model following is defined; they are called *phi-objects*. They must have an interior ("main part") and a boundary (frontier). Each phi-object is required to be the closure[2] of its interior. This requirement rules out such elements as isolated points, one-dimensional curves, etc.; they do not occur in realistic applications. Figure 15.3a shows an invalid phi-object — it has one-dimensional "rays," isolated points, and punctured interior points.

The smaller objects $T_1, \ldots, T_n$ always have finite size, in mathematical terms they are *bounded*. The larger object T_0 may be bounded or unbounded (it is common in applications that the container is a strip or a cylindrical tube of infinite length).

In addition, phi-objects should not have self-intersections along their boundary, as shown in Fig. 15.3b, c, because this may lead to confusion. For example, Fig. 15.3c shows a dark domain whose two ends touch each other; this must be prohibited. The reason is also demonstrated in that same figure: a similar object (the light grey "figure eight") is placed so that the two intersect each other only in their boundaries, which is generally allowed, but in this particular case we cannot position these objects as shown because one "cuts" through the other.

Mathematically, the above requirement can be stated as follows: a phi-object and its interior must have the same homotopy type (the same number of connected components, the same number of interior holes, etc.). Alternatively, one may require

[2] In general topology, closed sets that are closures of their interior are said to be *canonically closed*; this is what phi-objects are.

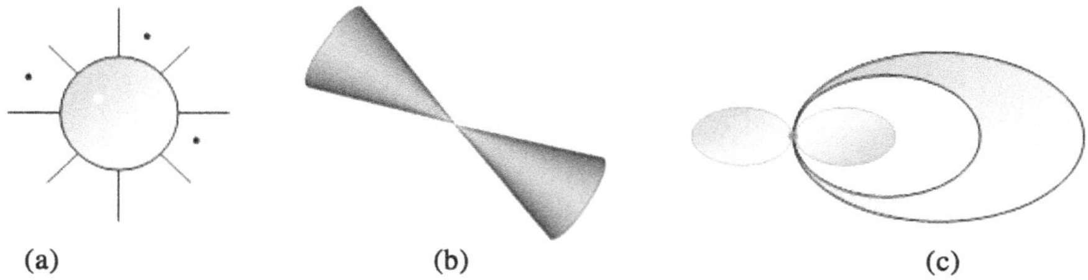

Fig. 15.3 Invalid phi-objects. By Prof. Dr. Tatiana Romanova, Copyright ©2020

that for any point z on the boundary $\mathrm{fr}(T)$ of a phi-object T, there exists an open neighborhood U_z of z such that $U_z \cap (\mathrm{int}\, T)$ is a connected set. These requirements may sound too abstract, but their practical meaning should be clear from the above example.

An important property of phi-objects is that if T is a phi-object, then $T^* = \mathbb{R}^d \setminus \mathrm{int}T)$, where $d = 2, 3$, is a phi-object too.

In most applications, the boundaries of 2D phi-objects are made by simple contours: straight lines and circular arcs. Likewise, the boundaries of 3D phi-objects mostly consist of flat sides, spherical, cylindrical, and conical surfaces.

15.2.2 Primary and Composed Phi-objects

Any phi-object in $\mathbb{R}^2$ is called a *phi-polygon* if its frontier is shaped by straight lines, rays, and line segments. An ordinary polygon is a phi-polygon, but there are also unbounded phi-polygons — half-plane, a sector bounded by two intersecting lines, etc.

A phi-object in $\mathbb{R}^3$ is called a *phi-polytope* if its boundary is shaped by phi-polygons. Other objects can be approximated by polygons and polytopes, which is a common practice, but we handle some curvilinear objects directly.

We call a *primary* phi-object in $\mathbb{R}^2$ a *circle*, *rectangle*, or *convex polygon*. In 3D, a primary object is a *sphere*, *cuboid*, *right circular cylinder*, *circular cone*, or *convex polytope*. In addition, if A is a 2D or 3D primary object, then the closure of its complement (in $\mathbb{R}^2$ or $\mathbb{R}^3$, respectively), denoted by A^*, is regarded a primary object, too (see some illustrations in [68]). Thus the list of primary objects is not limited to bounded or convex figures.

Note that convex polygons formally include rectangles but in practice it is convenient to treat the latter separately, as they can be handled more efficiently (e.g., compare (15.3.9) and (15.3.15) below).

More complex objects can be constructed from primary objects (by methods similar to those used in constructive solid geometry). We say that a phi-object T is

Fig. 15.4 Examples of composed phi-objects in 2D. By Prof. Dr. Tatiana Romanova, Copyright ©2020

composed if it is obtained by forming unions and intersections of primary objects, i.e.,

$$T = T_1 \circ_1 T_2 \circ_2 \cdots \circ_{k-1} T_k, \tag{15.2.1}$$

where T_i are primary objects, each $\circ_i$ denotes either a union ($\cup$) or an intersection ($\cap$), and the order in which these operations are executed can be specified by a set of parentheses, for example $T = T_1 \cup (T_2 \cap T_3)$. Composed phi-objects may be very complex, see some example in Fig. 15.4.

Fact 1 *In 2D, composed phi-objects are exactly those whose frontier is formed by straight lines, rays, line segments, and circular arcs.*

Indeed, every phi-object with such a boundary can be represented by unions and/or intersections of primary objects, in the sense of the formula (15.2.1). Similarly, 3D composed phi-objects have frontiers made by flat (planar) faces, parts of spheres, and parts of cylindrical and conical surfaces, see Fig. 15.5.

15.2.3 Geometric Parameters of Phi-objects

The shape of a phi-object can be specified in many ways. For a simple (primary) object, its type is named and its metric dimensions are listed. For example, a circle can be specified by a pair (C, r), where C is the type ("circle") and $r > 0$ is its radius, i.e.,

$$(C, r) = \{(x, y) : x^2 + y^2 \leq r^2\}.$$

Fig. 15.5 Examples of composed phi-objects in 3D. By Prof. Dr. Tatiana Romanova, Copyright ©2020

A sphere can be specified by a pair (S, r), where S is the type ("sphere") and $r > 0$ is its radius, i.e.,

$$(S, r) = \{(x, y, z) \colon x^2 + y^2 + z^2 \le r^2\}.$$

For a rectangle, a triple (R, a, b) can be used, where R is the type ("rectangle") and $a, b > 0$ are half-sides:

$$(R, a, b) = \{(x, y) \colon |x| \le a \text{ and } |y| \le b\}. \tag{15.2.2}$$

Similarly, a cuboid P (a 3D box) can be described:

$$(P, a, b, c) = \{(x, y, c) \colon |x| \le a, \quad |y| \le b, \quad |z| \le c\}.$$

For a cylinder, we use a triple (C, r, h), where C is the type, r is the radius of the base, and h is the half-height:

$$(C, r, h) = \{(x, y, z) \colon x^2 + y^2 \le r^2 \text{ and } |z| \le h\}. \tag{15.2.3}$$

All objects above are centrally symmetric. In such cases, the origin of the coordinate system is always placed at the center of symmetry to simplify the formulas. This explains the use of "half-sides," "half-heights," etc.

For a convex m-gon, its type is denoted by K and its shape is specified by a set of inequalities $\alpha_i x + \beta_i y \le \gamma_i$ for $i = 1, \dots, m$; it is convenient to assume that $(0, 0)$ belongs to the polygon (see Sect. 15.2.4); then $\gamma_i \ge 0$. It is also convenient to choose α_i and β_i so that $\alpha_i^2 + \beta_i^2 = 1$, this simplifies subsequent computations. Thus the convex m-gon is described by

$$\big(K, (\alpha_1, \beta_1, \gamma_1), \dots, (\alpha_m, \beta_m, \gamma_m)\big). \tag{15.2.4}$$

The (closure of the) complement of a primary object is specified similarly, except we add a star to its type; for example, the (closure of the) complement of a circle C can be defined as

$$(C^*, r) = \{(x, y) : x^2 + y^2 \geq r^2\}.$$

Recall that this is a primary object, too.

It is important that each primary object is specified by a set of *linear* or *quadratic* inequalities. Actually, quadratic formulas allow us to describe even more general shapes, such as ellipses, ellipsoids, and hyperboloids.

To represent a composed object, we can specify the primary objects used in its construction, their positions (in the way explained below, see Sect. 2.4), and the sequence of set-theoretic operations (unions and/or intersections) employed to produce the composed object from its primary constituents. The list of characteristics of a composed object may be quite long depending on the complexity of its shape.

15.2.4 Position Parameters of Phi-objects

One may notice that in the formulas which specify primary objects the origin (0, 0) plays a special role. It is called the *pole* of the primary object. If the object is centrally symmetric, then its center becomes the pole. Otherwise the choice of the pole may be quite arbitrary, for example in a generic polygon the pole can be placed at a vertex.

In addition, the orientation of the phi-object is usually fixed by its description, for example the sides of a rectangle may be aligned with the coordinate axes, see (15.2.2). Thus with each phi-object we associate not only a pole but also a coordinate frame originating at the pole. We call it the *eigen-* (own-) coordinate system of the object. The inequalities specifying a primary object are written in their eigen coordinates.

Next, in order to specify an arbitrary position of a 2D phi-object in $\mathbb{R}^2$, a translation vector $\nu = (\nu_1, \nu_2)$ and a rotation angle $\theta \in [0, 2\pi)$ are introduced. This means that the object is translated by ν, i.e., its pole moves to the point (ν_1, ν_2), and then the object is rotated around the pole by θ (for instance, counterclockwise).

The rotation parameter θ is optional. First of all, it is redundant for such objects as circles. Second, in many applications the objects cannot be rotated freely by their nature. In garment industry, which remains the largest field of applications of cutting and packing algorithms, free rotations are generally not allowed. One cuts pieces of predetermined shape from a long strip of fabric, and there are usually just two orientations in which the pieces can be placed: the original one and the one obtained by a 180° rotation (such a restriction is due to the existence of drawing patterns and to intrinsic characteristics of the fabric's weave). The cutting and packing problem both with and without rotation parameters will be analyzed here.

The position of a 3D phi-object in space $\mathbb{R}^3$ requires a translation vector $\nu = (\nu_1, \nu_2, \nu_3)$ and three (optional) rotation angles $\theta_1, \theta_2, \theta_3$.

To summarize, a composed phi-object on a plane or in space can be described by a list of characteristics that include

(i) Types of primary objects used in its construction and the rules of construction (the sequence of intersections and/or unions),
(ii) The metric dimensions of the constituent primary objects, and translation vectors and (optionally) rotation angle(s) that determine the position of the primary objects in the eigen coordinate system of the composed object
(iii) The translation vector and (optionally) rotation angle(s) that determine the position of the object in plane/space.

While the characteristics (i) and (ii) are fixed for every phi-object, those in (iii) are usually treated as variables by the optimization algorithms which try to arrange the objects in an optimal way.

15.2.5 Interaction of Phi-objects

In solving the C&P problems, it is very important to distinguish between different types of mutual location of two phi-objects (let us call them A and B):

- *Interior-intersection*: $\text{int}(A) \cap \text{int}(B) \neq \emptyset$.
- *Touching*: $\text{int}(A) \cap \text{int}(B) = \emptyset$ and $\text{fr}(A) \cap \text{fr}(B) \neq \emptyset$.
- *Non-intersection*: $A \cap B = \emptyset$.
- *Containment*: $A \subset B$, i.e., $\text{int}(A) \cap \text{int}(B^*) = \emptyset$.

Here B^* denotes the (closure of the) complement of B. Note that the containment is conveniently described by non-intersection of the interiors of A and B^*.

15.3 Phi-functions: Relating Phi-objects

To formalize the above relations between phi-objects, phi-functions are introduced by Stoyan and his co-workers [521, 523, 525]. As illustrated in Fig. 15.6, phi-functions serve two purposes: keeping phi-objects apart (left) and keeping phi-objects into a target container (right). The basic mathematical idea is that for any pair of phi-objects A and B with placement parameters u_A, u_B the phi-function $\Phi(u_A, u_B)$ must be positive for non-intersecting objects, zero for touching objects, and negative for objects with intersecting interiors (Fig. 15.7), i.e., $\Phi(u_A, u_B)$ must satisfy

$$\begin{cases} \Phi(u_A, u_B) > 0 \text{ if } A \cap B = \emptyset \\ \Phi(u_A, u_B) = 0 \text{ if } \text{int}(A) \cap \text{int}(B) = \emptyset \quad \text{and} \quad \text{fr}(A) \cap \text{fr}(B) \neq \emptyset \\ \Phi(u_A, u_B) < 0 \text{ if } \text{int}(A) \cap \text{int}(B) \neq \emptyset. \end{cases} \tag{15.3.5}$$

Fig. 15.6 Phi-functions in their positive mood serve two purposes: keeping objects apart (left) and keeping objects into a target container (right). Produced for this book by Diana Kallrath, Copyright ©2020

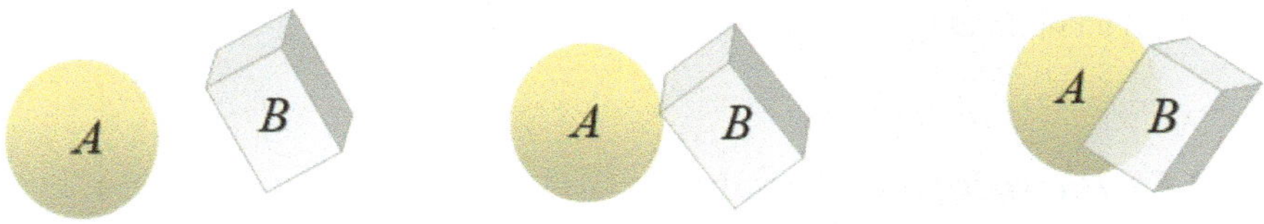

Fig. 15.7 Arrangements of two objects A and B

The phi-function is required to be *defined* and *continuous* for all values of its variables u_A, u_B.

Thus knowing the sign of $\Phi(u_A, u_B)$ for every pair of objects would allow us to distinguish between the basic types of their mutual location. The containment $A \subset B$, in particular, holds if and only if $\Phi(u_A, u_B)) \geq 0$ for the objects A and $B^* = \mathbb{R}^d \setminus \text{int}B$.

It is now clear that if $\Phi(u_A, u_B) > 0$, the objects are a certain distance apart; usually, decreasing $\Phi(u_A, u_B)$ brings them closer together. On the other hand, if $\Phi(u_A, u_B) < 0$, then the objects overlap, and increasing $\Phi(u_A, u_B)$ would force them apart. These features make the phi-functions instrumental for the performance of cutting and packing algorithms.

In some applications the metric dimensions of some objects should also be variable; then they can be included in the list of arguments of the phi-functions.

15.3.1 Construction of Phi-functions for Various Situations

While the sign of the phi-function plays a crucial role, its absolute value is not subject to any rigid requirements. In particular, if two objects A and B overlap, then $\Phi(u_A, u_B) < 0$ and the absolute value $|\Phi(u_A, u_B)|$ should just roughly measure the extent of overlap. For non-overlapping objects $\Phi(u_A, u_B) > 0$, and the value of $\Phi(u_A, u_B)$ may just roughly correspond to the distance between A and B.

In particular, one might set $\Phi(u_A, u_B) = \text{dist}(A, B)$ while interior nonintersecting of objects, where

$$\text{dist}(A, B) = \min_{X \in A,\ Y \in B} \text{dist}(X, Y) \tag{15.3.6}$$

denotes the geometric (Euclidean) distance between closed sets.

There is an issue of existence of the minimum provided by (15.3.6). For every pair of objects at least one has to be bounded; this property guarantees the existence of the above minimum.

In some cases, the geometric distance between objects is easy enough to compute and can be used as the value of the phi-function. But in many cases, the formula for the distance involves radicals, which makes it difficult for local optimization algorithms to use $\Phi(u_A, u_B)$ and its derivatives. In those cases, $\Phi(u_A, u_B)$ should be defined by a simpler formula that only roughly estimates the distance between the objects. Some examples are given below.

15.3.1.1 Phi-function for Two Circles

The phi-function for two circles C_i, $i = 1, 2$, with centers (x_i, y_i) and radii $r_i > 0$ can be defined more simply (see Fig. 15.8 for details):

$$\Phi^{CC} = (x_1 - x_2)^2 + (y_1 - y_2)^2 - r^2, \tag{15.3.7}$$

$$r = r_1 + r_2.$$

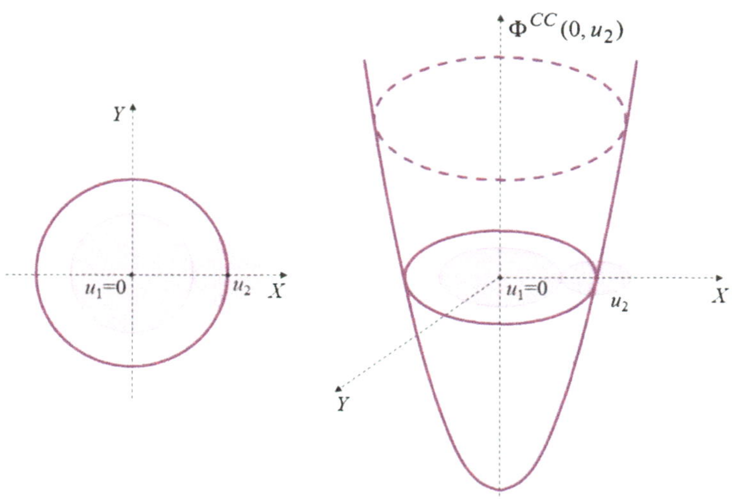

Fig. 15.8 Phi-function for two circles

Formula (15.3.7) allows avoiding square roots and uses only quadratic functions.

15.3.1.2 Phi-function for Two Spheres

Similarly, for two spheres S_i, $i = 1, 2$, with centers (x_i, y_i, z_i) and radii $r_i > 0$, the phi-function can be derived as follows:

$$\Phi^{SS} = (x_1 - x_2)^2 + (y_1 - y_2)^2 + (z_1 - z_2)^2 - r^2, \tag{15.3.8}$$

$$r = r_1 + r_2.$$

15.3.1.3 Phi-function for Two Rectangles

For two rectangles R_i, $i = 1, 2$, with centers (x_i, y_i) and half-sides $a_i, b_i > 0$ (assuming that the sides are aligned with the coordinate axes), the phi-function is defined by (see Fig. 15.9 for details)

$$\Phi^{RR} = \max_{i=1,\dots,4} \chi_i, \tag{15.3.9}$$

$$\chi_1 = -y - b,\ \chi_2 = -x - a,\ \chi_3 = y - b,\ \chi_4 = x - a,$$

$$x = x_2 - x_1,\ y = y_2 - y_1,\ a = a_1 + a_2, b = b_1 + b_2.$$

Observe that the above function (15.3.9) sometimes coincides with the geometric distance between the rectangles (if one is above the other or if they are placed side by side), but in general the distance involves square roots, while this formula is just a combination of linear functions.

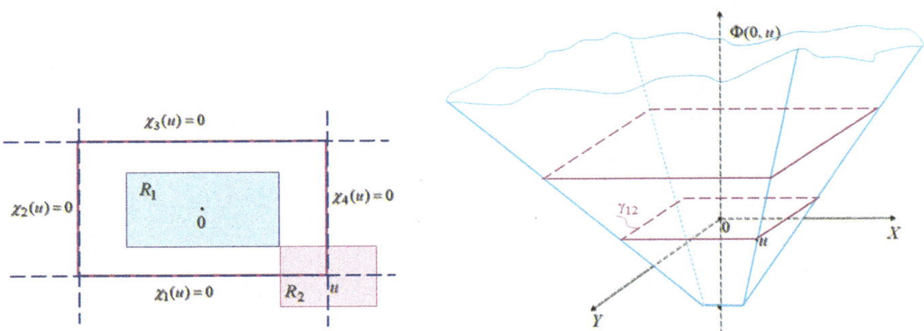

Fig. 15.9 Phi-function for two rectangles

15.3.1.4 Phi-function for Two Cuboids

Similarly, for two cuboids P_i, $i = 1, 2$, with centers (x_i, y_i, z_i) and half-sides $a_i, b_i, c_i > 0$, whose sides are aligned with the coordinate axes, the phi-function can be set

$$\Phi^{PP} = \max_{i=1,\dots,6} \chi_i,$$

$$\chi_1 = -x - a, \chi_2 = x - a, \chi_3 = y - b, \chi_4 = -y - b, \chi_5 = z - c, \chi_6 = -z - c$$

$$x = x_2 - x_1, y = y_2 - y_1, z = z_2 - z_1, a = a_1 + a_2, b = b_1 + b_2, c = c_1 + c_2.$$

15.3.1.5 Phi-function for Two Parallel Circular Cylinders

Now let C_i, $i = 1, 2$, be two cylinders with centers (x_i, y_i, z_i), radii of the bases r_i, and half-heights h_i, see (15.2.3). Assuming the axes of the cylinders are parallel to each other, the ϕ-function is derived as follows:

$$\Phi^{CC} = \max\{\chi_1, \chi_2, \omega\}, \tag{15.3.10}$$

$$\chi_1 = z - h, \chi_2 = -z - h, \omega = x^2 + y^2 - r^2,$$

$$x = x_2 - x_1, y = y_2 - y_1, z = z_2 - z_1, h = h_1 + h_2, \ r = r_1 + r_2.$$

15.3.1.6 Phi-function for Convex Polygons

Effectively, in (15.3.9) the distance between two vertices of the rectangles is replaced with the distance from a vertex of one rectangle to a side of the other (more precisely, to the line containing that side); and the distance from a point to a line is always given by a linear formula. This principle can be applied to any pair of convex phi-polygons.

The phi-function can be written explicitly for convex polygons (recall that those are primary phi-objects). Suppose

$$\left(K', (\alpha'_1, \beta'_1, \gamma'_1), \dots, (\alpha'_{m'}, \beta'_{m'}, \gamma'_{m'})\right) \tag{15.3.11}$$

and

$$\left(K'', (\alpha''_1, \beta''_1, \gamma''_1), \dots, (\alpha''_{m''}, \beta''_{m''}, \gamma''_{m''})\right) \tag{15.3.12}$$

are two convex polygons specified according to formula (15.2.4). Denote also by (x'_i, y'_i), $1 \leq i \leq m'$, the vertices of K' and by (x''_j, y''_j), $1 \leq j \leq m''$, the vertices of K''. As before, it is assumed that α_i's and β_i's satisfy $\alpha_i^2 + \beta_i^2 = 1$ for each polygon. Then the value $d = \alpha_i x + \beta_i y + \gamma_i$ is the "signed" distance from the point (x, y) to the ith edge of the polygon; the sign of d is automatically determined as follows: it is negative if the point (x, y) lies on the same side of the edge as the entire polygon and positive otherwise.

Now let

$$u_{ij} = \alpha'_i x''_j + \beta'_i y''_j + \gamma'_i \tag{15.3.13}$$

denote the "signed" distance from the jth vertex (x''_j, y''_j) of the polygon K'' to the ith edge of K' and

$$v_{ji} = \alpha''_j x'_i + \beta''_j y'_i + \gamma''_j \tag{15.3.14}$$

the "signed" distance from the ith vertex (x'_i, y'_i) of the polygon K' to the jth edge of K''. Now the phi-function can be defined in the form

$$\Phi^{K'K''} = \max\{ \max_{1 \leq i \leq m'} \min_{1 \leq j \leq m''} u_{ij}, \max_{1 \leq j \leq m''} \min_{1 \leq i \leq m'} v_{ji} \}. \tag{15.3.15}$$

This formula is based on two facts: The first one is a well-known geometric property of convex polygons: if two convex polygons are disjoint, then there is an edge E of one of them such that these polygons lie on the opposite sides of the line containing E. This property guarantees the basic features (15.3.5) of the function (15.3.15), in particular $\Phi^{K'K''} > 0$ whenever the polygons K', K'' are disjoint.

The second fact is a simple property of continuous functions: if f and g are continuous, then $\min\{f, g\}$ and $\max\{f, g\}$ are also continuous functions. This fact implies the continuity of Φ in (15.3.15).

Note that the restriction $\alpha_i^2 + \beta_i^2 = 1$ is no longer necessary as the phi-function need not represent actual distances.

If the polygons K' and K'' have fixed orientation, then their positions are completely specified by the coordinates of their poles; let us denote those by (x', y') and (x'', y''), respectively. These are the only variables in the formulas. It is easy to check that α_i's and β_i's are constants (independent of the coordinates of the pole), and γ_i's, x_i's, y_i's are just linear functions of the coordinates of the pole. Therefore the phi-function (15.3.15) is piecewise linear in its arguments (x', y') and (x'', y'').

The phi-function for convex polytopes can be found, for instance, Stoyan et al. (2002,[524]). The phi-function technique is compared with Lagrange multipliers approach in [375].

15.3.1.7 Phi-function for Non-convex Polygons

Suppose K' and K'' are non-convex phi-polygons (or polytopes), represented as unions $K' = K'_1 \cup \cdots \cup K'_p$ and $K'' = K''_1 \cup \cdots \cup K''_q$ of *convex* polygons (polytopes) K'_i and K''_j, are convex for $i = 1, \ldots, p$ and $j = 1, \ldots, q$. Then the phi-function for K' and K'' can be defined in the form

$$\Phi^{K'K''} = \min_{1 \le i \le p} \min_{1 \le j \le q} \Phi^{K'_i K''_j}.$$

The last formula illustrates a general principle. Suppose $A = A_1 \cup \cdots \cup A_p$ and $B = B_1 \cup \cdots \cup B_q$ are phi-objects, each of which is a union of some phi-objects A_i and B_j, respectively. These do not have to be disjoint unions, i.e., some A_i's may overlap, and so may some of B_j's. Then

$$\Phi^{AB} = \min_{1 \le i \le p} \min_{1 \le j \le q} \Phi^{A_i B_j}. \tag{15.3.16}$$

This fact can be verified by direct inspection, see also [68].

Now suppose K'' is a simply connected polygon and K' is a multiply connected polygon (i.e., polygon with holes) so that $K' = K'_1 \cap (K'_2 \cap \cdots \cap K'_p)$, where K'_1 is a simply connected convex phi-polygon and $K'_2, \cdots, K'_p$ are complements to simply connected phi-polygons (creating "holes"), then $\Phi^{K'K''}$ may be presented as follows:

$$\Phi^{K'K''} = \max_{1 \le i \le p} \Phi^{K'_i K''}. \tag{15.3.17}$$

Things may get more complicated when the frontiers of the objects are a mixture of arcs and line segments; then the constructions of phi-functions may require a degree of ingenuity; see next section.

15.3.1.8 Phi-function for a Rectangle and a Circle

Let R be a rectangle with center (x_1, y_1) and half-sides $a, b > 0$, and C be a circle with center (x_2, y_2) and radius $r > 0$. As can be seen in Fig. 15.10, the phi-function is defined by

$$\Phi^{RC} = \max\{\max_{i=1,\ldots,4} \chi_i, \min_{i=1,\ldots,4} \{\omega_i, \psi_i\}\} \tag{15.3.18}$$

$$\chi_1 = x - A, \ \chi_2 = y - B, \ \chi_3 = -x - A, \ \chi_4 = -y - B,$$

$$\omega_1 = (x + a)^2 + (y + b)^2 - r^2, \ \omega_2 = (x + a)^2 + (y - b)^2 - r^2,$$

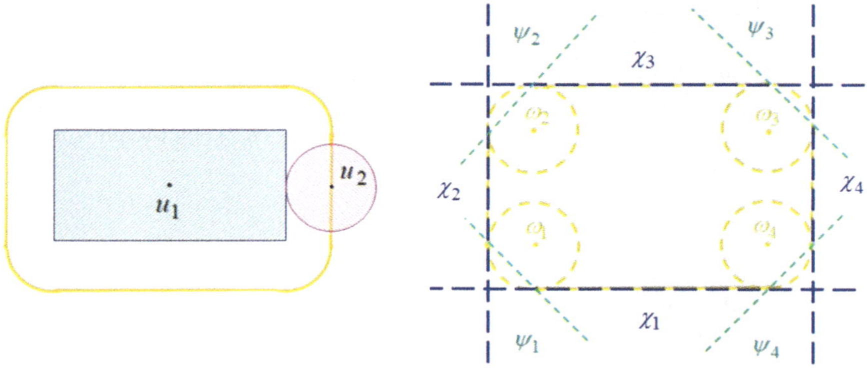

Fig. 15.10 Phi-function for a rectangle and a circle. By Prof. Dr. Tatiana Romanova, Copyright ©2020

$$\omega_3 = (x-a)^2 + (y-b)^2 - r^2,\ \omega_4 = (x-a)^2 + (y+b)^2 - r^2,$$

$$\psi_1 = -x - y - s,\ \psi_2 = -x + y - s,$$

$$\psi_3 = x + y - s,\ \psi_4 = x - y - s,\ s = a + b + r$$

$$A = a + r,\ B = b + r,\ x = x_2 - x_1,\ y = y_2 - y_1.$$

The reader may check by direct inspection that this Φ is continuous in x_1, y_1, x_2, y_2 and satisfies (15.3.5). Note that the phi-function is quadratic in its arguments (x_1, y_1) and (x_2, y_2).

15.3.1.9 Phi-function for a Convex Polygon and a Circle

Generalizing the above example, let K be a convex polygon with vertices (x_i, y_i), $1 \le i \le m$, and sides given by the equations $\alpha_i x + \beta_i y + \gamma_i = 0$ defined in Sect. 15.2, and subject to $\alpha_i^2 + \beta_i^2 = 1$. It is assumed that the vertices and sides are numbered clockwise and the ith side joins the ith and $(i+1)$st vertices. Let C be a circle with center (x_c, y_c) and radius r. Then the phi-function can be defined in the form

$$\Phi^{KC} = \max_{1 \le i \le m} \max \left\{ \alpha_i x_c + \beta_i y_c + \gamma - r, \Psi_i \right\}, \tag{15.3.19}$$

where

$$\Psi_i = \min \left\{ (x_c - x_i)^2 + (y_c - y_i)^2 - r^2, \right.$$
$$\left. (\beta_{i-1} - \beta_i)(x_c - x_i) - (\alpha_{i-1} - \alpha_i)(y_c - y_i) + r(\alpha_{i-1}\beta_i - \alpha_i\beta_{i-1}) \right\}.$$

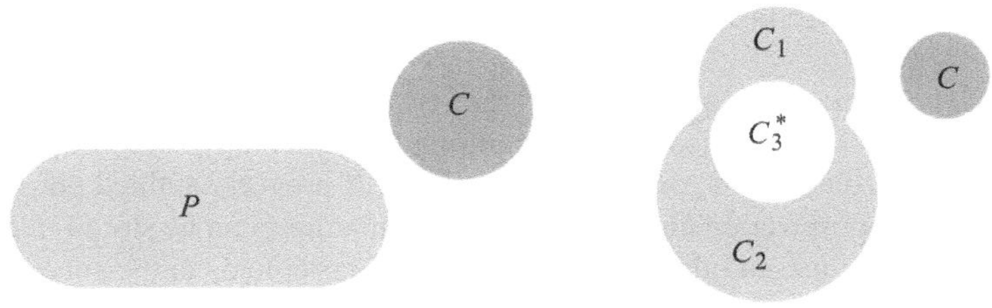

Fig. 15.11 Composed object and a circle. By Prof. Dr. Tatiana Romanova, Copyright ©2020

This formula generalizes (15.3.18).

As we see below, the construction of phi-functions for some composed objects may turn out rather simple.

15.3.1.10 Phi-function for a Composed Object and a Circle

Let P be the composed object defined by the form $P = R \cup C_1 \cup C_2$, where R is a rectangle, and C_i, $i = 1, 2$ are circles as displayed in Fig. 15.11a.

The phi-function for the composed object and a circle C can be defined by

$$\Phi^{PC} = \min\{\Phi^{RC}, \Phi^{C_1,C}, \Phi^{C_2,C}\},$$

where Φ^{RC}, $\Phi^{C_1,C}$, $\Phi^{C_2,C}$ as introduced in the previous sections. This follows from (15.3.16).

Let T be the composed object defined in the form $T = (C_1 \cup C_2) \cap C_3^*$, where C_i, $i = 1, 2$ are circles, C_3^* is the complement of the circle C_3 ($C_3^* = R^2 \backslash int C_3$), see Fig. 15.11b.

The phi-function for the composed object and a circle C can be defined by

$$\Phi^{PC} = \max\{\min\{\Phi^{C_1,C}, \Phi^{C_2,C}\}, \Phi^{C_3^*,C}\}.$$

This follows from (15.3.16) and (15.3.17).

Here, $\Phi^{C_1,C}$, $\Phi^{C_2,C}$ are defined as above and $\Phi^{C_3^*,C}$ can be given in the form

$$\Phi^{CC^*} = r^2 - (x_3 - x_c)^2 - (y_3 - y_c)^2,$$

$$r = r_3 - r_c.$$

15.3.1.11 Phi-functions for More General Objects

While the construction of phi-functions may be elaborate, it only needs to be done once for every pair of objects. In any cutting and packing problem with known shapes of available objects, one can prepare a set of properly defined phi-functions for the use by optimization algorithms. The phi-functions can be stored in advance, "offline," in a library, and then each instance of the problem can be solved quickly by calling the ready-to-use phi-functions from that library.

It is interesting to describe pairs of phi-objects for which one can find a radical-free phi-function expressed only by linear and quadratic formulas.

Fact 2 *If A and B are 2D composed objects (i.e., their frontiers are made by straight lines, rays, line segments, and circular arcs; recall Fact 1) fixing their orientations (i.e., exclude rotation angles), there exists a radical-free phi-function Φ^{AB}, the formula of which only involves linear and quadratic expressions.*

15.3.1.12 Phi-functions with Rotational Angles

If the orientations of the composed objects A and B are not fixed, then the formula for Φ^{AB} is obtained by changing variables that correspond to translating and rotating the coordinate system. It can be demonstrated by one example; the other cases are treated similarly.

Let K' and K'' be two convex polygons that are defined by (15.3.11)–(15.3.12). Suppose they are rotated about their poles by angles θ' and θ'' and then translated by vectors (u', v') and (u'', v''), respectively. Now let (x_i', y_i') be the coordinates of a vertex V_i' of K' in its eigen coordinate system. Then the coordinates of V_i' in the eigen system of K'' are

$$\begin{bmatrix} \tilde{x}_i' \\ \tilde{y}_i' \end{bmatrix} = \begin{bmatrix} c'' & s'' \\ -s'' & c'' \end{bmatrix} \left(\begin{bmatrix} c' & -s' \\ s' & c' \end{bmatrix} \begin{bmatrix} x_i' \\ y_i' \end{bmatrix} + \begin{bmatrix} u' \\ v' \end{bmatrix} - \begin{bmatrix} u'' \\ v'' \end{bmatrix} \right), \tag{15.3.20}$$

where the common notation $c' = \cos\theta'$ and $s' = \sin\theta'$ is used — the same procedure is applied to θ''. Similarly, if (x_j'', y_j'') are the coordinates of a vertex V_i'' of K'' in its eigen coordinate system, the coordinates of V_j'' in the eigen system of K' are

$$\begin{bmatrix} \tilde{x}_j'' \\ \tilde{y}_j'' \end{bmatrix} = \begin{bmatrix} c' & s' \\ -s' & c' \end{bmatrix} \left(\begin{bmatrix} c'' & -s'' \\ s'' & c'' \end{bmatrix} \begin{bmatrix} x_j'' \\ y_j'' \end{bmatrix} + \begin{bmatrix} u'' \\ v'' \end{bmatrix} - \begin{bmatrix} u' \\ v' \end{bmatrix} \right). \tag{15.3.21}$$

Now Eqs. (15.3.13)–(15.3.14) are modified as follows:

$$u_{ij} = \alpha_i' \tilde{x}_j'' + \beta_i' \tilde{y}_j'' + \gamma_i'$$
$$v_{ji} = \alpha_j'' \tilde{x}_i' + \beta_j'' \tilde{y}_i' + \gamma_j''.$$

Then the phi-functions $\Phi^{K'K''}$ are defined by the same formula (15.3.15) as before. This example shows how rotational angles (along with translation vectors) can be incorporated into the expressions for phi-functions.

Note that if the phi-function Φ^{AB} for two objects with a fixed orientation is radical-free, then including the rotational parameter θ brings the factors $\sin\theta$ and $\cos\theta$ into the formula, but it remains radical-free.

15.3.1.13 Normalized Phi-function

Some applications involve explicit restrictions on the allowable distances between certain pairs of objects (or between the objects and the walls of the container), i.e., some upper and/or lower limits on those distances may be set. In such cases, one may need to compute exact distances between the phi-objects to meet those requirements.

Thus there may be a use for phi-functions $\tilde{\Phi}^{AB}$ whose values equal dist(A, B) in case $A \cap B = \emptyset$. These phi-functions are called *normalized* phi-functions. The computation of geometric distances between primary and composed objects may involve rather complicated formulas with radicals, see a variety of examples detailed in [68], but they all can be done by using elementary geometry.

15.3.1.14 Normalized Phi-function for Two Circles

The normalized phi-function for two circles C_i, $i = 1, 2$, with centers (x_i, y_i) and radii $r_i > 0$ can be defined in the form

$$\tilde{\Phi}^{CC} = \sqrt{(x_1 - x_2)^2 + (y_1 - y_2)^2} - r,$$

$$r = r_1 + r_2.$$

Avoiding radicals is possible even in this case, provided that the restrictions on the distances between objects are known in advance, see the following section.

15.3.2 *Properties of Phi-functions*

Suppose the objects T_1 and T_2 have fixed metric characteristics and no rotation angles. Then the phi-function $\Phi(\nu_1, \nu_2)$ only depends on the two translation vectors ν_1 and ν_2. As Φ is determined by the relative position of two objects, it follows that

$$\Phi(\nu_1, \nu_2) = \Phi(\nu_1 - \nu_2, 0) = \Phi(0, \nu_2 - \nu_1).$$

Thus, to describe the phi-function, it is enough to fix the position of one object and only translate the other. Then the zero level of the phi-function, i.e.,

$$\gamma_{12} = \{\nu \in \mathbb{R}^d : \Phi(0, \nu) = 0\}$$

(here $d = 2, 3$) plays a special role: it describes all the translations of T_2 so that it touches T_1. This set is congruent ($\simeq$) to the *frontier* fr of the Minkowski sum of the two objects, i.e., $\gamma_{12} \simeq \text{fr}\, T_{12}(v)$, where $T_{12}(v) = T_1(0) \oplus -T_2(v)$ is the Minkowski sum of $T_1(0)$ and $-T_2(v)$. The Minkowski sum of two sets A and B is defined by

$$A \oplus B = \{X + Y \in \mathbb{R}^d : X \in A, Y \in B\}. \tag{15.3.22}$$

The set $\gamma_{12} \simeq \text{fr}\, T_{12}$ is also called *shape envelope* [34] or *hodograph* [521]. Note that $\gamma_{21} \simeq -\gamma_{12}$.

Most studies of the C&P problem in 2D are restricted to polygons (other shapes are simply approximated by polygons) and their orientation is usually fixed, thus no rotation angles are allowed. In that case γ_{12} is also a polygon; it is called the *No-Fit Polygon* (NFP). It bounds the region where the pole of T_2 should not be placed to avoid the overlap of T_2 with T_1.

Today, the No-Fit Polygon is the most common tool used in cutting and packing applications, and it remains the main object of study in modern literature on the subject. A number of efficient procedures have been developed for the construction of No-Fit Polygons; the first one was the orbiting algorithm (or sliding algorithm) of [380]. There are alternative algorithms, see [9, 67, 103, 216, 362].

Note that the No-Fit Polygon coincides with the zero level set of the phi-function in the absence of rotation angles and when one object is fixed. Thus the No-Fit Polygon is a special case of the broader theory of phi-functions [66].

15.4 Mathematical Optimization Model

In terms of phi-functions, the cutting and packing problem can be formulated as a constrained optimization problem suitable to be solved by general methods of mathematical programming.

First, for each object T_i a vector u_i of its variable parameters is defined; these may include

1. The translation vector ν_i,
2. The rotation angle(s) θ_i, and
3. Some metric dimensions if those are not fixed.

Thus $u_0, u_1, \ldots, u_n$ constitute the variables in the model.

15.4.1 Objective Function

The container T_0 is a special object. In most cases, it is not necessary to translate or rotate it, thus it is assumed that $\nu_0 = 0$ and $\theta_0 = 0$ and these parameters are excluded from the list of variables. On the other hand, the metric characteristics of the container are usually treated as variables, as some of those (for example, the length, perimeter, area, volume of the container) precisely are to be minimized. Thus the general goal is to minimize a certain objective function

$$\min F(u_0, u_1, \dots, u_n),$$

which may depend on some (or all) variables; though in most cases F only depends on the metric characteristics of T_0, i.e., $F = F(u_0)$.

15.4.2 Constraints

Next, all relevant constraints are listed. First, small objects T_i for $i = 1, \dots, n$ must be placed in the container, i.e.,

$$\Phi^{T_0^* T_i}(u_0, u_i) \geq 0 \qquad \text{for } i = 1, \dots, n,$$

where T_0^* denotes the (closure of the) complement of T_0.

Second, the small objects should not overlap, i.e.,

$$\Phi^{T_i T_j}(u_i, u_j) \geq 0 \qquad \text{for } 1 \leq i < j \leq n.$$

Third, there may be restrictions on the minimal and/or maximal distance between certain objects; in that case additional constraints have to be met:

$$\rho_{ij}^- \leq \tilde{\Phi}^{T_i T_j}(u_i, u_j) \leq \rho_{ij}^+$$

for some $1 \leq i < j \leq n$; here ρ_{ij}^- denotes the minimal allowable distance and ρ_{ij}^+ the maximal allowable distance. In this case the normalized phi-function $\tilde{\Phi}$ is used as the distances must be computed precisely. (But one can still avoid normalized phi-functions, see below.)

Fourth, there may be restrictions on the minimal and/or maximal distance from certain objects to the walls of the container, i.e.,

$$\rho_{0i}^- \leq \tilde{\Phi}^{T_0^* T_i}(u_0, u_i) \leq \rho_{0i}^+$$

for some $1 \leq i \leq n$. Lastly, there might be constraints on the rotation angles in the form $\theta_{\min} \leq \theta \leq \theta_{\max}$. This completes the list of constraints.

It is emphasized that (i) all placement constraints are defined by inequalities, and (ii) all phi-functions (except the optional constraints involving maximum and minimum distances) are fairly simple — they are continuous piecewise smooth functions expressed by linear and/or quadratic formulas. The objective function F is usually simple, too (for example, it may be just the length of the container).

15.4.3 Simplifying Distance Constraints

The distance constraints, as stated above, involve normalized phi-functions which may add unwanted radicals to the model. However, the formulas can be further simplified by eliminating radicals as follows. Suppose the minimal allowable distance ρ_{ij}^{-} for a pair of objects T_i, T_j is specified. An *adjusted* phi-function $\Phi(u_i, u_j)$ can be constructed such that

$$\Phi(u_i, u_j) = 0 \quad \text{if and only if} \quad \tilde{\Phi}(u_i, u_j) = \rho_{ij}^{-}$$

and so that the sign of $\Phi(u_i, u_j)$ coincides with that of $\tilde{\Phi}(u_i, u_j) - \rho_{ij}^{-}$. Since only the zero level set of the new function $\Phi(u_i, u_j)$ is rigidly specified, it can be defined by simpler formulas than those involved in the normalized phi-function $\tilde{\Phi}(u_i, u_j)$, i.e., via linear and quadratic formulas only. Now the minimal distance constraint $\tilde{\Phi}(u_i, u_j) \geq \rho_{ij}^{-}$ can be replaced with a simpler one:

$$\Phi(u_i, u_j) \geq 0.$$

In this way, all minimal and maximal distance constraints with inequalities can be replaced based on adjusted phi-functions and radicals are eliminated altogether.

For primary and composed objects, such a simplification is always possible. To see this, suppose A and B are primary or composed objects and the constraint reads $\text{dist}(A, B) \geq \rho^{-}$. Let $A_{\rho^-} = A \oplus (C, \rho^{-})$, where (C, ρ^{-}) denotes a sphere with radius ρ^{-} centered at the origin and $\oplus$ is the Minkowski sum. The object A_{ρ^-} consists of points that are either in A or at a distance $\leq \rho^{-}$ from A, and it is clearly a composed object, too.

Now the original constraint $\text{dist}(A, B) \geq \rho^{-}$ can be replaced with an equivalent one: $\Phi^{A_{\rho^-}, B} \geq 0$ (see Fig. 15.12). Due to Fact 2 there exists a phi-function $\Phi^{A_{\rho^-}, B}$ which can be constructed without radicals.

Example Suppose the constraint $\text{dist}(R_1, R_2) \geq \rho^{-}$ is given for two rectangles R_i, $i = 1, 2$, with centers (x_i, y_i) and half-sides $a_i, b_i > 0$ (assuming their sides are aligned with the coordinate axes). Then a phi-function for $R_1^{\rho^-} = R_1 \oplus (C, \rho^{-})$ and R_2 may be derived in the following radical-free form:

$$\Phi = \min\{ \min_{1 \leq m \leq 2} \Phi^{R_{1m} R_2}, \min_{1 \leq k \leq 4} \Phi^{C_{1k} R_2} \},$$

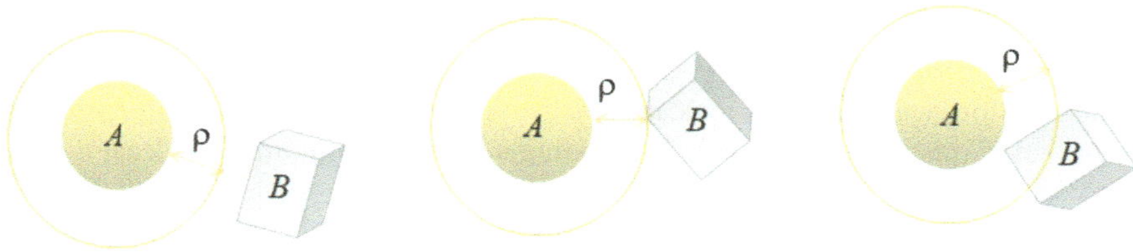

Fig. 15.12 Simplifying distance constraints for A and B

where

$$R_{11} = \{|x - x_1| \leq a_1 + \rho^-, \ |y - y_1| \leq b\}$$

$$R_{12} = \{|x - x_1| \leq a_1, \ |y - y_1| \leq b + \rho^-\}$$

and C_{1k} are circles of radius ρ^- centered on the vertices of the rectangle R_1.

15.4.4 General Remarks

All phi-function constraints define a feasible region W in the space of all the variables $u_0, u_1, \ldots, u_n$. The region W is also called the *solution space*. Let us consider a few characteristics of the model:

1. The solution space W is often a disconnected set. Each connected component of W may have a complicated structure, and in particular it may have multiple internal holes, "through" holes, and cavities.
2. The frontier of W is usually made of nonlinear surfaces containing valleys, ravines, etc.
3. The solution space W can be naturally represented as $W = \cup_{j=1}^{J} W_j$, where each W_j is specified by a system of inequalities of smooth functions *extracted* from the phi-function inequalities. It should be noted that J (the number of W_j's) may be huge, even larger than $n!$. Since each W_j is a non-convex set, the number of local extrema may be at least J.
4. The constraint optimization problem is multi-extremal and NP-hard.

Various solutions of this optimization problem are outlined in the next section.

15.5 Solving the Optimization Problem

In this section, we discuss various approaches for solving the optimization problem described in the previous section, i.e., finding a global minimum (or at least a good approximation to it) of the objective function F.

This task is treated as a mathematical minimization problem. Given an initial approximation, i.e., a point $U = (u_0, u_1, \ldots, u_n)$ in the solution space W, the algorithm performs a local search, i.e., it moves (modifies) the point $U \in W$ attempting to find a local minimum of F.

A point $U \in W$ corresponds to a particular layout of all the objects $T_1, \ldots, T_n$ inside T_0, and moving the point U through W is a *simultaneous* motion of all the objects $T_1, \ldots, T_n$ in T_0. This is where the algorithm allows a simultaneous motion of all the objects.

It is able to move all the objects at once, i.e., perform a local search in the multidimensional solution space W, because of using phi-functions. The phi-functions are continuous and piecewise smooth, and in most practical cases, they are conveniently defined by simple (linear and quadratic) formulas. These features are essential for smooth performance of local minimization schemes.

Thus, given an initial point $U_1 = (u_0, u_1, \ldots, u_n) \in W$, the algorithm finds a local minimum of the objective function. More precisely, given $U_1 \in W$ it forms a natural subset $W_{j,1} \subset W$ of the solution space W containing U_1, i.e., $U_1 \in W_{j,1}$. Then it finds $U_2 \in W_{j,1}$ so that $F(U_2) \leq F(U)$ for all $U \in W_{j,1}$ (or at least in a vicinity of U_2), and in addition $F(U_2) < F(U_1)$. Then, given U_2, the algorithm forms another natural subset $W_{j,2} \subset W$ containing U_2, i.e., $U_2 \in W_{j,2}$, and finds a point U_3 such that $F(U_3) \leq F(U)$ for all $U \in W_{j,2}$ (or at least in a vicinity of U_3) and $F(U_3) < F(U_2)$. This procedure is repeated until a local minimum of the objective function is found.

To find a global minimum of F over the whole space W, one would apply an exhaustive search, i.e., a search over *every* subset $W_j \subset W$, which is an unrealistic task in most cases, because the number of those subsets may be of the order 10^5 or 10^{10}. In practice, only a few (well chosen) initial points $U_1, \ldots, U_k \in W$ may be examined so that the task of choosing *good initial layouts* becomes of paramount importance.

In many industrial applications, experienced workers "manually" (with the help of CAD systems) build a high quality layout, cf. [229], which can then be followed by a quick run of a computer optimization program to improve the manual layout as much as (locally) possible.

In many other applications, however, there are no "expert layouts" available, and one has to rely on computer generated initial arrangements. In such situations various heuristics (and "metaheuristics") are used, the simplest and most popular perhaps being the *bottom-left placement procedure*. It places objects, one by one, in the most bottom-left vacant corner of the container. When positioning an object, the procedure takes into account the previously placed objects, first to avoid overlaps, and then (in some implementations) to fill holes left empty at earlier stages. Gomes

& Oliveira [228] also propose a randomized version of this method, where at each step the object to be placed next is selected randomly with probability proportional to its length.

Many authors then use various heuristics to (globally) alter the initial layout to obtain other layouts (and thus reach different components of the solution space W). One can swap two randomly chosen objects, or apply more sophisticated strategies so as "tabu search algorithms" [17, 65] or simulated annealing [229, 425], or various genetic algorithms [160].

In some implementations, objects are (temporarily) allowed to overlap and move through one another, so that the algorithm can perform a wider search over the solution space W. In that case, one needs to estimate (and penalize) the degree of overlap of objects so that the algorithm will gradually separate them and arrive at a feasible layout (with no overlaps) in the end. With respect to this phi-functions may be useful, too, as they provide such a feature as an estimate of the degree of overlap. Other authors develop different tools to penalize overlap; see [65, 229, 362].

Good initial layouts can be generated as follows, see [526]. First, the container T_0 and objects $T_1, \ldots, T_n$ are approximated by rectangular polygons (cuboids) $P_0, P_1, \ldots, P_n$ with sides parallel to two fixed coordinate axes. Then the polygonal (polyhedra) figures $P_1, \ldots, P_n$ are placed into P_0 consecutively, according to an object sequence $P_{i_1}, \ldots, P_{i_n}$ generated by a modification of the decremental neighborhood method. This procedure employs a probabilistic search and is designed to find the most promising ones. The latter will correspond to some points $U_1, \ldots, U_k$ in W. This time-consuming probabilistic search for local minima of F is only used to obtain the best initial points $U_1, \ldots, U_k$, and it produces k local minima $U_1^*, \ldots, U_k^*$ of F. In the end, the local minimum of F is chosen where the value of F is smaller than at the other local minima, i.e., $U^* = U_m^*$, where $m = \operatorname{argmin}\{F(U_i^*), 1 \le i \le k\}$.

Although the construction of an initial layout employs polygonal (polyhedra) approximations (and thus seems to be similar to many other techniques based on pixel and square representations, cf. [208]), strips as simpler enclosing shapes can be also applied, see [526, 528] and thus achieve a high speed in choosing an initial layout.

15.6 Numerical Examples

15.6.1 Arranging Two Triangles

To illustrate the usage of phi-functions in the context of deterministic global optimization, we consider two triangles and the problem of arranging them so that the convex hull perimeter becomes minimal. For this problem, Kallrath et al. (2021,[317]) have derived an analytic solution. From their general model for minimal perimeter convex hulls of polygons, we construct a simplified model dealing with two triangles T_1 and T_2 only. The input data are the x- and y-coordinates $(V_a^{\mathrm{x}}, V_a^{\mathrm{y}})$ and $(V_b^{\mathrm{x}}, V_b^{\mathrm{y}})$ of the vertices $(A_1, A_2, A_3) = (\mathrm{A}_1, \mathrm{B}_1, \mathrm{C}_1)$ of T_1 and $(B_1, B_2, B_3) = (\mathrm{A}_2, \mathrm{B}_2, \mathrm{C}_2)$ of T_2 describing both triangles in their own local coordinate system:

$$\mathrm{T}_1 : A_1 = (V_{A_1}^{\mathrm{x}}, V_{A_1}^{\mathrm{y}}) = (0,0),\ B_1 = (V_{A_2}^{\mathrm{x}}, V_{A_2}^{\mathrm{y}}) = (14,0),\ C_1 = (V_{A_3}^{\mathrm{x}}, V_{A_3}^{\mathrm{y}}) = (10,-5)$$

$$\mathrm{T}_2 : A_2 = (V_{B_1}^{\mathrm{x}}, V_{B_1}^{\mathrm{y}}) = (0,0),\ B_2 = (V_{B_2}^{\mathrm{x}}, V_{B_2}^{\mathrm{y}}) = (8,0),\ C_2 = (V_{B_3}^{\mathrm{x}}, V_{B_3}^{\mathrm{y}}) = (6,4).$$

We use the index sets $a \in \mathscr{A} = \{A_1, A_2, A_3)$ and $b \in \mathscr{B} = \{B_1, B_2, B_3\}$ to refer to the vertices of triangles T_1=A and T_2=B. The vertices A_a and B_b, *resp.*, are the origin of a straight line from A_a to A_{a++1} and B_b to B_{b++1} representing the sides $\mathbf{a}_a$ and $\mathbf{b}_b$ of the triangles. For two given triangles, we name them so that T_1 has the largest side (14 in our example for side $\mathbf{a}_1$).

As the triangle sides $\mathbf{a}_a$ and $\mathbf{b}_b$ are parts of straight lines we represent them exploiting the Hessian normal form

$$\mathbf{n}^{\mathrm{T}}\mathbf{x} = d.$$

For triangle T_1 and one of its sides $\mathbf{a}_a$ the coefficients for $\mathbf{n} = (\widetilde{\alpha}, \widetilde{\beta})$ and $d = -\gamma$ are given by

$$(n_x, n_y)(\mathbf{a}_a) = \frac{1}{\|\mathbf{a}_a\|}(V_{a++1}^{\mathrm{y}} - V_a^{\mathrm{y}}, V_a^{\mathrm{x}} - V_{a++1}^{\mathrm{x}}) \quad , \qquad d(\mathbf{a}_a) = n_x(\mathbf{a}_a)V_a^{\mathrm{x}} + n_y(\mathbf{a}_a)V_a^{\mathrm{y}},$$

i.e., for the given example

$$(\mathbf{n}_{\mathbf{a}_1}; d) = (-0.447213595499958, -0.894427190999916; 0)$$

$$(\mathbf{n}_{\mathbf{a}_2}; d) = (+0.780868809443030, -0.624695047554424; +10.932163)$$

$$(\mathbf{n}_{\mathbf{a}_3}; d) = (0, 1; 0).$$

Given a translation vector $(x_{\mathrm{T}}, y_{\mathrm{T}})$ and a rotation angle θ_{T}, each point $(\widetilde{x}, \widetilde{y}) \in T(0,0,0)$ in the local coordinate system of triangle T is transformed into point (x, y) as follows:

$$x = +\widetilde{x}\cos\theta_{\mathrm{T}} + \widetilde{y}\sin\theta_{\mathrm{T}} + x_{\mathrm{T}},$$

$$y = -\widetilde{x}\sin\theta_{\mathrm{T}} + \widetilde{y}\cos\theta_{\mathrm{T}} + y_{\mathrm{T}}.$$

Each straight line

$$\widetilde{L} = \{(x, y) \in R^2 \left| \widetilde{\alpha}x + \widetilde{\beta}y + \widetilde{\gamma} = 0, \widetilde{\alpha}^2 + \widetilde{\beta}^2 = 1 \right.\}$$

is transformed into the straight line

$$L = \{(x, y) \in R^2 \,|\alpha x + \beta y + \gamma = 0\},$$

where

$$\begin{aligned}
\alpha &= \widetilde{\alpha} \cos\theta_{\mathrm{T}} + \widetilde{\beta} \sin\theta_{\mathrm{T}} \\
\beta &= -\widetilde{\alpha} \sin\theta_{\mathrm{T}} + \widetilde{\beta} \cos\theta_{\mathrm{T}} \\
\gamma &= \widetilde{\gamma} - \alpha x_{\mathrm{T}} - \beta y_{\mathrm{T}},
\end{aligned}$$

and θ_{T} is a rotation parameter, $(x_{\mathrm{T}}, y_{\mathrm{T}})$ is a translation vector applied to T. Note that for the transformed straight line $\alpha^2 + \beta^2 = 1$ also holds.

The maximum number of sides e_s, $s \in \mathscr{S} := \{1, \ldots, m\}$, of the convex hull Ω is $m = 4$. As the fundamental variables we have the placement-rotation variables $\mathbf{u}_a = (x_a, y_a; \theta_a)$ and $\mathbf{u}_b = (x_b, y_b; \theta_b)$ of both triangles T_1 and T_2, respectively, and the side-information variables $\mathbf{e}_s = (x_s, y_s, \omega_s, \ell_s)$, where s refers to side s of Ω, and ℓ_s is the length of side s. To break symmetry we assume that triangle T_1 is fixed and triangle T_2 can be translated and rotated freely; therefore we have $\mathbf{u}_a = (x_a, y_a; \theta_a) = (0, 0; 0)$ leaving only $\mathbf{u}_b = (x_b, y_b; \theta_b)$ and the vertex coordinates of the convex hull as variables.

The objective function to be minimized is the linear term

$$\ell = \sum_s \ell_s .$$

Non-overlap of both triangles is enforced by

$$n_{A_3}^{\mathrm{x}} w_b^{\mathrm{x}} + n_{A_3}^{\mathrm{y}} w_b^{\mathrm{y}} - d_{A_3} \geq 0 \quad , \quad \forall b \tag{15.6.23}$$

with the translated and rotated vertex coordinates w_b^{x} and w_b^{y} of triangle B

$$\begin{aligned}
w_b^{\mathrm{x}} &= x_b + V_b^{\mathrm{x}} \cos\theta_b - V_b^{\mathrm{y}} \sin\theta_b \quad , \quad \forall b \\
w_b^{\mathrm{y}} &= y_b + V_b^{\mathrm{x}} \sin\theta_b + V_b^{\mathrm{y}} \cos\theta_b \quad , \quad \forall b.
\end{aligned}$$

Note that (15.6.23) is similar to the separation line or half-space approach in Sect. 13.3.1.2. All vertices of triangle T_2=B are above side $\mathbf{a}_1$ of the fixed triangle T_1=A. Geometrically, this means triangle B is attached to the largest side of A.

All vertices of the fixed triangle T_1 belong to the convex hull, i.e.,

$$-[V_a^{\mathrm{y}} - x_s] \sin\omega_s + [V_a^{\mathrm{y}} - y_s] \cos\omega_s \geq 0 \quad , \quad \forall s.$$

For the free triangle T_2, the inequalities describing that all its vertices are within the convex hull are

$$-[w_b^x - x_s]\sin\omega_s + [w_b^y - y_s]\cos\omega_s \geq 0 \quad , \quad \forall s.$$

Closing the convex hull, i.e., glueing the sides to each other, is enforced by

$$x_{s++1} = x_s + \ell_s \cos\omega_s \quad , \quad \forall s \tag{15.6.24}$$

$$y_{s++1} = y_s + \ell_s \sin\omega_s \quad , \quad \forall s. \tag{15.6.25}$$

To avoid degeneration of the sides of the convex hull, we add the inequalities

$$-(x_{s++2} - x_s)\sin\omega_s + (y_{s++2} - y_s)\cos\omega_s \geq \varepsilon \quad , \quad \forall s, \tag{15.6.26}$$

where ε is a small number, e.g., $\varepsilon = 0.00125$.

Degeneration caused by naming and counting the convex hull vertices is eliminated by fixing the first convex hull vertex S_1. This fixation requires some care. For the example below we can safely fix the first convex hull vertex S_1 to vertex A_3 of triangle T_1 as A_3 is always a convex hull vertex. With this fixation $(x_{S_1}, x_{S_2}) = (10, -5)$, the gap falls below 10^{-8} in 50 s using BARON.

For the following example of two triangles with vertices

$$T_1 : A_1 = (0,0), B_1 = (14,0), C_1 = (10,-5)$$

$$T_2 : A_2 = (0,0), B_2 = (8,0), C_2 = (6,4)$$

the model is contained in MCOL as *CH-2TRI.gms*; it produces the solution $u_* = 7$, $\alpha_2 = 33^\circ.6900675$, $x_{B_1} = x = 1$, and $\ell = 33.7079796215289$; see also Fig. 15.13.

15.6.2 Arranging Two Irregular Objects

The broad range of situations in which phi-functions and phi objects can be used successfully has been demonstrated by Bennell et al. (2015,[69]). From that paper, we reproduce a few examples:

1. Two irregular objects A and B are to be arranged so that various objective functions are minimized (Fig. 15.14). Both objects can be translated and rotated.
2. Minimal convex hulls of two objects A and B. Figure 15.15a displays the convex hull solution of two convex polygons, while Fig. 15.15b shows a locally optimal solution of two non-convex objects.
3. Minimal convex hulls for two non-convex objects A and B. A locally optimal solution, i.e., only an approximation, of the convex hull of minimal area is shown

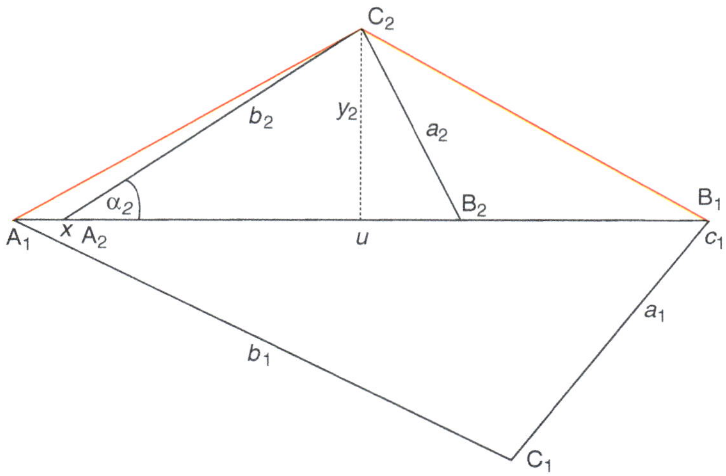

Fig. 15.13 Two triangles attached to each other and its convex hull

in Fig. 15.16a. An approximation of the convex hull of minimal perimeter for the two non-convex objects is displayed in Fig. 15.16b.

These and more examples in Bennell et al. (2015,[69]) demonstrate the potential of the algorithm to work with a large number of irregular objects in 2D and 3D and achieve tight packing arrangements that are hard to find otherwise, especially by manual work. More examples from industry with technical details and illustrations one can find in, cf. [466, 522, 533, 534].

15.7 Conclusions

In this chapter we have demonstrated how the combined use of phi-functions and mathematical programming can improve the performance of cutting and packing algorithms. Phi-functions have the following properties:

- They can be applied to 2D and 3D objects of very general types (phi-objects); these include disconnected objects, non-convex objects, regions with holes and cavities, etc.
- They take into account continuous translations and rotations of objects.
- They may take into account variable metric characteristics of objects.
- They take into account possible restrictions on the (minimal and/or maximal) distances between objects and from the objects to the walls of the container.
- They are useful when dealing with overlapping objects, as they measure the degree of overlap.

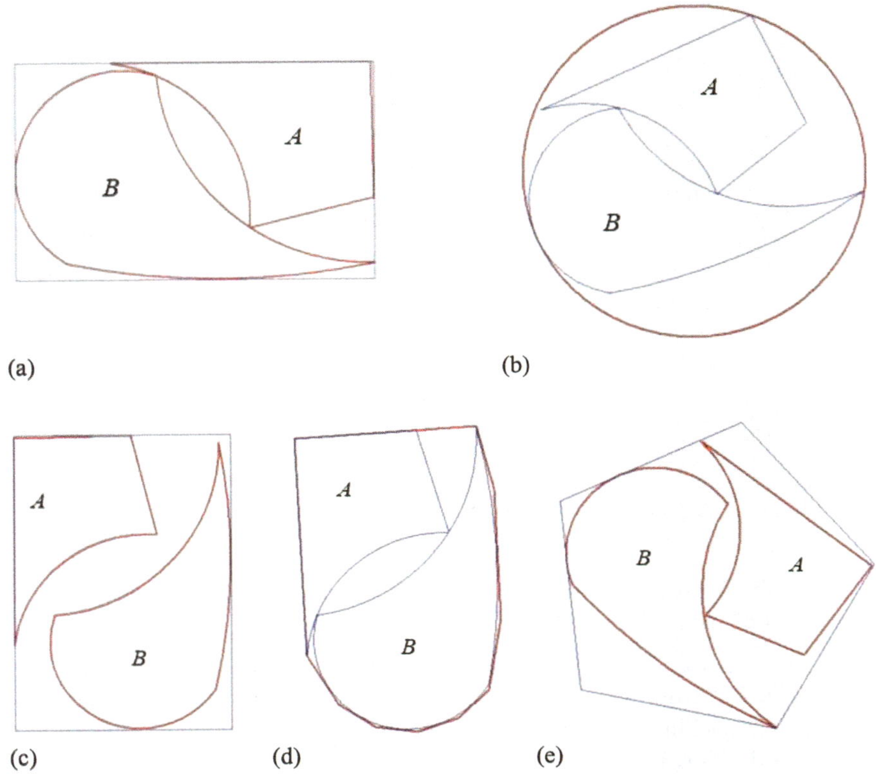

Fig. 15.14 Arrangement of two objects A and B as described in Example 1: (**a**) minimal enclosing rectangle, (**b**) minimal enclosing circle, (**c**) minimal enclosing rectangle taking into account distance constraints causing the objects not to touch each other, (**d**) minimal enclosing m-polygon, (**e**) minimum homothetic coefficient. Reprinted by permission from Springer Nature, Journal of Global Optimization, Bennell et al. (2015,[69]), Fig. 6

- In most practical cases, phi-functions (unlike geometric distances) are defined by simple (linear and quadratic) formulas, which allows us to use optimization algorithms of mathematical programming.
- Overall, phi-functions allow enlarging the class of optimization placement problems that can be effectively solved.

15.8 Summary and Recommended Bibliography

In this chapter we have presented phi-function techniques for solving cutting and packing problems. Thus the reader should now be familiar with:

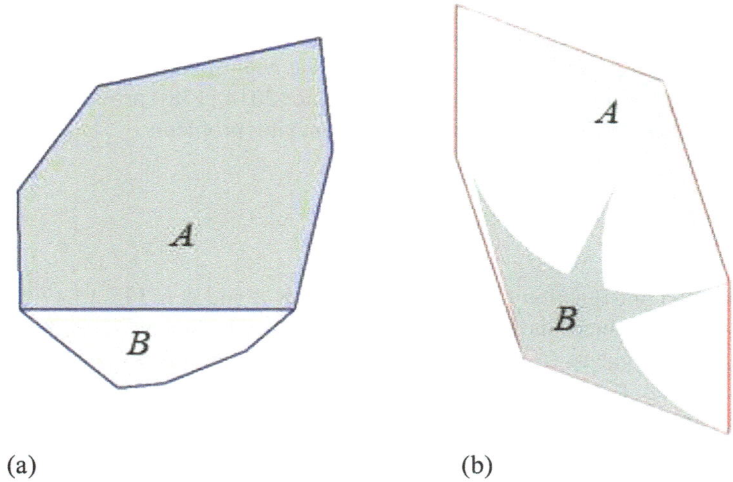

Fig. 15.15 Convex hull for two objects A and B: (**a**) two convex polygons, (**b**) two non-convex objects. Reprinted by permission from Springer Nature, Journal of Global Optimization, Bennell et al. (2015,[69]), Fig. 9

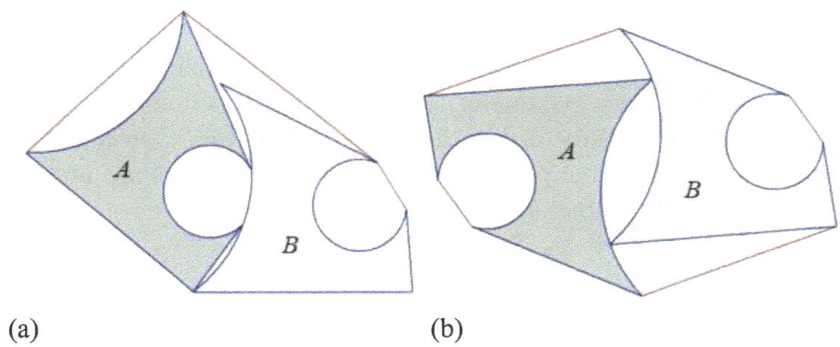

Fig. 15.16 An approximation of the m-polygonal convex hull of two non-convex objects of Example 3: (**a**) area of the convex hull; (**b**) perimeter of the convex hull. Reprinted by permission from Springer Nature, Journal of Global Optimization, Bennell et al. (2015,[69]), Fig. 11

- Phi-functions;
- Phi-objects;
- Benefits of phi-functions;
- General solution approaches for cutting and packing problems; and
- Constructing NLP models of cutting and packing problems using phi-functions.

Recommendations for further reading: A comprehensive overview of the most important and frequently considered optimization problems concerning cutting and packing are provided in the book *Introduction to Cutting and Packing Optimization - Problems, Modeling Approaches, Solution Methods* by Scheithauer (2018,[488]).

The paper *Irregular packing problems: A Review of Mathematical Models* by Leao et al. (2019,[357]) reviews mathematical models of nesting problems, highlighting differences and similarities among them. *Solving Non-Standard Packing Problems by Global Optimization and Heuristics* by Fasano (2014,[178]) provides an interesting overview on various types of uncommon packing problems.

15.9 Exercises

1. Construct a phi-function for the two phi-objects A and B in Fig. 15.17. Object A is a composed object and object B is a circle. *Hint*: The first object is composed of four primary objects C_1, C_2^*, C_3^*, R_4^* using the operations *union* and *intersection*; C_1 is a circle, C_2^* and C_3^* are two circular holes, R_4^* is a rectangular hole.
2. Construct a mathematical model and solve the following optimization problem. Place non-overlapping circles A, B, C of variable radii inside a convex polygon with vertices $\{(2,0),(0,2),(-2,0),(-2,-3)\}$ using the phi-function technique, so that the sum of the radii of the circles will be maximal. Use an available NLP solver to solve the nonlinear programming problem.
3. Construct a mathematical model and solve the following optimization problem. Let there be a polygonal region with vertices $\{(4,0),(1,3),(-1,3),(-4,0),(0,-4)\}$ and the polygon A with vertices $\{(2,-1),(0,2),(-2,0)\}$. Find the maximum size of the polygon A with respect to the value of the homothetic coefficient β (or scaling parameter), using the phi-function technique, so that βA will be arranged fully inside the polygonal region. Consider two cases: continuous rotations of the polygon are allowable; rotations are not allowable. Use an available NLP solver to solve the NLP problem.

Fig. 15.17 Two phi-objects A and B for which a phi-function should be constructed in Exercise 1

Chapter 16
The Impact and Implications of Optimization

In this chapter many issues are touched upon which are part of more general operational research concerns, particularly when these are amplified by using mathematical programming. It includes a discussion of the possibilities of parallel optimization.

16.1 Benefits of Mathematical Programming to Users

The process of modeling is a very rigorous one, in the sense that deficiencies in model *and* data are exposed by solvers. If one has forgotten a class of constraints, it is likely that an unbounded solution will result. If one has not been careful enough in collecting data, then the result is often infeasible models.

Though the first reaction of the modeler and the end user is usually one of frustration after the euphoria of building a syntactically correct model, the early exposure of deficiencies in data or logical thought is a huge benefit of optimization. All the work done in Quality Improvement Technologies has shown that the earlier a defect is caught, the less costly the rectification process, and this discovery is just as true in the context of business modeling.

Bringing together different parts of the organization into one model is a further benefit of optimization. Too often in companies, individual sections or groups act to maximize their own contribution, neglecting the effect of this local objective on overall profitability. And this effect can occur even if transfer prices have been set.

Optimization will specify a single overall objective function and will automatically achieve the coordination required to yield the best objective.

Since the optimization model will draw into one place all the required data, it is the ideal place to test data consistency and accuracy. Users of optimization frequently report that the results of this data "cleaning" and clarification are as, if not more, important than the precise results of optimization runs.

J. Kallrath, *Business Optimization Using Mathematical Programming*,
International Series in Operations Research & Management Science 307,
https://doi.org/10.1007/978-3-030-73237-0_16

It can be argued that optimization is too hard a taskmaster, that it will not be possible to achieve in practice what the model states is possible. This statement is, of course, true, but another benefit of optimization is that the decision maker has an objective measure (in both senses of the word objective!) of the best that could be done, and so is capable of making a rational assessment of what was actually achieved. There may be considerable satisfaction in knowing that even if you did not do the best, you did within, say, 2% of that best.

So, to summarize, users of mathematical programming benefit in three major ways when models are solved. First, there is the gain through the greater understanding of the problem. The model described in Sect. 10.4 is of this type. The very act of working through a model formulation with its builder can be of a considerable benefit to a client. Secondly, there is the production of a decision support system using the model and its solution capability. The airline model by Subramanian et al. (1994,[536]) cited in Sect. 7.4.5 provides an example of this and an indication of substantial savings made. Thirdly, there is the availability of a model for future experimental purposes. The model in Sect. 10.5 provides an example of this, where it is possible to try out ideas for future planning on the model that could not be conducted on the actual processes themselves. Thus the gains to clients are considerable. The prospect of saving even a few percent of a very large cost is exciting for a client or the prospect of devising a new business process which will make an organization more competitive can be very encouraging and justifies the operational research approach.

16.2 Implementing and Validating Solutions

As discussed in the large cases introduced in Chap. 10, regular communication with the client is important for model validation. Emphasis switches from validation by the modeler to validation by the end user. Once solutions are proposed the validation process continues and the modeler must continue to work with the client. The model would normally be tried out on test data to ensure that the model is robust, is predictively valid (produces predictions that are in line with existing possibilities), and is replicatively valid (produces working solutions). Thus communication and feedback between client and modeler continues. When, finally, a proposed solution is considered for implementation, checking will continue as the process of implementation may not be straightforward. It will also be important for the modeler to stress what assumptions have been made in the modeling and the potential shelf-life of the model. The modeler and users will need to monitor the model in the future to see if any breaches of the assumptions are made or any aspects of the decision support system pass their expiry date.

16.3 Communicating with Management

Even with a seemingly precise process such as mathematical programming there will be a stage once modeling has been undertaken when the modeler has to "sell the solution" to the client. This will be well before implementation is even contemplated. The results of a modeling exercise may produce solutions which are perhaps unexpected, e.g., indicative of inefficient[1] current practices and such results may prove unpopular with some individuals (but highly popular with others). Thus the modeler has to convince the client that the model is performing according to the conditions laid down by the client and to establish that no stage has been omitted or information misinterpreted. Thus it is important that a comfortable relationship of "mutual trust" exists throughout the lifetime of a mathematical programming project between modeler and the client for whom the model is being built.

16.4 Keeping a Model Alive

A number of difficulties exist for the users of mathematical programming models.

Firstly, there are the typical difficulties with the use of any sophisticated software system. The user of the model may not follow the "rules" envisaged by its developer and may try to use data that are not appropriate or to use the model for purposes for which it was not designed. This difficulty is particularly marked when users employ a model to investigate different parts of a company e.g. an organization with many outlets or branches. The model may have been tested on many such parts of the company, but there may still be some parts which produce unexpected and confusing results. Problems may turn out to be infeasible or unbounded and the user may not spot this and may go ahead and use what seems to be the optimal solution from such models. The remedies here lie with validation, as was discussed in Chap. 2, and the modeler must have made clear the limitations of the model and the assumptions under which it operates. However, once a model has been implemented there may be no real checks as to whether the user follows the rules of the game, especially when a model has been used for a considerable time. All that can be done is to provide accurate documentation, including what to do when things go wrong, and to encourage periodic health checks on users, data, and model.

Secondly, there is the difficulty of the total reliance on the model to produce the answers required. As the approach is a black box one, the user still needs to question what results the model produces and see if they make sense. The results may be remarkable, but they must still be reasonable. One of us was involved with an inventory model which had been built at an engineering company ten years earlier.

---

[1]The term *inefficient* is a relative term and should give nobody a bad conscience. Before the light bulb was invented a candle produced sufficient light for Homer to write the Iliad and Copernicus to prove Earth was not the center of the universe.

All the people who had been involved in the building of the model had long since left the organization and no one knew what the model really did. However, inventory decision making relied on it totally. Eventually consultants were called in to try to establish what the model actually did, find out whether it was still valid, and make appropriate changes. The elapsed period was clearly too long and users should have questioned earlier whether the model had passed its "sell by" date, and should not have continued to rely absolutely on it.

Thirdly, the converse to the above, there will be a danger that as time moves on new users will emerge who will feel that as they were not involved in the original development of the model then they can have no faith in it. This may have to be remedied by the involvement of the developers of the model from time to time.

The above three points suggest that even if a model is developed from outside an organization, that organization must plan to acquire or buy in the expertise for later validation of the model, even if it is continuing to produce apparently sensible results year after year. In the cases discussed in Chap. 10, communication with and feedback from the client was regularly mentioned, and this leads to a valid model being built.

16.5 Mathematical Optimization in Small and Medium Size Business

In many of the case studies, the reader might have got the impression that mathematical optimization is something particularly useful for larger companies or organizations. It is certainly true that at present small and medium size businesses make less use of it. Although governments in most western countries support small business development, these businesses usually do not have the know-how or even awareness of operational research benefits.

To increase the awareness and acceptance of mathematical methods the analyst and modeler should have a few things in mind when trying to model problems in small or medium size companies. It is not sufficient just to scale down models and approaches used in large companies. The modeler must ensure that he adjusts himself to the world the small business operates in. In contrast to large companies, which usually have some special service groups with strong academic backgrounds providing OR consulting to the company, the situation is quite different in small and medium size companies. While large companies very often seek solutions for complex integrated production networks, small and medium size business might concentrate on small and limited aspects changing frequently. The modeler has to be very flexible and to be prepared to solve completely different but relatively small problems. The mathematical challenge might be smaller but the modeler's social skills, e.g., ability to communicate and to convince people, become even more important.

Last but not least, the budget frame in small and medium size business is certainly significantly smaller than that of large companies. It is important to have that in mind because it helps to understand people and their motives.

16.6 Online Optimization by Exploiting Parallelism?

In some application areas, solution speed is critical for the success of the method and the acceptance of technology. Scheduling problems, for instance, must be solved in minutes when sudden changes occur in a factory and personnel has to be reallocated to machines. That is a sort of online optimization. To get a solver suitable for this task we need technological progress related to algorithms, software, and hardware.

In fluid mechanics, astrophysics, or quantum chemistry a considerable runtime improvement can be achieved by parallelization. Can this not also play a role in the optimization environment? The answer is certainly yes, whereas a distinction has to be made between parallelization efforts of the combinatorial part of the algorithms (especially non-deterministic aspects are to be expected here) and the core optimization problem (LP, MILP, NLP, MINLP).

For compatibility, we keep the material from the first edition of this book in the section below now named *Parallel Optimization: Status and Perspectives in 1997*. Most of the conceptual issues are still valid while some aspects may appear interesting from a historical point of view. In Sect. 16.6.2 we follow up with the state of the art and perspectives in 2020.

16.6.1 Parallel Optimization: Status and Perspectives in 1997

For mixed integer linear or nonlinear optimization problems, commercial software packages now offer standard parallel versions for PC networks or multiprocessor systems with almost linear behavior in the number of processors exploiting multi-thread techniques. While parallel algorithms and their implementation in the late 1970s were still rather a topic for specialists, they are now part of everyday life. In the field of nonlinear optimization, this may be a little different at the moment. However, this has more to do with the fact that the algorithms in this area are younger and less mature and have not yet been tested against many really big problems. Especially with mixed integer optimization, however, one should always bear in mind that with parallel architectures, even with linear behavior, the exponential growth of the B&B methods cannot be controlled. In this sense, parallel optimization software helps push back the limits of what is still solvable in benign problems; but it is only a limited means to solve structurally difficult problems, e.g., scheduling problems. In mixed integer nonlinear optimization, if you are not already using methods of global optimization, you can assume different starting points exploiting multi-start techniques exploiting several threads of your computer.

16.6.1.1 Algorithmic Components Suitable for Parallelization

The exact methods briefly described in Chap. 3 for solving mixed integer linear problems offer two different ways for parallelization: the combinatorial part of the algorithm and the linear program algorithm.

The combinatorial part is either a B&B or a B&C algorithm. In both cases, it is necessary to solve many LP subproblems. Obviously, the evaluation of the subproblems may be performed by a network of parallel processors or workstations. The subproblems are more or less decoupled from each other and allow a simple parallelization with coarse granularity. Positive results have been achieved (Ashford et al. 1992,[38]) on a transputer system with 8 slave- and one master processor. It was possible to get an almost linear speed-up [see Fig. 16.1].

The linear optimization kernel is much more difficult to optimize. As described in Chap. 3 commercial software uses two methods to solve linear programs: revised Simplex algorithm and interior-point methods. There exist attempts to parallelize the Simplex algorithm, but they only obtain a low speed-up. Therefore, there is more optimism toward the parallelization of interior-point methods. The major numerical work of solving IPMs is to solve nonlinear systems of equations. Linearization in combination with Newton's method leads to linear systems of equations. At that level, broad experience with parallelization is available.

In the next two subsections let us briefly reflect on some consequences parallelism has on the combinatorial part of the algorithm.

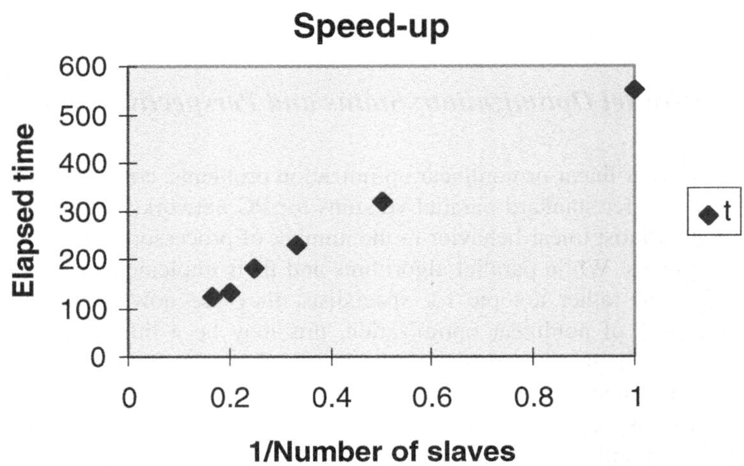

Fig. 16.1 Speed-up achieved with eight slaves

16.6.1.2 Non-determinism in Parallel Optimization

It comes as a surprise to many users to find that runs of exactly the same optimizer on the same input data can yield different optimal solutions. How can this happen?

A parallel MIP solver will typically work by sending LP relaxations to other processors for solution. Setting up the communication channel to send the data will generally take some random time, which will depend on the traffic on the network, or what other jobs are being run on the target processors. Thus the (clock) time before the LP solution is received back is a random variable. It is likely that several processors will be solving different LP relaxations at the same time, so in one run processor A might get its solution back first, while in another run processor B will get in first. The processor that controls the node selection will have different sets to choose from in the two runs, and may well choose a different node to work on next. As soon as that happens we have a completely different B&B tree, and integer solutions may be found in a different order.

Even worse, as we may find integer solutions in a different order, and if we are only trying to find the best IP solution within a certain tolerance, we may cut-off a slightly superior node that would have been encountered in a different run. Thus we stop, perfectly legitimately, with non-identical solutions. Parallel optimization is not unique in this initially worrying behavior, but it demonstrates the phenomenon frequently enough that the optimization specialist has to be prepared to explain this to the concerned end user.

16.6.1.3 Platforms for Parallel Optimization Software

There are at least two major platform types which can be recommended for parallel MILP software. The first of these consists of relatively small PC networks, or clusters of workstations (of the order of 10 rather than 100 machines), often comprising workstations from more than one manufacturer. These workstations are invariably networked, usually by a relatively slow Ethernet connection.

The second hardware platform is represented by quite cheap, closely coupled parallel hardware such as that manufactured by Parsytec GmbH (Aachen, Germany). Here the inter-processor communications are generally much faster, but there are restriction on the algorithms which can be implemented, as there is no disk storage attached to each processor.

Another design criterion requires the software to be easily portable between these two very different hardware bases. Furthermore, the software must be easy to support on heterogeneous workstations. Almost inevitably this means basing the inter-processor communication on PVM [PVM stands for *parallel virtual machines* and is the most frequently used platform for networked parallel computing (Geist et al. 1994,[210])], the *de facto* standard in the 1990s. However, PVM has the advantage that it addresses some more esoteric hardware platforms, which may be of use in some specialized applications. In 2020, MPI (Message Passing Interface) is the standard.

Given the very wide hardware base that is addressed, it is vital that the communications software is as simple as possible and ideally isolated in one module.

One consequence of supporting different hardware platforms means it cannot be assumed that different processing elements have sufficient memory to store information about all the nodes solved on that element, because not all hardware platforms have local disk storage. This limitation rules out several initially attractive architectures in which the B&B is effectively cut into small pieces and farmed out a priori, possibly with later load balancing, to identical slaves all doing their own local B&B. This approach has been adopted by IBM's Parallel OSL, which only addresses an architecture where each processor has its own local disk storage.

It is also important to ensure that a parallel version of MILP software is simple to use for existing users and involves little or no re-learning.

16.6.1.4 Design Decisions

In XPRESS-OPTIMIZER the design criteria set out above, coupled with the different targeted platforms, have dictated that a master–slave architecture be adopted. This is one of the classical paradigms of parallel computing, and its effectiveness on limited trials on a small array of transputers was demonstrated some years ago by Ashford et al. (1992,[38]).

The processors are partitioned into two sets, one containing a single master processor connected to local disk storage and logically connected to the rest of the processors (the slaves) which form the second disjoint set. The master processor has the following tasks:

1. Accept the problem specification from the host. In the case of a network of workstations, this task is done directly from disk storage local to the master. This is called the base LP.
2. Solve the first linear programming relaxation.
3. Farm out identical copies of the base LP to the remaining slave processors.
4. Maintain all global data, in particular data on the tree structure and summary node information.
5. Supply the modifications to the base LP to a currently free slave when specifying the problem to be solved.
6. Receive solution information from the slaves and update the global node information.
7. Notify the host when the B&B is complete (not necessary on workstations).
8. Deliver the solution to the host (again, not necessary on workstations).

In contrast the slaves' jobs are much simpler:

1. Receive the base LP.
2. Receive modifications that define which particular relaxation is to be solved. At the same time, receive an advanced basis so that re-optimization is faster.
3. Solve the LP.

4. If the relaxation is feasible, not cut-off and not integer, determine the branching entity and the branching direction. If the LP is either infeasible, integer, or has been cut-off (dominated), then just inform the master.

In summary, the master processor takes the role of central coordinator with the slaves acting as mere LP solvers with the added task of exploiting the LP solution to generate branching decisions.

16.6.1.5 Implementation

A centrally coordinated scheme may at first sight look vulnerable to difficulties in scaling, because all information has to be funneled through the master processor. To see whether this is so in practice it is necessary to carefully consider the following factors:

1. Average work at master, 2. Average work at each slave, 3. Quantity of data passed from master to slave, and from slave to master, 4. Latency in message passing.

Simple experiments on large problems show that the work at the master, which consists of node selection and B&B tree maintenance, is very light. However, the time required at each node to solve the LP relaxations associated with large MILP problems is very much larger, often running to several minutes.

A careful analysis of the data transfer needs on serial XPRESS-MP shows that by exploiting the detailed structure of the most common integer entities the number of bytes that had to be passed between master and slave could be greatly reduced. Consequentially, the transfers to and from disk could be reduced.

The data transfers grow linearly with the problem size, whereas the work at each node grows faster than linearly. As a result, the ratio of data transfer to computation time tends to be favorable in circumstances where more processors need to be added to reduce total elapsed time. This is not necessarily the situation in combinatorial problems, where the solution time at each node is often a very slowly growing function of problem size.

A parallel version of XPRESS-MP is available. The implementation was laborious but straightforward. A less sophisticated node selection routine at the master was required, and the results reported below were obtained using this new routine. The new routine tends to search more nodes in the B&B tree than the old serial version did, even when using just one slave, when it is directly comparable to the serial version. Work has begun on a third version which is believed to be as efficient as the serial version in terms of the number of nodes explored.

16.6.1.6 Performance

Early experiments on the limited hardware available at Dash (Ashford et al. 1992,[38]) have shown that speed-ups on the target problems were linear, or nearly so. It must be stressed, however, that it is very hard to make definite statements when only three workstations were available at Dash. The best test of performance came in early 1995 when the first commercial copy of Parallel XPRESS-MP passed its acceptance tests.

One commercial client had some medium- to large-scale long-term scheduling problems, with a need to run many scenarios. Preliminary results on problems run at night with Parallel XPRESS-MP on unused RS/6000s convinced the client to purchase 6 top-end RS/6000s to be dedicated to solving MILP problems. One of these workstations was a dedicated master, the other five were slaves.

The client reported achieving a factor of better than 4 speed-up, on average, compared with the current production serial XPRESS-MP solver, including the initial LP solution, on the 6 processors. Since only five of these were being used for solving LP relaxations, this represented a substantial speed-up over serial times. Failure to achieve a perfect linear speed-up can be explained by the observation that the parallel version typically explores more nodes.

16.6.1.7 Acceptability

As anticipated, the client encountered most difficulties in that the initial workstation cluster was on a heavily used local area network, and it was found that network traffic was causing delays. However, as a result of the efforts Dash had put into reducing the amount of data exchanged, the overall solution time was not significantly degraded.

The client and problem class can be considered typical of those who make up the expected market for parallel MILP solvers, with perhaps a bias toward the more challenging end of the spectrum. There was a heavily loaded network giving poor communication throughput, and very fast workstations meaning that the work at each node did not take long. Both of these factors have mitigated against our design performing as well as it might.

The client found no difficulties in understanding the small extensions to the parallel solver control language. However, it was a surprise that the client has not objected to the non-determinism [see Sect. 16.6.1.2 for what causes non-determinism] of the parallel solver.

Besides all technological progress, one should have in mind that the acceptance of fast optimization software also depends on how these techniques can overcome existing cultural, social, and psychological barriers.[2] Thus, besides technological efforts there should be a strong investment in improving the awareness and acceptance of mathematical optimization applied to real-world problems.

16.6.2 Parallel Optimization: Status and Perspectives in 2020

In Sect. 16.6.1 we have already addressed some of the conceptual issues related to parallel optimization. The state of the art and perspective of parallel optimization

[2] See page 547 for further details on this issue.

has been summarized in the book by Censor & Zenios (1997,[111]) — shortly after the publication of our first edition. During the time in between 1997 to 2020, there has been tremendous progress in the theory, algorithms, and applications of parallel optimization. We identify three major areas where parallel optimization approaches have been developed or used:

1. Parallel algorithmic techniques (*concurrent*, *concurrent-distributed-concurrent*, *distributed*) within the solvers CPLEX, GUROBI, XPRESS NONLINEAR, and SCIP.
2. Parallel metaheuristics, and
3. Machine learning and hyper-parameter optimization.

Before we discuss three areas in detail, we encourage the reader to consult Trelles & Rodriguez (2005,[554], pp. 522) for a very useful taxonomy of parallel architectures. The advantages of using multi-core platforms versus clusters of computer are discussed by Alba et al. (2013,[14], pp. 13).

16.6.2.1 Parallel Algorithms and Solver Worlds

Usually one finds parallelization techniques deep in solver technology, cf. Laundy (1999,[355]), Shinano et al. (2003,[503]), Baravykaité & Žilinskas (2006,[46]), Shinano et al. (2008,[504]), Shinano et al. (2016,[506]), Berthold et al. (2018,[74]), Shinano et al. (2018,[508]) or Shinano et al. (2018,[509]), or exploiting multiple threads (cf. Heipcke (2012,[260]) or Shinano et al. (2016,[507]) when implementing B&B based methods. A good overview on architectures of parallel algorithms and solvers for mixed integer linear optimization is by Ralphs et al. (2018,[450]). Most approaches for parallelization focus on B&B and B&C. While interior-point methods and their linear algebra part are more suitable to parallelization and have been parallelized already in the 1990s (cf. de Silva & Abramson(1998,[150]), the Simplex algorithm is inherently a sequential algorithm. Nevertheless, there are now also some promising attempts to parallelize the Simplex algorithm itself; cf. Huangfu & Hall (2018,[274]) or Coutinho et al. (2018,[133]), and reference therein.

Running the same problem with different algorithms, parameters, *etc.* and choosing the fastest one, also known as *concurrent optimization*, is one way — actually the easiest one, called *embarrassingly parallel* or *perfectly parallel* by Herligy & Shavit (2012,[267]) — to utilize the parallel computing power, i.e., controlling and tuning parameters of solvers, where *solver* refers to commercial MILP solvers CPLEX [278], GUROBI [250] or XPRESS-OPTIMIZER [260], or NLP/MINLP solvers such as BARON [215], ANTIGONE [407], or LINDO [491] to name a few.

All commercial MILP solvers allow *concurrent* runs with various flavors: *Concurrent*, *concurrent-distributed-concurrent*, *distributed*. *Concurrent* optimization for MILP can be understood as the simplest realization of the GEA and is available in CPLEX, GUROBI, or XPRESS-OPTIMIZER. The next level is *concurrent-distributed-concurrent* which allows communication and interaction between parallel runs on cores or threads. *Distributed MILP* means: Each B&B search is started with different parameter settings, a permutation of the columns/rows, or just another random seed. The best one wins and the job is done, or one even allows re-starts on that best one and only continues with those settings that perform best so far. CPLEX, for instance, offers *distributed* with the following tasks: (i) work on the lower bound on one thread; (ii) work on the primal bound (heuristics!) on the other, and (iii) have a third thread to manage the search tree. Impressive results are provided by Shinano et al. (2016,[506]) using a parallel enhanced version of the solver SCIP (cf. Shinano et al. (2010,[505]) or Gleixner et al. (2018,[222])) using 80,000 cores simultaneously on the Titan supercomputer to solve 12 previously unsolved MILP problems from the MIPLIB benchmark set.

Colombani & Heipcke (2004,[129]) and Heipcke (2012,[260]) present possibilities for problem decomposition and concurrent solving from a modeling point of view with example implementations in `Mosel` that show the handling of multiple models, multiple problems within a model, and as a new feature, distributed computation using a heterogeneous network of computers. In 2004 and 2012, the XPRESS-OPTIMIZER module *mmjobs* probably has focused on solving to MILP problems or solving NLP problems with multi-start techniques. This module allows one to determine on the modeling level what to parallelize and how to distribute jobs (whole model or submodels).

A very active and productive group is at ZIB in Berlin with Shinano (2018,[502]) reporting on seven years of progress in parallelizing B&B. When developing and implementing parallel algorithms, one has to keep in mind that it is very difficult to come up with and implement a parallel framework correctly leading to true parallelism, in the sense that investing more CPUs results in linear speed-ups. This may take many years. Shinano et al. (2020,[511]) have solved previously unsolved MIP Instances with ParaSCIP, a parallel version of SCIP,[3] on supercomputers by using up to 80,000 cores. Going back in time, see also Shinano et al. (2020,[510]), Munguia et al. (2019,[413]) for solving stochastic mixed integer programs, and Shinano et al. (2019,[509]) and their development of `FiberSCIP`, a shared memory parallelization of SCIP.

When it comes to developing parallel nonlinear optimization algorithms and software, some additional problems show up as NLP or MINLP problems can have very different structural properties such as being sparse, dense, mostly linear, convex, pseudoconvex, large, small, or parametric. In every case, there are different optimal ways of exploiting parallel hardware, but building a piece of software that implements all of these ways is difficult, which is why the solvers IPOPT, SCIP, KNITRO, and `NLPAROPT` probably have very different inner architectures. In other words, parallelism in nonlinear optimization is tricky from a design point of view because the choice of allocating parallel resources depends on the real-life problems the solver is designed to solve. There are three main applications of parallelism for

[3]See Gamrath et al. (2020,[207]).

MINLPs:

1. Calculating derivatives in parallel (important in very dense problems),
2. Implementing the factorization step in parallel (important in very large problems), and
3. Solving the problem many times in parallel, e.g., parallel B&B or multi-start (important in MINLP problems).

Solvers can usually only really do one of these in parallel, and that design choice determines what problems that solver can solve well. Even if solvers such as IPOPT or SCIP have been developed at research institutions, at least part of their motivation may stem from real-world problems influencing the inner architecture of the software.

Finally, we comment on determinism when using B&B or parallel solvers: In 2020, some solvers provide more determinism than others and have control parameters for such features! This also becomes obvious when switching from Linux to Windows using 4 cores to 16 cores, but turning on the control parameters allows for deterministic optimization runs; at a slight increase in run time, of course. A unique feature available in CPLEX is enforced by switching from CPU time limits to limits expressed in *ticks*. With this feature turned on, even when the Branch&Bound process is terminated prematurely, the solutions are identical. This is very helpful when developing and testing polylithic modeling and solutions approaches.

16.6.2.2 Parallel Metaheuristics

Parallel metaheuristics is another area in which parallel techniques are used; cf. Alba (2005,[11]), Alba et al. (2005,[13]), Alba & Luque (2005,[12]), various chapters in [11] about parallel versions of genetic algorithms, simulated annealing, and tabu search, the early work by Pardalos (1995,[437]), Gendreau & Potvin (2010,[211]), or Crainic (2019,[134]). If we follow Alba (2005,[11]) in his book *Parallel Metaheuristics* on p. 112, in many cases, pPMSA falls into the class of *independent run models*.

There exists a vast body of literature related to parallel techniques for solving multi-objective optimization problems. This requires to construct a set of solutions called the Pareto front. Figueira et al. (2010,[183]) favor evolutionary algorithms for this. Jozefowiez et al. (2002,[294]) have constructed a specially defined parallel tabu search applied to the Pareto front reached by an evolutionary algorithm.

To give an example, Lančinskas et al. (2015,[350]) have developed a stochastic search optimization algorithm and have applied it to solve a bi-objective competitive facility location problem for firm expansion. The parallel versions of the developed algorithm for shared- and distributed-memory parallel computing systems approximate the Pareto front and have almost linear speed-up when solving competitive facility location problems of different scope reasonable for practical applications.

16.6.2.3 Machine Learning and Hyper-Parameter Optimization

A different community and field where parallel solution approaches have an impact is machine learning and hyper-parameter optimization in the context of Bayesian optimization. In machine learning, hyper-parameter optimization or tuning, the goal is to select a set of optimal hyper-parameters for a learning algorithm. Hyper-parameters are those parameters the values of which are used to control the learning process, while the values of other parameters (usually node weights) are learned. Grid search and random search (cf. Bergstra & Benglio (2012,[72])) allow for easy implementation to parallel approaches. Bergstra et al. (2011,[73]) let a Gaussian process algorithm and a tree-structured Parzen estimator run asynchronously in order to make use of multiple computer nodes and to avoid wasting time while waiting for the trial evaluations to complete.

16.6.2.4 Parallel Optimization in the Real World

In 2020, real-world optimization is still struggling with making use of today's and tomorrow's multi-core computing architecture or exploiting graphical processing units (GPUs). Not every small or mid-size company has a cluster of 1,000 computers available. The good news: Parallel optimization is possible for practitioners using the inherent parallel algorithms of solvers offering parallel algorithms and — with some programming effort — using AMLs.

An easy task is to program a multi-start approach for solving NLP or MINLP problems. This requires a random number generator, a local NLP solver, and some features in an AML or other programming language to run a certain number of instances in parallel.

Beyond this, i.e., the development of one's own parallel algorithms and software, life becomes complicated due to the lack of good tools to implement, test, and deploy distributed algorithms and software. One has to keep in mind that parallel computing applications tend to target time-critical or large problems. Problem size has implications on data structures and distributing memory, and actually requires to taking precautions for all kinds of solver details to avoid bottlenecks, which are non-issues for smaller problems.

While usually one finds parallelization techniques deep in solver technology as discussed above, Kallrath et al. (2020,[314]) use a parallel PMSA (pPMSA), i.e., parallelization at a higher level of the application itself; cf. the example at the end of Sect. 14.2.1.1. This can be understood as follows: Running the same problem with different algorithms, parameters, *etc.* and choosing the fastest one, also known as *concurrent optimization* (cf. CPLEX User Manual) is one way to utilize the parallel computing power. A parallel PMSA takes this one step further by applying the multi-grid approach on the level of the application itself by exploiting the control parameters of the PMSA. This approach has been used both on multi-core platforms of up to 32 cores, and on clusters of up to 1,000 computers — it is very suitable and relevant for the following two real-world situations: Situation 1: One relevant practical requirement is that we have a limit on the time available for returning a solution back to the user, i.e., we usually cannot solve the problem to optimality; this is especially true for scheduling problems. In this situation, we want to get the best solution within the available time. Situation 2: Multi-criteria optimization problems with the following property: It is difficult for the problem owner to quantify what is a *good solution* to him. Therefore, we want to offer various solutions enabling the user to select by inspection the *best* solution: an example for this is the cutting stock problem with two conflicting objectives, the minimization of trimloss and the number of patterns. Note that both situations (time limit and multi-criteria objectives) can also show up in combination.

The previous paragraphs show the potential for parallel optimization. It is all about computing time and solution quality within a given time limit. Most academic literature is concerned about linear speed-ups. If a parallelized algorithm on n identical CPUs produces the optimal result within time T_n, what is the finishing times T_{kn} on kn identical CPUs? Ideally, one expects $kT_{kn} = T_n$. Related follow-up questions are:

1. For which range do we have linear or almost linear speed-up? This may depend strongly on hardware, communication between the CPUs, software and also on the problem itself.
2. If linear speed-up depends on hardware, software, and also on the problem as suggested above, is the concept of linear speed-up suitable for real-world optimization problems? LP solvers, and in 2020, also MILP solver are generic to a reasonable level. Deterministic global optimization solvers can already do a lot, but they are still limited to a certain problem size.
3. Are there applications areas for which parallel optimization is especially suitable? The energy sector with the need to evaluate thousands or millions of independent scenarios may be a good candidate.

An interesting project is *BEAM-ME* (Dec 2019, http://www.beam-me-projekt.de). This research project supported by the German *Bundesministerium für Wirtschaft und Energy* aims toward utilizing the full potential of parallel computing on HPC with decentralized memory architecture for the application to optimizing energy system models. A detailed description of project results is on https://gitlab.com/beam-me/best-practice-guide.

16.7 Summary

In this chapter, we have considered optimization as an approach to problem solving which is not without limitation. As with other techniques used in operational research, care and attention to many matters of good practice are required, especially in the area of communication. Mathematical programming is a powerful technique but needs to be used sensibly to encourage confidence in clients. The modeler must move steadily but carefully through systematic procedures that require constant modeler/client feedback in order to ensure the final implementation of the benefits of the modeling process. Optimization is an area under continual development and we are convinced it has much to offer for the future.

Chapter 17
Concluding Remarks and Outlook

In this chapter, we provide reflections on the material presented in this book. The chapter ends with the authors' view of future developments and suggests some conclusions.

17.1 Learnings from the Examples and Models

With the opening of markets and borders and a globalization of world economy, problems like the production network planning system described in Sect. 10.4 will increase both in number and in complexity. The ability to produce complete, accurate, and optimal solutions to larger problems offers the potential for enormous reductions in costs, huge increases in efficiency, and careful handling of resources. Some industry specialists estimate that savings would average 3–4% of turnover and could well be much higher. In particular, it becomes possible to exploit the inherent advantages and synergies in highly connected production networks (Kallrath & Schreieck, 1995,[320]). Thus, the reader can see the potential for business optimization illustrated by the case studies.

Some of case studies are scaled down examples of real problems solved in industry. The real cases usually contained additional subtleties which were either confidential or not appropriate for didactic reasons. The model formulations presented in this book provide the modeler with building blocks for developing comprehensive models, e.g., in *supply chain management*. Being able to calculate complete, accurate, and optimal solutions to large planning problems offers the potential to save enormous costs, increase efficiency, and carefully manage limited resources. Experiences with the use of mathematical optimization lead to cautious savings estimates of about 3–4%. The case studies can only hint at this potential, as for didactic reasons or for reasons of confidentiality, the models are frequently

J. Kallrath, *Business Optimization Using Mathematical Programming*,
International Series in Operations Research & Management Science 307,
https://doi.org/10.1007/978-3-030-73237-0_17

simplified and reduced in size. Nevertheless, the reader should have benefited from reading the book in the following way: the case studies

- offer a wide range of real-world problems from different application areas and industries and hopefully have given an impression of what problems can be addressed with optimization methods;
- contain some special formulations, subtleties, and "tricks-of-the-trade," which improve the model formulation considerably and which should be available in the modeler's box of tricks;
- show the need for careful, sometimes very specific, tailor-made solution techniques adapted to the problem at hand;
- contain structures and substructures relevant to other problems; and
- can be building blocks or starting points for more complex models.

The examples, especially, the bigger case studies in Chap. 10, should illustrate that modeling is not necessarily a straightforward process, but rather an iterative process with some reformulations. Sometimes, the question arises in practice whether one should develop a model precisely to the customer's needs and then possibly get problems with its solution, or whether one should already consider during the modeling process what one considers solvable and reduce the fidelity accordingly immediately. Here, as the author of this book, I would like to position myself very clearly: because of the acceptance and closeness to reality, mapping the reality is indispensable and should have the highest priority. In Chap. 14, ways were shown to solve also very difficult and large problems. If these possibilities are still not sufficient to solve the problem, only then, you may think about reducing the degree of mapping the reality.

17.2 Future Developments

We have seen that some large real-world problems have been successfully solved at large companies in chemical, airline, refining and other industries and that the use of optimization, even in smaller companies, demonstrates huge potential for reducing costs, increasing efficiency and flexibility, and generally contributing to the effective management of the enterprise.

17.2.1 Pushing the Limits

However, we should not be lulled into a false sense of security by a handful of success stories. Since its computational beginnings, optimization has always tested the available computer hardware and software to its limits. In a certain light-hearted way, the practical optimization specialist must be likened to a pole vaulter at an athletics competition. Ultimately, the pole vaulter goes home having failed — the competition may have been won, but the goal to clear the last jump is missed.

The same thing faces the optimization specialist, though not for the same reason of failing at some absolute target of achievement.

The professional optimization specialist always faces a moving target in that once a problem has been solved for the user, the user will inevitably come back with a tougher problem. For instance, the user will increase the number of time periods in a multi-time period model or perhaps disaggregate some process that has been modeled fairly crudely into its separate components, thus rendering the problem bigger and generally harder to solve. In our experience, this is not too much of a problem with pure linear programming problems as though the computing times increase with problem size, they do not do so at an exponential rate. If a ten-period model takes a certain time to solve, then one would expect that a twenty-period model might take perhaps four times as long, and it would be very unlikely to take twenty times as long.

In practice, users seem to think about solution times in several possible bands. The first acceptable band is where the solution time for the problem is of the order of 12 h, so one overnight run can be done per day. There is, in practice, little benefit to be gained from reducing this to 8 h as probably there will still only be one run possible per day. The next band of solution times is of the order of 1 h where, if we allow some time for users to inspect their results and decide upon another scenario to analyze, probably two or three runs per day are possible. The next band covers solution times of the order of a minute or so, at which point several benefits start to accrue. The first is that now we can start to contemplate optimization almost "on line," i.e., to use it to adapt to the situation as data changes. The second benefit is that we can rapidly analyze many scenarios and start to get a good understanding of how the solution changes as parameters change.

The final band of solution times is where the optimization takes the order of one second, rather like recalculating a medium-size spreadsheet. At this point, we have achieved a much desired objective of being able to deliver instantaneous optimization to the end user. The implications of having this technology are very profound because we can have a guarantee of optimality in many situations where currently we are forced to descend to heuristics because of the need to have an implementable solution within some restricted time limit. We really have online optimization.

The power of computers and high quality optimization software has increased significantly in the last decade, but we should not see this as a panacea. It seems unlikely that there will be an order of magnitude increase in performance of either Simplex or interior-point algorithms (for instance, a typical quality interior-point solver only takes the order of 30 iterations to converge to an optimal solution. It is extremely unlikely that this can be reduced to 15 iterations and at the same time improve the efficiency of those iterations by a factor of 5 which would be necessary to get a tenfold reduction in solution times.). Improvements in computational speed are increasingly being limited by access time of memory, and it is unfortunate that the Simplex algorithm tends to access memory in a rather random way, which means that various caching schemes do not perform very well.

17.2.2 Cloud Computing

Parallelization, as described in Sect. 16.6, is one option to push the limits. Another approach to consider could be *cloud computing*, i.e., the use of IT infrastructure and services that are not maintained locally on local computers but are rented as a service and accessed via a network (e.g., the Internet). Given the raise of the cloud, it looks very promising that we will able to solve larger models — although there is a lot of computational overhead. Kurschl et al. (2014,[347]) are convinced that mathematical optimization is one of the domains, which benefit from cloud computing by the use of additional computing power for optimization problems to reduce the calculation time (stronger hardware and higher degree of parallelism). The supportive arguments for cloud computing are to have hardware (memory and computers) available whenever needed, stability, reduced local IT dependence, and variable costs instead of investment and maintenance costs. The disadvantages are virtual computing power instead of real hardware under one's own control, security issues, and reliance on Internet access as well as the stability and availability of the Internet. Due to the memory on demand and machine on demand, larger problems can be solved than with personal or local hardware, and highly parallel structures can be used.

So, in summary, we cannot rely on technology alone to deal with the expectations of users, but there is a real need to solve ever larger or more complicated problems. How can we proceed? In the next section, we try to give an answer.

17.2.3 The Importance of Modeling

In earlier chapters, we have seen that *modeling* is vital to practical optimization. It is just about possible to get away with poor modeling if the problem is a pure LP. As long as the model is correct, the implications of a poor (larger and redundant) model are likely just to be longer running times by a factor of perhaps 2 or 3; not desirable, but not disastrous. But when we move to MILP problems, the difference between a good and a poor formulation may not be small factors 2 or 3, but perhaps *5 or 6 orders of magnitude* increase in solution times. The problem in the worst case cannot be solved in reasonable time at all.

We have found that the analysis and hard work by experienced and expert modelers very often yield several orders of magnitude improvements in solution times for MILP problems. Even when the problem is just too hard to solve, the insights obtained by this analysis often give very good heuristics.

But even with growing hardware and software capabilities, the importance of the experienced modeler cannot be underestimated. The opposite is true. *Analysts will become more important.* When hardware and software capabilities grow, there is demand for more complex and realistic models because clients will ask for more details in the model. It is the modelers' task and responsibility to bring clients' demands and mathematical programming reality to a fruitful liaison. In addition to better hardware, the modeler will be facing more intelligent algorithms implemented in commercial software, e.g., providing efficient B&C routines, which requires that modelers really keep themselves up to date. Not only integer programming will be more or less restricted to linear problems (as it is now), but also quadratic or mixed integer nonlinear programming in general will become tractable. Last but not least, modelers will have more flexible modeling tools at their finger tips: modeling tools supporting dynamic cutting planes, formulating optimizations problems from graphically designed network flow problems, providing links to complete different solution algorithms, and allowing to switch between different solvers. Despite the success observed when applying mixed integer linear programming, the support given to expert decision making and scenario generation by mathematical models and methods is still far from being widely accepted. Very often, analysts experience great reservations when talking to people working in production, logistics, or marketing. There is a psychological and/or cultural barrier. Experts are used to decision making based on experience and heuristic rules that are difficult to express explicitly. The approach to achieve objective solutions which can be controlled on a quantitative basis is new. It may create unconscious fears and may in addition require a huge effort to explain the problem of interest to a non-specialist with the appropriate degree of completeness and accuracy. Indeed, on the one hand, the mathematical kernel of the application operates as a black box usually difficult to understand for non-mathematicians. On the other hand, experts are afraid to lose influence and acknowledgment when outsiders, in this case mathematicians, can produce solutions that prove to be better in terms of costs, contribution margin, utilization rate, or some other valuable quantity, when compared to their own solutions. As a rule of thumb, usually at least 50% of all efforts and time spent during project work trying to solve a real-world problem using mathematical optimization methods is related to psychology, i.e., talking to clients in order to increase the acceptance of the solution techniques or removing reservations and fears against mathematics. Thus, besides technological efforts, there should be a strong investment in improving the awareness and acceptance of mathematical optimization applied to real-world problems.

The sceptics can be calmed down: mathematical methods and techniques cannot and will never replace human inventiveness or decision making, but they can successfully provide a quantitative basis for these decisions and allow analysts and decision makers to cope most readily with complex problems. Thereby, in this sense, mathematical optimization can successfully contribute to a safer and better world in which risks are decreased, quality is improved, and resources are used more efficiently.

The very rapid performance gains achieved in the last decade by general purpose MILP solvers have had a slightly negative influence on the development of special purpose algorithms for optimization. It is somewhat dispiriting for the developer of a special purpose algorithm to be beaten on solution times by general purpose software, even though it is probably because the latter has had several man-years of careful development and tuning. The recent availability of modular high performance optimization subroutine libraries has meant that if LP or MILP (or some parts thereof) are required by the special purpose algorithm, then this can be extracted from the library and will have top performance. The libraries will continue to develop over the next decade as algorithm developers demand extra functionality.

17.2.4 Tools Around Optimization

Originally, the kernel of optimization projects was to formulate the optimization problem at hand, possibly implementing it into an algebraic modeling language and to solve it by some mathematical algorithm implemented in a programming language or provided by the optimization software. Nowadays, these kernel components *implementing the model* and *solving it* are accompanied by various tools:

- automatic documentation of the model in the document preparation system LaTeX (frequently, used by mathematicians, physicists, and people in Operations Research),
- graphical representation of the solution output,
- advanced business analytic tools applied to cover the specifics of the solution and explain it in the language of the end user,
- additional statistical tools for data preprocessing or result analysis,
- support tools for scenario analysis, and
- connectivity to other programming languages such as `Python` (which seems to win the popularity competition).

We will not discuss these tools any further. Instead, we want to stress that they extend the functionality of modeling and optimization software and reach out to larger user communities. For modeling tools, we provide more background information and indicate some ideas for the presentation of results.

It is our belief that modeling tools will also continue to grow in power and functionality. These tools seem to bifurcate into enhanced *algebraic modelers* and *visual modeling support*.

A modeling language used to implement mathematical optimization models supports the expressions and symbols used in the community of mathematical optimization. Therefore, it is natural that algebraic modeling languages (AMLs) support the concept of structuring a model by separating *data* (what is given), *variables* (what we want to know), *constraints* (restrictions, bounds, etc.), and *objective function* (what we want to maximize or minimize). Those entities are connected not only by the algebraic operations $(+,-,\cdot)$ but also by nonlinear functional relationships. AMLs — the earliest, `GAMS`, `LINGO`, and `mp-model`, appeared in the late 1970s and early 1980s — are declarative languages for implementing optimization problems. They keep the relations (equalities and inequalities) and restrictions among the variables and connect data and the mathematical objects to a solver; they do not contain information on *how* to solve the optimization problem.

Since the early 1980s, AMLs have played and still play an important role in the world of mathematical optimization and optimization used in industry. In the 1950s and 1960s, Assembler and Fortran coded LP models were mostly replaced by IBM's matrix generators MPS that established the standard of industrial model formulation. Models in those days were LP models, many of them solved by IBM's LP solver MPSX. At that time, there was no market for AMLs, but there was no real support for NLP problems, and this was a niche for AMLs as they enabled the user to formulate NLP problems and supported automatic differentiation, i.e., they symbolically generated the first and second derivative information. Another line of development was triggered by the advent of personal computers (PCs). Dash Optimization provided a tool to PC users rather than mainframes in 1984 with their solver XPRESS-OPTIMIZER and their modeling language `mp-model`. Thus, after a while, AMLs also became superior in implementing LP models and succeeded MPS. Nowadays, academic research models (developed by scientists) are used for developing and testing solvers or constructing efficient model reformulations. Domain expert models (developed by analysts) are used within consulting projects or feasibility studies. And finally, AMLs often host the models for black box model users doing their operational planning. AMLs ensure robustness, stability, and data checks needed in industrially stable software. Furthermore, AMLs accelerate the development and improvement of solvers ranging from Linear Programming to Mixed Integer Nonlinear Programming and even Global Optimization techniques. If a user has an NLP problem implemented in an AML using a local solver to compute its local optimum, it is only a matter of minutes to switch from a local solver to a global solver such as BARON, ANTIGONE, or LINDOGLOBAL. Thus, there is a significantly reduced development risk for the user. But the solver developers can also count on a much larger market when their solver is embedded in an AML. The solver technology, in some sense, is now a commodity which allows the users to switch, for instance, from one MILP solver to another one or play around and collect experience with the free Coin-OR solvers. The implementation of polylithic modeling and solution approaches described in Kallrath (2011,[311]) is possible without huge development efforts. And, last but not least, the development of Microsoft Windows and improved hardware technology has lead to integrated development environments (IDEs)[1] such as Xpress Workbench for `Mosel`, `GAMSIDE`, or `GAMS Studio` in `GAMS` or systems such as `AIMMS` and `MPL`. This increases the efficiency of working with AMLs and contributes greatly to the fact that AMLs reduce the project time, making maintenance easier and increasing the lifetime of optimization software.

In contrast to the algebraic modeling approach, there exists the *visual modeling* approach. With these visual modeling tools, it is possible to lay out the design of the screen and from visual objects derive the various bits of programs. The extension to visual modeling is seemingly obvious — it involves the connecting of the various

[1] Not to be confused with standard "user interfaces" like reports in a browser, in Tableau, or Xpress Insight.

objects with arrows to show flows and the generating of equations that describe material balance, transformations, blending, etc. from the icons that represent the objects. This "model of modeling" seems to be attractive, but its application to anything other than logistics and network modeling soon raises difficulties. For instance, how does one denote that an arrow (denoting a flow) really represents a set of flows and how does one indicate which set this flow belongs to? How does one declare that certain data are required and need to be collected? If there are five units of a particular type, does one represent this with five objects, or in some way in which, e.g., one object stands for all five? If we do not adopt the former approach, the screen rapidly gets cluttered, whereas if we adopt the latter, a new set of notations has to be developed and learned.

Visual modeling can be readily applied if a problem at hand can be modeled as network flows, where a visual interpretation is easy. If modeling is to be used by relatively unsophisticated end users, a visual presentation of the model is far more attractive than an algebraic one. End users might not be able to build the model initially, but it is very likely they will be able to modify an existing model, and even more likely that they will find the visual model easier to understand and thus will be more likely to accept it.

Regardless of whether or not a model is implemented taking the algebraic or visual route, visualization of results becomes more and more important. We discuss this in detail in the next section.

17.2.5 Visualization of Input Data and Output Results

Algorithm and model developers regularly create and evaluate statistics of optimization runs in order to determine which parameter settings work best for a specific benchmark set. There is a trade-off between runtime and solution quality where solution quality can be a mixture of multiple parts of the objective function based on different units of measure. Such statistics are often presented in tabular format.

Furthermore, it is of similar importance to validate the data input and the output for feasibility during development phase. This already requires additional data and certain visualization capabilities. Validation often needs to be done by the domain experts, i.e., the planners or decision makers need to be involved, not only the modeling expert. To make this more tangible, we illustrate it with a few examples as the type of visualizations depend on the use case as well. It is key to review the information collected in a model in the right context to understand the results.

Example 1 A supply chain manager responsible for a production–distribution problem as in Sect. 10.4 benefits from drawing the supply and distribution network on a map as in Fig. 17.1 or the supply per supplier (Fig. 17.2). Financial institutes and sales departments need to understand the impact of various buyer types and geospatial influence on churn rates to make better decisions, see Fig. 17.3.

Fig. 17.1 A typical side-by-side scenario comparison of a supply chain network using a map. Screenshot taken from a FICO Xpress Insight demo app *Supply Chain Network Designer*

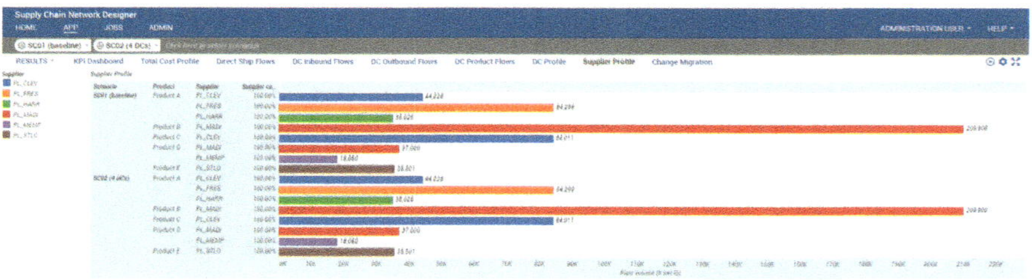

Fig. 17.2 A horizontal bar chart showing supplier profiles of two scenarios in the same chart. Screenshot taken from a FICO Xpress Insight demo app *Supply Chain Network Designer*

Example 2 Operations managers want to see the production plan, for instance, on a horizontal axis (e.g., via Gantt charts) and the machine utilization (see Fig. 17.4).

Example 3 Tour planners, e.g., for school bus routing, pickup and delivery services, benefit from seeing the routes proposed as in Fig. 17.5.

Visualization capabilities become even more important when models are handed over to a decision maker, such as an operations manager (cf. the scheduling example in Sect. 10.5), a supply chain manager (as in Sect. 10.4), marketing analyst, and so on. Visualization capabilities, in particular, are extremely important in the context of Decision Support Tools. Here, it is key to show all the data fed in and the results as close as possible in the business context. This is crucial for acceptance of the software solution. To base a decision on any solution proposed by a software tool, the users need to trust in the data quality, the algorithmic logic, and the feasibility of the results. This gain of trust can be supported by the way data and results are visualized.

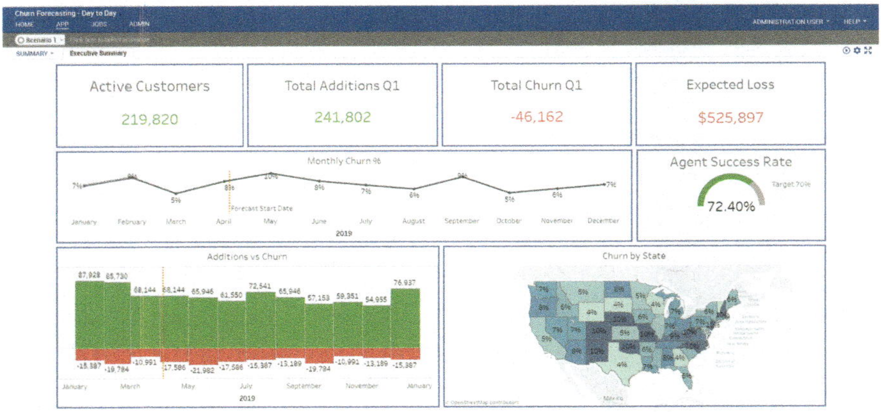

Fig. 17.3 Churn rates. Screenshot taken from a FICO Xpress Insight demo app *Churn Forecasting*

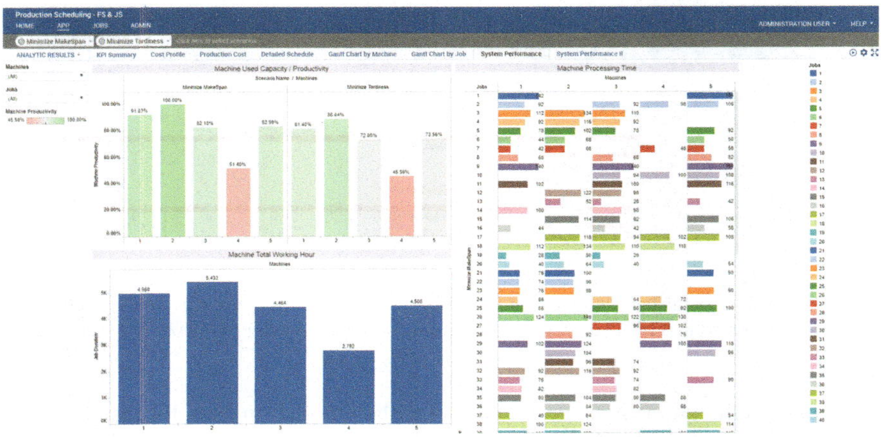

Fig. 17.4 A production plan. Screenshot taken from a FICO Xpress Insight demo app *Production Scheduling FS & JS*

Decision makers have several key needs which we first describe generically. Yet, figuring out how to incorporate these needs into software that gets accepted by the user poses another challenge that is discussed afterward.

Among the typical *user visualization needs* are understanding what input data are available, if the data are up to date, and results displayed in a way that users can quickly identify key decisions, the potential bottlenecks, and main drivers for the decision. Finally, users like to understand and to have an answer to the

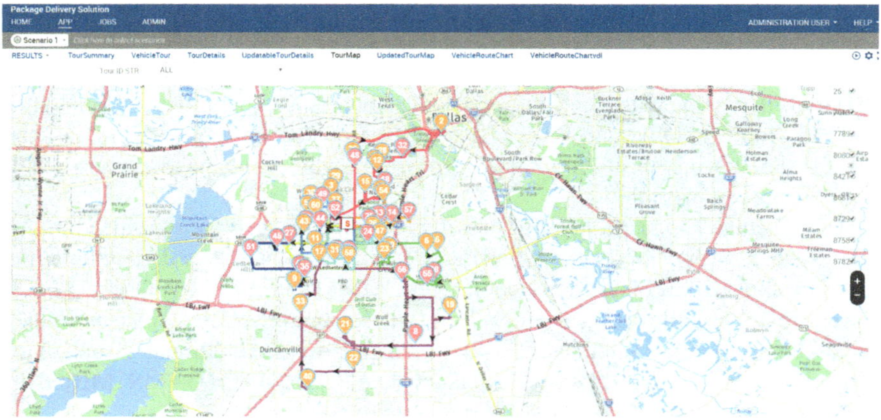

Fig. 17.5 A map showing delivery points and an optimal delivery route. Screenshot taken from a FICO Xpress Insight demo app *Package Delivery Solution*

question: why there is a deviation from the planned scenario? Given these needs, a tool needs to provide various visualization capabilities among them:

1. Multi-layer approach: high-level key performance indices (KPIs), drilling down into the main decisions and further down to each single entity.
2. The way data and solution proposals are displayed must fit the business context. Typical and prominent examples are Gantt charts for scheduling problems and maps for network optimization.
3. Drilling down on data requires standard spreadsheet functionality provided across reports, in particular, searching and filtering across multiple tables and charts, the so-called Dashboards.
4. Interactive Dashboards give the impression of control — while the algorithmic and model part may be a black box to the user, interacting with the model by adjusting some parameters and getting a response with reasonable results generates trust.
5. Scenario comparison in the same charts to explore trade-offs between different strategies.
6. Collaboration features such as the ability of sharing scenarios, data, and results to collaboratively develop and evaluate business strategies.
7. Traceability of decisions plays a major role. What were the major triggers to decide for a specific strategy and reject others?
8. Comparison of plan and actual figures to evaluate historical decisions and reevaluate the decision making process, KPIs, and input factors in a sustainable manner — fundamentally, to enable a learning curve.

Overall, we should keep in mind that good users become even better users when powerful and effective tools are available to them.

17.2.5.1 Tools and Software

The needs described above are partly very well covered by Business Intelligence (BI) tools that come with rich charting and dashboarding capabilities. Prominent BI tools are Tableau, Qlik, and Power BI from Microsoft. Yet, there are crucial gaps such as the missing ability to embed optimization algorithms or to support whole business process workflows.

There actually is just a handful of software tools tailored for optimization that offer such a variety of features partly by hooking up dashboard functionalities from the above BI Tools: `AIMMS` comes with AIMMS Pro, and FICO offers FICO Xpress Insight, while IBM provides the Decision Optimization Center. Formerly, stand-alone modeling languages have started to provide additional visualization capabilities as well: `AMPL` with Quandec and `GAMS` with GAMS MIRO are prominent examples.

Let us now focus on the *tool development process*. To develop sustainable optimization software, data and results must be demonstrated in their business context. Yet, a modeling expert or a solution developer is usually no business expert. On the other hand, a business expert is most likely unable to set up a requirement document that encompasses all needs.

To maximize the probability of overall success, it is recommended to develop optimization models, the visualization, and the overall tool in an iterative and collaborative manner. Hence, the framework used must support rapid application development. This is highly needed to get down to the guts and needs for a software solution and to unite the technical in-love-with-the-model perspective of the modeling experts with the needs of IT-affine decision makers.

The main advantages of a collaborative and agile tool development approach Are as follows:

1. The modeling expert gets acquainted with business terminology; the developer and the domain expert start speaking the same language.
2. The requirements to be developed get refined iteratively with a step-wise increasing joint understanding. This holds for the model and for the visualization.
3. Gaps in the model and input data can be identified early.
4. An inherent trust into the model and the data is build up, which will help in tool acceptance.

To get mathematical models into the hands of decision makers, and enable them to quantitatively assess their strategies, is a challenge on its own. Luckily, tools that closely integrate models with visualizations that can be customized to the specific use case are available. But visualizing results is just the start. Decision support tools require a rich amount of features, e.g., for collaboration and workflows. Given the success of WYSIWYG editors for webpages, the future may have some more surprises for model developers to ease their life when creating such tools.

17.2.5.2 The Broader Company Picture: IT

Another angle to consider is the requirements from IT departments. Software tools need to be embedded into existing IT infrastructure, with the company's user authentication management, matching its roles and authorities. Network security becomes important when decision support tools are no longer stand-alone local computer programs, but users can share data and scenarios by connecting to internal and external data sources. Value-based optimization tools bring together plain quantity figures (historical or forecasted production amounts and capacities) with cost and price figures. Such data are no longer classified as internal data or as just confidential, but usually as strictly or highly confidential. As this leads to the highest business impact, security measures such as access control and monitoring need to become part of the software.

17.2.5.3 Summary

Building sustainable optimization models and tools requires close collaboration of the developer, the users, and companies' IT departments. From the IT point of view, security, data base connectivity, authentication, User, and role management need to be taken into account. For Operations Researchers, this poses quite some challenges but also the possibility to broaden their horizon. It is key to collaboratively develop models and visualizations in order to find gaps in the data or model early, to be able to design the right User Interface and to get acceptance of the tool. This way, a decision maker is willing to trust the tool and rely on the proposals generated by an optimization-based Decision Support Tool.

17.2.6 Increasing Problem Size and Complexity

At the risk of spreading despondency, there is one more topic that we shall have to discuss before we finish. The theory of computational complexity has been very successful in dividing problems into two classes. Very roughly, it says that problems can either be solved in a time proportional to a polynomial of their size S (for instance, S^2 or S^7) or belong to the class of problems called $\mathcal{NP}$ where the solution time is proportional to an exponential of the size (2^S). (We apologize to expert readers for this gross simplification of computational complexity theory: it is not the solution time; it is a proof what type of polynomial/exponential worst case algorithms exists. But instances may be easy to solve. There is also weak NP-hardness.)

LP falls into the class of polynomial algorithms in practice (interestingly, the Simplex algorithm is not polynomial in theory but is so in practice, while some interior-point methods are polynomial both in theory and in practice, and the most widely used interior-point method has not been proven to be polynomial in theory).

On the other hand, IP is in the $\mathcal{NP}$ class, where solution time can grow as 2^S. The implications of this can be catastrophic. Suppose we can solve a problem of size $S = 1{,}000$ in 1 h. Then, we cannot expect to be able to solve a problem of size $S = 1{,}005$ in less than 32 h. In other words, increasing the problem size by half of 1% moves us from a problem that can be solved several times a day to one that requires well over a day. And, parallelism (see Sect. 16.6) is not the answer: we need 32 processors just to get the time back to 1 h. It is as if our poor pole vaulter has to double the run-up speed every time an extra 1cm is to be cleared.

So, in the long run, if end users keep on insisting on increasing the size of their problems and there is no huge breakthrough in mathematics which would move MILP into the class of polynomial problems, we are going to have to give up trying to solve MILP problems of a certain size.[2] Even if computer speeds double, we can only solve a problem of size one more than we could before. At that level, the last resort may have to be the use of heuristics to produce approximate solutions.

Many of the problems described and their model formulations show that very large linear problems can now be solved. We find examples of this not only in process industry, refineries, and airlines, i.e., mostly large organizations, but also in the medium-sized and smaller companies; there is a worthwhile potential for cost reduction and efficiency increase through the use of mixed integer optimization. As a result of faster processors and improved algorithms, planning problems that 10 years ago might have required a night's computing time on a company computer can now be solved in less than an hour, or perhaps even a few minutes, so that interactive work is possible. It is expected that in a few years MINLP and Global Optimization Techniques for non-convex, nonlinear optimization problems will play a similar role to today's MILP.

Standard software, e.g., in supply chain optimization or vehicle routing, provides the user with powerful tools for clearly defined problems. It is still necessary for these programs to be used by personnel with knowledge of optimization and modeling, but it is conceivable that the models and algorithms will become so robust that automated solutions will be possible for certain subtasks. The creation of models and solution methods in this quality requires a lot of experience, as well as the handling of newer solution techniques like B&C for solving MILP problems, methods for solving MINLP problems, or the possibility to use several solution methods in combination.

In addition to robustness, it is to be expected that the improved possibilities will lead to models becoming more detailed and realistic; this is also a broad field for the modeler, whose importance will increase even further as a result of the extended range of methods. It is the modeler's task to distinguish between useful details and exaggerated precision that is not consistent with the quality of the input data. It would often be useful to consider operational aspects when making strategic decisions, such as process design or purchasing equipment.

---

[2]It is sometimes possible to solve very large problems. But there are small problem instances leading to runtime issues.

Improved solution methods should also be used to make the models more realistic in terms of data quality. Except for Sect. 11.3, in this book, deterministic models were discussed throughout. However, as the data are subject to uncertainties, which may be the subject to probability distributions, robust solutions can be found for LP problems with uncertain data [63] or stochastic optimization — also called stochastic programming — as the means of choice. Even for MILP problems, solution methods exist to consider data with stochastic uncertainties [cf. Schultz (1995,[494]), Carøe & Schultz (1999,[108])]. Although these approaches are still subject to restrictions and were mostly implemented only in the form of university software, there are successful applications of these techniques, e.g., in the energy industry [Gollmer et al. (2000,[227])] and the chemical industry [Sand et al. (2000,[481]) and Engell et al. (2001,[174])].

In our considerations, we took little account of the complexity of the problems considered in the sense of complexity theory. This theory, cf. (2000,[209]), has proved very useful in classifying problems into two classes: those of easy and those of hard problems.[3] Problems solved in polynomial runtime belong to the light class, called $\mathscr{P}$. The effort to invert a matrix with n rows and columns increases with n^3,[4] so this problem is in class $\mathscr{P}$. The proof of a problem belonging to class $\mathscr{P}$ is provided if an algorithm can be constructed which solves this problem in polynomial runtime. For problems that are not solved in polynomial time, the computational effort usually increases exponentially, e.g., with 2^n or with $n!$. Klee & Minty (1972,[334]) have shown that there are LP problems for which the Simplex algorithm requires at least exponentially many iteration steps; most practical LP problems, on the other hand, can be solved with the Simplex algorithm in polynomial time. Khachian (1979,[332]) succeeded in proving that LP problems belong to class $\mathscr{P}$ by showing that some interior-point methods solve LP problems in polynomial runtime.[5]

However, for the most commonly used methods in practice, a corresponding theoretical proof is still lacking. This shows that it is not so easy to prove that a problem does not belong to $\mathscr{P}$. If no algorithm has been constructed for a problem that solves this problem in polynomial time, this does not mean that this will not be possible in the future. In the class $\mathscr{N}\mathscr{P}$ — this abbreviation stands for non-

[3] A classification into the class of nice hard and hopelessly hard would perhaps be more appropriate, because the computing time for the allegedly mild problems solvable in polynomial time can become quite large depending on the coefficients and powers occurring in this polynomial context. In addition, one should bear in mind that complexity theory deals with the determination of the runtime behavior as a function of the size of the problem. However, there are many problems, including those classified as $\mathscr{N}\mathscr{P}$-complete or $\mathscr{N}\mathscr{P}$ difficult, which occur in practice in small instances and can easily be solved with B&B or B&C methods. If the problem is hopelessly serious, the situation is not hopeless, depending on the size of the problem.

[4] Prefactors or lower order polynomials are not considered here.

[5] However, it should again be noted here that complexity theory makes statements about the scalability of problems with regard to their solution behavior. In practice, the Simplex algorithm often performs better than the interior-point methods.

deterministic polynomial — we find problems that could not be solved with any available (deterministic) algorithm in polynomial time. If the proof of membership to $\mathscr{P}$ is still lacking for a problem and can at the same time be shown in polynomial runtime and it can be determined that a proposed solution is actually a solution of the problem, then it belongs to the class $\mathscr{N}\mathscr{P}$. It is useful to note that $\mathscr{N}\mathscr{P}$ problems have certain relationships with respect to the required computational effort: they can be solved in the same time or with the same runtime behavior except for constants.

This leads to the definition of $\mathscr{N}\mathscr{P}$-complete problems. A problem is called $\mathscr{N}\mathscr{P}$-complete,[6] if the existence of a polynomial algorithm to solve this problem implies that all problems in $\mathscr{N}\mathscr{P}$ can be solved in polynomial time. Examples are the backpack problem from Sect. 7.1.1, the traveling salesman problem, or the satisfiability problem described in Sect. 7.5, for which $\mathscr{N}\mathscr{P}$-completeness was first proven. Also the mixed integer optimization in all its forms — MILP, MINLP, and also mixed binary optimization — falls into the class $\mathscr{N}\mathscr{P}$ (Karp 1972,[327]); often an exponential growth of computing time is observed. It may be argued that the classification of a problem as belonging to a complexity class is of more theoretical interest. If you have a certain problem with data instances of non-varying size and can actually solve it in a satisfactory computing time, you could agree with this argumentation. However, if one wants to know something about the scaling behavior of the problem or if one simply cannot solve it in the desired computation time, the classification or the proof that a problem is $\mathscr{N}\mathscr{P}$-complete or $\mathscr{N}\mathscr{P}$-hard[7] can help to avoid certain aberrations and to find the right means to approach the problem at least with efficient heuristics and search strategies. Especially in connection with scheduling, it becomes clear that the mixed integer optimization meets insurmountable barriers due to the complexity.

17.2.7 The Future of Planning and Scheduling

What are likely scenarios for planning and scheduling in the near future? It is already evident that there are a growing number of software packages that solve planning problems with accurate procedures. Mixed integer optimization is increasingly becoming an accepted standard for planning and design; this is very obvious in the process industry [304]. This is not yet the case for scheduling problems and heuristic methods can usually be found here. But a trend is recognizable, which lets the mathematical optimization community and the community of those preferring

[6]Furthermore, there are the $\mathscr{N}\mathscr{P}$-hard problems. Every problem to which we can map or transform an NP-complete problem in polynomial time and which, regardless of whether or not this problem belongs to the class $\mathscr{N}\mathscr{P}$, has the property that it cannot be solved in polynomial time, unless $\mathscr{P} = \mathscr{N}\mathscr{P}$, means $\mathscr{N}\mathscr{P}$-hard. The $\mathscr{N}\mathscr{P}$-hard problems are therefore at least as hard as the $\mathscr{N}\mathscr{P}$-complete problems.

[7]Usually, optimization problems are NP-hard (objective function); and the underlying decision version is NP-complete.

Constraint Programming grow closer together; cf. Heipcke (1999,[258]). Hybrid methods [cf. Harjunkoski et al. (2000,[254]), Hooker (2000,[270]), or Jain & Grossmann (2001,[284])] are being developed which combine language elements and algorithmic components of both worlds. It is to be expected that this will have great effects on supply chain optimization and scheduling problems. In 1999, the European Union decided to support the `LISCOS` (*Large Integrated Supply Chain Optimization Software*) project with several million Euros. This project (http://www.liscos.fc.ul.pt), initiated by the *Scientific Computing* group of BASF Aktiengesellschaft and carried out jointly with 8 consortium partners, aimed at the development of MIP-CP hybrid processes. Timpe (2002,[545]) describes a successful application of this approach to solve a planning and scheduling problem in chemical industry. At about the same time, the EU supported the project `COCONUT` [86] for the solution of non-convex optimization problems with methods from Global Optimization and Constraint Programming [86]. Another approach is continuous-time model formulations; here are especially the work [252, 279–281, 364] and [366] of Floudas and employees very promising.

17.2.8 Simultaneous Operational Planning & Design and Strategic Optimization

In industry, it often happens that a customer wants a planning tool or scheduling software for a production network that has just been built or expanded. In scheduling problems, especially, it often turns out that there are bottlenecks for certain products or situations. The situation could be significantly improved if the planning and scheduling aspects were included during the design phase. This problem may seem mathematically very complex, as scheduling problems are very difficult in themselves. However, scheduling problems are often only very difficult when some critical resources (e.g., raw materials, machine availability, or personnel) become particularly scarce. But this could be avoided by simultaneous analysis of design and planning problems. Simultaneous analysis requires reasonably realistic and detailed demand forecasts to be available and the affected operational and planning departments to cooperate; unfortunately, this is often a problem. The site analysis and design of the topology of a reactor system described in Kallrath (1999,[301]) are successful examples.

The problem described in Kallrath (2001b,[302]; 2002c,[303]) demonstrates the usefulness of a model that contains operational and strategic-tactical aspects simultaneously. The company wants to acquire additional sites or facilities, install new ones with improved technology, or close older facilities. In industry, there may be a logical link between different plants, i.e., if one plant is shut down, another may also have to be shut down. The strategic aspects of the problem are given by the cost of purchasing a site or commissioning or shutting down a plant, whereby the investment costs should prove to be reasonable over a period of several years.

For the plant types in question, data must be provided as required for plants already used in operational planning. The degrees of freedom of decision include the times of commissioning or shutdown. Since these decisions have long-term consequences for a production network, it is important that the optimality of the solution can be proven, but at least the quality of the solution can be evaluated in terms of upper and lower bounds, and then the solution is robust against market fluctuations — this can be achieved, for example, with the help of stochastic optimization. In Kallrath (2009,[309]), these aspects are discussed in detail.

Another integration of operational and strategic optimization is found in a work by Klosterhalfen et al. (2019,[336]). The authors develop a generic and innovative MP solution approach to derive the financial key performance indices (KPIs) required for future decision making at the upper management level. In a first step, they formulate an optimization model that finds the optimal product mix for the *entire* value chain such that the *contribution margin (CM) is maximized*. In a second step, they use another nonlinear optimization model [referred to as *cost allocation model* (CAM)] to allocate the costs of the optimized product mix to the individual products. Due to the existence of byproducts in chemical value chains, the CAM formulation becomes a nonlinear program (NLP). However, it is possible in this case, to decompose the problem into two linear programs (LPs), which can be efficiently solved sequentially, in order to obtain the optimal CAM solution for a truly CM-optimal product ranking. Based on this detailed product-specific solution, all KPIs can be calculated subsequently on any level of aggregation. Their approach allows both the strategic accounting and also the product and supply chain management (P&SCM) department to work with one and the same data model and calculation logic ensuring cross-functional consistency and transparency. Thus, they bridge the gap between the P&SCM and the accounting communities. It becomes obvious that pure isolated business unit analyses can lead to wrong assessments of future capacity investments in our practical use case because capacity bottlenecks in the value chain cannot be properly detected. The detailed cost split into different cost categories facilitates the identification of the major cost drivers in the different planning scenarios. By adopting this integrated approach that considers the entire value chain as a whole, companies can identify a large improvement potential over the common practice that splits the value chain according to organizational boundaries and calculates isolated solutions.

Simultaneous operational planning and design/strategic optimization is very demanding in terms of the data structure and data availability. For all design or strategic objects, the same data are required as for the existing objects, for instance, in the supply network structure. Therefore, the setup of simultaneous strategic and tactical planning, or simultaneous tactical and operational planning, i.e., scheduling, is not an easy task. As more as one goes down from strategic into tactical and operational planning, there is a lot of information one needs to collect and validate as input data. For the existing plants, these data are usually available to the degree required. For the design objects, e.g., a whole plant site to integrate into an existing network, this can be challenging as this affects data from different sources in procurement, production, and logistics like contracts, purchase orders, price matrix formulas, price forecasts, exchange rates, bill of materials, manufacturing costs, regional and local capacity in production and warehouses, freight costs, taxes and customs, markups, trade restrictions, IBC reimbursement, packaging and filling costs, and in most cases many additional data on top of that list. In strategic optimization, it is helpful to work out the optimal solution for a set of scenarios covering, for instance, the future of price and demand forecasts as well as exchange rates. These scenarios could be analyzed on their own or in the sense of an expected value approach.

A recent software package, `PlanNow`,[8] integrates the ongoing collection of procurement, production and logistics data into different scenarios. Transformation, simulation and change of planning attributes are executed within the scenarios of `PlanNow`. After implementation of the numerical and mathematical optimization model strategic, tactical and operational planning could be combined and executed in communicated via the collaboration elements of `PlanNow`. Scenario solution can be visualized and analytically compared also based on different assumptions. Team and stakeholders could vote, agree or disagree, and approve plans. All information like data, models, team and stakeholder collaboration, and approval within each scenario are stored in `PlanNow` and can be retrieved at any time.

17.3 Mathematical Optimization for a Better World *

The title *Mathematical optimization for a better world* may need some explanations and justification in order not to appear immodest. In this book, mathematical optimization is presented as a modeling approach and a solution technique for problems that occur in the real world — primarily in industry, but indeed wherever decisions are required — and are accessible to quantification in a deterministic or probabilistic sense. Often, solutions are worked out in optimization projects, i.e., projects with a mathematical optimization kernel. The process of model building within an optimization project is, ideally, based on four key elements:

1. a binding agreement about the goal of the project (this include a formulation of the objective function and the constraints),
2. clarity, precision, and transparency,
3. logical consistency, and
4. measurability and quantification.

The result of the project could be an operative decision support system for production planning, cutting stock, or service selection to mention a few. It could also deliver support for a strategic decision to be made. Ideally, at the end, the successful project and tool implementation leads to a *win–win* situation to all people

[8]The webpage www.PlanNow.ai provides further details.

participating, involved or effected by the project. Thus, when talking about a better world, it is about *making good decisions*.

Decisions are made on Earth by humans for humans and other living beings with consequences for all life on this planet. Are the key elements identified above for mathematical optimization are really limited to optimization projects, or could they be applied much more comprehensively where conflicting interests meet in decision making processes? Mathematical optimization offers quantitative support, especially with regard to the robustness of decisions, and reveals inconsistencies in the assumptions and prerequisites of a model rather than, for example, decision support based on selective simulations. It provides a framework for multi-criteria optimization with conflicting goals. In individual optimization projects, it can lead to win–win situations and reduced usage of natural resources on a company level.

In a much broader sense — and that is meant by better world — it can also help the society, a country, or the world, as all decisions seem to contain the following elements: goal or goals, constraints which one is willing to accept or which one must keep for physical reasons, and possible degrees of freedom. Decisions based on an optimization project are usually better accepted if all people affected have been included in the modeling process. Good examples for this are production planning systems and personnel deployment plans, especially those that can easily switch from holiday planning to staff reduction. It is important for quality and acceptance to put decisions on a quantitative basis and being able to *measure and quantify the improvements* — mathematical optimization supports this. This becomes even more important when extending mathematical optimization by some concepts of game theory. Collective decision making processes have their pitfalls. They can, as proven in mathematical game theory and discussed in Eichner et al. (1996,[171]), lead to strange paradoxes. If more than three prioritized alternatives are available, it is possible that decisions will be made that no one actually wants.

If the decision is to be as good as possible, then all possible consequences (logical consistency) and aspects must be examined. We are not claiming that they have to be considered in the model to be developed — but we propose that they can only be omitted based on a detailed consideration and safe error estimation. Thus, for example, Newton's theory of gravity is sufficient for the construction of an elevator under earthly gravitational conditions within the required constructional inaccuracy, and the general theory of relativity does not need to be applied. Not considering certain aspects, however, can be a bad mistake and lead to acceptance problems (see page 10 and the truck drivers concerned).

It helps in the modeling process and that anyone who can make a factually competent contribution to this without taking their status into account and without consequences for their career is also allowed to do so — for example, involving some colleagues from the production floor during the design or modeling phase could be useful. So during the modeling phase, each participant should be completely independent. This increases the chance that *unusual ideas* can survive the early stage and so can lead to unbiased results. The most successful projects in my career were those with new employees who had only recently joined the company, or those close to their retirement; both groups were hard to beat in terms of motivation, openness, objectivity, and commitment — and courage.

If a decision problem can be formulated in whole or in part as a mathematical optimization problem, and if it is to be solved as such, then both the client people and the modeling people are needed who want to solve the problem as well as possible and are not bound by too many other things in their minds and freedom of decision. The basic idea of optimization does not let you rest until you have been able to derive the optimal solution or at least a certain statement. Optimization is not only a solution technique, but in the good sense rather an *attitude toward life*. Good doctors, who still obey the *oath of Hippocrates*[9] and care about a patient who is seriously ill, will not rest[10] until they either saved the patient or can be sure they have tried everything. Fire-fighters looking for survivors in a burning building will have a similar attitude — why else would they risk their lives? The 80% mentality and reasoning, *80% are good enough, everything else is not reasonable in cost,* is a completely different approach, which, if well justified, is appropriate — only, usually the justification is not good, because (a) often the total potential is not known at all: it is often only known after a rigorous mathematical investigation (otherwise it is unclear: 80% of what?) and (b) the additional potential is not set in relation to the costs. The above-mentioned attitude to life has essentially also to do with what is accepted as an argument and reasoning structure. Another aspect of this 80% mentality is that independent or innocent people suffer from the consequences of this 80% mentality and decisions. It was fair if the decision maker would take 100% responsibility, but usually this is not what happens. Would they still resort to this 80% responsibility if they were 100% liable for all consequences of their decisions?

As far as the modeler is concerned, the success of the project depends on the modeler's technical knowledge of mathematical optimization, but perhaps even more on motivation — sportive or altruistic elements can lead to great achievements. A well-executed project that contains mathematical aspects of optimization will, if the complexity of the underlying problem allows it — although this often only becomes apparent in the course of the project for mathematical reasons — leads to success with high probability and to a situation in which all participants win — on their own, self-chosen scale accepted by all participants. But it requires the participation of all and their motivation, as well as an atmosphere that allows to talk about goals in an objective, open and trusting way, and to reach a real consensus. Based on the observation "*Tell me your goals and I will tell you how you will behave*," the goals for the members of the project team, and also objective

[9]Physician of antiquity, who lived from about 460 to 377 BC and laid the base of ancient Greek medicine. He formulated the Hippocratic Oath, which contains moral commandments still valid and binding today for a true physician. For natural scientists the equivalent of the Hippocratic oath would probably be "Measured values must not be faked"; in the case of persons with decision making responsibility, especially managers and politicians, perhaps the moral commandment "name all goals and intentions" openly and completely would make sense.

[10]In the truest sense of the word *rest*, such doctors will not worry about the end of the working day, weekends or other things that are often anchored in company agreements or labor law.

function of the optimization problem, should not be too narrowly defined but must be formulated in such a way that it implies a consistent and desirable behavior. It is now worthwhile to examine which elements were characteristic of successful projects:

1. The project results and the planning tools based on mathematical optimization were really wanted by all participants. And no interested party was excluded.
2. The results — and this was usually foreseeable early — led to a situation in which everyone profited — by their own yardsticks. But here, again it becomes clear how important openness, trust, and a tolerant, constructive culture of error are; otherwise, not everyone will necessarily disclose their goals and standards.
3. The projects usually had strong personalities as project leaders who acted and decided in the team's spirit but clearly had the responsibility as individuals. It is probably one of the most questionable ideas in politics, economics, and administration of the past 20 years to distribute decision making authority among teams, committees or the like — just imagine this in nature: a crowd of hunters, for example, who debate a lot during the hunt; not to mention the debates after the hunt, when it comes to distributing the prey, if this system is successful at all. The analysis of successful projects shows that projects led by individuals perform much better than others.
4. The project teams were very carefully composed, taking into account the task at hand. Particular attention was paid to the *spectrum* of the team members.
5. The number of people involved in successful projects has always been quite small, usually no more than five. This does not necessarily mean that projects with more participants only bring bad luck. But then, apart from the technical skills to run such projects, there are certainly even higher demands on objectivity, trust and openness to be met; project leaders must then really be very good leaders, who motivate their team, respond to the peculiarities of the team members, and understand how to use their knowledge and skills.
6. The best clients or customers were employees who had only been with the company for a few years, or those who had been on the verge of retirement for a short time, i.e., not more than two years. They just did what they thought is reasonable — they were neither afraid of negative effects on their careers nor strictly aligned with bureaucratic rules.

Mathematical optimization is very closely linked to one's view of life — and *vice versa*. It leads to higher precision and objectivity, because inconsistencies always raise new questions to be clarified — but also to improved leadership and leadership style due to the transparency. If openness, honesty, and professional competence take precedence over politics in a corporate culture, if there is a good mix of people with different backgrounds who accept, respect, and cooperate with their respective professional backgrounds, it should be obvious and self-evident that decision making processes are supported by mathematical optimization. It does not always have to be mathematical optimization, but it is often the only adequate means to safely and quantitatively support decisions in complex systems with many degrees of freedom or many restrictions, to use existing resources more efficiently, or to make robust strategic decisions with uncertain input data. Especially, in times when the willingness to take risks is rather modest, one should expect that mathematical optimization is highly valued as it enables us to make reliable statements about the feasibility and quality of decisions. There are not always spectacular, career-accelerating results, rather mathematical optimization can contribute in small steps, especially in large companies — a physicist would probably say "in an *adiabatic* way" — to improve the technical, procedural, discussion-argumentative basis of a company, as long as it desires and allows this process.

Large companies with a large number of employees are, for statistical reasons, a smaller reflection of society, in fact, of the state as a whole. However, as their employees are also members of state and society and partly contribute the internalized corporate culture to the social culture, it becomes clear that mathematical optimization not only improves the decision making processes and economic situation of a company or its internal culture but can also contribute far beyond that to a better world for everyone.

Appendix A
Software Related Issues

In this appendix we briefly address how to access data from external files using algebraic modeling systems. For convenience, we list the model files in MCOL.

A.1 Accessing Data from Algebraic Modeling Systems

There is not just one approach for data access and people naturally prefer various techniques depending on their background or objectives. The oldest, easiest, and most transparent way (W1) is to read *flat text files*. Widespread is another way (W2) to extract data from spreadsheets or databases using various data connectors — these data connectors can be very elaborated and powerful tools. Finally, many people nowadays (W3) do all data reading and preprocessing using programming languages such as `Matlab`, `Python`, or `R`. Finally, there are Extract, Transform, Load (ETL) tools (W4), e.g., `Alteryx`, `FICO DataPipelines`, `Kettle Pentaho Data Integration` (open source), and `RapidMiner Turbo Prep`. Extraction means extract the relevant data from different data sources, Transformation stands for transform the data into the structure and format of the target database, and Load is to load the data into the target database. Most importantly, ETL tools define a clear cut of ETL tasks which prepare data, e.g., for optimization, take care of streaming, batch jobs. They provide the clean and (partly) validated data input an optimization model requires. Partly, these can call out to optimization and analytic services for some preprocessing tasks.

The balance between the advantages and disadvantages of these approaches W1 to W4 depends on personal preferences or company requirements. My personal preference are time-independent techniques and only reading and writing flat text files (it best fits my ideal of autarky or self-sufficiency). This has enabled me to have operative optimization-based decision support systems running for more than 20 years without any interaction and maintenance work surviving many migrations

J. Kallrath, *Business Optimization Using Mathematical Programming*,
International Series in Operations Research & Management Science 307,
https://doi.org/10.1007/978-3-030-73237-0

of the operating system or interfaces to SAP without my involvement. I think the best approach for the customer is to try not to depend on IT support as this avoids after sales costs — ideally: buy it, and then use it forever.

IT departments more often recommend W2 to W4 as they often go with what is up to date. It is like visiting car dealers: It is more likely that they recommend a fancy modern car, maybe even one operating on electric energy — rather than recommending to buy a (used) 1992 Scirocco. Guess, why?

The good news is: Algebraic modeling systems allow you to incorporate data from external files in *all* forms and flavors W1 to W4 mentioned above. They all have their syntax for reading text files, accessing spreadsheets or databases, and tools for connecting to `Python` or `R`. They are loyal to their customers and keep supporting these different data connectors — and add new ones from time to time.

A.2 List of Case Studies and Model Files

Filename	*Problem description*	*Section*
ABSVAL	Modeling absolute value terms	6.5
BENCH101	Solution to Exercise 10.1 in Chap. 10	
BENCH102	Solution to Exercise 10.2 in Chap. 10	
BLENDX	Ore blending problem (`XPRESS-MP` manual)	2.7.1
BOATDUAL	Dual of the “Boat Renting” problem (Ex. 3.3)	
BREWERY	Brewery planning	8.3
BURGAP	Knapsack exercise (Exercise 7.3)	7.10
BURGLAR	Knapsack problem	7.1.1
BUSCREW	Bus crew scheduling	7.8.4
CALVES	Calves and pigs problem	3.3.1
CARTON	Carton scheduling problem	10.3
CH-2TRI∗	Minimal perimeter convex hull for two triangles	10.3
COUPLES	Solution to Exercise 6.5 in Chap. 6	
DEA	Data envelopment analysis (Exercise 5.2)	5.3
DUAL	Solution to Exercise 3.1b) in Chap. 3	3.5.1
dynBigM*	Dynamic computation of big-M coefficients	14.1.2.1
USDO	Solution to Exercise 7.5 in Chap. 7	
FLOWSHOP	Solution to Exercise 7.7 in Chap. 7	
fracProg*	Fractional programming example	11.1
GAP	Generalized assignment problem (Exercise 7.2)	
goalProg*	Example exploiting a hierarchy of goals	5.4.3
lagRel*	Lagrange relaxation applied to the GAP	14.1.3.3
LIM1	Solution to Exercise 5.1 in Chap. 5	

LIM2	Solution to Exercise 6.7 in Chap. 6	
MULTK	Exercise on multiple knapsack (Exercise 7.4)	7.10
NETWORK	Network problem (Exercise 4.3)	
newsVendor*	Newsvendor problem — two stage stochastic programming	11.3.2.1
NPV	Solution to Exercise 6.8 in Chap. 6	
Portfolio*	EEV or generalized quantities for multi-stage problems	11.3.2.5
optGrid*	Optimal breakpoints for piecewise linear approximations	14.2.3
PRDX	Simple production planning exercise	2.5.2
PRIMAL	Primal problem (Exercise 3.1b)	
PROJSCHD	Project scheduling case study	10.2.3
QUADRAT	Quadratic programming example	11.4
SET	Set covering problem (Exercise 7.6)	
SIMPLE1	Solution to Exercise 2.2 in Chap. 2	
SIMPLE2	Solution to Exercise 2.3 in Chap. 2	
SLAB	Solution to Exercise 6.6 in Chap. 6	
SLUDGE	Example illustrating recursion	11.2.1
TRIM1	Trimloss problem (Exercise 4.1)	4.1.1
TRIM2	Trimloss problem (Exercise 4.2)	4.1.2
TrimMINLP*	Trimloss problem formulated as a MINLP problem	13.3
TSP	Traveling salesman problem (Exercise 7.1)	
vehRoute*	Heating oil delivery (*e4delivr.mos*)	7.2.3
YLDMNGMT	Yield management, financial modeling	8.4.2

Appendix B
Glossary

The terms used in this book are also defined here in this glossary for the purpose of subsequent reference. Within this glossary all terms written in boldface are explained in the glossary.

Algebraic modeling language (AML): A high-level programming language similar to the mathematical notation of optimization problems. External solvers can be accessed via the AML to obtain a solution of the implemented model formulation.

Algorithm: A systematic procedure organized into a series of steps, mathematically or otherwise.

Arc: An object within a **graph**. Arcs, sometimes also called edges, usually represent roads, pipelines, or similar paths along which some material can flow. Often arcs have a capacity. Arcs connect the **nodes** in a graph.

Basic variables: Those variables in optimization problems whose values, in non-degenerate cases, are away from their **bounds** and are uniquely determined from a system of equations.

Basis (Basic feasible solution): In an LP problem with constraints $\mathsf{A}\mathbf{x} = \mathbf{b}$ and $\mathbf{x} \geq 0$ the set of m linearly independent columns of the $m \times n$ system matrix A of an LP problem with m constraints and n variables forming a regular matrix $\mathscr{B}$. The vector $\mathbf{x}_B = \mathscr{B}^{-1}\mathbf{b}$ is called a basic solution. $\mathbf{x}_B$ is called a basic feasible solution if $\mathbf{x}_B \geq 0$.

Bilevel programming: Bilevel programming (BLP) problems are mathematical programming problems that contain an optimization problem in the constraints. Alternatively, one might say, it is an optimization problem constrained by another optimization problem.

Bound: Bounds on variables are special constraints. A bound involves only one variable and a constant which fixes the variable to that value, or serves as a lower or upper limit.

J. Kallrath, *Business Optimization Using Mathematical Programming*, International Series in Operations Research & Management Science 307,
https://doi.org/10.1007/978-3-030-73237-0

Branch & Bound: An implicit enumeration **algorithm** for solving combinatorial problems. A general Branch & Bound algorithm for **MILP** problems operates by solving an **LP** relaxation of the original problem and then performing a systematic search for an optimal solution among sub-problems formed by branching on a variable which is not currently at an integer value to form a sub-problem, resolving the sub-problems in a similar manner.
Branch & Cut: An **algorithm** for solving mixed integer linear programming problems. It operates by solving a linear program (step 1), which is a **relaxation** of the original problem, and then performing a systematic search (step 2) for an optimal solution. During this search, a series of valid constraints (cuts) is added to the relaxation or to sub-problems generated from the relaxation. These cuts must be satisfied by the integer constraints of the original problem. The problem or sub-problems are resolved likewise (step 3).
Constraint: A relationship that implicitly or explicitly limits the values of the variables in a model. Usually, constraints are formulated as inequalities or equations representing conditions imposed on a problem, but other types of relations exist, e.g., set membership relations.
Continuous relaxation: An optimization problem in which the requirements that certain variables take integer or discrete values have been removed.
Convex region: A region in multi-dimensional space where a line segment joining any two points lying in the region remains completely in the space.
Cutting-planes: Additional valid inequalities that are added to **MILP** problems to improve their LP relaxation when all variables are treated as continuous variables.
Duality: A useful concept in optimization theory connecting the (primal) optimization problem and its dual.
Duality gap: For feasible points of the primal and dual optimization problem the difference between the primal and dual objective function values. In LP the duality gap of the optimal solution is zero.
Dual problem: An optimization problem closely related to the original problem which is called the primal problem. The dual of an LP problem is obtained by exchanging the objective function and the right-hand side constraint vector and transposing the constraint matrix.
Dual values: A synonym for shadow prices. The dual values are the dual variables, i.e., the variables in the dual optimization problem.
Feasible point (feasible problem): A point (or vector) to an optimization problem that satisfies all the constraints of the problem. (A problem for which at least one feasible point exists.)
Global optimum: A feasible point $\mathbf{x}^*$ to an optimization problem that gives the optimal value of the objective function $f(\mathbf{x})$. In a minimization problem, we have the relation $f(\mathbf{x}) \geq f(\mathbf{x}^*)$ for all other points of the feasible region.
Goal programming: A method of formulating a multi-objective optimization problem by expressing each objective as a goal with a hypothetical attainment level, modeled as a constraint, and using an expression which will minimize deviation from goals as the objective function. The goals are also called targets.

Graph: A mathematical object consisting of **nodes** and **arcs**, useful in describing network flow problems. The structure and properties of graphs are analyzed in graph theory, a mathematical discipline.
Heuristic solution: A feasible point of an optimization problem which is not necessarily optimal and has been found by a constructive technique which could not guarantee the optimality of the solution.
Infeasible problem: A problem for which no **feasible point** exists.
Integrality gap: The difference between the objective function value of the continuous relaxation of an integer, mixed integer, or discrete programming problem and its optimal objective function value.
Kuhn–Tucker conditions: Generalization of the necessary and sufficient conditions for steady points in nonlinear optimization problems involving equalities and inequalities.
Linear combination: A linear combination of vectors $\mathbf{v}_1, \ldots, \mathbf{v}_n$ is the vector $\sum_i a_i \mathbf{v}_i$ with real valued numbers a_i. The trivial linear combination is generated by multiplying all vectors by zero and then adding them up, i.e., $a_i = 0$ for all i.
Linear function: A function $f(\mathbf{x})$ of a vector $\mathbf{x}$ which has a constant gradient $\nabla f(\mathbf{x}) = \mathbf{c}$. In that case $f(\mathbf{x})$ is of the form $f(\mathbf{x}) = \mathbf{c}^T \mathbf{x} + \alpha$ for some fixed scalar α.
Linear independence: A set of vectors is linearly independent if there exists no non-trivial linear combination which represents the zero vector. The trivial linear combination is the only linear combination which generates the zero vector.
Linear Programming (LP): A technique for solving optimization problems containing only continuous variables appearing in linear constraints and in a linear objective function.
Local optimum: A feasible point $\mathbf{x}^*$ to an optimization problem that gives the optimal value of the objective function in the neighborhood of that point $\mathbf{x}^*$. In a minimization problem, we have the relation $f(\mathbf{x}) \geq f(\mathbf{x}^*)$ for all other points of that neighborhood. Contrast with **global optimum**.
Matrix: A rectangular array of elements such as symbols or numbers arranged in rows and columns. A matrix may have operations such as addition, subtraction, or multiplication associated with it, if these are valid for the matrix elements.
Metaheuristics: A metaheuristic in the context of mathematical optimization is a generic, high-level procedure to generate (near-optimal) feasible solutions to an optimization problem. Metaheuristics cannot guarantee global optimality.
Mixed Integer Linear Programming (MILP): An extension of LP problems which allows some of the variables to take on binary, integer, semi-continuous, or partial-integer values.
Mixed Integer Nonlinear Programming (MINLP): A technique for solving optimization problems which allows some of the variables to take on binary, integer, semi-continuous, or partial-integer values and allows nonlinear constraints and objective functions.

Model: A mathematical representation of a real-world problem using variables, constraints, objective functions, and other mathematical objects.
Modeling system: In the context of mathematical optimization, a software system for formulating an optimization problem. The optimization problem can be formulated in an algebraic language or can be represented by a visual model. In the modeling system one brings together the structure of the problem and the data.
Network: A representation of a problem as a series of points (nodes), some of which are connected by lines or curves (arcs), which may or may not have a direction characteristic and a capacity characteristic. The network is usually represented by a **graph**.
Node: An object within a **graph**. Nodes usually represent plants, depots, or a point in a network. Nodes can be connected by **arcs**.
Non-basic variables: Those variables in optimization problems which are independently fixed to one of their bounds.
Nonlinear function: Any function $f(\mathbf{x})$ of a vector $\mathbf{x}$ which has a non-constant gradient $\nabla f(\mathbf{x})$.
Nonlinear Programming (NLP): Optimization problems containing only continuous variables and nonlinear constraints and objective functions.
$\mathcal{NP}$ **completeness**: Characterization of how difficult it is to solve a certain class of optimization problems. The computational requirements increase exponentially with some measure of the problem size.
Objective (objective function): An expression in an optimization problem that has to be maximized or minimized.
Optimization: The process of finding the best solution, according to some criterion, or objective function, respectively, of an unconstrained or constrained optimization problem.
Optimum (optimal solution): A feasible point of an optimization problem that cannot be improved in terms of the objective function without violating the constraints of the problem.
Outer approximation: Algorithm for solving **MINLP** problems based on an equivalent representation of the feasible region by an intersection of hyperplanes and linearization so as to bound the objective function.
Pivot: An element in a matrix used to divide a set of other elements. In the context of solving systems of linear equations the pivot element is chosen with respect to numerical stability. In linear programming the pivot element is selected by pricing and the minimum ratio rule. In that context, the linear algebra step calculating the new basis inverse, although not explicitly, is sometimes called the pivoting step.
Parametric programming: Investigation of the effects of significant changes to problem coefficients on the optimal solution in an optimization problem.
Post-optimality (Post-optimal analysis): Investigation of the effect of marginal changes in the problem's coefficients on the optimal solution.
Presolve: An algorithm for modifying an optimization problem before solving it, whereby redundant features are removed and valid additional features may be added.

Ranging: Investigation of the limits of changes in coefficients in an optimization problem which will not fundamentally affect the optimal solution.
Reduced cost: The price (or the gain) for moving a non-basic variable away from the bound it is fixed to.
Relaxation: An optimization problem created from another where some of the constraints have been removed or weakened.
Report writer: A software system used to take the mathematical results from solving an optimization problem and turn them into a report in business terms.
Scaling: Reducing the variability in the size of the elements in a matrix (e.g., an LP matrix) by a series of row or column operations.
Sensitivity analysis: The analysis of how an optimal solution of an optimization problem changes if some input data used in the problem are slightly changed.
Shadow price: The marginal change to the objective function value of an optimal solution of an optimization problem caused by making a marginal change to the right-hand side value of a constraint of the problem. Shadow prices are also termed **dual values**.
Simplex algorithm: Algorithm for solving LP problems that investigate vertices of polyhedra.
Slack variables: Variables inserted into $\leq$ inequalities with positive unit coefficients to convert them into equalities.
Solver: A software (library) with a set of implemented algorithms like Simplex and Branch & Bound capable of solving (various types of) optimization problems. Examples are CPLEX and BARON.
Special ordered set: An ordered set of variables in which at most a fixed number of the variables may be non-zero, and if more than one is allowed to be non-zero, they must be adjacent in the ordering.
Stackelberg game: A strategic game in game theory named after the German economist Heinrich Freiherr von Stackelberg who introduced this hierarchical game in 1934 in his book *Marktform und Gleichgewicht* (engl.: *Market Structure and Equilibrium*). In a Stackelberg game, the players of the game compete with each other according to the rule: The leader makes the first move, and then the follower reacts optimally to the leader's action. In this asymmetric hierarchical game the leader and the follower cannot be interchanged.
Stochastic optimization: A technique to solve optimization problems in which some of the input data are random or subject to fluctuations.
Successive Linear Programming (SLP): Algorithm for NLP problems containing a modest number of nonlinear terms in constraints and objective function.
Surplus variables: Variables inserted into $\geq$ inequalities with negative unit coefficients to convert them into equalities.
Traveling salesman problem: A very hard MILP optimization problem which consists of finding the cheapest closed tour in a graph subject to the constraint that all nodes must be visited exactly once.

Unbounded problem (LP,MILP): A problem in which no optimal solution exists because the objective function tends to increase to infinity or to decrease to minus infinity. Often missing bounds on variables or missing constraints cause a problem to be unbounded.

Unbounded problem (NLP,MINLP): A problem in which a convex objective function to be maximized or a concave objective function to be minimized tends to increase to infinity or to decrease to minus infinity due to missing bounds on variables or missing constraints. Another reason is that the objective function has points $\mathbf{x}_p$ in the feasible region with $\lim_{\mathbf{x}\rightarrow\mathbf{x}_p} f(\mathbf{x}) = \pm\infty$.

Unimodularity: A property of a matrix. A quadratic matrix is called unimodular if its determinant is +1. An LP matrix is called unimodular if all its sub-matrices have determinants with values +1, 0, or -1. If an LP matrix is unimodular, and the right-side constraint vector has only integer entries, then all basic feasible solutions to the LP take integer values.

Variable: An algebraic symbol used to represent a decision or other varying quantity. Variables are also called "unknowns" or just "columns."

Vector: A single-row or single-column matrix of variables.

Appendix C
Mathematical Foundations: Linear Algebra and Calculus

In this appendix — a first draft has been provided by Dominik Schweisgut (Heidelberg University) — we summarize mathematical principles connected to mathematical optimization, model building, and solution algorithms. We introduce different topics by giving definitions, mathematical statements, and examples.

In particular, we present some elementary aspects of matrices and sets of equations as well as the basic of one-dimensional calculus to improve the readers' understanding of the techniques contained in this book. In this appendix, we want to get the reader closer to a more formalized mathematical language by introducing the terminology used when reading or writing mathematical documents on a higher scientific level.

The material provided in this appendix should not be a complete introduction into higher mathematics needed when working on optimization problems but it should provide some fundamental and formalized knowledge that can be useful when one wants to get a deeper understanding of the mathematics of an optimization problem.

C.1 Sets and Quantifiers

First of all, we introduce the concept of sets already introduced in Chap. 1. Straightforwardly, one can say that a set is a collection of objects that often have one or more properties in common. As an example we can look at the set of all banknotes in USD. Let us give this set the name $\mathscr{B}$ (for banknotes) and we call the \$1 USD banknote "1USD" and so on. So we have

$$\mathscr{B} = \{1\text{USD}, 5\text{USD}, 10\text{USD}, 20\text{USD}, 50\text{USD}, 100\text{USD}\}.$$

J. Kallrath, *Business Optimization Using Mathematical Programming*, International Series in Operations Research & Management Science 307,
https://doi.org/10.1007/978-3-030-73237-0

Important to recognize is that there is no order in a set. Therefore,

$$\hat{\mathscr{B}} = \{10\text{USD}, 5\text{USD}, 100\text{USD}, 1\text{USD}, 20\text{USD}, 50\text{USD}\}$$

is the same set as $\mathscr{B}$. The entries contained in a set are called "objects" or "elements," and elements of a set do not appear twice in a set. Sets without elements are called *empty sets* usually denoted by the symbol $\emptyset$. In mathematics, there are various sets of numbers denoted by special symbols and we introduce them as further examples. For doing so, we write the set symbol followed by ":=" followed by a list of set elements:

1. $\mathbb{N} := \{1, 2, 3, \ldots\}$ (set of natural numbers) as in Sect. 1.9.1,
2. $\mathbb{N}_0 := \{0, 1, 2, 3, \ldots\}$ (set of natural numbers containing the zero),
3. $\mathbb{Z} := \{\ldots - 3, -2, -1, 0, 1, 2, \ldots\}$ (set of integers),
4. $\mathbb{Q} :=$ Set of fractions (all rational numbers),
5. $\mathbb{R} :=$ Set of rational and irrational numbers (set of real numbers).

Before we introduce some definitions and show some properties that sets can have, we introduce the so-called *quantifiers* and have a look at some principles of propositional logic. Why would we use quantifiers and propositional logic?

First of all using quantifiers and propositional logic, it is easy to express mathematical definitions, theorems, or propositions short and precise. We make this clear by giving examples where appropriate. When we look at propositional logic, we always have the Boolean algebra in mind. This helps us because a mathematical proposition can only take two values, *true* and *false*.[1] We now introduce some fundamental operations frequently used in mathematics.

First we introduce the very easy negation "¬". We make this clear by looking at the Boolean operation table for this operation:

A	¬A
w	f
f	w

Example: Instead of $x \neq y$ (x is not equal to y) we can now write $\neg(x = y)$.

Next we introduce the *and*-operation:

A	B	A ∧ B
w	w	w
w	f	f
f	w	f
f	f	f

[1] We do not go into the subtleties of undecidable propositions. Readers interested in this are referred to *Gödel's incompleteness theorem*.

where "A∧B" reads "A and B." As an example we can look at the following set: $\mathscr{S} = \{2, 4, 6, 8, 10, \ldots\}$ of positive and even integers. This enables us to claim the following proposition: "If an integer α is an element of $\mathscr{S}$, then α is even and positive" or if we use the and-operation: "If an integer α is an element of $\mathscr{S}$ then: α is even $\wedge$ α is positive." Notice that both properties have to be true.

Now we introduce the *or*-operator:

A	B	A ∨ B
w	w	w
w	f	w
f	w	w
f	f	f

,

where "A∨B" reads as "A or B." As an example we could look at the following set: $\mathscr{S}_1 = \{\ldots, -4, -2, 0, 1, 2, 3, \ldots\}$. Now we can say: "An integer α is element of $\mathscr{S}_1$ if α is even or α is positive" or, if we use the *or*-operator: "An integer α is element of $\mathscr{S}_1$ if: α is even $\vee$ α is positive."

Last we introduce the useful operators of *implication* and *equivalence* (implication: "⇒", equivalence: "⇔")

A	B	A ⇒ B
w	w	w
w	f	f
f	w	w
f	f	w

,

A	B	A ⇔ B
w	w	w
w	f	f
f	w	f
f	f	w

.

Looking only at these tables this may seem a little bit abstract so we give some examples: α is even $\wedge$ α is positive $\Rightarrow$ α is element of the set $\mathscr{S}$. But also the equivalence is true: α is even $\wedge$ α is positive $\Leftrightarrow$ α is element of the set $\mathscr{S}$.

There are two things to keep in mind: First, the implication $A \Rightarrow B$ requires great care when expressing mathematical propositions, because if A is wrong, B is always true. Second, to show that two propositions A and B are equivalent, is equivalent to showing (A⇒B) ∧ (B⇒A).

To express propositions shorter and to express more interesting propositions, we introduce now some quantifiers. With these, it is easier to specialize propositions if one wants to make propositions about special elements that have some properties in common.

First we introduce the universal quantifier "∀" (reads "for all") already listed in Sect. 1.9.1. Given a set $\mathscr{X}$, the universal quantifier is used to express a proposition about all elements of $\mathscr{X}$. *Example*: ($\mathscr{X} = \mathbb{N}$). $\forall x \in \mathbb{N} : x$ is positive. This could also be formulated as: If $x \in \mathbb{N}$, then $x > 0$; or $x \in \mathbb{N} \Rightarrow x > 0$.

Another useful symbol is "∈" (is element of) used when one wants to express that x is element of $\mathbb{N}$. Furthermore ":" means: For all x in $\mathbb{N}$ it holds that x is positive.

Notice that if $\mathscr{X} = \emptyset$ every proposition: $\forall x \in \emptyset : A(x)$ is true for every proposition $A(x)$, so one has to be careful about the sets one works with.

Next we introduce the existence quantifier or exists operator "∃" (reads "it exists a ...") that we use when we want to express that it exists a special element of a set that has specific properties. *Example*: $\exists n \in \mathbb{N}_0 : n = 3n$. This reads as: "It exists an element n of $\mathbb{N}_0$ such that $n = 3n$ is fulfilled." — in this example, there is only one element which fulfills the condition, namely, $n = 0$. *Exactly one* is linked to the existing operator, not to the conditional ":". Notice that the conditional ":" does not ask for "exactly one." The expression

$$\exists n \in \mathbb{N} : \frac{n}{2} \in \mathbb{N}$$

provides an example with infinite many elements fulfilling the condition. If we want to express "exactly one" we use the conditional ! and write

$$\exists! n \in \mathbb{N}_0 : n = 3n.$$

When using these quantifiers be careful with switching the position of the quantifiers when combining them. As an example: "For every woman in a class, there is one woman that is the mother of these women.". But: "There is one woman for every woman in a class, so that this woman is their mother." is very unlikely and for sure not the same proposition. So keep in mind that: $\forall x \exists y(x) : A(x, y)$ is not the same as $\exists y : \forall x : A(x, y)$ in general. Being now familiar with quantifiers and propositional logic allows us to introduce further concepts related to sets.

Definition Given a set X. A subset of X (we call it A), $A \subseteq X$, is a set with the property: $\forall x \in A : x \in A \Rightarrow x \in X$. Notice that $X = A$ is also possible.
For two given sets $A \subseteq X$ and $B \subseteq X$, the standard operations union "∪", intersection "∩", complement "A^c", and relative complement "$B \setminus A$" are commonly used. With the conditional | (read: *for that holds*) we have

$$\begin{aligned} A \cup B &:= \{x \mid x \in A \vee x \in B\} \\ A \cap B &:= \{x \mid x \in A \wedge x \in B\} \\ A^c &:= \{x \mid x \in X \wedge x \notin A\} \\ B \setminus A &:= \{x \mid x \in B \wedge x \notin A\}. \end{aligned}$$

Often useful is also the Cartesian product of two sets $A \times B = \{(a, b) \mid a \in A, b \in B\}$. Notice that the order is important here.

The operations union and intersection can be expanded to more than two sets. For given sets $A_1, \ldots, A_n$, their union is

$$\bigcup_{i=1,\ldots,n} A_i = \{x \mid x \in A_1 \vee x \in A_2 \vee \ldots \vee A_n\},$$

while the intersection is

$$\bigcap_{i=1,\ldots,n} A_i = \{x \mid x \in A_1 \wedge x \in A_2 \wedge \ldots \wedge A_n\}.$$

The following laws hold when working with sets. Given $A \subseteq X$, $B \subseteq X$, $C \subseteq X$:

$$\begin{aligned}
\text{Commutative law} &: A \cup B = B \cup A \text{ and } A \cap B = B \cap A \\
\text{Associative law} &: (A \cup B) \cup C = A \cup (B \cup C) \text{ and } (A \cap B) \cap C = A \cap (B \cap C) \\
\text{Distributive law} &: (A \cup B) \cap C = (A \cap C) \cup (B \cap C) \text{ and} \\
&\quad (A \cap B) \cup C = (A \cup C) \cap (B \cup C) \\
\text{De Morgan's laws} &: (A \cup B)^c = A^c \cap B^c \text{ and } (A \cap B)^c = A^c \cup B^c.
\end{aligned}$$

C.2 Absolute Value and Triangle Inequality

The absolute value and the triangle inequality become important when we talk about convergence and are always useful in the context of bounds.

Definition For given $x \in \mathbb{R}$, the absolute value $|x|$ of x is

$$|x| := \begin{cases} x, & x \geq 0 \\ -x, & x < 0. \end{cases}$$

Examples $|-6| = 6$, $|27| = 27$, $|0.4| = 0.4$. The absolute value function $x \to |x|$ obeys the *triangle inequality*

$$|x_1 + x_2| \leq |x_1| + |x_2| \quad , \quad x_1, x_2 \in \mathbb{R},$$

and the *inverse triangle inequality*

$$|x_1 - x_2| \geq ||x_1| - |x_2|| \quad , \quad x_1, x_2 \in \mathbb{R}.$$

We specify these inequalities here for the one-dimensional case, but it holds similarly in higher dimensions and vectors spaces. In Sect. 6.5 we have shown how to replace the absolute function by an equivalent MILP formulation.

C.3 Vectors in $\mathbb{R}^n$ and Matrices in $\mathcal{M}(m \times n, \mathbb{R})$

A (column) vector (sometimes, also called array) in the vector space $\mathbb{R}^n$ (formal definition of a vector space follows in the next section) is an ordered tuple of n real numbers

$$\mathbf{x} = \begin{pmatrix} x_1 \\ \cdot \\ \cdot \\ \cdot \\ x_n \end{pmatrix} = (x_1, \ldots, x_n)^{\mathrm{T}}$$

and these vectors are the elements of the vector space $\mathbb{R}^n$. The array $(x_1, \ldots, x_n)$ is called a row vector. Note that the transpose-operator T changes a row vector into an column vector, and *vice versa*. Given $\mathbf{x} \in \mathbb{R}^n$ as above, and $\mathbf{y} = (y_1, \ldots, y_n)^{\mathrm{T}} \in \mathbb{R}^n$, the vector addition is defined by adding up component by component, i.e.,

$$\mathbf{x} + \mathbf{y} := \begin{pmatrix} x_1 + y_1 \\ \cdot \\ \cdot \\ \cdot \\ x_n + y_n \end{pmatrix}.$$

Furthermore, we introduce the Euclidean scalar product

$$\mathbf{x} \cdot \mathbf{y} = \mathbf{x}^{\mathrm{T}}\mathbf{y} := \sum_{i=1}^{n} x_i y_i .$$

Definition (matrix) Let $m \in \mathbb{N}$, $n \in \mathbb{N}$. An $m \times n$ - matrix M (m rows, n columns, $m \times n$ entries) is a family of elements in the following form:

$$\mathsf{M} = \begin{pmatrix} a_{11} & \ldots & a_{1n} \\ \cdot & & \cdot \\ \cdot & & \cdot \\ \cdot & & \cdot \\ a_{m1} & \ldots & a_{mn} \end{pmatrix}.$$

The set of all $m \times n$- matrices with entries in $\mathbb{R}$ is called $\mathcal{M}(m \times n, \mathbb{R})$. Matrices allow the operations addition, multiplication, and scalar multiplication with scalars $\lambda \in \mathbb{R}$.

Addition Given $\mathsf{M}_1 \in \mathcal{M}(m \times n, \mathbb{R})$, $\mathsf{M}_2 \in \mathcal{M}(m \times n, \mathbb{R})$:

$$\mathsf{M}_1 + \mathsf{M}_2 = \begin{pmatrix} a_{11} \ldots a_{1n} \\ \cdot \qquad \cdot \\ \cdot \qquad \cdot \\ \cdot \qquad \cdot \\ a_{m1} \ldots a_{mn} \end{pmatrix} + \begin{pmatrix} b_{11} \ldots b_{1n} \\ \cdot \qquad \cdot \\ \cdot \qquad \cdot \\ \cdot \qquad \cdot \\ b_{m1} \ldots b_{mn} \end{pmatrix} = \begin{pmatrix} a_{11} + b_{11} \ldots a_{1n} + b_{1n} \\ \cdot \qquad \cdot \\ \cdot \qquad \cdot \\ \cdot \qquad \cdot \\ a_{m1} + b_{m1} \ldots a_{mn} + b_{mn} \end{pmatrix}.$$

Multiplication Given $\mathsf{M}_1 \in \mathcal{M}(m \times n, \mathbb{R})$, $\mathsf{M}_2 \in \mathcal{M}(n \times r, \mathbb{R})$:

$$\mathsf{M}_1 \cdot \mathsf{M}_2 = \begin{pmatrix} \sum_{j=1}^{n} a_{1j} b_{j1} \ldots \sum_{j=1}^{n} a_{1j} b_{jr} \\ \cdot \qquad \cdot \\ \cdot \qquad \cdot \\ \cdot \qquad \cdot \\ \sum_{j=1}^{n} a_{mj} b_{j1} \ldots \sum_{j=1}^{n} a_{mj} b_{jr} \end{pmatrix}.$$

Notice that $\mathsf{M}_1 \cdot \mathsf{M}_2 \in \mathcal{M}(m \times r, \mathbb{R})$, and notice that the number of columns of M_1 and the number of rows of M_2 have to be equal.

Further, instead of this formula it is easier to remember that the entry with index (i, k) is the i-th row times the k-th column.

Scalar Multiplication Given $\mathsf{M}_1 \in \mathcal{M}(m \times n, \mathbb{R})$, $\lambda \in \mathbb{R}$:

$$\lambda \cdot \mathsf{M}_1 = \begin{pmatrix} \lambda a_{11} \ldots \lambda a_{1n} \\ \cdot \qquad \cdot \\ \cdot \qquad \cdot \\ \cdot \qquad \cdot \\ \lambda a_{m1} \ldots \lambda a_{mn} \end{pmatrix}.$$

For matrices the following calculation rules apply: Given $\mathsf{M}_1, \mathsf{M}_2 \in \mathcal{M}(m \times n, \mathbb{R})$, $\mathsf{N}_1, \mathsf{N}_2 \in \mathcal{M}(n \times r, \mathbb{R})$, $\mathsf{P} \in \mathcal{M}(r \times s, \mathbb{R})$, $\lambda \in \mathbb{R}$:

$$\begin{aligned} \mathsf{M}_1 \cdot (\mathsf{N}_1 + \mathsf{N}_2) &= \mathsf{M}_1 \cdot \mathsf{N}_1 + \mathsf{M}_1 \cdot \mathsf{N}_2 \\ (\mathsf{M}_1 + \mathsf{M}_2) \cdot \mathsf{N}_1 &= \mathsf{M}_1 \cdot \mathsf{N}_1 + \mathsf{M}_2 \cdot \mathsf{N}_1 \\ \mathsf{M}_1 \cdot (\lambda \cdot \mathsf{N}_1) &= \lambda \cdot (\mathsf{M}_1 \cdot \mathsf{N}_1) \\ \mathsf{M}_1 \cdot (\mathsf{N}_1 \cdot \mathsf{P}) &= (\mathsf{M}_1 \cdot \mathsf{N}_1) \cdot \mathsf{P}. \end{aligned}$$

Notice that generally it is not $\mathsf{A} \cdot \mathsf{B} = \mathsf{B} \cdot \mathsf{A}$ for matrices A and B! Furthermore with the unit matrix $E_n \in \mathcal{M}(n \times n, \mathbb{R})$

$$E_n = \begin{pmatrix} 1 \dots 0 \\ \cdot \quad\quad \cdot \\ \cdot \quad\quad \cdot \\ \cdot \quad\quad \cdot \\ 0 \dots 1 \end{pmatrix}$$

(only ones on the main diagonal; the rest of the entries are zero) it holds that:

$$\mathsf{M}_1 \cdot E_n = \mathsf{M}_1 = E_n \cdot \mathsf{M}_1$$

for all quadratic matrices M_1 , i.e., $m = n$ (same number of columns and rows).

Definition (Invertible Matrix) We call $\mathsf{A} \in \mathcal{M}(n \times n, \mathbb{R})$ invertible if $\exists \mathsf{B} \in \mathcal{M}(n \times n, \mathbb{R})$ such that $\mathsf{A} \cdot \mathsf{B} = \mathsf{B} \cdot \mathsf{A} = E_n$ applies. We call $\mathsf{B} = \mathsf{A}^{-1}$ the inverse of A. Notice that $GL(n, \mathbb{R}) = \{\mathsf{A} \in \mathcal{M}(n \times n, \mathbb{R}) \mid \mathsf{A} \text{ is invertible}\}$ (| means "such that") is a group in terms of multiplication, i.e., every matrix $\mathsf{A} \in GL(n, \mathbb{R})$ is invertible, $E_n \in GL(n, \mathbb{R})$, the associative law applies and $\mathsf{A} \cdot \mathsf{B} \in GL(n, \mathbb{R})$ if $\mathsf{A}, \mathsf{B} \in GL(n, \mathbb{R})$.

Annotation: For $\mathsf{A}, \mathsf{B} \in GL(n, \mathbb{R})$ it is

$$(\mathsf{A} \cdot \mathsf{B})^{-1} = \mathsf{B}^{-1} \cdot \mathsf{A}^{-1}.$$

Definition (Transposed Matrix) Given $\mathsf{A} \in \mathcal{M}(m \times n, \mathbb{R})$ the transposed matrix A^T is given by

$$\mathsf{A}^T = \begin{pmatrix} a_{11} \dots a_{m1} \\ \cdot \quad\quad \cdot \\ \cdot \quad\quad \cdot \\ \cdot \quad\quad \cdot \\ a_{1n} \dots a_{mn} \end{pmatrix}, \quad \mathsf{A} = \begin{pmatrix} a_{11} \dots a_{1n} \\ \cdot \quad\quad \cdot \\ \cdot \quad\quad \cdot \\ \cdot \quad\quad \cdot \\ a_{m1} \dots a_{mn} \end{pmatrix}.$$

Example

$$\mathsf{A} = \begin{pmatrix} 1 & 2 \\ 3 & 4 \\ 5 & 6 \end{pmatrix}, \quad \mathsf{A}^T = \begin{pmatrix} 1 & 3 & 5 \\ 2 & 4 & 6 \end{pmatrix}.$$

Lemma *Given* $\mathsf{A}_1, \mathsf{A}_2, \mathsf{A}_3 \in \mathscr{M}(m \times n, \mathbb{R}), \mathsf{B} \in \mathscr{M}(n \times r, \mathbb{R}), \lambda \in \mathbb{R}$. *It applies*

$$\begin{aligned}(\mathsf{A}_1 + \mathsf{A}_2)^{\mathrm{T}} &= \mathsf{A}_1^{\mathrm{T}} + \mathsf{A}_2^{\mathrm{T}} \\ (\lambda \cdot \mathsf{A}_1)^{\mathrm{T}} &= \lambda \cdot \mathsf{A}_1^{\mathrm{T}} \\ (\mathsf{A}_1^{\mathrm{T}})^{\mathrm{T}} &= \mathsf{A}_1 \\ (\mathsf{A}_1 \cdot \mathsf{B})^{\mathrm{T}} &= \mathsf{B}^{\mathrm{T}} \cdot \mathsf{A}_1^{\mathrm{T}}.\end{aligned}$$

Finally in this chapter, we introduce the dyadic product. The dyadic product is a vector multiplication which generates a matrix based on the two input vectors. For vectors $\mathbf{x}, \mathbf{y} \in \mathbb{R}^n$*, the dyadic product* $\mathbf{x}\mathbf{y}^{\mathrm{T}}$ *is*

$$\mathbf{x}\mathbf{y}^{\mathrm{T}} := \begin{pmatrix} x_1 \\ \cdot \\ \cdot \\ \cdot \\ x_n \end{pmatrix} \begin{pmatrix} y_1 & \dots & y_n \end{pmatrix} = \begin{pmatrix} x_1 y_1 & x_1 y_2 & \dots & x_1 y_n \\ & \cdot & & \cdot \\ & \cdot & & \cdot \\ & \cdot & & \cdot \\ x_n y_1 & x_n y_2 & \dots & x_n y_n \end{pmatrix}.$$

C.4 Vector Spaces, Bases, Linear Independence, and Generating Systems

When we talked about $\mathbb{R}^n$ in Chaps. 3 or 12, we repeatedly talked informally about vector spaces. Now we have a closer look on vector spaces and some principles associated with them. Before we formally define a vector space, we introduce some other mathematical structures first. This will help us to understand the structure of a vector space. We start with introducing the concept of an (abelian) group.

Definition (Group) When we talk about a group, we talk about a tuple $(G, \star)$ containing a set G and a map

$$\star : G \times G \longrightarrow G; (g, h) \longmapsto g \star h,$$

such that the following axioms are fulfilled:

1. Associative law : $(a \star b) \star c = a \star (b \star c) \quad , \quad \forall a, b, c \in G$
2. Neutral element : $\exists! e \in G : e \star g = g \star e = g \quad , \quad \forall g \in G$
3. Inverse element : $\forall g \in G \exists! h \in G : g \star h = h \star g = e.$

G is called an abelian group $\Leftrightarrow g \star h = h \star g$ for all elements $g, h \in G$, i.e., the commutative law holds. The unique inverse element h is also called g^{-1}.

The next entity to be introduced is that of a field. This concept is more familiar to the reader, even though one does not notice. The well-known examples are $\mathbb{Q}$ or $\mathbb{R}$.

Definition (Field) A field is a tuple $(\mathbb{K}, \star, \bullet)$ containing a set $\mathbb{K}$ and two maps $\star$ and $\bullet$:

$$\star : \mathbb{K} \times \mathbb{K} \longrightarrow \mathbb{K};\ (g, h) \longmapsto g \star h$$
$$\bullet : \mathbb{K} \times \mathbb{K} \longrightarrow \mathbb{K};\ (g, h) \longmapsto g \bullet h.$$

For a short and precise definition we define two "subtuples" $(\mathbb{K}, \star)$ and $(\mathbb{K}, \bullet)$ and define: $\mathbb{K}$ is a field $\Longleftrightarrow$ $(\mathbb{K}, \star)$ *and* $(\mathbb{K}, \bullet)$ are both abelian groups.
When combining these maps the distributive laws have to hold:

$$(a \star b) \bullet c = (a \bullet c) \star (b \bullet c)$$
$$a \bullet (b \star c) = (a \bullet b) \star (a \bullet c).$$

Often in literature the neutral element of $(\mathbb{K}, \star)$ is called "0", and the map "$\star$" is called "+". Further the neutral element of $(\mathbb{K}, \bullet)$ is called "1" and the map $\bullet$ is called "$\cdot$". This may seem more familiar to the reader; especially, when looking at the distributive law these rules look similar as in $\mathbb{R}$. But we introduced this in a general way on purpose to show that there are more fields with sometimes abstract maps $\star$ and $\bullet$.

Now we are ready to introduce a vector space in a universal way.

Definition (Vector Space) We define a vector space over a field $\mathbb{K}$ as a set V together with two maps $\star$ and $\bullet$:

$$\star : V \times V \longrightarrow V;\ (v, w) \longmapsto v \star w$$
$$\bullet : \mathbb{K} \times V \longrightarrow V;\ (\lambda, v) \longmapsto \lambda \bullet v.$$

For the tuple $(V, \star)$, V is a vector space over a field $\mathbb{K}$, if the following aspects are fulfilled:

1. $(V, \star)$ is an abelian group. According to our annotation above we call the neutral element of this group "0."
2. $(\lambda + \mu) \bullet v = (\lambda \bullet v) \star (\mu \bullet w),\ \forall v, w \in V.$
3. $\lambda \bullet (v \star w) = (\lambda \bullet v) \star (\lambda \bullet w),\ \forall v, w \in V.$
4. $(\lambda \cdot \mu) \bullet v = \lambda \bullet (\mu \bullet v),\ \forall v, w \in V.$
5. $1 \bullet v = v,\ \forall v \in V.$

Notice that we used the terms for the field $\mathbb{K}$ (considering the neutral element and the maps) according to our annotation above. Doing this it is important to distinguish between the maps of the vector space and the maps of the field. In literature, the map "$\bullet$" is called "multiplication" and the map "$\star$" is called "addition" when talking about a field, and "$\bullet$" is called "scalar multiplication" when talking about a vector space. Examples for vector spaces are $\mathbb{R}^n$ over the field $\mathbb{R}$, or $\mathbb{C}$ over the field $\mathbb{R}$.

Next we introduce the vector space equivalent of a subset namely a vector subspace or simply subspace.

Definition (Subspace) Given the vector space introduced above. A non-empty subset $W \subseteq V$, $W \neq \emptyset$ is called a subspace if the following conditions hold:

$$
\begin{aligned}
&1 : 0 \in W\\
&2 : v, w \in W \Longrightarrow v \star w \in W\\
&3 : \lambda \in \mathbb{K}, v \in W \Longrightarrow \lambda \bullet v \in W.
\end{aligned}
$$

Another useful concept is the linear span of a subset $X \subseteq V$ of a vector space V over a field $\mathbb{K}$, written as $span(X)$. This is the smallest subspace of V that contains the subset X. We distinguish between a linear span of a finite set and a linear span of an infinite set. A linear combination of elements (or the so-called vectors) of V is defined as

Definition (Linear Combination) Let V be a vector space over a field $\mathbb{K}$ and $X_n = \{v_1, \ldots, v_n\} \subseteq V$ a finite subset of V. Then we call $v = \mu_1 v_1 + \ldots + \mu_n v_n$ a linear combination of $v_1, \ldots, v_n$. Notice that $v \in V$ as V is a vector space. Now the linear span follows as:

Definition (Linear Span) Let V be the vector space mentioned above, and let $X \subseteq V$ be the same finite subset as in the previous definition. Further, let $R \subseteq V$ be an infinite subset of the vector space V.

1. $span(X) = \{v \mid v = \mu_1 v_1 + \ldots + \mu_n v_n, \mu_1, \ldots, \mu_n \in \mathbb{K}\}$ is called the linear span of X.
2. $span(R) = \bigcup_{n \in \mathbb{N}} \{v \mid v = \sum_{i=1}^{n} \mu_i v_i, v_i \in V, \mu_i \in \mathbb{K}, (i = 1, \ldots, n)\}$ is called the linear span of R. This is the union $\bigcup_{n \in \mathbb{N}} span(X_n)$.

Notice that $\emptyset$ is a finite subset of V, too. To cover this case we define $span(\emptyset) = \{0\}$, where 0 is the neutral element of $(V, \star)$.

The next definition introduced is the concept of linear independence that is frequently used explicitly or implicitly in this book and is an important concept when talking about bases of vector spaces (we will introduce this term later on).

Definition (Linear Independent) Considering the vector space from above and the finite subset X we write also $X' = (v_1, \ldots, v_n)$ as an ordered tuple (sometimes called "vector family"). So now for example $v_1 = v_2$ is possible and the order is important. We call this tuple linear independent if the following implication holds

$$\mu_1 v_1 + \ldots + \mu_n v_n = 0 \Longrightarrow \mu_1 = \ldots = \mu_n = 0,$$

where $\mu_1, \ldots, \mu_n \in \mathbb{K}$. The definition in the case when X is infinite is analogous to the definition of the linear span over the finite subsets of the infinite set. Formulating this is a good task for the reader to get used to some maybe a little abstract definitions.

Before we get to some useful propositions in this topic, we have to introduce two more concepts that are absolutely necessary to know to work with vector spaces.

Definition (Generating System) Let V be the vector space over the field $\mathbb{K}$ and X' the vector family from the last definition.

1. X' is called a generating system of the vector space V if $V = span(X')$ holds. (Notice that we can define the span for this tuples, too. The definition does not change.)
2. We say that V is finally generated $\Leftrightarrow$ X' is finite (that is the case here).
3. We call X' a basis of the vector space $V \Leftrightarrow X'$ is linear independent and a generating system of V.
4. The dimension of V is defined by

$$dim_{\mathbb{K}}(V) = \begin{cases} n & \text{, V is finally generated and } \mid X' \mid = n \\ \infty & \text{, V is not finally generated.} \end{cases}$$

Notice that $|X'|$ is the cardinality of the family (or in more general terms this is defined for normal sets, too), i.e., the number of elements contained in the family (or the set). Sometimes $|X'|$ is called the length of the basis or the length of the generating system.

Let us finish this section by introducing some useful propositions associated to this topic.

Lemma C.1 *Let V be a vector space over a field $\mathbb{K}$ and I and index set. Then the following propositions hold (notice that "$0_{\mathbb{K}}$" defines the neutral element of field $\mathbb{K}$ considering the "addition" and "1" is the neutral element of $\mathbb{K}$ considering the "multiplication"):*

1. $v \bullet 0_{\mathbb{K}} = 0, \forall v \in V.$
2. $0 \bullet \mu = 0, \forall \mu \in \mathbb{K}.$
3. $v \bullet \mu = 0 \Rightarrow v = 0 \vee \mu = 0_{\mathbb{K}}, \forall v \in V, \forall \mu \in \mathbb{K}.$
4. $v \bullet (-1) = -v, \forall v \in V$ *(notice that "$-$" is used to mark the Inverse considering "$\star$" or "$+$", respectively).*
5. *Let $W_1, \ldots, W_n$ be subspaces of $V \Rightarrow \bigcap_{i \in I} W_i$ is a subspace, too.*
6. *The union of two or more subspaces is in general no subspace of V.*

Simple examples are subspaces of the vector space $\mathbb{R}^2$ established by lines through the origin, for example.

Lemma C.2 *Let the requirements of this lemma be the same as in Lemma C.1 and let $X = (v_1, \ldots, v_n)$ be a family of vectors. Then it holds:*
X is linear independent $\Leftrightarrow$ every $v \in span(X)$ can be expressed in unique way. An

analogous lemma holds for infinite families of vectors (The proposition is reduced to the finite case of every finite subset).

Lemma C.3 *Let the requirements of this lemma be the same as in Lemma C.1 and let $U \subseteq V$ be subspace of V. Then it holds*

1. *For every finally generated vector space V there is a basis B with finite dimension.*
2. *V has finite dimension $\Rightarrow$ U has finite dimension.*
3. *$dim_{\mathbb{K}}(U) \leq dim_{\mathbb{K}}(V)$ and $dim_{\mathbb{K}}(U) = dim_{\mathbb{K}}(V) \Leftrightarrow U = V$.*

C.5 Rank of Matrices, Determinant, and Criteria for Invertible Matrices

Maybe the reader thinks that this section would better fit after the section about matrices. It is placed here because it is highly linked to the previous section. First, we introduce a special kind of vector space.

Definition (Rank) Given $\mathsf{A} \in \mathscr{M}(m \times n, \mathbb{R})$ and

$$\mathsf{A} = \begin{pmatrix} a_{11} \dots a_{1n} \\ \cdot \quad\quad \cdot \\ \cdot \quad\quad \cdot \\ \cdot \quad\quad \cdot \\ a_{m1} \dots a_{mn} \end{pmatrix}.$$

Let us further define: $\mathbf{a}_1 = \left(a_{11} \dots a_{m1}\right)^{\mathrm{T}}$, $\mathbf{a}_2 = \left(a_{12} \dots a_{m2}\right)^{\mathrm{T}}$, $\dots$, $\mathbf{a}_n = \left(a_{1n} \dots a_{mn}\right)^{\mathrm{T}}$ the columns of the matrix A. Further we define $\tilde{A} = (\mathbf{a}_1, \dots, \mathbf{a}_n)$ the vector family containing the columns of A (notice that $\mathbf{a}_1, \dots, \mathbf{a}_n \in \mathbb{R}^m$).

1. We call the subspace of $\mathbb{R}^m$ $span(\tilde{A}) \subseteq \mathbb{R}^m$ the column space of the matrix A.
2. The dimension of the column space $span(\tilde{A})$ is called the column rank of A.
3. Analogously, it is possible to define the row space of the matrix A (subspace of $\mathbb{R}^n$).

An important function of a matrix $\mathsf{A} \in \mathscr{M}(n \times n, \mathbb{R})$ is its determinant written as $\det \mathsf{A}$, $\det(\mathsf{A})$, or $|\mathsf{A}|$ (notice that this is just a concept for quadratic matrices). The determinant could be called a scalar classification number for a matrix A. In a more abstract way, we can look at the determinant as the map $\det : \mathscr{M}(n \times n, \mathbb{R}) \longrightarrow \mathbb{R}; \mathsf{A} \longmapsto \det(\mathsf{A})$ that has some interesting properties and is very useful, for example, for deciding whether a matrix is invertible or not. The determinant plays an important role in the theory of eigenvalues and eigenvectors.

It even has some geometrical meaning but we do not want to show this here; notice that this concept can be used for more general fields $\mathbb{K}$ (corresponding to $\mathcal{M}(n \times n, \mathbb{K})$, too).

There are several ways to compute the determinant of a matrix A, but we only present two formulas for the case of 2×2 and 3×3 as this is enough to get an understanding of this concept; in practice calculations in higher dimensions without a computer are rare.

So let us define $\mathsf{A} \in \mathcal{M}(2 \times 2, \mathbb{R})$ and $\mathsf{B} \in \mathcal{M}(3 \times 3, \mathbb{R})$. The determinants of such matrices follow as

1. For $\mathsf{A} = \begin{pmatrix} a & b \\ c & d \end{pmatrix}$ the determinant of A is given by $\det \mathsf{A} = ad - bc$.
2. For $\mathsf{B} = \begin{pmatrix} a & b & c \\ d & e & f \\ g & h & i \end{pmatrix}$, the formula of Leibnitz gives us $\det \mathsf{B} = a(ei - fh) + c(dh - eg) - b(di - fg)$.

To close this section, we introduce some useful propositions for this topic.

Lemma C.4 *For quadratics matrices $\mathsf{A}, \mathsf{B} \in \mathcal{M}(n \times n, \mathbb{R})$ with column rank k holds: row rank = column rank (notice that this statement is true for $\mathsf{A} \in \mathcal{M}(m \times n, \mathbb{R})$) too, and we shortly say "rank" of a matrix). Furthermore, we have*

$$\det E_n = \det(\mathsf{A}) \cdot \det(\mathsf{A}^{-1}) \quad , \quad \det \mathsf{A}^{\mathrm{T}} = \det \mathsf{A}$$

$$\mathsf{A}^{-1}\text{exists (}\mathsf{A}\text{is "regular")} \Leftrightarrow \det \mathsf{A} \neq 0 \Leftrightarrow k = n$$

$$\det(\mathsf{A} \cdot \mathsf{B}) = \det \mathsf{A} \cdot \det \mathsf{B} \quad , \quad \det(\mu \cdot \mathsf{A}) = \mu^n \cdot \det \mathsf{A}, \forall \mu \in \mathbb{R}.$$

C.6 Systems of Linear Equations

The next thing we want to introduce is the concept of a system of linear equations that is frequently used in this book. First of all this is a set of linear equations, i.e., equations that consist of one or more variables with linear coefficients (no powers or exponential functions and so on). The goal is to find values for these variables such that all equations are fulfilled at the same time. We write such a system in a general way as follows:

$$a_{11}x_1 + a_{12}x_2 + \ldots + a_{1n}x_n = b_1$$
$$a_{21}x_1 + a_{22}x_2 + \ldots + a_{2n}x_n = b_2$$
$$\ldots$$
$$a_{m1}x_1 + a_{m2}x_2 + \ldots + a_{mn}x_n = b_m.$$

Alternatively we can express this in shorter terms in matrix form $\mathsf{A}\mathbf{x} = \mathbf{b}$ with

$$\mathsf{A} = \begin{pmatrix} a_{11} & \dots & a_{1n} \\ \cdot & & \cdot \\ \cdot & & \cdot \\ \cdot & & \cdot \\ a_{m1} & \dots & a_{mn} \end{pmatrix} \in \mathscr{M}(m \times n, \mathbb{R}),$$

and $\mathbf{x} = \begin{pmatrix} x_1 & \dots & x_n \end{pmatrix}^{\mathrm{T}} \in \mathbb{R}^n$ and $\mathbf{b} = \begin{pmatrix} b_1 & \dots & b_m \end{pmatrix}^{\mathrm{T}} \in \mathbb{R}^m$.
Another frequently used concept is the so-called extended coefficient matrix

$$(\mathsf{A}|\mathbf{b}) = \begin{pmatrix} a_{11} & \dots & a_{1n} & b_1 \\ \cdot & & \cdot & \cdot \\ \cdot & & \cdot & \cdot \\ \cdot & & \cdot & \cdot \\ a_{m1} & \dots & a_{mn} & b_m \end{pmatrix}.$$

Generally when trying to find a solution for such a system three cases can occur:

1. The system has no solution.
2. The system has a unique determined solution.
3. The system has an infinite number of solutions.

Mathematicians ideally want to know which of these three cases applies just by A and $\mathbf{b}$. Therefore, we give some criteria for identifying those cases:

1. $r =$ rank of $\mathsf{A} \neq c =$ rank of $(\mathsf{A}|\mathbf{b}) \Rightarrow$ case (1).
2. $r = c = n =$ number of variables in the system $\Rightarrow$ case (2).
3. $r = c < n \Rightarrow$ case (3).

Annotations:

1. $\mathsf{A} \in \mathscr{M}(n \times n, \mathbb{R})$ and $\det \mathsf{A} \neq 0 \Rightarrow$ case (2).
2. Further, it should be noted that there are several ways to determine the rank of matrices, and there are other ways to determine which case occurs.

Finally, we introduce the concept of elementary row transformations which occurred in this book often. The elementary row transformations have special properties, and we want to introduce them now (notice that when we talk about rows we always mean rows of $(A \mid b)$):

1. Adding one row to another.
2. Multiplication of $\mu \in \mathbb{R}$ with a row of $(\mathsf{A}|\mathbf{b})$ $(\mu \neq 0)$.
3. Adding μ-times one row to another row $(\mu \neq 0)$.

These row transformations help us to simplify the system $\mathsf{A}\mathbf{x} = \mathbf{b}$. But this is only possible due to Lemma C.5:

Lemma C.5 *Elementary row transformations do not change the rank of the matrix* $(\mathsf{A}|\mathbf{b})$, *and this implies that they do not change the number of solutions and their values of the system* $\mathsf{A}\mathbf{x} = \mathbf{b}$.

C.7 Some Facts on Calculus

In this last section of this appendix we cover the need of calculus in this book because in particular the theoretical propositions of this book are easier to understand with some knowledge of this topic.

So first of all for the rest of the section let $f, g : U \to \mathbb{R}$, $U \subset \mathbb{R}$ be two functions.

First of all we start with the elementary definition of an environment, especially for the real vector space $\mathbb{R}$:

Definition (Environment) An environment of a point $x \in \mathbb{R}$ is a subset $U \subseteq \mathbb{R}$ such that there exists an $\varepsilon \in \mathbb{R}$, $\varepsilon > 0$ with $U = (x - \varepsilon, x + \varepsilon)$.

We call $x_0 \in U$

1. a *local minimum* if $\exists\, \varepsilon > 0 : \forall x \in (x - \varepsilon, x + \varepsilon) \cap U : f(x_0) \leq f(x)$.
2. a *local maximum* if $\exists\, \varepsilon > 0 : \forall x \in (x - \varepsilon, x + \varepsilon) \cap U : f(x_0) \geq f(x)$.
3. a *global minimum* if $f(x_0) \leq f(x), \quad \forall x \in U$.
4. a *global maximum* if $f(x_0) \geq f(x), \quad \forall x \in U$.

Further we call a function f bounded upwards if $\exists C \in \mathbb{R}$ such that: $C \geq f(x)$, $\forall x \in U$. The definition of downwards bounded functions is similar. Further, we call a function bounded if it is bounded upwards and downwards. We start by introducing two special functions:

Definition (Maximum, Minimum)

1. The maximum of f and g is given by
$$u(x) := \max(f, g)(x) = \begin{cases} f(x), & \text{if } f(x) \geq g(x) \\ g(x), & \text{if } g(x) > f(x). \end{cases}$$
2. The minimum of f and g is given by
$$v(x) := \min(f, g)(x) = \begin{cases} f(x), & \text{if } f(x) \leq g(x) \\ g(x), & \text{if } g(x) < f(x). \end{cases}$$

Notice that $u(x)$ and $v(x)$ are functions.

Definition (Function Max, Min) $f^+ := max(f, 0)$ and $f^- := min(-f, 0)$, where "0" is the function that has constant value 0.

Now we summarize a few simple definitions and a very elementary proposition about convergence of sequences. When we talk about a sequence $(x_n)_{n\in\mathbb{N}}$ we mean an infinite list of elements that is ordered and the elements can occur more than once. A simple example is the sequence $(\frac{1}{n})_{n\in\mathbb{N}} = \frac{1}{1}, \frac{1}{2}, \frac{1}{3}, \ldots$

In this book we often talked about the convergence of algorithms and the number of iterations. One way to approach this topic is to have a look at the convergence of sequences.

Definitions (Convergence of a Sequence, Limit)

1. We say a sequence $(x_n)_{n\in\mathbb{N}}$ converges to a $x \in \mathbb{R}$, if $\forall\varepsilon > 0$ $(\varepsilon \in \mathbb{R})$ $\exists n_0 \in \mathbb{N}$ such that $\forall n \geq n_0 : | x_n - x | < \varepsilon$. To express this we write: $\lim_{n\to\infty} x_n = x$, and we say x_n is convergent to its limit x.
2. A real sequence $(x_n)_{n\in\mathbb{N}}$ is called bounded upwards if $\exists C \in \mathbb{R}$ such that $x_n \leq C$, $\forall n \in \mathbb{N}$ (bounded downwards is defined analogously).
3. A real sequence is called bounded if $\exists C \in \mathbb{R}$ such that $| x_n | \leq C$, $\forall n \in \mathbb{N}$.

The calculation of limits of real sequences is based on the following limit rules:

Lemma C.6 *Given two convergent sequences $(x_n)_{n\in\mathbb{N}}$ and $(y_n)_{n\in\mathbb{N}}$ with limits x and y, respectively. Then it holds*

$$\lim_{n\to\infty}(x_n \pm y_n) = x \pm y \quad ; \quad \lim_{n\to\infty}(cx_n) = cx, \quad c \in \mathbb{R})$$

$$\lim_{n\to\infty}(x_n y_n) = xy \quad ; \quad \lim_{n\to\infty}\frac{x_n}{y_n} = \frac{x}{y} \quad , \quad \text{if } (y_n \neq 0 \wedge y \neq 0)\,.$$

Let us transfer this concept to a topic central in this book: algorithms. We look at the starting problem named P_1, and let an algorithm operate on it leading to state P_2 of the problem and so on. The desired state of the problem is named P. This could be computing the root of a nonlinear equation or the optimum of an optimization problem. Based on the output x_n of the algorithm, and an error tolerance ε, we are able to quantify when the algorithm has approached the desired state P quantified by x close enough. Then we say the algorithm converges when it is, after finitely many steps, within the error tolerance ε and one can apply the mathematical notation introduced above.

In calculus, it is very important to know the domain of a function when analyzing it. For instance, in mathematical optimization, depending on the type of domain on which a function is defined, it is guaranteed that a continuous function takes it optimum.

Definitions (Open, Closed, Compact) Let $X \subseteq \mathbb{R}$ be a non-empty subset (we will talk about the empty set separately). Further let I be an index set.

1. We call X open if $\forall x \in X, \exists\varepsilon \in \mathbb{R}, \varepsilon > 0$ such that $(x - \varepsilon, x + \varepsilon) \subset X$.
2. We call X closed if X^c is an open set.
3. We call a closed set X compact, if for every open cover $\bigcup_{V_i, i\in I} V_i$ such that $X \subseteq \bigcup_{V_i, i\in I} V_i$ there is a finite subset $J \subset I$ such that $X \subseteq \bigcup_{V_j, j\in J} V_j$, where V_i is open $\forall i \in I$.

Annotation: The empty set is considered as open and closed (both!). Because there is nothing to check we can consider the empty set as open. But the definition for a closed set is fulfilled, too. This implies that $\mathbb{R}$ itself is open and closed, too. We call sets like this "closed open sets" and there are more of them. Additionally, there are also sets that are not open neither closed like $(a, b]$ for $a, b \in \mathbb{R}$ and $a < b$.

This last point may seem a little bit abstract. Fortunately, the theorem of Heine–Borel tells us that a set $X \subset \mathbb{R}$ is compact $\Leftrightarrow$ X is closed and bounded. Bounded means that $\exists C \in \mathbb{R}$ such that $|x| \leq C, \forall x \in X$.

Now we have a look at two important properties when talking about real functions: Let $U = (a, b), a < b$.

1. We call f continuous in a point $x_0 \in U$ if $\forall \varepsilon > 0, \exists \delta > 0$ such that $| f(x_0) - f(x) |< \varepsilon, \forall x$ with $| x - x_0 |< \delta$.
2. We call f continuous, if f is continuous in all points $x_0 \in U$.
3. We call f differentiable in a point $x_0 \in U$, if the limit $\lim\limits_{h \to 0} \frac{f(x_0+h)-f(x_0)}{h}$ exists. The limit is called derivative or in other terms $f'(x_0)$.
4. We call f differentiable, if f is differentiable in all points $x_0 \in U$.
5. We say that f is continuously differentiable, if f is differentiable and its derivative is continuous.

Now we have the concepts to introduce the extreme value theorem from Karl Weierstrass which is very essential in real calculus. This theorem tells us that every function $f : [a, b] \longrightarrow \mathbb{R}$ $(a < b)$ that is continuous in every point of $[a, b]$ takes its (global) minimum and maximum on the compact interval $[a, b]$.

The final thing to mention is a property of multi-variate functions $f : U \longrightarrow \mathbb{R}$, where $U \subset \mathbb{R}^n$, i.e., we have now n variables $x_1, \ldots, x_n$.

Definition (Partial Differentiable) We define f_{x_i} as the function we get when taking all other variables $x_1, \ldots, x_{i-1}, x_i, x_{i+1}, \ldots, x_n$ as fixed constants. Now we call f partial differentiable if every function f_{x_i} is differentiable in the x_i-th variable, and the i-th partial derivative is written as $\partial_{x_i} f$.

With this knowledge we finally define a special vector namely the gradient $\nabla f(x) = (\partial_{x_1} f(x), \ldots, \partial_{x_n} f(x))$. This concept turns out as very useful in many topics of mathematics and physics; it has been used in Chap. 12. Certainly, there is much more to say about the gradient of a function than what fits in this appendix.

References

1. Abadie, J.: The GRG method for nonlinear programming. In: Greenberg, H.J. (ed.) Design and Implementation of Optimization Software, pp. 335–363. Sijthoff and Noordhoff, Holland (1978)
2. Abadie, J., Carpenter, J.: Generalization of the Wolfe reduced gradient method to the case of nonlinear constraints. In: Fletcher, R. (ed.) Optimization, pp. 37–47. Academic Press, New York (1969)
3. Achterberg, T., Koch, T., Martin, A.: Branching rules revisited. Oper. Res. Lett. **33**(1), 42–54 (2005)
4. Achterberg, T., Berthold, T., Koch, T., Wolter, K.: Constraint integer programming: a new approach to integrate CP and MIP. In: Perron, L., Trick, M.A. (eds.) Integration of Artificial Intelligence (AI) and Operations Research (OR) Techniques in Constraint Programming for Combinatorial Problems, pp. 6–20. Springer, Berlin, Heidelberg (2008)
5. Achterberg, T., Bixby, R.E., Gu, Z., Rothberg, E., Weninger, D.: Presolve reductions in mixed integer programming. INFORMS J. Comput. **32**(2), 473–506 (2020)
6. Adjiman, C.S.J.: Global Optimization Techniques for Process Systems Engineering. PhD Dissertation, Dept. of Chemical Engineering, Princeton University, Princeton, NJ (1999)
7. Adjiman, C., Dallwig, S., Floudas, C., Neumaier, A.: A global optimization method, αBB, for general twice-differentiable constrained NLPs - I. Theoretical advances. Comput. Chem. Eng. **22**, 1137–1158 (1998)
8. Adjiman, C., Dallwig, S., Floudas, C., Neumaier, A.: A global optimization method, αBB, for general twice-differentiable constrained NLPs - II. Implementation and computational results. Comput. Chem. Eng. **22**, 1159–1179 (1998)
9. Agarwal, P.K., Flato, E., Halperin, D.: Polygon decomposition for efficient construction of Minkowski sums. Comp. Geom. Theory Appl. **21**, 29–61 (2002)
10. Aiyoshi, E., Shimizu, K.: Hierarchical decentralized systems and its new solution by a barrier method. IEEE Trans. Syst. Man Cybernet. **40**, 444–449 (1988)
11. Alba, E.: Parallel Metaheuristics: A New Class of Algorithms. Wiley-Interscience, New York, NY (2005)
12. Alba, E., Luque, G.: Measuring the performance of parallel metaheuristics. In: Alba, E. (ed.) Parallel Metaheuristics: A New Class of Algorithms. Wiley Series on Parallel and Distributed Computing, chap. 2, pp. 43–62. Wiley, New York (2005)
13. Alba, E., Talbi, E.G., Luque, G., Melab, N.: Metaheuristics and parallelism. In: E. Alba (ed.) Parallel Metaheuristics: A New Class of Algorithms. Wiley Series on Parallel and Distributed Computing, chap. 4, pp. 79–104. Wiley, New York (2005)

J. Kallrath, *Business Optimization Using Mathematical Programming*,
International Series in Operations Research & Management Science 307,
https://doi.org/10.1007/978-3-030-73237-0

14. Alba, E., Luque, G., Nesmachnow, S.: Parallel metaheuristics: recent advances and new trends. Int. Trans. Oper. Res. **20**(1), 1–48 (2013)
15. Al-Khayyal, F.A.: Jointly constrained bilinear programs and related problems: an overview. Comput. Math. Appl. **19**, 53–62 (1990)
16. Al-Khayyal, F.A., Falk, J.E.: Jointly constrained biconvex programming. Math. Ops. Res. **8**, 273–286 (1983)
17. Albano, A., Sapuppo, G.: Optimal allocation of two-dimensional irregular shapes using heuristic search methods. IEEE Transl. Syst. Man Cybernet. **10**, 242–248 (1980)
18. Anbil, R., Gelman, E., Patty, B., Tanga, R.: Recent advances in crew pairing optimization at American airlines. Interfaces **21**(1), 62–74 (1991)
19. Andersen, E.D.: On exploiting problem structure in a basis identifications procedure for linear programming. Department Publication 6, Department of Management Sciences, Odense Universitet, Odense (1996)
20. Andersen, E.D., Andersen, K.D.: Presolving in linear programming. Math. Program. **71**, 221–245 (1995)
21. Andersen, E.D., Ye, Y.: Combining Interior-Point and Pivoting Algorithms for Linear Programming. Technical report, Department of Management Sciences, The University of Iowa, Ames, Iowa (1994)
22. Andersen, E.D., Ye, Y.: On a Homogeneous Algorithm for the Monotone Complementary Problem. Technical report, Department of Management Sciences, The University of Iowa, Ames, Iowa (1995)
23. Andersen, E.D., Gonzio, J., Meszaros, C., Xu, X.: Implementation of Interior Point Methods for Large Scale Linear Programming. Department Publication 1, Department of Management Sciences, Odense Universitet, Odense (1996)
24. Andrade, R., Lisser, A., Maculan, N., Plateau, G.: BB strategies for stochastic integer programming. In: Spielberg, K., Guignard, M. (eds.) Special Volume of Annals of OR: State-of-the-Art IP and MIP (Algorithms, Heuristics and Applications). Kluwer Academic Publishers, Dordrecht (2005)
25. Andrei, N.: Nonlinear Optimization Applications Using the GAMS Technology. Springer, New York (2013)
26. Anthonisse, J.: An input system for linear programming problems. Stat. Nederland. **24**, 71–81 (1970)
27. Appa, G.M.: The transportation problem and its variants. Oper. Res. Quart. **24**, 79–99 (1973)
28. Applegate, D.L., Bixby, R.E., Chvatal, V., Cook, W.J.: The Traveling Salesman Problem: A Computational Study. Princeton Series in Applied Mathematics. Princeton University Press, Princeton, NJ (2007)
29. Arabeyre, J., Fearnley, J., Steiger, F., Teather, W.: The airline crew scheduling problem: a survey. Transport. Sci. **3**, 140–163 (1969)
30. Arbel, A.: Exploring Interior-Point Linear Programming Algorithms and Software. MIT Press, London (1994)
31. Arellano-Garcia, H., Martini, W., Wendt, M., Li, P., Wozny, G.: Chance-constrained batch distillation process optimization under uncertainty. In: Grossmann, I.E., McDonald, C.M. (eds.) Proceedings of the 4th International Conference on Foundations of Computer-Aided Process Operations (FOCAPO), pp. 609–612. OMNI Press, Wisconsin (2003)
32. Arellano-Garcia, H., Martini, W., Wendt, M., Wozny, G.: Robust optimization process design optimization under uncertainty. In: Floudas, C.A., Agrawal, R. (eds.) Proceedings of the 6th International of Conference on Foundations of Computer-Aided Process Design (FOCAPD), pp. 505–508. CACHE Corp., Austin, TX (2004)
33. Arntzen, B.C., Brown, G.C., Harrison, T.P., Trafton, L.L.: Global supply management at Digital Equipment Corporation. Interfaces **25**, 69–93 (1995)
34. Art, R.C.: An Approach to the Two-Dimensional Irregular Cutting Stock Problem. Tech. Report 36.008, IBM Cambridge Scientific Centre (1966)
35. Ashford, R.W., Daniel, R.C.: LP-MODEL XPRESS-LP's model builder. Inst. Math. Appl. J. Math. Manage. **1**, 163–176 (1987)

36. Ashford, R.W., Daniel, R.C.: Practical aspects of mathematical programming. In: Munford, A.G., Bailey, T.C. (eds.) Operational Research Tutorial Papers, pp. 105–122. Operational Research Society, Birmingham (1991)
37. Ashford, R.W., Daniel, R.C.: Some lessons in solving practical integer programming problems. J. Oper. Res. Soc. **43**, 425–433 (1992)
38. Ashford, R.W., Connard, P., Daniel, R.C.: Experiments in solving mixed integer programming problems on a small array of transputers. J. Oper. Res. Soc. **43**, 519–531 (1992)
39. Athanassopoulos, A.D., Thanassoulis, E.: Separating market efficiency from profitability and its implications for planning. J. Oper. Res. Soc. **46**, 20–34 (1995)
40. Audet, C., Brimberg, J., Hansen, P., Le Digabel, S., Mladenović, N.: Pooling problem: alternate formulations and solution methods. Manage. Sci. **50**, 761–776 (2004)
41. Baker, B., Baia, A.P.: Branch-and-bound algorithms for a regional water authority distribution problem. J. Oper. Res. Soc. **46**, 698–707 (1995)
42. Baker, E., Fisher, M.: Computational results for very large air crew scheduling problems. OMEGA **9**, 613–618 (1981)
43. Balas, E.: An additive algorithm for solving linear programs with zero-one variables. Oper. Res. **13**, 517–546 (1965)
44. Balas, E.: Disjunctive Programming. Springer Nature, Cham (2018)
45. Baldacci, R., Toth, P., Vigo, D.: Exact algorithms for routing problems under vehicle capacity constraints. Ann. OR **175**(1), 213–245 (2010)
46. Baravykaitė, M., Žilinskas, J.: Implementation of parallel optimization algorithms using generalized branch and bound template. In: Bogle, I.D.L., Žilinskas, J. (eds.) Computer Aided Methods in Optimal Design and Operations, chap. 3, pp. 21–28. World Scientific Publishing Co. Pte. Ltd., Singapore (2006)
47. Bard, J.F.: Convex two-level programming. Math. Program. **40**, 15–27 (1988)
48. Bard, J.F.: Practical Bilevel Optimization: Algorithms and Applications. Kluwer Academic Publishers, Dordrecht (1998)
49. Barnhart, C., Johnson, E.L., Nemhauser, G.L., Savelsberg, M.W.P., Vance, P.H.: Branch-and-price: column generation for solving huge integer programs. Oper. Res. **46**(3), 316–329 (1998)
50. Bartusch, M., Möhring, R.H., Radermacher, F.J.: Scheduling project networks with resource constraints and time windows. Ann. Oper. Res. **16**, 201–240 (1988)
51. Baston, V.J.D., Rahmouni, M.K., Williams, H.P.: The practical conversion of linear programmes to network flow models. Eur. J. Oper. Res. **50**, 325–334 (1991)
52. Bazaraa, M.S., Sherali, H.D.: On the choice of step sizes in subgradient optimization. Eur. J. Oper. Res. **7**, 380–388 (1981)
53. Bazaraa, M.S., Sherali, H.D., Shetty, C.M.: Nonlinear Programming Theory and Algorithms, 2nd edn. Wily-Interscience Series in Discrete Mathematics and Optimization. Wiley, New York (1993)
54. Bazaraa, M.S., Jarvis, J.J., Sherali, D.: Linear Programming and Network Flows, 4th edn. Wiley, New York (2010)
55. Beale, E.M.L.: Integer programming. In: Jacobs, D.A.H. (ed.) The State of the Art of Numerical Analysis, pp. 409–448. Academic Press, London (1977)
56. Beale, E.M.L., Daniel, R.C.: Chains of Linked Ordered Sets. Tech. Rep., Scicon, Wavendon Tower, Wavendon, Milton Keynes (1980)
57. Beale, E.M.L., Forrest, J.J.H.: Global optimization using special ordered sets. Math. Program. **10**, 52–69 (1976)
58. Beale, E.M.L., Tomlin, J.A.: Special facilities in a general mathematical programming system for non-convex problems using ordered sets of variables. In: J. Lawrence (ed.) OR '69: Proceedings of the 5th International Conference on Operational Research, pp. 447–454. Tavistock, London (1970)
59. Bellman, R.: Dynamic Programming. Princeton University Press, Princeton, NJ (1957)
60. Belotti, P., Kirches, C., Leyffer, S., Linderoth, J., Luedtke, J., Mahajan, A.: Mixed-integer nonlinear optimization. Acta Numer. **22**, 1–131 (2013)

61. Belotti, P., Bonami, P., Fischetti, M., Lodi, A., Monaci, M., Nogales-Gómez, A., Salvagnin, D.: On handling indicator constraints in mixed integer programming. Comp. Opt. Appl. **65**(3), 545–566 (2016)
62. Belton, V., Vickers, S.P.: Demystifying DEA - a visual interactive approach based on multiple criteria analysis. J. Oper. Res. Soc. **44**, 883–896 (1993)
63. Ben-Tal, A., Nemirovski, A.: Robust solutions of linear programming problems contaminated with uncertain data. Math. Program. **88**, 411–424 (2000)
64. Benders, J.F.: Partitioning procedures for solving mixed-variables programming problems. Numer. Math. **4**, 238–252 (1962)
65. Bennell, J.A., Dowsland, K.A.: Hybridising Tabu search with optimisation techniques for irregular stock cutting. Manage. Sci. **47**, 1160–1172 (2001)
66. Bennell, J.A., Oliveira, J.F.: The geometry of nesting problems: a tutorial. Eur. J. Oper. Res. **184**, 397–415 (2008)
67. Bennell, J.A., Song, X.: A comprehensive and robust procedure for obtaining the nofit polygon using Minkowski sums. Comput. Oper. Res. **35**, 267–281 (2008)
68. Bennell, J., Scheithauer, G., Stoyan, Y., Romanova, T.: Tools of mathematical modelling of arbitrary object packing problems. J. Ann. Oper. Res. **179**, 343–368 (2010)
69. Bennell, J., Scheithauer, G., Stoyan, Y., Romanova, T., Pankratov, A.: Optimal clustering of a pair of irregular objects. J. Global Optim. **61**, 497–524 (2015)
70. Benoist, T., Bourreau, E.: Fast global filtering for eternity II. Constraint Program. Lett. **3**, 35–50 (2008)
71. Benson, H.: Global optimization of nonlinear sums of ratios. J. Math. Anal. Appl. **263**, 301–315 (2001)
72. Bergstra, J., Bengio, Y.: Random search for hyper-parameter optimization. J. Mach. Learn. Res. **13**, 281–305 (2012)
73. Bergstra, J., Bardenet, R., Bengio, Y., Kégl, B.: Algorithms for hyper-parameter optimization. In: Proceedings of the 24th International Conference on Neural Information Processing Systems, NIPS'11, pp. 2546–2554. Curran Associates Inc., New York (2011)
74. Berthold, T., Farmer, J., Heinz, S., Perregaard, M.: Parallelization of the FICO Xpress-optimizer. Optim. Methods Softw. **33**(3), 518–529 (2018)
75. Bertsimas, D., Sim, M.: Robust discrete optimization and network flows. Math. Program. Ser. B **98**, 49–71 (2003)
76. Beyer, H.G., Sendhoff, B.: Robust optimization - a comprehensive survey. Comput. Methods Appl. Mech. Eng. **196**(33), 3190 – 3218 (2007)
77. Biegler, L.T., Grossmann, I.: Retrospective on optimization. Comput. Chem. Eng. **28**, 1169–1192 (2004)
78. Birge, J.R.: Stochastic programming computation and applications. INFORMS J. Comput. **9**, 111–133 (1997)
79. Birge, J.R., Louveaux, F.V.: Introduction to Stochastic Programming. Springer Series in Operations Research and Financial Engineering. Springer, New York (2000)
80. Bisschop, J., Meeraus, A.: On the development of a general algebraic modeling system in a strategic planning environment. In: Applications. Mathematical Programming Studies, vol. 20, pp. 1–29. Springer, Berlin, Heidelberg (1982)
81. Bisschop, J.J., Kuip, C.A.: Compound sets in mathematical programming modeling languages. Manage. Sci. **39**, 746–756 (1993)
82. Bixby, R.E., Cunningham, W.H.: Converting linear programs to network problems. Math. Oper. Res. **5**, 321–357 (1980)
83. Blackburn, R., Kallrath, J., Klosterhalfen, S.T.: Operations research in BASF's supply chain operations. Int. Trans. Oper. Res. **22**(3), 385–405 (2014)
84. Blais, J.Y., Lamont, J., Rousseau, J.M.: The HASTUS vehicle and manpower scheduling system at Societeè de Transport de la Communantè Urbaine de Montrèal. Interfaces **20**(1), 26–42 (1990)
85. Blazewicz, J., Ecker, K., Schmidt, G., Weglarz, J.: Scheduling in Computer and Manufacturing Systems. Springer, Berlin, Heidelberg (1993)

86. Bliek, C., Spellucci, P., Vicente, L., Neumaier, A., Granvilliers, L., Monfroy, E., Benhamouand, F., Huens, E., Hentenryck, P.V., Sam-Haroud, D., Faltings, B.: Algorithms for solving nonlinear constrained and optimization problems: The State of the Art. Report of the European Community funded project COCONUT, Mathematisches Institut der Universität Wien, http://www.mat.univie.ac.at/~neum/glopt/coconut/StArt.html (2001)
87. Bock, H.G.: Randwertproblemmethoden zur Parameteridentifizierung in Systemen nichtlinearer Differentialgleichungen. Preprint 142, Universität Heidelberg, SFB 123. Institut für Angewandte Mathematik, Heidelberg (1987)
88. Bock, H.G., Zillober, C.: Interior point methods and extrapolation. In: Kleinschmidt, P. (ed.) Symposium Operations Research (SOR'95): Program and Abstract of the Annual Conference of the DGOR, GMÖOR and ÖGOR, p. 39. University of Passau, Passau (1995)
89. Bodington, C.E., Baker, T.E.: A history of mathematical programming in the petroleum industry. Interfaces **20**(4), 117–127 (1990)
90. Bohoris, G.A., Thomas, J.M.: A heuristic for vehicle routing and depot staffing. J. Oper. Res. Soc. **46**, 1184–1191 (1995)
91. Bomze, I.M., Grossmann, W.: Optimierung – Theorie und Algorithmen. Wissenschaftsverlag, Mannheim (1993)
92. Bonami, P., Kilinç, M., Linderoth, J.: Algorithms and software for convex mixed integer nonlinear programs. In: Lee, J., Leyffer, S. (eds.) Mixed Integer Nonlinear Programming. The IMA Volumes in Mathematics and its Applications, vol. 154, pp. 1–39. Springer, New York (2012)
93. Bossel, H.: Modellbildung und Simulation, 2nd edn. Vieweg, Braunschweig (1994)
94. Boukouvala, F., Misener, R., Floudas, C.A.: Global optimization advances in mixed-integer nonlinear programming, MINLP, and constrained derivative-free optimization, CDFO. Eur. J. Oper. Res. **252**(3), 701–727 (2016)
95. Box, G.E.P., Draper, N.R.: Empirical Model-Building and Response Surfaces. Wiley, New York (1987)
96. Bracken, J., McGill, J.M.: Mathematical programs with optimization problems in the constraints. Oper. Res. **21**, 37–44 (1973)
97. Bradley, G., Brown, G.H., Graves, G.W.: Design and implementation of large-scale primal transshipment algorithms. Manage. Sci. **24**(1), 1–34 (1977)
98. Brearley, A.L., Mitra, G., Williams, H.P.: Analysis of mathematical programming problems prior to applying the simplex algorithm. Math. Program. **8**, 54–83 (1975)
99. Brockmüller, B., Günlük, O., Wolsey, L.A.: Designing private line networks - polyhedral analysis and computation. Trans. Oper. Res. **16**, 7–24 (1997)
100. Brooke, A., Kendrick, D., Meeraus, A.: GAMS: A User's Guide. The Scientific Press, Redwood City, CA (1988)
101. Brooke, A., Kendrick, D., Meeraus, A.: GAMS - A User's Guide (Release 2.25). Boyd & Fraser Publishing Company, Danvers, MA (1992)
102. Burer, S., Letchford, A.N.: Non-convex mixed-integer nonlinear programming: a survey. Surv. Oper. Res. Manage. Sci. **17**(2), 97–106 (2012)
103. Burke, E., Hellier, R., Kendall, G., Whitwell, G.: A new bottom-left-fill heuristic algorithm for the two-dimensional irregular packing problem. Oper. Res. **54**, 587–601 (2006)
104. Burkhard, R.E.: Methoden der Ganzzahligen Optimierung. Springer, Wien, New York (1972)
105. Burkhard, R.E., Derigs, U.: Assignment and Matching Problems: Solution Methods with Fortran-Programs. Springer, Berlin (1980)
106. Bussieck, M.R., Meeraus, A.: General algebraic modeling system (GAMS). In: Kallrath, J. (ed.) Modeling Languages in Mathematical Optimization, pp. 137–157. Kluwer Academic Publishers, Norwell, MA (2003)
107. Cacchiani, V., Bolton, C.C., Toth, P.: Models and algorithms for the traveling salesman problem with time-dependent service times. Eur. J. Oper. Res. **283**(3), 825–843 (2020)
108. Carœ, C.C., Schultz, R.: Dual decomposition in stochastic integer programming. Oper. Res. Lett. **24**, 37–45 (1999)

109. Carpaneto, G., Martello, S., Toth, P.: Algorithms and codes for the assignment problem. Ann. Oper. Res. **13**, 193–223 (1988)
110. Castillo, P., Castro, P., Mahalec, V.: Global optimization of nonlinear Blend-scheduling problems. Engineering **3**, 188–201 (2017)
111. Censor, Y., Zenios, S.: Parallel Optimization: Theory, Algorithms, and Applications. Oxford University Press, Oxford (1997)
112. Chakraborty, A., Malcom, A., Colberg, R.D., Linninger, A.A.: Optimal waste reduction and investment planning under uncertainty. Comput. Chem. Eng. **28**, 1145–1156 (2004)
113. Charnes, A., Cooper, W.W.: The stepping-stone method of explaining linear programming calculations in transportation problems. Manage. Sci. **1**, 49–69 (1954)
114. Charnes, A., Cooper, W.W.: Chance-constrained programming. Manage. Sci. **5**, 73–79 (1959)
115. Charnes, A., Cooper, W.W., Rhodes, E.: Measuring the efficiency of decision making units. Eur. J. Oper. Res. **2**, 429–444 (1978)
116. Chauhan, S.S., Eremeev, A.V., Romanova, A.A., Servakh, V.V.: Approximation of linear cost supply management problem with lower-bounded demands solutions. In: Kolokolov, A.A., Eremeev, A.V. (eds.) Proceedings of Discrete Optimization Methods in Production and Logistics (DOM 2004), OMSK, pp. 16–21 (2004)
117. Cheng, L., Subrahmanian, E., Westerberg, A.W.: Design and planning under uncertainty: issues on problem formulation and solution. Comput. Chem. Eng. **27**, 781–801 (2003)
118. Chernov, N., Stoyan, Y., Romanova, T.: Mathematical model and efficient algorithms for object packing problem. Comput. Geom. **43**, 535–553 (2010)
119. Chernov, N., Stoyan, Y., Romanova, T., Pankratov, A.: Phi-functions for 2D objects formed by line segments and circular arcs. Adv. Oper. Res. (2012). Article ID 346358
120. Cheshire, M.K., McKinnon, K.I.M., Williams, H.P.: The efficient allocation of private contractors to public works. J. Oper. Res. Soc. **35**, 705–709 (1984)
121. Choi, T.M. (ed.): Handbook of Newsvendor Problems: Models, Extensions and Applications. International Series in Operations Research & Management Science, vol. 176. Springer, New York (2012)
122. Christofides, N., Beasley, J.E.: A tree search algorithm for the p-Median problem. Eur. J. Oper. Res. **10**, 196–204 (1982)
123. Christofides, N., Alvarez-Valdes, R., Tamarit, J.M.: Project scheduling with resource constraints: a branch and bound approach. Eur. J. Oper. Res. **29**, 262–273 (1987)
124. Ciriani, T.A., Colombani, Y., Heipcke, S.: Embedding optimisation algorithms with Mosel. 4OR **1**(2), 155–168 (2003)
125. Clark, P.A., Westerberg, A.W.: Bilevel programming for steady state chemical process design: fundamentals and algorithms. Comput. Chem. Eng. **14**, 87–97 (1990)
126. Collatz, L., Wetterling, W.: Optimierungsaufgaben, 2nd edn. Springer, Berlin (1971)
127. Colombani, Y., Heipcke, S.: The Constraint Solver SchedEns. Tutorial and Documentation. Technical report, LIM Laboratoire d'Informatique Marseille, Marseille (1997)
128. Colombani, Y., Heipcke, S.: Mosel: an extensible environment for modeling and programming solutions. In: Jussien, N., Laburthe, F. (eds.) Proceedings of CP-AI-OR'02, pp. 277–290. Le Croisic (2002)
129. Colombani, Y., Heipcke, S.: Multiple Models and Parallel Solving with Mosel. Tech. rep., FICO Xpress Optimization, Birmingham (2004). http://www.fico.com/fico-xpress-optimization/docs/latest/mosel/mosel_parallel/dhtml
130. Colombani, Y., Daniel, B., Heipcke, S.: Mosel: a modular environment for modeling and solving problems. In: Kallrath, J. (ed.) Modeling Languages in Mathematical Optimization, pp. 211–238. Kluwer Academic Publishers, Norwell, MA (2004)
131. Conejo, A.J., Castillo, E., Mínguez, R., García-Bertrand, R.: Decomposition Techniques in Mathematical Programming: Engineering and Science Applications. Springer, Berlin (2006)
132. Cornuejols, G., Fisher, M.L., Nemhauser, G.L.: Location of bank accounts to optimize floats: an analytic study of exact and approximate algorithms. Manage. Sci. **23**, 789–810 (1977)
133. Coutinho, D., de Souza, S.X., Aloise, D.: A scalable shared-memory parallel simplex for large-scale linear programming (2018). CoRR **abs/1804.04737**

134. Crainic, T.G.: Parallel metaheuristics and cooperative search. In: Gendreau, M., Potvin, J.Y. (eds.) Handbook of Metaheuristics, pp. 419–451. Springer, New york (2019)
135. Crowder, H.E., Johnson, E.L., Padberg, M.W.: Solving large scale 0-1 linear programming problems. Oper. Res. **31**, 803–834 (1983)
136. Curtis, A.R., Reid, J.K.: On the automatic scaling of matrices for Gaussian elimination. J. Inst. Math. Appl. **10**, 118–124 (1972)
137. Dakin, R.J.: A tree search algorithm for mixed integer programming problems. Comput. J. **8**, 250–255 (1965)
138. Danna, E., Rothberg, E., Le Pape, C.: Exploring relation induced neighborhoods to improve MIP solutions. Math. Program. **102**, 71–90 (2005)
139. Dantzig, G.B.: Application of the simplex method to a transportation problem. In: T.C. Koopmans (ed.) Activity Analysis of Production and Allocation, pp. 359–373. Wiley, New York (1951)
140. Dantzig, C.B.: Linear programming under uncertainty. Manage. Sci. **1**, 197–206 (1955)
141. Dantzig, G.B.: Discrete variable extremum problems. Oper. Res. **5**, 266–277 (1957)
142. Dantzig, G.B.: Linear Programming and Extensions. Princeton University Press, Princeton, NJ (1963)
143. Dantzig, G.B., Wolfe, P.: Decomposition principle for linear programs. Oper. Res. **8**, 101–111 (1960)
144. Dantzig, B., Wolfe, P.: The decomposition algorithm for linear programming. Oper. Res. **8**, 101–111 (1960)
145. Dantzig, G.B., Fulkerson, D.R., Johnson, S.M.: Solution of a large-scale traveling salesman problem. Oper. Res. **2**, 393–410 (1954)
146. Dantzig, G.B., Dempster, M.A.H., Kallio, M.J. (eds.): Large-scale Linear Programming. IIASA Collaborative Proceedings Series, vol. CP-81-51. International Institute for Applied System Analysis, Laxenburg (1981)
147. Day, R.: MAGIC version 1.0: A Matrix Generator Convertor for Linear and Integer Programming Models. Tech. Rep., Research Report and Department of Business Studies and University of Edinburgh (1982)
148. de Farias Jr., I.R., Johnson, E.L., Nemhauser, G.L.: A generalized assignment problem with special ordered sets: a polyhedral approach. Math. Program. Ser. A **89**, 187–203 (2000)
149. de Farias Jr., I.R., Zhao, M., Zhao, H.: A special ordered set approach for optimizing a discontinuous separable piecewise linear function. Oper. Res. Lett. **36**, 234–238 (2008)
150. de Silva, A., Abramson, D.: A parallel interior point method and its application to facility location problems. Comput. Optim. Appl. **9**, 249–273 (1998)
151. Dempe, S.: Foundations of Bilevel Programming. Kluwer Academic Publisher, Dordrecht (2002)
152. Dempe, S., Kue, F.M.: Solving discrete linear bilevel optimization problems using the optimal value reformulation. J. Global Optim. **68**(2), 255–277 (2017)
153. Dempe, S., Kalashnikov, V., Pérez-Valdés, G.A., Kalashnykova, N.I.: Natural gas bilevel cash-out problem: convergence of a penalty function method. Eur. J. Oper. Res. **215**(3), 532–538 (2011)
154. Dempe, S., Khamisov, O.V., Kochetov, Y.: A special three-level optimization problem. J. Glob. Optim. **76**(3), 519–531 (2020)
155. Dennis, J.B.: A high-speed computer technique for the transportation problem. J. Assoc. Comput. Mach. **5**, 132–153 (1958)
156. Desaulniers, G., Desrosiers, J., Erdmann, A., Solomon, M.M., Soumis, F.: VRP with pickup and delivery. In: Toth, P., Vigo, D. (eds.) The Vehicle Routing Problem, pp. 225–242. Society for Industrial and Applied Mathematics, Philadelphia, PA (2001)
157. Desrochers, M., Desrosiers, J., Solomon, M.M.: A new optimization algorithm for the vehicle routing problem with time windows. Oper. Res. **40**(2), 342–354 (1992)
158. Desrosiers, J., Dumas, Y., Solomon, M.M., Soumis, F.: Time constrained routing and scheduling. In: Ball, M.E., Magnanti, T.L., Monma, C., Nemhauser, G.L. (eds.) Handbook in Operations Research and Management Science, pp. 35–140. Society for Industrial and Applied Mathematics, Philadelphia (1995)

159. Di Domenica, N., Lucas, C., Mitra, G., Valente, P.: Scenario generation for stochastic programming and simulation: a modelling perspective. IMA J. Manage. Math. **20**(1), 1–38 (2007)
160. Dighe, R., Jakiela, M.J.: Solving pattern nesting problems with genetic algorithms employing task decomposition and contact. Evolut. Comput. **3**, 239–266 (1996)
161. Dikin, I.I.: Iterative solution of problems of linear and quadratic programming. Sov. Math. Dokl. **8**, 674–675 (1967)
162. Dinter, J.V., Rebennack, S., Kallrath, J., Denholm, P., Newman, A.: The unit commitment model with concave emissions costs: a hybrid benders' decomposition with nonconvex master problems. Ann. Oper. Res. **210**, 361–386 (2013)
163. Diwekar, U.: Introduction to Applied Optimization, vol. 22. Springer, New York (2008)
164. Drud, A.S.: CONOPT - a large-scale GRG code. ORSA J. Comput. **6**(2), 207–218 (1994)
165. Duran, M.A., Grossmann, I.E.: An outer-approximation algorithm for a class of mixed-integer nonlinear programmes. Math. Program. **36**, 307–339 (1986)
166. Duriagina, Z., Lemishka, I., Litvinchev, I., Marmolejo, J., Pankratov, A., Romanova, T., Yaskov, G.: Optimized filling a given cuboid with spherical powders for additive manufacturing. J. Oper. Res. Soc. China **2020**, 1–16 (2020)
167. Dyson, R.G., Gregory, A.S.: The cutting stock problem in the glass industry. Oper. Res. Quart. **25**, 41–54 (1974)
168. Dyson, R.G., Thanassoulis, E.: Reducing weight flexibility in data envelopment analysis. J. Oper. Res. Soc. **39**, 563–576 (1988)
169. Edmonds, J., Johnson, E.L.: Matching and Euler tours and the Chinese postman problem. Math. Program. **5**, 88–124 (1973)
170. Eglese, R.W.: Routing winter gritting vehicles. Discr. Appl. Math. **48**, 231–244 (1994)
171. Eichner, T., Pfingsten, A., Wagener, A.: Strategisches Abstimmungsverhalten bei Verwendung der Hare-Regel. zfbv **48**, 466–473 (1996)
172. Eley, M.: A bottleneck assignment approach to the multiple container loading problem. OR Spectrum **25**, 113–130 (2003)
173. Emet, S., Westerlund, T.: Solving a dynamic separation problem using MINLP techniques. Appl. Numer. Math. **58**(12), 395–406 (2008)
174. Engell, S., Märkert, A., Sand, G., Schultz, R., Schulz, C.: Online scheduling of multiproduct batch plants under uncertainty. In: Online Optimization of Large Scale Systems, pp. 649–676. Springer, Berlin (2001)
175. Escudero, L., Garín, A., Merino, M., Pérez, G.: The value of the stochastic solution in multistage problems. TOP: An Off. J. Span. Soc. Stat. Oper. Res. **15**(1), 48–64 (2007)
176. Farley, A.A.: Planning the cutting of photographic color paper rolls for Kodak (Australasia) Pty. Ltd. Interfaces **21**(1), 96–106 (1991)
177. Farrell, M.J.: The measurement of productive efficiency. J. R. Stat. Soc. Ser. A **120**, 253–281 (1957)
178. Fasano, G.: Solving Non-Standard Packing Problems by Global Optimization and Heuristics. Springer, Cham (2014)
179. Ferrier, G., Lovell, K.: Measuring cost efficiency in banking: econometric and linear programming evidence. J. Econometr. **46**, 229–245 (1990)
180. Ferris, M.C., Dirkse, S.P., Jagla, J.H., Meeraus, A.: An extended mathematical programming framework. Comput. Chem. Eng. **33**, 87–97 (2009)
181. Fiacco, A.V., McCormick, G.P.: Nonlinear Programming. Sequential Unconstrained Minimization Techniques. Wiley, New York (1968)
182. Fieldhouse, M.: The pooling problem. In: Ciriani, T., Leachman, R.C. (eds.) Optimization in Industry: Mathematical Programming and Modeling Techniques in Practice, pp. 223–230. Wiley, Chichester (1993)
183. Figueira, J., Liefooghe, A., Talbi, E.G., Wierzbicki, A.: A parallel multiple reference point approach for multi-objective optimization. Eur. J. Oper. Res. **205**(2), 390–400 (2010)
184. Fischetti, M., Glover, F.: The feasibility pump. Math. Program. **104**, 91–104 (2005)
185. Fischetti, M., Lodi, A.: Local branching. Math. Program. **98**, 23–47 (2003)

186. Fisher, M.L.: An applications oriented guide to Lagrangian relaxation. Interfaces **15**, 10–21 (1985)
187. Fletcher, R.: Practical Methods of Optimization, 2nd edn. Wiley, Chichester (1987)
188. Flin, H., Liebling, T.M., Prodon, A.: Optimal subtrees and extensions. Ann. Discr. Math. **16**, 121–127 (1982)
189. Floudas, C.A.: Nonlinear and Mixed-Integer Optimization : Fundamentals and Applications. Oxford University Press, Oxford (1995)
190. Floudas, C.A.: Deterministic Global Optimization: Theory, Methods and Applications. Kluwer Academic Publishers, Dordrecht (2000)
191. Floudas, C.A., Gounaris, C.E.: A review of recent advances in global optimization. J. Glob. Optim. **45**, 3–38 (2009)
192. Floudas, C.A., Pardalos, P.M. (eds.): Frontiers in Global Optimization. Kluwer Academic Publishers, Dordrecht (2004)
193. Floudas, C.A., Pardalos, P.M., Adjiman, C.S., Esposito, W.R., Gümüs, Z.H., Harding, S.T., Klepeis, J.L., Meyer, C.A., Schweiger, C.A.: Handbook of Test Problems of Local and Global Optimization. Kluwer Academic Publishers, Dordrecht (1999)
194. Floudas, C.A., Akrotirianakis, I.G., Caratzoulas, S., Meyer, C.A., Kallrath, J.: Global optimization in the 21st century: advances and challenges for problems with nonlinear dynamics. Comput. Chem. Eng. **29**, 1185–1202 (2005)
195. Ford Jr., L.R., Fulkerson, D.R.: Flows in Networks. Princeton University Press, Princeton, NJ (1962)
196. Forrest, J.J.H., Hirst, J.P.H., Tomlin, J.A.: Practical solution of large mixed integer programming problems with UMPIRE. Manage. Sci. **20**, 736–773 (1974)
197. Fourer, R.: Modeling languages versus matrix generators for linear programming. ACM Trans. Math. Softw. **9**, 143–183 (1983)
198. Fourer, R., Gay, D.M., Kernighan, B.W.: AMPL: A Mathematical Programming Language. Tech. Rep. Computing Science Technical Report No. 133, AT&T Bell Laboratories, Murray Hill, NJ (1987). Revised June 1989
199. Fourer, R., Gay, D.M., Kernighan, H.W.: A modeling language for mathematical programming. Manage. Sci. **36**, 519–554 (1990)
200. Fourer, R., Gay, D.M., Kernighan, B.W.: Design principles and new developments in the AMPL modeling language. In: Kallrath, J. (ed.) Modeling Languages in Mathematical Optimization, pp. 137–135. Kluwer Academic Publishers, Norwell, MA (2003)
201. Freund, R.M., Mizuno, S.: Interior point methods: current status and future directions. Optima (Math. Program. Soc. Newslett.) **51**, 1–9 (1996)
202. Frisch, K.R.: The Logarithmic Potential Method for Convex Programming. Technical Report, University Institute of Economics, Oslo, Norway (1955)
203. Fulkerson, D., Wolfe, P.: An algorithm for scaling matrices. SIAM Rev. **4**, 142–147 (1962)
204. Furini, F., Traversi, E., Belotti, P., Frangioni, A., Gleixner, A., Gould, N., Liberti, L., Lodi, A., Misener, R., Mittelmann, H., Sahinidis, N.V., Vigerske, S., Wiegele, A.: QPLIB: a library of quadratic programming instances. Math. Program. Comput. **11**(2), 237–265 (2019)
205. Galati, M.: Decomposition Methods for Integer Linear Programming. PhD thesis, Lehigh University (Industrial Engineering), Lehigh (2009)
206. Gamrath, G., Koch, T., Martin, A., Miltenberger, M., Weninger, D.: Progress in presolving for mixed integer programming. Math. Program. Comput. **7**(4), 367–398 (2015)
207. Gamrath, G., Anderson, D., Bestuzheva, K., Chen, W.K., Eifler, L., Gasse, M., Gemander, P., Gleixner, A., Gottwald, L., Halbig, K., Hendel, G., Hojny, C., Koch, T., Bodic, P.L., Maher, S.J., Matter, F., Miltenberger, M., Mühmer, E., Müller, B., Pfetsch, M., Schlösser, F., Serrano, F., Shinano, Y., Tawfik, C., Vigerske, S., Wegscheider, F., Weninger, D., Witzig, J.: The SCIP Optimization Suite 7.0. Tech. Rep. 20-10, ZIB, Takustr. 7, 14195 Berlin (2020)
208. Gan, M., Gopinathan, N., Jia, X., Williams, R.A.: Predicting packing characteristics of particles of arbitrary shapes. KONA **22**, 2–93 (2004)
209. Garey, M.R., Johnson, D.S.: Computers and intractability - a guide to the theory of NP completeness, 22nd edn. W. H. Freeman and Company, New York (2000)

210. Geist, A., Beguelin, A., Dongarra, J., Jiang, W., Manchek, R., Sunderam, V.: PVM: Parallel Virtual Machines - A User's Guide and Tutorial for Networked Parallel Computing. The MIT Press, Cambridge, MA (1994)
211. Gendreau, M., Potvin, J.Y.: Handbook of Metaheuristics, 2nd edn. Springer Publishing Company, Incorporated, New York (2010)
212. Geoffrion, A.M.: Generalized benders decomposition. J. Optim. Theor. Appl. **10**, 237–260 (1972)
213. Geoffrion, A.M.: Lagrangian relaxation and its uses in integer programming. Math. Program. Study **2**, 82–114 (1974)
214. Gershkoff, I.: Optimizing flight crew schedules. Interfaces **19**(4), 29–43 (1989)
215. Ghildyal, V., Sahinidis, N.V.: Solving Global Optimization Problems with BARON. In: Migdalas, A., Pardalos, P., Värbrand, P. (eds.) From Local to Global Optimization, pp. 205–230. Kluwer Academic Publishers, Dordrecht (2001)
216. Ghosh, P.K.: An algebra of polygons through the notion of negative shapes. CVGIP: Image Understand. **54**, 119–144 (1991)
217. Gill, P.E., Murray, W., Wright, M.H.: Practical Optimization. Academic Press, London (1981)
218. Gill, P.E., Murray, W., Saunders, M.A., Tomlin, J.A., Wright, M.H.: On the projected Newton barrier methods for linear programming and an equivalence to Karmarkar's projective method. Math. Program. **36**, 183–209 (1986)
219. Gill, P.E., Murray, W., Saunders, M.A.: SNOPT: an SQP Algorithm for Large-scale Constrained Optimization. Numerical analysis report 97-2, Department of Mathematics, University of California, San Diego, San Diego, La Jolla, CA (1997)
220. Gilmore, P.C., Gomory, R.E.: A linear programming approach to the cutting stock problem. Oper. Res. **9**, 849–859 (1961)
221. Gilmore, P.C., Gomory, R.E.: A linear programming approach to the cutting stock problem, part II. Oper. Res. **11**, 863–888 (1963)
222. Gleixner, A., Bastubbe, M., Eifler, L., Gally, T., Gamrath, G., Gottwald, R.L., Hendel, G., Hojny, C., Koch, T., Lübbecke, M.E., Maher, S.J., Miltenberger, M., Müller, B., Pfetsch, M.E., Puchert, C., Rehfeldt, D., Schlösser, F., Schubert, C., Serrano, F., Shinano, Y., Viernickel, J.M., Walter, M., Wegscheider, F., Witt, J.T., Witzig, J.: The SCIP Optimization Suite 6.0. Technical report, Optimization Online (2018). http://www.optimization-online.org/DB_HTML/2018/07/6692.html
223. Glen, J.J.: Sustainable yield analysis in a multicohort single-species fishery: a mathematical programming approach. J. Oper. Res. Soc. **46**, 1052–1062 (1995)
224. Glover, F.: A new foundation for a simplified primal integer programming algorithm. Oper. Res. **16**, 727–740 (1968)
225. Glover, F., Klingman, D., Philips, N.: Netform modeling and applications. Interfaces **20**(4), 7–27 (1990)
226. Goel, V., Grossmann, I.E.: A stochastic programming approach to planning of offshore gas field developments under uncertainty in reserves. Comput. Chem. Eng. **28**(8), 1409–1429 (2004)
227. Gollmer, R., Nowak, M.P., Römisch, W., Schultz, R.: Unit commitment in power generation - a basic model and some extensions. Ann. Oper. Res. **96**, 167–189 (2000)
228. Gomes, A.M., Oliveira, J.F.: A 2-exchange heuristic for nesting problems. Eur. J. Oper. Res. **141**, 359–370 (2002)
229. Gomes, A.M., Oliveira, J.F.: Solving irregular strip packing problems by hybridising simulated annealing and linear programming. Eur. J. Oper. Res. **171**, 811–829 (2006)
230. Gomory, R.E.: Outline of an algorithm for integer solutions to linear programming. Bull. Am. Math. Soc. **64**, 275–278 (1958)
231. Gonzaga, C.L.: Path-following methods for linear programming. SIAM Rev. **34**, 167–224 (1992)
232. Gould, N.I.M., Reid, J.K.: New crash procedures for large systems of linear constraints. Math. Program. **45**, 475–503 (1989)

233. Granot, F., Hammer, P.: On the use of boolean functions in 0-1 programming. Methods Oper. Res. **12**, 154–184 (1972)
234. Greenberg, H.J.: How to analyse the results of linear programs - Part 1: preliminaries. Interfaces **23**(4), 56–67 (1993)
235. Greenberg, H.J.: How to analyse the results of linear programs - Part 2: price interpretation. Interfaces **23**(5), 97–114 (1993)
236. Greenberg, H.J.: How to analyse the results of linear programs - Part 3: infeasibility. Interfaces **23**(6), 120–139 (1993)
237. Greenberg, H.J.: How to analyse the results of linear programs - Part 4: forcing substructures. Interfaces **24**(1), 121–130 (1994)
238. Greenberg, H.J., Murphy, F.H.: A comparison of mathematical programming modeling systems. Ann. Oper. Res. **5**, 177 – 238 (1992)
239. Gregory, C., Darby-Dowman, K., Mitra, G.: Robust optimization and portfolio selection: the cost of robustness. Eur. J. Oper. Res. **212**(2), 417–428 (2011)
240. Grossmann, I.E.: Review of nonlinar mixed-integer and disjunctive programming techniques. Optim. Eng. **3**, 227–252 (2002)
241. Grossmann, I.E., Harjunkoski, I.: Process systems engineering: academic and industrial perspective. Comput. Chem. Eng. **126**, 474–484 (2019)
242. Grossmann, I.E., Trespalacios, F.: Systematic modeling of discrete-continuous optimization models through generalized disjunctive programming. AIChE J. **59**, 3276–3295 (2013)
243. Grötschel, M.: Discrete mathematics in manufacturing. In: O'Malley, R.E. (ed.) ICIAM 91: Proceedings of the Second International Conference on Industrial and Applied Mathematics, pp. 119–145. SIAM, Philadelphia (1992)
244. Grötschel, M., Lovasz, L.: Combinatorial optimization. In: Graham, R.L. (ed.) Handbook on Combinatorics, pp. 1541–1597. North-Holland, Amsterdam (1982)
245. Guéret, C., Heipcke, S., Prins, C., Sevaux, M.: Applications of Optimization with Xpress-MP. Dash Optimization, Blisworth (2002)
246. Guignard, M.: Lagrange relaxation. Sociedad de Estad ística e Investgaciøn Operativa **11**(2), 151–228 (2003)
247. Gupta, A., Maranas, C.D.: Managing demand uncertainty in supply chain planning. Comput. Chem. Eng. **27**, 1219–1227 (2003)
248. Gupta, O.K., Ravindran, V.: Branch and bound experiments in convex nonlinear integer programming. Manage. Sci. **31**, 1533–1546 (1985)
249. Gupta, A., Maranas, C.D., McDonald, C.M.: Mid-term supply chain planning under demand uncertainty: customer demand satisfaction and inventory management. Comput. Chem. Eng. **24**(12), 2613–2621 (2000)
250. Gurobi Optimization, L.: Gurobi Optimizer Reference Manual (2019). http://www.gurobi.com
251. Hales, T.C.: A proof of the Kepler conjecture. Ann. Math. **162**, 1065–1185 (2005)
252. Harding, S.T., Floudas, C.A.: Locating heterogeneous and reactive azeotropes. Ind. Eng. Chem. Res. **39**, 1576–1595 (2000)
253. Harjunkoski, I.: Application of MINLP Methods on a Scheduling Problem in the Paper Converting Industry. PhD Dissertation, Abo Akademi University, Abo (1997)
254. Harjunkoski, I., Jain, V., Grossmann, I.E.: Hybrid mixed-integer/constrained logic programming strategies for solving scheduling and combinatorial optimization problems. Comput. Chem. Eng. **24**, 337–343 (2000)
255. Harris, P.M.J.: Pivot selection methods of the Devex LP code. Math. Program. **5**, 1–28 (1973)
256. Heinz, S., Schlechte, T., Stephan, R.: Solving steel mill slab design problems. Constraints **17**, 39–50 (2012)
257. Heipcke, S.: Resource Constrained Job-Shop-Scheduling with Constraint Nets. Diplomarbeit, Katholische Universität Eichstätt, Mathem.-Geographische Fakultät, Universität Eichstätt, Eichstätt (1995)
258. Heipcke, S.: Comparing constraint programming and mathematical programming approaches to discrete optimisation. The change problem. J. Oper. Res. Soc. **50**(6), 581–595 (1999)

259. Heipcke, S.: Mosel: Modeling and Optimization. Dash Optimization, Blisworth (2002)
260. Heipcke, S.: Xpress-Mosel: multi-solver, multi-problem, multi-model, multi-node modeling and problem solving. In: Kallrath, J. (ed.) Algebraic modeling systems: Modeling and solving real world optimization problems, pp. 77–110. Springer, Heidelberg (2012)
261. Helber, S.: Operations Management Tutorial. S. Helber (2014). https://books.google.de/books?id=HUTioQEACAAJ
262. Held, M.H., Wolfe, P., Crowder, H.D.: Validation of subgradient optimization. Math. Program. **6**, 62–88 (1974)
263. Hendrix, E.M.T., G.-Tóth, B.: Introduction to Nonlinear and Global Optimization. Springer Optimization and Its Applications, vol. 37. Springer, New York, NY (2010)
264. Hendry, L.C., Fok, K.K., Shek, K.W.: A cutting stock and scheduling problem in the copper industry. J. Oper. Res. Soc. **47**, 38–47 (1996)
265. Henrion, R., Küchler, C., Römisch, W.: Discrepancy distances and scenario reduction in two-stage stochastic mixed-integer programming. J. Ind. Manage. Optim. **4**, 363–384 (2008)
266. Henrion, R., Küchler, C., Römisch, W.: Scenario reduction in stochastic programming with respect to discrepancy distances. Comput. Optim. Appl. **43**, 67–93 (2009)
267. Herlihy, M., Shavit, N.: The Art of Multiprocessor Programming, Revised Reprint, 1st edn. Morgan Kaufmann Publishers Inc., San Francisco, CA (2012)
268. Hertz, D.: The extreme eigenvalues and stability of real symmetric interval matrices. IEEE Trans. Autom. Control **37**, 532–535 (1992)
269. Hitchcock, F.L.: The distribution of a product from several sources to numerous localities. J. Math. Phys. **20**, 224–230 (1941)
270. Hooker, J.: Logic-Based Methods for Optimization: Combining Optimization and Constraint Satisfaction. Wiley, Chichester (2000)
271. Horst, R., Pardalos, P.M. (eds.): Handbook of Global Optimization. Kluwer Academic Publishers, Dordrecht (1995)
272. Horst, R., Pardalos, P.M., Thoai, N.V.: Introduction to Global Optimization, 2nd edn. Kluwer Academic Publishers, Dordrecht (2000)
273. Horst, R., Tuy, H.: Global Optimization: Deterministic Approaches, 3rd edn. Springer, New York (1996)
274. Huangfu, Q., Hall, J.A.J.: Parallelizing the dual revised simplex method. Math. Program. Comput. **10**(1), 119–142 (2018)
275. Hummeltenberg, W.H.: Implementations of special ordered sets in MP software. Eur. J. Oper. Res. **17**, 1–15 (1984)
276. Hung, M.S., Rom, W.O.: Solving the assignment problem by relaxation. Oper. Res. **28**, 969–982 (1980)
277. IBM: AIX Easy Modeler/6000 and User Guide. IBM World Trade Corporation, New York, Paris (1993)
278. IBM: IBM ILOG CPLEX Optimization Studio (2017) CPLEX Users Manual (2017). http://www.ibm.com
279. Ierapetriou, M.G., Floudas, C.A.: Effective continuous-time formulation for short-term scheduling. 1. Multipurpose batch processes. Ind. Eng. Chem. Res. **37**, 4341–4359 (1998)
280. Ierapetriou, M.G., Floudas, C.A.: Effective continuous-time formulation for short-term scheduling. 2. Continuous and semicontinuous processes. Ind. Eng. Chem. Res. **37**, 4360–4374 (1998)
281. Ierapetriou, M.G., Hene, T.S., Floudas, C.A.: Continuous time formulation for short-term scheduling with multiple intermediate due dates. Ind. Eng. Chem. Res. **38**, 3446–3461 (1999)
282. Ignizio, J.P.: Goal Programming and Extensions. Heath, Lexington, MA (1976)
283. Irnich, S., Toth, P., Vigo, D.: The family of vehicle routing problems. In: Toth, P., Vigo, D. (eds.) Vehicle Routing. MOS-SIAM Series on Optimization, vol. 18, pp. 1–33. SIAM, New York (2014)
284. Jain, V., Grossmann, I.E.: Algorithms for hybrid MILP/CP models for a class of optimization problems. IFORMS J. Comput. **13**, 258–276 (2001)

285. Janak, S.L., Floudas, C.A., Kallrath, J., Vormbrock, N.: Production scheduling of a large-scale industrial batch plant: I. Short-term and medium-term scheduling. Ind. Eng. Chem. Res. **45**, 8234–8252 (2006a)
286. Janak, S.L., Lin, X., Floudas, C.A.: A new robust optimization approach for scheduling under uncertainty - II. Uncertainty with known probability distribution. Comput. Chem. Eng. **31**, 171–195 (2007)
287. Jaulin, L., Kieffer, M., Didrit, O., Walter, E.: Applied Interval Analysis. Springer, London (2001)
288. Jeroslow, R.G., Lowe, J.K.: Modelling with integer variables. Math. Program. Study **22**, 167–184 (1984)
289. Jeroslow, R.G., Lowe, J.K.: Experimental results with the new techniques for integer programming formulations. J. Oper. Res. Soc. **36**, 393–403 (1985)
290. Jesson, D., Mayston, D., Smith, P.: Performance assessment in the education sector; educational and economic perspectives. Oxford Rev. Educ. **13**, 249–266 (1987)
291. Jia, Z., Ierapetritou, M.: Mixed-integer linear programming model for gasoline blending and distribution scheduling. Ind. Eng. Chem. Res. **42**, 825–835 (2003)
292. Johnson, E.L., Kostreva, M.M., Suhl, U.H.: Solving 0-1 integer programming problems arising from large scale planning models. Oper. Res. **33**, 803–819 (1985)
293. Jonker, R., Volgenat, T.: Improving the Hungarian assignment algorithm. Oper. Res. Lett. **5**, 171–175 (1986)
294. Jozefowiez, N., Semet, F., Talbi, E.G.: Parallel and hybrid models for multi-objective optimization: application to the vehicle routing problem. In: Guervós, J.J.M., Adamidis, P., Beyer, H.G., Schwefel, H.P., Fernández-Villacañas, J.L. (eds.) Parallel Problem Solving from Nature — PPSN VII, pp. 271–280. Springer, Berlin, Heidelberg (2002)
295. Jünger, M., Liebling, T.M., Naddef, D., Nemhauser, G.L., Pulleyblank, W.R., Reinelt, G., Rinaldi, G., Wolsey, L.A. (eds.): 50 Years of Integer Programming 1958–2008 - From the Early Years to the State-of-the-Art. Springer, Heidelberg (2010)
296. Jungnickel, D.: Graphs, Networks and Algorithms, 4th edn. Springer, Berlin (2012)
297. Kalan, J.E.: Aspects of large-scale in-core linear programming. In: ACM Annual Conference Proceedings 1971 and Chicago, pp. 304–313. ACM, New York (1971)
298. Kall, P.: Stochastic Linear Programming. Springer, Berlin (1976)
299. Kall, P., Wallace, S.W.: Stochastic Programming. Wiley, Chichester (1994)
300. Kallrath, J.: Diskrete Optimierung in der chemischen Industrie. In: Bachem, A., Jünger, M., Schrader, R. (eds.) Mathematik in der Praxis - Fallstudien aus Industrie, Wirtschaft, Naturwissenschaften und Medizin, pp. 173–195. Springer, Berlin (1995)
301. Kallrath, J.: Mixed-integer nonlinear programming applications. In: Ciriani, T.A., Gliozzi, S., Johnson, E.L., Tadei, R. (eds.) Operational Research in Industry, pp. 42–76. Macmillan, Houndmills, Basingstoke (1999)
302. Kallrath, J.: Combined strategic, design and operative planning - two success stories in MILP and MINLP. In: Bulatov, V., Baturin, V. (eds.) Proceedings of 12th Baikal International Conference: Optimization Methods and Their Applications, pp. 123–128. Institute of System Dynamics and Control Theory, Irkutsk (2001)
303. Kallrath, J.: Combined strategic and operational planning - an MILP success story in chemical industry. OR Spectrum **24**(3), 315–341 (2002)
304. Kallrath, J.: Planning and scheduling in the process industry. OR Spectrum **24**(3), 219–250 (2002)
305. Kallrath, J.: Exact computation of global minima of a nonconvex portfolio optimization problem. In: Floudas, C.A., Pardalos, P.M. (eds.) Frontiers in Global Optimization, pp. 237–254. Kluwer Academic Publishers, Dordrecht (2004)
306. Kallrath, J. (ed.): Modeling Languages in Mathematical Optimization. Kluwer Academic Publishers, Norwell, MA (2004)
307. Kallrath, J.: Modeling difficult optimization problems. In: Floudas, C.A., Pardalos, P.M. (eds.) Encyclopedia of Optimization, pp. 2284–2297. Springer, New York (2008)

308. Kallrath, J.: Pricing problems in the chemical process industry. Comput. Manage. Sci. **5**, 403–405 (2008)
309. Kallrath, J.: Combined strategic design and operative planning in the process industry. Comput. Chem. Eng. **33**, 1983–1993 (2009)
310. Kallrath, J.: Cutting circles and polygons from area-minimizing rectangles. J. Glob. Optim. **43**, 299–328 (2009)
311. Kallrath, J.: Polylithic modeling and solution approaches using algebraic modeling systems. Optim. Lett. **5**, 453–466 (2011). https://doi.org/10.1007/s11590-011-0320-4
312. Kallrath, J. (ed.): Algebraic Modeling Systems: Modeling and Solving Real World Optimization Problems. Springer, Heidelberg (2012)
313. Kallrath, J.: Packing ellipsoids into volume-minimizing rectangular boxes. J. Glob. Optim. **67**(1), 151–185 (2017)
314. Kallrath, J., Blackburn, R., Näumann, J.: Grid-enhanced polylithic modeling and solution approaches for hard optimization problems. In: Bock, H.G., Jäger, W., Kostina, E., Phu, H.X. (eds.) Modeling, Simulation and Optimization of Complex Processes HPSC 2018 – Proceedings of the 7th International Conference on High Performance Scientific Computing, Hanoi, March 19–23, 2018, pp. 1–15. Springer Nature, Cham (2020)
315. Kallrath, J., Frey, M.M.: Packing circles into perimeter-minimizing convex hulls. J. Glob. Optim. **73**(4), 723–759 (2019)
316. Kallrath, J., Maindl, T.I.: Real Optimization with SAP-APO. Springer, Heidelberg (2006)
317. Kallrath, J., Pankratov, A., Romanova, T., Litvinchev, I.: Minimal perimeter convex hulls of convex polygons. J. Glob. Optim. (2021, submitted (under review))
318. Kallrath, J., Rebennack, S.: Cutting ellipses from area-minimizing rectangles. J. Glob. Optim. **59**(2–3), 405–437 (2014)
319. Kallrath, J., Rebennack, S., Kallrath, J., Kusche, R.: Solving real-world cutting stock-problems in the paper industry: mathematical approaches, experience and challenges. Eur. J. Oper. Res. **238**, 374–389 (2014)
320. Kallrath, J., Schreieck, A.: Discrete optimization and real world problems. In: Hertzberger, B., Serazzi, G. (eds.) High-Performance Computing and Networking. Lecture Notes in Computer Science, vol. 919, pp. 351–359. Springer, Berlin, Heidelberg, New York (1995)
321. Kalvelagen, E.: Branch-and-Bound methods for an MINLP model with semi-continuous variables (2003, discontinued on http://www.gams.com)
322. Kantorovich, L.V.: Mathematical methods in the organization and planning of production. Transl. Manage. Sci. **6**, 366–422 (1960 (1939))
323. Karelathi, J.: Solving the Cutting Stock Problem in the Steel Industry. Master thesis, Helsinki University of Technology, Department of Engineering Physics and Mathematics, Helsinki (2002)
324. Karelathi, J., Vainiomäki, P., Westerlund, T.: Large scale production planning in the stainless steel industry. Ind. Eng. Chem. Res. **50**, 4893–4906 (2011)
325. Karmarkar, N.: A new polynomial time algorithm for linear programming. Combinatorica **4**, 375–395 (1984)
326. Karmarkar, U.S., Schrage, L.: The deterministic dynamic product cycling problem. Oper. Res. **33**, 326–345 (1985)
327. Karp, R.M.: Reducibility among combinatorial problems. In: Miller, R.E., Thatcher, J.W. (eds.) Complexity of Computer Computations, pp. 85–103. Plenum Press, New York (1972)
328. Karuppiah, R., Grossmann, I.E.: A Lagrangean based branch-and-cut algorithm for global optimization of nonconvex mixed-integer nonlinear programs with decomposable structures. J. Global Optim. **41**, 163–186 (2008)
329. Karush, W.: Minima of Functions of Several Variables with Inequalities as Side Constraints. Master thesis, Department of Mathematics, University of Chicago, Chicago (1939)
330. Kearfott, R.B.: Rigorous Global Search: Continuous Problems. Kluwer Academic Publishers, Dordrecht (1996)
331. Kelley, J.E.: The cutting plane method for solving convex programs. J. SIAM **8**(4), 703–712 (1960)

332. Khachian, L.G.: A polynomial algorithm in linear programming. Sov. Math. Dokl. **20**, 191–194 (1979)
333. Kilinç, M.R., Sahinidis, N.V.: State of the Art in mixed-integer nonlinear optimization, chap. 21, pp. 273–292. SIAM, Philadelphia (2017)
334. Klee, V., Minty, G.J.: How good is the simplex algorithm? In: Shisha, O. (ed.) Inequalities III, pp. 159–175. Academic Press, New York (1972)
335. Klein-Haneveld, W.K., van der Vlerk, M.H.: Stochastic integer programming: general models and algorithms. Ann. Oper. Res. **85**, 39–57 (1999)
336. Klosterhalfen, S.T., Kallrath, J., Frey, M.M., Schreieck, A., Blackburn, R., Buchmann, J., Weidner, F.: Creating cost transparency to support strategic planning in complex chemical value chains. Eur. J. Oper. Res. **279**, 605–619 (2019)
337. Klotz, E.: Identification, assessment, and correction of ill-conditioning and numerical instability in linear and integer programs. In: INFORMS TutORials in Operations Research. INFORMS, chap. 3, pp. 54–108 (2014)
338. Klotz, E., Newman, A.M.: Practical guidelines for solving difficult linear programs. Surv. Oper. Res. Manage. Sci. **18**(1), 1–17 (2013)
339. Klotz, E., Newman, A.M.: Practical guidelines for solving difficult mixed integer linear programs. Surv. Oper. Res. Manage. Sci. **18**(1), 18–32 (2013)
340. Koopmans, T.C.: Optimum utilization of the transport systems. Econometr. Suppl. **17**, 136–145 (1947)
341. Korte, B., Vygen, J.: Combinatorial Optimization: Theory and Algorithms. Algorithms and Combinatorics, vol. 21, 6th edn. Springer, New York (2018)
342. Kristjansson, B.: MPL Modeling System. Maximal Software, Arlington, VA (1992)
343. Kronqvist, J., Bernal, D.E., Lundell, A., Grossmann, I.E.: A review and comparison of solvers for convex MINLP. Optim. Eng. **20**(2), 397–455 (2019)
344. Kuhn, H.: Nonlinear programming: a historical view. In: Cottle, R., Lemke, C. (eds.) Nonlinear Programming. SIAM-AMS Proceedings, vol. 9, pp. 1–26. American Mathematical Society, Providence, RI (1976)
345. Kuhn, H.W.: The Hungarian method for the assignment problem. Naval Res. Logist. Quart. **2**, 83–97 (1955)
346. Kuhn, H.W., Tucker, A.W.: Nonlinear programming. In: Neumann, J. (ed.) Proceedings Second Berkeley Symposium on Mathematical Statistics and Probability, pp. 481–492. University of California, Berkeley, CA (1951)
347. Kurschl, W., Pimminger, S., Wagner, S., Heinzelreiter, J.: Concepts and requirements for a cloud-based optimization service. In: 2014 Asia-Pacific Conference on Computer Aided System Engineering (APCASE), pp. 9–18 (2014)
348. Lancia, G., Serafini, P.: Compact Extended Linear Programming Models. Springer International Publishing Company AG, Cham (2018)
349. Land, A.H., Doig, A.G.: An automatic method for solving discrete programming problems. Econometrica **28**, 497–520 (1960)
350. Lančinskas, A., Ortigosa, P.M., Žilinskas, J.: Parallel optimization algorithm for competitive facility location. Math. Modell. Anal. **20**(5), 619–640 (2015)
351. Laporte, G., Nickel, S., Saldanha da Gama, F.: Location Science. Springer, Cham (2015)
352. Laporte, G., Toth, P., Vigo, D.: Vehicle routing: historical perspective and recent contributions. EURO J. Transport. Logist. **2**(1-2), 1–4 (2013)
353. Lasdon, L.S., Waren, A.D.: Generalized reduced gradient method for linearly and nonlinearly constrained programming. In: Greenberg, H.J. (ed.) Design and Implementation of Optimization Software, pp. 363–397. Sijthoff and Noordhoff, Alphen aan den Rijn (1978)
354. Lasdon, L.S., Waren, A.D., Jain, A., Ratner, M.: Design and testing of a generalized reduced gradient code for nonlinear programming. ACM Trans. Math. Softw. **4**, 34–50 (1978)
355. Laundy, R.S.: Implementation of parallel Branch-and-Bound algorithms in Xpress-MP. In: Ciriani, T.A., Gliozzi, S., Johnson, E.L., Tadei, R. (eds.) Operational Research in Industry. MacMillan, London (1999)

356. Lawler, E.L., Lenstra, J.K., Rinnooy-Kan, A.H.G., Shmoys, D.B.: The Traveling Salesman Problem: A Guided Tour of Combinatorial Optimization. Wiley, Chichester (1985)
357. Leao, A.A.S., Toledo, F.M.B., Oliveira, J.F., Carravilla, M., Alvarez-Valdes, R.: Irregular packing problems: a review of mathematical models. Eur. J. Oper. Res. **282**(3), 803–822 (2019)
358. Lenstra, J.K., Kan, A.H.G.R., Brucker, P.: Complexity of machine scheduling problems. Ann. Discr. Math. **1**, 343–362 (1977)
359. Letchford, A.: Allocation of school bus contracts using integer programming. J. Oper. Res. Soc. **47**, 369–372 (1996)
360. Leung, J., Magnanti, T.L.: Valid inequalities and facets of the capacitated plant location problem. Math. Program. **44**, 271–291 (1989)
361. Leyffer, S.: Deterministic Methods for Mixed Integer Nonlinear Programming. PhD Thesis, Department of Mathematics and Computer Science, University of Dundee, Dundee (1993)
362. Li, Z., Milenkovic, V.: Compaction and separation algorithms for non-convex polygons and their applications. Eur. J. Oper. Res. **84**, 539–561 (1995)
363. Liberti, L., Maculan, N. (eds.): Global Optimization: From Theory to Implementation. Nonconvex Optimization and Its Applications, vol. 84, pp. 223–232. Springer, New York (2006)
364. Lin, X., Floudas, C.A.: Design, synthesis and scheduling of multipurpose batch plants via an effective continuous-time formulation. Comput. Chem. Eng. **25**, 665–674 (2001)
365. Lin, X., Floudas, C.A., Kallrath, J.: Global solution approaches for nonconvex MINLP problems in product portfolio optimization. J. Glob. Optim. **32**, 417–431 (2005)
366. Lin, X., Floudas, C.A., Modi, S., Juhasz, N.M.: Continuous-time optimization approach for medium-range production scheduling of a multiproduct batch plant. Ind. Eng. Chem. Res. **41**, 3884–3906 (2002)
367. Lin, X., Janak, S.L., Floudas, C.A.: A new robust optimization approach for scheduling under uncertainty - I. bounded uncertainty. Comput. Chem. Eng. **28**, 1069–1085 (2004)
368. Linderoth, J., Savelsbergh, M.W.: A computational study of search strategies for mixed integer programming. INFORMS J. Comput. **11**(2), 173–187 (1999)
369. Little, J.D.C., Murty, K.G., Sweeney, D.W., Karel, C.: An algorithm for the traveling salesman problem. Oper. Res. **11**, 972–989 (1963)
370. Litvinchev, I.: Decomposition-aggregation method for convex programming problems. Optimization **22**(1), 47–56 (1991)
371. Litvinchev, I., Espinosa, E.L.O.: Solving the two-stage capacitated facility location problem by the Lagrangian heuristic. In: Hu, H., Shi, X., Stahlbock, R., Voß, S. (eds.) Computational Logistics. ICCL 2012. Lecture Notes in Computer Science, vol. 7555, pp. 92–103 (2012)
372. Litvinchev, I., Rangel, S.: Localization of the optimal solution and a posteriori bounds for aggregation. Comput. Oper. Res. **26**(10-11), 967–988 (1999)
373. Litvinchev, I., Tsurkov, V.: Aggregation in large-scale optimization. Appl. Optim. **83**, 291 (2003)
374. Litvinchev, I., Mata, M., Rangel, S., Saucedo, J.: Lagrangian heuristic for a class of the generalized assignment problems. Comput. Math. Appl. **60**(4), 1115–1123 (2010)
375. Litvinchev, I., Romanova, T., Corrales-Diaz, R., Esquerra-Arguelles, A., Martinez-Noa, A.: Lagrangian approach to modeling placement conditions in optimized packing problems. Mob. Netw. Appl. **25**, 2126–2133 (2020) (2020)
376. Lübbecke, M.E., Desrosiers, J.: Selected topics in column generation. Oper. Res. **53**(6), 1007–1023 (2005)
377. Lundell, A., Westerlund, T.: Solving global optimization problems using reformulations and signomial transformations. Comput. Chem. Eng. **116**, 122–134 (2018)
378. Lustig, I.J., Marsten, R.E., Shanno, D.F.: Computational experience with a primal-dual interior point method for linear programming. Linear Algebra Appl. **152**, 191–222 (1991)
379. Lustig, I.J., Marsten, R.E., Shanno, D.F.: On implementing Mehrotra's predictor-corrector interior-point method for linear programming. SIAM J. Optim. **2**, 435–449 (1992)

380. Mahadevan, D.A.: Optimization in Computer-Aided Pattern Packing. Ph.D. Thesis, North Carolina State University (1984)
381. Main, R.A.: Large recursion models: practical aspects of recursion techniques. In: Ciriani, T., Leachman, R.C. (eds.) Optimization in Industry: Mathematical Modeling Techniques in Practice, pp. 241–249. Wiley, Chichester (1993)
382. Makhorin, A.: GNU Linear Programming Kit - Reference Manual (2009). Version 4.37, 2009. http://www.gnu.org/software/glpk
383. Maniezzo, V., Stützle, T., Voß, S. (eds.): Matheuristics: Hybridizing Metaheuristics and Mathematical Programming. Springer, Heidelberg (2009)
384. Maranas, C.D., Floudas, C.A.: Global minimum potential energy confirmations of small molecules. J. Global Optim. **4**, 135–170 (1994)
385. Margot, F.: Symmetry in integer linear programming. In: Jünger, M., Liebling, T.M., Naddef, D., Nemhauser, G.L., Pulleyblank, W.R., Reinelt, G., Rinaldi, G., Wolsey, L.A. (eds.) 50 Years of Integer Programming 1958–2008 - From the Early Years to the State-of-the-Art, chap. 17, pp. 647–686. Springer, Heidelberg (2010)
386. Maros, I., Mitra, G.: Finding better starting bases for the simplex method. In: Kleinschmidt, P. (ed.) Operations Research Proceedings 1995. Springer, Berlin (1996)
387. Martello, S., Toth, P.: Knapsack Problems: Algorithms and Computer Implementations. Wiley, Chichester (1990)
388. Martin, R.K.: Large Scale Linear and Integer Optimization – A Unified Approach. Kluwer, Dortrecht (1999)
389. McCormick, G.P.: Computation of global solutions to factorable nonconvex programs: part I - convex underestimations problems. Math. Program. **10**, 147–175 (1976)
390. McKinnon, K.I.M., Williams, H.P.: Constructing integer programming models by the predicate calculus. Ann. Oper. Res. **21**, 227–246 (1989)
391. McMullen, P.: The maximum number of faces of convex polytopes. Mathematika **17**, 179–184 (1970)
392. Meeraus, A.: An algebraic approach to modeling. J. Econ. Develop. Control **5**(1), 81–108 (1983)
393. Mehrotra, S.: On the implementation of a primal-dual interior point method. SIAM J. Optim. **2**(4), 575–601 (1992)
394. Mei-Ko, K.: Graphic programming using odd or even points. Chin. Math. **1**, 273–277 (1962)
395. Meyer, M.: Applying linear programming to the design of ultimate pit limits. Manage. Sci. **16**, 121–135 (1969)
396. Meyn, S.P.: Stability, performance evaluation, and optimization. In: Handbook of Markov Decision Processes. International Series in Operations Research & Management Science, vol. 40, pp. 305–346. Kluwer Academic Publishers, Boston, MA (2002)
397. Michalewicz, Z., Fogel, D.B.: How to Solve It: Modern Heuristics. Springer, Berlin (2000)
398. Migdalas, A., Pardalos, P.M., Värbrand, P. (eds.): Multilevel Optimization: Algorithms and Applications. Kluwer Academic Publishers, Boston, MA (1998)
399. Miliotis, P.A.: Data envelopment analysis applied to electricity distribution districts. J. Oper. Res. Soc. **43**, 549–555 (1992)
400. Miller, C.E.: The simplex method for local separable programming. In: Graves, R.L., Wolfe, P.L. (eds.) Recent Advances in Mathematical Programming, pp. 311–317. McGraw-Hill, London (1963)
401. Miller, C.E., Tucker, A.W., Zemlin, R.A.: Integer programming formulation of traveling salesman problems. ACM **7**, 326–329 (1960)
402. Mingozzi, A., Roberti, R., Toth, P.: An exact algorithm for the multitrip vehicle routing problem. INFORMS J. Comput. **25**(2), 193–207 (2013)
403. Mirrlees, J.A.: The theory of moral hazard and unobservable behaviour: part I. Rev. Econ. Stud. **66**, 3–21 (1999)
404. Misener, R., Floudas, C.A.: Advances for the pooling problem: modeling, global optimization, and computational studies survey. Appl. Computat. Math. **8**, 3–22 (2009)

405. Misener, R., Floudas, C.A.: Piecewise-linear approximations of multidimensional functions. J. Optim. Theor. Appl. **145**, 120–147 (2010)
406. Misener, R., Floudas, C.: GloMIQO: global mixed-integer quadratic optimizer. J. Glob. Optim. 1–48 (2012). http://dx.doi.org/10.1007/s10898-012-9874-7
407. Misener, R., Floudas, C.: ANTIGONE: algorithms for coNTinuous/Integer Global Optimization of Nonlinear Equations. J. Glob. Optim. **59**, 503–526 (2014)
408. Mitchell, G.: The Practice of Operational Research. Wiley, Chichester (1993)
409. Mitra, G., Darby-Dowman, K., Lucas, C., Smith, J.W.: Maritime scheduling using discrete optimization and artificial intelligence techniques. In: Sciomachen, A. (ed.) Optimization in Industry 3: Mathematical Programming Techniques in Practice, pp. 1–17. Wiley, Chichester (1995)
410. Mitra, G., Poojari, C., Sen, S.: Strategic and tactical planning models for supply chain: an application of stochastic mixed integer programming. In: Aardal, I., Nemhauser, G.L., Weismantel, R. (eds.) Handbook of Discrete Optimization. Elsevier, North-Holland (2004)
411. Mulvey, J.M.: Testing of a large-scale network optimization program. Math. Program. **15**, 291–315 (1978)
412. Mulvey, J.M., Beck, M.P.: Solving capacitated clustering problems. Eur. J. Oper. Res. **18**, 339–348 (1984)
413. Munguia, L.M., Oxberry, G., Rajan, D., Shinano, Y.: Parallel PIPS-SBB: multi-level parallelism for stochastic mixed-integer programs. Comput. Optim. Appl. (2019). Epub ahead of print
414. Muñoz, J., Gutierrez, G., Sanchis, A.: Evolutionary techniques in a constraint satisfaction problem: puzzle eternity II. In: Proceedings 2009 IEEE Congress on Evolutionary Computation, pp. 2985–2991 (2009)
415. Murtagh, B.A., Saunders, M.A.: Large-scale linearly constrained optimization. Math. Program. **14**, 41–72 (1978)
416. Murtagh, B.A., Saunders, M.A.: A projected Lagrangian algorithm and its implementation for sparse nonlinear constraints. Math. Program. Study (Algorithm for Constrained Minimization of Smooth Nonlinear Function) **16**, 84–117 (1982)
417. Mutapcic, A., Boyd, S.: Cutting-set methods for robust optimization with pessimizing oracles. Optim. Methods Softw. **24**, 381–406 (2009)
418. Muts, P., Nowak, I., Hendrix, E.M.T.: The decomposition-based outer approximation algorithm for convex mixed-integer nonlinear programming. J. Glob. Optim. **77**(1), 75–96 (2020)
419. Nelder, J.A., Mead, R.: A simplex method for function minimization. Comp. J. **7**, 308–313 (1965)
420. Nemhauser, G.L.: The age of optimization: solving large-scale real world-problems. Oper. Res. **42**, 5–13 (1994)
421. Nemhauser, G.L., Wolsey, L.A.: Integer and Combinatorial Optimization. Wiley, New York (1988)
422. Neumann, K., Morlock, M.: Operations Research. Carl Hanser, München, Wien (1993)
423. Nowak, I.: Relaxation and Decomposition Methods for Mixed Integer Nonlinear Programming. Birkhäuser, Basel (2005)
424. Nowak, I., Muts, P., Hendrix, E.M.T.: Multi-tree decomposition methods for large-scale mixed integer nonlinear optimization. In: Velásquez-Bermúdez, J.M., Khakifirooz, M., Fathi, M. (eds.) Large Scale Optimization in Supply Chains and Smart Manufacturing: Theory and Applications, pp. 27–58. Springer International Publishing, Cham (2019)
425. Oliveira, J.F., Ferreira, J.S.: Algorithms for nesting problems. In: Vidal, R. (ed.) Applied Simulated Annealing. Lecture Notes in Economics and Mathematical Systems, vol. 396, pp. 255–274. Springer, Heidelberg (1993)
426. Orchard-Hays, W.: Advanced Linear Programming Computing Techniques. McGraw-Hill, New York (1969)
427. Orçun, S., Altinel, I.K., Hortaçsu, O.: Scheduling of batch processes with operational uncertainties. Comput. Chem. Eng. **20**, S1215–S1220 (1996)
428. Orden, A.: LP from the '40s to the '90s. Interfaces **23**(5), 2–12 (1993)

429. Osman, I.H., Christofides, N.: Capacitated clustering problems by hybrid simulated annealing and tabu search. Int. Trans. Oper. Res. **1**, 317–336 (1994)
430. Ostrowski, J., Linderoth, J.T., Rossi, F., Smriglio, S.: Orbital branching. Math. Program. **126**(1), 147–178 (2011)
431. Padberg, M.: Linear Optimization and Extensions. Springer, Berlin, Heidelberg (1996)
432. Padberg, M.W., Rinaldi, G.: Optimization of a 532-city traveling salesman problem by branch and cut. Oper. Res. Lett. **6**, 1–6 (1987)
433. Pankratov, A., Romanova, T., Litvinchev, I.: Packing ellipses in an optimized convex polygon. J. Glob. Optim. **75**(2), 495–522 (2019)
434. Pankratov, A., Romanova, T., Litvinchev, I.: Packing ellipses in an optimized rectangular container. Wirel. Netw. **26**(7), 4869–4879 (2020)
435. Pankratov, A., Romanova, T., Litvinchev, I.: Packing oblique 3D objects. Mathematics **8**(7) (2020)
436. Papadimitriou, C.H., Steiglitz, K.: Combinatorial Optimization: Algorithms and Complexity. Prentice Hall, Englewood Cliffs, NJ (1982)
437. Pardalos, P.M., Pitsoulis, L.S., Mavridou, T.D., Resende, M.G.C.: Parallel search for combinatorial optimization: genetic algorithms, simulated annealing, tabu search and GRASP. In: Parallel Algorithms for Irregularly Structured Problems, Second International Workshop, IRREGULAR '95, Lyon, September 4–6, 1995, Proceedings, pp. 317–331 (1995)
438. Patterson, J.H., Slowinski, R., Talbot, F.B., Weglarz, J.: An algorithm for a general class of precedence and resource constrained scheduling problems. In: Slowinski, R., Weglarz, J. (eds.) Advances in Project Scheduling, pp. 1–26. Elsevier, Amsterdam (1989)
439. Pidd, M.: Computer Simulation in Management Science, 3rd edn. Wiley, Chichester (1992)
440. Pillo, G.D., Lucidi, S., Rinaldi, F.: An approach to constrained global optimization based on exact penalty functions. J. Glob. Optim. **54**, 251–260 (2012)
441. Pochet, Y., Wolsey, L.A.: Production Planning by Mixed Integer Programming. Springer, New York (2006)
442. Poljak, B.T.: A general method of solving extremum problems. Sov. Math. Dokl. **8**, 593–597. Transl. Dokl. Akad. Nauk. SSSR **174**, 1967 (1967)
443. Polya, G.: Vom Lernen und Lösen mathematischer Aufgaben. Einsicht und Entdeckung. Lernen und Lehren. Birkhäuser Verlag, Basel (1979)
444. Popper, K.R.: The Logic of Scientific Discovery, 10th edn. Hutchinson, London (1980)
445. Potts, C.N., Wassenhove, L.N.V.: Integrating scheduling with batching and lot-sizing: a review of algorithms and complexity. J. Oper. Res. Soc. **43**, 395–406 (1992)
446. Prékopa, A.: Stochastic Programming. Kluwer Academic Publishers, Dordrecht (1995)
447. Press, W.H., Flannery, B.P., Teukolsky, S.A., Vetterling, W.T.: Numerical Recipes - The Art of Scientific Computing, 2nd edn. Cambridge University Press, Cambridge (1992)
448. Pritsker, A.B., Watters, L.J., Wolfe, P.M.: Multiproject scheduling with limited resources: a zero-one programming approach. Manage. Sci. **16**, 93–108 (1969)
449. Rahimian, H., Mehrotra, S.: Distributionally robust optimization: a review (2019). arXiv:1908.05659 [math.OC]
450. Ralphs, T., Shinano, Y., Berthold, T., Koch, T.: Parallel solvers for mixed integer linear optimization. In: Hamadi, Y., Sais, L. (eds.) Handbook of Parallel Constraint Reasoning, pp. 283 – 336. Springer, Cham (2018)
451. Ratschek, H., Rokne, J.: Experiments using interval analysis for problem solving a circuit design problem. J. Glob. Optim. **3**, 501–518 (1993)
452. Ratschek, H., Rokne, J.: Interval methods. In: Horst, R., Pardalos, P.M. (eds.) Handbook of Global Optimization, pp. 751–828. Kluwer Academic Publishers, Dordrecht (1995)
453. Ravindran, A., Phillips, D.T., Solberg, J.J.: Operations Research. Principles and Practice. Wiley, New York (1987)
454. Rebennack, S., Kallrath, J.: Continuous piecewise linear delta-approximations for bivariate and multivariate functions. J. Optim. Theor. Appl. **167**, 102–117 (2015)
455. Rebennack, S., Kallrath, J.: Continuous piecewise linear delta-approximations for univariate functions: computing minimal breakpoint systems. J. Optim. Theor. Appl. **167**, 617–643 (2015)

456. Rebennack, S., Kallrath, J., Pardalos, P.M.: Column enumeration based decomposition techniques for a class of non-convex MINLP problems. J. Glob. Optim. **43**, 277–297 (2009)
457. Rebennack, S., Nahapetyan, A., Pardalos, P.M.: Bilinear modeling solution approach for fixed charged network flow problems. Optim. Lett. **3**, 347–355 (2009)
458. Rebennack, S., Oswald, M., Theis, D.O., Seitz, H., Reinelt, G., Pardalos, P.M.: A branch and cut solver for the maximum stable set problem. J. Combin. Optim. **21**, 434–457 (2011)
459. Rebennack, S., Reinelt, G., Pardalos, P.M.: A tutorial on branch&cut algorithms for the maximum stable set problem. Int. Trans. Oper. Res. **19**, 161–199 (2012)
460. Reinfeld, N.V., Vogel, W.R.: Mathematical Programming. Prentice-Hall International, Englewood Cliffs, NJ (1958)
461. Robinson, S.M.: A quadratically convergent algorithm for general nonlinear for programming problems. Math. Program. **3**, 145–156 (1972)
462. Robinson, E.P., Gao, L.L., Muggenborg, S.D.: Designing an integrated distribution system at DowBrands Inc. Interfaces **23**(3), 107–117 (1993)
463. Romanova, T., Bennell, J., Stoyan, Y., Pankratov, A.: Packing of concave polyhedra with continuous rotations using nonlinear optimisation. Eur. J. Oper. Res. **268**, 37–53 (2018)
464. Romanova, T., Pankratov, A., Litvinchev, I., Pankratova, Y., Urniaieva, I.: Optimized packing clusters of objects in a rectangular container. Math. Probl. Eng. **2019** (2019). https://doi.org/10.1155/2019/4136430
465. Romanova, T., Stetsyuk, P., Chugay, A., Shekhovtsov, S.: Parallel computing technologies for solving optimization problems of geometric design. Cybernet. Syst. Anal. **55**(6), 894–904 (2019)
466. Romanova, T., Stoyan, Y., Pankratov, A., Litvinchev, I., Avramov, K., Chernobryvko, M., Yanchevskyi, I., Mozgova, I., Bennell, J.: Optimal layout of ellipses and its application for additive manufacturing. Int. J. Prod. Res. **59**(2), 560–575 (2019)
467. Romanova, T., Litvinchev, I., Grebennik, I., Kovalenko, A., Urniaieva, I., Shekhovtsov, S.: Packing convex 3D objects with special geometric and balancing conditions. In: Vasant, P., Zelinka, I., Weber, G.W. (eds.) Modeling an Optimization in Space Engineering Applications. Intelligent Computing and Optimization. Advances in Intelligent Systems and Computing, vol. 1072, pp. 273–281. Springer, Cham (2020)
468. Romanova, T., Litvinchev, I., Pankratov, A.: Packing ellipses in an optimized cylinder. Eur. J. Oper. Res. **285**, 429–443 (2020)
469. Romanova, T., Pankratov, A., Litvinchev, I., Plankovskyy, S., Tsegelnyk, Y., Shypul, O.: Sparsest packing of two-dimensional objects. Int. J. Prod. Res. (2020). https://doi.org/10.1080/00207543.2020.1755471
470. Romanova, T., Stoyan, Y., Pankratov, A., Litvinchev, I., Plankovskyy, S., Tsegelnyk, Y., Shypul, O.: Sparsest balanced packing of irregular 3d objects in a cylindrical container. Eur. J. Oper. Res. **285**(2), 429–443 (2020)
471. Romero, C.: Handbook of Critical Issues in Goal Programming. Pergamon Press, Oxford (1991)
472. Rommelfanger, H.: Fuzzy Decision Support-Systeme - Entscheiden bei Unschärfe, 2nd edn. Springer, Heidelberg (1993)
473. Roslöf, J., Harjunkoski, I., Westerlund, T., Isaksson, J.: Solving a large-scale industrial scheduling problem using MILP combined with a heuristic procedure. Eur. J. Oper. Res. **138**(1), 29–42 (2002)
474. Ruszczyński, A., Shapiro, A.: Stochastic programming. In: Handbooks in Operations Research and Management Science, vol. 10. Elsevier, North-Holland (2003)
475. Ryan, D.M.: The solution of massive generalized set partitioning problems in aircrew rostering. J. Oper. Res. Soc. **43**, 459–468 (1992)
476. Ryu, J., Lee, M., Kim, D., Kallrath, J., Sugihara, K., Kim, D.S.: VOROPACK-D: real-time disk packing algorithm using Voronoi diagram. Appl. Math. Comput. **375**, 125076 (2020)
477. Sahinidis, N.V.: Optimization under uncertainty: state-of-the-art and opportunities. Comput. Chem. Eng. **28**, 971–983 (2004)

478. Sahinidis, N.V.: Mixed-integer nonlinear programming 2018. Optim. Eng. **20**, 301–306 (2018)
479. Salkin, H.M., Kluyver, C.A.D.: The knapsack problem: a survey. Naval Res. Logist. Quart. **24**, 127–144 (1975)
480. Salveson, M.E.: The assembly line balancing problem. J. Ind. Eng. **6**, 18–25 (1955)
481. Sand, G., Engell, S., Märkert, A., Schultz, R., Schulz, C.: Approximation of an ideal online scheduler for a multiproduct batch plant. Comput. Chem. Eng. **24**, 361–367 (2000)
482. Santos, M.O., Almada-Lobo, B.: Integrated pulp and paper mill planning and scheduling. Comput. Ind. Eng. **63**(1), 1–12 (2012)
483. SAS Institute Inc.: SAS/OR 15.1 User's Guide: Mathematical Programming Examples. SAS Institute Inc., Cary, NC (2018)
484. Savelsbergh, M.W.P.: Preprocessing and probing techniques for mixed integer programming problems. ORSA J. Comput. **6**, 445–454 (1994)
485. Savelsbergh, M.W.P.: A branch-and-price algorithm for the generalized assignment problem. Oper. Res. **6**, 831–841 (1997)
486. Savelsbergh, M.W.P.: Branch-and-price: integer programming with column generation. In: Floudas, C.A., Pardalos, P. (eds.) Encyclopedia of Optimization, pp. 218–221. Kluwer Academic Publishers, Dordrecht, Holland (2001)
487. Savelsbergh, M.W.P., Sismondi, G.C., Nemhauser, G.L.: Functional description of MINTO and a mixed INTeger optimizer. Oper. Res. Lett. **8**, 119–124 (1994)
488. Scheithauer, G.: Introduction to Cutting and Packing Optimization - Problems, Modeling Approaches, Solution Methods. International Series in Operations Research and Management Science, vol. 263. Springer, Cham (2018)
489. Schindler, S., Semmel, T.: Station staffing at Pan American world airways. Interfaces **23**(3), 91–94 (1993)
490. Schniederjans, M.J.: Goal Programming: Methodology and Applications. Kluwer Academic Publishers, Boston, MA (1995)
491. Schrage, L.: LindoSystems: LindoAPI (2004)
492. Schrage, L.: Optimization Modeling with LINGO. LINDO Systems, Inc., Chicago, IL (2006)
493. Schrijver, A.: Combinatorial Optimization: Polyhedra and Efficiency. Springer Science & Business Media, Heidelberg (2003). Journal of Computer and System Sciences B
494. Schultz, R.: On structure and stability in stochastic programs with random technology matrix and complete integer recourse. Math. Program. **70**, 73–89 (1995)
495. Schultz, R.: Stochastic programming with integer variables. Math. Program. Ser. B **97**, 285–309 (2003)
496. Schweiger, C.A., Rojnuckarin, A., Floudas, C.A.: MINOPT: a software package for mixed-integer nonlinear optimization. Dept. of Chemical Engineering, Princeton University, Princeton, NJ (1996)
497. Sen, S.: Algorithms for stochastic mixed-integer programming models. In: Aardal, I., Nemhauser, G.L., Weismantel, R. (eds.) Handbook of Discrete Optimization. Elsevier, North-Holland (2004)
498. Sen, S., Higle, J.L.: An introductory tutorial on stochastic linear programming models. Interfaces **29**(2), 33–61 (1999)
499. Sexton, T.R., Sleeper, S., Taggart Jr., R.E.: Improving pupil transportation in North Carolina. Interfaces **24**(1), 87–103 (1994)
500. Shapiro, A., Dentcheva, D., Ruszczynski, A.: Lectures on stochastic programming: Modeling and theory. MPS-SIAM Series on Optimization, vol. 9. Society for Industrial and Applied Mathematics, Philadelphia, PA (2009)
501. Sharda, R.: Linear and discrete optimization and modeling software. Unicom in Association with Lionheart Publishing Inc, Atlanta, GA (1993)
502. Shinano, Y.: The ubiquity generator framework: 7 years of progress in parallelizing branch-and-bound. In: Operations Research Proceedings 2017, pp. 143–149 (2018)
503. Shinano, Y., Fujie, T., Kounoike, Y.: Effectiveness of parallelizing the ILOG-CPLEX mixed integer optimizer in the PUBB2 framework. In: Kosch, H., Böszörményi, L., Hellwagner, H.

(eds.) Euro-Par 2003 Parallel Processing. Euro-Par 2003. Lecture Notes in Computer Science, vol. 2790, pp. 770–779 (2003)
504. Shinano, Y., Achterberg, T., Fujie, T.: A dynamic load balancing mechanism for new ParaLEX. In: 2008 14th IEEE International Conference on Parallel and Distributed Systems, pp. 455–462 (2008)
505. Shinano, Y., Achterberg, T., Berthold, T., Heinz, S., Koch, T.: ParaSCIP: a parallel extension of SCIP. In: Competence in High Performance Computing 2010 - Proceedings of an International Conference on Competence in High Performance Computing, Schloss Schwetzingen, June 2010, pp. 135–148 (2010)
506. Shinano, Y., Achterberg, T., Berthold, T., Heinz, S., Koch, T., Winkler, M.: Solving open MIP instances with ParaSCIP on supercomputers using up to 80,000 cores. In: 2016 IEEE International Parallel and Distributed Processing Symposium (IPDPS), pp. 770–779 (2016)
507. Shinano, Y., Berthold, T., Heinz, S.: A first implementation of ParaXpress: combining internal and external parallelization to solve MIPs on supercomputers. In: International Congress on Mathematical Software, pp. 308–316. Springer, New York (2016)
508. Shinano, Y., Berthold, T., Heinz, S.: ParaXpress: an experimental extension of the FICO Xpress-optimizer to solve hard MIPs on supercomputers. Optim. Methods Softw. **33**(3), 530–539 (2018)
509. Shinano, Y., Heinz, S., Vigerske, S., Winkler, M.: FiberSCIP - a shared memory parallelization of SCIP. INFORMS J. Comput. **30**(1), 11–30 (2018)
510. Shinano, Y., Rehfeldt, D., Gally, T.: An easy way to build parallel state-of-the-art combinatorial optimization problem solvers: a computational study on solving Steiner tree problems and mixed integer semidefinite programs by using ug[SCIP-*,*]-libraries. In: Proceedings of the 9th IEEE Workshop Parallel/Distributed Combinatorics and Optimization, pp. 530–541 (2019)
511. Shinano, Y., Achterberg, T., Berthold, T., Heinz, S., Koch, T., Winkler, M.: Solving Previously Unsolved MIP Instances with ParaSCIP on Supercomputers by using up to 80,000 Cores. Tech. Rep. 20-16, ZIB, Berlin (2020)
512. Skjäl, A., Westerlund, T.: New methods for calculating αBB-type underestimators. J. Glob. Optim. **58**(3), 411–427 (2014)
513. Skjäl, A., Westerlund, T., Misener, R., Floudas, C.A.: A generalization of the classical αBB convex underestimation via diagonal and nondiagonal quadratic terms. J. Optim. Theory Appl. **154**(2), 462–490 (2012)
514. Smith, P., Mayston, D.: Measuring efficiency in the public sector. OMEGA **15**, 181–189 (1987)
515. Sousa, J.P., Wolsey, L.A.: A time indexed formulation of non-preemptive single machine scheduling problems. Math. Program. **54**, 353–367 (1992)
516. Spelluci, P.: Numerische Verfahren der nichtlinearen Optimierung. Birkhäuser, Basel (1993)
517. Spencer(III), T., Brigandi, A.J., Dargon, D.R., Sheehan, M.J.: AT&Ts telemarketing site selection system offers customer support. Interfaces **20**(1), 83–96 (1990)
518. Spendley, W., Hext, G.R., Himsworth, F.R.: Sequential application of simplex designs in optimisation and evolutionary operation. Technometrics **4**, 441–461 (1962)
519. Stigler, G.J.: The cost of subsistence. J. Farm Econ. **27**, 303–314 (1945)
520. Stoer, J.: Foundations of recursive quadratic programming methods for solving nonlinear programs. In: Schittkowski, K. (ed.) Computational Mathematical Programming. NATO ASI Series, vol. 15. Springer, Heidelberg (1985)
521. Stoyan, Y.: Mathematical methods for geometric design. In: Advances in CAD/CAM, Proceedings of PROLAMAT82 (Leningrad, USSR, May 1982), pp. 67–86. North-Holland, Amsterdam (1983)
522. Stoyan, Y., Romanova, T.: Mathematical models of placement optimisation: two- and three-dimensional problems and applications. In: Fasano, G., Pinter, J.D. (eds.) Modeling and Optimization in Space Engineering. Lecture Notes in Economics and Mathematical Systems, vol. 73, pp. 363–388. Springer, New York (2013)

523. Stoyan, Y., Terno, J., Scheithauer, G., Gil, N., Romanova, T.: Phi-functions for primary 2D-objects. Stud. Inf. Univ. **2**, 1–32 (2001)
524. Stoyan, Y., Gil, M., Terno, J., Romanova, T., Scheithauer, G.: Construction of a Phi-function for two convex polytopes. Appl. Math. **2**, 199–218 (2002)
525. Stoyan, Y., Scheithauer, G., Gil, N., Romanova, T.: Φ-functions for complex 2D-objects. 4OR: Quart. J. Belg. French Ital. Oper. Res. Soc. **2**, 69–84 (2004)
526. Stoyan, Y., Gil, N.I., Scheithauer, G., Pankratov, A., Magdalina, I.: Packing of convex polytopes into a parallelepiped. Optimization **54**, 215–235 (2005)
527. Stoyan, Y., Romanova, T., Scheithauer, G., Krivulya, A.: Covering a polygonal region by rectangles. Comput. Optim. Appl. **48**(3), 675–695 (2011)
528. Stoyan, Y., Romanova, T., Pankratov, A., Chugay, A.: Optimized object packings using quasi-phi-functions. In: Fasano, G., Pinter, J.D. (eds.) Optimized Packings with Applications, pp. 265–293. Springer International Publishing, Cham (2015)
529. Stoyan, Y., Romanova, T., Pankratov, A., Kovalenko, A., Stetsyuk, P.: Modeling and optimization of balance layout problems. In: Fasano, G., Pinter, J.D. (eds.) Space Engineering. Modeling and Optimization with Case Studies. Optimization and its Applications, vol. 114, pp. 369–400. Springer, New York (2016)
530. Stoyan, Y., Pankratov, A., Romanova, T.: Cutting and packing problems for irregular objects with continuous rotations: mathematical modelling and non-linear optimization. J. Oper. Res. Soc. **67**, 786–800 (2016)
531. Stoyan, Y., Pankratov, A., Romanova, T.: Quasi-phi-functions and optimal packing of ellipses. J. Glob. Optim. **65**, 283–307 (2016)
532. Stoyan, Y., Pankratov, A., Romanova, T.: Placement problems for irregular objects: Mathematical modeling, optimization and applications. In: Butenko, S., Pardalos, P.M., Shylo, V. (eds.) Optimization Methods and Applications, pp. 521–559. Springer International Publishing, Cham (2017)
533. Stoyan, Y., Pankratov, A., Romanova, T., Fasano, G., Pinter, J.D., Stoian, Y.E., Chugay, A.: Optimized packings in space engineering applications: part I. In: Fasano, G., Pinter, J.D. (eds.) Modeling and Optimization in Space Engineering. Springer Optimization and its Applications, vol. 144, pp. 395–437. Springer, Cham (2019)
534. Stoyan, Y., Grebennik, I., Romanova, T., Kovalenko, A.: Optimized packings in space engineering applications: Part II. In: Fasano, G., Pinter, J.D. (eds.) Modeling and Optimization in Space Engineering. Springer Optimization and its Applications, vol. 144, pp. 439–457. Springer, Cham (2019)
535. Stoyan, Y., Yaskov, G., Romanova, T., Litvinchev, I., Yakovlev, S., Velarde Cantu, J.M.: Optimized packing multidimensional hyperspheres: a unified approach. Math. Biosci. Eng. **17**, 6601–6630 (2020)
536. Subramanian, R., Scheff(Jr.), R.P., Quinlan, J.D., Wiper, D.S., Marsten, R.E.: Coldstart: fleet assignment at delta air lines. Interfaces **24**(1), 104–120 (1994)
537. Suhl, L., Mellouli, T.: Optimierungssysteme: Modelle, Verfahren, Software, Anwendungen. Springer, Berlin, Heidelberg (2007)
538. Talbot, F.B., Patterson, J.H.: An efficient integer programming algorithm with network cuts for solving resource-constrained scheduling problems. Manage. Sci. **24**, 1163–1174 (1978)
539. Tavares, L.V.: A review on the contributions of operational research to project management. In: Proceedings of the 14th European Conference on Operational Research. The Hebrew University, Jerusalem (1995)
540. Tawarmalani, M., Sahinidis, N.V.: Convexification and global optimization in continuous and mixed-integer nonlinear programming: theory, algorithms, software, and applications. In: Nonconvex Optimization And Its Applications, vol. 65. Kluwer Academic Publishers, Dordrecht (2002)
541. Tawarmalani, M., Sahinidis, N.V.: The pooling problem. In: Sahinidis, N.V., Tawarmalani, M. (eds.) Convexification and Global Optimization in Continuous and Mixed-Integer Nonlinear Programming. Nonconvex Optimization and Its Applications, vol. 65, pp. 253–283. Springer, Boston, MA (2002)

542. Tawarmalani, M., Sahinidis, N.V.: Global optimization of mixed integer nonlinear programs: a theoretical and computational study improve MIP solutions. Math. Program. **99**, 563–591 (2004)
543. Thanassoulis, E., Dyson, R.G., Foster, M.J.: Relative efficiency assessments using data envelopment analysis: an application to data on rates departments. J. Oper. Res. Soc. **38**, 397–412 (1987)
544. Thue, A.: Über die dichteste Zusammenstellung von kongruenten Kreisen in einer Ebene. Christiana Vid. Selsk. Skr. **1**, 1–9 (1910)
545. Timpe, C.: Solving mixed planning & scheduling problems with mixed branch & bound and constraint programming. OR Spectrum **24**, 431–448 (2002)
546. Tomlin, J.A.: On scaling linear programming problems. Math. Program. **4**, 146–166 (1975)
547. Tomlin, J.A., Welch, J.S.: A pathological case in the reduction of linear programs. Oper. Res. Lett. **2**, 53–57 (1983)
548. Tomlin, J.A., Welch, J.S.: Formal optimisation of some reduced linear programming problems. Math. Program. **27**, 232–240 (1983)
549. Toregas, C., Swain, R., Revelle, C., Bergman, L.: The location of emergency service vehicles. Oper. Res. **19**, 1363–1373 (1971)
550. Toth, P., Vigo, D.: 2. Branch-and-bound algorithms for the capacitated VRP. In: Toth, P., Vigo, D. (eds.) The Vehicle Routing Problem, pp. 29–51. SIAM, Philadelphia, PA (2002)
551. Toth, P., Vigo, D.: Models, relaxations and exact approaches for the capacitated vehicle routing problem. Discr. Appl. Math. **123**(1–3), 487–512 (2002)
552. Toth, P., Vigo, D.: The Vehicle Routing Problem. SIAM, Philadelphia, PA (2002)
553. Toth, P., Vigo, D. (eds.): Vehicle Routing. MOS-SIAM Series on Optimization, vol. 18. SIAM, Philadelphia (2014)
554. Trelles, O., Rodriguez, A.: Bioinformatics and parallel metaheuristics. In: Alba, E. (ed.) Parallel Metaheuristics: A New Class of Algorithms. Wiley Series on Parallel and Distributed Computing, chap. 21, pp. 517–549. Wiley, Hoboken (2005)
555. Trespalacios, F., Grossmann, I.E.: Review of mixed-integer nonlinear and generalized disjunctive programming methods. Chem. Ing. Tech. **86**, 991–1012 (2014)
556. Vajda, S.: Mathematical Programming. Adison-Wesley, Reading, MA (1961)
557. Valente, C., Mitra, G., Sadki, M., Fourer, R.: Extending algebraic modelling languages for stochastic programming. INFORMS J. Comput. **21**(1), 107–122 (2009)
558. van den Akker, M.: LP-Based Solution Methods for Single-Machine Scheduling Problems. PhD Thesis, Eindhoven University of Technology, Eindhoven (1994)
559. Vanderbeck, F., Wolsey, L.A.: An exact algorithm for IP column generation. Oper. Res. Lett. **19**, 151–160 (1996)
560. Vanderbei, R.J.: Linear Programming - Foundations and Extensions. Kluwer, Dordrecht (1996)
561. Vanderbei, R.J.: Linear Programming - Foundations and Extensions, 4th edn. Springer, New York (2014)
562. VanRoy, T.J., Wolsey, L.A.: Solving mixed integer programs by automatic reformulation. Oper. Res. **35**(1), 45–57 (1987)
563. Vasquez-Marques, A.: American airlines arrival slot allocation system (ASAS). Interfaces **21**(1), 42–61 (1991)
564. Vielma, J.P., Nemhauser, G.: Modeling disjunctive constraints with a logarithmic number of binary variables and constraints. Math. Program. Ser. A **128**, 49–72 (2011)
565. Vielma, J.P., Ahmed, S., Nemhauser, G.: Mixed-integer models for nonseparable piecewise-linear optimization: unifying framework and extensions. Oper. Res. **53**, 303–315 (2009)
566. Vishnoi, N.: Algorithms for Convex Optimization. Cambridge University Press, Cambridge (2021)
567. Viswanathan, J., Grossmann, I.E.: A combined penalty function and outer-approximation method for MINLP optimization. Comp. Chem. Eng. **14**(7), 769–782 (1990)
568. Wächter, A., Biegler, L.T.: On the implementation of a primal-dual interior point filter line search algorithm for large-scale nonlinear programming. Math. Program. **106**, 25–57 (2006)

569. Wallace, S.W.: Decision making under uncertainty: is sensitivity analysis of any use? Oper. Res. **48**, 20–25 (2000)
570. Wenger, K.M.: Generic Cut Generation Methods for Routing Problems. Ph.D. thesis, University of Heidelberg, Institute of Computer Science, Im Neuenheimer Feld 368, D-69120 Heidelberg (2003). ISBN 3-8322-2545-5, Shaker Verlag, http://www.shaker.de
571. Werner, J.: Numerische Mathematik. Vieweg, Wiesbaden, Deutschland (1992)
572. Westerlund, T., Isaksson, J.: Some efficient formulations for the simultaneous solution of trim-loss and scheduling problems in the paper-converting industry. Chem. Eng. Res. Des. **76**, 677–684 (1998)
573. Westerlund, T., Pörn, R.: Solving pseudo-convex mixed integer problems by cutting plane techniques. Optim. Eng. **3**, 253–280 (2002)
574. Westerlund, T., Petterson, F.: An extended cutting plane method for solving convex MINLP problems. Comput. chem. Eng. Sup. **19**, S131–136 (1995)
575. Westerlund, T., Petterson, F., Grossmann, I.E.: Optimization of pump configuration problems as a MINLP problem. Comput. Chem. Eng. **18**(9), 845–858 (1994)
576. Westerlund, T., Skrifvars, H., Harjunkoski, I., Pörn, R.: An extended cutting plane method for solving a class of non-convex MINLP problems. Comput. Chem. Eng. **22**, 357–365 (1998)
577. Westerlund, J., Hästbacka, M., Forssell, S., Westerlund, T.: A mixed-time MILP scheduling model. Ind. Eng. Chem. Res. **46**, 2781–2796 (2007)
578. Westerlund, J., Papageorgiou, L.G., Westerlund, T.: A MILP model for N-dimensional allocation. Comput. Chem. Eng. **31**(12), 1702–1714 (2007)
579. Westerlund, T., Eronen, V., Mäkelä, M.M.: On solving generalized convex MINLP problems using supporting hyperplane techniques. J. Glob. Optim. **71**(4), 987–1011 (2018)
580. Williams, H.P.: Model Building in Mathematical Programming, 3rd edn. Wiley, Chichester (1993)
581. Williams, H.P.: The Dual of a Logical Linear Programme. Research paper, Mathematical Sciences, University of Southampton, Southhampton (1995)
582. Williams, H.P., Redwood, A.C.: A structured linear programming model in the food industry. Oper. Res. Quart. **25**, 517–528 (1974)
583. Wilson, J.M.: Alternative formulations of a flow-shop scheduling problem. J. Oper. Res. Soc. **40**, 395–399 (1989)
584. Wilson, J.M.: Generating cuts in integer programming with families of special ordered sets. Eur. J. Oper. Res. **46**, 101–108 (1990)
585. Wilson, E.J.G., Willis, R.G.: Scheduling of telephone betting operators - a case study. J. Oper. Res. Soc. **33**, 991–998 (1983)
586. Wolsey, L.A.: Group-theoretic results in mixed integer programming. Oper. Res. **19**, 1691–1697 (1971)
587. Wolsey, L.A.: Uncapacitated lot-sizing problems with start-up costs. Oper. Res. **37**, 741–747 (1989)
588. Wolsey, L.A.: Valid inequalities for 0-1 knapsacks and MIPS with generalised upper bound constraints. Discr. Appl. Math. **29**, 251–261 (1990)
589. Wright, M.M.: Speeding up the Hungarian algorithm. Comput. Oper. Res. **17**, 95–96 (1990)
590. Wright, S.: Primal-Dual Interior-Point Methods. Society for Industrial and Applied Mathematics, Philadelphia, PA (1996)
591. Xiao, X., Floudas, C.: Integrated gasoline blending and order delivery operations: part I. Short-term scheduling and global optimization for single and multi-period operations. AIChE J. **62**(6), 2043–2070 (2016)
592. Young, R.D.: A simplified primal (all-integer) integer programming algorithm. Oper. Res. **16**, 750–782 (1968)
593. Young, R.E., Baumol, W.J.: Integer programming and pricing. Econometrica **28**, 520–550 (1960)
594. Zenios, S.A. (ed.): Financial Optimization. Cambridge University Press, Cambridge (1993)
595. Zhang, J., Kim, N., Lasdon, L.: An improved successive linear programming algorithm. Manage. Sci. **31**, 1312–1331 (1985)

596. Zimmermann, H.J.: Fuzzy Set Theory and Its Applications, 2nd edn. Kluwer Academic Publishers, Boston, MA (1987)
597. Zimmermann, H.J.: Fuzzy Sets, Decision Making, and Expert Systems. Kluwer Academic Publishers, Boston, MA (1987)
598. Zimmermann, H.J.: An application-oriented view of modeling uncertainty. Eur. J. Oper. Res. **122**, 190–198 (2000)

Index

J. Kallrath, *Business Optimization Using Mathematical Programming*, International Series in Operations Research & Management Science 307,
https://doi.org/10.1007/978-3-030-73237-0

D

E

F

The manufacturer's authorised representative in the EU is Springer Nature Customer Service Centre GmbH, Europaplatz 3, 69115 Heidelberg, Germany. If you have any concerns regarding our products, please contact ProductSafety@springernature.com

Printed and bound by CPI Group (UK) Ltd, Croydon, CR0 4YY
30/09/2026
02225287-0018